P9-APA-939

INTRODUCTION TO PHYSICAL METALLURGY

Second Edition

SIDNEY H. AVNER
Professor
New York City Community College
City University of New York

McGRAW-HILL BOOK COMPANY

New York
St. Louis
San Francisco
Düsseldorf
Johannesburg
Kuala Lumpur
London
Mexico
Montreal
New Delhi
Panama
Rio de Janeiro
Singapore
Sydney
Toronto

Library of Congress Cataloging in Publication Data

Avner, Sidney H
Introduction to physical metallurgy.

Includes bibliographies.
I. Physical metallurgy. I. Title.
TN690.A86 1974 669'.9 74-807
ISBN 0-07-002499-5

7 8 9 10 KPKP 83210

The editors for this book were Robert Buchanan and Myrna W. Breskin, the designer was Edward Butler, and its production was supervised by Patricia Ollague. It was set in Helvetica by Black Dot, Inc.
It was printed and bound by Kingsport Press, Inc.

CONTENTS

Preface v
Introduction vii
1. Tools of the Metallurgist 1
2. Metal Structure and Crystallization 65
3. Plastic Deformation 107
4. Annealing and Hot Working 129
5. Constitution of Alloys 147
6. Phase Diagrams 155
7. The Iron-Iron Carbide Equilibrium Diagram 225
8. The Heat Treatment of Steel 249
9. Alloy Steels 349
10. Tool Steels 387
11. Cast Iron 423
12. Nonferrous Metals and Alloys 461
13. Metals at High and Low Temperatures 547
14. Wear of Metals 567
15. Corrosion of Metals 583
16. Powder Metallurgy 605
17. Failure Analysis 633

Appendix: Temperature-conversion Table 666
Glossary 667
Index 689

To Judy, Kenny, and Jeffrey,
in whose hands
the future lies

PREFACE

The emphasis of the second edition of this text remains on the basic concepts and applications of physical metallurgy.

The level of this edition is also essentially unchanged. The text is still considered appropriate for the teaching of physical metallurgy to students who are not majors in metallurgy as well as to engineering students as an introductory course. It has also proved useful for technician training programs in industry. The fundamental concepts are still presented in a simplified form yet as accurately as possible. The only background required is an elementary course in physics.

During the past decade, the first edition of this text was found to be quite effective, and many favorable comments were received from both students and faculty members. However, advances in certain areas and suggestions from users have necessitated a revision of the first edition.

The following is a summary of the most notable improvements in the second edition:

In Chapter 1, which covers some of the important tools and tests in the field of metallurgy, the section on nondestructive testing has been expanded to include eddy current testing and holography. The latest ASTM code has been used in the section on hardness testing.

Some changes have been made in Chapter 2 in order to make the simplified explanation of atomic and metal structure more understandable. A brief explanation of x-ray diffraction and grain size measurement was added.

Chapters 3 and 4 which cover the fundamentals of plastic deformation and the effect of heat on cold-worked materials remain essentially the same except for an expanded discussion of dislocations and fracture.

Chapter 6, on binary phase diagrams, now includes diffusion, a more detailed explanation of the theory of age hardening, and more actual phase diagrams as illustrations.

Chapter 7 which considers the iron-iron carbide equilibrium diagram in some detail now also discusses wrought iron and the effect of small quantities of other elements on the properties of steel.

The section on case hardening of steel in Chapter 8 has been expanded to include a more detailed explanation of nitriding, flame hardening and induction hardening. A section on hardenable carbon steels has also been added.

In Chapter 9, the portion on stainless steels now encompasses new sections on precipitation-hardening stainless steels, maraging steels, and ausforming.

In Chapter 10 the section on cemented carbide tools has been expanded and a new section on ceramic tools has been added.

Chapter 11 now covers only cast iron and has been enhanced by additional diagrams.

The numerous additional photomicrographs added to Chapter 12 illustrate various nonferrous microstructures. An entire section on titanium and titanium alloys has been included because of their increased commercial importance.

The chapter on wear of metals has been moved next to the one on corrosion of metals to improve the continuity of subject matter.

Chapter 15 now discusses the corrosion of metals in greater detail.

A brief discussion of the powder metallurgy processing techniques has been added to Chapter 16.

There are two major changes in this edition as compared to the first edition.

1. The replacement of Chapter 17 on extractive metallurgy by an entirely new chapter on failure analysis. It was felt that extractive metallurgy was not really part of physical metallurgy and that a chapter on failure analysis would be of greater interest and value to technicians and engineers.
2. The addition of a glossary of terms related to physical metallurgy.

There is very little on the details of operation of heat-treating and testing equipment since they are covered in the laboratory course which is taken in conjunction with the theory course.

Numerous photomicrographs have been used to illustrate typical structures. Many tables have been included to present representative data on commercial alloys.

The aid received from the following people in reading portions of the manuscript or in preparations of photomicrographs for the first edition is gratefully acknowledged: J.E. Krauss, G. Cavaliere, A. Dimond, A. Smith, A. Cendrowski, J. Sadofsky, C. Pospisil, T. Ingraham, J. Kelch, and O. Kammerer. Many companies have contributed generously from their publications and credit is given where possible.

I make no particular claim for originality of material. The information of other authors and industrial companies has been drawn upon. The only justification for this book, then, lies in the particular topics covered, their sequence, and the way in which they are presented.

I would like to express my appreciation to Miss Barbara Worth for typing most of the first edition manuscript, to Mrs. Helen Braff and Mrs. Lillian Schwartz for typing the second edition material, and finally to my wife, without whose patience and understanding this book could never have been written.

Sidney H. Avner

INTRODUCTION

Metallurgy is the science and technology of metals. It is beyond the scope of this text to cover the development of metallurgy as a science. Only certain highlights will be mentioned here for the purpose of orientation.

The worker of metals is mentioned in the Bible and in Greek and Norse mythology. Metallurgy as an art has been practiced since ancient times. Ancient man knew and used many native metals. Gold was used for ornaments, plates, and utensils as early as 3500 B.C. The art of smelting, refining, and shaping metals was highly developed by both the Egyptians and the Chinese. The ancient Egyptians knew how to separate iron from its ore and that steel had the ability to harden, but iron was not used widely before 1000 B.C. Iron was not popular with ancient people because of its tendency to rust, and they preferred working with gold, silver, copper, brass, and bronze.

Knowledge of dealing with metals was generally passed directly from master to apprentice in the Middle Ages, leading to an aura of superstition surrounding many of the processes. Very little was written on metallurgical processes until Biringuccio published his "Pirotechnia" in 1540, followed by Agricola's "De Re Metallurgica" in 1556. In succeeding years, much knowledge was added to the field by people trying to duplicate the composition and etched structure of Damascus steel.

Until the beginning of the last quarter of the nineteenth century, most investigations of metal structure had been macroscopic (by eye) and superficial. The science of the structure of metals was almost nonexistent. The situation was ripe for the detailed attention of individuals whose background was more scientific than practical. The individual most responsible for the period of rapid development that followed was Henry Clifton Sorby.

Sorby was an amateur English scientist who started with a study of meteorites and then went on to study metals.

In September 1864, Sorby presented a paper to the British Association for the Advancement of Science in which he exhibited and described a number of microscopical photographs of various kinds of iron and steel. This paper marks the beginning of metallography, the field concerned with the use of the microscope to study the structure of metals. It seems that while many people appreciated the value of Sorby's studies at the time they were done, none of them had sufficient interest to develop the technique independently, and metallography lay dormant for almost twenty years.

Additional work by Martens in Germany (1878) revived Sorby's interest in metallurgical problems, and in 1887 he presented a paper to the Iron and Steel Institute which summarized all his work in the field. Considerable attention was now generated by both scientists and industrial metallurgists in other countries. In the early part of the twentieth century, Albert Sauveur convinced American steel companies that the microscope was a practical tool to aid in the manufacture and heat treatment of steel.

About 1922, more knowledge of the structure and properties of metals was added by the application of x-ray diffraction and wave mechanics.

Metallurgy is really not an independent science since many of its fundamental concepts are derived from physics, chemistry, and crystallography.

The metallurgist has become increasingly important in modern technology. Years ago, the great majority of steel parts were made of cheap low-carbon steel that would machine and fabricate easily. Heat treatment was reserved largely for tools. Designers were unable to account for structural inhomogeneity, surface defects, etc., and it was considered good practice to use large factors of safety. Consequently, machines were much heavier than they should have been, and the weight was considered a mark of quality. This attitude has persisted, to some extent, to the present time but has been discouraged under the leadership of the aircraft and automotive industries. They have emphasized the importance of the strength-weight ratio in good design, and this has led to the development of new high-strength, lightweight alloys.

New technical applications and operating requirements pushed to higher levels have created a continued need for the development of new alloys. For example, an exciting development has been the Wankel rotary engine—an internal combustion engine of unusual design that is more compact, lighter, and mechanically far simpler than the ordinary reciprocating piston motor of equivalent horsepower. A particularly bothersome problem has been the seals between the rotor and the metal wall. Originally, the seals were made of carbon and seldom lasted more than 20,000 miles. Research developed a new sintered titanium-carbide alloy seal which has given lifetimes of up to 100,000 miles.

The metallurgical field may be divided into two large groups:

1. Process or extractive metallurgy—the science of obtaining metals from their ores, including mining, concentration, extraction, and refining metals and alloys.

2. Physical metallurgy—the science concerned with the physical and mechanical characteristics of metals and alloys. This field studies the properties of metals and alloys as affected by three variables:

- **a.** Chemical composition—the chemical constituents of the alloy
- **b.** Mechanical treatment—any operation that causes a change in shape such as rolling, drawing, stamping, forming, or machining
- **c.** Thermal or heat treatment—the effect of temperature and rate of heating and cooling

REFERENCES

Hoover and Hoover: "Georgius Agricola's *De Re Metallurgica*," Dover Publications, New York, 1912.

Howe, H. M.: The Metallurgy of Steel, *The Engineering and Mining Journal*, 1st ed., New York, 1890.

Rickard, Thomas: "Man and Metals," McGraw-Hill Book Company, New York, 1932.

Sauveur, Albert: "The Metallography and Heat Treatment of Iron and Steel," 4th ed., McGraw-Hill Book Company, New York, 1935.

Smith and Gnudi: "Pirotechnia of Vannoccia Biringuccio," American Institute of Mining and Metallurgical Engineers, New York, 1943.

Smith, Cyril Stanley: "A History of Metallography," University of Chicago Press, 1960.

Sullivan, F.: "The Story of Metals," American Society for Metals, Metals Park, Ohio, 1951.

1 TOOLS OF THE METALLURGIST

The purpose of this chapter is to give the student an understanding of some of the common tools and tests that are used in the metallurgical field.

1·1 Temperature Scales In scientific research and in most foreign countries, the standard temperature-measuring scale is the centigrade scale. However, in American industrial plants, the Fahrenheit scale is used almost exclusively. Therefore, all references to temperature in this book will be in terms of the Fahrenheit scale since this is the one most likely to be encountered by the industrial technician. Conversion from one scale to the other may be made by the following equations:

$$°C = 5/9\ °F - 32 \qquad (1\cdot1)$$

$$°F = 9/5\ °C + 32 \qquad (1\cdot2)$$

The accuracy with which temperatures are measured and controlled will determine the successful operation of some metallurgical processes such as casting, smelting, refining, and heat treatment. It will also have a profound effect on the strength properties of many metals and alloys.

TEMPERATURE MEASUREMENT

In order to understand the effect of thermal treatment on the properties, it is necessary to have some knowledge of how temperature is measured.

Pyrometry deals with the measurement of elevated temperatures, generally above 950°F, and instruments used for this purpose are known as *pyrometers.*

Thermometry deals with the measurement of temperatures below 950°F, and instruments for this purpose are known as *thermometers.*

1·2 Temperature Measurement by Color One of the simplest methods of estimating the temperature of a metal is by noting the color of the hot body. There is an apparent correlation between the temperature of a metal and

TABLE 1·1 Variation of Color with Temperature

COLOR	TEMP., °F
Faint red	950
Dark red	1150
Dark cherry	1175
Cherry red	1300
Bright cherry	1475
Dark orange	1650
Orange	1750
Yellow	1800

its color, as shown by Table 1·1. Except when applied by an experienced observer, this method will give only rough temperature estimates. The principal difficulty is that judgment of color varies with the individual. Other sources of error are that the color may not be uniform and may vary somewhat with different materials.

If a continuous indication or recording of temperature is required, then the instruments in use may be divided into two general classifications: (1) mechanical systems that deal essentially with the expansion of a metal, a liquid, a gas or vapor; and (2) electrical systems which deal with resistance, thermocouple, radiation, and optical pyrometers.

1·3 Metal-expansion Thermometers Most metals expand when heated, and the amount of expansion will depend upon the temperature and the coefficient of expansion. This principle is incorporated in the bimetallic strip which is used in the common thermostat. The bimetallic strip is made by bonding a high-expansion metal on one side with a low-expansion metal on the other. As a result of small temperature changes, the strip will curve and therefore make or break an electrical circuit which will control the heating of a house.

When it is used as an industrial temperature indicator, the bimetallic strip is usually bent into a coil, one end of which is fixed so that on expansion a rotary motion is automatically obtained (Fig. 1·1).

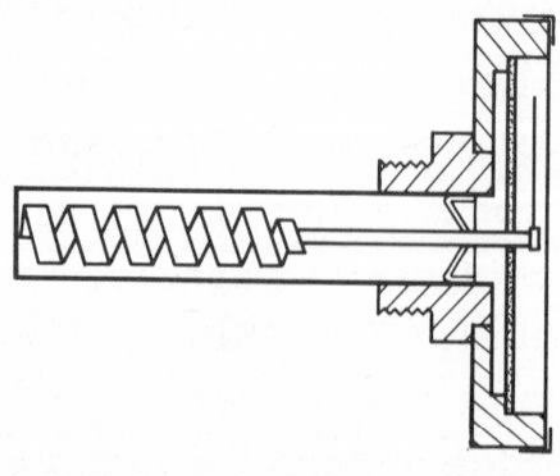

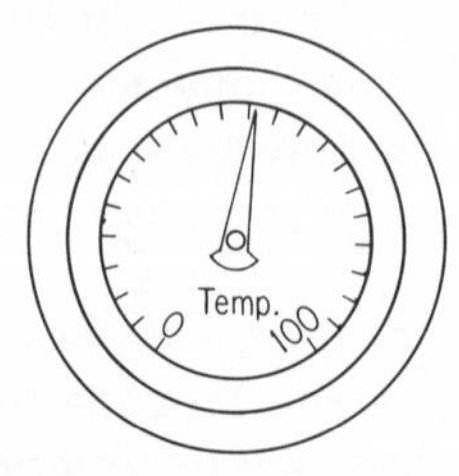

Fig. 1·1 Industrial temperature indicator with a helical bimetallic element. (By permission from P. J. O'Higgins, "Basic Instrumentation," McGraw-Hill Book Company, New York, 1966.)

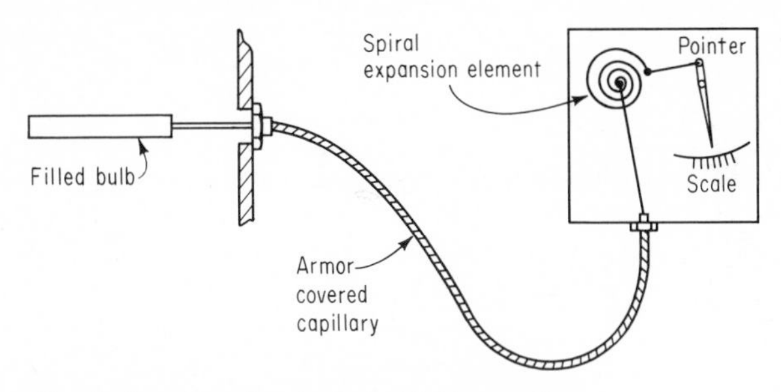

Fig. 1·2 Simple thermal system for industrial temperature measurement. (By permission from P. J. O'Higgins, "Basic Instrumentation," McGraw-Hill Book Company, New York, 1966.)

Most bimetallic strips have Invar as one metal, because of its low coefficient of expansion, and yellow brass as the other metal for low temperatures or a nickel alloy for higher temperatures. They can be used in the range of −100 to 1000°F, are very rugged, and require virtually no maintenance. Their main disadvantage is that, owing to the necessity for enclosing the element in a protecting tube, the speed of response may be lower than that of other instruments.

1·4 Liquid-expansion Thermometers The remainder of the mechanical system temperature-measuring instruments, whether liquid-expansion or gas- or vapor-pressure, consist of a bulb exposed to the temperature to be measured and an expansible device, usually a Bourdon tube, operating an indicating pointer or a recording pen. The bulb and Bourdon tube are connected by capillary tubing and filled with a suitable medium (Fig. 1·2).

The liquid-expansion thermometer has the entire system filled with a suitable organic liquid or mercury. Changes in bulb temperature cause the liquid to expand or contract, which in turn causes the Bourdon tube to expand or contract. Temperature changes along the capillary and at the case also cause some expansion and contraction of the liquid, and some form of compensation is therefore required. Figure 1·3 shows a fully compensated liquid-expansion thermometer using a duplicate system, less bulb, arranged so that motions are subtracted. Some of the liquids used and the temperatures covered by them are:

Mercury	−35 to +950°F
Alcohol	−110 to +160°F
Pentane	+330 to + 85°F
Creosote	+20 to +400°F

1·5 Gas- or Vapor-pressure Thermometers In the vapor-pressure thermometer, a volatile liquid partially fills the bulb. Different temperatures of the bulb cause corresponding pressure variations in the saturated vapor above the liquid surface in the bulb. These pressure variations are transmitted to the

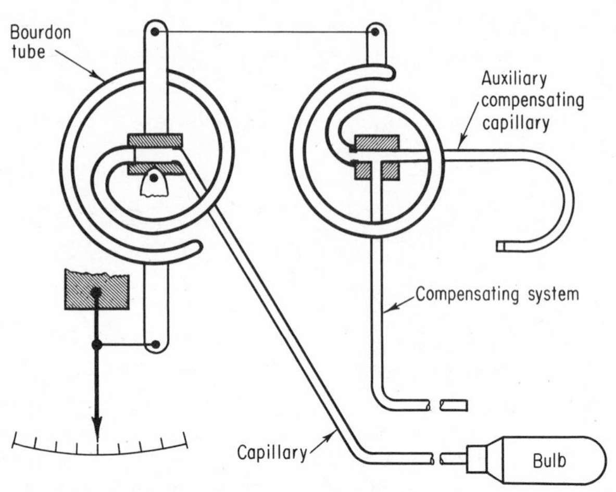

Fig. 1·3 A fully compensated liquid-expansion thermometer. (From "Temperature Measurement," American Society for Metals, 1956.)

Bourdon tube, the pressure indications acting as a measure of the temperature in the bulb. By suitable choice of volatile liquid, almost any temperature from −60 to +500°F can be measured. Some liquids used are methyl chloride, ether, ethyl alcohol, and toluene.

The gas-pressure thermometer is similar to the vapor-pressure thermometer except that the system is filled with a gas, usually nitrogen. The range of temperature measured by the gas-pressure thermometer is from −200 to +800°F.

Filled-system thermometers are used primarily for low-temperature applications such as plating and cleaning baths, degreasers, cooling water and oil temperatures, and subzero temperatures in the cold treatment of metals. These instruments are relatively inexpensive but are not used where quick repair or exceptionally high accuracy is required.

1·6 Resistance Thermometer The principle of the resistance thermometer depends upon the increase of electrical resistance with increasing temperature of a conductor. If the temperature-resistance variations of a metal are calibrated, it is possible to determine the temperature by measuring its electrical resistance. The resistance coil is mounted in the closed end of a protecting tube and the leads are extended to a suitable resistance-measuring instrument, usually a Wheatstone bridge.

Resistance coils are usually made of copper, nickel, or platinum. Nickel and copper are most satisfactory for temperatures between 150 and 500°F, whereas platinum may be used between −350 and +1100°F. The resistance thermometer is very accurate and is of great importance in the laboratory. However, its industrial use is limited because it is fragile and requires many precautions in use.

1·7 Thermoelectric Pyrometer This is the most widely used method for metallurgical temperature measurement and control; it will perform satisfactorily up to about 3000°F.

The simple thermoelectric pyrometer, shown in Fig. 1·4, consists of the following units:

1 The *thermocouple*, composed of two different metals or alloys
2 The *junction block*, just outside the furnace
3 The *extension leads*
4 The *indicating instrument* or recorder

The operation of this pyrometer is based upon two principles:

Peltier Effect If two dissimilar metallic wires are brought into electrical contact, an emf will exist across the point of contact. The magnitude of the emf developed will be determined by the chemical composition of the wires and the temperature of the junction point.

Thomson Effect If there is a temperature difference between the ends of a single homogeneous wire, an emf will exist between the ends of the wire. The magnitude of the emf developed will be determined by the composition, the chemical uniformity of the wire, and the temperature difference.

The total emf in a thermoelectric pyrometer, sometimes called the *Seebeck effect*, is therefore the algebraic sum of four emf's, two Peltier emf's at the hot and cold junctions and two Thomson emf's along each of the wires.

The cold junction, or reference junction, must be kept at a constant temperature. This is usually 0°C, or 32°F. At the indicating instrument this is usually done by means of a cold-junction compensating coil which changes its resistance with fluctuations in ambient temperature, always keeping the instrument at 32°F. If the cold junction, or reference junction, is kept at constant temperature, then the measured emf in the pyrometer circuit will be a definite function of the temperature of the hot junction. By suitable

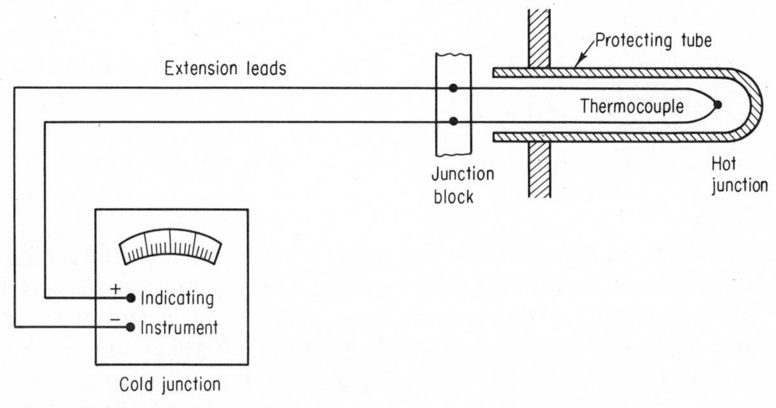

Fig. 1·4 A simple thermoelectric pyrometer.

TABLE 1·2 Temperature vs. Electromotive Force*
Emf in millivolts; cold junction 32° F

TEMP, °F	PT + 10%RH VS. PLATINUM	CHROMEL VS. ALUMEL	IRON VS. CONSTANTAN	COPPER VS. CONSTANTAN
32	0.0	0.0	0.0	0.0
100	0.221	1.52	1.94	1.517
200	0.595	3.82	4.91	3.967
300	1.017	6.09	7.94	6.647
400	1.474	8.31	11.03	9.525
500	1.956	10.57	14.12	12.575
600	2.458	12.86	17.18	15.773
700	2.977	15.18	20.25	19.100
800	3.506	17.53	23.32	
900	4.046	19.89	26.40	
1000	4.596	22.26	29.52	
1100	5.156	24.63	32.72	
1200	5.726	26.98	36.01	
1300	6.307	29.32	39.43	
1400	6.897	31.65	42.96	
1500	7.498	33.93	46.53	
1600	8.110	36.19	50.05	
1700	8.732	38.43		
1800	9.365	40.62		
1900	10.009	42.78		
2000	10.662	44.91		
2100	11.323	47.00		
2200	11.989	49.05		
2300	12.657	51.05		
2400	13.325	53.01		
2500	13.991	54.92		
2600	14.656			
2700	15.319			
2800	15.979			
2900	16.637			
3000	17.292			

*By permission from P. H. Dike, "Thermoelectric Thermometry," p. 82, Leeds and Northrup Company, 1954.

calibration, it is possible to determine an exact relationship between the developed emf and the true temperature of the hot junction (Table 1·2).

Another useful thermoelectric law states that, if a third metal is introduced into the circuit, the total emf of the circuit will not be affected if the temperature of this third metal is uniform over its entire length.

The purpose of the extension leads is to move the reference junction to a point where the temperature will not vary. Thermocouple wires are usually not long enough nor well enough insulated to run directly to the instrument. The extension leads are usually made of the same material as the thermocouple wires and are placed in a duplex cable with the individual covering color-coded for identification. Copper extension leads may be used in some cases, but then the cold junctions are at the junction block instead of the instrument and may be more difficult to maintain at constant temperatures.

1·8 Thermocouple Materials Theoretically, any two dissimilar metallic wires will develop an emf when there is a temperature difference between their junction points. Industrially, however, only a few combinations are actually used for thermocouples. These were chosen primarily for their thermoelectric potential, reasonable cost, grain-size stability, linearity of the temperature-emf curve, and melting points higher than the temperature to be measured. The first material in the combination is always connected to the positive terminal.

Chromel-Alumel Chromel (90 percent nickel, 10 percent chromium) vs. Alumel (94 percent nickel, 3 percent manganese, 2 percent aluminum, 1 percent silicon) is one of the most widely used industrial combinations. It has a fairly linear calibration curve and good resistance to oxidation. It is most useful in the range from 1200 to 2200°F.

Iron-Constantan Constantan is an alloy containing approximately 54 percent copper and 46 percent nickel. This combination may be used in the range from 300 to 1400°F. The primary advantages are its comparatively low cost, high thermoelectric power, and adaptability to different atmospheres.

Copper-Constantan The constantan alloy used with copper differs slightly from that used with iron and may contain small amounts of manganese and iron. This combination is most useful for measuring low temperatures, down to −420°F. The upper limit is approximately 600°F.

The above combinations are known as *base-metal* thermocouples.

Platinum, 10 percent Rhodium–Platinum This is a "noble-metal" thermocouple. It is used for measuring temperatures which are too high for base-metal thermocouples and where radiation or optical pyrometers are not satisfactory. It is suitable for continuous use in the range from 32 to 3000°F but deteriorates rapidly in a reducing atmosphere.

Thermocouples are manufactured by cutting off suitable lengths of the two wires; the ends are carefully twisted together for about two turns, or sometimes butted together, and welded to form a smooth well-rounded head (Fig. 1·5*a*).

The thermocouple wires should be in electrical contact only at the hot junction, since contact at any other point will usually result in too low a

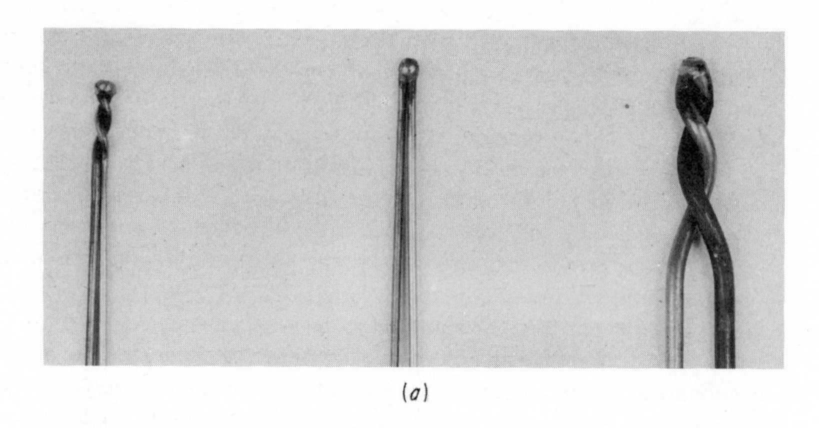

(a)

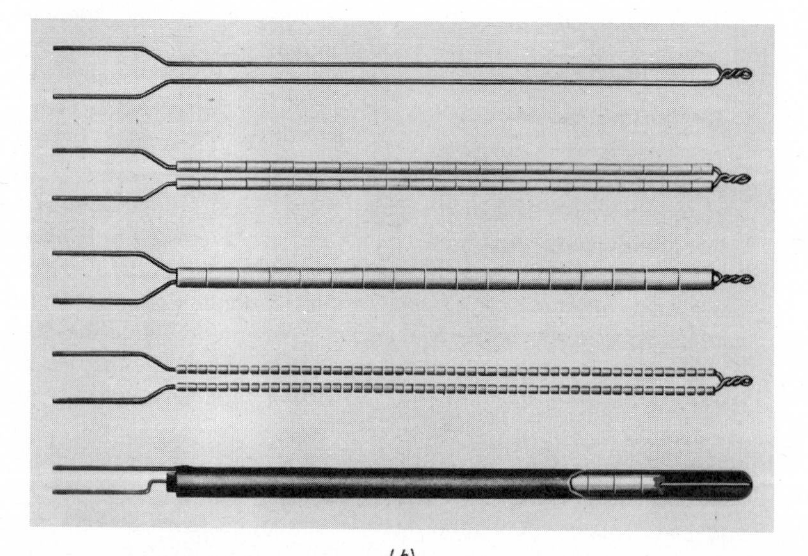

(b)

Fig. 1·5 (a) Examples of properly welded thermocouples. (b) Different types of porcelain separators. (Leeds & Northrup Company.)

measured emf. The two wires are insulated from each other by porcelain beads or ceramic tubes (Fig. 1·5b).

In most cases, thermocouples are enclosed in protecting tubes. The protecting tubes may be either ceramic or metallic materials. The tube guards the thermocouple against mechanical injury and prevents contamination of the thermocouple materials by the furnace atmosphere. A variety of metallic protecting tubes are available, such as wrought iron or cast iron (up to 1300°F); 14 percent chrome iron (up to 1500°F); 28 percent chrome iron, or Nichrome (up to 2000°F). Above 2000°F, porcelain or silicon carbide protecting tubes are used.

1·9 Measurement of Emf The temperature of the hot junction is determined by measuring the emf generated in the circuit. A potentiometer is one of the most accurate instruments available for measuring small emfs. Essentially the emf developed by a thermocouple is balanced against a known emf and is measured in terms of this standard. The slide-wire scale may be calibrated in millivolts or directly in temperature. In the latter case, the instrument should be used only with the type of thermocouple for which it is calibrated. This information is usually stamped on the dial face of the instrument.

A simple direct-indicating potentiometer circuit is shown in Fig. 1·6. Current from the dry cell is passed through a main circuit consisting of a slide-wire and an adjustable resistance *R*. The slide-wire *AB* is a uniform resistance wire which may be considered as divided into an equal number of divisions. With the polarity of the dry cell as shown, there is a drop of potential along the slide-wire from *A* to *B*, the magnitude of which depends upon the current flowing through it from the dry cell. Since the slide-wire is of uniform resistance, there are equal drops of potential across each division. In order to standardize the drop between *A* and *B* to correspond to the fixed markings on the indicating dial, a standard cell of known and fixed voltage is connected into the circuit by moving the switch to the standard cell (S.C.) position. Notice that the polarity of the standard cell is such that the current flowing from it opposes the current flowing from the dry cell. The resistance *R* is adjusted so that these currents are made equal, with the net result that no current flows through the circuit—as indicated by zero deflection of the galvanometer. The circuit is now standardized so that the potential drop across each division of the slide-wire corresponds to a definite amount of millivolts.

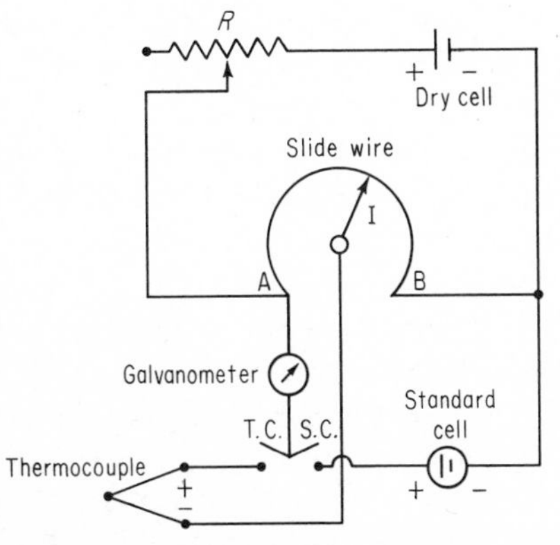

Fig. 1·6 A simple direct-indicating potentiometer. (Leeds & Northrup Company.)

When the emf of the thermocouple is to be measured, it replaces the standard cell in the circuit by moving the switch to the thermocouple (T.C.) position. The thermocouple must be properly connected so that the current flowing from it opposes the flow of current from the dry cell. The circuit is balanced, not by adjusting the resistance R, but by adjusting the resistance of that portion of the slide-wire which is contained in the thermocouple circuit. This adjustment is made by turning the indicator dial until the galvanometer reads zero. At this point, the drop of potential through the slide-wire up to the point of contact is equal to the emf of the thermocouple, and the millivolts may be read directly on the slide-wire scale. Reference to a suitable calibration table, such as Table 1·2, for the particular thermocouple being used will allow the conversion of millivolts to temperature, or the temperature may be read directly if the dial is so calibrated.

Since the cold junction at the instrument is usually higher than the standard cold junction (32°F), it is necessary to compensate for this variation. The compensation may be made manually, or automatically by a temperature-sensitive resistor called a cold-junction compensator. In contrast to most materials, the cold-junction compensator has a negative temperature resistance coefficient. This means that its resistance decreases with increasing temperature. It will, therefore, maintain the cold junction at a constant temperature by balancing any change in resistance as the instrument temperature varies.

1·10 Recording and Controlling Pyrometer In most industrial installations, the instrument is required to do more than simply indicate temperature. The pointer of the potentiometer may be replaced by a pen that moves over a traveling chart to obtain a complete record of the temperature. This is called a *recording pyrometer.* The instrument, through the use of electric circuits, may also be used to control the flow of gas to the burners or electricity to the heating elements, and thereby maintain a constant predetermined furnace temperature. This is called a *controlling pyrometer.* It is possible to design the instrument to record and control the temperature from one or more thermocouples.

1·11 Radiation Pyrometer The basic principles of the operation of the radiation pyrometer involve a standard radiating source known as a *blackbody.* A blackbody is a hypothetical body that absorbs all the radiation that falls upon it. Such a body radiates energy at a higher rate than any other body at the same temperature. Radiation pyrometers are generally calibrated to indicate blackbody or true temperatures. The Stefan-Boltzmann law, which is the basis for the temperature scale of radiation pyrometers, shows that the rate of radiant energy from a blackbody is proportional to the fourth power of its absolute temperature.

$$W = KT^4 \tag{1·3}$$

where W = rate at which energy is emitted by a blackbody
K = proportionality constant
T = absolute temperature of blackbody

The apparent temperature measured from non-blackbody materials will always be lower than the true temperature. This is due to the emissivity of the material, which is defined as the ratio of the rate at which radiant energy is emitted from the non-blackbody material to the rate of that emitted from a blackbody at the same temperature. Hence

$$W = Ke_tT^4 = KT_a^4 \tag{1·4}$$

or

$$T_a^4 = e_tT^4 \tag{1·5}$$

where T_a = apparent absolute temperature of non-blackbody measured by pyrometer
e_t = total emissivity of non-blackbody

Therefore, knowing the total emissivity of the material, the indicated pyrometer temperature may be easily corrected to the true absolute temperature that would be read by the pyrometer under blackbody conditions.

Figure 1·7 shows a cross section of a mirror-type radiation pyrometer. Radiation from the target passes through window *A* to mirror *B* and is focused to form an image of the target in the plane of the internal diaphragm *J*. This image is then focused by mirror *D* upon a group of thermocouples called a *thermopile E*. By viewing hole *C* through lens *H* it can be determined whether the image of the target is sufficiently large to cover the hole and whether the pyrometer is properly aimed. The rise in temperature of the thermopile is approximately proportional to the rate at which radiant energy impinges on it, and the emf is therefore proportional to T^4. In actual practice, however, not all the radiant energy reaches the thermocouple since some will be absorbed by the atmosphere and optical parts of the instrument. Therefore, the Stefan-Boltzmann law is not followed very closely, and the relation between the temperature of the radiating source and the emf of the thermocouple may be expressed empirically as

$$E = KT^b \tag{1·6}$$

The constants K and b must be determined experimentally by calibration at two standardization points.

The radiation pyrometer does not require direct contact with the hot body, and therefore the upper temperature range is not limited by the ability of the pyrometer itself to withstand high temperatures. By using suitable stops in the optical system, there is no upper temperature limit. The lower temperature limit is approximately 1000°F.

1·12 Optical Pyrometer The instrument described in the previous section which

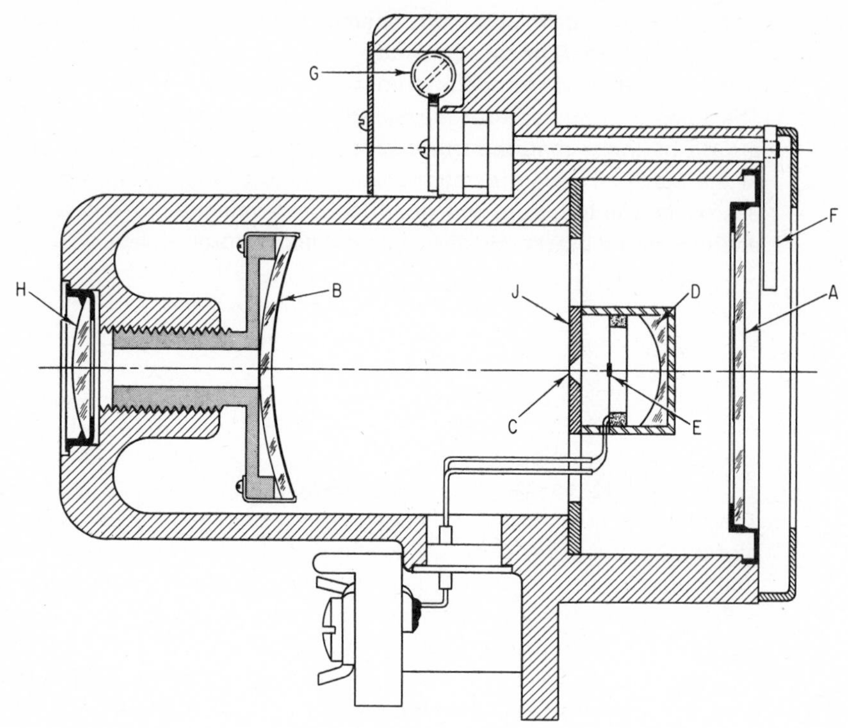

Fig. 1·7 A mirror-type radiation pyrometer. (Leeds & Northrup Company.)

responded to all wavelengths of radiation is known as a total-radiation pyrometer. While the general principles on which the optical pyrometer is based are the same as for the radiation pyrometer, they differ in that the optical pyrometer makes use of a single wavelength or a narrow band of wavelengths in the visible part of the spectrum. The optical pyrometer measures temperature by comparing the brightness of light emitted by the source with that of a standard source. To make the color comparison easier, a red filter is used which restricts the visible radiation to only the wavelength of red radiation.

The type most widely used in industry is the *disappearing-filament* type. This pyrometer consists of two parts, a telescope and a control box. The telescope (Fig. 1·8*a*) contains a red-glass filter mounted in front of the eyepiece and a lamp with a calibrated filament upon which the objective lens focuses an image of the body whose temperature is being measured. It also contains a switch for closing the electric circuit of the lamp, and an absorbing screen for changing the range of the pyrometer.

The control box contains the main parts of the measuring circuit shown in Fig. 1·8*b*. These include dry cells to provide the current to illuminate the lamp, a rheostat R to adjust filament current, and a potentiometer slide-wire, with associated standard cell and galvanometer, to measure the filament current accurately. This current is manually adjusted by rotating R_1 until the filament matches the brightness of the image of the object sighted upon and the filament seems to disappear (Fig. 1·8*c*). Accurate balance is then obtained by rotating P_1 until the galvanometer reads zero. A scale attached to the potentiometer contact P indicates the temperature directly.

The temperature range of the optical pyrometer described is from 1400 to about 2400°F. This upper limit is due partly to danger of deterioration of the filament at higher temperatures and partly to the dazzling effect on the eye of the brightness at elevated temperatures. The temperature range may be extended upward by use of an absorbing screen between the objective lens and the filament, thus permitting brightness matches to be secured at lower filament temperatures. The pyrometer can then be calibrated for the higher temperature range by using the lower filament temperatures. Thus, by using various absorbing screens, the upper limit of the optical pyrometer can be extended to 10,000°F or higher.

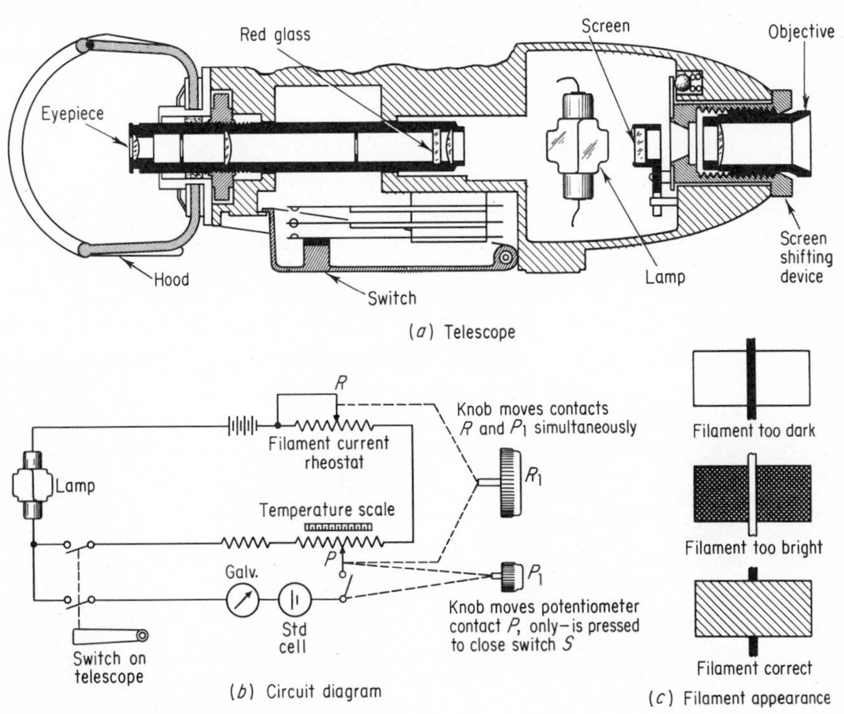

Fig. 1·8 The disappearing-filament type of optical pyrometer. (*a*) Telescope; (*b*) circuit diagram; (*c*) filament appearance. (Leeds & Northrup Company.)

Some advantages of the optical and radiation pyrometers are:

1 Measurement of high temperature.
2 Measurement of inaccessible bodies.
3 Measurement of moving or small bodies.
4 No part of the instrument is exposed to the destructive effects of heating.

The principal disadvantages are:

1 Errors introduced because the photometric match is a matter of individual judgment.
2 Errors introduced by smoke or gases between the observer and the source.
3 Uncertainty as to the amount of departure from blackbody conditions.

METALLOGRAPHY

1·13 Introduction Metallography or microscopy consists of the microscopic study of the structural characteristics of a metal or an alloy. The microscope is by far the most important tool of the metallurgist from both the scientific and technical standpoints. It is possible to determine grain size and the size, shape, and distribution of various phases and inclusions which have a great effect on the mechanical properties of the metal. The microstructure will reveal the mechanical and thermal treatment of the metal, and it may be possible to predict its expected behavior under a given set of conditions.

Experience has indicated that success in microscopic study depends largely upon the care taken in the preparation of the specimen. The most expensive microscope will not reveal the structure of a specimen that has been poorly prepared. The procedure to be followed in the preparation of

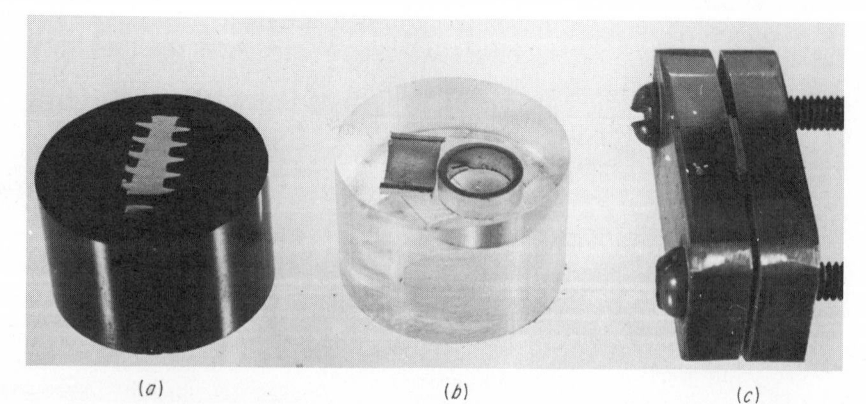

(*a*) (*b*) (*c*)

Fig. 1·9 (*a*) Specimen mounted in Bakelite, enlarged 2X. (*b*) Specimen mounted in Lucite, enlarged 2X. (*c*) Specimen held in metal clamp, enlarged 2X.

a specimen is comparatively simple and involves a technique which is developed only after constant practice. The ultimate objective is to produce a flat, scratch-free, mirrorlike surface. The steps required to prepare a metallographic specimen properly are covered in Secs. 1·14 to 1·19.

1·14 Sampling The choice of a sample for microscopic study may be very important. If a failure is to be investigated, the sample should be chosen as close as possible to the area of failure and should be compared with one taken from the normal section.

If the material is soft, such as nonferrous metals or alloys and non-heat-treated steels, the section may be obtained by manual hacksawing. If the material is hard, the section may be obtained by use of an abrasive cutoff wheel. This wheel is a thin disk of suitable cutting abrasive, rotating at high speed. The specimen should be kept cool during the cutting operation.

1·15 Rough Grinding Whenever possible, the specimen should be of a size that is convenient to handle. A soft sample may be made flat by slowly moving it up and back across the surface of a flat smooth file. The soft or hard specimen may be rough-ground on a belt sander, with the specimen kept cool by frequent dropping in water during the grinding operation. In all grinding and polishing operations the specimen should be moved perpendicular to the existing scratches. This will facilitate recognition of the stage when the deeper scratches have been replaced by shallower ones characteristic of the finer abrasive. The rough grinding is continued until the surface is flat and free of nicks, burrs, etc., and all scratches due to the hacksaw or cutoff wheel are no longer visible. (The surface after rough grinding is shown in Fig. 1·10*a*.)

1·16 Mounting Specimens that are small or awkwardly shaped should be mounted to facilitate intermediate and final polishing. Wires, small rods, sheet metal specimens, thin sections, etc., must be appropriately mounted in a suitable material or rigidly clamped in a mechanical mount.

Synthetic plastic materials applied in a special mounting press will yield mounts of a uniform convenient size (usually 1 in., 1.25 in., or 1.5 in. in diameter) for handling in subsequent polishing operations. These mounts, when properly made, are very resistant to attack by the etching reagents ordinarily used. The most common thermosetting resin for mounting is Bakelite, Fig. 1·9*a*. Bakelite molding powders are available in a variety of colors, which simplifies the identification of mounted specimens. The specimen and the correct amount of Bakelite powder, or a Bakelite preform, are placed in the cylinder of the mounting press. The temperature is gradually raised to 150°C, and a molding pressure of about 4,000 psi is applied simultaneously. Since Bakelite is set and cured when this temperature is reached, the specimen mount may be ejected from the molding die while it is still hot.

Lucite is the most common thermoplastic resin for mounting. Lucite is completely transparent when properly molded, as shown in Fig. 1·9*b*. This transparency is useful when it is necessary to observe the exact section that is being polished or when it is desirable for any other reason to see the entire specimen in the mount. Unlike the thermosetting plastics, the thermoplastic resins do not undergo curing at the molding temperature; rather they set on cooling. The specimen and a proper amount of Lucite powder are placed in the mounting press and are subjected to the same temperature and pressure as for Bakelite (150°C and 4,000 psi). After this temperature has been reached, the heating coil is removed, and cooling fins are placed around the cylinder to cool the mount to below 75°C in about 7 min while the molding pressure is maintained. Then the mount may be ejected from the mold. Ejecting the mount while still hot or allowing it to cool slowly in the molding cylinder to ordinary temperature before ejection will cause the mount to be opaque.

Small specimens may be conveniently mounted for metallographic preparation in a laboratory-made clamping device as shown in Fig. 1·9*c*. Thin sheet specimens, when mounted in such a clamping device, are usually alternated with metal "filler" sheets which have approximately the same hardness as the specimens. The use of filler sheets will preserve surface irregularities of the specimen and will prevent, to some extent, the edges of the specimen from becoming rounded during polishing.

1·17 Intermediate Polishing After mounting, the specimen is polished on a series of emery papers containing successively finer abrasives. The first paper is usually No. 1, then 1/0, 2/0, 3/0, and finally 4/0.

The surface after intermediate polishing on 4/0 paper is shown in Fig. 1·10*b*. The intermediate polishing operations using emery paper are usually done dry; however, in certain cases such as the preparation of soft materials, silicon carbide abrasive may be used. As compared to emery paper, silicon carbide has a greater removal rate and, as it is resin-bonded, can be used with a lubricant. Using a lubricant prevents overheating the sample, minimizes smearing of soft metals, and also provides a rinsing action to flush away surface removal products so the paper will not become clogged.

1·18 Fine Polishing The time consumed and the success of fine polishing depend largely upon the care that was exercised during the previous polishing steps. The final approximation to a flat scratch-free surface is obtained by use of a wet rotating wheel covered with a special cloth that is charged with carefully sized abrasive particles. A wide range of abrasives is available for final polishing. While many will do a satisfactory job, there appears to be a preference for the gamma form of aluminum oxide for ferrous and copper-based materials, and cerium oxide for aluminum, magnesium, and their alloys. Other final polishing abrasives often used are diamond paste, chromium oxide, and magnesium oxide.

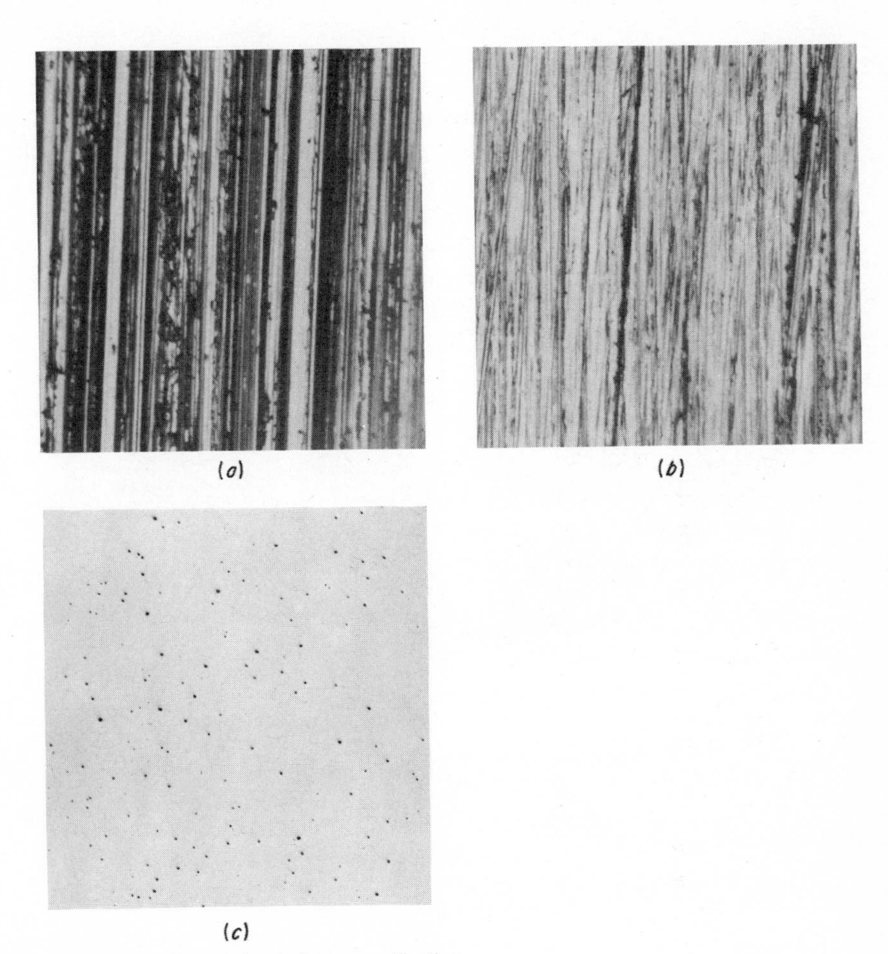

Fig. 1·10 (*a*) Surface after rough grinding, magnification 100X. (*b*) Surface after intermediate polishing on 4/0 paper, magnification 100X. (*c*) Scratch-free surface after final polishing, magnification 50X. Black spots are oxide impurities.

The choice of a proper polishing cloth depends upon the particular material being polished and the purpose of the metallographic study. Many cloths are available of varying nap or pile, from those having no pile, such as silk, to those of intermediate pile, such as broadcloth, billiard cloth, and canvas duck, and finally to a deep pile, such as velvet. Synthetic polishing cloths are also available for general polishing purposes, of which two, under the trade names of Gamal and Microcloth, are most widely used. A properly polished sample will show only the nonmetallic inclusions and will be scratchfree (Fig. 1·10*c*).

1·19 Etching The purpose of etching is to make visible the many structural characteristics of the metal or alloy. The process must be such that the various parts of the microstructure may be clearly differentiated. This is accomplished by use of an appropriate reagent which subjects the polished surface to chemical action.

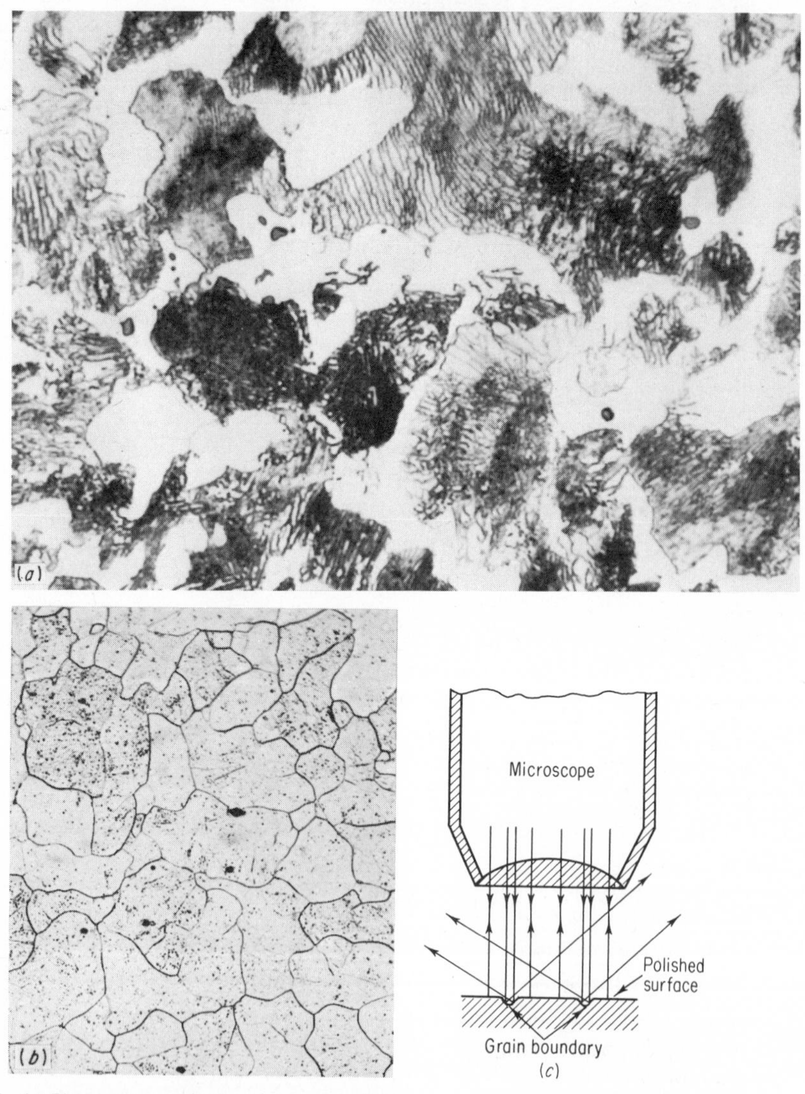

Fig. 1·11 (*a*) Photomicrograph of a mixture revealed by etching. (*b*) Photomicrograph of pure iron. (The International Nickel Company.) (*c*) Schematic illustration of the microscopic appearance of grain boundaries as dark lines.

In alloys composed of two or more phases, the components are revealed during etching by a preferential attack of one or more of these constituents by the reagent, because of difference in chemical composition of the phases (Fig. 1·11*a*). In uniform single-phase alloys or pure metals, contrast is obtained and grain boundaries are made visible because of differences in the rate at which various grains are attacked by the reagent (Fig. 1·11*b*). This difference in the rate of attack is mainly associated with the angle of the different grain sections to the plane of the polished surface. Because of chemical attack by the etching reagent, the grain boundaries will appear as valleys in the polished surface. Light from the microscope hitting the side of these valleys will be reflected out of the microscope, making the grain boundaries appear as dark lines. This is illustrated schematically in Fig. 1·11*c*.

The selection of the appropriate etching reagent is determined by the metal or alloy and the specific structure desired for viewing. Table 1·3 lists some of the common etching reagents.

1·20 Metallurgical Microscopes At this point it is appropriate to discuss briefly the principles of the metallurgical microscope. In comparison with a biological type, the metallurgical microscope differs in the manner by which the specimen is illuminated. Since a metallographic sample is opaque to light, the sample must be illuminated by reflected light. As shown in Fig. 1·12, a horizontal beam of light from some light source is reflected, by means of a plane-glass reflector, downward through the microscope objective onto the surface of the specimen. Some of this incident light reflected from the specimen surface will be magnified in passing through the lower lens system, the objective, and will continue upward through the plane-glass reflector and be magnified again by the upper lens system, the eyepiece. The initial magnifying power of the objective and the eyepiece is usually engraved on the lens mount. When a particular combination of objective and eyepiece is used at the proper tube length, the total magnification is equal to the product of the magnifications of the objective and the eyepiece. Figure 1·13*a* shows a table-type metallurgical microscope.

It is possible to mount a camera bellows above the eyepiece and use the table-type microscope for photomicrography. However, the bench-type metallograph illustrated in Fig. 1·13*b*, which is specifically designed for both visual examination and permanent recording of metallographic structures by photographic methods, will give superior photomicrographs.

The maximum magnification obtained with the optical microscope is about 2,000×. The principal limitation is the wavelength of visible light, which limits the resolution of fine detail in the metallographic specimen. The magnification may be extended somewhat by the use of shorter-wavelength radiation, such as ultraviolet radiation, but the sample preparation technique is more involved.

The greatest advance in resolving power was obtained by the electron

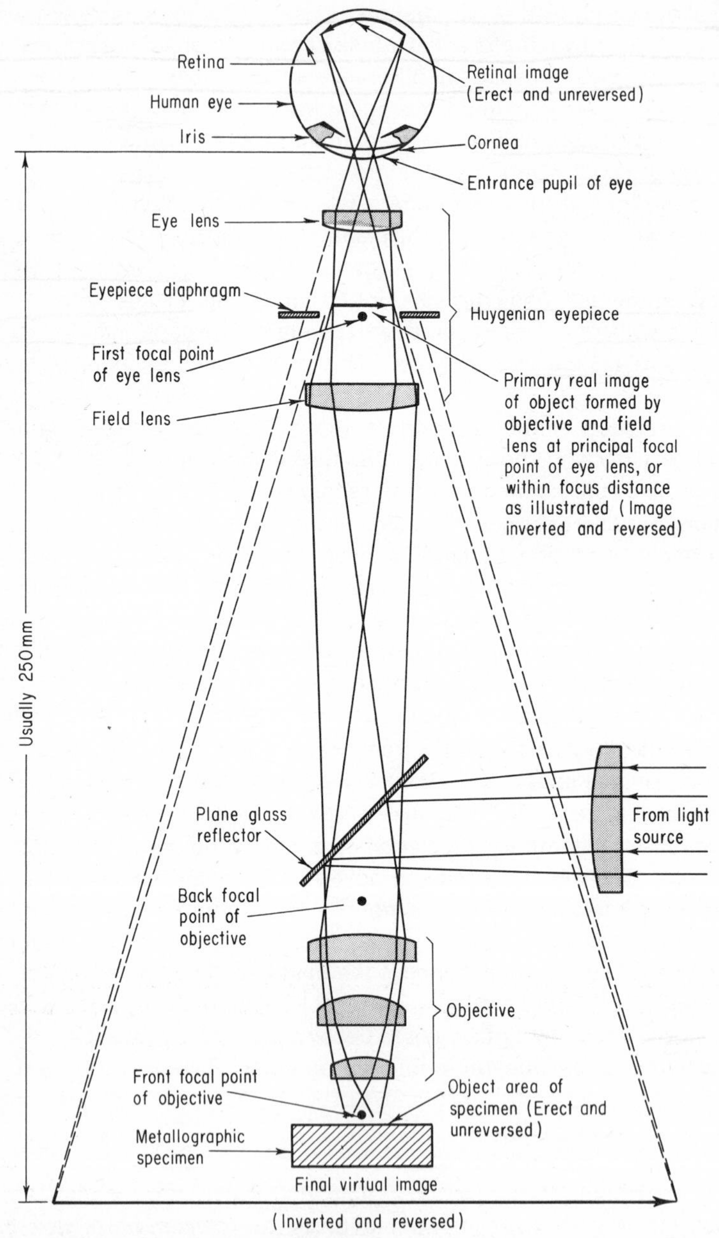

Fig. 1·12 Illustrating the principle of the metallurgical compound microscope and the trace of rays through the optical system from the object field to the final virtual image. (By permission from G. L. Kehl, "Principles of Metallographic Laboratory Practice," 3d ed., McGraw-Hill Book Company, New York, 1949.)

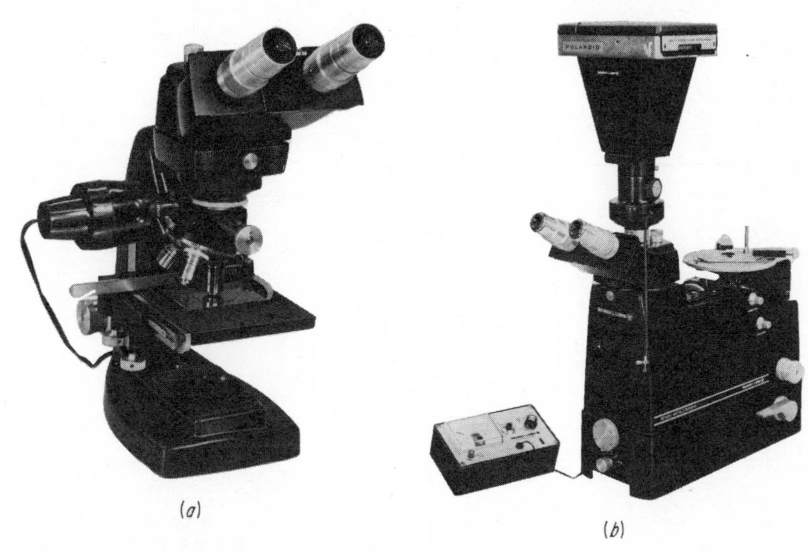

Fig. 1·13 (*a*) Metallurgical microscope. (*b*) Bench-type metallograph. (Bausch & Lomb, Inc.)

microscope. Under certain circumstances, high-velocity electrons behave like light of very short wavelength. The electron beam has associated with it a wavelength nearly 100,000 times smaller than the wavelength of visible light, thus increasing the resolving power tremendously. An electron microscope is shown in Fig. 1·14*a*.

Although in principle the electron microscope is similar to the light microscope (Fig. 1·14*b*), its appearance is very much different. It is much larger because of the highly regulated power supplies that are needed to produce and control the electron beam. The entire system must be kept pumped to a high vacuum since air would interfere with the motion of the electrons.

The lenses of the electron microscope are the powerful magnetic fields of the coils, and the image is brought into focus by changing the field strength of the coils while the coils remain in a fixed position. In the optical microscope the image is brought into focus by changing the lens spacing.

Since metallographic specimens are opaque to an electron beam, it is necessary to prepare, by special techniques, a thin replica of the surface to be studied. The specimen is polished and etched following normal metallographic practice. It is then placed on a hot plate with a small pellet of suitable plastic on the etched surface. As the temperature rises, the plastic begins to flow and pressure is applied to ensure intimate contact between the plastic and the surface. After cooling, the replica is carefully peeled off. To improve contrast, a thin coating of carbon or tungsten is evaporated onto the replica at an angle and from one side. Since the shadowed replica is fragile, it is supported on a disk of very fine copper-

TABLE 1·3 Etching Reagents for Microscopic Examination*

ETCHING REAGENT	COMPOSITION		USES	REMARKS
Nitric acid (nital)	White nitric acid Ethyl or methyl alcohol (95% or absolute) (also amyl alcohol)	1–5 ml 100 ml	In carbon steels: (1) to darken pearlite and give contrast between pearlite colonies, (2) to reveal ferrite boundaries, (3) to differentiate ferrite from martensite	Etching rate is increased, selectivity decreased, with increasing percentages of HNO_3. Reagent 2 (picric acid) usually superior Etching time a few seconds to 1 min
Picric acid (picral)	Picric acid Ethyl or methyl alcohol (95% or absolute)	4 g 100 ml	For all grades of carbon steels: annealed, normalized, quenched, and tempered, spheroidized, austempered. For all low-alloy steels attacked by this reagent	More dilute solutions occasionally useful. Does not reveal ferrite grain boundaries as readily as nital Etching time a few seconds to 1 min or more
Ferric chloride and hydrochloric acid	Ferric chloride Hydrochloric acid Water	5 g 50 ml 100 ml	Structure of austenitic nickel and stainless steels	
Ammonium hydroxide and hydrogen peroxide	Ammonium hydroxide Water Hydrogen peroxide	5 parts 5 parts 2–5 parts	Generally used for copper and many of its alloys	Peroxide content varies directly with copper content of alloy to be etched Immersion or swabbing for about 1 min. Fresh peroxide for good results
Ammonium persulfate	Ammonium persulfate Water	10 g 90 ml	Copper, brass, bronze, nickel silver, aluminum bronze	Use either cold or boiling; immersion
Palmerton reagent	Chromic oxide Sodium sulfate Water	200 g 15 g 1,000 ml	General reagent for zinc and its alloys	Immersion with gentle agitation
Ammonium molybdate	Molybdic acid (85%) Ammonium hydroxide (sp gr 0.9) Water Filter and add to nitric acid (sp gr 1.32)	100 g 140 ml 240 ml 60 ml	Rapid etch for lead and its alloys; very suitable for removing thick layer of worked metal	Alternately swab specimen and wash in running water
Hydrofluoric acid	Hydrofluoric acid (conc) H_2O	0.5 ml 99.5 ml	General microscopic for aluminum and its alloys	Swab with soft cotton for 15 s

* From "Metals Handbook," 1948 ed., American Society for Metals, Metals Park, Ohio.

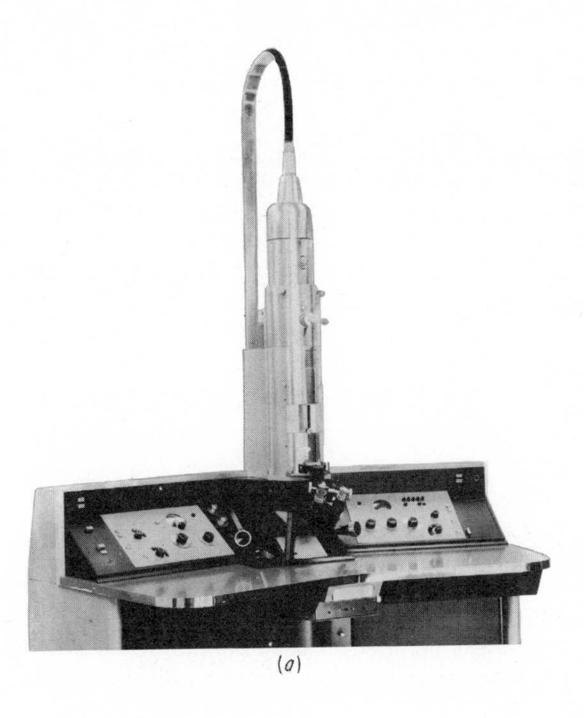

(a)

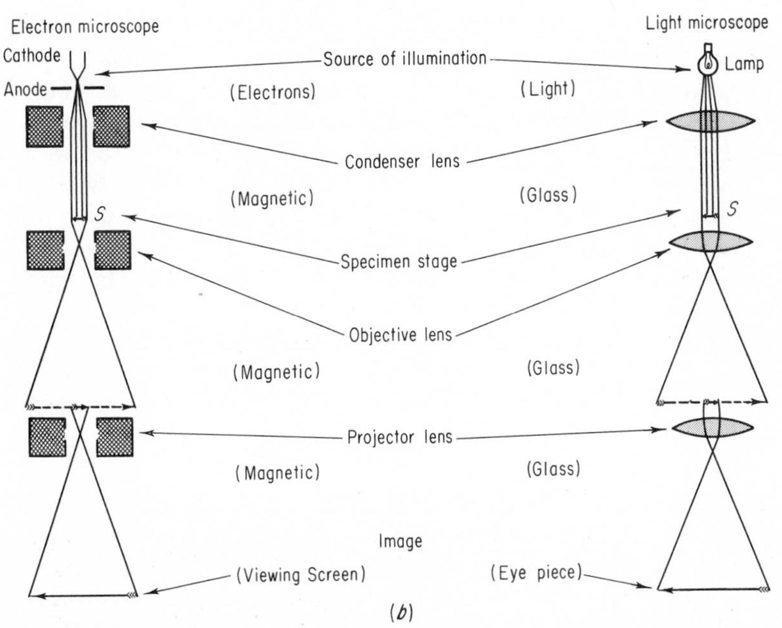

(b)

Fig. 1·14 (a) The electron microscope. (b) Similarity of light and electron microscopes. (Radio Corporation of America.)

wire mesh. The disk is then placed over the opening in the specimen holder, which is inserted in the column of the instrument.

The electrons emitted by a hot tungsten-filament cathode are accelerated, to form a high-velocity beam, by the anode. This beam is concentrated on the replica by the condensing lens. Depending upon the density and thickness of the replica at each point, some of the electrons are absorbed or scattered while the remainder pass through. The magnetic field of the objective lens focuses and enlarges the electron beam that has passed through the replica. Some of the electrons in this image are brought into a second focus on a fluorescent screen by the projector lens. The electron microscope shown in Fig. 1·14*a* has a basic magnification range of 1,400 to 32,000×, which may be extended to 200,000× with accessory lenses.

TESTS FOR MECHANICAL PROPERTIES

1·21 Hardness The property of "hardness" is difficult to define except in relation to the particular test used to determine its value. It should be observed that a hardness number or value cannot be utilized directly in design, as can a tensile strength value, since hardness numbers have no intrinsic significance.

Hardness is not a fundamental property of a material but is related to the elastic and plastic properties. The hardness value obtained in a particular test serves only as a comparison between materials or treatments. The test procedure and sample preparation are usually simple, and the results may be used in estimating other mechanical properties. Hardness testing is widely used for inspection and control. Heat treatment or working usually results in a change in hardness. When the hardness resulting from treating a given material by a given process is established, it affords a rapid and simple means of inspection and control for the particular material and process.

The various hardness tests may be divided into three categories:

Elastic hardness

Resistance to cutting or abrasion

Resistance to indentation

1·22 Elastic Hardness This type of hardness is measured by a scleroscope (Fig. 1·15), which is a device for measuring the height of rebound of a small diamond-tipped hammer after it falls by its own weight from a definite height onto the surface of the test piece. The instrument usually has a self-indicating dial so that the height of rebound is automatically indicated. When the hammer is raised to the starting position, it has a certain amount

Fig. 1·15 Scleroscope hardness tester. (The Shore Instrument & Manufacturing Company.)

of potential energy. When it is released, this energy is converted to kinetic energy until it strikes the surface of the test piece. Some of the energy is now absorbed in forming the impression, and the rest is returned to the hammer for its rebound. The height of rebound is indicated by a number on an arbitrary scale such that the higher the rebound, the larger the number and the harder the test piece. This test is really a measure of the resilience of a material, that is, the energy it can absorb in the elastic range.

1·23 Resistance to Cutting or Abrasion

Scratch Test This test was developed by Friedrich Mohs. The scale consists of 10 different standard minerals arranged in order of increasing hardness. Talc is No. 1, gypsum No. 2, etc., up to 9 for corundum, 10 for diamond. If an unknown material is scratched noticeably by No. 6 and not by No. 5, the hardness value is between 5 and 6. This test has never been used to any great extent in metallurgy but is still used in mineralogy. The primary disadvantage is that the hardness scale is nonuniform. When the hardness of the minerals is checked by another hardness-test method, it is found that the values are compressed between 1 and 9, and there is a large gap in hardness between 9 and 10.

File Test The test piece is subjected to the cutting action of a file of known hardness to determine whether a visible cut is produced. Comparative tests with a file depend upon the size, shape, and hardness of the file; the speed, pressure, and angle of filing during the test; and the composition and heat treatment of the material under test. The test is generally used industrially as one of acceptance or rejection.

In many cases, particularly with tool steels, when the steel is properly heat-treated it will be hard enough so that if a file is run across the surface it will not cut the surface. It is not unusual to find heat-treating specifications which simply say "heat-treat until the material is file-hard." By running a file across the surface an inspector may rapidly check a large number of heat-treated parts to determine whether the treatment has been successful.

1·24 Resistance to Indentation This test is usually performed by impressing into the specimen, which is resting on a rigid platform, an indenter of fixed and known geometry, under a known static load applied either directly or by means of a lever system. Depending on the type of test, the hardness is expressed by a number that is either inversely proportional to the depth of indentation for a specified load and indenter, or proportional to a mean load over the area of indentation. The common methods of making indentation hardness tests are described below.

Brinell Hardness Test The Brinell hardness tester usually consists of a hand-operated vertical hydraulic press, designed to force a ball indenter into the test specimen (Fig. 1·16*a*). Standard procedure requires that the test be made with a ball of 10 mm diameter under a load of 3,000 kg for ferrous metals, or 500 kg for nonferrous metals. For ferrous metals the loaded ball is pressed into the test specimen for at least 10 s; for nonferrous metals the time is 30 s. The diameter of the impression produced is measured by means of a microscope containing an ocular scale, usually graduated in tenths of a millimeter, permitting estimates to the nearest 0.05 mm.

The Brinell hardness number (HB) is the ratio of the load in kilograms to the impressed area in square millimeters, and is calculated from the following formula:

$$\mathrm{HB} = \frac{L}{(\pi D/2)(D - \sqrt{D^2 - d^2})} \tag{1·7}$$

where L = test load, kg

D = diameter of ball, mm

d = diameter of impression, mm

Calculation is usually unnecessary because tables are available for convert-

(a) (b)

Fig. 1·16 (a) Brinell hardness tester. (Ametek/Testing Equipment Systems, East Moline, Ill.) (b) Rockwell hardness tester. (Wilson Mechanical Instrument Division, American Chain & Cable Company.)

ing the observed diameter of impression to the Brinell hardness number (see Table 1·4).

The Brinell hardness number followed by the symbol HB without any suffix numbers denotes standard test conditions using a ball of 10 mm diameter and a load of 3,000 kg applied for 10 to 15 s. For other conditions, the hardness number and symbol HB are supplemented by numbers indicating the test conditions in the following order: diameter of ball, load, and duration of loading. For example, 75 HB 10/500/30 indicates a Brinell hardness of 75 measured with a ball of 10 mm diameter and a load of 500 kg applied for 30 s.

The Brinell hardness number using the standard ball is limited to approximately 500 HB. As the material tested becomes harder, there is a tendency for the indenter itself to start deforming, and the readings will not be accurate. The upper limit of the scale may be extended by using a tungsten carbide ball rather than a hardened steel ball. In that case, it is possible to go to approximately 650 HB.

TABLE 1·4 Approximate Hardness Relations for Steel*

BRINELL, 3,000 KG			VICKERS DIAMOND PYRAMID	ROCKWELL, USING BRALE				SCLEROSCOPE	MOHS	TENSILE STRENGTH, 1,000 PSI
DIAMETER, MM	STANDARD BALL	TUNGSTEN CARBIDE BALL		C 150 KG	D 100 KG	A 60 KG	SUPERFICIAL 30 N			
2.35	. . .	682	737	61.7	72.0	82.2	79.0	84		
2.40	. . .	653	697	60.0	70.7	81.2	77.5	81		
2.45	. . .	627	667	58.7	69.7	80.5	76.3	79	8.0	323
2.50	. . .	601	640	57.3	68.7	79.8	75.1	77	. . .	309
2.55	. . .	578	615	56.0	67.7	79.1	73.9	75	. . .	297
2.60	. . .	555	591	54.7	66.7	78.4	72.7	73	7.5	285
2.65	. . .	534	569	53.5	65.8	77.8	71.6	71	. . .	274
2.70	. . .	514	547	52.1	64.7	76.9	70.3	70	. . .	263
2.75	495	. . .	539	51.6	64.3	76.7	69.9	. .	. . .	259
	. . .	495	528	51.0	63.8	76.3	69.4	68	. . .	253
2.80	477	. . .	516	50.3	63.2	75.9	68.7	. .	. . .	247
	. . .	477	508	49.6	62.7	75.6	68.2	66	. . .	243
2.85	461	. . .	495	48.8	61.9	75.1	67.4	. .	. . .	237
	. . .	461	491	48.5	61.7	74.9	67.2	65	. . .	235
2.90	444	. . .	474	47.2	61.0	74.3	66.0	. .	7.0	226
	. . .	444	472	47.1	60.8	74.2	65.8	63	. . .	225
2.95	429	429	455	45.7	59.7	73.4	64.6	61	. . .	217
3.00	415	415	440	44.5	58.8	72.8	63.5	59	. . .	210
3.05	401	401	425	43.1	57.8	72.0	62.3	58	. . .	202
3.10	388	388	410	41.8	56.8	71.4	61.1	56	. . .	195
3.15	375	375	396	40.4	55.7	70.6	59.9	54	6.5	188
3.20	363	363	383	39.1	54.6	70.0	58.7	52	. . .	182
3.25	352	352	372	37.9	53.8	69.3	57.6	51	. . .	176
3.30	341	341	360	36.6	52.8	68.7	56.4	50	. . .	170
3.35	331	331	350	35.5	51.9	68.1	55.4	48	. . .	166
3.40	321	321	339	34.3	51.0	67.5	54.3	47	. . .	160
3.45	311	311	328	33.1	50.0	66.9	53.3	46	. . .	155
3.50	302	302	319	32.1	49.3	66.3	52.2	45	6.0	150
3.55	293	293	309	30.9	48.3	65.7	51.2	43	. . .	145
3.60	285	285	301	29.9	47.6	65.3	50.3	42	. . .	141
3.65	277	277	292	28.8	46.7	64.6	49.3	41	. . .	137
3.70	269	269	284	27.6	45.9	64.1	48.3	40	. . .	133
3.75	262	262	276	26.6	45.0	63.6	47.3	39	. . .	129
3.80	255	255	269	25.4	44.2	63.0	46.2	38	. . .	126
3.85	248	248	261	24.2	43.2	62.5	45.1	37	5.5	122
3.90	241	241	253	22.8	42.0	61.8	43.9	36	. . .	118
3.95	235	235	247	21.7	41.4	61.4	42.9	35	. . .	115
4.00	229	229	241	20.5	40.5	60.8	41.9	34	. . .	111

* Adapted from H. E. Davis, G. E. Troxell, and C. T. Wiskocil, "The Testing and Inspection of Engineering Materials," 2d ed., McGraw-Hill Book Company, New York, 1955; based on "Metals Handbook," 1948 ed., American Society for Metals, Metals Park, Ohio. See ASTM E 140 for additional relations.

TABLE 1·4 Approximate Hardness Relations for Steel *(Continued).*

BRINELL, 3,000 KG		ROCKWELL						SCLEROSCOPE	MOHS	TENSILE STRENGTH, 1,000 PSI
		BRALE				SUPERFICIAL				
DIAMETER, MM	STANDARD BALL	D 100 KG	A 60 KG	B 100 KG 1/16-IN. BALL	E 100 KG 1/8-IN. BALL	30 N BRALE	30 T 1/16-IN. BALL			
4.05	223	40	60	97	. . .	41	80.5	33	. . .	108
4.10	217	39	60	96	. . .	40	80.0	32	. . .	105
4.15	212	38	59	95	. . .	39	79.0	31	. . .	102
4.20	207	37	59	94	. . .	38	78.5	31	. . .	100
4.25	202	37	58	93	110	37	78.0	30	. . .	98
4.30	197	36	58	92	110	36	77.5	29	. . .	96
4.35	192	35	57	91	109	35	77.0	28	5.0	94
4.40	187	34	57	90	109	34	76.0	28	. . .	92
4.45	183	34	56	89	109	33	75.5	27	. . .	90
4.50	179	33	56	88	108	32	75.0	27	. . .	88
4.55	174	33	55	87	108	31	74.5	26	. . .	86
4.60	170	32	55	86	107	30	74.0	26	. . .	84
4.65	166	32	54	85	107	30	73.5	25	. . .	82
4.70	163	31	53	84	106	29	73.0	25	. . .	81
4.75	159	31	53	83	106	28	72.8	24	. . .	79
4.80	156	30	52	82	105	27	71.5	24	. . .	77
4.85	153	. .	. .	81	105	. .	71.0	23	. . .	76
4.90	149	. .	. .	80	104	. .	70.0	23	4.5	75
4.95	146	. .	. .	79	104	. .	69.5	22	. . .	74
5.00	143	. .	. .	78	103	. .	69.0	22	. . .	72
5.05	140	. .	. .	76	103	. .	68.0	21	. . .	71
5.10	137	. .	. .	75	102	. .	67.0	21	. . .	70
5.15	134	. .	. .	74	102	. .	66.0	21	. . .	68
5.20	131	. .	. .	73	101	. .	65.0	20	. . .	66
5.25	128	. .	. .	71	100	. .	64.0	. .	. . .	65
5.30	126	. .	. .	70	100	. .	63.5	. .	. . .	64
5.35	124	. .	. .	69	99	. .	62.5	. .	. . .	63
5.40	121	. .	. .	68	98	. .	62	. .	. . .	62
5.45	118	. .	. .	67	97	. .	61	. .	. . .	61
5.50	116	. .	. .	65	96	. .	60	. .	. . .	60
5.55	114	. .	. .	64	95	. .	59	. .	. . .	59
5.60	112	. .	. .	63	95	. .	58	. .	. . .	58
5.65	109	. .	. .	62	94	. .	58	. .	. . .	56
5.70	107	. .	. .	60	93	. .	57	. .	. . .	55
5.75	105	. .	. .	58	92	. .	55	. .	. . .	54
5.80	103	. .	. .	57	91	. .	54	. .	. . .	53

Rockwell Hardness Test This hardness test uses a direct-reading instrument based on the principle of differential depth measurement (Fig. 1·16*b*). The test is carried out by slowly raising the specimen against the indenter until a fixed minor load has been applied. This is indicated on the dial gauge. Then the major load is applied through a loaded lever system. After the dial pointer comes to rest, the major load is removed and, with the minor load still acting, the Rockwell hardness number is read on the dial gauge. Since the order of the numbers is reversed on the dial gauge, a shallow impression on a hard material will result in a high number while a deep impression on a soft material will result in a low number.

There are two Rockwell machines, the normal tester for relatively thick sections, and the superficial tester for thin sections. The minor load is 10 kg on the normal tester and 3 kg on the superficial tester.

A variety of indenters and loads may be used, and each combination determines a particular Rockwell scale. Indenters include hard steel balls 1/16, 1/8, 1/4, and 1/2 in. in diameter and a 120° conical diamond (brale) point.

TABLE 1·5 The Rockwell Hardness Scales*

SCALE	MAJOR LOAD, KG	TYPE OF INDENTER	TYPICAL MATERIALS TESTED
A	60	Diamond cone	Extremely hard materials, tungsten carbides, etc.
B	100	1/16" ball	Medium hard materials, low- and medium-carbon steels, brass, bronze, etc.
C	150	Diamond cone	Hardened steels, hardened and tempered alloys
D	100	Diamond cone	Case-hardened steel
E	100	1/8" ball	Cast iron, aluminum and magnesium alloys
F	60	1/16" ball	Annealed brass and copper
G	150	1/16" ball	Beryllium copper, phosphor bronze, etc.
H	60	1/8" ball	Aluminum sheet
K	150	1/8" ball	Cast iron, aluminum alloys
L	60	1/4" ball	Plastics and soft metals such as lead
M	100	1/4" ball	Same as L scale
P	150	1/4" ball	Same as L scale
R	60	1/2" ball	Same as L scale
S	100	1/2" ball	Same as L scale
V	150	1/2" ball	Same as L scale

* Ametek Testing Equipment Systems, East Moline, Ill.

Major loads are usually 60, 100, and 150 kg on the normal tester and 15, 30, and 45 kg on the superficial tester.

The most commonly used Rockwell scales are the B ($^1/_{16}$-in. ball indenter and 100-kg load) and the C (diamond indenter and 150-kg load), both obtained with the normal tester. Because of the many Rockwell scales, the hardness number must be specified by using the symbol HR followed by the letter designating the scale and preceded by the hardness numbers. For example, 82 HRB means a Rockwell hardness of 82 measured on the B scale ($^1/_{16}$-in. ball and 100-kg load). The Rockwell hardness scales and some typical applications are given in Table 1·5.

The performance of the machine should be checked frequently with standard test blocks supplied by the manufacturer. The operating crank should be returned gently to its starting position; snapping the crank to remove the major load may cause an error of several points in the dial indication. Care must be taken to seat the anvil and indenter firmly. Any vertical movement at these points results in additional depth being registered on the gauge and, therefore, a false hardness reading.

Vickers Hardness Test In this test, the instrument uses a square-based diamond-pyramid indenter with an included angle of 136° between opposite faces (see Fig. 1·17). The load range is usually between 1 and 120 kg. The Vickers hardness tester operates on the same basic principle as the Brinell tester, the numbers being expressed in terms of load and area of the impression. As a result of the indenter's shape, the impression on the surface of the specimen will be a square. The length of the diagonal of the square is measured through a microscope fitted with an ocular micrometer that contains movable knife-edges. The distance between knife-edges is indicated on a counter calibrated in thousandths of a millimeter. Tables

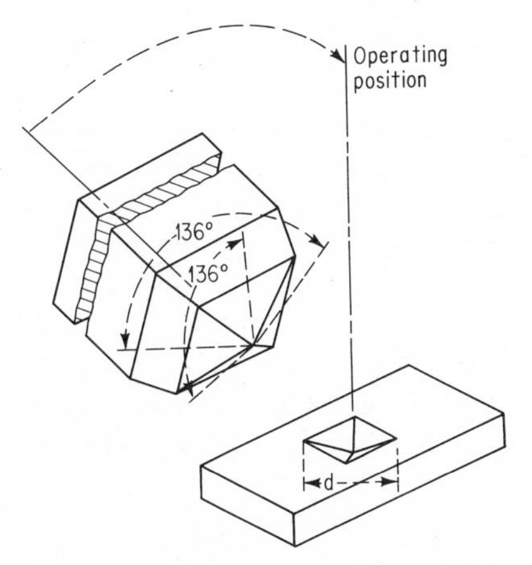

Fig. 1·17 The Vickers diamond-pyramid indenter.

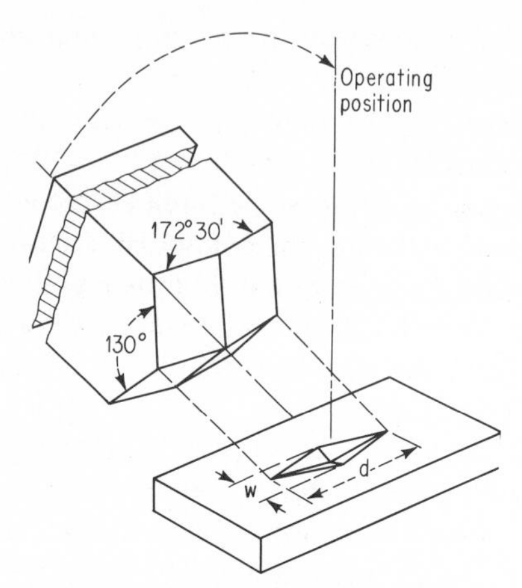

Fig. 1·18 The Knoop diamond-pyramid indenter.

are usually available to convert the measured diagonal to Vickers pyramid hardness number (HV), or the following formula may be used:

$$HV = \frac{1.854L}{d^2} \tag{1·8}$$

where L = applied load, kg

d = diagonal length of square impression, mm

As a result of the latitude in applied loads, the Vickers tester is applicable to measuring the hardness of very thin sheets as well as heavy sections.

Microhardness Test This term, unfortunately, is misleading, as it could refer to the testing of small hardness values when it actually means the use of small indentations. Test loads are between 1 and 1,000 gm. Two types of indenters are used for microhardness testing: the 136° square-base Vickers diamond pyramid described previously, and the elongated Knoop diamond indenter.

The Knoop indenter (Fig. 1·18) is ground to pyramidal form that produces a diamond-shaped indentation having long and short diagonals of approximate ratio of 7:1. The pyramid shape employed has included longitudinal angles of 172°30′ and transverse angles of 130°. The depth of indentation is about 1/30 of its length. As in the Vickers test, the long diagonal of the impression is measured optically with a filar micrometer eyepiece. The Knoop hardness number is the load divided by the area of the impression. Tables are usually available to convert the measured diagonal to Knoop hardness number (HK), or the following formula may be used:

$$HK = \frac{14.229L}{d^2} \tag{1·9}$$

where L = applied load, kg

d = length of long diagonal, mm

The Tukon microhardness tester is shown in Fig. 1·19. Some typical applications of indentation hardness tests are given in Table 1·6.

1·25 Accuracy of Any Indentation Hardness Test Some of the factors that influence the accuracy of any indentation hardness test are:

Condition of the Indenter Flattening of a steel-ball indenter will result in errors in the hardness number. The ball should be checked frequently

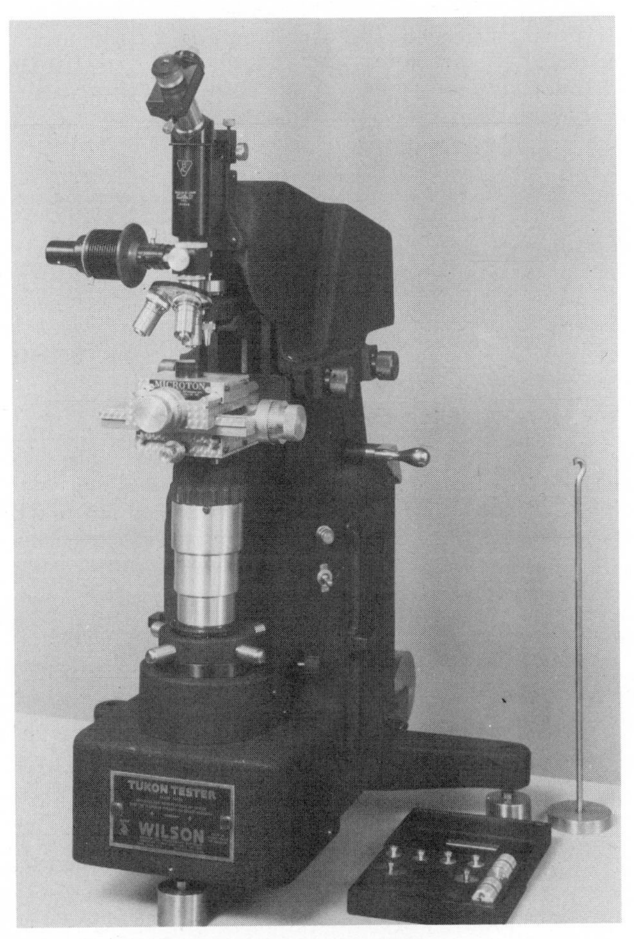

Fig. 1·19 The Tukon microhardness tester. (Wilson Mechanical Instrument Division, American Chain & Cable Company.)

TABLE 1·6 Typical Applications of Indentation Hardness Tests*

BRINELL	ROCKWELL	ROCKWELL SUPERFICIAL	VICKERS	MICROHARDNESS
Structural steel and other rolled sections Most castings, including steel, cast iron, and aluminum Most forgings	Finished parts, such as bearings, bearing races, valves, nuts, bolts, gears, pulleys, rolls, pins, pivots, stops, etc. Cutting tools, such as saws, knives, chisels, scissors Forming tools Small castings and forgings Sheet metal Large-diameter wire Electrical contacts Plastic sheet or parts Case-hardened parts Cemented carbides	Same as standard Rockwell except where shallower penetration is necessary, as in: Thin case-hardened parts, to .010 in. Thin materials down to .006 in. Cemented carbides Powdered metals	Same as Rockwell and Rockwell Superficial except where higher accuracy or shallower penetration is necessary, as in: Thin case-hardened parts, .005 to .010 in. Thin materials down to .005 in. Highly finished parts to avoid a removal operation Thin sections, such as tubing Weak structures Plating thickness	Plated surfaces Coatings, such as lacquer, varnish, or paint Foils and very thin materials down to .0001 in. To establish case gradients Bimetals and laminated materials Very small parts or areas, such as watch gears, cutting tool edges, thread crests, pivot points, etc. Very brittle or fragile materials (Knoop indenter), such as silicon, germanium, glass, tooth enamel Opaque, clear, or translucent materials Powdered metals To investigate individual constituents of a material To determine grain or grain boundary hardness

* Ametek/Testing Equipment Systems, East Moline, Ill.

for permanent deformation and discarded when such deformation occurs. Diamond indenters should be checked for any sign of chipping.

Accuracy of Load Applied The tester should apply loads in the stated range with negligible error. Loads greater than the recommended amount should not be used for accurate testing.

Impact Loading Besides causing inaccurate hardness readings, impact loading may damage diamond indenters. The use of a controlled oil dashpot will ensure smooth, steady operation of the loading mechanism.

Surface Condition of the Specimen The surface of the specimen on which the hardness reading is to be taken should be flat and representative of sound material. Any pits, scale, or grease should be removed by grinding or polishing.

Thickness of Specimen The specimen should be thick enough so that no bulge appears on the surface opposite that of the impression. The recommended thickness of the specimen is at least ten times the depth of the impression.

Shape of the Specimen The greatest accuracy is obtained when the test surface is flat and normal to the vertical axis of the indenter. A long specimen should be properly supported so that it does not tip. A flat surface should be prepared, if possible, on a cylindrical-shaped specimen, and a V-notch anvil should be used to support the specimen unless parallel flats are ground, in which case a flat anvil may be used. If a Rockwell hardness test is made on a round specimen less than 1 in. in diameter without grinding a flat, the observed reading must be adjusted by an appropriate correction factor (Table 1·7).

Location of Impressions Impressions should be at least $2\frac{1}{2}$ diameters from the edge of the specimen and should be at least 5 diameters apart for ball tests.

Uniformity of Material If there are structural and chemical variations in the material, the larger the impression area the more accurate the average-hardness reading. It is necessary to take many readings if the impression area is small to obtain a true average hardness for the material.

1·26 Advantages and Disadvantages of Different Types of Tests The selection of a hardness test is usually determined by ease of performance and degree of accuracy desired. Since the Brinell test leaves a relatively large impression, it is limited to heavy sections.

This is an advantage, however, when the material tested is not homogeneous. The surface of the test piece when running a Brinell test does not have to be so smooth as that for smaller impressions; however, using a microscope to measure the diameter of the impression is not so convenient as reading a dial gauge. Because of deformation of the steel ball, the Brinell test is generally inaccurate above 500 HB. The range may be extended to about 650 HB with a tungsten carbide ball.

The Rockwell test is rapid and simple in operation. Since the loads and

TABLE 1·7 Wilson Cylindrical Correction Chart*

Cylindrical work corrections (approximate only) to be added to observed Rockwell number

DIAMOND BRALE INDENTER							
C, D, A SCALES	DIAMETER OF SPECIMEN, IN.						
	1/4	3/8	1/2	5/8	3/4	7/8	1
80	0.5	0.5	0.5	0	0	0	0
70	1.0	1.0	0.5	0.5	0.5	0	0
60	1.5	1.0	1.0	0.5	0.5	0.5	0.5
50	2.5	2.0	1.5	1.0	1.0	0.5	0.5
40	3.5	2.5	2.0	1.5	1.0	1.0	1.0
30	5.0	3.5	2.5	2.0	1.5	1.5	1.0
20	6.0	4.5	3.5	2.5	2.0	1.5	1.5
1/16-IN. BALL INDENTER							
B, F, G SCALES	DIAMETER OF SPECIMEN, IN.						
	1/4	3/8	1/2	5/8	3/4	7/8	1
100	3.5	2.5	1.5	1.5	1.0	1.0	0.5
90	4.0	3.0	2.0	1.5	1.5	1.5	1.0
80	5.0	3.5	2.5	2.0	1.5	1.5	1.5
70	6.0	4.0	3.0	2.5	2.0	2.0	1.5
60	7.0	5.0	3.5	3.0	2.5	2.0	2.0
50	8.0	5.5	4.0	3.5	3.0	2.5	2.0
40	9.0	6.0	4.5	4.0	3.0	2.5	2.5

* Courtesy of Wilson Mechanical Instrument Division, American Chain & Cable Co.

indenters are smaller than those used in the Brinell test, the Rockwell test may be used on thinner specimens, and the hardest as well as the softest materials can be tested.

The Vickers tester is the most sensitive of the production hardness testers. It has a single continuous scale for all materials, and the hardness number is virtually independent of load. Because of the possibility of using light loads, it can test thinner sections than any other production test, and the square indentation is the easiest to measure accurately.

The microhardness test is basically a laboratory test. The use of very light loads permits testing of very small parts and very thin sections. It can be used to determine the hardness of individual constituents of the microstructure. Since the smaller the indentation, the better the surface finish must be, a great deal more care is required to prepare the surface for microhardness testing. The surface is usually prepared by the technique of metallographic polishing described in Secs. 1·14 to 1·19.

The principal advantages of the scleroscope are the small impressions that remain, the rapidity of testing, and portability of the instrument. However, results tend to be inaccurate unless proper precaustions are taken.

The tube must be perpendicular to the test piece, thin pieces must be properly supported and clamped, the surface to be tested must be smoother than for most other testing methods, and the diamond tip should not be chipped or cracked.

1·27 Hardness Conversion The approximate hardness conversion between the various hardness-test machines is shown in Table 1·4. These data are generally applicable to steel and have been derived by extensive hardness tests on carbon and alloy steels, mainly in the heat-treated condition.

1·28 Stress and Strain When an external force is applied to a body which tends to change its size or shape, the body resists this external force. The internal resistance of the body is known as *stress* and the accompanying changes in dimensions of the body are called *deformations* or *strains*. The total stress is the total internal resistance acting on a section of the body. The quantity usually determined is the intensity of stress or unit stress, which is defined as the stress per unit area. The unit stress is usually expressed in units of pounds per square inch (psi), and for an axial tensile or compressive load it is calculated as the load per unit area.

The total deformation or total strain in any direction is the total change of a dimension of the body in that direction, and the unit deformation or unit strain is the deformation or strain per unit of length in that direction.

1·29 The Tensile Test Next to the hardness test, the tensile test is most frequently performed to determine certain mechanical properties. A specifically prepared sample is placed in the heads of the testing machine and an

Fig. 1·20 A tensile sample with an extensometer attached.

axial load is placed on the sample through a hydraulic or mechanical lever loading system. The force is indicated on a calibrated dial. If the original cross-sectional area of the specimen is known, the stress developed at any load may be calculated. The deformation or strain is measured at a fixed length, usually 2 in., by a dial gauge called an *extensometer* (see Fig. 1·20). The unit strain may then be determined by dividing the measured elongation by the gauge length used. In some cases, an electrical strain gauge may be used to measure the total strain.

The relation between unit stress s and unit strain ϵ, found experimentally, is represented by the stress-strain graph in Fig. 1·21 for a ductile material and by the graph in Fig. 1·22 for a brittle material.

1·30 Tensile Properties The properties which may be determined by a tension test follow.

Proportional Limit It is found for many structural materials that the early part of the stress-strain graph may be approximated by a straight line OP in Figs. 1·21 and 1·22. In this range, the stress and strain are proportional to each other, so that any increase in stress will result in a proportionate increase in strain. The stress at the limit of proportionality point P is known as the *proportional limit*.

Elastic Limit If a small load on the test piece is removed, the extensometer needle will return to zero, indicating that the strain, caused by the load, is elastic. If the load is continually increased, then released after each increment and the extensometer checked, a point will be reached at which the extensometer needle will not return to zero. This indicates that

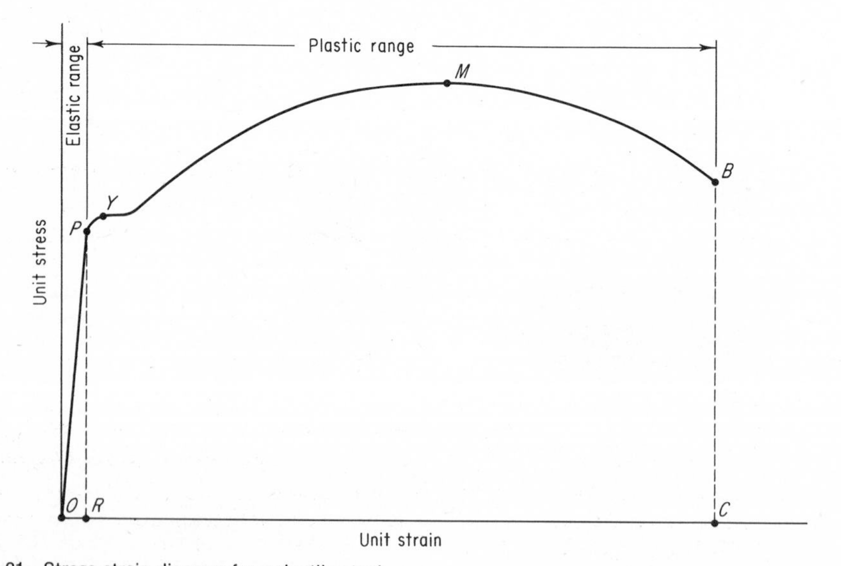

Fig. 1·21 Stress-strain diagram for a ductile steel.

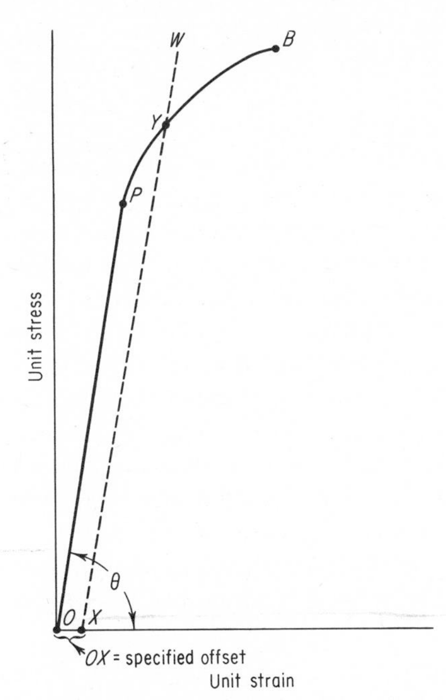

Fig. 1·22 Stress-strain diagram for a brittle material.

the material now has a permanent deformation. The elastic limit may therefore be defined as the minimum stress at which permanent deformation first occurs. For most structural materials the elastic limit has nearly the same numerical value as the proportional limit.

Yield Point As the load in the test piece is increased beyond the elastic limit, a stress is reached at which the material continues to deform without an increase in load. The stress at point *Y* in Fig. 1·21 is known as the *yield point.* This phenomenon occurs only in certain ductile materials. The stress may actually decrease momentarily, resulting in an upper and lower yield point. Since the yield point is relatively easy to determine and the permanent deformation is small up to yield point, it is a very important value in the design of many machine members whose usefulness will be impaired by considerable permanent deformation. This is true only for materials that exhibit a well-defined yield point.

Yield Strength Most nonferrous materials and the high-strength steels do not possess a well-defined yield point. For these materials, the maximum useful strength is the yield strength. The yield strength is the stress at which a material exhibits a specified limiting deviation from the proportionality of stress to strain. This value is usually determined by the "offset method." In Fig. 1·22, the specified offset *OX* is laid off along the strain axis. Then *XW* is drawn parallel to *OP*, and thus *Y*, the intersection of *XW* with the stress-strain diagram, is located. The value of the stress at point *Y* gives the yield strength. The value of the offset is generally between 0.10 and 0.20 percent of the gauge length.

Ultimate Strength As the load on the test piece is increased still further, the stress and strain increase, as indicated by the portion of the curve *YM* (Fig. 1·21) for a ductile material, until the maximum stress is reached at point *M*. The ultimate strength or the tensile strength is therefore the maximum stress developed by the material based on the original cross-sectional area. A brittle material breaks when stressed to the ultimate strength (point *B* in Fig. 1·22), whereas a ductile material will continue to stretch.

Breaking Strength For a ductile material, up to the ultimate strength, the deformation is uniform along the length of the bar. At the maximum stress, localized deformation or necking occurs in the specimen, and the load falls off as the area decreases. This necking elongation is a nonuniform deformation and occurs rapidly to the point of failure (Fig. 1·23). The breaking strength (point *B*, Fig. 1·21), which is determined by dividing the breaking load by the original cross-sectional area, is always less than the ultimate

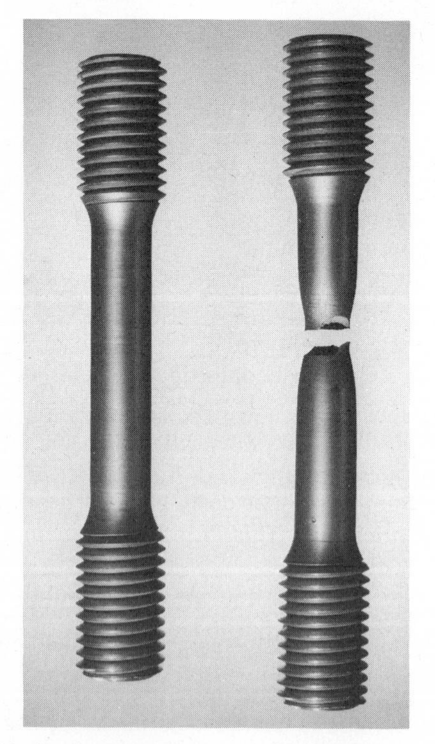

Fig. 1·23 Tension sample before and after failure.

strength. For a brittle material, the ultimate strength and breaking strength coincide.

Ductility The ductility of a material is indicated by the amount of deformation that is possible until fracture. This is determined in a tension test by two measurements:

Elongation This is determined by fitting together, after fracture, the parts of the specimen and measuring the distance between the original gauge marks.

$$\text{Elongation (percent)} = \frac{L_f - L_0}{L_0} \times 100 \tag{1·10}$$

where L_f = final gauge length

L_0 = original gauge length, usually 2 in.

In reporting percent elongation, the original gauge length must be specified since the percent elongation will vary with gauge length.

Reduction in Area This is also determined from the broken halves of the tensile specimen by measuring the minimum cross-sectional area and using the following formula:

$$\text{Reduction in area (percent)} = \frac{A_0 - A_f}{A_0} \times 100 \tag{1·11}$$

where A_0 = original cross-sectional area

A_f = final cross-sectional area

Modulus of Elasticity, or Young's Modulus Consider the straight-line portion of the stress-strain curve. The equation of a straight line is $y = mx + b$, where y is the vertical axis, in this case *stress*, and x is the horizontal axis, in this case *strain*. The intercept of the line on the y axis is b, and in this case it is zero since the line goes through the origin. The slope of the line is m. When the equation is solved for m, the slope is equal to y/x. Therefore, the slope of the line may be determined by drawing any right triangle and finding the tangent of the angle θ (Fig. 1·22), which is equal to y/x or *stress/strain*. The slope is really the constant of proportionality between stress and strain below the proportional limit and is known as the *modulus of elasticity* or *Young's modulus*.

The modulus of elasticity, which is an indication of the stiffness of a material, is measured in pounds per square inch. For example, the modulus of elasticity of steel is approximately 30 million psi, while that of aluminum is 10 million psi. Therefore, steel is approximately three times as stiff as aluminum. The modulus of elasticity is a very useful engineering property and will appear in formulas dealing with the design of beams and columns where stiffness is important.

1·31 True Stress-strain The conventional tensile test described will give valuable information up to the point of yielding. Beyond this point, the stress

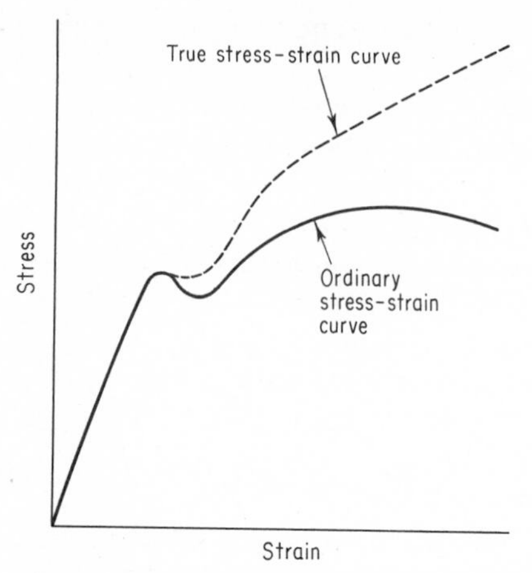

Fig. 1·24 True stress-strain and conventional stress-strain diagrams for mild steel.

values are fictitious since the actual cross-sectional area will be considerably reduced. The true stress is determined by the load divided by the cross-sectional area at that moment of loading. The true strain is determined by the change in length divided by the immediately preceding length. The true stress-strain diagram (Fig. 1·24) yields useful information regarding plastic flow and fracture of metals.

1·32 Resilience and Toughness It is possible to divide the stress-strain diagram into two parts as shown in Fig. 1·21. The part to the left of the elastic limit may be called the *elastic range* and that to the right of the elastic limit the *plastic range.* The area under the curve in the elastic range (area *OPR*) is a measure of the energy per unit volume which can be absorbed by the material without permanent deformation. This value is known as the *modulus of resilience.* The energy per unit volume that can be absorbed by a material (the area under the entire stress-strain diagram) up to the point of fracture is known as *toughness.* This is mainly a property of the plastic range, since only a small part of the total energy absorbed is elastic energy that can be recovered when the stress is released.

1·33 Impact Test Although the toughness of a material may be obtained by the area under the stress-strain diagram, the impact test will give an indication of the relative toughness.

Generally, notch-type specimens are used for impact tests. Two general types of notches are used in bending impact tests, the keyhole notch and the V notch. Two types of specimens are used, the Charpy and the Izod, shown in Fig. 1·25. The Charpy specimen is placed in the vise so that it is a simple beam supported at the ends. The Izod specimen is placed in the vise so that one end is free and is therefore a cantilever beam.

The ordinary impact machine has a swinging pendulum of fixed weight which is raised to a standard height depending upon the type of specimen tested (see Fig. 1·26). At that height, with reference to the vise, the pendulum has a definite amount of potential energy. When the pendulum is released, this energy is converted to kinetic energy until it strikes the specimen. The Charpy specimen will be hit behind the V notch, while the Izod specimen, placed with the V notch facing the pendulum, will be hit above the V notch. In either case, some of the energy of the pendulum will be used to rupture the specimen so that the pendulum will rise to a height lower than the initial height on the opposite side of the machine. The weight of the pendulum times the difference in heights will indicate the energy, usually in foot-pounds, absorbed by the specimen, or the notched impact strength.

From the description of the test, it is apparent that the notched-bar impact test does not yield the true toughness of a material but rather its behavior with a particular notch. The results are useful, however, for comparative purposes. The notched-bar test is used by the aircraft and automotive industries, which have found by experience that high impact

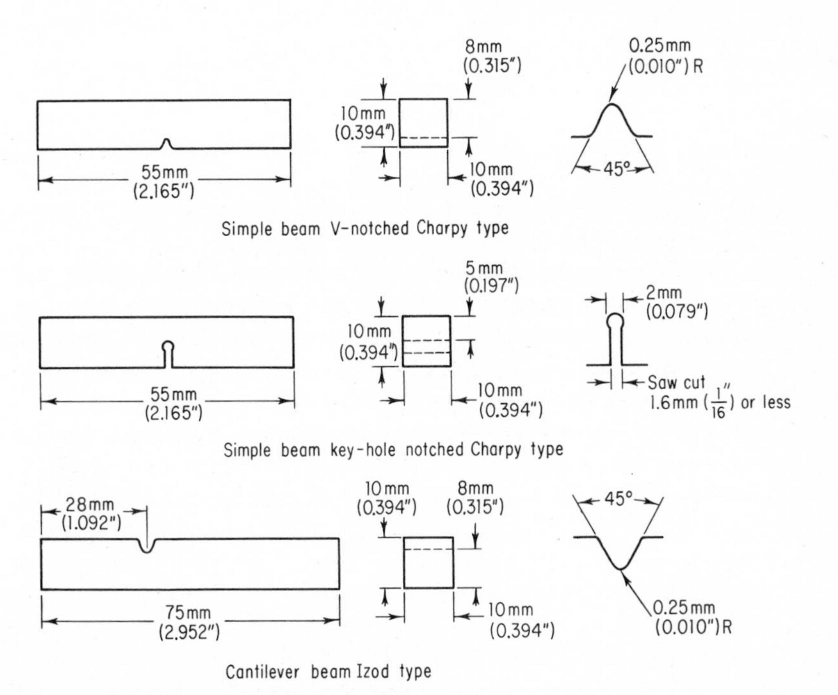

Fig. 1·25 Notched-bar impact test specimens. (By permission from "Tentative Methods for Notched-Bar Impact Testing of Metallic Materials," ASTM Designation E23-56T.)

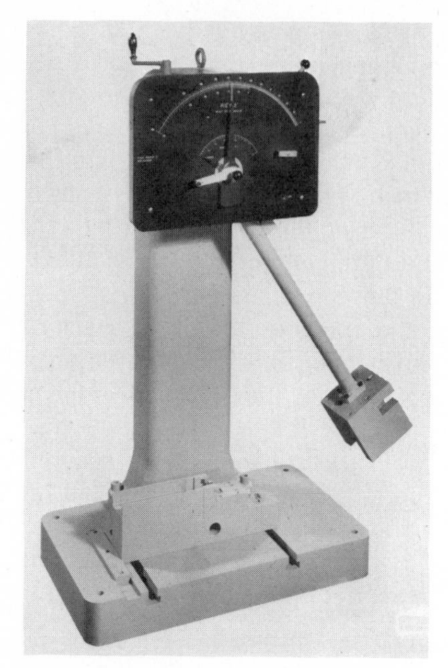

Fig. 1·26 The impact testing machine. (Ametek/Testing Equipment Systems, East Moline, Ill.)

strength by test generally will give satisfactory service where shock loads are encountered.

1·34 Fatigue Tests The fatigue test is a dynamic type of test which determines the relative behavior of materials when subjected to repeated or fluctuating loads. It attempts to simulate stress conditions developed in machine parts by vibration of cycling loads. The magnitude of the stress may be changed on the machine, and the type of stress (tension, compression, bending, or torsion) is determined by the machine and the type of specimen tested. The stress placed on the specimen during test continually alternates between two values, the maximum of which is usually lower than the yield strength of the material. The cycles of stress are applied until failure of the specimen or until a limiting number of cycles has been reached.

Those results are then plotted on a semilogarithmic scale with the stress S as the ordinate and the number of cycles N, to cause failure, as the abscissa. The "endurance limit" of any material is defined as the limiting stress below which the metal will withstand an indefinitely large number of cycles of stress without fracture. At that point on the S-N curve, the curve becomes parallel to the abscissa. For steel this will occur at approximately 10^7 cycles of stress. For some nonferrous alloys, however, the curve does not become horizontal, and the term *endurance limit* is often applied to the stress corresponding to some specific number of cycles. A typical *S-N* plot for alloy steel heat-treated, medium carbon steel heat-treated, aluminum-copper alloy, and gray cast iron is shown in Fig. 1·27.

Fatigue tests are widely used to study the behavior of materials not only for type and range of fluctuating loads but also for the effect of corrosion, surface conditions, temperature, size, and stress concentration.

1·35 Creep Tests The creep test determines the continuing change in the deformation of a material at elevated temperature when stressed below the yield strength. The results are important in the design of machine parts which are exposed to elevated temperatures. Creep behavior will be discussed in greater detail in Chap. 13.

NONDESTRUCTIVE TESTING

1·36 Introduction A nondestructive test is an examination of an object in any manner which will not impair the future usefulness of the object. Although in most cases nondestructive tests do not provide a direct measurement of mechanical properties, they are very valuable in locating material defects that could impair the performance of a machine member when placed in service. Such a test is used to detect faulty material before it is formed or machined into component parts, to detect faulty components before assembly, to measure the thickness of metal or other materials, to determine level of liquid or solid contents in opaque containers, to identify and sort materials, and to discover defects that may have developed during processing or use. Parts may also be examined in service, permitting their removal before failure occurs.

Nondestructive tests are used to make products more reliable, safe, and economical. Increased reliability improves the public image of the manu-

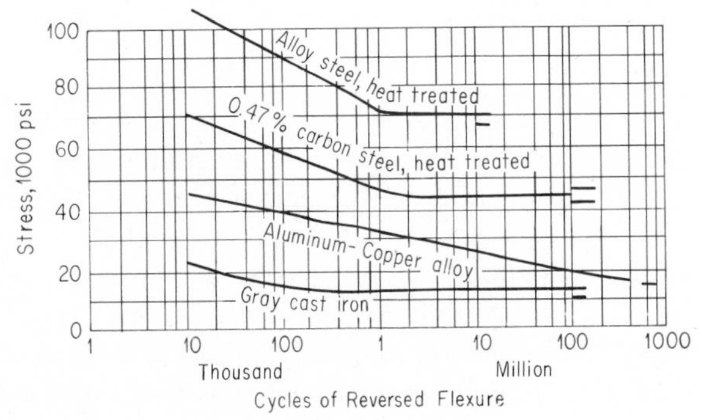

Fig. 1·27 Typical S-N (stress-cycle) diagrams. (From "Metals Handbook," 1948 ed., American Society for Metals, Metals Park, Ohio.)

facturer, which leads to greater sales and profits. In addition, manufacturers use these tests to improve and control manufacturing processes.

Before Wold War II, nondestructive testing was not urgent because of the large safety factors that were engineered into almost every product. Service failures did take place, but the role of material imperfections in such failures was not then fully recognized, and, therefore, little concentrated effort was made to find them. During, and just after, World War II the significance of imperfections to the useful life of a product assumed greater importance. In aircraft design, in nuclear technology, and in space exploration, high hazards and costs have made maximum reliability essential. At the same time, there has been extensive growth of all inspection methods in industrial and scientific applications.

There are five basic elements in any nondestructive test.

1 Source A source which provides some probing medium, namely, a medium that can be used to inspect the item under test.

2 Modification This probing medium must change or be modified as a result of the variations or discontinuities within the object being tested.

3 Detection A detector capable of determining the changes in the probing medium.

4 Indication A means of indicating or recording the signals from the detector.

5 Interpretation A method of interpreting these indications.

While there are a large number of proven nondestructive tests in use, this section will concentrate on the most common methods and on one recent development. The most common methods of nondestructive testing or inspection are:

Radiography

Magnetic-particle inspection

Fluorescent-penetrant inspection

Ultrasonic inspection

Eddy current inspection

1·37 Radiography of Metals The radiography of metals may be carried out by using x-rays or gamma rays—short-wavelength electromagnetic rays capable of going through relatively large thicknesses of metal. Gamma rays may be obtained from a naturally radioactive material such as radium or a radioactive isotope such as cobalt-60. Gamma radiation is more penetrating than that of x-rays, but the inferior sensitivity limits its application. There is no way that the source may be regulated for contrast or variable thickness, and it usually requires much longer exposure times than the x-ray method.

X-rays are produced when matter is bombarded by a rapidly moving stream of electrons. When electrons are suddenly stopped by matter, a

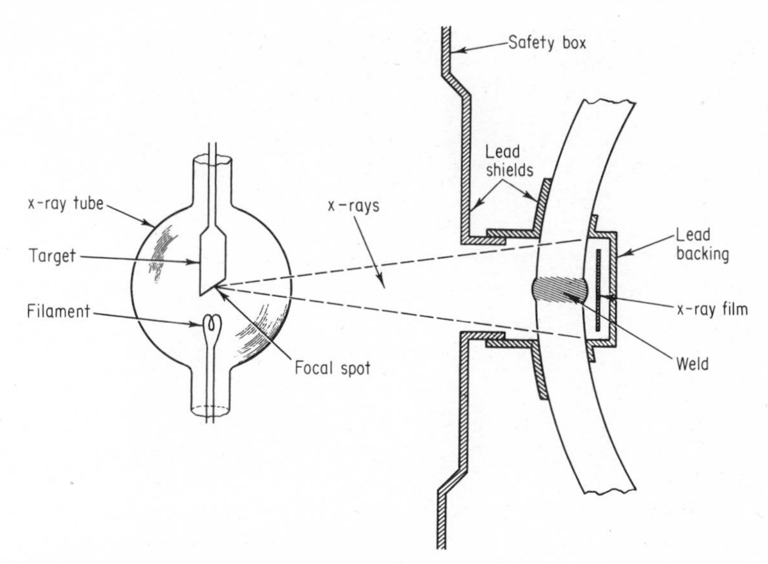

Fig. 1·28 Schematic representation of the use of x-rays for examination of a welded plate. (From "Basic Metallurgy," vol. 2, American Society for Metals, Metals Park, Ohio, 1954.)

part of their kinetic energy is converted to energy of radiation, or x-rays. The essential conditions for the generation of x-rays are (1) a filament (cathode) to provide the source of electrons proceeding toward the target, (2) a target (anode) located in the path of electrons, (3) a voltage difference between the cathode and anode which will regulate the velocity of the electrons striking the target and thus regulate the wavelength of x-rays produced, and (4) a means of regulating tube current to control the number of electrons striking the target. The first two requirements are usually incorporated in an x-ray tube. The use of x-rays for the examination of a welded plate is shown schematically in Fig. 1·28. X-rays are potentially dangerous, and adequate safeguards must be employed to protect operating personnel.

A radiograph is a shadow picture of a material more or less transparent to radiation. The x-rays darken the film so that regions of lower density which readily permit penetration appear dark on the negative as compared with regions of higher density which absorb more of the radiation. Thus a hole or crack appears as a darker area, whereas copper inclusions in aluminum alloy appear as lighter areas (see Fig. 1·29).

While the radiography of metals has been used primarily for the inspection of castings and welded products, it may also be used to measure the thickness of materials. Fig. 1·30 shows a simple radiation thickness gauge.

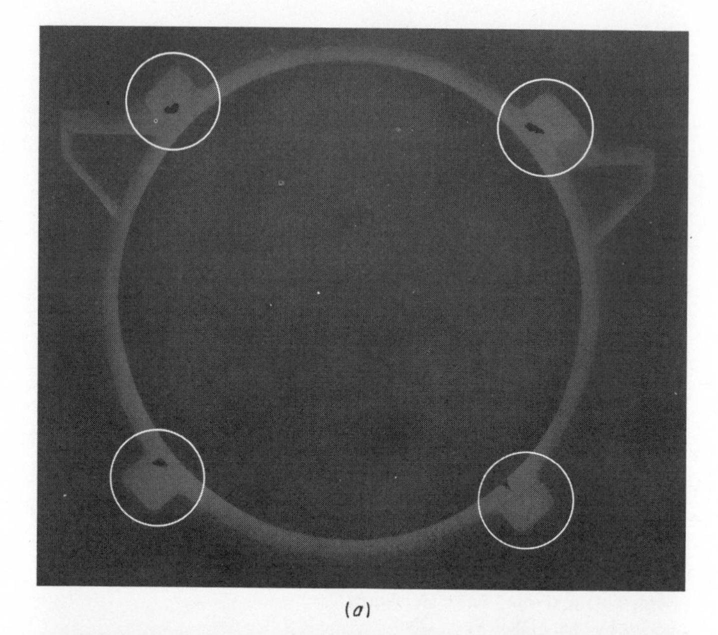

(a)

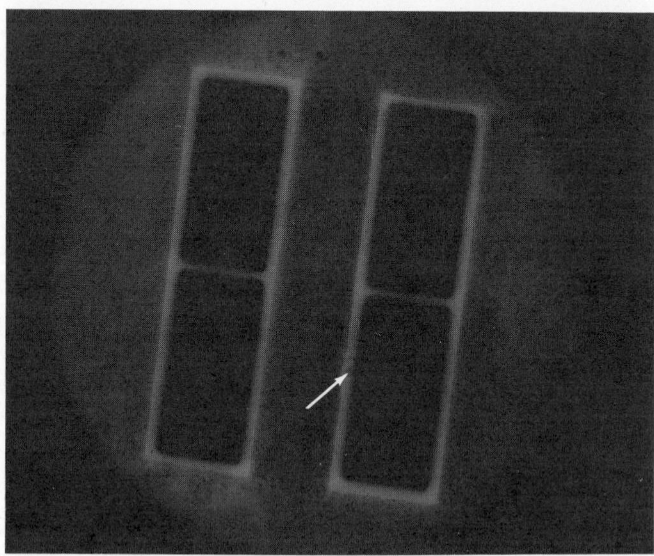

(b)

Fig. 1·29 (a) Radiograph of a stainless steel casting; dark spots are shrinkage voids. (b) Radiograph of a brass sand casting; numerous black spots indicate extensive porosity.

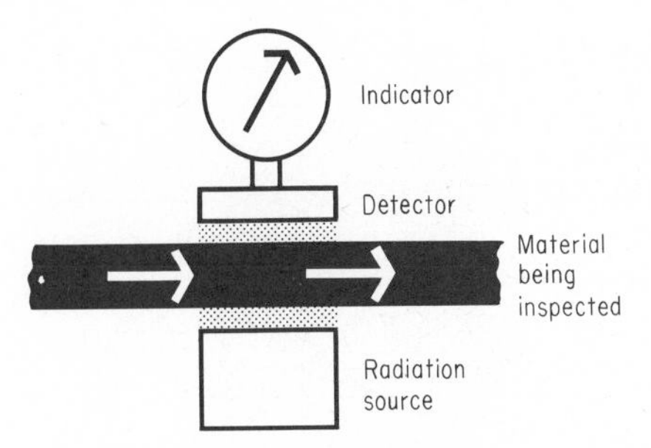

Fig. 1·30 A simple radiation thickness gauge.

The radiation from the source is influenced by the material being tested. As the thickness increases, the radiation intensity reaching the detector decreases. If the response of the detector is calibrated for known thicknesses, the detector reading can be used to indicate the thickness of the inspected material. With a suitable feedback circuit, the detector may be used to control the thickness between predetermined limits.

1·38 Magnetic-particle Inspection (Magnaflux) This is a method of detecting the presence of cracks, laps, tears, seams, inclusions, and similar discontinuities in ferromagnetic materials such as iron and steel. The method will detect surface discontinuities too fine to be seen by the naked eye and will also detect discontinuities which lie slightly below the surface. It is not applicable to nonmagnetic materials.

Magnetic-particle inspection may be carried out in several ways. The piece to be inspected may be magnetized and then covered with fine magnetic particles (iron powder). This is known as the residual method. Or, the magnetization and application of the particles may occur simultaneously. This is known as the continuous method. The magnetic particles may be held in suspension in a liquid that is flushed over the piece, or the piece may be immersed in the suspension (wet method). In some applications, the particles, in the form of a fine powder, are dusted over the surface of the workpiece (dry method). The presence of a discontinuity is shown by the formation and adherence of a particle pattern on the surface of the workpiece over the discontinuity. This pattern is called an *indication* and assumes the approximate shape of the surface projection of the discontinuity. The Magnaglo method developed by the Magnaflux Corporation is a variation of the Magnaflux test. The suspension flowed over the magnetized workpiece contains fluorescent magnetic particles. The workpiece is then viewed under black light, which makes the indications stand out more clearly.

When the discontinuity is open to the surface, the magnetic field leaks out to the surface and forms small north and south poles that attract the

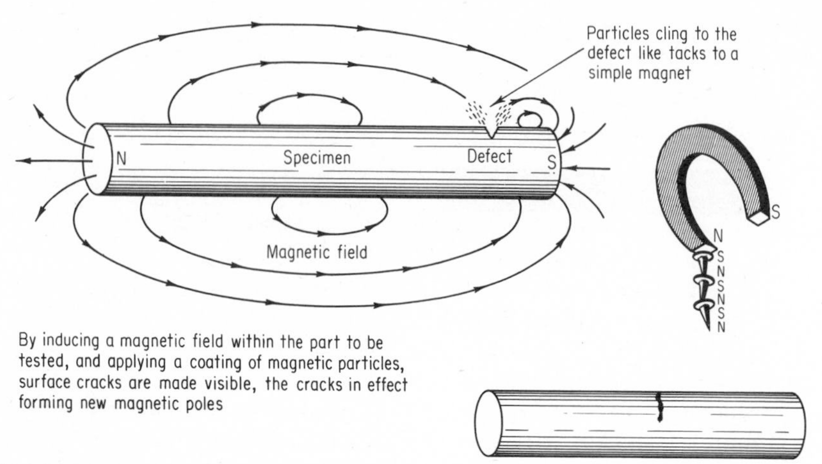

Fig. 1·31 Principle of the Magnaflux test. (Magnaflux Corporation, Chicago, Ill.)

magnetic particles (see Fig. 1·31). When fine discontinuities are under the surface, some part of the field may still be deflected to the surface, but the leakage is less and fewer particles are attracted, so that the indication obtained is much weaker. If the discontinuity is far below the surface, no leakage of the field will be obtained and consequently no indication. Proper use of magnetizing methods is necessary to ensure that the magnetic field set up will be perpendicular to the discontinuity and give the clearest indication.

As shown in Fig. 1·32 for longitudinal magnetization, the magnetic field may be produced in a direction parallel to the long axis of the workpiece by placing the piece in a coil excited by an electric current so that the long axis of the piece is parallel to the axis of the coil. The metal part then becomes the core of an electromagnet and is magnetized by induction from the magnetic field created in the coil. Very long parts are magnetized in steps by moving the coil along the length. In the case of circular magnetization, also shown in Fig. 1·32, a magnetic field transverse to the long axis of the workpiece is readily produced by passing the magnetizing current through the piece along this axis.

Direct current, alternating current, and rectified alternating current are all used for magnetizing purposes. Direct current is more sensitive than alternating current for detecting discontinuities that are not open to the surface. Alternating current will detect discontinuities open to the surface and is used when the detection of this type of discontinuity is the only interest. When alternating current is rectified, it provides a more penetrating magnetic field.

The sensitivity of magnetic-particle inspection is affected by many factors, including strength of the indicating suspension, time in contact with the suspension, time allowed for indications to form, time subject to magnetizing current, and strength of the magnetizing current. Some examples of cracks detectable by Magnaflux or Magnaglo are shown in Fig. 1·33.

All machine parts that have been magnetized for inspection must be put through a demagnetizing operation. If these parts are placed in service without demagnetizing, they will attract filings, grindings, chips, and other steel particles which may cause scoring of bearings and other engine parts. Detection of parts which have not been demagnetized is usually accomplished by keeping a compass on the assembly bench.

1·39 Fluorescent-penetrant Inspection (Zyglo) This is a sensitive nondestructive method of detecting minute discontinuities such as cracks, shrinkage, and porosity that are open to the surface. While this method may be applied to both magnetic and nonmagnetic materials, its primary application

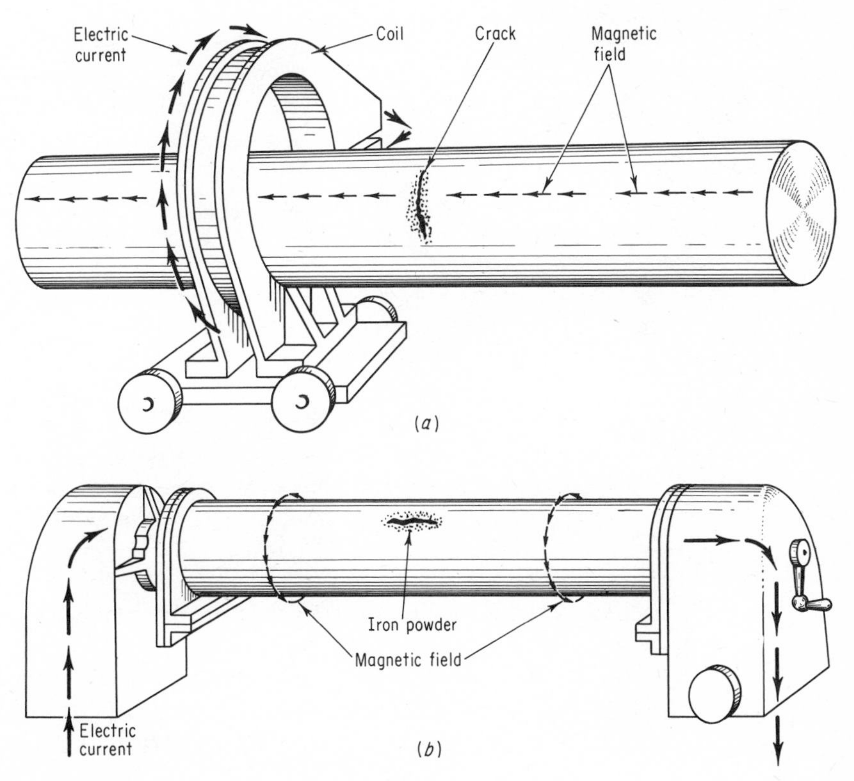

Fig. 1·32 Illustrating two kinds of magnetization: (a) Longitudinal magnetization; (b) circular magnetization. (Magnaflux Corporation, Chicago, Ill.)

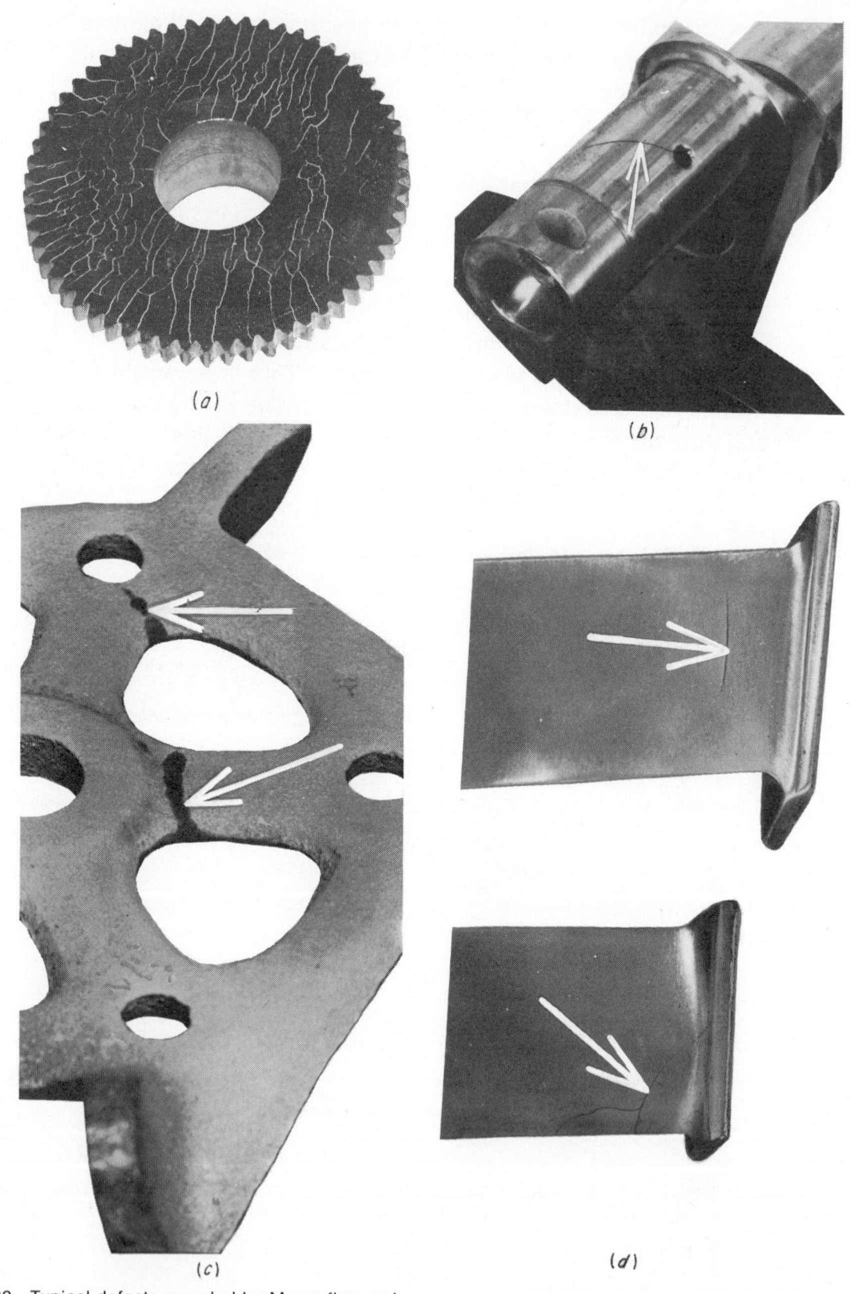

Fig. 1·33 Typical defects revealed by Magnaflux and Magnaglo. (*a*) Grinding cracks; (*b*) fatigue crack in an aircraft crankshaft; (*c*) casting cracks in a lawnmower casting; (*d*) cracks in critical jet engine blades. (Magnaflux Corporation, Chicago, Ill.)

is for nonmagnetic materials. Penetrant techniques can be used for inspecting any homogeneous material that is not porous, such as metals, glass, plastic, and some ceramic materials.

Parts to be tested are first treated with a penetrant. Penetrants are usually light, oil-like liquids which are applied by dipping, spraying, or brushing, or in some other convenient manner. The liquid penetrant is drawn into cracks and other discontinuities by strong capillary action. After the penetrant has had time to seep in, the portion remaining on the surface is removed by wiping or washing. This leaves the penetrant in all surface-connected discontinuities. The test part is now treated with a dry powder or a suspension of powder in a liquid. This powder or developer acts like a sponge drawing the penetrant from the defect and enlarging the size of the area of penetrant indication. In order for the inspection process to be completed, the penetrant must be easily observed in the developing powder. One method is to use contrasting colors for the penetrant and developer. A combination of white developer and red penetrant is very common.

Another method is to use a fluorescent penetrant. The major steps in fluorescent penetrant inspection are shown in Fig. 1·34. The steps are exactly the same as described previously except that the penetrating liquid contains a material that emits visible light when it is exposed to ultraviolet radiation. Lamps that emit ultraviolet are called black lights, because the visible light they might normally emit is stopped by a filter, making them appear black or dark purple. When the part to be inspected is viewed under black light, the defect appears as a bright fluorescing mark against a black background. Figure 1·35 shows a nonmagnetic stainless steel valve body being tested by fluorescent penetrant.

Fluorescent penetrant inspection is used to locate cracks and shrinkage in castings, cracks in the fabrication and regrinding of carbide tools, cracks and pits in welded structures, cracks in steam- and gas-turbine blading,

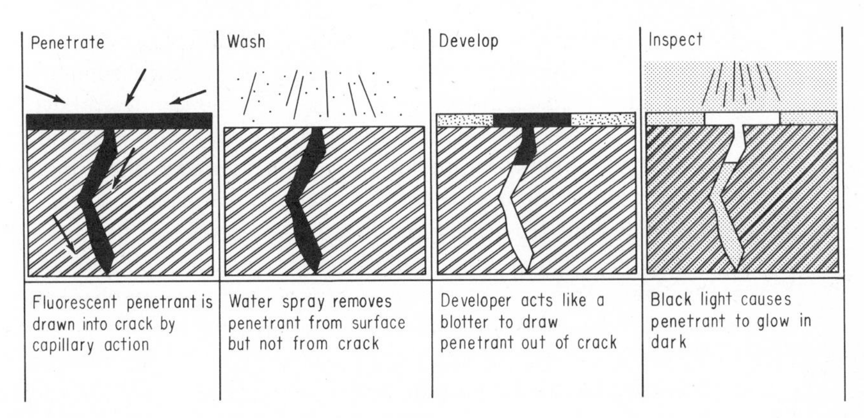

Fig. 1·34 Major steps in fluorescent-penetrant inspection.

Fig. 1·35 Nonmagnetic stainless steel valve body being inspected by fluorescent penetrant. (Magnaflux Corporation, Chicago, Ill.)

and cracks in ceramic insulators for spark plugs and electronic applications.

1·40 Ultrasonic Inspection The use of sound waves to determine defects is a very ancient method. If a piece of metal is struck by a hammer, it will radiate certain audible notes, of which the pitch and damping may be influenced by the presence of internal flaws. However, this technique of hammering and listening is useful only for the determination of large defects.

A more refined method consists of utilizing sound waves above the audible range with a frequency of 1 to 5 million Hz (cycles per second)—hence the term ultrasonic. Ultrasonics is a fast, reliable nondestructive testing method which employs electronically produced high-frequency sound waves that will penetrate metals, liquids, and many other materials at speeds of several thousand feet per second. Ultrasonic waves for nondestructive testing are usually produced by piezoelectric materials. These materials undergo a change in physical dimension when subjected to an electric field. This conversion of electrical energy to mechanical energy is known as the *piezoelectric* effect. If an alternating electric field is applied to a piezoelectric crystal, the crystal will expand during the first half of the cycle and contract when the electric field is reversed. By varying the

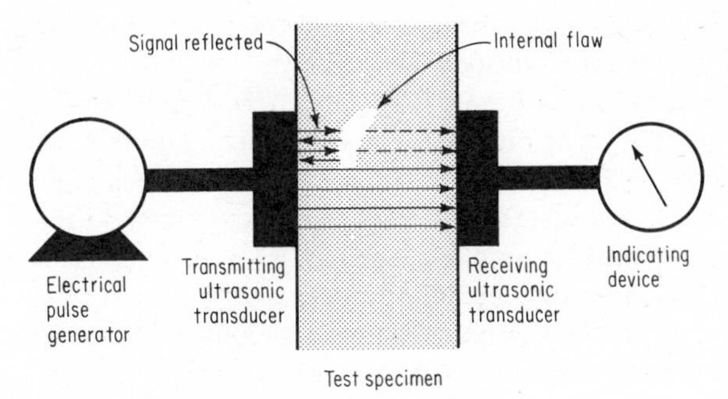

Fig. 1·36 The through-transmission and pulse-echo methods of ultrasonic inspection.

frequency of the alternating electric field, we can vary the frequency of the mechanical vibration (sound wave) produced in the crystal. Quartz is a widely used ultrasonic transducer. A transducer is a device for converting one form of energy to another.

Two common ultrasonic test methods, the through-transmission and the pulse-echo methods, are illustrated in Fig. 1·36. The through-transmission method uses an ultrasonic transducer on each side of the object being inspected. If an electrical pulse of the desired frequency is applied to the transmitting crystal, the ultrasonic waves produced will travel through the specimen to the other side. The receiving transducer on the opposite side receives the vibrations and converts them into an electrical signal that can be amplified and observed on the cathode-ray tube of an oscilloscope, a meter, or some other indicator. If the ultrasonic wave travels through the specimen without encountering any flaw, the signal received is relatively large. If there is a flaw in the path of the ultrasonic

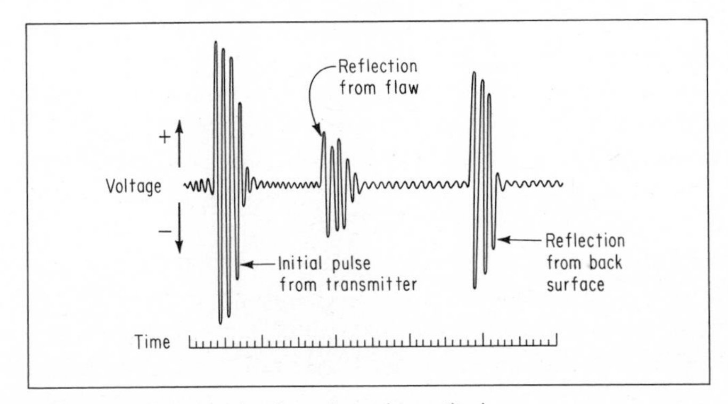

Fig. 1·37 Oscilloscope pattern for the pulse-echo method of ultrasonic inspection.

wave, part of the energy will be reflected and the signal received by the receiving transducer will be reduced.

The pulse-echo method uses only one transducer which serves as both transmitter and receiver. The pattern on an oscilloscope for the pulse-echo method would look similar to that shown in Fig. 1·37. As the sound wave enters the material being tested, part of it is reflected back to the crystal where it is converted back to an electrical impulse. This impulse is amplified and rendered visible as an indication or pip on the screen of the oscilloscope. When the sound wave reaches the other side of the material, it is reflected back and shows as another pip on the screen farther to the right of the first pip. If there is a flaw between the front and back surfaces of the material, it will show as a third pip on the screen between the two indications for the front and back surfaces. Since the indications on the oscilloscope screen measure the elapsed time between reflection of the pulse

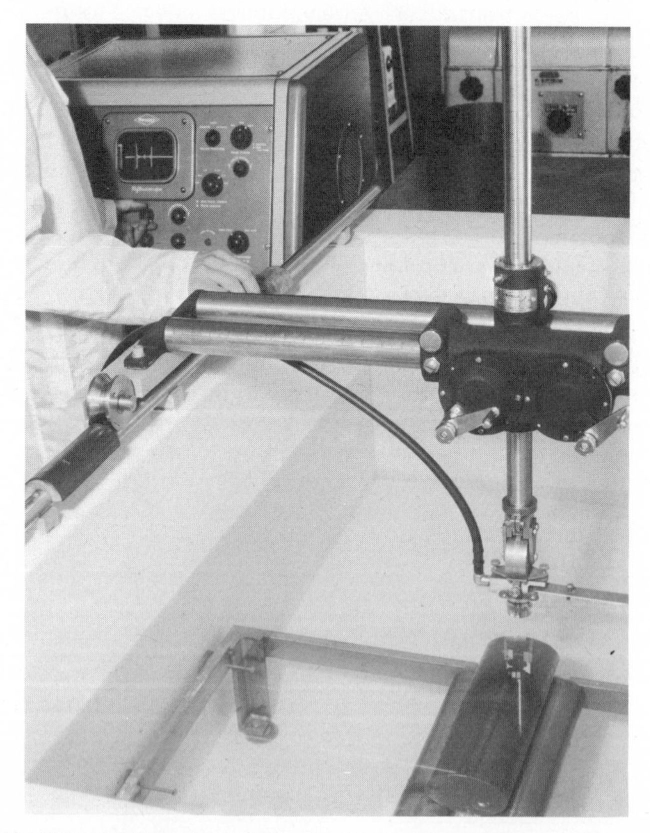

Fig. 1·38 Ultrasonic inspection by immersion in a water tank. (Fansteel Metallurgical Corporation)

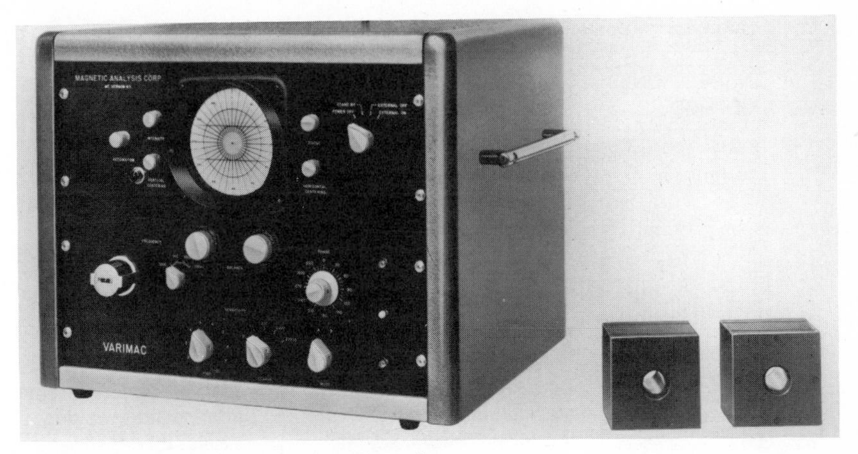

Fig. 1·39 An eddy current tester and two encircling coils. (Magnetic Analysis Corporation, Mount Vernon, N.Y.)

from the front and back surfaces, the distance between indications is a measure of the thickness of the material. The location of a defect may therefore be accurately determined from the indication on the screen.

In general, smooth surfaces are more suitable for the higher-frequency testing pulse and thereby permit detection of smaller defects. Proper transmission of the ultrasonic wave has a great influence on the reliability of the test results. For large parts, a film of oil ensures proper contact between the crystal searching unit and the test piece. Smaller parts may be placed in a tank of water, oil, or glycerin. The crystal searching unit transmits sound waves through the medium and into the material being examined (Fig. 1·38). Close examination of the oscilloscope screen in this picture shows the presence of three pips. The left pip indicates the front of the piece, the right pip the back of the piece, and the smaller center pip is an indication of a flaw.

Ultrasonic inspection is used to detect and locate such defects as shrinkage cavities, internal bursts or cracks, porosity, and large nonmetallic inclusions. Wall thickness can be measured in closed vessels or in cases where such measurement cannot otherwise be made.

1·41 Eddy Current inspection Eddy current techniques are used to inspect electrically conducting materials for defects, irregularities in structure, and variations in composition. In eddy current testing, a varying magnetic field is produced if a source of alternating current is connected to a coil. When this field is placed near a test specimen capable of conducting an electric current, eddy currents will be induced in the specimen. The eddy currents, in turn, will produce a magnetic field of their own. The detection unit will measure this new magnetic field and convert the signal into a volt-

TABLE 1·8 Major Nondestructive Testing Methods

INSPECTION METHOD	WHEN TO USE	WHERE TO USE
Eddy current	Measuring variations in wall thickness of thin metals or coatings; detecting longitudinal seams or cracks in tubing; determining heat treatments and metal compositions for sorting.	Tubing and bar stock, parts of uniform geometry, flat stock, or sheets and wire.
Radiography: x-rays	Detecting internal flaws and defects; finding welding flaws, cracks, seams, porosity, holes, inclusions, lack of fusion; measuring variations in thickness.	Assemblies of electronic parts, casting, welded vessels; field testing of welds; corrosion surveys; components of nonmetallic materials.
Gamma x-rays	Detecting internal flaws, cracks, seams, holes, inclusions, weld defects; measuring thickness variations.	Forgings, castings, tubing, welded vessels; field testing welded pipe; corrosion surveys.
Magnetic particle	Detecting surface or shallow subsurface flaws, cracks, porosity, nonmetallic inclusions, and weld defects.	Only for ferromagnetic materials; parts of any size, shape, composition, or heat treatment.
Penetrant	Locating surface cracks, porosity, laps, cold shuts, lack of weld bond, fatigue, and grinding cracks.	All metals, glass, and ceramics, castings, forgings, machined parts, and cutting tools; field inspections.
Ultrasonic pulse echo	Finding internal defects, cracks, lack of bond, laminations, inclusions, porosity; determining grain structure and thicknesses.	All metals and hard nonmetallic materials; sheets, tubing, rods, forgings, castings; field and production testing; inservice part testing; brazed and adhesive-bonded joints.

* Metals Progress Data Sheet, August 1968, American Society for Metals, Metals Park, Ohio.

ADVANTAGES	LIMITATIONS
High speed, noncontact, automatic.	False indications result from many variables; only good for conductive materials; limited depth of penetration.
Provides permanent record on film; works well on thin sections; high sensitivity; fluoroscopy techniques available; adjustable energy level.	High initial cost; power source required; radiation hazard; trained technicians needed.
Detects variety of flaws; provides a permanent record; portable; low initial cost; source is small (good for inside shots); makes panoramic exposures.	One energy level per source; radiation hazard; trained technicians needed; source loses strength continuously.
Economical, simple in principle, easy to perform; portable (for field testing); fast for production testing.	Material must be magnetic; demagnetizing after testing is required; power source needed; parts must be cleaned before finishing.
Simple to apply, portable, fast, low in cost; results easy to interpret; no elaborate setup required.	Limited to surface defects; surfaces must be clean.
Fast, dependable, easy to operate; lends itself to automation, results of test immediately known; relatively portable, highly accurate, sensitive.	Requires contact or immersion of part; Interpretation of readings requires training.

age that can be read on a meter or a cathode-ray tube. Properties such as hardness, alloy composition, chemical purity, and heat treat condition influence the magnetic field and may be measured directly by a single coil.

An important use for eddy current testing is sorting material for heat treat variations or composition mix-ups. This application requires the use of two coils (see Fig. 1·39). A standard piece is placed in one coil and the test piece in the other coil. Acceptance or rejection of the test piece may be determined by comparing the two patterns on the oscilloscope screen.

Eddy current testing may be used to detect surface and sub-surface defects, plate or tubing thickness, and coating thickness.

A summary of the major nondestructive testing methods is given in Table 1·8.

1·42 Recent Developments The most interesting of the recent developments in nondestructive testing methods is the use of holography,[1] a unique method of recording on film visual data about a three-dimensional object and recreating a three-dimensional image of the object. Whereas conventional photography shows the image of an object on film, the holographic process records interference patterns which are used to reconstruct the image.

A simplified setup for making and viewing holograms is shown in Fig. 1·40. The narrow, single-wavelength light beam from the laser passes through a lens-pinhole assembly to increase its area of coverage. The diverged beam strikes a mirror and the object. The reference beam reflected from the mirror and the light reflected from the object both strike a photographic plate. Since no lens is used to focus the light from the object, the film records not an image but the interference pattern resulting

[1]Much of the following description was taken from publications of GC Optronics, Inc., Ann Arbor, Mich.

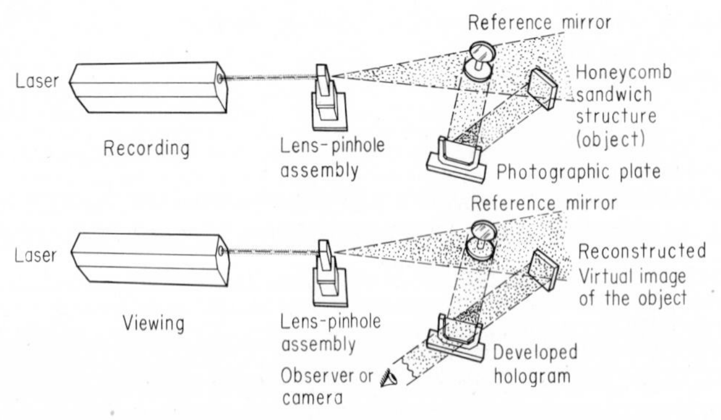

Fig. 1·40 A simplified setup for making and viewing holograms. (GC Optronics, Inc., Ann Arbor, Mich.)

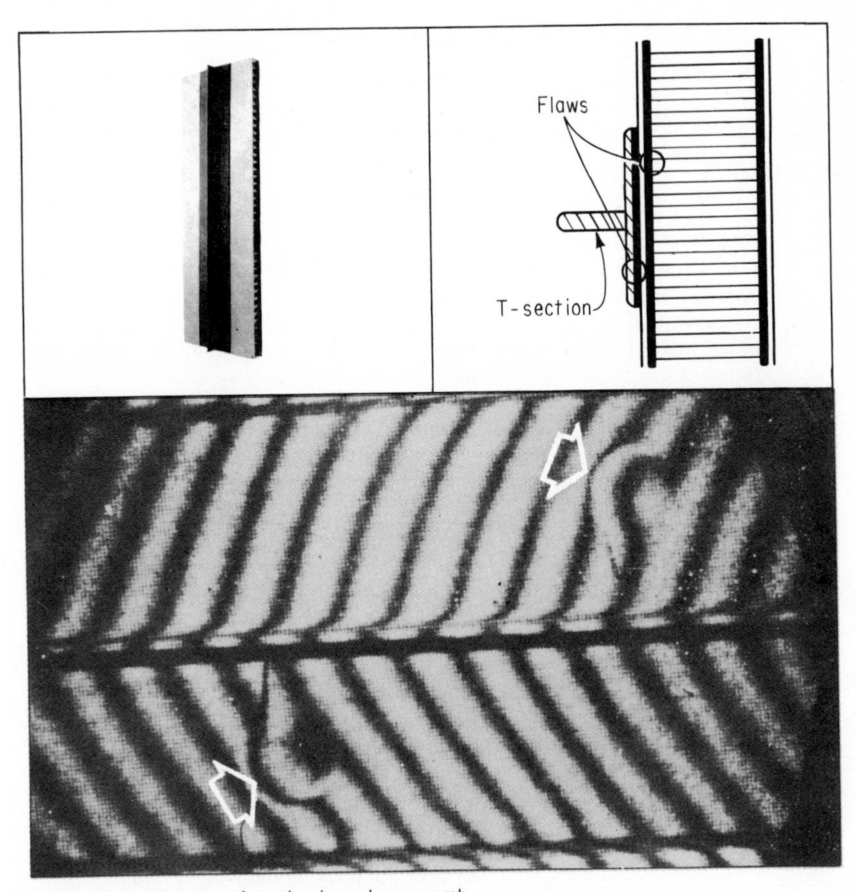

Fig. 1·41 A real-time hologram of an aluminum honeycomb-sandwich structure with a T-extrusion. Two debonds are revealed after mild stressing by heat. (GC Optronics, Inc., Ann Arbor, Mich.)

from a mixture of these two beams. After exposure, the film is developed and the hologram is ready for viewing. In viewing, the laser light, similar to the reference beam used during the exposure, illuminates the hologram. An observer, looking through the hologram as though it were a window, sees a reconstructed, three-dimensional image of the original object, apparently in the exact position occupied by the object during exposure.

In the real-time method, the original hologram is viewed so that the reconstructed virtual image is superimposed on the object. The object is then stressed. Any stress-induced deformation of the object will appear as fringe patterns on the picture and will reveal voids, flaws, unbonded areas, and inclusions. Fig. 1·41 shows a hologram of an aluminum honeycomb-sandwich structure with a T-extrusion bonded to the surface. The structure was mildly stressed by heat, and the real-time hologram reveals two debonds under the T-section.

QUESTIONS

1·1 How are thermocouples calibrated?
1·2 What factors may lead to errors in a thermoelectric circuit?
1·3 Aside from being able to measure high temperatures, what is another advantage of the optical pyrometer?
1·4 Assuming that bare, not fused wires of copper and constantan are put into a liquid metal below the melting point of copper, will the thermocouple measure the temperature of the liquid metal?
1·5 Assume that the thermocouple wires are reversed when connected to the potentiometer; how may this be detected?
1·6 How is "true stress" calculated?
1·7 Differentiate between resilience and toughness.
1·8 Which property in a tension test is an indication of the stiffness of a material?
1·9 Which properties in a tension test indicate the ductility of a material?
1·10 How will the speed of testing affect the yield strength and ultimate strength?
1·11 On a stress-strain graph, for a load beyond the yield strength that is suddenly removed, show the elastic strain and the plastic strain.
1·12 Why is the yield strength usually determined rather than the elastic limit?
1·13 What is the difference between the proportional limit and the elastic limit?
1·14 Why are impact specimens notched?
1·15 Discuss the effect of the type of notch and velocity of the hammer on the results of the impact test.
1·16 What limits the range of hardness in the Brinell machine?
1·17 What are the units for the Brinell hardness number?
1·18 Why is it possible to obtain the approximate tensile strength of steel by 500 times the Brinell hardness number?
1·19 Is there a unit associated with the Rockwell hardness number? Explain.
1·20 Why is a correction factor necessary for Rockwell readings on a specimen less than 1 in. in diameter?
1·21 Is the correction factor in Question 1·20 to be added or subtracted from the observed readings? Explain.
1·22 What is the minimum thickness of the specimen if a reading is to be taken in the range of Rockwell C 60? (Refer to "Metals Handbook," 1948 edition.)
1·23 If the specimen in Question 1·22 is to be checked with the Brinell test, what should its minimum thickness be?
1·24 If the specimen in Question 1·22 is to be checked on the Rockwell 15 N scale, what should its minimum thickness be?
1·25 How may one determine whether the specimen was too thin to be checked with a particular Rockwell scale?
1·26 List three factors that contribute to the inaccuracy of a scleroscope reading.
1·27 What factors may be varied in taking a radiograph with x-rays?
1·28 In a radiograph, what will be the difference in appearance of gas cavities, cracks, and impurities?
1·29 What are the limitations of magnetic-particle inspection?
1·30 What are the limitations of fluorescent-penetrant inspection?

1·31 What are the limitations of ultrasonic inspection?

1·32 Explain the difference between through-transmission and pulse-echo methods of ultrasonic inspection.

1·33 What is a transducer?

1·34 Name and explain three transducers.

1·35 Which nondestructive testing method is best suited to determine the wall thickness at the bottom of a steel tank?

1·36 Which nondestructive testing method should be used to sort out bars of mixed steel?

REFERENCES

American Society for Metals: "Metals Handbook," 1948 ed., Metals Park, Ohio.

———: "Nondestructive Testing for Management," Metals Park, Ohio, 1963.

———: "Temperature Measurement," Metals Park, Ohio, 1956.

American Society for Testing and Materials: "Annual Book of Standards," Part 31, Philadelphia, Pa., 1971.

Betz, C. E.: "Principles of Penetrants," Magnaflux Corporation, Chicago, Ill., 1963.

Carlin, B.: "Ultrasonics," McGraw-Hill Book Company, New York, 1960.

Coxon, W. F.: "Temperature Measurement and Control," The Macmillan Company, New York, 1960.

Davis, H. E., G. E. Troxell, and C. T. Wiskocil: "The Testing and Inspection of Engineering Materials," 2d ed., McGraw-Hill Book Company, New York, 1955.

Dike, P. H.: "Thermoelectric Thermometry," Leeds & Northrup Company, Philadelphia, Pa., 1954.

Doane, F. B.: "Principles of Magnaflux Inspection," Magnaflux Corporation, Chicago, Ill., 1940.

Eastman Kodak Company: "Radiography in Modern Industry," Rochester, N.Y., 1957.

Kehl, G. L.: "Principles of Metallographic Laboratory Practice," 3d ed., McGraw-Hill Book Company, New York, 1949.

Lysaght, V. E.: "Indentation Hardness Testing," Van Nostrand Reinhold Company, New York, 1949.

McMaster, R. C.: "Nondestructive Testing Handbook," The Ronald Press Company, New York, 1959.

Williams, S. R.: "Hardness and Hardness Measurement," American Society for Metals, Metals Park, Ohio, 1942.

2 METAL STRUCTURE AND CRYSTALLIZATION

2·1 Introduction All matter is considered to be composed of unit substances known as *chemical elements*. These are the smallest units that are distinguishable on the basis of their chemical activity and physical properties. The elements are composed of atoms which have a distinct structure characteristic of each element.

2·2 Atomic Structure While each chemical element is composed of atoms, it is the difference in atomic structure that gives the element its characteristic properties. It is useful to think of the free atom as being composed of three elementary particles: (1) *electrons*, tiny particles of negative electricity, (2) *protons*, particles of positive electricity, and (3) electrically neutral particles, called *neutrons*. Almost the entire mass of the atom is concentrated in the nucleus, which contains the protons and neutrons. The mass of the proton is approximately 1.673×10^{-24} g, and the neutron is just slightly heavier, approximately 1.675×10^{-24} g, while the mass of the electron is much lighter, approximately 9.11×10^{-28} g. Therefore, the electron is approximately 1/1800 the mass of a proton. The diameter of the nucleus is of the order of 10^{-12} cm and is very small compared with the atomic diameter, which is approximately 10^{-8} cm. The atom consists of a minute positively charged nucleus made up of protons and neutrons surrounded by a sufficient number of electrons to keep the atom as a whole electrically neutral. Since the electron and proton have equal but opposite electrical charge, the neutral atom must contain an equal number of electrons and protons.

The discovery of the atomic particles started in 1874, when William Crooks, an Englishman, while experimenting with electrical discharges inside a partially evacuated tube, discovered that the so-called cathode rays could turn a small paddle wheel placed in the tube and would cast a shadow of any solid object placed in their path. This convinced him that the rays were material in nature. From the direction that they were deflected in a magnetic field he knew that the particles possessed negative charges. It was suggested by his colleagues that these rays consisted of negatively charged particles. In 1897, his countryman, J. J. Thomson, made another discovery. From the extent to which the rays were deflected by a positively charged object and by a magnet, Thompson determined that the particles of which the rays were composed were traveling at high speeds and that each one had the mass of about 1/1800 of the hydrogen atom. Further investigations showed that these little particles had identical properties, regardless of the kind of material used as the electrodes. Thus it was evident that these small negatively charged particles, called electrons, were parts of all atoms. In 1911, fourteen years after Thomson's discovery of electrons, Ernest Rutherford bombarded light elements, such as aluminum, with high-speed alpha particles. The targets emitted a new kind of particle, positive in charge, and of a mass almost identical with that of the hydrogen atom. These particles were named protons. The bombardment of light elements, notably beryllium, with alpha particles resulted in the emission of a third kind of atomic building unit. Chadwick, in 1932, named this particle the neutron.

It is sometimes difficult for the mind to conceive of such tiny particles. If the simplest atom, hydrogen, were magnified so that its diameter would be $1/2$ mile, its nucleus would be the size of a baseball, and its electron, $1/4$ of a mile distant, would be the size of a softball. The first conception of atomic structure that met with wide acceptance was that advanced by the American scientists Gilbert Lewis and Irving Langmuir about 1916.

In their theory, the protons and neutrons constituted a central dense nucleus. Each electron was supposed to occupy a fixed position outside the nucleus. These positions, beginning with the third electron, were arranged in concentric cubes. The inner cube could hold eight electrons, the next cube could hold eight electrons, and even more in cases of elements of large atomic weight. Successive cubes showed an increasing complexity of electronic structure. Figure 2.1(*a*) shows the Lewis-Langmuir model of the sodium atom. The nucleus, with eleven protons and twelve neutrons, is placed in the center, and the electrons, represented by dots, are placed at the corners of hypothetical stationary concentric cubes.

Shortly thereafter a dynamic model of the atom was developed by the Danish scientist Niels Bohr. It possessed fewer defects than the Lewis-Langmuir model and is substantially the working model now in use. The Bohr concept of the sodium atom is shown in Figure 2.1(*b*). According to this concept, the nucleus of an atom contains all the neutrons and protons. Hence the mass or weight of an atom is centered in its nucleus. The electrons revolve in circular and elliptical orbits about the nucleus somewhat as the planets and comets of our solar system revolve about the sun.

Each unit of atomic weight is supplied by a proton or a neutron. Thus an

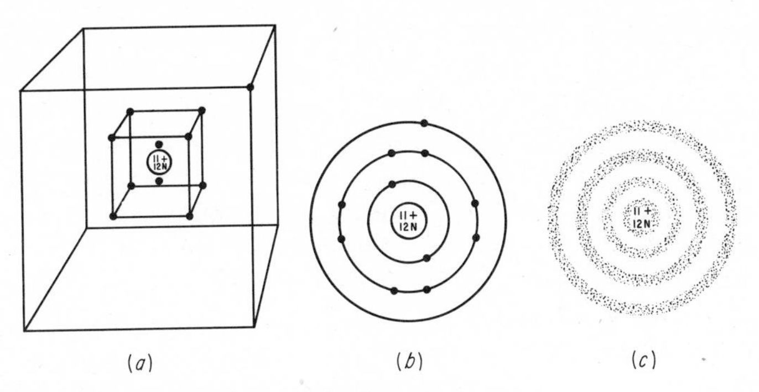

Fig. 2·1 Changing concepts of atomic structure—the sodium atom. (*a*) Lewis-Langmuir model; (*b*) Bohr concept; (*c*) wave mechanics model.

atom of oxygen with an atomic weight of 16 could contain 16 protons or 16 neutrons or a total of 16 of the two. Which of these three possibilities is correct?

To determine the answer to this question, you must use the concept of atomic number. If the elements are arranged in order of their increasing atomic weight, the number of nuclear protons increases in the same order. In other words, hydrogen, our lightest atom, has one proton in the nucleus. The next heavier element, helium, has two. The third, lithium, has three. Uranium, the 92nd element, has 92 protons. Now since oxygen has an atomic number of 8, and its atomic weight is 16, we know that the nucleus of the atom is composed of eight protons and eight neutrons. Therefore the atomic number of an element is equal to the number of protons, or since the atom has to be electrically neutral, the atomic number is also equal to the number of electrons.

The electrons, spinning on their own axes as they rotate around the nucleus, are arranged in definite shells. The maximum number of electrons that can fit in each shell is $2n^2$, where n is the shell number. Therefore the maximum number of electrons that will fit in the first shell is two, the second eight, the third eighteeen, the fourth thirty-two, and so on. Each shell is further subdivided into energy states or levels. According to the Pauli Exclusion Principle, no more than two electrons can fit on any one energy level; and if two are to fit on the same level, they must be of opposite spin. The number of energy levels increases with distance from the nucleus, and electrons tend to occupy the lowest energy levels. Therefore, the inner shells tend to fill up first, but this is not always true for some of the heavier elements. It is possible for the highest energy level of an inner shell to have more energy than the lowest energy level of the next shell, as will be illustrated later.

TABLE 2·1 Atomic Number, Atomic Weight, and Chemical Symbols of the Elements*

ELEMENT	SYMBOL	AT. NO.	AT. WT.	ELEMENT	SYMBOL	AT. NO.	AT. WT.
Actinium	Ac	89	227	Mercury	Hg	80	200.61
Aluminum	Al	13	26.98	Molybdenum	Mo	42	95.95
Americium	Am	95	243	Neodymium	Nd	60	144.27
Antimony	Sb	51	121.76	Neon	Ne	10	20.183
Argon	A	18	39.944	Neptunium	Np	93	237
Arsenic	As	33	74.91	Nickel	Ni	28	58.69
Astatine	At	85	211	Niobium	Nb	41	92.91
Barium	Ba	56	137.36	(Columbium)	(Cb)		
Berkelium	Bk	97	247	Nitrogen	N	7	14.008
Beryllium	Be	4	9.013	Nobelium	No	102	254
Bismuth	Bi	83	209.00	Osmium	Os	76	190.2
Boron	B	5	10.82	Oxygen	O	8	16.000
Bromine	Br	35	79.916	Palladium	Pd	46	196.7
Cadmium	Cd	48	112.41	Phosphorus	P	15	30.975
Calcium	Ca	20	40.08	Platinum	Pt	78	195.23
Californium	Cf	98	251	Plutonium	Pu	94	242
Carbon	C	6	12.011	Polonium	Po	84	210
Cerium	Ce	58	140.13	Potassium	K	19	39.100
Cesium	Cs	55	132.91	Praseodymium	Pr	59	140.92
Chlorine	Cl	17	35.457	Promethium	Pm	61	145
Chromium	Cr	24	54.01	Protactinium	Pa	91	231
Cobalt	Co	27	58.94	Radium	Ra	88	226.05
Copper	Cu	29	63.54	Radon	Rn	86	222
Curium	Cm	96	247	Rhenium	Re	75	186.31
Dysprosium	Dy	66	162.46	Rhodium	Rh	45	102.91
Einsteinium	E	99	245	Rubidium	Rb	37	85.48
Erbium	Er	68	167.2	Ruthenium	Ru	44	101.1
Europium	Eu	63	152.0	Samarium	Sm	62	150.43
Fermium	Fm	100	253	Scandium	Sc	21	44.96
Fluorine	F	9	19.00	Selenium	Se	34	78.96
Francium	Fr	87	233	Silicon	Si	14	28.09
Gadolinium	Gd	64	156.9	Silver	Ag	47	107.880
Gallium	Ga	31	69.72	Sodium	Na	11	22.991
Germanium	Ge	32	72.60	Strontium	Sr	38	87.63
Gold	Au	79	197.0	Sulfur	S	16	32.066
Hafnium	Hf	72	178.6	Tantalum	Ta	73	180.95
Helium	He	2	4.003	Technetium	Te	43	98
Holmium	Ho	67	164.94	Tellurium	Te	52	127.61
Hydrogen	H	1	1.008	Terbium	Tb	65	158.93
Indium	In	49	114.76	Thallium	Tl	81	204.39
Iodine	I	53	126.91	Thorium	Th	90	232.05
Iridium	Ir	77	192.2	Thulium	Tm	69	168.94
Iron	Fe	26	55.85	Tin	Sn	50	118.70
Krypton	Kr	36	83.80	Titanium	Ti	22	47.90
Lanthanum	La	57	138.92	Tungsten	W	74	183.92
Lawrencium	Lw	103†	257	Uranium	U	92	238.07
Lead	Pb	82	207.21	Vanadium	V	23	50.95
Lithium	Li	3	6.940	Xenon	Xe	54	131.3
Lutecium	Lu	71	174.99	Ytterbium	Yb	70	173.04
Magnesium	Mg	12	24.32	Yttrium	Y	39	88.92
Manganese	Mn	25	54.94	Zinc	Zn	30	65.38
Mendelevium	Mv	101	256	Zirconium	Zr	40	91.22

* From "Metals Handbook," 1961 edition, American Society for Metals, Metals Park, Ohio.
† Elements 104 Hahnium (Ha) and 105 Kurchatorium (Ku) have also been found.

The modified Bohr theory is our nearest approach to a satisfactory mechanical model of the atom, but it contains certain defects. It is generally agreed that the electron has the nature of both a particle and a wave. In the early 1920s, through the work of Heisenberg and Schroedinger and with the development of wave mechanics, the modern concept of the atom is best expressed by mathematical equations. The electron shells may be visualized as concentric halos of negative electricity with its greatest densities concentrated as in the Bohr shells, Fig. 2·1(*c*).

It is not possible to determine the exact orbit of an electron, but rather its position is determined by the probability that it will be found in a given region of the atom. This probability is represented mathematically by a certain wave function, and the solution of the wave equation leads to four quantities known as the *quantum numbers* n, l, m_l, and m_s. Of these, n is the principal quantum number related to the total energy of the electron in a particular state and may have values $n = 1, 2, 3$, etc. The second quantum number l is a measure of the angular momentum of the electron and may have values from 0 to $(n - 1)$. The letters *s*, *p*, *d*, *f*, *g*, and *h* have been introduced to signify $l = 0, 1, 2, 3, 4, 5$, so that the energy level corresponding to $n = 1$ and $l = 0$ is called the (1*s*) level, that corresponding to $n = 2$ and $l = 1$ is called the (2*p*) level, etc.

The quantum number m_l is related to the component of the angular momentum in a specified direction and may have values of $+l$ to $-l$, including zero. The fourth quantum number m_s, related to the spin of the electron on its own axis, may have the value of $\pm\frac{1}{2}$ depending upon the direction of the spin. Figure 2·2 schematically shows some of the possible

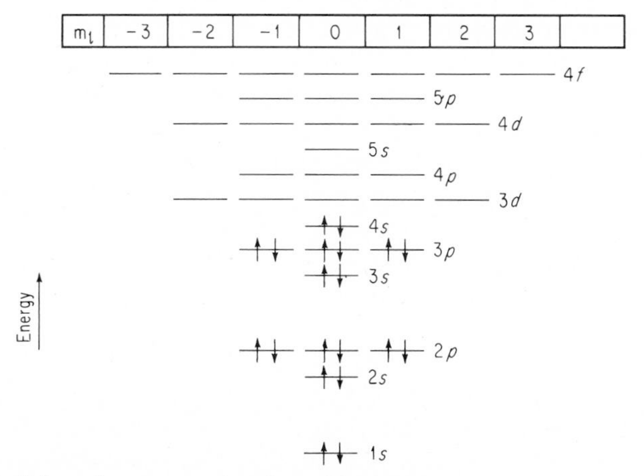

Fig. 2·2 Schematic representation of possible electron states in an atom. (From M. M. Eisenstadt, "Introduction to Mechanical Properties of Materials," The Macmillan Company, New York, 1971.)

electron states in an atom. Each horizontal line may be occupied by two electrons of opposite spin, which are shown in the first few levels as vertical arrows. Notice the overlap of energy levels at larger values of n. The $(4s)$ level is below the $(3d)$ level, the $(5s)$ level is below the $(4d)$ level, etc. This will have an interesting effect on the periodic table which will be explained in the next section. The state of an electron is completely specified by the four quantum numbers, and no two electrons in an atom may have the same four quantum numbers.

The electron configuration of an atom may be indicated by a number representing the principal shell number n followed by the letter representing the subshell (value of l) and finally the number of electrons in the subshell as a superscript. For example, the element lithium, which contains three electrons, would be represented as $(1s)^2\ (2s)^1$; the element oxygen, containing eight electrons, as $(1s)^2\ (2s)^2\ (2p)^4$.

The atomic weight of an element is the weight of the particular atom relative to the atomic weight of oxygen, which is taken to be 16.000. On this scale, the atomic weight of hydrogen is 1.008. The atomic number of an element is equal to the number of electrons or the number of protons. The atomic number and atomic weight of the elements are given in Table 2·1.

2·3 The Periodic Table Several scientists in the first half of the 1800s recognized that certain elements showed a similarity in physical and chemical properties, and they tried various groupings. This led to the tabulation in 1869, by a Russian scientist, Mendeleev, of what is called the *periodic table* or periodic classification of the elements. The periodic table is the best classification device for the elements and is one of the most fundamental concepts in science.

Elements of any given family which show a similarity in chemical properties were arranged in the same column or group. For Example, Group VII: fluorine, chlorine, bromine, and iodine are all similar in properties. The elements in each horizontal row or period increase from left to right in atomic weight, with few exceptions. Using all the elements that were known in 1869 naturally left many blank spaces in the first table. So satisfactory was this classification of the elements that Mendeleev was able to make certain predictions and formulate generalizations that have been exactly verified by subsequent research. Each blank space in the table predicted the existence of an element and predicted its approximate atomic weight and physical and chemical properties.

In the years that have elapsed since Mendeleev formulated his table, all of his missing elements have been discovered, and twelve elements beyond uranium have been prepared in the course of atomic-nuclear research. The complete periodic table is shown in Table 2·2. The modified form of the periodic table is shown in Table 2·3. The lines running from top to bottom connect elements of similar chemical properties. In each horizontal row

TABLE 22 Periodic Table of the Elements

PERIODS	GROUPS																	
	LIGHT METALS		HEAVY METALS										NONMETALS					INERT GASES
	IA	IIA	IIIB	IVB	VB	VIB	VIIB		VIIIB		IB	IIB	IIIA	IVA	VA	VIA	VIIA	O
1	1 H 1.008																	2 He 4.003
2	3 Li 6.94	4 Be 9.013							METALLOIDS →				5 B 10.82	6 C 12.011	7 N 14.008	8 O 16.000	9 F 19.00	10 Ne 20.183
3	11 Na 22.991	12 Mg 24.32			BRITTLE				DUCTILE			Low Melt- ing	13 Al 26.98	14 Si 28.09	15 P 30.975	16 S 32.066	17 Cl 35.457	18 Ar 39.944
4	19 K 39.100	20 Ca 40.08	21 Sc 44.96	22 Ti 47.90	23 V 50.95	24 Cr 52.01	25 Mn 54.94	26 Fe 55.85	27 Co 58.94	28 Ni 58.71	29 Cu 63.54	30 Zn 65.38	31 Ga 69.72	32 Ge 72.60	33 As 74.91	34 Se 78.96	35 Br 79.916	36 Kr 83.80
5	37 Rb 85.48	38 Sr 87.63	39 Y 88.92	40 Zr 91.22	41 Nb 92.91	42 Mo 95.95	43 Tc 99	44 Ru 101.1	45 Rh 102.91	46 Pd 106.4	47 Ag 107.88	48 Cd 112.41	49 In 114.82	50 Sn 118.70	51 Sb 121.76	52 Te 127.61	53 1 126.91	54 Xe 131.30
6	55 Cs 132.91	56 Ba 137.36	Lan- thanide Series	72 Hf 178.50	73 Ta 180.95	74 W 183.86	75 Re 186.22	76 Os 190.2	77 Ir 192.2	78 Pt 195.09	79 Au 197.0	80 Hg 200.61	81 Tl 204.39	82 Pb 207.21	83 Bi 209.00	84 Po 210	85 At 210*	86 Rn 222
7	87 Fr 223*	88 Ra 226.05	Acti- nide Series															
LANTHANIDE SERIES				57 La 138.92	58 Ce 140.13	59 Pr 140.92	60 Nd 144.27	61 Pm 145*	62 Sm 150.35	63 Eu 152.0	64 Gd 157.26	65 Tb 158.93	66 Dv 162.51	67 Ho 164.94	68 Er 167.27	69 Tm 168.94	70 Yb 173.04	71 Lu 174.99
ACTINIDE SERIES				89 Ac 227	90 Th 232.05	91 Pa 231	92 U 238.07	93 Np 237*	94 Pu 242*	95 Am 243*	96 Cm 245*	97 Bk 249*	98 Cf 249*	99 Es 255	100 Fm 255	101 Md 256*	102 No 254*	103 Lw 257*

* Mass number of the isotope of longest known half-life.

the brackets designated (1*s*), (2*s*), (2*p*) and so on denote the filling of subshells of electrons. The period number refers to values of the principal shell, or the first quantum number *n*. Table 2·4 shows the placement of the electrons in the principal shell and subshells of the first 26 elements.

Some understanding of the arrangement of the elements may now be obtained. The subsequent discussion may be followed by referring to Tables 2·2, 2·3, and particularly 2·4. Since the first principal shell can contain the maximum of only two electrons, the first period contains only two elements: hydrogen, written as $(1s)^1$ and helium, $(1s)^2$. In the next element, lithium, the third electron enters the second principal shell, and the electron configuration is written therefore $(1s)^2\ (2s)^1$. The sum of the superscripts indicates the total number of electrons, which in the case of lithium is 3. Both hydrogen and lithium have similar electron configurations, that is, one electron in the (*s*) subshell, and therefore are listed in the same group. The electrons in the unfilled shells are known as *valence electrons* and are largely responsible for the chemical behavior of the elements. Therefore the elements in the same group will have the same chemical valence.

In going from lithium to neon (atomic number 10), the second shell is filled to its maximum of eight electrons. In this process the (2*s*) subshell is filled first with beryllium. Then from boron to neon the electrons fill the (2*p*) subshell to its maximum of 6 electrons. The outer group of 8 electrons, when the (*s*) and (*p*) subshells are filled (or 2, in the case of helium), is a very stable group, and whenever this occurs, the element shows very little chemical activity. Helium, neon, and the rest of the elements in the same group are known as *inert gases*.

In the next element, sodium (atomic number 11), the last electron has to enter the third principal shell. The electron configuration for sodium, therefore, is $(1s)^2\ (2s)^2\ (2p)^6\ (3s)^1$. Notice that the electron configuration is similar to that of hydrogen and lithium. Therefore sodium is placed in the same group. From sodium to argon the process is the same as that occurring between lithium and neon. The (3*s*) and (3*p*) subshells are now filled, and a very stable grouping is obtained. Therefore argon is an inert gas.

According to the $2n^2$ rule, the third shell can contain a maximum of 18 electrons, and you would think that the next electron should enter the (3*d*) subshell. However, the (4*s*) subshell has less energy, that is, it is closer to the nucleus than the (3*d*) subshell, and therefore it is filled first, so the next electron goes into the (4*s*) subshell, and we have the element potassium (atomic number 19), with an electron configuration similar to that of hydrogen, lithium, and sodium, and it is placed in the same group. This is shown schematically in Figs. 2·2 and 2·3.

TABLE 2·3 Modified Periodic Table*

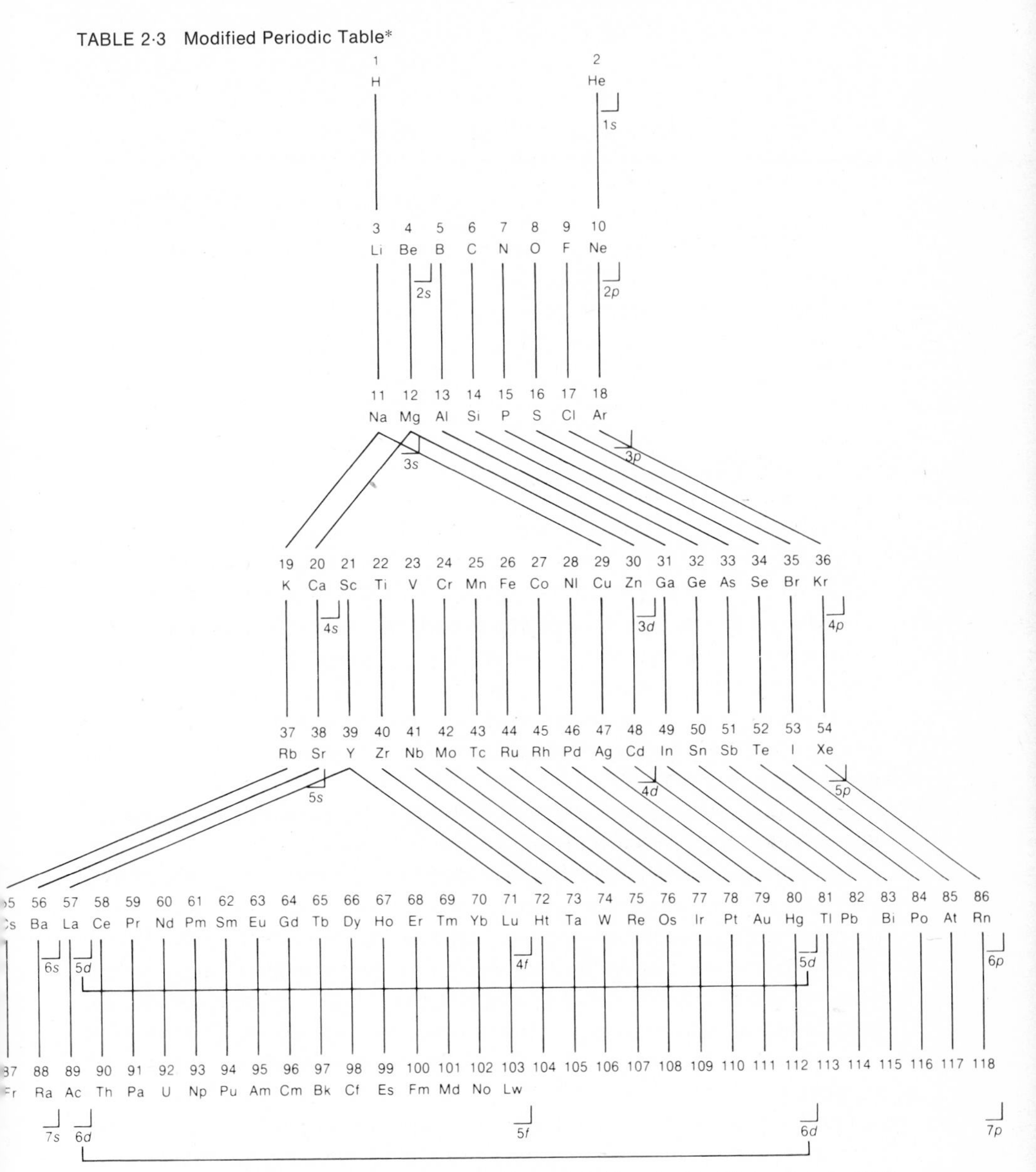

*From Glenn T. Seaborg and Justin L. Bloom, The Synthetic Elements, *Scientific American*, April 1969.

After the (4*s*) subshell is filled with calcium, the next electrons enter the third subshell (3*d*), expanding the third shell from a group of 8 electrons to

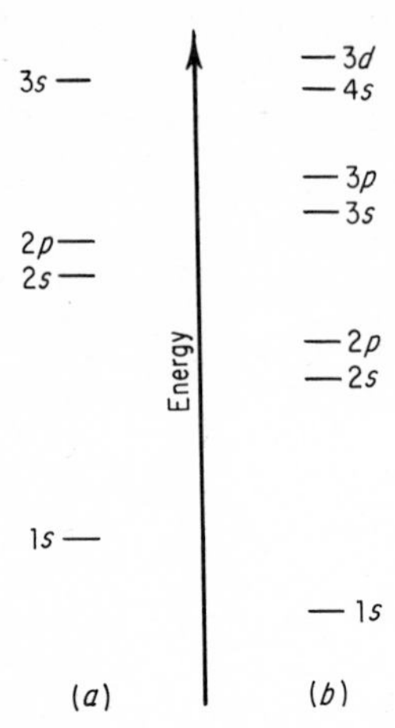

Fig. 2·3 Relative electronic energies in two free atoms: (*a*) magnesium, (*b*) iron. (From A. G. Guy, "Elements of Physical Metallurgy," 2d ed., Addison-Wesley Publishing Company, Inc., Reading, Mass., 1959.)

a maximum of 18. These elements are known as the transition elements and are all of variable valence. After copper (atomic number 29), the electrons fill the (4*p*) subshell in a normal manner, reaching a stable grouping of eight electrons in the fourth shell with krypton, atomic number 26, an inert gas. The electron configuration of the elements in the first four periods is given in Table 2·4.

Transition processes also occur in the later periods, and a similar process gives rise to the rare-earth metals, atomic numbers 58 to 71.

The periodic table is constantly referred to in the development of new alloys for specific purposes. Since elements in the same group have similar electron configurations, they often may adequately replace each other in alloys. Tungsten is added to tool steels to improve their softening resistance at elevated temperatures, and molybdenum or chromium is sometimes used as a substitute. Sulfur is used to improve the machinability of steel; selenium and tellurium are added to stainless steels for the same purpose.

2·4 Isotopes It is possible for the nucleus of an element to have more or less than the normal number of neutrons. Since the number of protons or electrons has not changed, the atomic number will remain the same. However, the atomic weight will be different. The atoms of varying atomic weight are called the isotopes of the element. Most of the elements are mixtures of two or more naturally occurring isotopes. Therefore the atomic weights which are measured are usually not whole numbers. Hydrogen, for example, is a mixture of three isotopes. Most of its atoms contain one proton,

TABLE 2·4 Electron Configuration of the Elements in the First Four Periods*

ATOMIC NUMBER AND ELEMENT	PRINCIPAL SHELL AND SUBSHELLS									
	$n = 1$	2		3			4			
	$l = s$	s	p	s	p	d	s	p	d	f
1 H	1									
2 He	2									
3 Li	2	1								
4 Be	2	2								
5 B	2	2	1							
6 C	2	2	2							
7 N	2	2	3							
8 O	2	2	4							
9 F	2	2	5							
10 Ne	2	2	6							
11 Na	2	2	6	1						
12 Mg	2	2	6	2						
13 Al	2	2	6	2	1					
14 Si	2	2	6	2	2					
15 P	2	2	6	2	3					
16 S	2	2	6	2	4					
17 Cl	2	2	6	2	5					
18 A	2	2	6	2	6					
19 K	2	2	6	2	6		1			
20 Ca	2	2	6	2	6		2			
21 Sc†	2	2	6	2	6	1	2			
22 Ti†	2	2	6	2	6	2	2			
23 V†	2	2	6	2	6	3	2			
24 Cr†	2	2	6	2	6	5	1			
25 Mn†	2	2	6	2	6	5	2			
26 Fe†	2	2	6	2	6	6	2			
27 Co†	2	2	6	2	6	7	2			
28 Ni†	2	2	6	2	6	8	2			
29 Cu	2	2	6	2	6	10	1			
30 Zn	2	2	6	2	6	10	2			
31 Ga	2	2	6	2	6	10	2	1		
32 Ge	2	2	6	2	6	10	2	2		
33 As	2	2	6	2	6	10	2	3		
34 Se	2	2	6	2	6	10	2	4		
35 Br	2	2	6	2	6	10	2	5		
36 Kr	2	2	6	2	6	10	2	6		

* From W. Hume-Rothery, "Atomic Theory for Students of Metallurgy," The Institute of Metals, London, 1955.
† Transition elements.

but a few contain one proton and one neutron, and still fewer contain one proton and two neutrons. The double-weight form of hydrogen is known as deuterium, the prime constituent of heavy water. The triple-weight form is known as tritium. In ordinary hydrogen these masses of one, two, and three are mixed in such proportions as to give an average atomic weight of 1.008. Nickel has isotopes of atomic weights 58, 60, 61, and 62, producing an average atomic weight of 58.71. Many artificially prepared isotopes, such as cobalt 60, are radioactive and are used for industrial and medical applications.

2·5 Classification of Elements The chemical elements may be roughly classified into three groups, metals, metalloids, and nonmetals. Elements considered to be metals are distinguished by several characteristic properties: (1) in the solid state they exist in the form of crystals; (2) they have relatively high thermal and electrical conductivity; (3) they have the ability to be deformed plastically; (4) they have relatively high reflectivity of light (metallic luster). The metals are on the left side of the periodic table and constitute about three-fourths of the elements (see Table 2·2).

Metalloids resemble metals in some respects and nonmetals in others. Generally they have some conductivity but little or no plasticity. Examples of metalloids are carbon, boron, and silicon.

The remaining elements are known as *nonmetals.* This includes the inert gases, the elements in Group VIIA, and N, O, P, and S.

2·6 Atom Binding It is characteristic of the solid state that all true solids exhibit a crystal structure which is a definite geometric arrangement of atoms or molecules. Some materials, such as glass or tar, that are rigid at room temperature do not have a regular arrangement of molecules but rather the random distribution that is typical of the liquid state. These materials are not true solids but rather supercooled liquids.

The question now arises as to what holds the atoms or molecules of a solid together. There are four possible types of bonds:

1 Ionic bond
2 Covalent or homopolar bond
3 Metallic bond
4 Van der Waals forces

2·7 Ionic Bond As was pointed out earlier, the electron structure of atoms is relatively stable when the outer shells contain eight electrons (or two in the case of the first shell). An element like sodium with one excess electron will readily give it up so that it has a completely filled outer shell. It will then have more protons than electrons and become a positive ion (charged atom) with a +1 charge. An atom of chlorine, on the other hand, with seven electrons in its outer shell, would like to accept one electron. When it does, it will have one more electron than protons and become a negative ion with a −1 charge. When sodium and chlorine atoms are placed together, there is a transfer of electrons from the sodium to the chlorine atoms, resulting in a strong electrostatic attraction between the positive sodium ions and the negative chlorine ions (Fig. 2·4) and forming the compound sodium chloride, which is ordinary table salt. The fact that this compound has its own properties, not necessarily related to sodium or chlorine, demonstrates that the ionic bond is a very strong bond. This is fortunate, since sodium is a highly reactive metal and chlorine is a poisonous gas; yet table salt is something we use every day without any harm. This explains the strong attraction between paired ions typical of the gas or liquid state. In the solid state, however, each sodium ion is surrounded by six negative chlorine ions, and vice versa, so that the attraction is equal in all directions.

2·8 Covalent Bond Atoms of some elements may attain a stable electron structure by sharing one or more electrons with adjacent atoms. As shown in Fig. 2·5, nitrogen (atomic number 7) has 5 electrons in the outer shell and needs 3 more to complete that shell. Hydrogen has 1 electron in the outer shell. To attain a stable structure, nitrogen and hydrogen behave differently than do sodium and chlorine. A nitrogen atom shares the electrons of three hydrogen atoms and in turn shares three of its electrons with the three hydrogen atoms to form the compound ammonia (NH_3). In this case ions are not formed; instead, the strong bond is due to the attraction of the shared electrons by the positive nuclei. The three hydrogen atoms are united to the nitrogen atom by three pairs of electrons, each atom furnishing one electron of each pair. This is known as the covalent or homopolar bond. The covalent bond is typical of most gas molecules.

2·9 Metallic Bond The lack of oppositely charged ions in the metallic structure and the lack of sufficient valence electrons to form a true covalent bond necessitate the sharing of valence electrons by more than two atoms. Each of the atoms of the metal contributes its valence electrons to the formation of a negative electron "cloud." These electrons are not associated with a particular ion but are free to move among the positive metallic ions in

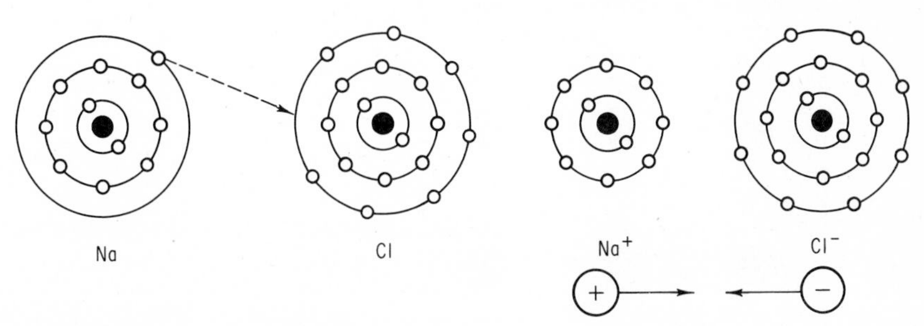

Fig. 2·4 Electron transfer in NaCl formation. (From L. H. Van Vlack, "Elements of Materials Science," Addison-Wesley Publishing Company, Inc., Reading, Mass., 1959.)

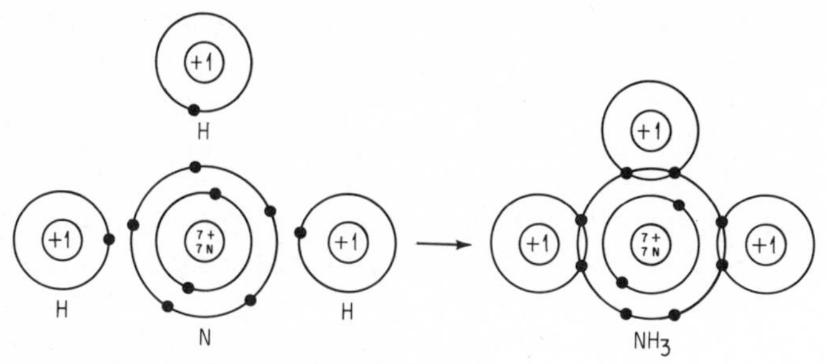

Fig. 2·5 Covalent bond in the formation of ammonia.

definite energy levels. The metallic ions are held together by virtue of their mutual attraction for the negative electron cloud. This is illustrated schematically in Fig. 2·6. The metallic bond may be thought of as an extension of the covalent bond to a large number of atoms.

2·10 Van der Waals Forces This type of bond arises in neutral atoms such as the inert gases. When the atoms are brought close together there is a separation of the centers of positive and negative charges, and a weak attractive force results. It is of importance only at low temperatures when the weak attractive force can overcome the thermal agitation of the atoms.

METAL STRUCTURE

2·11 Atomic Diameter When atoms of a metal approach each other, two opposing forces influence the internal energy, an attractive force between the electrons and both positive nuclei, and a repulsive force between the posi-

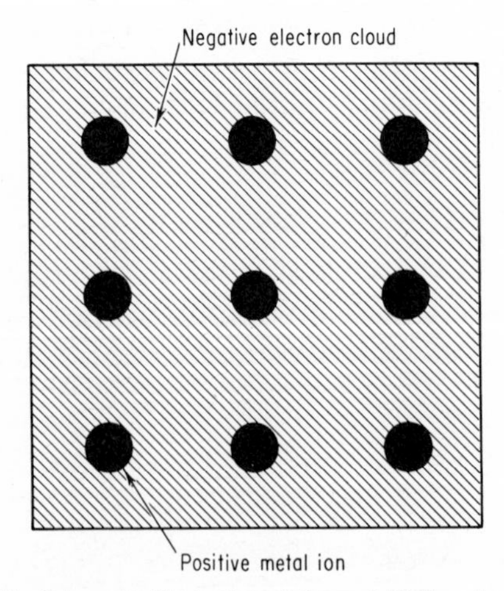

Fig. 2·6 Schematic illustration of the metallic bond.

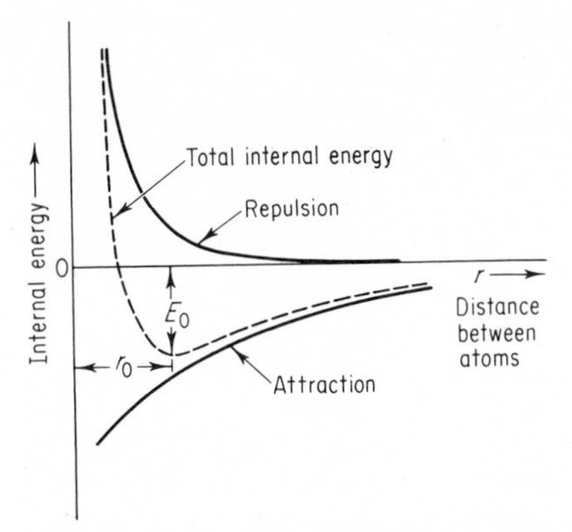

Fig. 2·7 Internal energy in relation to distance between atoms.

tive nuclei and also between the electrons. The first force tends to decrease the internal energy and the second force tends to increase it. At some distance these two forces will just balance each other and the total internal energy E_0 will be a minimum, corresponding to an equilibrium condition (Fig. 2·7). The equilibrium distance r_0 is different for each element and is determined by measuring the distance of closest approach of atoms in the solid state. If the atoms are visualized as spheres just touching at equilibrium, then the distance between centers of the spheres may be taken as the approximate atomic diameter. The atomic diameter increases as the number of occupied shells increases (Table 2·5) and decreases as the number of valence electrons increases (Table 2·6).

TABLE 2·5 Atomic Diameter of the Elements in Groups IA and IIA of the Periodic Table*

ELEMENT	ATOMIC NUMBER	ATOMIC DIAMETER, ANGSTROMS†
	GROUP IA	
Lithium	3	3.03
Sodium	11	3.71
Potassium	19	4.62
Rubidium	37	4.87
Cesium	55	5.24
	GROUP IIA	
Beryllium	4	2.22
Magnesium	12	3.19
Calcium	20	3.93
Strontium	38	4.30
Barium	56	4.34

* From "Metals Handbook," 1948 ed., American Society for Metals, Metals Park, Ohio.
† One angstrom = 10^{-8} cm.

TABLE 2·6 Atomic Diameter and Valence of Some Elements in the Fourth Period*

ELEMENT	VALENCE	ATOMIC DIAMETER, ANGSTROMS
Potassium	1	4.62
Calcium	2	3.93
Scandium	3	3.20
Titanium	4	2.91
Vanadium	5	2.63
Chromium	Variable	2.49
Manganese	Variable	2.37
Iron	Variable	2.48
Cobalt	Variable	2.50
Nickel	Variable	2.49

* From "Metals Handbook," 1948 ed., American Society for Metals, Metals Park, Ohio.

2·12 Crystal Structure Since atoms tend to assume relatively fixed positions, this gives rise to the formation of crystals in the solid state. The atoms oscillate about fixed locations and are in dynamic equilibrium rather than statically fixed. The three-dimensional network of imaginary lines connecting the atoms is called the *space lattice*, while the smallest unit having the full symmetry of the crystal is called the *unit cell.* The specific unit cell for each metal is defined by its parameters (Fig. 2·8), which are the edges of the unit cell a, b, c and the angles α (between b and c), β (between a and c), and γ (between a and b).

There are only 14 possible types of space lattices, and they fall into seven crystal systems listed in Table 2·7.

Fortunately, most of the important metals crystallize in either the cubic or hexagonal systems, and only three types of space lattices are commonly encountered: the b.c.c. (body-centered cubic), the f.c.c. (face-centered cubic), and the c.p.h. (close-packed hexagonal). Unit cells of these are shown schematically in Figs. 2·9 to 2·11. In each case the atom is represented as a point (left) and more accurately as a sphere (right).

2·13 Body-centered Cubic If the atoms are represented as spheres, the center atom touches each corner atom but these do not touch each other. Since each corner atom is shared by eight adjoining cubes and the atom in the center cannot be shared by any other cube (see Fig. 2·12*a*), the unit cell of the b.c.c. structure contains:

$$\begin{aligned} 8 \text{ atoms at the corners} \times 1/8 &= 1 \text{ atom} \\ 1 \text{ center atom} &= \underline{1 \text{ atom}} \\ \text{Total} &= 2 \text{ atoms} \end{aligned}$$

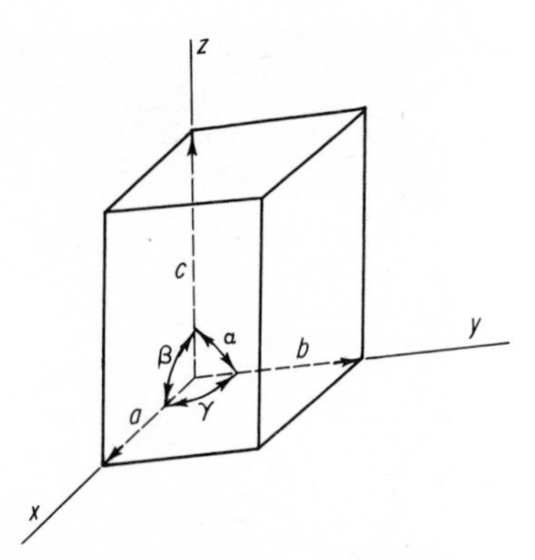

Fig. 2·8 Space lattice illustrating lattice parameters.

Examples of metals that crystallize in the b.c.c. structure are chromium, tungsten, alpha (α) iron, delta (δ) iron, molybdenum, vanadium, and sodium.

2·14 Face-centered Cubic In addition to an atom at each corner of the cube, there is one in the center of each face, but none in the center of the cube. Each face atom touches its nearest corner atom. Since each corner atom

TABLE 2·7 The Crystal Systems*

In this table $\neq$ means "not necessarily equal to"

1. Triclinic	Three unequal axes, no two of which are perpendicular $a \neq b \neq c \quad \alpha \neq \beta \neq \gamma \neq 90°$
2. Monoclinic	Three unequal axes, one of which is perpendicular to the other two $a \neq b \neq c \quad \alpha = \gamma = 90° \neq \beta$
3. Orthorhombic	Three unequal axes, all perpendicular $a \neq b \neq c \quad \alpha = \beta = \gamma = 90°$
4. Rhombohedral (trigonal)	Three equal axes, not at right angles $a = b = c \quad \alpha = \beta = \gamma \neq 90°$
5. Hexagonal	Three equal coplanar axes at 120° and a fourth unequal axis perpendicular to their plane $a = b \neq c \quad \alpha = \beta = 90° \quad \gamma = 120°$
6. Tetragonal	Three perpendicular axes, only two equal $a = b \neq c \quad \alpha = \beta = \gamma = 90°$
7. Cubic	Three equal axes, mutually perpendicular $a = b = c \quad \alpha = \beta = \gamma = 90°$

* From C. S. Barrett, "Structure of Metals," McGraw-Hill Book Company, Inc., New York, 1952.

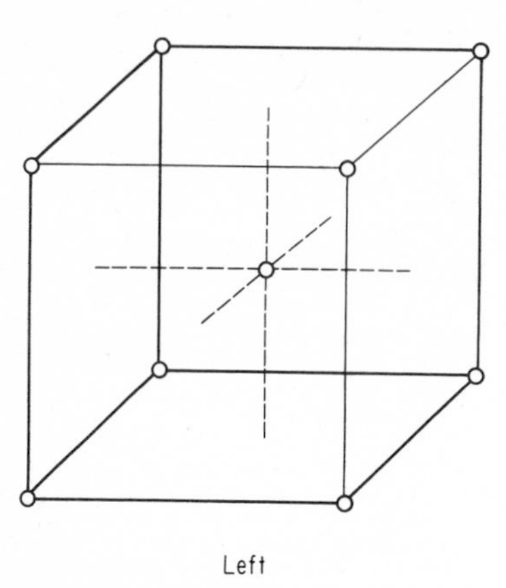

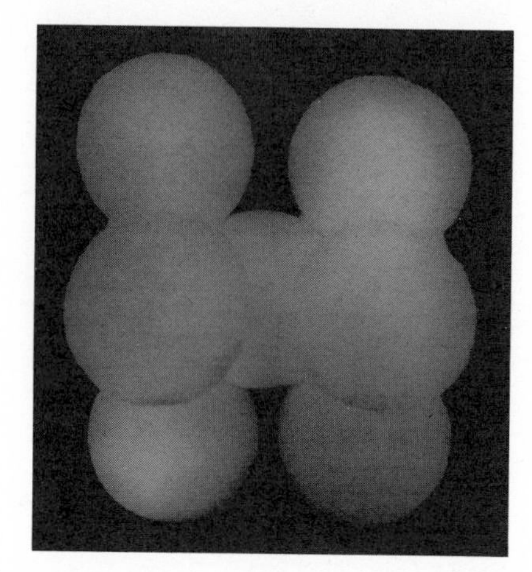

Fig. 2·9 Unit cell of the b.c.c. structure represented by points (left) and as spheres (right).

is shared by eight adjoining cubes and each face atom is shared by only one adjacent cube (see Fig. 2·12*b*), the unit cell contains:

$$\begin{aligned} 8 \text{ atoms at the corners} \times 1/8 &= 1 \text{ atom} \\ 6 \text{ face-centered atoms} \times 1/2 &= \underline{3 \text{ atoms}} \\ \text{Total} &= 4 \text{ atoms} \end{aligned}$$

This indicates that the f.c.c. structure is more densely packed than the b.c.c. structure. Another way to show the difference in packing is to calculate the fraction of the volume of a f.c.c. cell that is occupied by atoms and

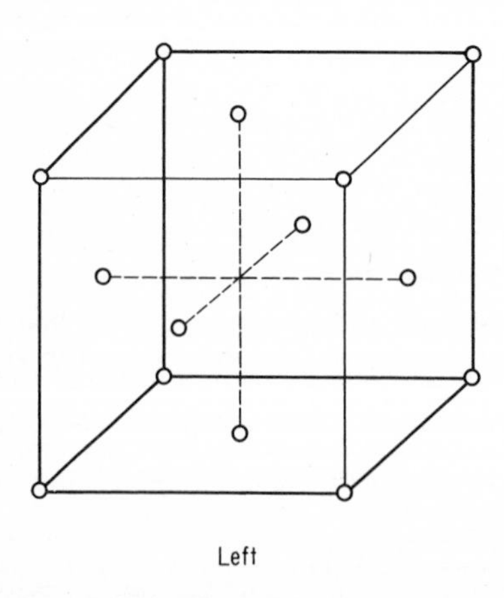

Fig. 2·10 Unit cell of the f.c.c. structure represented by points (left) and as spheres (right).

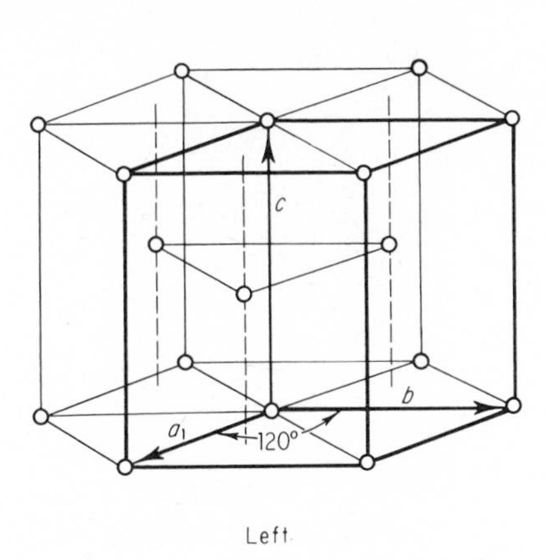

Fig. 2·11 The c.p.h. structure—as points, with the unit cell shown in heavy lines (left), and as spheres (right).

compare it to that of a b.c.c. cell. Since there are four atoms per unit cell and each atom is a sphere of radius r_a, then

$$V_{\text{atoms}} = 4(\tfrac{4}{3}\pi r_a^3) = \frac{16}{3}\pi r_a^3$$

and

$$V_{\text{cell}} = a^3$$

where a is the lattice parameter. It is now necessary to find the cell volume in terms of r_a. Consider a cube face as shown:

$$a\sqrt{2} = 4r_a \qquad \text{or } a = 2\sqrt{2}r_a$$

$$\text{Packing factor} = \frac{V_{\text{atoms}}}{V_{\text{cell}}} = \frac{\frac{16}{3}\pi r_a^3}{(2\sqrt{2}r_a)^3} = \frac{\pi}{3\sqrt{2}} = 0.74$$

It is left as an exercise for the student to show that the packing factor for the b.c.c. structure turns out to be $\pi\sqrt{3}/8$, or 0.68.

Examples of metals that crystallize in the f.c.c. lattice are aluminum, nickel, copper, gold, silver, lead, platinum, and gamma (γ) iron.

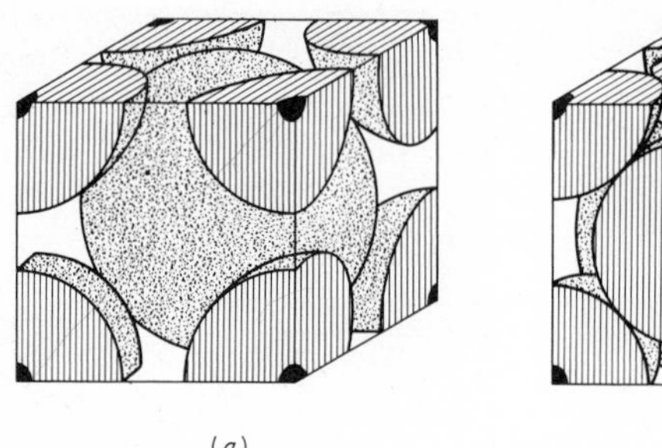

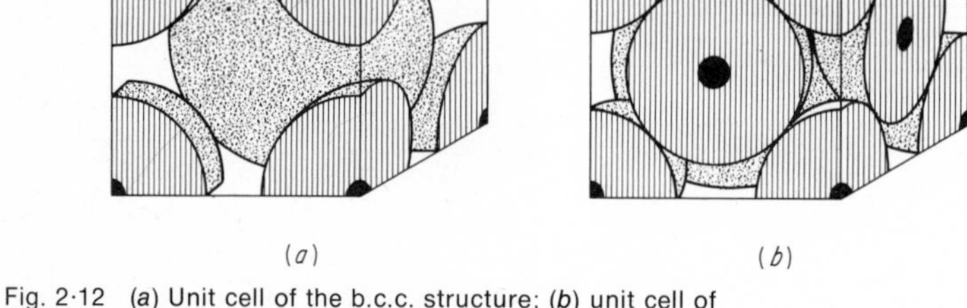

Fig. 2·12 (*a*) Unit cell of the b.c.c. structure; (*b*) unit cell of the f.c.c. structure. (From "Basic Metallurgy," American Society for Metals, Metals Park, Ohio, 1967.)

2·15 Close-packed Hexagonal The usual picture of the close-packed hexagonal lattice shows two basal planes in the form of regular hexagons with an atom at each corner of the hexagon and one atom at the center. In addition, there are three atoms in the form of a triangle midway between the two basal planes. If the basal plane is divided into six equilateral triangles, the additional three atoms are nestled in the center of alternate equilateral triangles (Fig. 2·11).

The parallel repetition of this hexagonal prism will not build up the entire lattice. The true unit cell of the hexagonal lattice is in fact only the portion shown by heavy lines in Fig. 2·11. The hexagonal prism, therefore, contains two whole unit cells and two halves.

It may not be readily apparent from the unit cell why the structure is called hexagonal. It remains to show that, if a number of these unit cells are packed together with axes parallel to one another, as in a space lattice, a hexagonal prism may be carved out of them.

Figure 2·13 shows many unit cells, with the open circles representing atoms in the plane of the paper and the filled circles representing atoms halfway above and below. The hexagonal lattice derived from the unit cells is shown by means of either the filled or the open circles. In each case, there are seven atoms representing the basal plane and three atoms in the form of a triangle in the center of alternate equilateral triangles.

Since each atom at the corner of the unit cell is shared by eight adjoining cells and one atom inside the cell cannot be shared, the c.p.h. unit cell contains two atoms. Examples of metals that crystallize in this type of structure are magnesium, beryllium, zinc, cadmium, and hafnium.

The unit cell of the cubic system may be specified by a single lattice parameter a, but the hexagonal cell requires the width of the hexagon a and the distance between basal planes c. These determine the axial ratio c/a, which is sometimes given. It may be shown mathematically that the

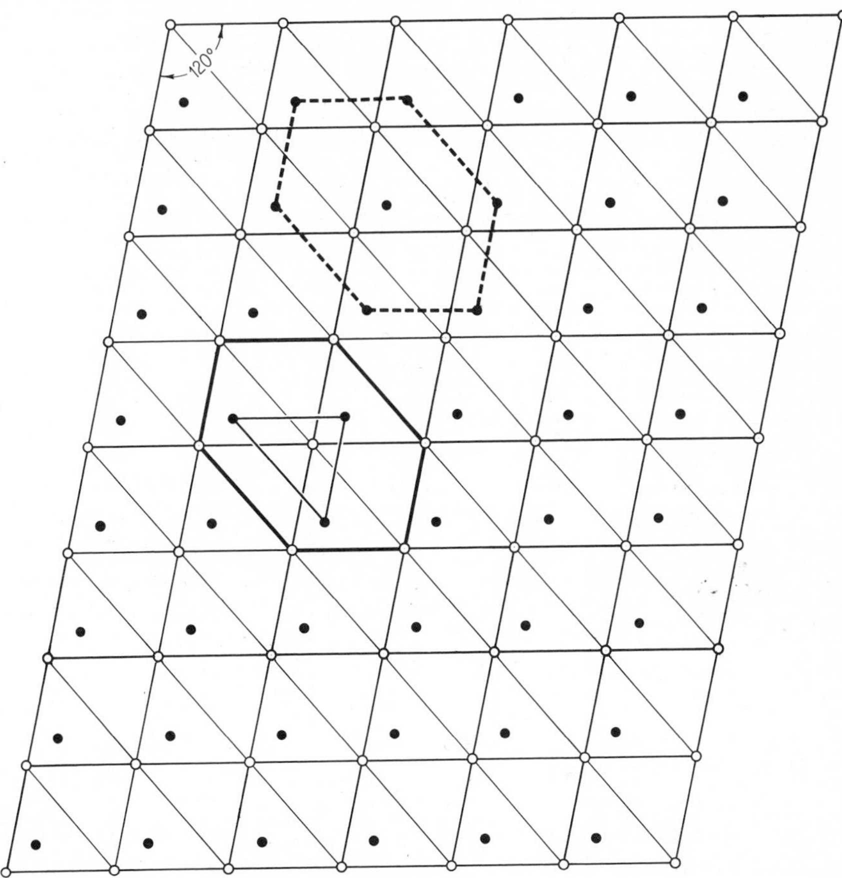

Fig. 2·13 Derivation of the hexagonal lattice from many unit cells.

axial ratio of a c.p.h. structure formed of spheres in contact is 1.633. In reality, metals of this structure have axial ratios that vary from 1.58 for beryllium to 1.88 for cadmium. Therefore, if the atoms are still considered to be in contact, they must be spheroidal in shape rather than spherical.

The types of crystal structure, lattice parameters, and other physical property data of the common metals are given in Table 2·8.

2·16 Polymorphism and Allotropy Polymorphism is the property of a material to exist in more than one type of space lattice in the solid state. If the change in structure is reversible, then the polymorphic change is known as *allotropy*. At least fifteen metals show this property, and the best-known example is iron. When iron crystallizes at 2800°F it is b.c.c. (δ Fe), at 2554°F the structure changes to f.c.c. (γ Fe), and at 1670°F it again becomes b.c.c. (α Fe).

TABLE 2·8 Physical Property Data of Some Common Metals*

METAL	SYMBOL	DENSITY AT 68°F, LB PER CU IN.	MELTING PT, °F	BOILING PT, °F	CRYSTAL STRUCTURE	LATTICE PARAMETER ANGSTROMS†	CLOSEST APPROACH OF ATOMS, ANGSTROMS†
Aluminum	Al	0.0975	1220.4	4442	f.c.c.	4.0491	2.862
Antimony	Sb	0.239	1166.9	2516	Rhombohedral	4.5065	2.904
Beryllium	Be	0.067	2332	5020	c.p.h.	$a = 2.2858$ $c = 3.5842$	2.221
Bismuth	Bi	0.354	520.3	2840	Rhombohedral	4.7457	3.111
Cadmium	Cd	0.313	609.6	1409	c.p.h.	$a = 2.9787$ $c = 5.617$	2.972
Carbon (graphite)	C	0.081	6740	8730	Hexagonal	$a = 2.4614$ $c = 6.7041$	1.42
Chromium	Cr	0.260	3407	4829	b.c.c.	2.884	2.498
Copper	Cu	0.324	1981.4	4703	f.c.c.	3.6153	2.556
Gold	Au	0.698	1945.4	5380	f.c.c.	4.078	2.882
Iron (α)	Fe	0.284	2797.7	5430	b.c.c.	2.8664	2.4824
Lead	Pb	0.4097	621.3	3137	f.c.c.	4.9489	3.499
Magnesium	Mg	0.0628	1202	2025	c.p.h.	$a = 3.2088$ $c = 5.2095$	3.196
Manganese	Mn	0.270	2273	3900	Cubic (complex)	8.912	2.24
Molybdenum	Mo	0.369	4730	10,040	b.c.c.	3.1468	2.725
Nickel	Ni	0.322	2647	4950	f.c.c.	3.5238	2.491
Platinum	Pt	0.775	3217	8185	f.c.c.	3.9310	2.775
Silicon	Si	0.084	2570	4860	Diamond cubic	5.428	2.351
Silver	Ag	0.379	1760.9	4010	f.c.c.	4.086	2.888
Tin	Sn	0.2637	449.4	4120	b.c. tetragonal	$a = 5.8314$ $c = 3.1815$	3.016
Titanium	Ti	0.164	3035	5900	c.p.h.	$a = 2.9503$ $c = 4.683$	2.91
Tungsten	W	0.697	6170	10,706	b.c.c.	3.1585	2.734
Vanadium	V	0.220	3450	6150	b.c.c.	3.039	2.632
Zinc	Zn	0.258	787	1663	c.p.h.	$a = 2.6649$ $c = 4.9470$	2.6648

* "Metals Handbook," 1961 ed., American Society for Metals, Metals Park, Ohio.
† One angstrom = 10^{-8} cm.

2·17 Crystallographic Planes The layers of atoms or the planes along which atoms are arranged are known as *atomic* or *crystallographic planes.* The relation of a set of planes to the axes of the unit cell is designated by *Miller indices.* One corner of the unit cell is assumed to be the origin of the space coordinates, and any set of planes is identified by the reciprocals of its intersections with these coordinates. The unit of the coordinates is the lattice parameter of the crystal. If a plane is parallel to an axis, it intersects it at infinity.

In Fig. 2·14, or the cubic system, the crosshatched plane *BCHG* intersects the *Y* axis at one unit from the origin and is parallel to the *X* and *Z* axes or intersects them at infinity. Therefore,

	X	Y	Z
Intersection	∞	1	∞
Reciprocal	$\frac{1}{\infty}$	$\frac{1}{1}$	$\frac{1}{\infty}$
Miller indices	0	1	0

The illustrated plane has Miller indices of (010). If a plane cuts any axis on the negative side of the origin, the index will be negative and is indicated by placing a minus sign above the index, as $(h\bar{k}l)$. For example, the Miller indices of the plane *ADEF* which goes through the origin (point *A*) cannot be determined without changing the location of the origin. Any point in the cube may be selected as the origin. For convenience, take point *B*. The

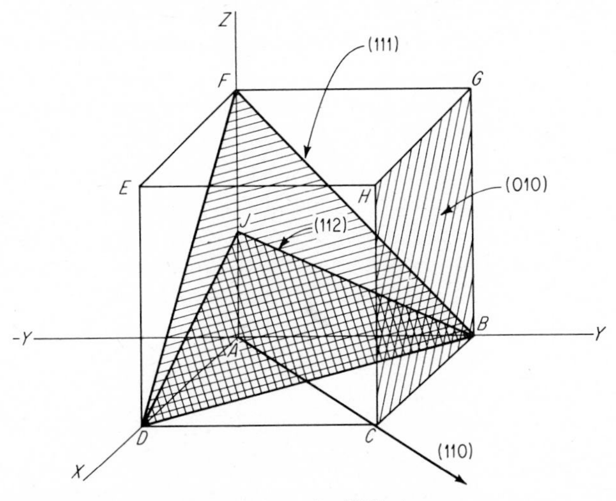

Fig. 2·14 Determination of Miller indices: the (010) plane, the (111) plane, and the (112) plane.

plane *ADEF* is parallel to the *X* axis (*BC*) and the *Z* axis (*BG*) but intersects the *Y* axis at -1. The plane has Miller indices of $(0\bar{1}0)$.

As another illustration, the Miller indices of the plane *BDJ* (Fig. 2·14) may be determined as follows:

	X	*Y*	*Z*
Intersection	1	1	$\frac{1}{2}$
Reciprocal	$\frac{1}{1}$	$\frac{1}{1}$	$\frac{1}{\frac{1}{2}}$
Miller indices	1	1	2

This plane has Miller indices of (112).

If the Miller indices of a plane result in fractions, these fractions must be cleared. For example, consider a plane that intersects the *X* axis at 1, the *Y* axis at 3, and the *Z* axis at 1. Taking reciprocals gives indices of 1, $\frac{1}{3}$, and 1. Multiplying through by 3 to clear fractions results in Miller indices of (313) for the plane.

All parallel planes have the same indices. Parentheses (*hkl*) around Miller indices signify a specific plane or set of parallel planes. Braces signify a family of planes of the same "form" (which are equivalent in the crystal), such as the cube faces of a cubic crystal: $\{100\} = (100) + (010) + (001) + (\bar{1}00) + (0\bar{1}0) + (00\bar{1})$.

Reciprocals are not used to determine the indices of a direction. In order to arrive at a point on a given direction, consider that starting at the origin it is necessary to move a distance *u* times the unit distance *a* along the *X* axis, *v* times the unit distance *b* along the *Y* axis, and *w* times the unit distance *c* along the *Z* axis. If *u, v,* and *w* are the smallest integers to accomplish the desired motion, they are the indices of the direction and are enclosed in square brackets [*uvw*]. A group of similar directions are enclosed in angular brackets $\langle uvw \rangle$. For example, in Fig. 2·14, to determine the direction *AC,* starting at the origin (point *A*), it is necessary to move one unit along the *X* axis to point *D* and one unit in the direction of the *Y* axis to reach point *C*. The direction *AC* would have indices of [110]. In a cubic crystal, a direction has the same indices as the plane to which it is perpendicular.

An approximate idea of the packing of atoms on a particular plane may be obtained by visualizing a single unit cell of the b.c.c. and f.c.c. structure. Considering the atoms as the lattice points, the number of atoms on a particular plane would be:

PLANE	b.c.c.	f.c.c.
(100)	4	5
(110)	5	6
(111)	3	6
(120)	2	3
(221)	1	1

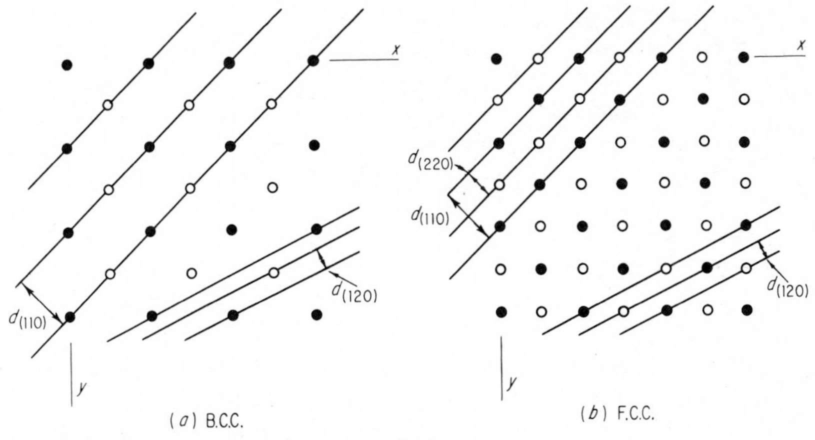

Fig. 2·15 Projection of the lattice on a plane perpendicular to the Z axis to illustrate interplanar spacing. Filled circles are in the plane of the paper.

An infinite number of planes may be taken through the crystal structure, but most are just geometrical constructions and have no practical importance. Remembering that each complete set of parallel planes must account for all the atoms, the most important planes are the ones of high atomic population and largest interplanar distance. In the b.c.c. structure these are the $\{110\}$ planes, and in the f.c.c. structure these are the $\{111\}$ planes (see Figs. 2·15 and 2·16).

2·18 X-ray Diffraction One may wonder at this point how it is possible to measure lattice dimensions since atomic spacings are in the order of only a few angstrom units. The most useful tool for studying crystal structure is x-ray diffraction, and a brief introduction will be given here.

Since x-rays have a wavelength about equal to the distance separating

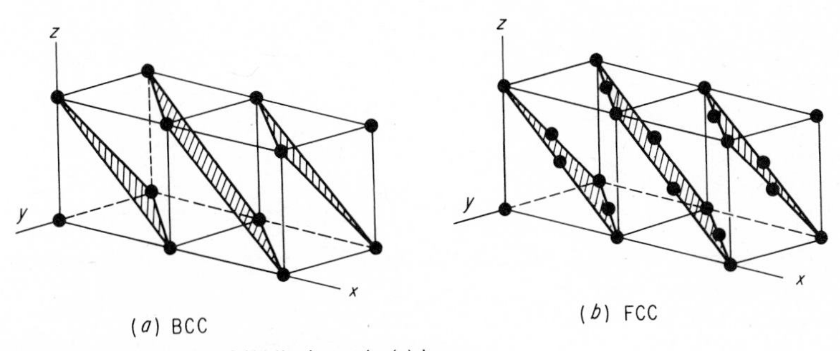

Fig. 2·16 Interplanar spacing of (111) planes in (*a*) b.c.c. and (*b*) f.c.c. structures. (By permission from L. H. Van Vlack, "Elements of Materials Science," Addison-Wesley Publishing Company, Inc., Reading, Mass., 1959.)

the atoms in solids, when x-rays are directed at a crystalline material they will be diffracted by the planes of atoms in the crystal. If a beam of x-rays strikes a set of crystal planes at some arbitrary angle, there will usually be no reflected beam because the rays reflected from the crystal planes must travel different lengths and will tend to be "out of phase" or to cancel each other out. However, at a particular angle, known as the Bragg angle, θ, the reflected rays will be in phase because the distance traveled will be an integral number of wavelengths. Consider the parallel planes of atoms in Fig. 2·17 from which a wave is diffracted. The waves may be reflected from an atom at H or H' and remain in phase at K; however, the reflected rays from atoms in subsurface planes, such as H'', must also be in phase at K. In order for this to occur, the distance $MH''P$ must equal one or more integral wavelengths. If d is the spacing between planes and θ is the angle of incidence, the distance MH'' is equal to $d \sin \theta$ and the distance $MH''P$ is equal to $2d \sin \theta$ or

$$n\lambda = 2d \sin \theta$$

where n can have values of 1, 2, 3, etc. This is known as the Bragg equation. Without going into the details of equipment, for a given wavelength λ, θ can be measured and d calculated.

2·19 The States of Matter Three states of matter are distinguishable: gas, liquid, and solid. In the gaseous state the metal atoms occupy a great deal of space because of their rapid motion. Their motion is entirely random, and as they travel they collide with each other and the walls of the container. The combination of all the collisions with the wall is the pressure of the gas on the wall. The atoms move independently and are usually widely separated so that the attractive forces between atoms are negligible. The arrangement of atoms in a gas is one of complete disorder.

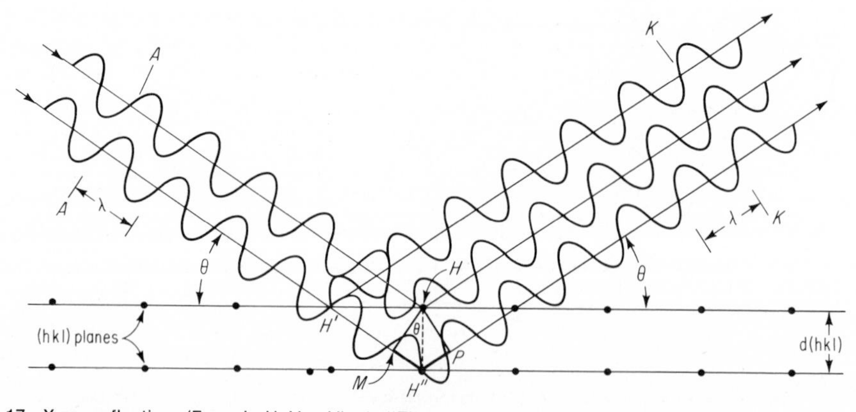

Fig. 2·17 X-ray reflection. (From L. H. Van Vlack, "Elements of Materials Science," Addison-Wesley Publishing Company, Inc., Reading, Mass., 1959.)

At some lower temperature, the kinetic energy of the atoms has decreased so that the attractive forces become large enough to bring most of the atoms together in a liquid. Not all the atoms are in the liquid. Atoms remain in the vapor above the liquid, and there is a continual interchange of atoms between the vapor and liquid across the liquid surface. In a confined vessel, at a definite temperature, the interchange of atoms will reach equilibrium and there will be a constant value of vapor pressure of the gas above the liquid. If the vapor is free to escape, equilibrium will not be reached and more atoms will leave the liquid surface than are captured by it, resulting in evaporation. Evidence of attractive forces between atoms in a liquid may be demonstrated by the application of pressure. A gas may be easily compressed into a smaller volume, but it takes a high pressure to compress a liquid. There is, however, still enough free space in the liquid to allow the atoms to move about irregularly.

As the temperature is decreased, the motions are less vigorous and the attractive forces pull the atoms closer together until the liquid solidifies. Most materials contract upon solidification, indicating a closer packing of atoms in the solid state. The atoms in the solid are not stationary but are vibrating around fixed points, giving rise to the orderly arrangement of crystal structures discussed previously.

2·20 Mechanism of Crystallization Crystallization is the transition from the liquid to the solid state and occurs in two stages:

1 Nuclei formation
2 Crystal growth

Although the atoms in the liquid state do not have any definite arrangement, it is possible that some atoms at any given instant are in positions exactly corresponding to the space lattice they assume when solidified (Fig. 2·18). These chance aggregates or groups are not permanent but continually break up and reform at other points. How long they last is determined by the temperature and size of the group. The higher the temperature, the greater the kinetic energy of the atoms and the shorter the life of the group. Small groups are very unstable since they are formed of only a small number of atoms and the loss of only one atom may destroy the group. When the temperature of the liquid is decreased, the atom movement decreases, lengthening the life of the group, and more groups will be present at the same time.

Atoms in a material have both kinetic and potential energy. Kinetic energy is related to the speed at which the atoms move and is strictly a function of temperature. The higher the temperature, the more active are the atoms and the greater is their kinetic energy. Potential energy, on the other hand, is related to the distance between atoms. The greater the average distance between atoms, the greater is their potential energy.

Now consider a pure metal at its freezing point where both the liquid

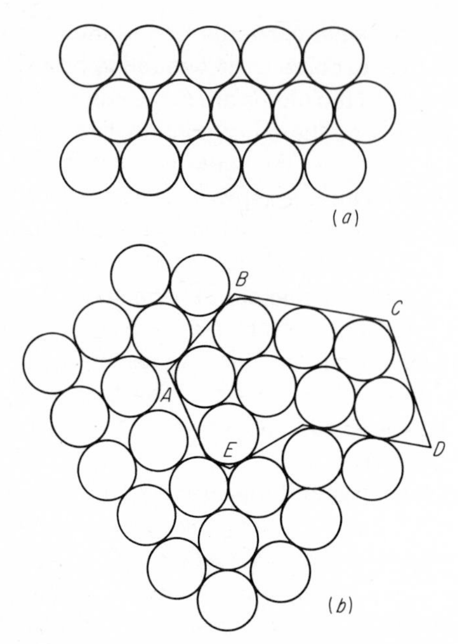

Fig. 2·18 Schematic diagram of structures of (*a*) crystal and (*b*) liquid. Area *ABCDE* in liquid is identical in arrangement as in crystal. (From Chalmers, "Physical Metallurgy," John Wiley & Sons, Inc., New York, 1959.)

and solid states are at the same temperature. The kinetic energy of the atoms in the liquid and the solid must be the same, but there is a significant difference in potential energy. The atoms in the solid are much closer together, so that solidification occurs with a release of energy. This difference in potential energy between the liquid and solid states is known as the *latent heat of fusion.* However, energy is required to establish a surface between the solid and liquid. In pure materials, at the freezing point, insufficient energy is released by the heat of fusion to create a stable boundary, and some undercooling is always necessary to form stable nuclei. Subsequent release of the heat of fusion will raise the temperature to the freezing point (Fig. 2·19). The amount of undercooling required may be reduced by the presence of solid impurities which reduce the amount of surface energy required.

When the temperature of the liquid metal has dropped sufficiently below its freezing point, stable aggregates or nuclei appear spontaneously at various points in the liquid. These nuclei, which have now solidified, act as centers for further crystallization. As cooling continues, more atoms tend to freeze, and they may attach themselves to already existing nuclei or form new nuclei of their own. Each nucleus grows by the attraction of atoms from the liquid into its space lattice. Crystal growth continues in

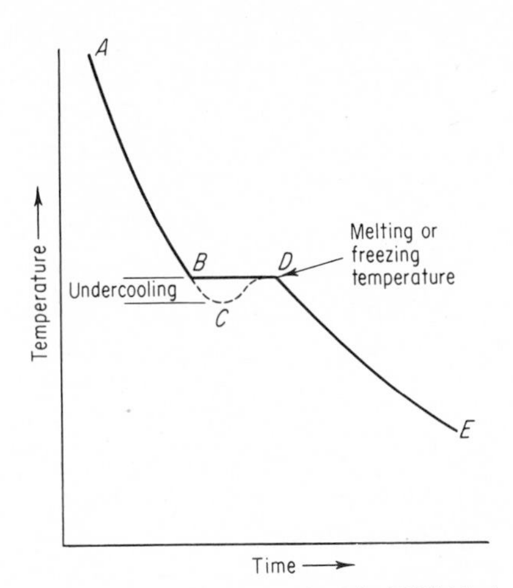

Fig. 2·19 Cooling curve for a pure metal; *ABDE* ideal, *ABCDE* actual.

three dimensions, the atoms attaching themselves in certain preferred directions, usually along the axes of the crystal. This gives rise to a characteristic treelike structure which is called a *dendrite* (Fig. 2·20). Since each nucleus is formed by chance, the cyrstal axes are pointed at random and the dendrites growing from them will grow in different directions in each crystal. Finally, as the amount of liquid decreases, the gaps between the arms of the dendrite will be filled and the growth of the dendrite will be

Fig. 2·20 Magnesium dendrites growing from liquid.

mutually obstructed by that of its neighbors. This leads to a very irregular external shape. The crystals found in all commercial metals are commonly called *grains* because of this variation in external shape. The area along which crystals meet, known as the *grain boundary*, is a region of mismatch (Fig. 2·21). This leads to a noncrystalline (amorphous) structure at the grain boundary with the atoms irregularly spaced. Since the last liquid to solidify is generally along the grain boundaries, there tends to be a higher concentration of impurity atoms in that area. Figure 2·22 shows schematically the process of crystallization from nuclei to the final grains.

2·21 Crystal Imperfections It is apparent from the preceding section that most materials when solidified consist of many crystals or grains. It is possible under carefully controlled conditions to manufacture a single crystal. The so-called metal "whiskers," which in some cases are made directly from the vapor, are nearly perfect single crystals. Figure 2·23 shows tin whiskers growing from a copper substrate made with the scanning electron microscope at a magnification of 20,000 times.

Single crystals may also be made by withdrawing a crystal fragment or seed at a carefully controlled speed from a melt which is held at just above the freezing point. In any case, single crystals approach a nearly perfect

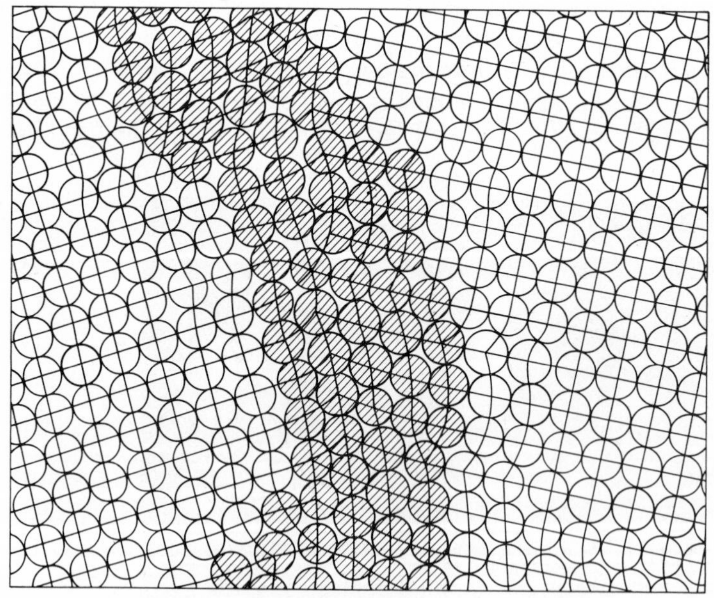

Fig. 2·21 Schematic representation of a grain boundary between two crystals. The cross-hatched atoms are those which constitute the boundary material. (From L. F. Mondolfo and O. Zmeskal, "Engineering Metallurgy," McGraw-Hill Book Company, New York, 1955.)

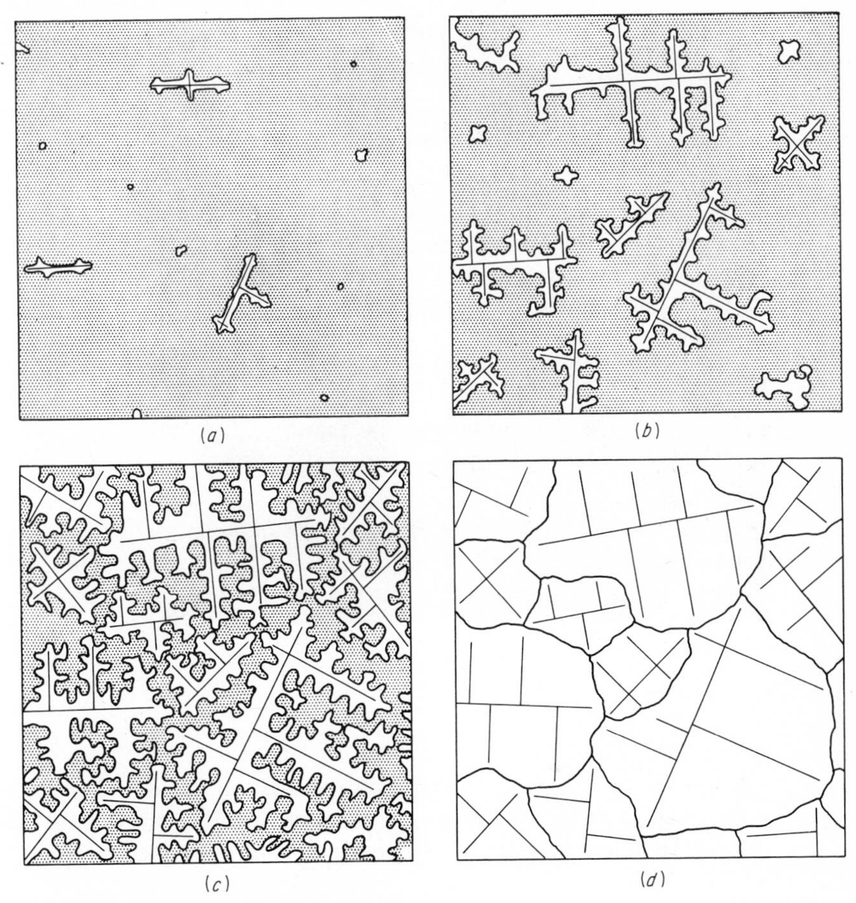

Fig. 2·22 Schematic representation of the process of crystallization by nucleation and dendritic growth. (From L. F. Mondolfo and O. Zmeskal, "Engineering Metallurgy," McGraw-Hill Book Company, New York, 1955.)

lattice structure. It is possible to calculate the theoretical strength of a metal by the force required to separate the bond between adjoining atoms. This turns out to be several million pounds per square inch and is approached by the strength of single crystals or metal whiskers. However, the ordinary strength of metals is 100 to 1,000 times less. Why is there such a difference? The answer is found in the occurrence of defects in the crystal structure.

It is interesting to realize the amount of activity that is occurring on the surface of a crystal during growth. A very slow growth rate, such as 1 mm per day, requires the deposition of about one hundred layers of atoms per

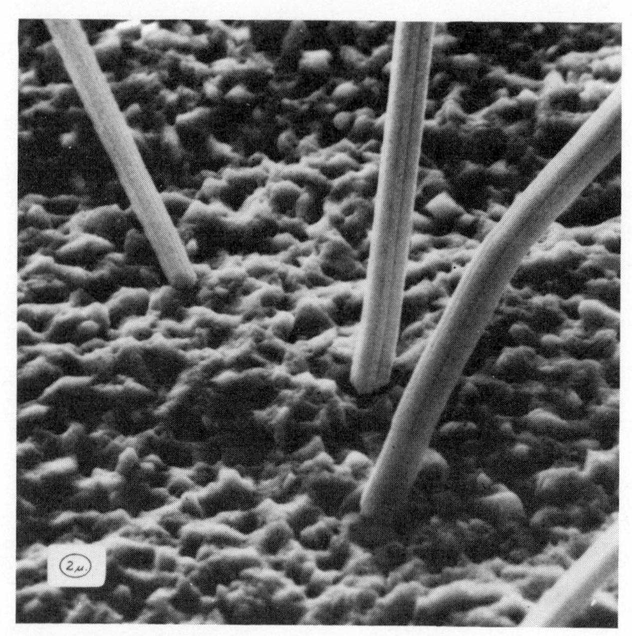

Fig. 2·23 Tin whiskers growing from a copper substrate. Made with the scanning electron microscope at a magnification of 20,000×.

second on the surface. All these atoms must be laid down in exactly the right sort of order for the crystal to be perfect. It is therefore not surprising that few crystals are perfect and that imperfections exist on an atomic scale.

The most important crystal imperfections are vacancies, interstitials, and dislocations. Vacancies are simply empty atom sites (Fig. 2·24*a*). It may be shown by thermodynamic reasoning that lattice vacancies are a stable feature of metals at all temperatures above absolute zero. By successive jumps of atoms, just like playing Chinese checkers, it is possible for a vacancy to move in the lattice structure and therefore play an important part in diffusion of atoms through the lattice. Notice that the atoms surrounding a vacancy tend to be closer together, distorting the lattice planes.

It is possible, particularly in lattice structures that are not close-packed and in alloys between metals that have atoms widely different in atomic diameters, that some atoms may fall into interstitial positions or in the spaces of the lattice structure (Fig. 2·24*b*). Interstitials tend to push the surrounding atoms farther apart and also produce distortion of the lattice planes. Vacancies are not only present as a result of solidification but can

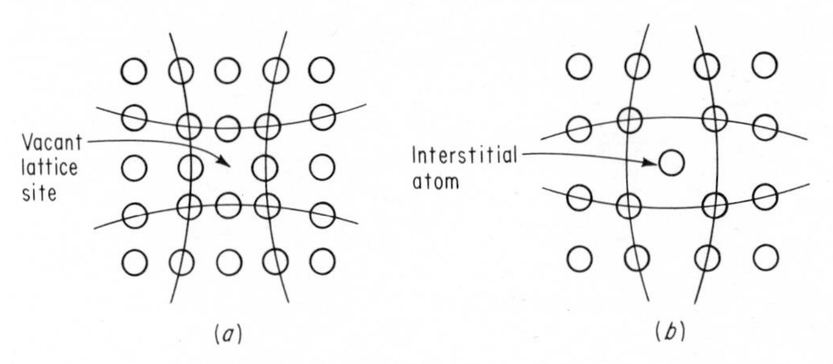

Fig. 2·24 Vacancy and interstitial crystal defects.

be produced by raising the temperature or by irradiation with fast-moving nuclear particles. Interstitial atoms may be produced by the severe local distortion during plastic deformation as well as by irradiation.

A dislocation may be defined as a disturbed region between two substantially perfect parts of a crystal. Two simple kinds of dislocation are illustrated schematically in Fig. 2·25. The edge dislocation consists of an extra half plane of atoms in the crystal. The screw dislocation is so named because of the spiral surface formed by the atomic planes around the screw-dislocation line. The dislocation line produces compressive stress below the dislocation and tensile stresses above it and a disturbed region in the lattice structure. Where the mismatch between neighboring grains is not too great, the grain boundary may be represented by an array of parallel edge dislocations (Fig. 2·26). The creation, multiplication, and interaction between dislocations are very useful in explaining many of the properties of metals.

The student should refer to the references at the end of this chapter for a more complete description of the types and theory of dislocations.

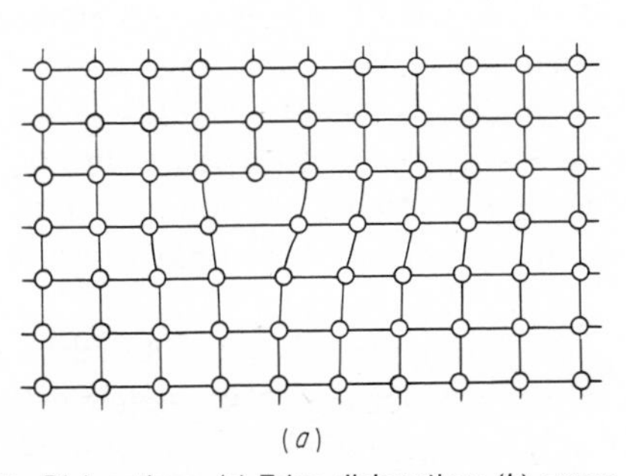

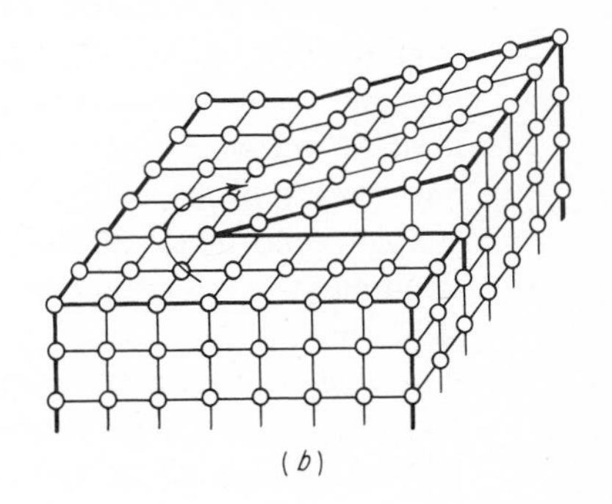

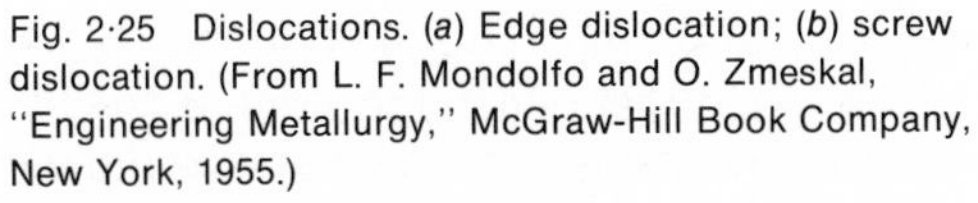
Fig. 2·25 Dislocations. (*a*) Edge dislocation; (*b*) screw dislocation. (From L. F. Mondolfo and O. Zmeskal, "Engineering Metallurgy," McGraw-Hill Book Company, New York, 1955.)

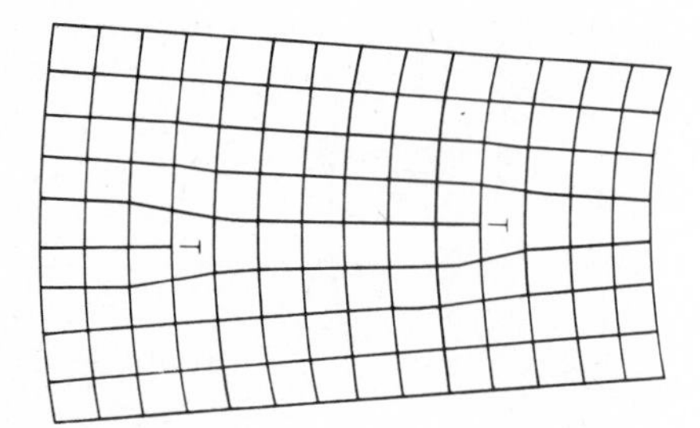

Fig. 2·26 A small angle boundary composed of edge dislocations, indicated at the T symbols. (By permission from C. S. Barrett, "Structure of Metals," 2d ed., McGraw-Hill Book Company, New York, 1952.)

2·22 Macrodefects in Castings The preceding section discussed defects on an atomic scale that arise from solidification. Other defects that may result from solidification are large enough to be visible to the naked eye. These are known as *macrodefects.* The most common macrodefects are shrinkage cavities and porosity.

Liquid metals, with few exceptions, undergo a contraction in volume due to solidification. This decrease in volume may be as much as 6 percent. In a properly designed mold, with provision for liquid supply to the portion that solidifies last, the contraction in volume presents no serious problem. If, however, the entire exterior of the casting should solidify first, the decrease in volume of the interior during solidification will result in a large *shrinkage cavity* at the mid-section, as shown in Fig. 2·27. In the solidification of steel ingots, the shrinkage cavity, called *pipe*, is usually concentrated in the top central portion of the ingot. This portion is cut off and discarded before working.

The ideal solidification would be that in which the metal first freezes at the bottom of the mold and continues upward to a riser at the top; however, heat is dissipated more rapidly from the top of the mold. To minimize the formation of shrinkage cavities, abrupt changes in thickness and combinations of heavy and light sections should be avoided. If the casting does have heavy sections, they should be designed with risers at the top to supply liquid metal during solidification. Heavy sections should be cast uppermost in the mold, and chills may be used in the sand adjacent to the slow-cooling parts.

Porosity or *blowholes* occur whenever gases are trapped in the casting. They are usually more numerous and smaller than shrinkage cavities and may be distinguished by their rounded form (Fig. 2·28). Air may be entrapped in the casting by the sudden rush of metal during pouring. Since gases are generally more soluble in liquid metal than the solid, dissolved gases may be liberated during solidification. Gases may also be produced

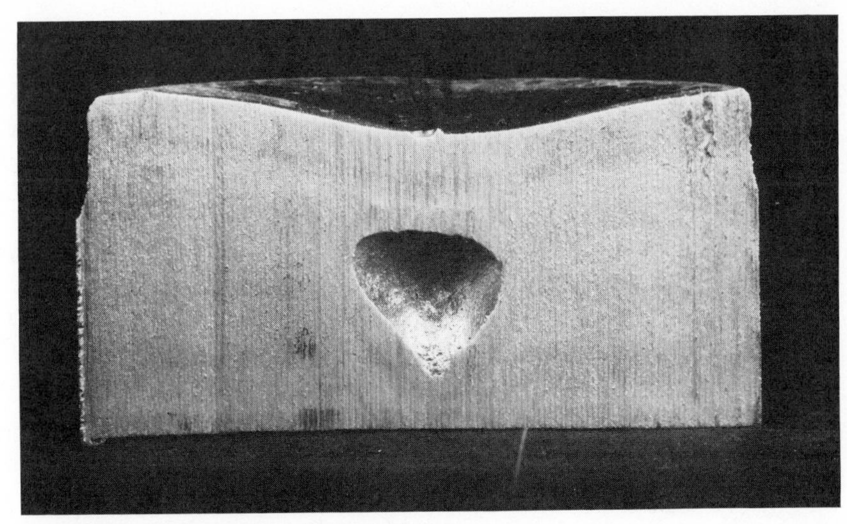

Fig. 2·27 Shrinkage cavity in a niobium (columbium) alloy billet. (Fansteel Metallurgical Corporation.)

by reaction of the liquid metal with volatile substances, such as moisture, in the mold. Porosity may be greatly reduced by proper venting of the mold, and by not unduly compacting the sand.

Hot tears are cracks due to heavy shrinkage strains set up in the solid casting just after solidification. A common cause is the failure of the sand mold to collapse and allow the casting to contract. Hot tears may also result from the same nonuniform cooling conditions that give rise to shrink-

Fig. 2·28 Gas pockets and porosity in a tool steel casting, enlarged 2×.

age cavities. Proper design of the casting will minimize the danger of hot tears. Fig. 2·29 shows a hot tear in a stainless steel casting.

2·23 Grain Size The size of grains in a casting is determined by the relation between the rate of growth *G* and the rate of nucleation *N*. If the number of nuclei formed is high, a fine-grained material will be produced, and if only a few nuclei are formed, a coarse-grained material will be produced. The rate of cooling is the most important factor in determining the rate of nucleation and therefore the grain size. Rapid cooling (chill cast) will result in a large number of nuclei formed and fine grain size, whereas in slow cooling (sand cast or hot mold) only a few nuclei are formed and they will have a chance to grow, depleting the liquid before more nuclei can form.

Other factors that increase the rate of nucleation, thus promoting the formation of fine grain, are:

1 Insoluble impurities such as aluminum and titanium that form insoluble oxides in steel.

2 Stirring the melt during solidification which tends to break up the crystals before they have a chance to grow very large.

The rate of growth relative to the rate of nucleation is greatest at or just under the freezing point. If the liquid is kept accurately at the freezing temperature and the surface is touched by a tiny crystal (seed), the crystal will grow downward into the liquid. If it is withdrawn slowly, a single crystal can be produced.

In general, fine-grained materials exhibit better toughness or resistance to shock (Fig. 2·30). They are harder and stronger than coarse-grained material.

In industrial casting processes, where a hot liquid is in contact with an originally cool mold, a temperature gradient (difference in temperature)

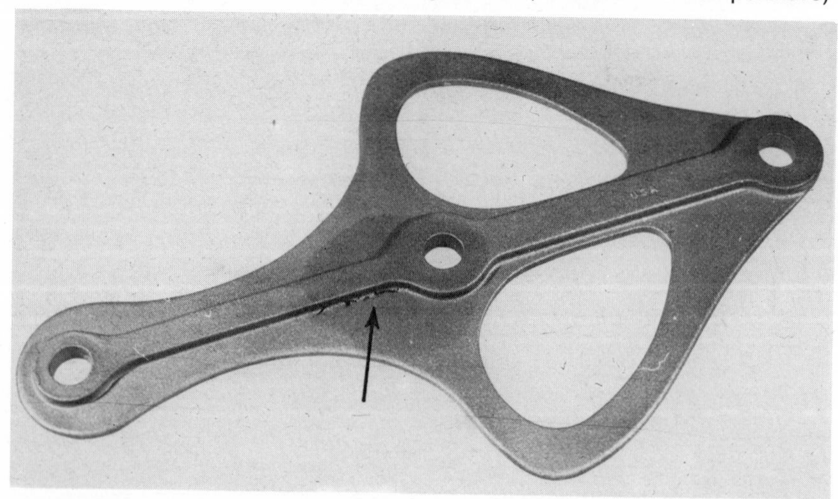

Fig. 2·29 Hot tear (arrow) in a stainless steel casting.

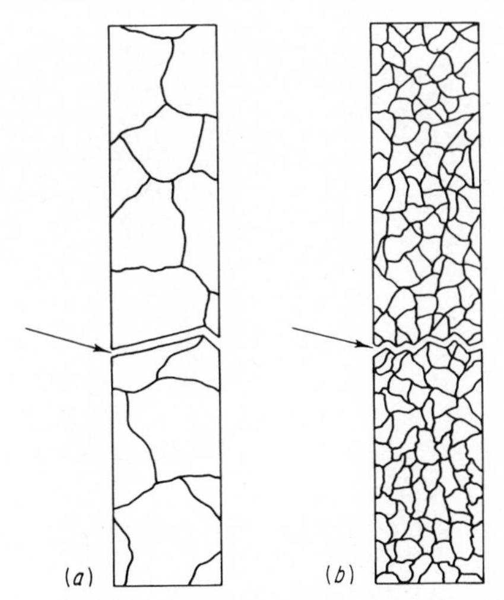

Fig. 2·30 (*a*) Brittleness of a coarse-grained metal under shock; (*b*) frequent change of direction of force in rupture of a fine-grained metal. (By permission from Doan and Mahla, "Principles of Physical Metallurgy," McGraw-Hill Book Company, New York, 1941.)

will exist in the liquid. The outside is at a lower temperature than the center and therefore starts to solidify first. Thus many nuclei are formed at the mold wall and begin to grow in all directions. They soon run into the side of the mold and each other, so that the only unrestricted direction for growth is toward the center. The resulting grains are elongated columnar ones, perpendicular to the surface of the mold. This is illustrated for high-purity lead, as cast, in Fig. 2·31. Next to the mold wall, where the cooling rate is fast, the grains are small, while toward the center, where the cooling rate is much slower, the grains are larger and elongated.

If the mold has sharp edges, a plane of weakness will develop from this corner because both gaseous and solid impurities tend to concentrate along this plane. Such castings may cause internal rupture during rolling or forging operations. To avoid this plane, it is good casting design to provide the mold with rounded corners.

2·24 Grain Size Measurement The three basic methods for grain size estimation recommended by the ASTM are:

1 Comparison method
2 Intercept (or Heyn) method
3 Planimetric (or Jeffries) method

Comparison Method The specimen is prepared and etched according to the metallographic procedure described in Chap. 1. The image of the microstructure projected at a magnification of 100×, or a photomicrograph of the structure at the same magnification, is compared with a series of

Fig. 2·31 High-purity lead, as cast. Magnified 2×. (National Lead Co. Research Laboratory.)

graded standard grain-size charts (ASTM E112–63). By trial and error a match is secured, and the grain size of the metal is then designated by a number corresponding to the index number of the matching chart. Metals showing a mixed grain size are rated in a similar manner, and it is customary in such cases to report the grain size in terms of two numbers denoting the approximate percentage of each size present. The comparison method is most convenient and sufficiently accurate for specimens consisting of equiaxed grains. The ASTM grain-size number n may be obtained as follows:

$$N = 2^{n-1}$$

where N is the number of grains observed per square inch at 100× magnification (see Table 2·9).

Intercept (or Heyn) Method The grain size is estimated by counting on a ground-glass screen, or photomicrograph, or on the specimen itself, the number of grains intersected by one or more straight lines. Grains touched by the end of the line count only as half grains. Counts are made on at

TABLE 2·9 ASTM grain-size Ranges, $N = 2^{n-1}$

GRAIN SIZE NO.	GRAINS PER SQ IN. AT 100×	
	MEAN	RANGE
$n =$ 1	$N =$ 1	------------
2	2	1.5–3
3	4	3–6
4	8	6–12
5	16	12–24
6	32	24–48
7	64	48–96
8	128	96–192
9	256	192–384
10	512	384–768

least three fields to assure a reasonable average. The length of the line in millimeters divided by the average number of grains intersected by it gives the average intercept length or grain "diameter." The intercept method is recommended particularly for grains that are not equiaxed.

Planimetric (or Jeffries) Method A circle or a rectangle of known area (usually 5,000 sq mm) is inscribed on a photomicrograph or on the ground glass of the metallograph. A magnification should be selected which will give at least 50 grains in the field to be counted. The sum of all the grains included completely within the known area plus one-half the number of grains intersected by the circumference of the area gives the total number or equivalent whole grains within the area. Knowing the magnification of the specimen, the number of grains per square millimeter is determined by multiplying the equivalent number of whole grains by the corresponding magnification factor (Jeffries' multiplier) f given in Table 2·10.

TABLE 2·10 Relationship between Magnification Used and Jeffries' Multiplier f for an Area of 5,000 sq. mm

MAGNIFICATION USED	f
1	0.002
25	0.125
50	0.5
75	1.125
100	2.0
200	8.0
300	18.0
500	50.0
1000	200.0

Thus, if the equivalent number of whole grains is found to be 75 at a magnification of 100×, the number of grains per square millimeter is equal to 75 × 2.0, or 150.

In case of dispute, the planimetric method is preferred over the comparison method for equiaxed grains.

It is important to realize that in using any method to determine grain size, the estimation is not a precise measurement. A metal structure is a mixture of three-dimensional crystals of varying sizes and shapes. Even if all the crystals were identical in size and shape, the cross sections of the grains on the polished surface would show varying areas depending upon where the plane cuts each individual crystal. Therefore, no two fields of observation can be exactly the same.

QUESTIONS

2·1 Differentiate between atomic number and atomic weight.

2·2 What is meant by an *isotope*?

2·3 Explain the arrangement of the elements in the periodic table.

2·4 What may be said about all the elements in the same group of the periodic table?

2·5 Why are some elements known as *transition elements*?

2·6 Using the system explained in Sec. 2·2, write the electron configuration of the elements in the fourth period.

2·7 Define a *solid*. Glass is not considered a true solid. Why?

2·8 How does the metallic bond differ from the ionic and covalent bonds?

2·9 Describe an experiment to show that the atoms of a solid are in motion.

2·10 Explain the existence of attractive and repulsive forces between atoms.

2·11 What is *atomic diameter* and how may its value be approximated?

2·12 How does the atomic diameter change within one group of the periodic table? Why?

2·13 How does the atomic diameter change within a period of the periodic table? Why?

2·14 Give three examples of allotropic metals. For each example, give the temperatures and the changes in crystal structure.

2·15 What are the Miller indices of a plane that intersects the X axis at 2 and the Y axis at $1/2$ and is parallel to the Z axis? The structure is cubic.

2·16 What are the Miller indices of a plane in the cubic structure that goes through $Y = 1/2$, $Z = 1$, and is parallel to the X axis? Draw the cubic structure and crosshatch this plane.

2·17 What are the Miller indices of a plane in the cubic structure that intersects the X axis at $1/2$, the Y axis at 1, and is perpendicular to the XY plane?

2·18 When a gas liquefies, energy is released as the heat of vaporization. What is this energy due to?

2·19 Is there any difference in the kinetic energy of the atoms in the liquid and the gas at the boiling point? Explain.

2·20 Differentiate between a crystal, a dendrite, and a grain.

2·21 Why is the grain boundary irregular?

2·22 List three factors that tend to promote fine grain in a casting.
2·23 Why is it important to avoid sharp corners in castings?
2·24 Describe a method of obtaining a uniform grain size in a casting which has a thick and thin section.
2·25 In the cubic system, the interplanar spacing d_{hkl} measured at right angles to the planes is given by the following formula:

$$d_{hkl} = \frac{a}{\sqrt{h^2 + k^2 + l^2}}$$

where (hkl) are the Miller indices and a is the lattice parameter. Calculate the interplanar spacing for the (110), (111), (120), (221), and (123) planes of copper. Which of the above planes has the greatest interplanar spacing?
2·26 In the cubic system, the angle ϕ between the plane $(h_1k_1l_1)$ and the plane $(h_2k_2l_2)$ may be found by the following formula:

$$\cos\phi = \frac{h_1h_2 + k_1k_2 + l_1l_2}{\sqrt{(h_1^2 + k_1^2 + l_1^2)(h_2^2 + k_2^2 + l_2^2)}}$$

Calculate the angle between the (100) plane and the (110) plane; the (100) plane and the (111) plane; the (110) plane and the (111) plane.
2·27 An x-ray diffraction analysis of a crystal is made with x-rays having a wavelength of 0.58 Å. A reflection is observed at an angle of 6.45°. What is the interplanar spacing?

Ans. 2.575 Å

2·28 Density calculations may be used to check the validity of x-ray diffraction analysis.

$$\text{Density} = \frac{\text{weight/unit cell}}{\text{volume/unit cell}}$$

where weight/unit cell = (no. of atoms/unit cell) (weight of atom) and the weight of atom is the atomic weight/Avogadro's number (6.02×10^{23} atoms).

(*a*) Copper is f.c.c. and has a lattice parameter of 3.61 Å(10^{-8}cm). Calculate its density in gm/cm³, and check with the density value obtained in a handbook.
(*b*) Do the same with iron (b.c.c. and lattice parameter of 2.86 Å).

2·29 If copper is f.c.c. and has an atomic diameter of 2.556 Å, calculate the lattice parameter.
2·30 If iron is b.c.c. and has a lattice parameter of 2.86 Å, calculate the atomic diameter.
2·31 The equation for the equilibrium fraction of atom sites that are vacant is:

$$N = e^{-\Delta H/RT}$$

where ΔH = molar heat of reaction accompanying the formation of vacancies
R = gas content
T = absolute temperature in °K
Using ΔH = 20,000 cal/mol for aluminum
and R = 2 cal/mol-deg
calculate the fraction of lattice sites that are vacant at 300, 500, 1000, and 1300°K.
2·32 Using the planimetric method for measuring grain size, the number of equivalent grains at 200× was 62. Calculate the number of grains per square millimeter. What is the equivalent ASTM grain size? (Refer to Table II, ASTM E112–63).

REFERENCES

American Society for Metals: "Atom Movements," Metals Park, Ohio, 1951.

Barrett, C. S.: "Structures of Metals," 2d ed., McGraw-Hill Book Company, New York, 1952.

Cottrell, A. H.: "Dislocations and Plastic Flow in Crystals," Oxford University Press, Fair Lawn, N.J., 1956.

Cullity, B. D.: "Elements of X-ray Diffraction," Addison-Wesley Publishing Company, Inc., Reading, Mass., 1956.

Eisenstadt, M. M.: "Introduction to Mechanical Properties of Materials," The Macmillan Company, New York, 1971.

Guy, A. G.: "Elements of Physical Metallurgy," 2d ed., Addison-Wesley Publishing Company, Inc., Reading, Mass., 1959.

Hume-Rothery, W.: "Atomic Theory for Students of Metallurgy," The Institute of Metals, London, 1955.

——— and G. V. Raynor: "The Structure of Metals and Alloys," The Institute of Metals, London, 1969.

Mason, C. W.' "Introductory Physical Metallurgy," American Society for Metals, Metals Park, Ohio, 1947.

Mondolfo, L. F., and O. Zmeskal: "Engineering Metallurgy," McGraw-Hill Book Company, New York, 1955.

Rogers, B. A.: "The Nature of Metals," American Society for Metals, Metals Park, Ohio, 1951.

Van Vlack, L. H.: "Elements of Materials Science," Addison-Wesley Publishing Company, Inc., Reading, Mass., 1958.

———: "Materials Science for Engineers," Addison-Wesley Publishing Company, Reading, Mass., 1970.

Weertman, J., and J. R. Weertman: "Elementary Dislocation Theory," The Macmillan Company, New York, 1964.

3 PLASTIC DEFORMATION

3·1 Introduction When a material is stressed below its elastic limit, the resulting deformation or strain is temporary. Removal of the stress results in a gradual return of the object ot its original dimensions. When a material is stressed beyond its elastic limit, plastic or permanent deformation takes place, and it will not return to its original shape by the application of force alone. The ability of a metal to undergo plastic deformation is probably its most outstanding characteristic in comparison with other materials. All shaping operations such as stamping, pressing, spinning, rolling, forging, drawing, and extruding involve plastic deformation of metals. Various machining operations such as milling, turning, sawing, and punching also involve plastic deformation. The behavior of a metal under plastic deformation and the mechanism by which it occurs are of essential interest in perfecting the working operation.

Much information regarding the mechanism of plastic deformation may be obtained by studying the behavior of a single crystal under stress and later applying this knowledge to a polycrystalline material.

Plastic deformation may take place by slip, twinning, or a combination of both methods.

3·2 Deformation by Slip If a single crystal of a metal is stressed in tension beyond its elastic limit, it elongates slightly, a step appears on the surface indicating relative displacement of one part of the crystal with respect to the rest, and the elongation stops. Increasing the load will cause movement on another parallel plane, resulting in another step. It is as if neighboring thin sections of the crystal had slipped past one another like sliding cards on a deck. Each successive elongation requires a higher stress and results in the appearance of another step, which is actually the intersection of a slip plane with the surface of the crystal. Progressive increase of the load eventually causes the material to fracture.

Investigations showed that sliding occurred in certain planes of atoms in the crystal and along certain directions in these planes. The mechanism by which a metal is plastically deformed was thus shown to be a new type of flow, vastly different from the flow of liquids or gases. It is a flow that depends upon the perfectly repetitive structure of the crystal which allows the atoms in one face of a slip plane to shear away from their original neighbors in the other face, to slide in an organized way along this face carrying their own half of the crystal with them, and finally to join up again with a new set of neighbors as nearly perfect as before.

It was pointed out in the preceding chapter that parallel planes of high atomic density and corresponding large interplanar spacing exist in the crystal structure. Any movement in the crystal takes place either along these planes or parallel to them.

Investigation of the orientation of the slip plane with respect to the applied stress indicates that slip takes place as a result of simple shearing stress. Resolution of the axial tensile load F in Fig. 3·1 gives two loads. One is a shear load ($F_s = F \cos \theta$) along the slip plane and the other a normal tensile load ($F_n = F \sin \theta$) perpendicular to the plane. The area of the slip plane is $A/\sin \theta$, where A is the cross-sectional area perpendicular to F.

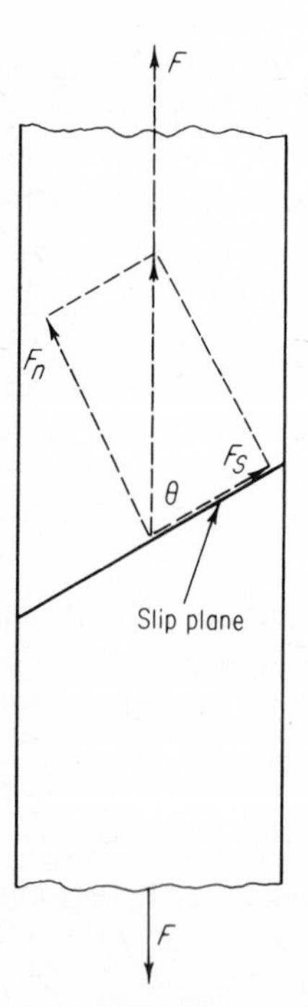

Fig. 3·1 Components of force on a slip plane.

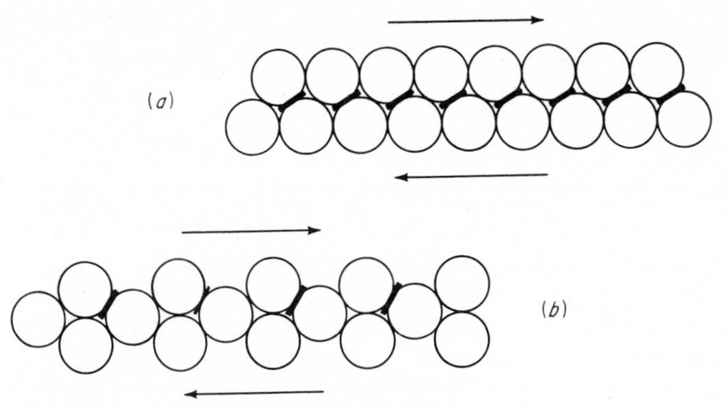

Fig. 3·2 Plastic flow occurs when planes of atoms slip past one another. Close-packed planes do this more easily (*a*) than planes aligned in another direction (*b*).

The resulting stresses are

$$\text{Shear stress } S_s = \frac{F \cos\theta}{A/\sin\theta} = \frac{F}{A}\cos\theta \sin\theta = \frac{F}{2A}\sin 2\theta \tag{3·1}$$

$$\text{Normal stress } S_n = \frac{F \sin\theta}{A/\sin\theta} = \frac{F}{A}\sin^2\theta \tag{3·2}$$

From Eq. (3·1), it is evident that the shear stress on a slip plane will be maximum when $\theta = 45°$.

A more important factor in determining slip movement is the direction of shear on the slip plane. Slip occurs in directions in which the atoms are most closely packed, since this requires the least amount of energy. It was pointed out in the previous chapter that close-packed rows are farther apart from each other than rows that are not close-packed, therefore they can slip past each other with less interference. We might also expect, since the atoms are not bonded directly together but are merely held together by the free electron gas, that these close-packed rows of atoms could slide past each other particularly easily without coming apart. Referring to Fig. 3·2, the atoms in a row in (*a*) are closer together and the rows are farther apart vertically than they are in (*b*), so that less force is required for a given horizontal displacement, as suggested by the slope of the black bars between the atoms. In addition, less displacement is required to move the atoms into unstable positions from which they will be pulled forward into stable ones, when these stable positions are closer together as in Fig. 3·2(*a*).

Figure 3·3 shows the packing of atoms on a slip plane. There are three directions in which the atoms are close-packed, and these would be the easy slip directions. The shear stress S_s on the slip plane, which was de-

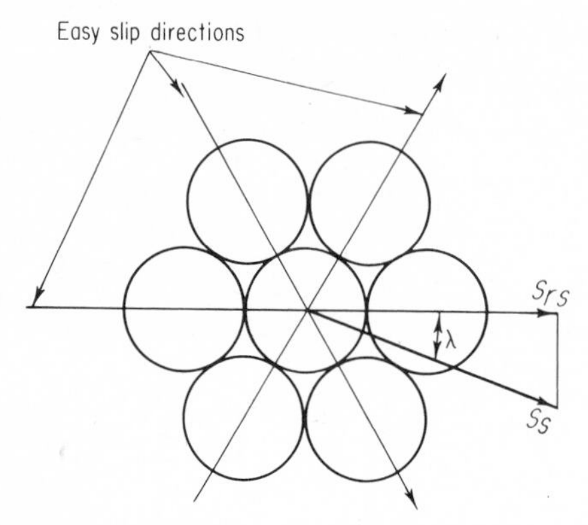

Fig. 3·3 Resolution of shear stress into a slip direction.

rived earlier (Eq. 3·1), may not coincide with one of these easy slip directions and has to be resolved into the nearest slip direction to determine the resolved shear stress S_{rs}. As the diagram shows, the stresses are related by the cosine of the angle λ, and the resolved shear stress is

$$S_{rs} = \frac{F}{2A} \sin 2\theta \cos \lambda \tag{3·3}$$

Investigation has shown that differently oriented crystals of a given metal will begin to slip when different axial stresses are applied but that the critical resolved shear stress, that is, the stress required to initiate slip, is always the same.

If the slip planes are either parallel or perpendicular to the direction of applied stress, slip cannot occur, and either the material deforms by twinning or it fractures. As deformation proceeds and the tensile load remains axial, both the plane of slip and the direction of slip tend to rotate into the axis of tension (Fig. 3·4).

3·3 Mechanism of Slip Portions of the crystal on either side of a specific slip plane move in opposite directions and come to rest with the atoms in nearly equilibrium positions, so that there is very little change in the lattice orientation. Thus the external shape of the crystal is changed without destroying it. Sensitive x-ray methods show that some bending or twisting of the lattice planes has occurred and that the atoms are not in exactly normal positions after deformation. Slip is illustrated schematically in Figs. 3·5 and 3·6 in an f.c.c. (face-centered cubic) lattice.

The (111) plane (Fig. 3·5), which is the plane of densest atomic population, intersects the $(00\bar{1})$ plane in the line *ac*. When the $(00\bar{1})$ plane is assumed to be the plane of the paper and many unit cells are taken together (Fig. 3·6), slip is seen as a movement along the (111) planes in the close-

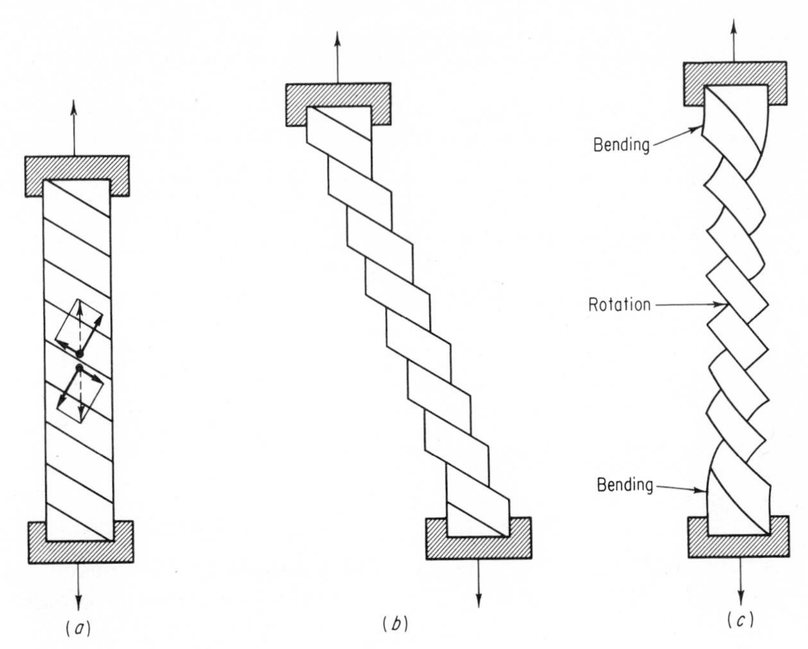

Fig. 3·4 Schematic representation of slip in tension. (*a*) Before straining; (*b*) with ends not constrained; (*c*) ends constrained. (From B. D. Cullity, "Elements of X-ray Diffraction," Addison-Wesley Publishing Company, Inc., Reading, Mass., 1956.)

packed $[\bar{1}10]$ direction, a distance of one lattice dimension or multiple of that dimension. The series of steps formed will generally appear under the microscope as a group of approximately parallel lines (Figs. 3·7 and 3·8). In Fig. 3·7, a single vertical line was scribed on the surface before

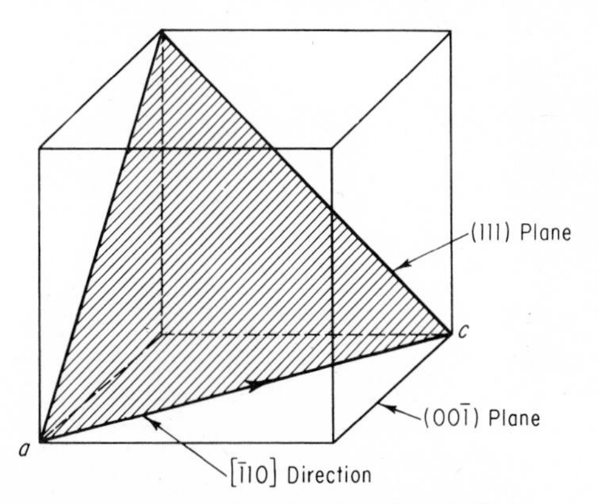

Fig. 3·5 Slip plane and slip direction in an f.c.c. lattice.

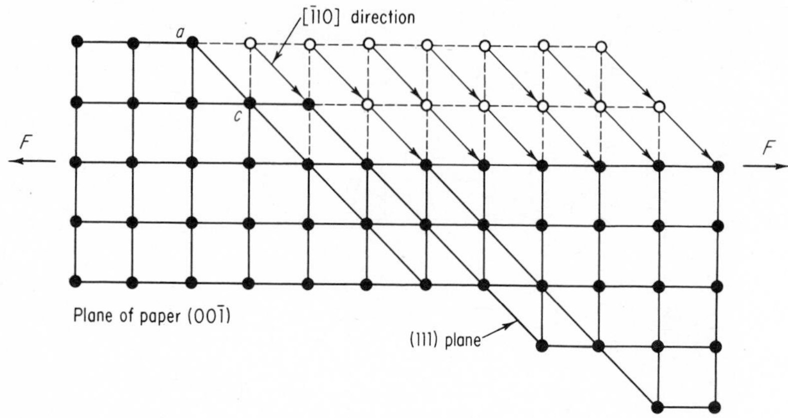

Fig. 3·6 Schematic diagram of slip in an f.c.c. crystal. (By permission from Doan and Mahla, "Principles of Physical Metallurgy," McGraw-Hill Book Company, New York, 1941.)

straining. After straining, the slip lines appear as parallel lines or steps at an angle of approximately 45°. The vertical line is no longer straight, and each step amounts to a movement of about 700 to 800 atoms.

Figure 3·8 shows slip lines in two grains of copper. Notice that the slip lines are parallel inside each grain but because of the different orientation of the unit cells in each grain, they have to change direction at the grain boundary. The grain boundary starts in the lower left corner and runs upward and to the right.

From the schematic picture of slip in Fig. 3·6, one may assume, at first, that the motion consists of a simultaneous movement of planes of atoms

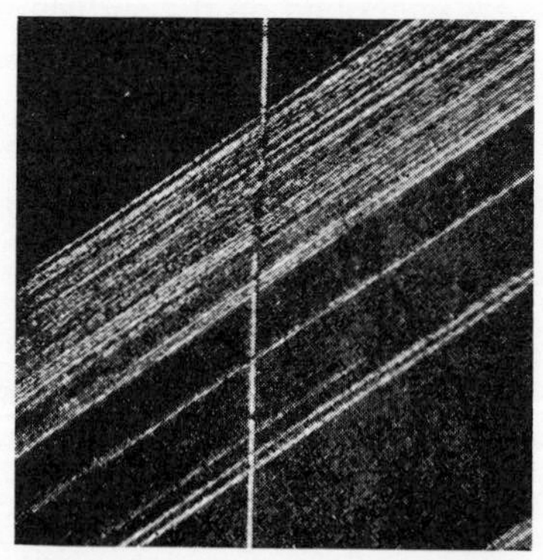

Fig. 3·7 Single crystal of brass strained in tension, 200×. (By permission from R. M. Brick and A. Phillips, "Structure and Properties of Alloys," 2d ed., McGraw-Hill Book Company, New York, 1949.)

Fig. 3·8 Slip lines in copper. Specimen polished, etched, and then strained. 100×.

across each other. This, however, requires that the shearing force must have the same value over all points of the slip plane. The vibrations of the atoms and the difficulties of applying a uniformly distributed force make this condition unattainable. A more reasonable assumption is that the atoms slip consecutively, starting at one place or at a few places in the slip plane, and then move outward over the rest of the plane.

Sir Neville Mott has likened slip to the sliding of a large heavy rug across a floor. If you try to slide the entire rug as one piece, the resistance is too great. What you can do instead is to make a wrinkle in the rug and then slide the whole thing a little at a time by pushing the wrinkle along, thereby enlarging the slipped region behind it at the expense of the unslipped region in front of it. A similar analogy to the wrinkle in the rug is the movement of the earthworm (see Fig. 3·9). Examination of Fig. 3·9 shows that, by application of the shear force, an extra plane of atoms (called a dislocation) has been formed above the slip plane. This dislocation moves across the slip plane and leaves a step when it comes out at the surface of the crystal. Each time the dislocation moves across the slip plane, the crystal moves one atom spacing. Since the atoms do not end up in exactly normal positions after the passage of the dislocation, subsequent movement of the dislocation across the same slip plane encounters greater resistance. Eventually, this resistance or distortion of the slip plane becomes great enough to lock the dislocation in the crystal structure, and the movement stops. Further deformation will require movement on another slip plane. Although

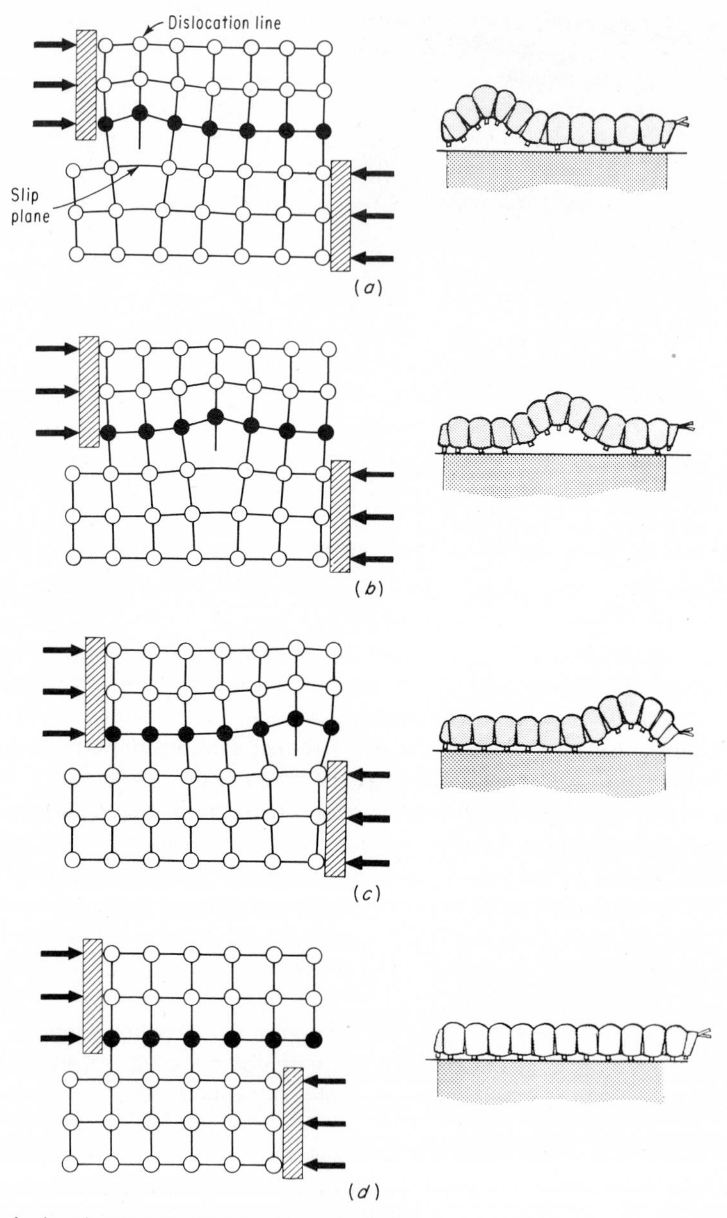

Fig. 3·9 Analogy between the movement of a dislocation through a crystal and the movement of an earthworm as it arches its back while going forward. (Courtesy Westinghouse Electric Corp.)

the distortion is greatest on the active slip plane, its effect is felt throughout the lattice structure, and the applied load must be increased to cause movement on another slip plane.

The stress required to initiate slip in a perfect crystal, that is, the stress required to move one atom over another, may be calculated for a given metal. This result, however, is 100 to 1,000 times larger than the experimentally observed critical resolved shear stress for slip in single crystals. The much lower observed critical resolved shear stress is not surprising since it was pointed out in Sec. 2·21 that dislocations already exist in the crystal structure as a result of solidification. It is therefore not necessary to create a dislocation, but simply to start an existing one moving on the slip plane. This theory suggests that one method of attaining high strength in metals would be to manufacture more nearly perfect crystals without imperfections. The strength of metal "whiskers," which are nearly perfect crystals, has approached the theoretical strength and lends support to the dislocation theory.

3·4 Slip in Different Lattice Structures The combination of a slip plane and a slip direction is known as a *slip system.* The slip direction is always the one of densest atomic packing in the slip plane and is the most important factor in the slip system.

In f.c.c. materials there are four sets of (111) planes and three close-packed ⟨110⟩ directions (Fig. 3·3) in each plane, making a total of 12 possible slip systems. These slip systems are well distributed in space; therefore, it is almost impossible to strain an f.c.c. crystal and not have at least one {111} plane in a favorable position for slip. As expected, the critical resolved shear stress for slip would be low (Table 3·1), and metals with this type of lattice structure (silver, gold, copper, aluminum) are easily deformed.

TABLE 3·1 Critical Resolved Shear Stress for Several Metals at Room Temperature*

METAL	STRUCTURE	SLIP PLANE	SLIP DIRECTION	CRITICAL RESOLVED SHEAR STRESS (PSI)
Silver	f.c.c.	{111}	$\langle 1\bar{1}0 \rangle$	54
Copper	f.c.c.	{111}	$\langle 1\bar{1}0 \rangle$	71
Aluminum	f.c.c.	{111}	$\langle 1\bar{1}0 \rangle$	114
Magnesium	c.p.h.	{0001}	$\langle 11\bar{2}0 \rangle$	64
Cobalt	c.p.h.	{0001}	$\langle 11\bar{2}0 \rangle$	960
Titanium	c.p.h.	$\{10\bar{1}0\}$	$\langle 11\bar{2}0 \rangle$	1,990
Iron	b.c.c.	{110}	$\langle \bar{1}11 \rangle$	3,980
Columbium	b.c.c.	{110}	$\langle \bar{1}11 \rangle$	4,840
Molybdenum	b.c.c.	{110}	$\langle \bar{1}11 \rangle$	10,400

* Values were tabulated from various sources by W. J. M. Tegart, "Elements of Mechanical Metallurgy," The Macmillan Company, New York, 1966, p. 106.

The c.p.h. (close-packed hexagonal) metals (cadmium, magnesium, cobalt, titanium) have only one plane of high atomic population, the (0001) plane (or basal plane) and three close-packed $\langle 11\bar{2}0 \rangle$ in that plane (see Fig. 3·10). This structure does not have so many slip systems as the f.c.c. lattice, and the critical resolved shear stress is higher than for f.c.c. materials (Table 3·1). While the number of slip systems is limited, deformation by twinning helps to bring more slip systems into proper position, thereby approaching the plasticity of the f.c.c. structure and surpassing that of b.c.c. metals.

Since b.c.c. (body-centered cubic) metals have fewer atoms per unit cell, they do not have a well-defined slip system and do not have a truly close-packed plane. The slip direction is the close-packed $\langle 111 \rangle$ direction. Slip lines in b.c.c. metals are wavy and irregular, often making identification of a slip plane extremely difficult. The $\{110\}$, $\{112\}$, and $\{123\}$ planes have all been identified as slip planes in b.c.c. crystals. Studies have indicated that any plane that contains a close-packed $\langle 111 \rangle$ direction can act as a slip plane. In further agreement with the lack of a close-packed plane is the relatively high critical resolved shear stress for slip (Table 8·1). Therefore, b.c.c. metals [molybdenum, alpha (α) iron, tungsten] do not show a high degree of plasticity.

3·5 Deformation by Twinning In certain materials, particularly c.p.h. metals, twinning is a major means of deformation. This may accomplish an extensive change in shape or may bring potential slip planes into a more favorable position for slip. Twinning is a movement of planes of atoms in the lattice parallel to a specific (twinning) plane so that the lattice is divided into two symmetrical parts which are differently oriented. The amount of movement of each plane of atoms in the twinned region is proportional to its distance from the twinning plane, so that a mirror image is formed

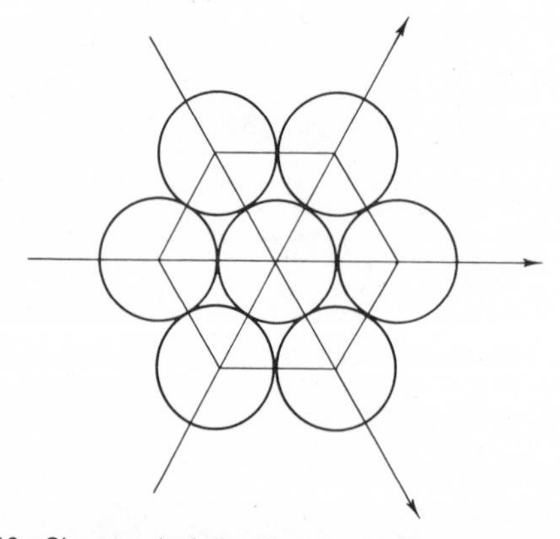

Fig. 3·10 Close-packed directions in the basal plane of the hexagonal lattice.

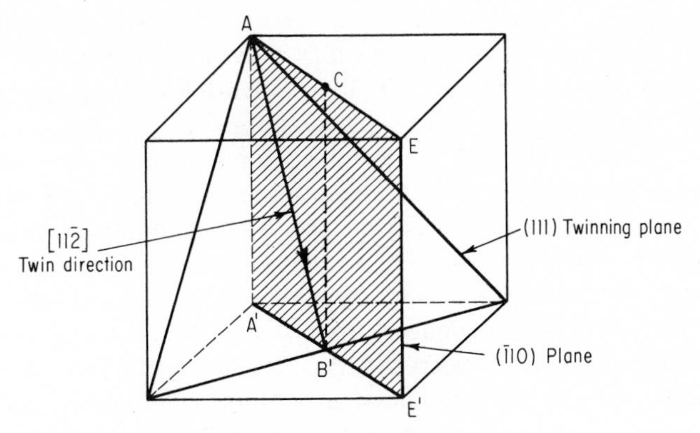

Fig. 3·11 Diagram of a twin plane and twin direction in an f.c.c. lattice.

across the twin plane. Twinning is illustrated schematically in an f.c.c. lattice in Figs. 3·11 and 3·12.

In Fig. 3·11, the (111) twinning plane intersects the ($\bar{1}$10) plane along the line AB', which is the twin direction. The mechanism of twinning is shown in Fig. 3·12. The plane of the paper is the ($\bar{1}$10) plane, and many unit cells

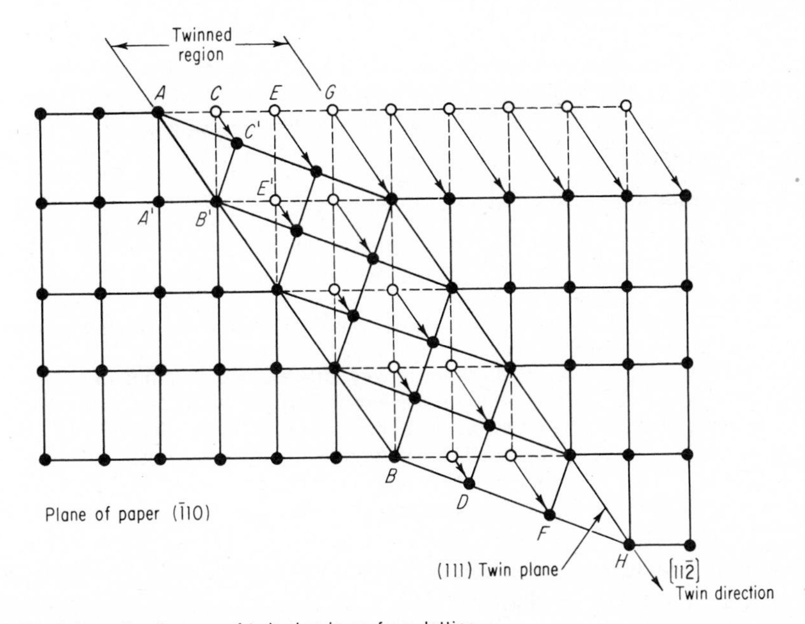

Fig. 3·12 Schematic diagram of twinning in an f.c.c. lattice. (By permission from G. E. Doan, "Principles of Physical Metallurgy," 3d ed., McGraw-Hill Book Company, New York, 1953.)

are taken together. Each (111) plane in the twin region moves in shear in the [11$\bar{2}$] direction. The first one, *CD*, moves one-third of an interatomic distance; the second one, *EF*, moves two-thirds of an interatomic distance; and the third one, *GH*, moves an entire spacing.

If a line is drawn perpendicular to the twin plane (AB′) from atom A′, notice that we have another atom, C′, exactly the same distance away from the twinned plane but on the other side. The same thing is true for all the atoms in the twinned region, so what we really have here is a mirror image in the twinned region of the untwinned portion of the crystal. Since the atoms end up in interatomic spaces, the orientation of the atoms, or the distance between atoms, has been changed. Generally, the twinned region involves the movement of a large number of atoms, and usually it appears microscopically as a broad line or band, as shown in Fig. 3·13. This picture shows twin bands in zinc; notice how the bands change direction at the grain boundary.

The twinning plane and direction are not necessarily the same as those for slip. In f.c.c. metals the twin plane is the (111) plane, and the twin direction is the [11$\bar{2}$] direction; in b.c.c. it is the (112) plane and the [111] direction.

Two kinds of twins are of interest to the metallurgist:

1 Deformation or mechanical twins, most prevalent in c.p.h. metals (magnesium, zinc, etc.) and b.c.c. metals (tungsten, α iron, etc.).
2 Annealing twins, most prevalent in f.c.c. metals (aluminum, copper, brass, etc.). These metals have been previously worked and then reheated. The twins are formed because of a change in the normal growth mechanism.

3·6 Slip vs. Twinning Slip and twinning differ in:

1 Amount of movement: in slip, atoms move a whole number of interatomic spacings, while in twinning the atoms move fractional amounts depending on their distance from the twinning plane.
2 Microscopic appearance: slip appears as thin lines, while twinning appears as broad lines or bands.
3 Lattice orientation: in slip there is very little change in lattice orientation, and the steps are visible only on the surface of the crystal. If the steps are removed by polishing, there will be no evidence that slip has taken place. In twinning, however, since there is a different lattice orientation in the twinned region, removal of the steps by surface polishing will not destroy the evidence of twinning. Proper etching solutions, sensitive to the differences in orientation, will reveal the twinned region.

3·7 Fracture Fracture is the separation of a body under stress into two or more parts. The failure is characterized as either *brittle* or *ductile*. Brittle fracture generally involves rapid propagation of a crack with minimal energy absorption and plastic deformation. In single crystals, brittle fracture occurs by cleavage along a particular crystallographic plane [for example,

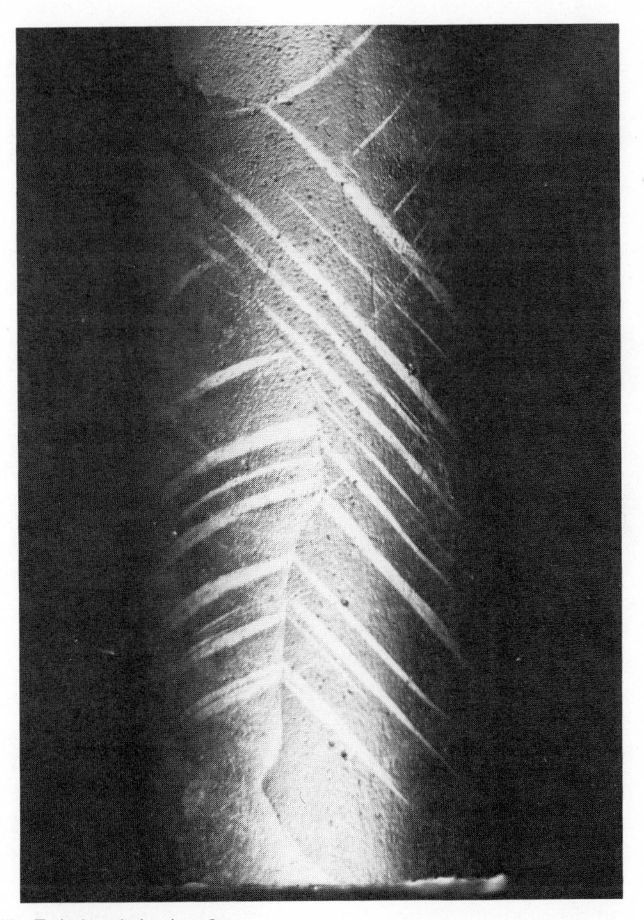

Fig. 3·13 Twin bands in zinc. 2×.

the (100) plane in iron]. In polycrystalline materials, the brittle-fracture surface shows a granular appearance (Fig. 3·14*b*) because of the changes in orientation of the cleavage planes from grain to grain.

As with plastic deformation, the difference between the theoretical fracture strength and the actual fracture strength is due to structural irregularities. Freshly drawn glass fibers have strengths approaching theoretical values, but anything that can give rise to surface irregularities, such as nicks or cracks, weakens them. The first explanation of this discrepancy was given by A. A. Griffith in 1921. He theorized that failure in brittle materials was caused by many fine elliptical, submicroscopic cracks in the metal. The sharpnesss at the tip of such cracks will result in a very high stress concentration which may exceed the theoretical fracture strength

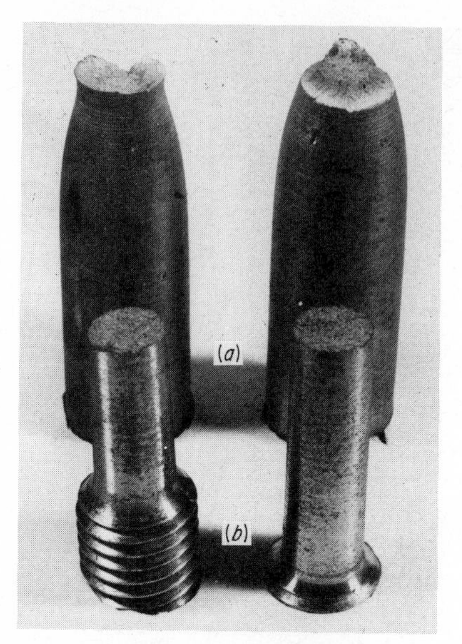

Fig. 3·14 (*a*) Ductile cup-and-cone fracture in low-carbon steel; (*b*) brittle fracture in high-carbon steel.

at this localized area and cause the crack to propagate even when the body of the material is under fairly low applied tensile stress.

It is possible that microcracks may exist in the metal due to the previous history of solidification or working. However, even an initially sound material may develop cracks on an atomic scale. As shown in Fig. 3·15*a*, the pile-up of dislocations at a barrier which might be of grain boundary or included particle may result in a microcrack. Another method (Fig. 3·15*b*) is for three unit dislocations to combine into a single dislocation. From the above explanation, it is apparent that any method that will increase the mobility of dislocations will tend to reduce the possibility of brittle fracture.

Ductile fracture occurs after considerable plastic deformation prior to failure. The failure of most polycrystalline ductile materials occurs with a *cup-and-cone fracture* associated with the formation of a neck in a tensile specimen. See Fig. 3·14*a*. The fracture begins by the formation of cavities in the center of the necked region. In most commercial metals, these internal cavities probably form at nonmetallic inclusions. This belief is supported by the fact that extremely pure metals are much more ductile than those of slightly lower purity. Under continued applied stress, the cavities coalesce to form a crack in the center of the sample. The crack proceeds outward toward the surface of the sample in a direction perpendicular to the applied stress. Completion of the fracture occurs very rapidly along a surface that makes an angle of approximately 45° with the tensile axis.

The final stage leaves a circular lip on one half of the sample and a bevel on the surface of the other half. Thus one half has the appearance of a shallow cup, and the other half resembles a cone with a flattened top, giving rise to the term cup-and-cone fracture.

3·8 Slip, Twinning, and Fracture The amount of deformation that can occur before fracture is determined by the relative values of the stresses required for slip, twinning, and cleavage. There is a critical resolved shear stress for slip which is increased by alloying, decreasing temperature, and prior deformation. There is a critical resolved shear stress for twinning which is also increased by prior deformation. There is also a critical normal stress for cleavage on a particular plane which is not sensitive to prior deformation and temperature. When a stress is applied to a crystal, which process takes place depends upon which critical stress is exceeded first. If the critical resolved shear stress for slip or twinning is reached first, the crystal will slip or twin and show some ductility. If, however, the critical normal stress is reached first, the crystal will cleave along the plane concerned with little or no plastic deformation.

3·9 Polycrystalline Material The preceding discussion described plastic deformation in single crystals. Commercial material, however, is always made up of polycrystalline grains, whose crystal axes are oriented at random. When a polycrystalline material is subjected to stress, slip starts first in those grains in which the slip system is most favorably situated with respect to the applied stress. Since contact at the grain boundaries must be main-

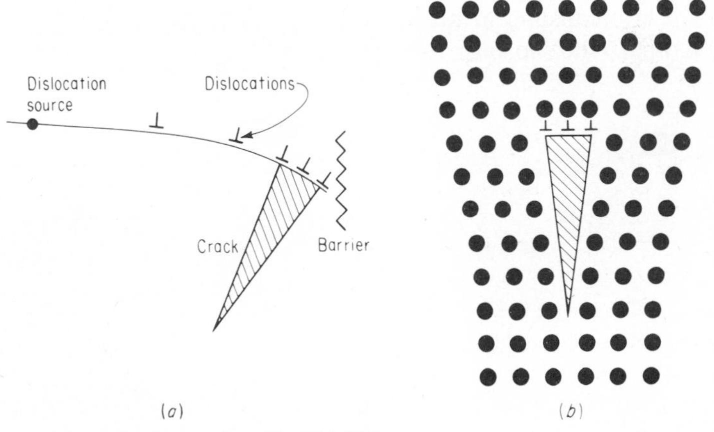

Fig. 3·15 A mechanism of crack formation. (*a*) Edge dislocations piling up at a barrier, (*b*) three piled-up dislocations form an incipient crack within the crystal lattice. (From A. G. Guy, "Elements of Physical Metallurgy," 2d ed., Addison-Wesley Publishing Company, Inc., Reading, Mass., 1959.)

tained, it may be necessary for more than one slip system to operate. The rotation into the axis of tension brings other grains, originally less favorably oriented, into a position where they can now deform. As deformation and rotation proceed, the individual grains tend to elongate in the direction of flow (Fig. 3·16). After a certain amount of deformation, most grains will have a particular crystal plane in the direction of deformation. The material now shows *preferred orientation*, which will result in somewhat different properties, depending upon the direction of measurement.

A fine-grained metal in which the grains are oriented at random will possess identical properties in all directions, but a metal with preferred orientation of grains will have directional properties. This may be troublesome—for example, in the deep drawing of sheet metal. Preferred orientation is also of prime importance in the manufacture of steel for electrical instruments because the magnetic properties will be different depending upon the direction of working. If the deformation is severe, the grains may be fragmented or broken.

Not all the work done in deformation is dissipated in heat; part of it is stored in the crystal as an increase in internal energy. Since the crystal axes of adjacent grains are randomly oriented, the slip planes and twinning planes must change direction in going from grain to grain (Figs. 3·8 and 3·13). This means that more work is done at the grain boundaries and more internal energy will exist at those points.

Figure 3·17 shows the microstructure of polycrystalline brass after being deformed slightly in a vise. Notice that the thin, parallel slip lines change direction at the grain boundaries. The grain in the lower portion of the picture is interesting in that it illustrates both slip and twinning in the same grain. The slip lines, running vertically and to the right, become hori-

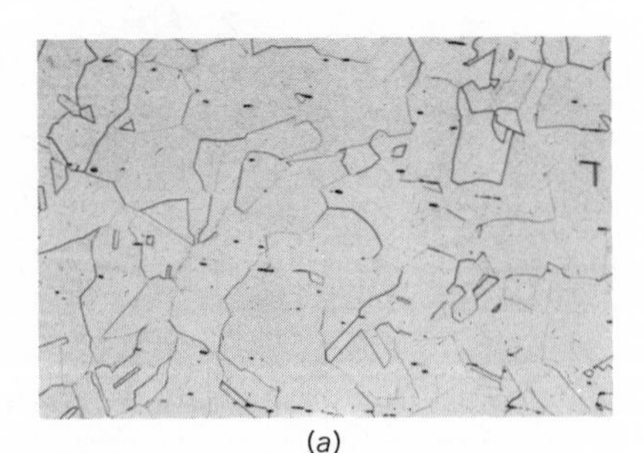

(*a*)

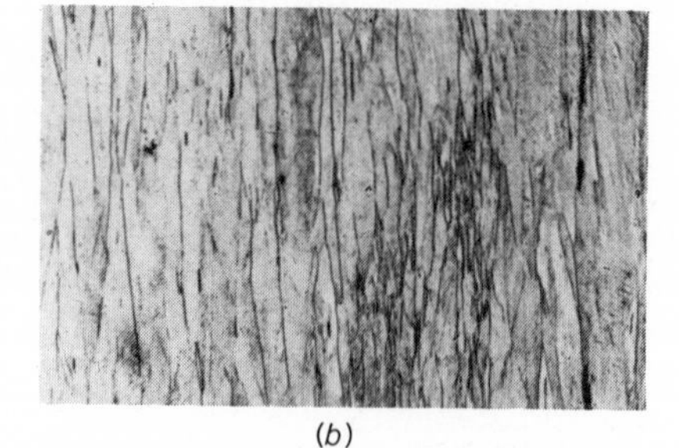

(*b*)

Fig. 3·16 The microstructure of a nickel-copper alloy, Monel, showing: (*a*) the equiaxed grains in the unworked condition, hardness BHN 125, etched in cyanide persulfate, 100×, and (*b*) cold-drawn condition showing the grains elongated in the direction of drawing, hardness BHN 225, etched in Carapellas' reagent, 100×. (The International Nickel Company.)

Fig. 3·17 Polycrystalline brass, polished, etched, and then deformed slightly in a vise. 100×. (By permission from R. M. Brick and A. Phillips, "Structure and Properties of Alloys," 2d ed., McGraw-Hill Book Company, New York, 1949.)

zontal when they cross the twinning band and then resume their original direction on the other side. Since the slip lines have the same direction on either side of the twinning band, this indicates that the deformation is occurring in the same grain.

When a crystal deforms, there is some distortion of the lattice structure. This distortion is greatest on the slip planes and grain boundaries and increases with increasing deformation. This is manifested by an increase in resistance to further deformation. The material is undergoing *strain hardening* or *work hardening*. One of the remarkable features of plastic deformation is that the stress required to initiate slip is lower than that required to continue deformation on subsequent planes. Aside from distortion of the lattice structure, the pile-up of dislocations against obstacles (such as grain boundaries and foreign atoms) and the locking of dislocations on intersecting slip planes increase the resistance to further deformation.

In reality, crystals usually contain complex networks of interconnected dislocation lines, as well as other defects and impurities in the crystal lattice. When the dislocations begin to move, their ends remain tied to other parts of the network, or to other defects. Because the ends are anchored, the active slip planes can never get rid of their slip dislocations; in fact, the dislocations in the plane multiply when the plane slips. Since the ease with which a dislocation moves across the slip plane is an indication of the ductility of the material, it suggests that materials may be made harder by putting various obstacles in the way of the dislocations. Since dislocations pile up at grain boundaries, metals can, to some extent, be hardened by reducing the size of the grains.

Alloying introduces foreign atoms that distort the crystal locally around themselves, and these local distortions offer resistance to the movement of a nearby dislocation. If the alloy atoms are gathered together in clumps, their effect is enhanced, and this can be accomplished, as will be explained later in the section on age-hardening, by heat treatment. In the hardening that is produced by various processes of plastic working, such as hammering or rolling, the obstacles are paradoxically the dislocations themselves. When the number of dislocations in the worked metal becomes large enough, those moving along intersecting slip planes obstruct one another's movement, an effect readily appreciated by anyone who has been held up at a road junction in dense traffic.

It is important to remember that whenever there is distortion of the lattice structure, whether it is a result of plastic deformation, heat treatment, or alloying, there will be an increase in the strength and hardness of the material.

3·10 Effect of Cold Working on Properties A material is considered to be cold-worked if its grains are in a distorted condition after plastic deformation is completed. All the properties of a metal that are dependent on the lattice structure are affected by plastic deformation or cold working. Tensile strength, yield strength, and hardness are increased, while ductility, as represented by percent elongation, is decreased (Table 3·2). Although both strength and hardness increase, the rate of change is not the same. Hardness generally increases most rapidly in the first 10 percent reduction,

TABLE 3·2 Effect of Plastic Deformation on the Tensile Properties of 70:30 Brass*

REDUCTION BY COLD ROLLING, PERCENT	TENSILE STRENGTH, PSI	ELONGATION, % IN 2 IN.	HARDNESS ROCKWELL X†
0	43,000	70	12
10	48,000	52	62
20	53,000	35	83
30	60,000	20	89
40	70,900	12	94
50	80,000	8	97
60	90,000	6	100

* From R. M. Brick and A. Phillips, "Structure and Properties of Alloys," McGraw-Hill Book Company, New York, 1942.
† Rockwell × = 1/16-in. ball indenter, 75 kg load.

whereas the tensile strength increases more or less linearly. The yield strength increases more rapidly than the tensile strength, so that, as the amount of plastic deformation is increased, the gap between the yield and tensile strengths decreases (Fig. 3·18). This is important in certain forming operations where appreciable deformation is required. In drawing, for example, the load must be above the yield point to obtain appreciable deformation but below the tensile strength to avoid failure. If the gap is narrow, very close control of the load is required.

Ductility follows a path opposite to that of hardness, a large decrease in

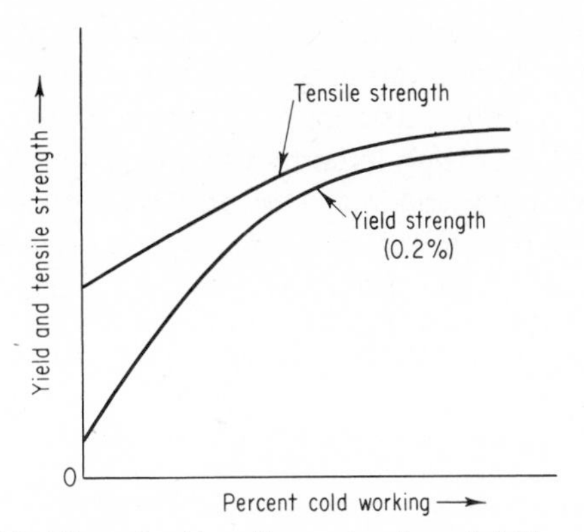

Fig. 3·18 Effect of cold working on tensile and yield strength of copper.

the first 10 percent reduction and then a decrease at a slower rate.

Distortion of the lattice structure hinders the passage of electrons and decreases electrical conductivity. This effect is slight in pure metals but is appreciable in alloys (Fig. 3·19).

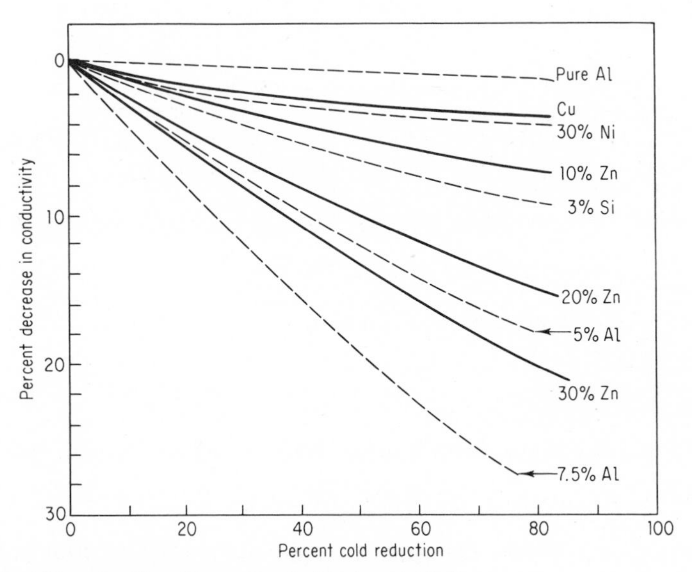

Fig. 3·19 Effect of cold working on the electrical conductivity of pure aluminum; pure copper; Cu + 30 percent Ni; Cu + 3 percent Si; Cu + 5 percent and 7.5 percent Al; Cu + 10 percent, 20 percent, and 30 percent Zn. (By permission from R. M. Brick and A. Phillips, "Structure and Properties of Alloys," 2d ed., McGraw-Hill Book Company, New York, 1949.)

The increase in internal energy, particularly at the grain boundaries, makes the material more susceptible to intergranular corrosion (see Fig. 12·4), thereby reducing its corrosion resistance. Known as *stress corrosion*, this is an acceleration of corrosion in certain environments due to residual stresses resulting from cold working. One of the ways to avoid stress corrosion cracking is to relieve the internal stresses by suitable heat treatment after cold working and before placing the material in service.

QUESTIONS

3·1 Describe an experiment to determine the increase in internal energy of a crystal as a result of deformation.

3·2 The movement in slip is sometimes described as analogous to simple gliding. Why is this a poor analogy?

3·3 Is it possible to have both slip and twinning occur in the same grain? Explain.

3·4 How may one distinguish between slip and twinning if the width of the twin band is of the same order as a slip line?

3·5 What is the effect of the rate of deformation on the mechanical properties?

3·6 How would a difference in grain size affect the change in mechanical properties due to deformation?

3·7 Which properties would be affected by preferred orientation and why?

3·8 In Fig. 3·19, why does the addition of 30 percent nickel to copper have less effect on electrical conductivity than the addition of 5 percent aluminum to copper?

3·9 Using Eqs. (3·1) and (3·2), plot a curve showing the variation of the shear stress and the normal stress as θ varies from 0 to 90°. Assume F is constant.

3·10 The critical resolved shear stress for slip in copper is 142 psi. Using Eq. (3·3) and assuming λ to remain constant at 10°, plot a curve showing the change in the axial load F with change in θ.

3·11 Explain the reason for the increase in ductility of most metals as the temperature is raised.

3·12 Why is there a greater tendency for brittle failure to occur at high rates of straining?

3·13 What is the difference between brittle fracture and ductile fracture?

REFERENCES

Barrett, C. S.: "Structure of Metals," 2d ed., McGraw-Hill Book Company, New York, 1952.

Boas, W.: "An Introduction to the Physics of Metals and Alloys," John Wiley & Sons, Inc., New York, 1947.

Brick, R. M., and A. Phillips: "Structure and Properties of Alloys," 2d ed., McGraw-Hill Book Company, New York, 1949.

Cottrell, A. H.: "Dislocations and Plastic Flow in Crystals," Oxford University Press, Fair Lawn, N.J., 1956.

Guy, A. G.: "Elements of Physical Metallurgy," 2d ed., Addison-Wesley Publishing Company, Inc., Reading, Mass., 1959.

Mondolfo, L. F., and O. Zmeskal: "Engineering Metallurgy," McGraw-Hill Book Company, New York, 1955.

Rogers, B. A.: "The Nature of Metals," American Society for Metals, Metals Park, Ohio, 1951.

Van Vlack, L. H.: "Elements of Materials Science," Addison-Wesley Publishing Company, Inc., Reading, Mass., 1959.

———: "Materials Science For Engineers," Addison-Wesley Publishing Company, Inc., Reading, Mass., 1970.

Weertman, J., and J. R. Weertman: "Elementary Dislocation Theory," The Macmillan Company, New York, 1964.

Wulff, J.: "The Structure and Properties of Materials," vol. 3: Mechanical Behavior, John Wiley & Sons, Inc., New York, 1965.

Wulpi, D. J.: "How Components Fail," American Society for Metals, Metals Park, Ohio, 1967.

4 ANNEALING AND HOT WORKING

FULL ANNEALING

4·1 Introduction In the previous chapter, the mechanism of plastic deformation by slip and twinning and the effect of cold working on the properties of the metal were studied. As a result of cold working, the hardness, tensile strength, and electrical resistance increased, while the ductility decreased. There was also a large increase in the number of dislocations, and certain planes in the crystal structure were severely distorted. It was emphasized that while most of the energy used to cold-work the metal was dissipated in heat, a finite amount was stored in the crystal structure as internal energy associated with the lattice defects created by the deformation. The stored energy of cold work is that fraction, usually from 1 to approximately 10 percent of the energy put into a material while producing a cold-worked state, which is retained in the material. Figure 4·1 shows the relationship between stored energy and the amount of deformation in high-purity copper.

Full annealing is the process by which the distorted cold-worked lattice structure is changed back to one which is strain-free through the application of heat. This process is carried out entirely in the solid state and is usually followed by slow cooling in the furnace from the desired temperature. The annealing process may be divided into three stages: recovery, recrystallization, and grain growth.

4·2 Recovery This is primarily a low-temperature process, and the property changes produced do not cause appreciable change in the microstructure. The principal effect of recovery seems to be the relief of internal stresses due to cold working. This is shown in Fig. 4·2. At a given temperature, the rate of decrease in residual strain hardening is fastest at the beginning and drops off at longer times. Also, the amount of reduction in residual stress that occurs in a practical time increases with increasing temperature.

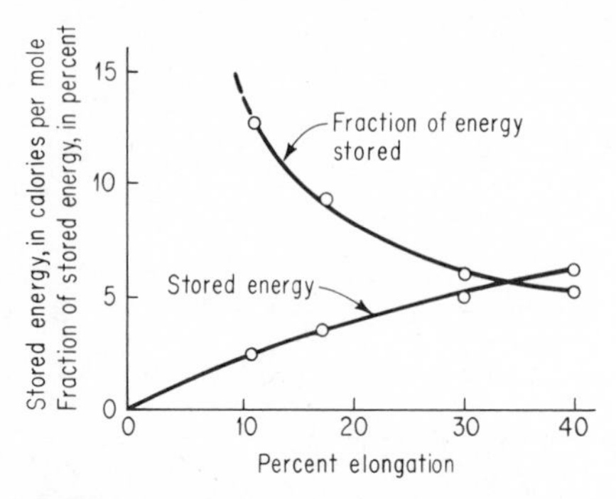

Fig. 4·1 Stored energy of cold work and fraction of the total work of deformation remaining as stored energy for high-purity copper plotted as functions of tensile elongation. (From data of P. Gordon, *Trans. AIME*, vol. 203, p. 1043, 1955.)

When the load which has caused plastic deformation in a polycrystalline material is released, all the elastic deformation does not disappear. This is due to the different orientation of the crystals, which will not allow some of them to move back when the load is released. As the temperature is increased, there is some springback of these elastically displaced atoms which relieves most of the internal stress. In some cases there may be a slight amount of plastic flow, which may result in a slight increase in hardness and strength. Electrical conductivity is also increased appreciably during the recovery stage.

Since the mechanical properties of the metal are essentially unchanged, the principal application of heating in the recovery range is in stress-relieving cold-worked alloys to prevent stress-corrosion cracking or to minimize the distortion produced by residual stresses. Commercially, this low-temperature treatment in the recovery range is known as *stress-relief annealing.*

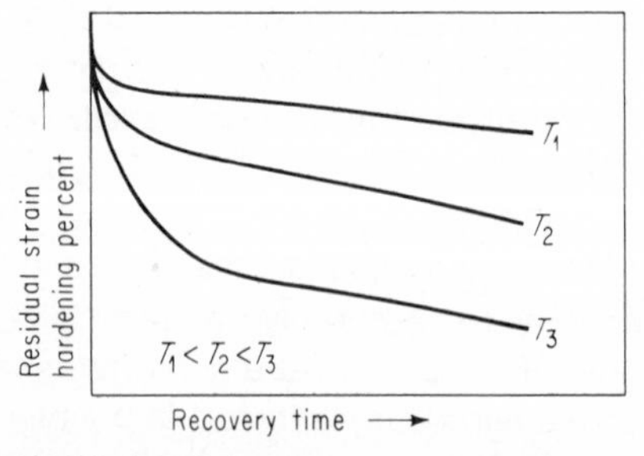

Fig. 4·2 Residual strain hardening vs. recovery time at three constant annealing temperatures.

4·3 Recrystallization As the upper temperature of the recovery range is reached, minute new crystals appear in the microstructure. These new crystals have the same composition and lattice structure as the original underformed grains and are not elongated but are approximately uniform in dimensions (equiaxed). The new crystals generally appear at the most drastically deformed portions of the grain, usually the grain boundaries and slip planes. The cluster of atoms from which the new grains are formed is called a nucleus. Recrystallization takes place by a combination of nucleation of strain-free grains and the growth of these nuclei to absorb the entire cold-worked material.

Figure 4·3 shows a typical recrystallization curve. This is a plot of the percent of the material recrystallized versus the time of annealing at constant temperature for a fixed composition and degree of cold working. This curve is typical of any process that occurs by nucleation and growth. Initially, there is an incubation period in which enough energy is developed in order to start the process going. In this case, the incubation period is to allow the strain-free nuclei to reach a visible microscopic size. It is important to realize that the growth of recrystallized embryos is irreversible. During the study of crystallization in Chap. 2, it was pointed out that solidification from the liquid would start when a group of atoms had reached a critical size to form a stable cluster. Embryos, that is, clusters of less than critical size, would redissolve or disappear. However, since there is no simple way to recreate the distorted, dislocation-filled structure, the recrystallization embryo cannot redissolve. Therefore, these embryos merely wait for additional energy to attract more atoms into their lattice structure. Eventually the critical size is exceeded, and visible recrystallization begins. The incubation period corresponds to the irreversible growth of the embryos.

Exactly how recrystallization takes place is not yet clearly understood; however, some idea of the process may be obtained by examining it in

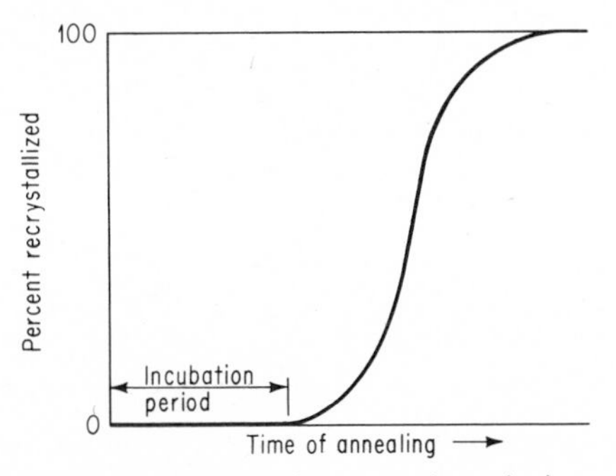

Fig. 4·3 A typical recrystallization curve at constant temperature.

terms of the energy of the lattice. In the discussion of plastic deformation, it was emphasized that the slip planes and grain boundaries were localized points of high internal energy as a result of the pile-up of dislocations. Because of the nature of strain hardening, it is not possible for the dislocation or the atoms to move back to form a strain-free lattice from the distorted lattice. A simplified analogy is shown in Fig. 4·4. Consider that some atoms, at the grain boundaries or slip planes, have been pushed up an energy hill to a value of E_1 above the internal energy of atoms in the undeformed lattice. The energy required to overcome the rigidity of the distorted lattice is equal to E_2. The atoms cannot reach the energy of the strain-free crystal by the same path they went up the hill; instead, they must get over the top, from which they are able to roll down easily. This difference in energy, $E_2 - E_1$, is supplied by heat. When the temperature is reached at which these localized areas have an energy content equal to E_2, they give up part of their energy as *heat of recrystallization* and form nuclei of new strain-free grains. Part of this heat of recrystallization is absorbed by surrounding atoms so that they have sufficient energy to overcome the rigidity of the distorted lattice and be attracted into the lattice structure of the strain-free grains, initiating grain growth. The number and energy content of these high-energy points depend to a large extent on the amount of prior deformation, the number increasing with increasing deformation.

4·4 Recrystallization Temperature The term *recrystallization temperature* does not refer to a definite temperature below which recrystallization will not occur but refers to the approximate temperature at which a highly cold-

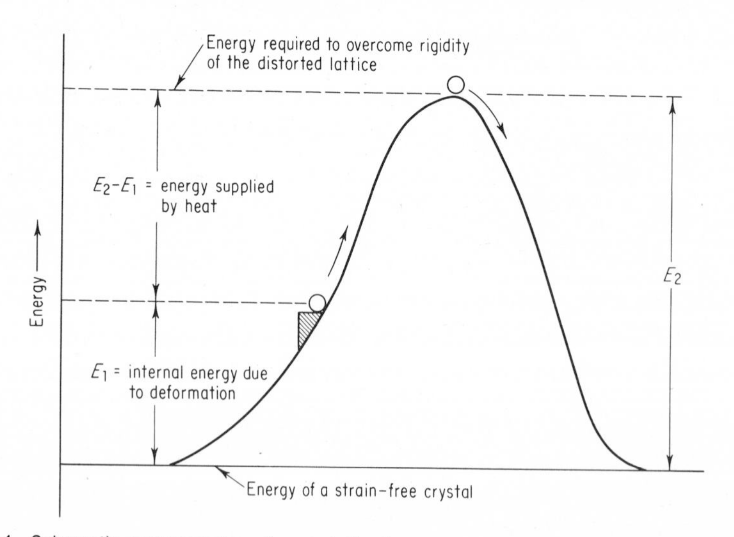

Fig. 4·4 Schematic representation of recrystallization.

worked material completely recrystallizes in 1 hr. The recrystallization temperature of several metals and alloys is listed in Table 4·1.

Notice that very pure metals seem to have low recrystallization temperatures as compared with impure metals and alloys. Zinc, tin, and lead have recrystallization temperatures below room temperature. This means that these metals cannot be cold-worked at room temperature since they recrystallize spontaneously, reforming a strain-free lattice structure.

The greater the amount of prior deformation, the lower the temperature for the start of recrystallization (Fig. 4·5), since there will be greater distortion and more internal energy left.

Increasing the annealing time decreases the recrystallization temperature. The recrystallization process is far more sensitive to changes in temperature than to variations in time at constant temperature. The influence of time and temperature on the tensile strength of highly cold-worked copper is shown in Fig. 4·6. Recrystallization is indicated by the sharp drop in tensile strength. A tensile strength of 40,000 psi may be obtained by heating for 12 hr at 300°F, 6 hr at 320°F, 2 hr at 340°F, 1 hr at 370°F, or 1/2 hr at 390°F.

For equal amounts of cold-working, more strain hardening is introduced into initially fine-grained metal than into initially coarse-grained metal. Therefore, the finer the initial grain size the lower the recrystallization

TABLE 4·1 Approximate recrystallization temperatures for several metals and alloys*

MATERIAL	RECRYSTALLIZATION TEMP, °F
Copper (99.999%)	250
Copper, 5% zinc	600
Copper, 5% aluminum	550
Copper, 2% beryllium	700
Aluminum (99.999%)	175
Aluminum (99.0%+)	550
Aluminum alloys	600
Nickel (99.99%)	700
Monel metal	1100
Iron (electrolytic)	750
Low-carbon steel	1000
Magnesium (99.99%)	150
Magnesium alloys	450
Zinc	50
Tin	25
Lead	25

* By permission from A. G. Guy, "Elements of Physical Metallurgy," 2d ed., Addison-Wesley Publishing Company, Inc., Reading, Mass., 1959.

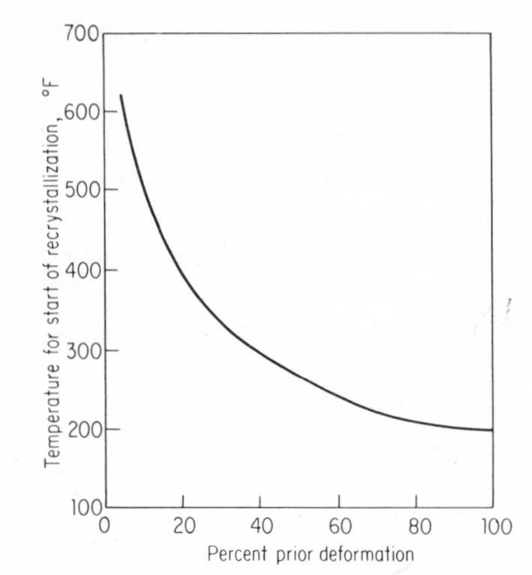

Fig. 4·5 Effect of prior deformation on the temperature for the start of recrystallization of copper.

temperature. By the same reasoning, the lower the temperature of cold working, the greater the amount of strain introduced, effectively decreasing the recrystallization temperature for a given annealing time.

A certain minimum amount of cold working (usually 2 to 8 percent) is

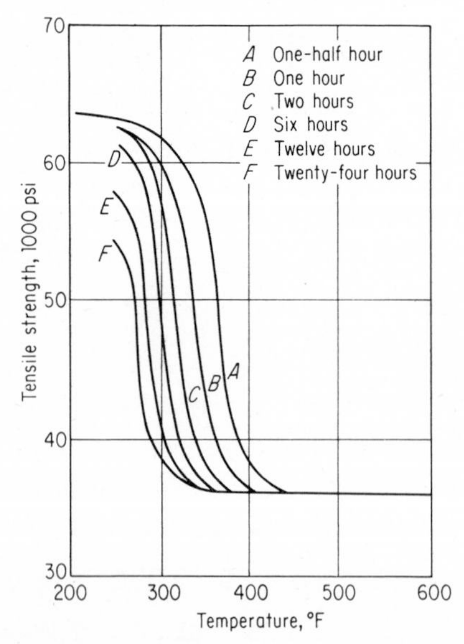

Fig. 4·6 Effect of time and temperature on annealing. (From "Metals Handbook," 1948 ed., American Society for Metals, Metals Park, Ohio.)

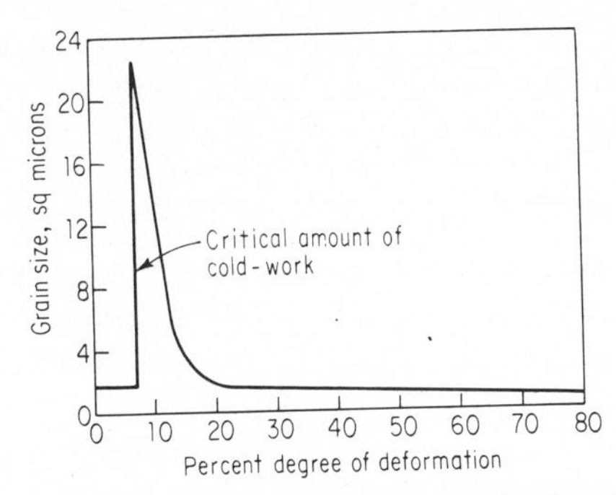

Fig. 4·7 Effect of cold working on grain size developed in a low-carbon steel after annealing at 1740° F. (From "Metals Handbook," 1948 ed., American Society for Metals, Metals Park, Ohio.)

necessary for recrystallization. In Fig. 4·7, it is seen that a deformation of approximately 7 percent is required before any change in grain size occurs. This is known as the *critical deformation*. At degrees of deformation smaller than this, the number of recrystallization nuclei becomes very small.

4·5 Grain Growth Large grains have lower free energy than small grains. This is associated with the reduction of the amount of grain boundary. Therefore, under ideal conditions, the lowest energy state for a metal would be as a single crystal. This is the driving force for grain growth. Opposing this force is the rigidity of the lattice. As the temperature increases, the rigidity of the lattice decreases and the rate of grain growth is more rapid. At any given temperature there is a maximum grain size at which these two effects are in equilibrium (Fig. 4·8).

It is therefore theoretically possible to grow very large grains by holding a specimen for a long time high in the grain-growth region. The very large

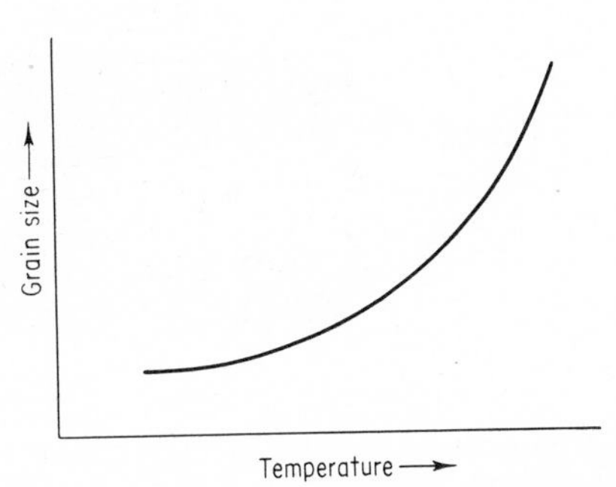

Fig. 4·8 Effect of temperature on recrystallized grain size.

grains shown in Fig. 4·9 were obtained by this method. The specimen was held at a temperature just under the melting point of this alloy. Notice that some melting has occurred in the lower left corner because of temperature fluctuation in the furnace.

4·6 Grain Size Since annealing involves nucleation and grain growth, factors that favor rapid nucleation and slow growth will result in fine-grained material, and those which favor slow nucleation and rapid growth will result in coarse-grained material. The factors that govern the final recrystallized grain size are:

Degree of Prior Deformation This is the most important factor. Increasing the amount of prior deformation favors nucleation and decreases the final grain size (see Fig. 4·7). It is interesting to note that, at the critical deformation, the grains will grow to a very large size upon annealing. The formation of large grains during recrystallization with the minimum deformation is due to the very few recrystallization nuclei that are formed during the time available for recrystallization. If the deformation is carefully controlled at the critical amount, subsequent annealing will result in very large grains or single crystals. This is the basis for the strain-anneal method of producing single crystals. With increasing degrees of deformation, an increasing number of points of high stress or high energy are present, leading to recrystallization from a greater number of nuclei, and finally to a greater number of grains, and thus to a continually smaller grain size.

Time at Temperature Increasing the time at any temperature above the recrystallization temperature favors grain growth and increases the final grain size. The progress of recrystallization of cold-worked brass is shown in the group of photomicrographs of Fig. 4·10. The first one (*a*) shows the alloy after the rolling operation in which the reduction was 33 percent. There are numerous slip lines and several dark twinning bands. The remaining samples were reheated at 1075°F in a lead bath for increasing periods of time. The second one (*b*) shows new grains beginning to form

Fig. 4·9 Large grains in a titanium-vanadium alloy formed by holding for a long time high in the grain-growth region. Magnification 2×.

along the slip lines and at the grain boundaries, which are points of high internal energy. After 8 min at 1075°F (photomicrograph *c*), recrystallization is just about complete, there being no evidence of the old distorted structure. Beyond this point, increasing the time at 1075°F (photomicrographs *d* to *g*) merely serves to increase the grain size. Notice the presence of annealing twin bands, which are found in wrought and annealed brasses. They arise during annealing by a change in the normal growth mechanism due to previous straining.

Annealing Temperature The lower the temperature *above* the recrystallization temperature, the finer the final grain size (see Fig. 4·8).

Heating Time The shorter the time heating to the annealing temperature, the finer the final grain size. Slow heating will form few nuclei, favoring grain growth and resulting in coarse grain.

Insoluble Impurities The greater the amount and the finer the distribution of insoluble impurities the finer the final grain size. They not only increase nucleation but act as barriers to the growth of grains. It was shown in Table 4·1 that adding alloying elements (such as zinc in copper) or soluble impurities raised the recrystallization temperature. Insoluble impurities, such as Cu_2O in copper, do not noticeably affect the recrystallization temperature but decrease the recrystallized grain size. This latter effect is used commercially to obtain fine-grained structures in annealed metals.

The amount of strain hardening introduced by a given amount of elongation increases as the grain size decreases. If both coarse- and fine-grained material are given the same amount of strain hardening, their annealing behavior will be very similar.

The rate of cooling from the annealing temperature has a negligible effect on final grain size. This factor will be of interest only if the material has been heated far in the grain-growth range and slow-cooled. During slow cooling the material may have enough energy to continue grain growth, and some coarsening may result.

4·7 Effect on Properties Since full annealing restores the material to a strain-free lattice structure, it is essentially a softening process. Property changes produced by plastic deformation are removed, and the material returns very nearly to its original properties. Therefore, during annealing, the hardness and strength decrease, whereas the ductility increases. The change in properties is shown schematically in Fig. 4·11 and for 70–30 brass in Table 4·2.

HOT WORKING

Hot working is usually described as working a material above its recrystallization temperature. The above definition, however, does not take into account the rate of working.

(a) (b)

(c) (d)

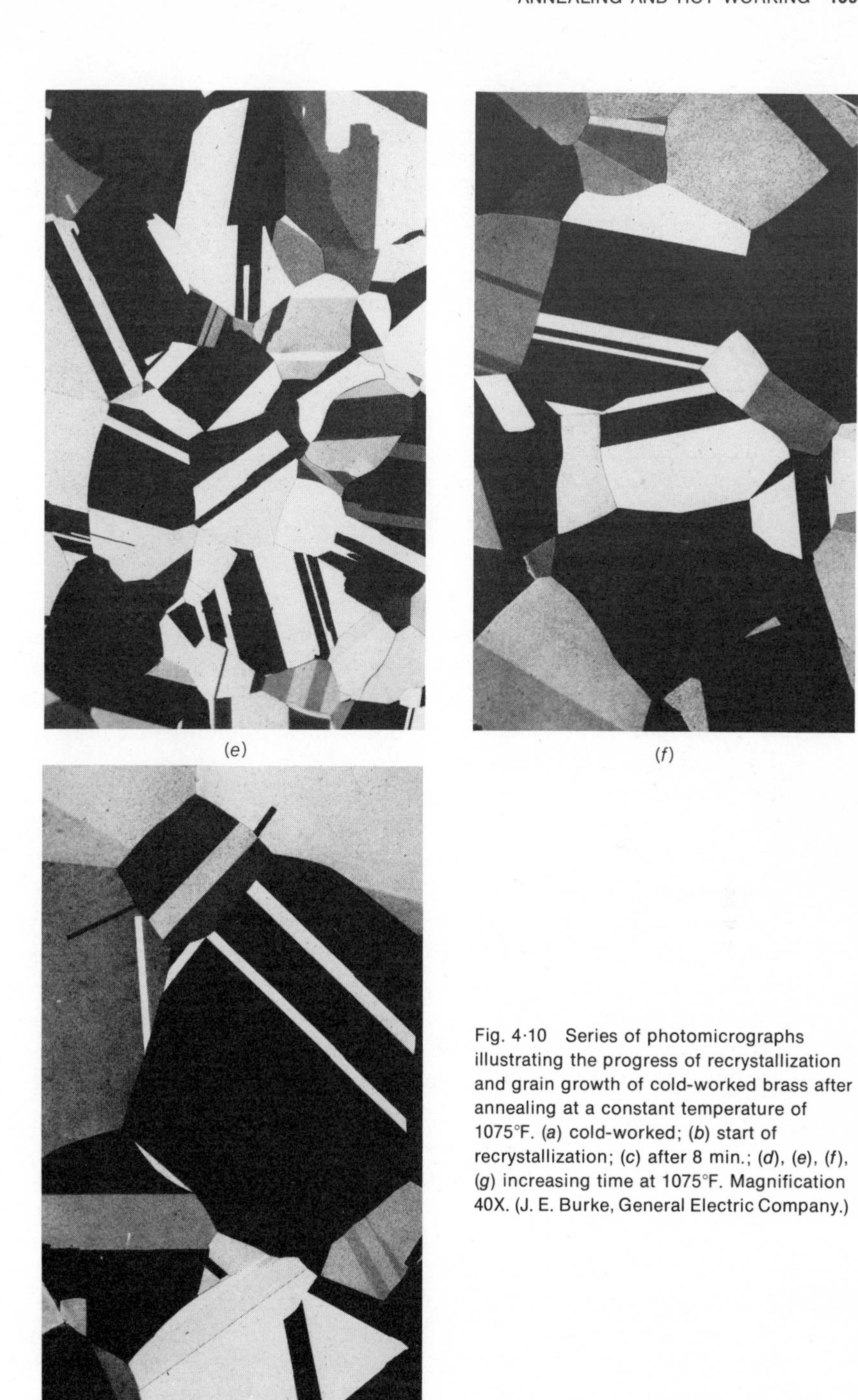

(e)

(f)

(g)

Fig. 4·10 Series of photomicrographs illustrating the progress of recrystallization and grain growth of cold-worked brass after annealing at a constant temperature of 1075°F. (*a*) cold-worked; (*b*) start of recrystallization; (*c*) after 8 min.; (*d*), (*e*), (*f*), (*g*) increasing time at 1075°F. Magnification 40X. (J. E. Burke, General Electric Company.)

TABLE 4·2 Annealing of 70-30 brass after 50 percent cold reduction with time constant at 30 min*

ANNEALING TEMP, °F	HARDNESS, ROCKWELL X†	TENSILE STRENGTH, PSI	ELONGATION, % IN 2 IN.
None (cold work)	97	80,000	8
300	98	81,000	8
392	100	82,000	8
482	101	82,000	8
572	98	76,000	12
662	80	60,000	28
842	58	46,000	51
1112	34	44,000	66
1292	14	42,000	70

* From R. M. Brick and A. Phillips, "Structure and Properties of Alloys," 2d ed., McGraw-Hill Book Company, New York, 1949.
† Rockwell X = 1/16-in. ball penetrator, 75-kg load.

4·8 Dividing Line between Hot and Cold Working When a material is plastically deformed, it tends to become harder, but the rate of work hardening decreases as the temperature is increased. When a material is plastically deformed at an elevated temperature, two opposing effects take place at the same time—a hardening effect due to plastic deformation and a softening effect due to recrystallization. For a given rate of working, there must be some temperature at which these two effects will just balance. If the material is worked above this temperature, it is known as *hot working*; below this temperature it is known as *cold working*. The hardening of copper under slow deformation in a tension test at various temperatures is shown in Fig. 4·12. At about 750°F, the rate of softening will equal the rate of hardening and the material can be continuously deformed without an increase in load. If the rate of deformation is increased considerably, such as in hammer forging, the temperature will have to be increased to about 1475°F before the two rates will be equal. The effect of working temperature on hardness with varying rate of working is shown schematically in Fig. 4·13.

The terms *hot* and *cold* as applied to working do not have the same significance that they ordinarily have. For example, lead and tin, whose recrystallization temperature is below room temperature, may be hot-worked at room temperature; but steel, with a high recrystallization temperature, may be cold-worked at 1000°F.

4·9 Hot Working vs. Cold Working Most of the metal shapes are produced from cast ingots. To manufacture sheet, plate, rod, wire, etc., from this ingot, the most economical method is by hot working. However, in the case of steel, the hot-worked material reacts with oxygen as it cools down to room temperature and forms a characteristic dark oxide coating called *scale*.

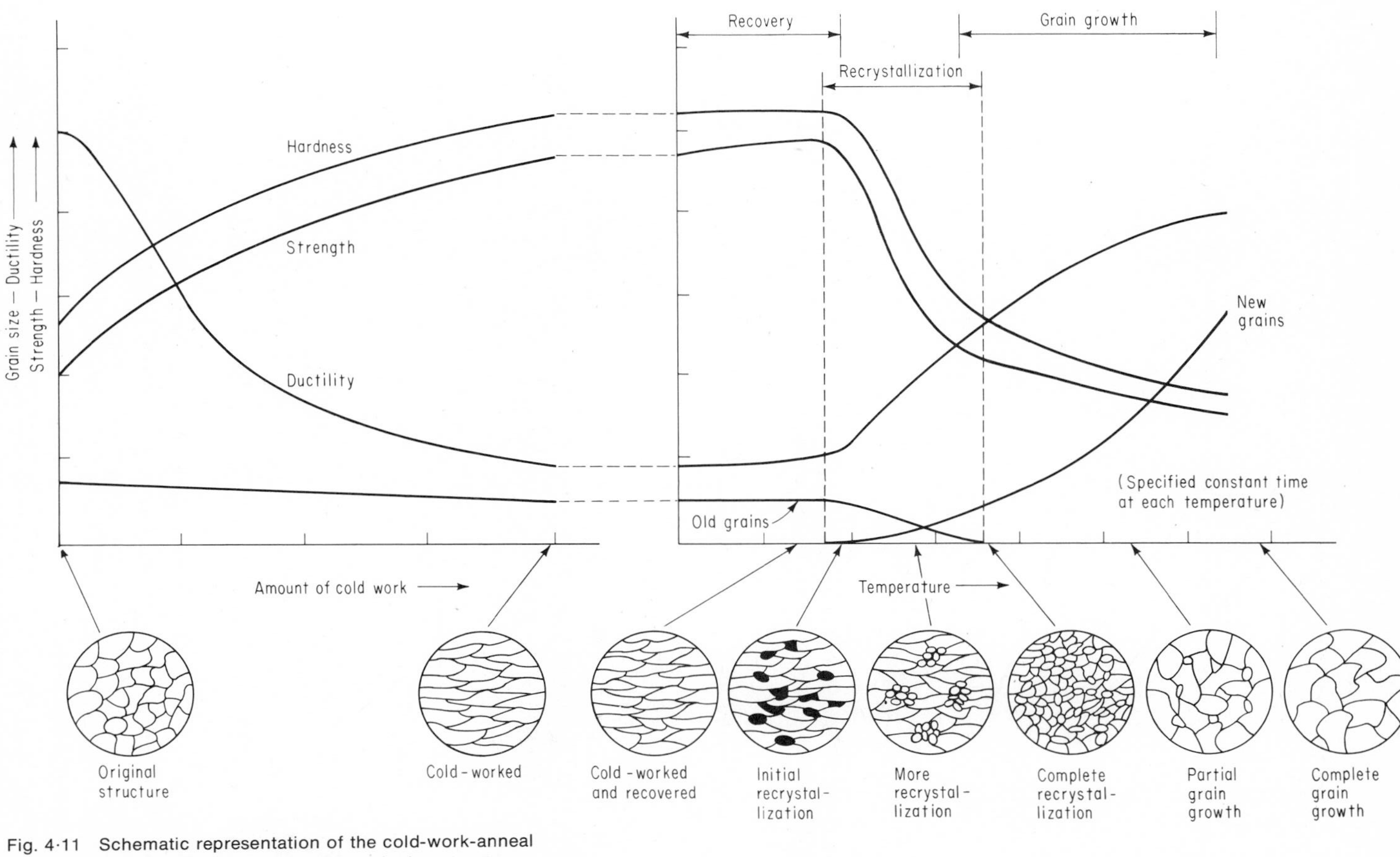

Fig. 4·11 Schematic representation of the cold-work-anneal cycle showing the effects on properties and microstructure. (From C. O. Smith, "The Science of Engineering Materials," Prentice-Hall Inc., Englewood Cliffs, N.J., 1969.)

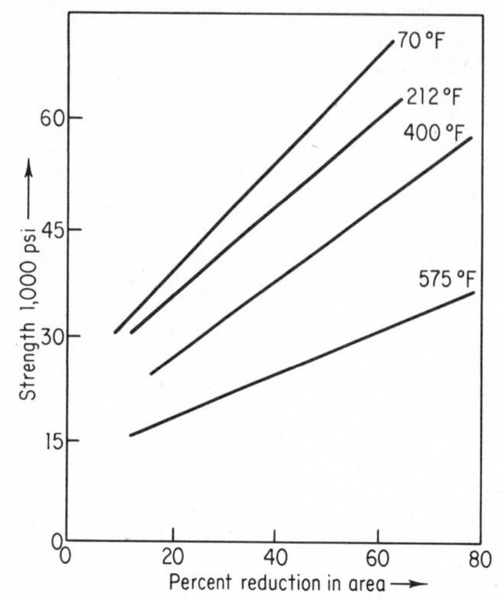

Fig. 4·12 Effect of the amount of cold working on the strength of copper determined by tension tests at various temperatures.

Occasionally, this scale may give difficulty during machining or forming operations.

It is not possible to manufacture hot-worked material to exact size because of dimensional changes that take place during cooling. Cold-worked material, on the other hand, may be held to close tolerances. It is free of surface scale but requires more power for deformation and is therefore more expensive to produce. Commercially, the initial reductions are car-

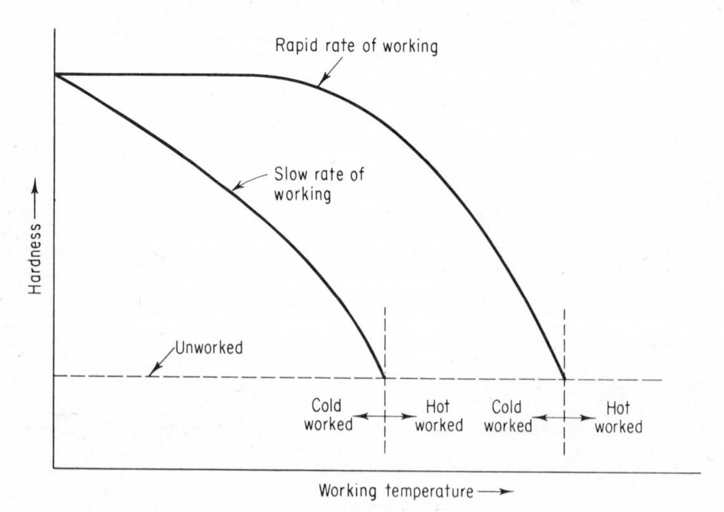

Fig. 4·13 Schematic illustration of the effect of working temperature on hardness with varying rate of working.

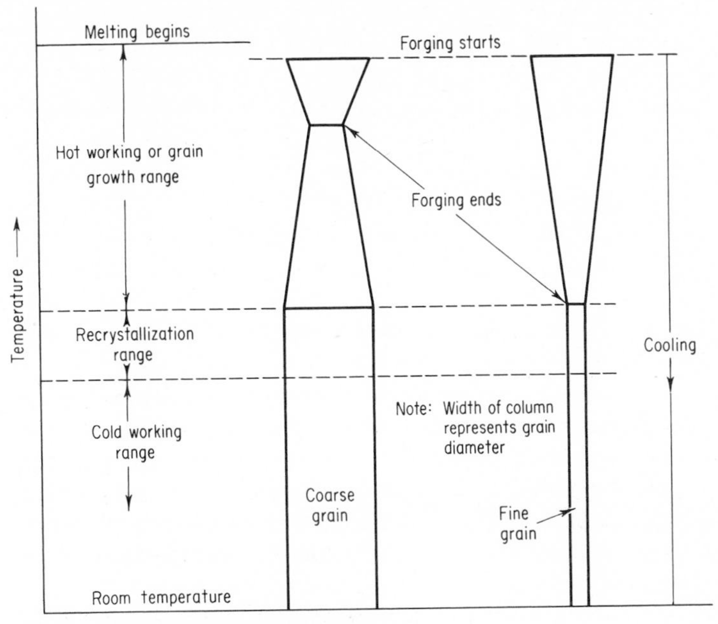

Fig. 4·14 Effect of finishing temperature in forging on grain size.

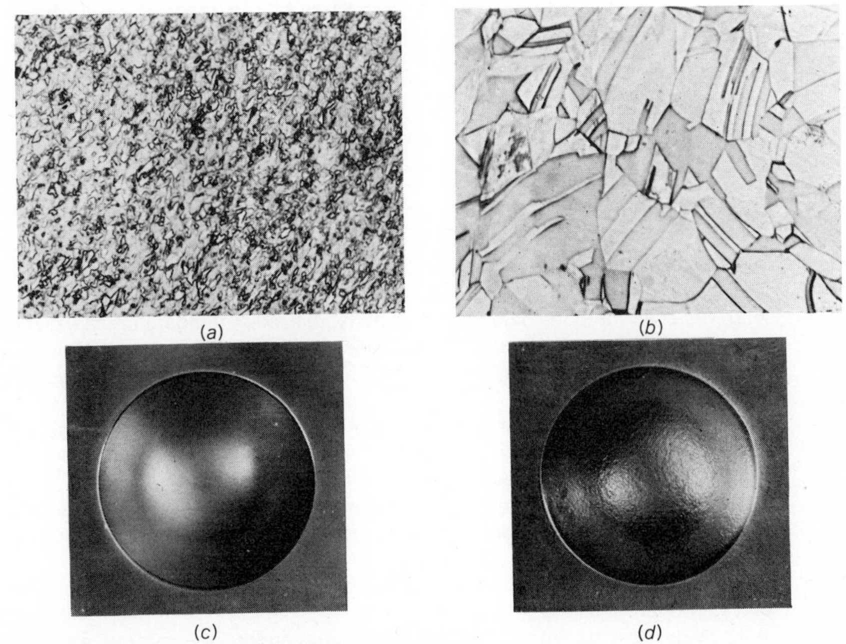

Fig. 4·15 Effect of grain size on the surface appearance of cold-drawn 70–30 brass sheet. (H. L. Burghoff, Chase Brass and Copper Company.)

TABLE 4·3 Recommended Grain Sizes in Brass for Cold-forming Operations*

GRAIN SIZE, MM	TYPE OF COLD-FORMING OPERATION
0.015	Slight forming operations
0.025	Shallow drawing
0.035	For best average surface combined with drawing
0.050	Deep drawing
0.100	Heavy drawing on thick sheet

* From "Metals Handbook," 1948 ed., table II, p. 879, American Society for Metals, Metals Park, Ohio.

ried out with the material at an elevated temperature, and the final reductions are done cold to take advantage of both processes.

The finishing temperature in hot working will determine the grain size that is available for further cold working. Higher temperatures are used initially to promote uniformity in the material, and the resulting large grains allow more economical reduction during the early working operation. As the material cools and working continues, the grain size will decrease, becoming very fine just above the recrystallization temperature. This is illustrated schematically in Fig. 4·14.

Proper control of annealing temperature will approximate the final grain size required for subsequent cold working. Although coarse-grained material has better ductility, the nonuniformity of deformation from grain to grain creates a problem in surface appearance. Figure 4·15 shows the "orange-peel" surface on coarse-grained material that is subjected to severe deformation. The choice of grain size is therefore a compromise determined by the particular cold-forming operation (Table 4·3).

QUESTIONS

4·1 Explain the importance of heating in the recovery range for some industrial applications.

4·2 Give two methods of lowering the recrystallization temperature of a given metal.

4·3 Assume that a tapered piece of copper has been stressed in tension beyond its yield point and then annealed. Explain how the grain size will change along the taper.

4·4 Suppose a bullet hole is made in a plate of aluminum. How will the grain size from the hole outward vary if the plate is annealed?

4.5 Describe two methods of making a single crystal.

4·6 Why does the recrystallization temperature vary with different metals?

4·7 Why does the addition of alloying elements change the recrystallization temperature?

4·8 Cite an industrial application which requires periodic annealing between cold-working operations.

4·9 Many heat-treating processes, including annealing, involve nucleation and growth. The relation between the rate of the process and the temperature may be expressed, in general, as

$$\text{Rate} = Ae^{-B/T}$$

where A and B are constants, and T is the absolute temperature in degrees Kelvin. For 50 percent recrystallization of copper, $A = 10^{12}$ min^{-1} and $B = 15{,}000$. The time for 50 percent recrystallization may be taken as the reciprocal of the rate. Calculate the time for 50 percent recrystallization at 100, 150, 200, 250, and 275°F.

4·10 Plot the results obtained in Question 4·9 on semilog paper with time on the log scale and temperature as the ordinate. This plot should be a straight line.

4·11 Extrapolate the line obtained in Question 4·10 and determine the temperature at which copper will be 50 percent recrystallized after 15 years.

REFERENCES

American Society for Metals: "Metals Handbook," 1948 ed., Metals Park, Ohio.

Brick, R. M., and A. Phillips: "Structure and Properties of Alloys," 2d ed., McGraw-Hill Book Company, New York, 1949.

Byrne, J. E.: "Recovery, Recrystallization and Grain Growth," The Macmillan Company, New York, 1965.

Guy, A. G.: "Elements of Physical Metallurgy," Addison-Wesley Publishing Company, Inc., Reading, Mass., 1959.

Mason, C. W.: "Introductory Physical Metallurgy," American Society for Metals, Metals Park, Ohio, 1947.

Reed-Hill, R. E.: "Physical Metallurgy Principles," Van Nostrand Reinhold Company, New York, 1964.

Rogers, B. A.: "The Nature of Metals," American Society for Metals, Metals Park, Ohio, 1951.

Smith, C. O.: "The Science of Engineering Materials," Prentice-Hall, Inc., Englewood Cliffs, N.J., 1969.

5 CONSTITUTION OF ALLOYS

5·1 Introduction An alloy is a substance that has metallic properties and is composed of two or more chemical elements, of which at least one is a metal.

An alloy system contains all the alloys that can be formed by several elements combined in all possible proportions. If the system is made up of two elements, it is called a *binary alloy system*; three elements, a *ternary alloy system*; etc. Taking only 45 of the most common metals, any combination of two gives 990 binary systems. Combinations of three give over 14,000 ternary systems. However, in each system, a large number of different alloys are possible. If the composition is varied by 1 percent, each binary system will yield 100 different alloys. Since commercial alloys often contain many elements, it is apparent that the number of possible alloys is almost infinite.

Alloys may be classified according to their structure, and complete alloy systems may be classified according to the type of their equilibrium or phase diagram. The basic types of phase diagrams will be studied in Chap. 6.

5·2 Classification of Alloys Alloys may be homogeneous (uniform) or mixtures. If the alloy is homogeneous it will consist of a single phase, and if it is a mixture it will be a combination of several phases. A phase is anything which is homogeneous and physically distinct. The uniformity of an alloy phase is not determined on an atomic scale, such as the composition of each unit lattice cell, but rather on a much larger scale. Any structure which is visible as physically distinct microscopically may be considered a phase. For most pure elements the term *phase* is synonymous with *state*. There is, therefore, for pure elements, a gaseous, liquid, and solid phase. Some metals are allotropic in the solid state and will have different solid phases. When the metal undergoes a change in crystal structure, it undergoes a phase change since each type of crystal structure is physically distinct.

In the solid state there are three possible phases: (1) pure metal, (2) intermediate alloy phase or compound, and (3) solid solution.

If an alloy is homogeneous (composed of a single phase) in the solid state, it can be only a solid solution or a compound. If the alloy is a mixture, it is then composed of any combination of the phases possible in the solid state. It may be a mixture of two pure metals, or two solid solutions, or two compounds, or a pure metal and a solid solution, and so on. The mixture may also vary in degree of fineness.

5·3 Pure Metal The characteristics of a pure metal have been discussed in detail in an earlier chapter. However, one property is worth repeating. Under equilibrium conditions, all metals exhibit a definite melting or freezing point. The term *under equilibrium conditions* implies conditions of extremely slow heating and cooling. In other words, if any change is to occur, sufficient time must be allowed for it to take place. If a cooling curve is plotted for a pure metal, it will show a horizontal line at the melting or freezing point (Fig. 5·1).

5·4 Intermediate Alloy Phase or Compound Because the reason for referring to this type of solid phase as an *intermediate alloy phase* will be more apparent during the study of phase diagrams, it will be simpler at this point to call it a *compound.*

It is now necessary to obtain some understanding of compounds in general. Most ordinary chemical compounds are combinations of positive and negative valence elements. The various kinds of atoms are combined in a definite proportion, which is expressed by a chemical formula. Some typical examples are water, H_2O (two atoms of hydrogen combined with one

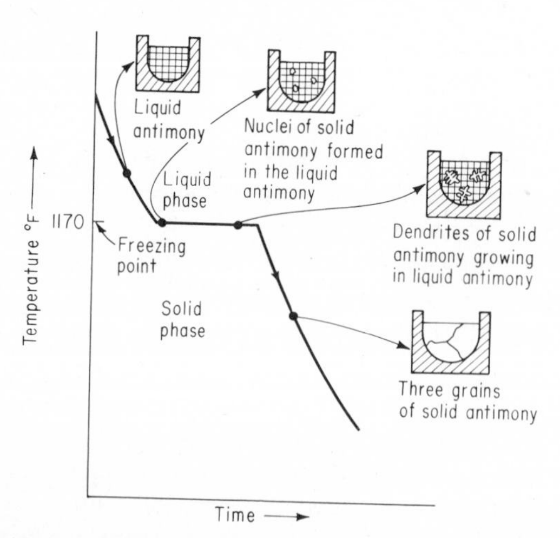

Fig. 5·1 Time-temperature cooling curve for the solidification of a small crucible of liquid antimony.

atom of oxygen), and table salt, NaCl (one atom of sodium combined with one atom of chlorine). The atoms that are combined to form the molecule, which is the smallest unit that has the properties of the compound, are held together in a definite bond. Various types of atomic bonding have been discussed in Chap. 2. The bond is generally strong, and the atoms are not easily separated. Most students are familiar with the classical high school chemistry demonstration of the electrolysis of water. By passing an electric current through water it is possible to separate the hydrogen and oxygen atoms.

When a compound is formed, the elements lose their individual identity and characteristic properties to a large extent. A good example is table salt (NaCl). Sodium (Na) is a very active metal that oxidizes rapidly and is usually stored under kerosene. Chlorine (Cl) is a poisonous gas. Yet one atom of each combines to give the harmless and important compound, table salt. Water (H_2O) is composed of elements that are normally gases at room temperature, yet the compound is a liquid at room temperature. What exists then is not the individual elements but rather the combination or compound. The compound will have its own characteristic physical, mechanical, and chemical properties.

Most compounds, like pure metals, also exhibit a definite melting point within narrow limits of temperature. Therefore, the cooling curve for a compound is similar to that for a pure metal (see Fig. 5·1). It is then referred to as a *congruent melting phase.* In reference to equilibrium diagrams, the intermediate alloy phases are phases whose chemical compositions are intermediate between the two pure metals and generally have crystal structures different from those of the pure metals.

The three most common intermediate alloy phases are:

Intermetallic Compounds or Valency Compounds These are generally formed between chemically dissimilar metals and are combined by following the rules of chemical valence. Since they generally have strong bonding (ionic or covalent), their properties are essentially nonmetallic. They usually show poor ductility and poor electrical conductivity and may have a complex crystal structure. Examples of valency compounds are CaSe, Mg_2Pb, Mg_2Sn, and Cu_2Se.

Interstitial Compounds These compounds formed between the transition metals such as scandium (Sc), titanium (Ti), tantalum (Ta), tungsten (W), and iron (Fe), with hydrogen, oxygen, carbon, boron, and nitrogen. The word *interstitial* means between the spaces, and the latter five elements have relatively small atoms that fit into the spaces of the lattice structure of the metal. These same five elements also form interstitial solid solutions, which will be described shortly. The interstitial compounds are metallic, may have a narrow range of composition, high melting points, and are extremely hard. Examples are TiC, TaC, Fe_4N, Fe_3C, W_2C, CrN, and TiH.

Many of these compounds are useful in hardening steel and in cemented-carbide tools.

Electron Compounds A study of the equilibrium diagrams of the alloys of copper, gold, silver, iron, and nickel with the metals cadmium, magnesium, tin, zinc, and aluminum shows striking similarities. A number of

TABLE 5·1 Examples of Electron Compounds

ELECTRON-ATOM RATIO 3:2 (B.C.C. STRUCTURE)	ELECTRON-ATOM RATIO 21:13 (COMPLEX CUBIC)	ELECTRON-ATOM RATIO 7:4 (C.P.H. STRUCTURE)
AgCd	Ag_5Cd_8	$AgCd_3$
AgZn	Cu_9Al_4	Ag_5Al_3
Cu_3Al	$Cu_{31}SN_8$	$AuZn_3$
AuMg	Au_5Zn_8	Cu_3Si
FeAl	Fe_5Zn_{21}	$FeZn_7$
Cu_5Sn	Ni_5Zn_{21}	Ag_3Sn

intermediate phases are formed in these systems with similar lattice structures. Hume-Rothery first pointed out that these intermediate phases are found to exist at or near compositions in each system that have a definite ratio of valence electrons to atoms and are therefore called *electron compounds.* Some examples are given in Table 5·1. For example, in the compound AgZn, the atom of silver has one valence electron while that of zinc has two valence electrons so that the two atoms of the compound will have three valence electrons, or an electron-to-atom ratio of 3:2. In the compound Cu_9Al_4, each atom of copper has one valence electron and each atom of aluminum three valence electrons, so that the 13 atoms that make up the compound have 21 valence electrons, or an electron-to-atom ratio of 21:13. For the purpose of calculation, the atoms of iron and nickel are assumed to have zero valence.

Many electron compounds have properties resembling those of solid solutions, including a wide range of composition, high ductility, and low hardness.

5·5 Solid Solutions Any solution is composed of two parts: a solute and a solvent. The solute is the minor part of the solution or the material which is dissolved, while the solvent constitutes the major portion of the solution. It is possible to have solutions involving gases, liquids, or solids as either the solute or the solvent. The most common solutions involve water as the solvent, such as sugar or salt dissolved in water.

The amount of solute that may be dissolved by the solvent is generally a function of temperature (with pressure constant) and usually increases with increasing temperature.

There are three possible conditions for a solution: unsaturated, saturated, and supersaturated. If the solvent is dissolving less of the solute than it could dissolve at a given temperature and pressure, it is said to be *unsaturated.* If it is dissolving the limiting amount of solute, it is *saturated.* If it is dissolving more of the solute than it should, under equilibrium conditions, the solution is *supersaturated.* The latter condition may be accomplished by doing work on the solution, such as stirring, or preventing equilibrium conditions by rapidly cooling the solution. The supersaturated condition is an unstable one, and given enough time or a little energy, the solution tends to become stable or saturated by rejecting or precipitating the excess solute.

A solid solution is simply a solution in the solid state and consists of two kinds of atoms combined in one type of space lattice. There is usually a considerable difference in the solubility of the solute in the liquid and solid states of the solution. The solute is generally more soluble in the liquid state than in the solid state. Moreover, when solidification of the solution starts, the temperature may be higher or lower than the freezing point of the pure solvent. Most solid solutions solidify over a range in temperature. Figure 5·2 shows the cooling curve for a solid solution alloy containing 50 percent Sb (antimony) and 50 percent Bi (bismuth). Compare this cooling curve with the one shown in Fig. 5·1. Notice that this alloy begins to solidify at a temperature lower than the freezing point of pure antimony (1170°F) and higher than the freezing point of pure bismuth (520°F). The process of solidification and the composition of the solid solution alloy and the liquid solution during freezing will be explained

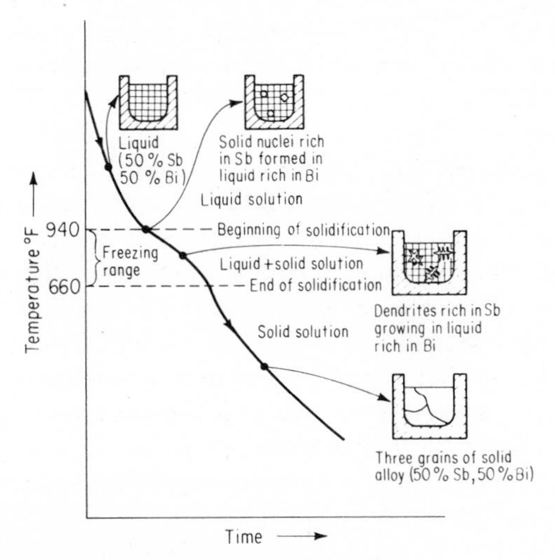

Fig. 5·2 Time-temperature cooling curve for the solidification of a small crucible of 50 percent antimony, 50 percent bismuth alloy.

in the next chapter. There are two types of solid solutions, substitutional and interstitial.

5·6 Substitutional Solid Solution In this type of solution, the atoms of the solute substitute for atoms of the solvent in the lattice structure of the solvent. For example, silver atoms may substitute for gold atoms without losing the f.c.c. (face-centered cubic) structure of gold, and gold atoms may substitute for silver atoms in the f.c.c. lattice structure of silver. All alloys in the silver-gold system consist of an f.c.c. lattice with silver and gold atoms distributed at random through the lattice structure. This entire system consists of a continuous series of solid solutions.

Several factors are now known, largely through the work of Hume-Rothery, that control the range of solubility in alloy systems.

Crystal-structure Factor Complete solid solubility of two elements is never attained unless the elements have the same type of crystal lattice structure.

Relative-size Factor The size factor is favorable for solid-solution formation when the difference in atomic radii is less than about 15 percent. If the relative size factor is greater than 8 percent but less than 15 percent, the alloy system usually shows a minimum. If the relative-size factor is greater than 15 percent, solid-solution formation is very limited. For example, silver and lead are both f.c.c., and the relative-size factor is about 20 percent. The solubility of lead in solid silver is about 1.5 percent, and the solubility of silver in solid lead is about 0.1 percent. Antimony and bismuth are completely soluble in each other in all proportions. They have the same type of crystal structure (rhombohedral) and differ in atomic radii by about 7 percent. However, the solubility of antimony in f.c.c. aluminum is less than 0.1 percent, although the relative-size factor is only about 2 percent.

Chemical-affinity Factor The greater the chemical affinity of two metals, the more restricted is their solid solubility and the greater is the tendency toward compound formation. Generally, the farther apart the elements are in the periodic table, the greater is their chemical affinity.

Relative-valence Factor If the solute metal has a different valence from that of the solvent metal, the number of valence electrons per atom, called the electron ratio, will be changed. Crystal structures are more sensitive to a decrease in the electron ratio than to an increase. In other words, a metal of lower valence tends to dissolve more of a metal of higher valence than vice versa. For example, in the aluminum-nickel alloy system, both metals are face-centered cubic. The relative-size factor is approximately 14 percent. However, nickel is lower in valence than aluminum, and in accord with the relative-valence factor solid nickel dissolves 5 percent aluminum, but the higher valence aluminum dissolves only 0.04 percent nickel.

By considering the above four factors, some estimate of the solid solubility of one metal in another can be determined. It is important to note that an unfavorable relative-size factor alone is sufficient to limit solubility to a low value. If the relative-size factor is favorable, then the other three factors should be considered in deciding on the probable degree of solid solubility. While the Hume-Rothery rules are a very good guide to solid solubility, there are exceptions to these rules.

The lattice structure of a solid solution is basically that of the solvent with slight changes in lattice parameter. An expansion results if the solute atom is larger than the solvent atom and a contraction if the solute atom is smaller.

5·7 Interstitial Solid Solutions These are formed when atoms of small atomic radii fit into the spaces or interstices of the lattice structure of the larger solvent atoms. Since the spaces of the lattice structure are restricted in size, only atoms with atomic radii less than 1 angstrom are likely to form interstitial solid solutions. These are hydrogen (0.46), boron (0.97), carbon (0.77), nitrogen (0.71), and oxygen (0.60).

This type of solution differs from interstitial compounds in that the amount of smaller atoms required to form the compound is always greater than the amount that may be dissolved interstitially. When a small amount of solute is added to the solvent and the difference in atomic radii is great enough, an interstitial solid solution is formed. In this condition, the solute atoms have considerable mobility and may move in the interstitial spaces of the lattice structure. More solute atoms may be dissolved interstitially until the solution becomes saturated at that temperature. Increasing the amount of solute atoms beyond this limit severely restricts the mobility of these atoms in a particular area and the interstitial compound of fixed composition starts to form. The interstitial compound, showing a narrow range of composition, is expressed by a chemical formula, but the interstitial solution, being of variable composition, cannot be represented by a chemical formula. The lattice structure always shows an expansion when this type of solution is formed.

Interstitial solid solutions normally have very limited solubility and generally are of little importance. Carbon in iron is a notable exception and forms the basis for hardening steel, which will be discussed in Chap. 8. Carbon dissolves in iron interstitially. The maximum solubility of carbon in γ iron (f.c.c.) is 2 percent at 2065°F, while the maximum solubility of carbon in α iron (b.c.c.) is only 0.025 percent at 1333°F.

Both types of solid solutions are illustrated in Fig. 5·3. Distortion of the lattice structure will exist in the region of the solute atom. This distortion will interfere with the movement of dislocations on slip planes and will therefore increase the strength of the alloy. This is the primary basis for the strengthening of a metal by alloying.

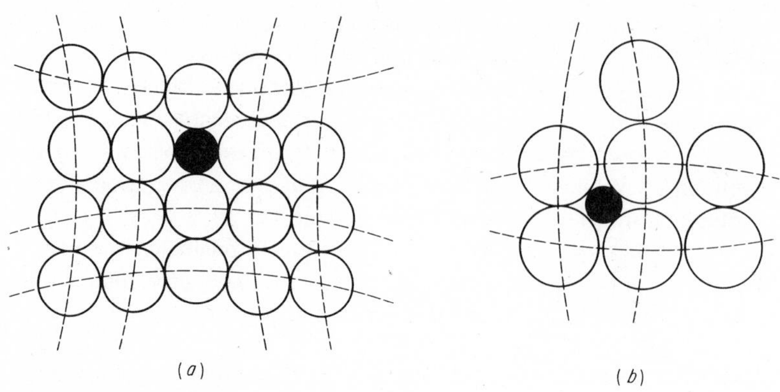

Fig. 5·3 Schematic representation of both types of solid solutions. (*a*) Substitutional; (*b*) interstitial.

In contrast to intermetallic and interstitial compounds, solid solutions in general are easier to separate, melt over a range in temperature, have properties that are influenced by those of the solvent and solute, and usually show a wide range of composition so that they are not expressed by a chemical formula.

A summary of the possible alloy structures is shown in Fig. 5·4.

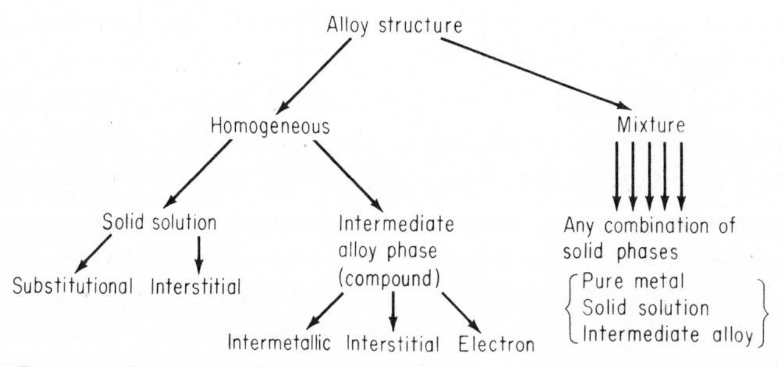

Fig. 5·4 Possible alloy structures.

6 PHASE DIAGRAMS

6·1 Introduction In the previous chapter it was indicated that there were many possibilities for the structure of an alloy. Since the properties of a material depend to a large extent on the type, number, amount, and form of the phases present, and can be changed by altering these quantities, it is essential to know (1) the conditions under which these phases exist and (2) the conditions under which a change in phase will occur.

A great deal of information concerning the phase changes in many alloy systems has been accumulated, and the best method of recording the data is in the form of *phase diagrams*, also known as *equilibrium diagrams* or constitutional diagrams.

In order to specify completely the state of a system in equilibrium, it is necessary to specify three independent variables. These variables, which are externally controllable, are temperature, pressure, and composition. With pressure assumed to be constant at atmospheric value, the equilibrium diagram indicates the structural changes due to variation of temperature and composition. The diagram is essentially a graphical representation of an alloy system.

Ideally, the phase diagram will show the phase relationships under equilibrium conditions, that is, under conditions in which there will be no change with time. Equilibrium conditions may be approached by extremely slow heating and cooling, so that if a phase change is to occur, sufficient time is allowed. In actual practice, phase changes tend to occur at slightly higher or lower temperatures, depending upon the rate at which the alloy is heated or cooled. Rapid variation in temperature, which may prevent phase changes that would normally occur under equilibrium conditions, will distort and sometimes limit the application of these diagrams.

It is beyond the scope of this text to cover all the possible conditions of equilibrium between phases in binary alloys. Only the most important ones will be considered, and they may be classified as follows:

1 Components completely soluble in the liquid state
  - **a** Completely soluble in the solid state (Type I)
  - **b** Insoluble in the solid state: the eutectic reaction (Type II)
  - **c** Partly soluble in the solid state: the eutectic reaction (Type III)
  - **d** Formation of a congruent-melting intermediate phase (Type IV)
  - **e** The peritectic reaction (Type V)

2 Components partly soluble in the liquid state: the monotectic reaction (Type VI)

3 Components insoluble in the liquid state and insoluble in the solid state (Type VII)

4 Transformations in the solid state
  - **a** Allotropic change
  - **b** Order-disorder
  - **c** The eutectoid reaction
  - **d** The peritectoid reaction

A study of these diagrams will illustrate basic principles which may be applied to understand and interpret more complex alloy systems.

6·2 Coordinates of Phase Diagrams Phase diagrams are usually plotted with temperature, in degrees centigrade or Fahrenheit, as the ordinate and the alloy composition in weight percentage as the abscissa. It is sometimes more convenient for certain types of scientific work to express the alloy composition in atomic percent. The conversion from weight percentage to atomic percentage may be made by the following formulas:

$$\text{Atomic percent of } A = \frac{100X}{X + Y(M/N)} \tag{6·1}$$

$$\text{Atomic percent of } B = \frac{100Y(M/N)}{X + Y(M/N)} \tag{6·2}$$

where M = atomic weight of metal A
N = atomic weight of metal B
X = weight percentage of metal A
Y = weight percentage of metal B

Regardless of the scale chosen for temperature or composition, there will be no difference in the form of the resulting phase diagram.

6·3 Experimental Methods The data for the construction of equilibrium diagrams are determined experimentally by a variety of methods, the most common being:

Thermal Analysis This is by far the most widely used experimental method. As was shown in Chap. 5, when a plot is made of temperature vs. time, at constant composition, the resulting cooling curve will show a a change in slope when a phase change occurs because of the evolution of heat by the phase change. This method seems to be best for determining the initial and final temperature of solidification. Phase changes occurring solely in the solid state generally involve only small heat changes, and other methods give more accurate results.

Metallographic Methods This method consists in heating samples of an alloy to different temperatures, waiting for equilibrium to be established, and then quickly cooling to retain their high-temperature structure. The samples are then examined microscopically.

This method is difficult to apply to metals at high temperatures because the rapidly cooled samples do not always retain their high-temperature structure, and considerable skill is then required to interpret the observed microstructure correctly. This method is best suited for verification of a diagram.

X-ray diffraction Since this method measures lattice dimensions, it will indicate the appearance of a new phase either by the change in lattice dimension or by the appearance of a new crystal structure. This method is simple, precise, and very useful in determining the changes in solid solubility with temperature.

6·4 Type I—Two Metals Completely Soluble in the Liquid and Solid States

Since the two metals are completely soluble in the solid state, the only type of solid phase formed will be a substitutional solid solution. The two metals will generally have the same type of crystal structure and differ in atomic radii by less than 8 percent.

The result of running a series of cooling curves for various combinations or alloys between metals *A* and *B*, varying in composition from 100 percent *A* 0 percent *B* to 0 percent *A* 100 percent *B*, is shown in Fig. 6·1. In order to see the relationship between the cooling curves, they have been plotted on a single set of axes. However, the student should realize that each cooling curve has its own coordinates. In other words, each cooling curve is a separate experiment starting from zero time. The cooling curves for the pure metals A and B show only a horizontal line because the beginning and end of solidification take place at a constant temperature. However, since intermediate compositions form a solid solution, these cooling curves show two breaks or changes in slope. For an alloy containing 80*A* and 20*B*, the first break is at temperature T_1, which indicates the beginning of solidification, and the lower break at T_2 indicates the end of solidification. All intermediate alloy compositions will show a similar type of cooling curve. The sense of the phase diagram, or some idea of its form, may be obtained by drawing a line connecting all the points that show the beginning of solidification, the upper dotted line in Fig. 6·1, and another line connecting all the points that show the end of solidification, which is the lower dotted line in Fig. 6·1.

It is now possible to determine the actual phase diagram by plotting temperature vs. composition. The appropriate points are taken from the series of cooling curves and plotted on the new diagram. For example, in Fig. 6·2, since the left axis represents the pure metal *A*, T_A is plotted along this line. Similarly, T_B is plotted. Since all intermediate compositions are percent-

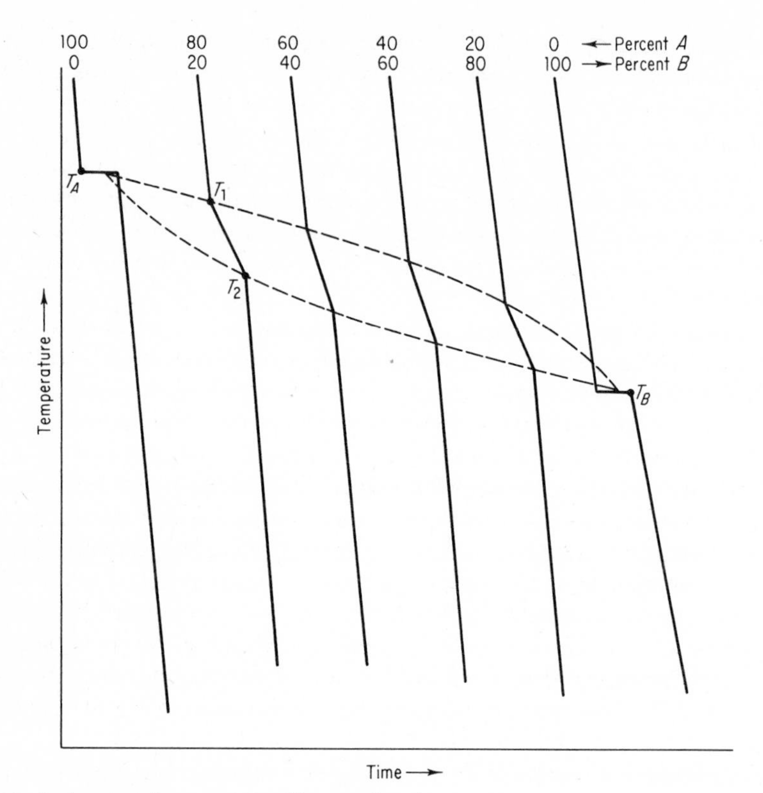

Fig. 6·1 Series of cooling curves for different alloys in a completely soluble system. The dotted lines indicate the form of the phase diagram.

ages of *A* and *B*, for simplicity the percent sign will be omitted. A vertical line representing the alloy 80*A*-20*B* is drawn, and T_1 and T_2 from Fig. 6·1 are plotted along this line. The same procedure is used for the other compositions.

The phase diagram consists of two points, two lines, and three areas. The two points T_A and T_B represent the freezing points of the two pure metals.

The upper line, obtained by connecting the points showing the beginning of solidification, is called the *liquidus* line; and the lower line, determined by connecting the points showing the end of solidification, is called the *solidus* line. The area above the liquidus line is a single-phase region, and any alloy in that region will consist of a homogeneous liquid solution. Similarly, the area below the solidus line is a single-phase region, and any alloy in this region will consist of a homogeneous solid solution. It is common practice, in the labeling of equilibrium diagrams, to represent solid solutions and sometimes intermediate alloys by Greek letters. In this case, let us label the solid solution alpha (α). Uppercase letters such as *A* and *B* will be used to represent the pure metals. Between the liquidus and solidus lines there exists a two-phase region. Any alloy in this region will consist of a mixture of a liquid solution and a solid solution.

Specification of temperature and composition of an alloy in a two-phase region indicates that the alloy consists of a mixture of two phases but does not give any information regarding this mixture. It is sometimes desirable to know the actual chemical composition and the relative amounts of the two phases that are present. In order to determine this information, it is necessary to apply two rules.

6·5 Rule I—Chemical Composition of Phases To determine the actual chemical composition of the phases of an alloy, in equilibrium at any specified temperature in a two-phase region, draw a horizontal temperature line, called a *tie line*, to the boundaries of the field. These points of intersection are dropped to the base line, and the composition is read directly.

In Fig. 6·3, consider the alloy composed of 80*A*-20*B* at the temperature *T*. The alloy is in a two-phase region. Applying Rule I, draw the tie line *mo* to the boundaries of the field. Point *m*, the intersection of the tie line with the solidus line, when dropped to the base line, gives the composition of the phase that exists at that boundary. In this case, the phase is a solid solution α of composition 90*A*-10*B*. Similarly, point *o*, when dropped to the base line, will give the composition of the other phase constituting the mixture, in this case the liquid solution of composition 74*A*-26*B*.

6·6 Rule II—Relative Amounts of Each Phase To determine the relative amounts of the two phases in equilibrium at any specified temperature in a two-

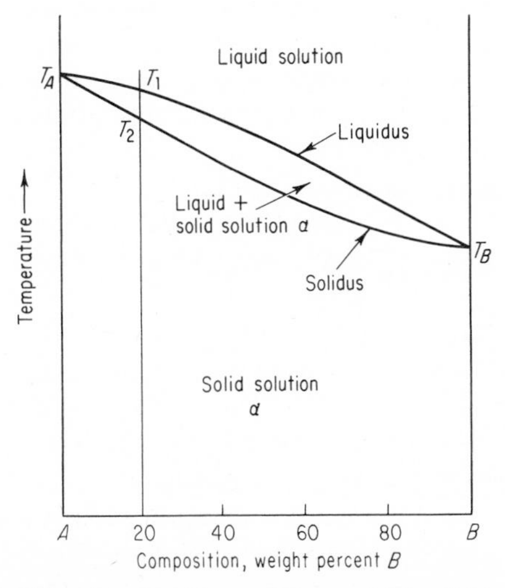

Fig. 6·2 Phase diagram of two metals completely soluble in the liquid and solid states.

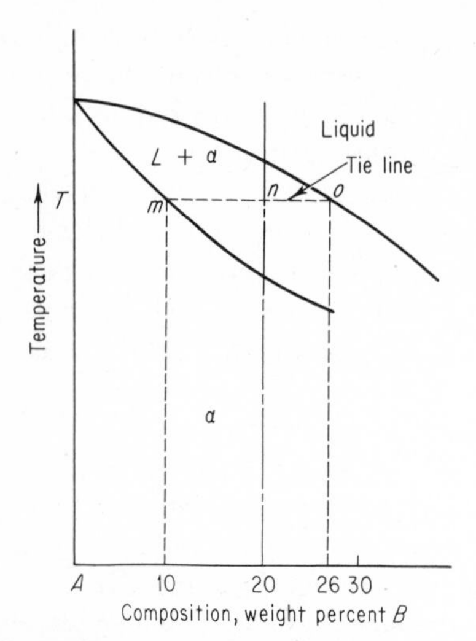

Fig. 6·3 Diagram showing the tie line *mo* drawn in the two-phase region at temperature *T*.

phase region, draw a vertical line representing the alloy and a horizontal temperature line to the boundaries of the field. The vertical line will divide the horizontal line into two parts whose lengths are inversely proportional to the amount of the phases present. This is also known as the *lever rule.* The point where the vertical line intersects the horizontal line may be considered as the fulcrum of a lever system. The relative lengths of the lever arms multiplied by the amounts of the phases present must balance.

In Fig. 6·3, the vertical line, representing the alloy 20B, divides the horizontal tie line into two parts, *mn* and *no*. If the entire length of the tie line *mo* is taken to represent 100 percent, or the total weight of the two phases present at temperature *T*, the lever rule may be expressed mathematically as

$$\text{Liquid (percent)} = \frac{mn}{mo} \times 100$$

$$\alpha \text{ (percent)} = \frac{no}{mo} \times 100$$

If the tie line is removed from the phase diagram and the numerical values are inserted, it will appear as shown in Fig. 6·4. Applying the above equations,

$$\text{Liquid (percent)} = \frac{10}{16} \times 100 = 62.5 \text{ percent}$$

$$\alpha \text{ (percent)} = \frac{6}{16} \times 100 = 37.5 \text{ percent}$$

$$\text{Liquid (percent)} = \frac{\alpha_2 T_2}{\alpha_2 L_2} \times 100 = \frac{20}{35} \times 100 = 57 \text{ percent}$$

$$\alpha_2 \text{ (percent)} = \frac{T_2 L_2}{\alpha_2 L_2} \times 100 = \frac{15}{35} \times 100 = 43 \text{ percent}$$

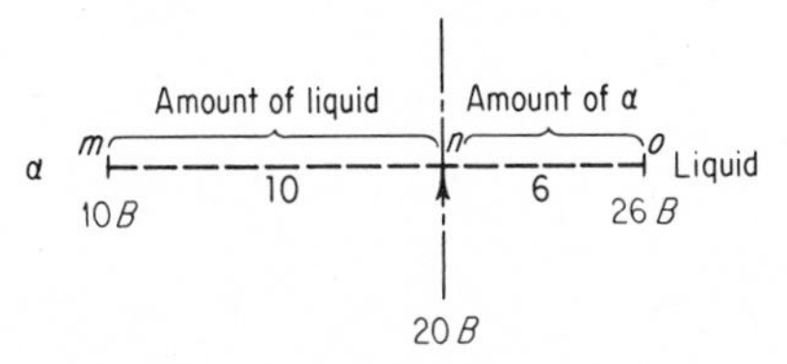

Fig. 6·4 The tie line *mo* removed from Fig. 6·3 to illustrate application of the lever rule.

To summarize both rules, the alloy of composition 80*A*-20*B* at the temperature *T* consists of a mixture of two phases. One is a liquid solution of composition 74*A*-26*B* constituting 62.5 percent of all the material present and the other a solid solution of composition 90*A*-10*B* making up 37.5 percent of all the material present.

6·7 Equilibrium Cooling of a Solid-Solution Alloy The very slow cooling, under equilibrium conditions, of a particular alloy 70*A*-30*B* will now be studied to observe the phase changes that occur (see Fig. 6.5). This alloy at temperature T_o is a homogeneous single-phase liquid solution (Fig. 6.5*a*) and remains so until temperature T_1 is reached. Since T_1 is on the liquidus line, freezing or solidification now begins. The first nuclei of solid solution to form, α_1, will be very rich in the higher-melting-point metal *A* and will be composed of 95*A*-5*B* (Rule I). Since the solid solution in forming takes material very rich in *A* from the liquid, the liquid must get richer in *B*. Just after the start of solidification, the composition of the liquid is approximated as 69*A*-31*B* (Fig. 6·5*b*).

When the lower temperature T_2 is reached, the liquid composition is at L_2. The only solid solution in equilibrium with L_2 and therefore the only solid solution forming at T_2 is α_2. Applying Rule I, α_2 is composed of 10B. Hence, as the temperature is decreased, not only does the liquid composition become richer in *B* but also the solid solution. At T_2, crystals of α_2 are formed surrounding the α_1 composition cores and also separate dendrites of α_z (Fig. 6·6). In order for equilibrium to be established at T_2, the entire solid phase must be a composition α_2. This requires diffusion of *B* atoms to the *A*-rich core not only from the solid just formed but also from the liquid. This is possible only if the cooling is extremely slow so that diffusion may keep pace with crystal growth (Fig. 6·5*c*).

At T_2, the relative amounts of the liquid and solid solution may be determined by applying Rule II:

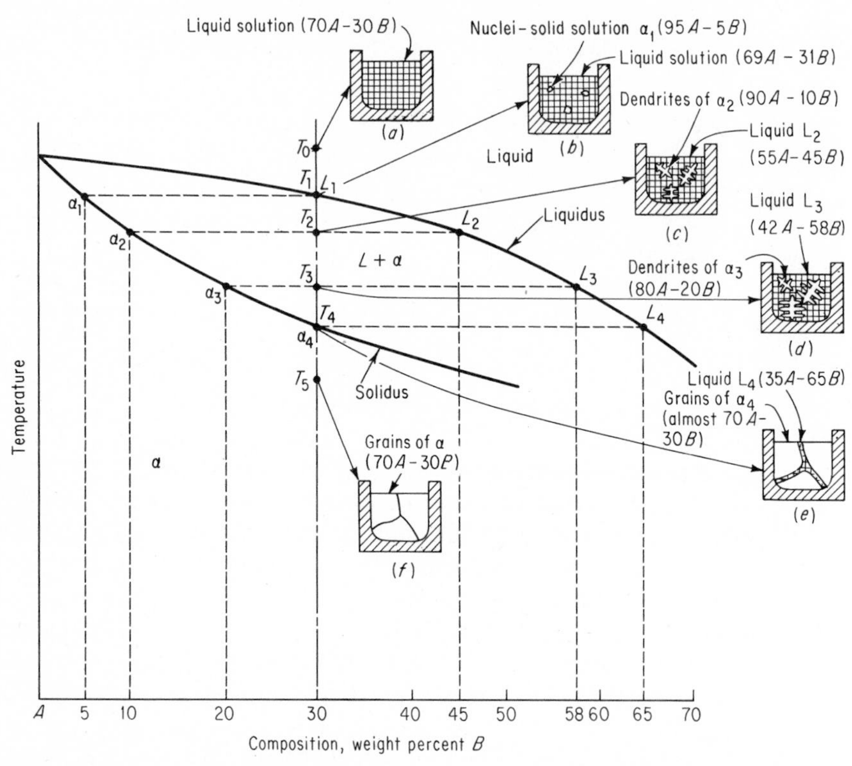

Fig. 6·5 The slow cooling of a 70A-30B alloy, showing the microstructure at various points during solidification.

As the temperature falls, the solid solution continues to grow at the expense of the liquid. The composition of the solid solution follows the solidus line while the composition of liquid follows the liquidus line, and both phases are becoming richer in *B*. At T_3 (Fig. 6·5*d*), the solid solution will make up approximately three-fourths of all the material present. The student should apply the lever rule at T_3 and determine the relative quantities of α_3 and L_3. Finally, the solidus line is reached at T_4 and the last liquid L_4, very rich in *B*, solidifies primarily at the grain boundaries (Fig. 6·5*e*). However, diffusion will take place and all the solid solution will be of uniform composition α(70*A*-30*B*), which is the overall composition of the alloy (Fig. 6·5*f*). Figure 6·7 shows the microstructure of a slow-cooled solid solution alloy. There are only grains and grain boundaries. There is no evidence of any difference in chemical composition inside the grains, indicating that diffusion has made the grain homogeneous.

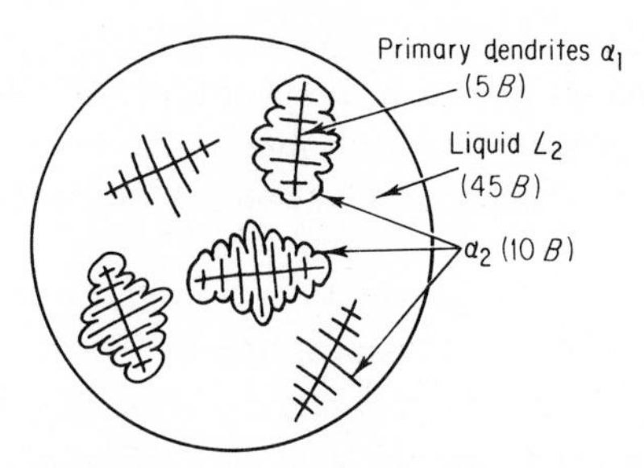

Fig. 6·6 Schematic picture of the alloy 30B at temperature T_2 before diffusion.

6·8 Diffusion It was pointed out in the previous section that diffusion, or the movement of atoms, in the solid state was an important phenomenon. Through the mechanism of diffusion under slow cooling the dendritic structure disappeared, and the grain became homogeneous. The purpose of this section is to explain briefly how diffusion in solids may occur.

Diffusion is essentially statistical in nature, resulting from many random movements of individual atoms. While the path of an individual atom may be zigzag and unpredictable, when large numbers of atoms make such

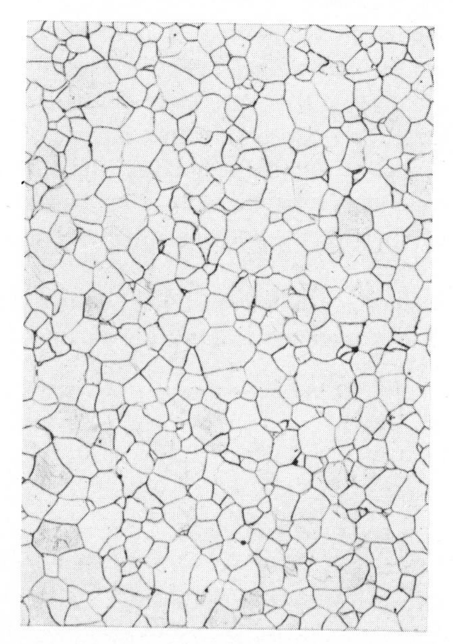

Fig. 6·7 Microstructure of a solid solution iron alloy; magnification 100X. (Courtesy of the Research Laboratory, U.S.Steel.)

movements they can produce a systematic flow.

There are three methods by which diffusion in substitutional solid solutions may take place; the vacancy mechanism, interstitial mechanism, and atom interchange mechanism. These are illustrated schematically in Fig. 6·8.

In Chap. 2, under crystallization, it was pointed out that vacancies and interstitial sites were a normal feature of a crystal structure. These imperfections greatly facilitate diffusion, or the jumping of adjoining atoms. Figure 6·8*a* shows how a solute atom might move one atomic spacing to the left by jumping into a vacancy. It is equally probable, of course, that any one of the other atoms neighboring the vacancy could have made the jump. The vacancy has moved to the right to occupy the position of the previous atom and is now ready for another random interchange. The interstitial mechanism is illustrated in Fig. 6·8*b*, where an atom in normal position moves into an interstitial space, and the vacated spot is taken by the interstitial atom. As shown in the same figure, diffusion may occur by an interstitial atom wandering through the crystal, but this method is more likely in interstitial solid solutions.

It is possible for movement to take place by a direct interchange between two adjacent atoms, as shown in Fig. 6·8*c*, or by a four-atom ring interchange, as in Fig. 6·8*d*. However, these would probably occur only under special conditions, since the physical problem of squeezing between closely packed neighboring atoms would increase the barrier for diffusion. Experimental evidence has indicated that the use of vacancies is the primary method of diffusion in metals. The rate of diffusion is much greater in a rapidly cooled alloy than in the same alloy slow-cooled. The difference is due to the larger number of vacancies retained in the alloy by fast cooling. Vacancy migration also has a lower activation energy when compared with the other methods.

The rate of diffusion of one metal in another is specified by the diffusion coefficient, which is given in units of square centimeters per second. While

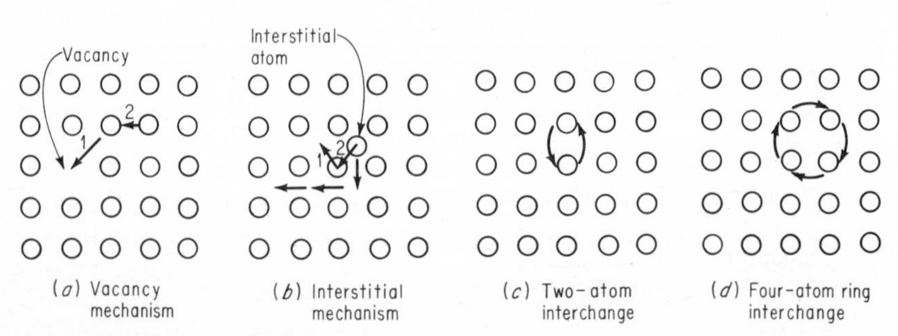

Fig. 6·8 Schematic diffusion mechanisms.

the diffusion coefficient is a function of many variables, the most important is temperature. As a general rule, it may be stated that the diffusion coefficient doubles for every 20-deg (centigrade) rise in temperature. This is not surprising, since all atoms are constantly vibrating about their equilibrium positions in the lattice, and since the amplitude of vibration increases with increasing temperature. The energy associated with these thermal vibrations, often referred to as the thermal energy, is sufficient to cause an atom to jump out of its lattice position under suitable conditions. Therefore, temperature is obviously an important factor in determining whether jumping or diffusion is likely to occur. An alloy has the lowest free energy when it is in a homogeneous condition, and this is the driving force for diffusion.

6·9 Nonequilibrium Cooling—Origin of Coring In actual practice it is extremely difficult to cool under equilibrium conditions. Since diffusion in the solid state takes place at a very slow rate, it is expected that with ordinary cooling rates there will be some difference in the conditions as indicated by the equilibrium diagram. Referring again to the alloy 30B (Fig. 6·9), solidification starts at T_1, forming a solid solution of composition α_1. At T_2 the liquid is at L_2 and the solid solution now forming is of composition α_2 (see Fig. 6·6). Since diffusion is too slow to keep pace with crystal growth, not enough time will be allowed to achieve uniformity in the solid, and the average composition will be between α_1 and α_2, say α'_2. As the temperature drops, the average composition of the solid solution will depart still further from equilibrium conditions. It seems that the composition of the solid solution is following a "nonequilibrium" solidus line α_1 to α'_5, shown dotted in Fig. 6·9. The liquid, on the other hand, has essentially the composition given by the liquidus line, since diffusion is relatively rapid in liquid. At T_3 the average solid solution will be of composition α'_3 instead of α_3. Under equilibrium cooling, solidification should be complete at T_4; however, since the average composition of the solid solution α'_4 has not reached the composition of the alloy, some liquid must still remain. Applying the lever rule at T_4 gives

$$\alpha'_4 \text{ (percent)} = \frac{T_4L_4}{\alpha'_4L_4} \times 100 \approx 75 \text{ percent}$$

$$L_4 \text{ (percent)} = \frac{\alpha'_4T_4}{\alpha'_4L_4} \times 100 \approx 25 \text{ percent}$$

Solidification will therefore continue until T_5 is reached. At this temperature the composition of the solid solution α'_5 coincides with the alloy composition, and solidification is complete. The last liquid to solidify, L_5, is richer in B than the last liquid to solidify under equilibrium conditions. It is apparent from a study of Fig. 6·9 that the more rapidly the alloy is cooled

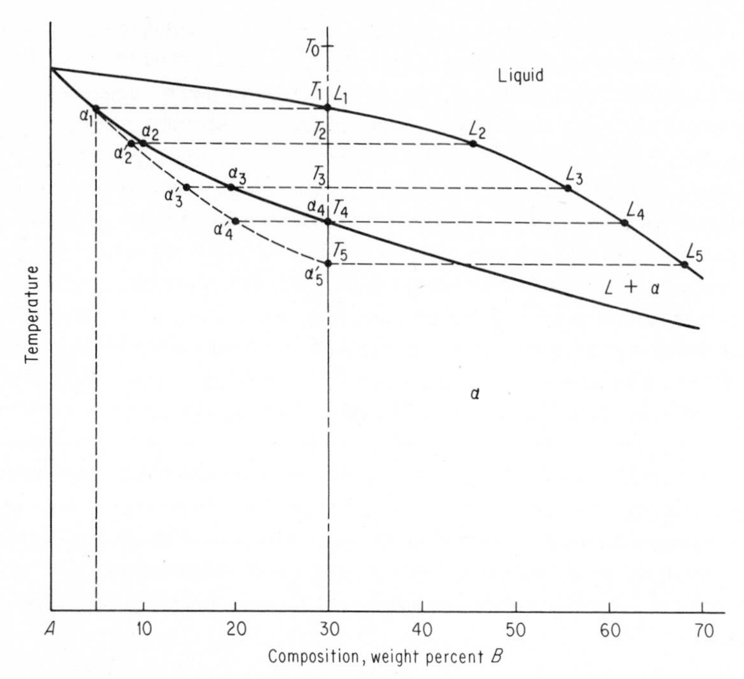

Fig. 6·9 Nonequilibrium cooling; the origin of coring.

the greater will be the composition range in the solidified alloy. Since the rate of chemical attack varies with composition, proper etching will reveal the dendritic structure microscopically (Fig. 6·10). The final solid consists of a "cored" structure with a higher-melting central portion surrounded by the lower-melting, last-to-solidify shell. The above condition is referred to as *coring* or *dendritic segregation.*

To summarize, nonequilibrium cooling results in an increased temperature range over which liquid and solid are present; final solidification occurs at a lower temperature than predicted by the phase diagram; the final liquid to solidify will be richer in the lower-melting-point metal; and since diffusion has not kept pace with crystal growth, there will be a difference in chemical composition from the center to the outside of the grains. The faster the rate of cooling, the greater will be the above effects.

6·10 Homogenization Cored structures are most common in as-cast metals. From the above discussion of the origin of a cored structure, it is apparent that the last solid formed along the grain boundaries and in the interdendritic spaces is very rich in the lower-melting-point metal. Depending upon the properties of this lower-melting-point metal, the grain boundaries may act as a plane of weakness. It will also result in a serious lack of uniformity in mechanical and physical properties and, in some cases, increased susceptibility to intergranular corrosion because of preferential attack by a corrosive medium. Therefore, for some applications, a cored structure is objectionable.

There are two methods for solving the problem of coring. One is to prevent its formation by slow freezing from the liquid, but this results in large grain size and requires a very long time. The preferred method industrially is to achieve equalization of composition or homogenization of the cored structure by diffusion in the solid state.

At room temperature, for most metals, the diffusion rate is very slow; but if the alloy is reheated to a temperature below the solidus line, diffusion will be more rapid and homogenization will occur in a relatively short time.

Figure 6·11 shows the actual equilibrium diagram for the copper-nickel system, and the alloy 85Cu-15Ni is shown as a dotted line. The effect of homogenization on the cored structure of an 85Cu-15Ni alloy is illustrated by the series of photomicrographs in Fig. 6·12. The first picture of this sequence shows the microstructure of the alloy as chill-cast. As the equi-

Fig. 6·10 Fine-cored dendrites in a copper-lead alloy; magnification 100X. (Research Laboratories, National Lead Company.)

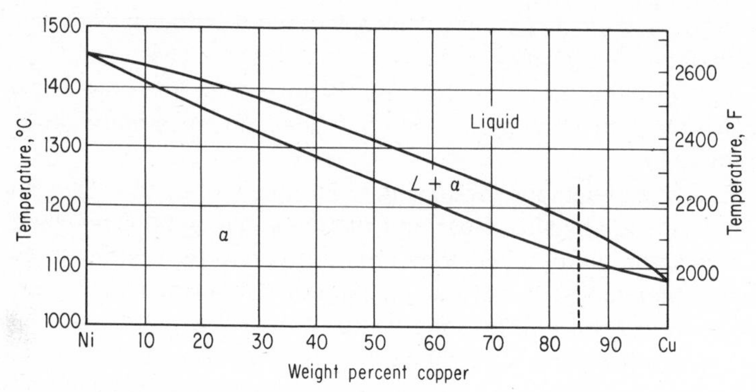

Fig. 6·11 Copper-nickel equilibrium diagram. (From "Metals Handbook," 1948 ed., p. 1198, American Society for Metals, Metals Park, Ohio.)

librium diagram predicts, the first solid to be formed in the central axes of the dendrites is rich in nickel. Because of rapid cooling, there is a great difference in nickel content between the central axes of the dendrites and the interdendritic spaces. This difference is revealed by suitable etching. The next figure shows the same sample after heating at 1382°F for 3 h. Counterdiffusion of nickel and copper atoms between the nickel-rich cores and the copper-rich fillings has reduced the composition differences somewhat. The microstructure of the same sample heated to 1742°F for 9 h is shown in the third figure. The composition is completely equalized, and the dendrites have disappeared. The grain boundaries are clearly evident. Black particles are copper oxide or nickel oxide inclusions. The fourth figure illustrates the same alloy slowly cooled by casting in a hot mold. The dendritic structure is coarser than that of the chill-cast alloy. The last figure shows this same sample heated 15 h at 1742°F. The structure is now completely homogenized. Despite the smaller initial composition differences across the coarse dendrites as compared with the fine dendrites, it took a longer time for equalization because of the greater distance through which the copper and nickel atoms had to diffuse in the coarse structure. Extreme care must be exercised in this treatment not to cross the solidus line; otherwise liquation of the grain boundaries will occur, impairing the shape and physical properties of the casting (Fig. 6·13).

6·11 Properties of Solid-solution Alloys In general, in an alloy system forming a continuous series of solid solutions, most of the property changes are caused by distortion of the crystal lattice of the solvent metal by additions of the solute metal. The effect of composition on some physical and mechanical properties of annealed alloys in the copper-nickel system is given in Table 6·1. Electrical resistivity depends upon distortion of the lattice structure. Since the distortion increases with the amount of solute metal

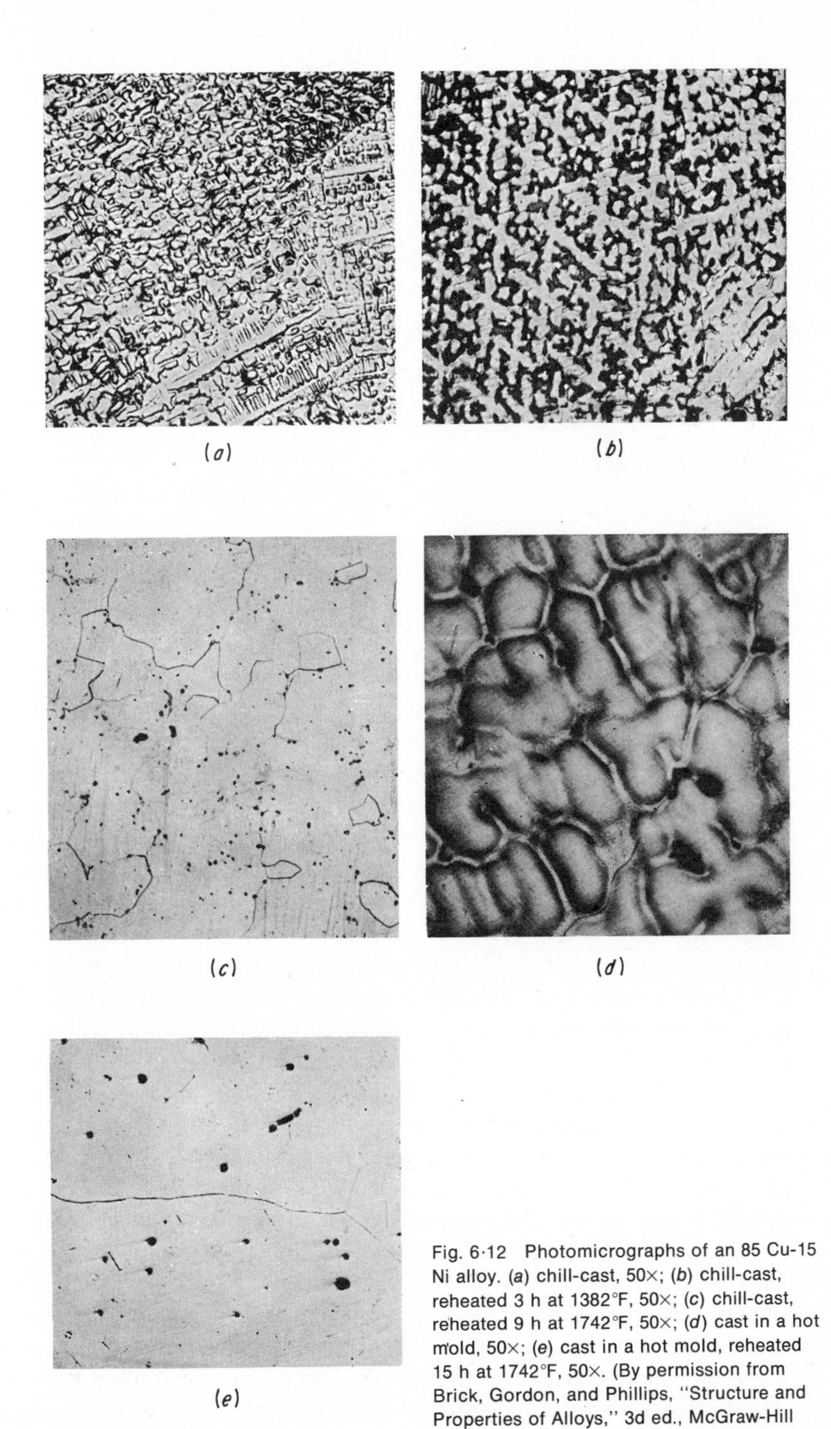

Fig. 6·12 Photomicrographs of an 85 Cu-15 Ni alloy. (*a*) chill-cast, 50×; (*b*) chill-cast, reheated 3 h at 1382°F, 50×; (*c*) chill-cast, reheated 9 h at 1742°F, 50×; (*d*) cast in a hot mold, 50×; (*e*) cast in a hot mold, reheated 15 h at 1742°F, 50×. (By permission from Brick, Gordon, and Phillips, "Structure and Properties of Alloys," 3d ed., McGraw-Hill Book Company, New York, 1965.)

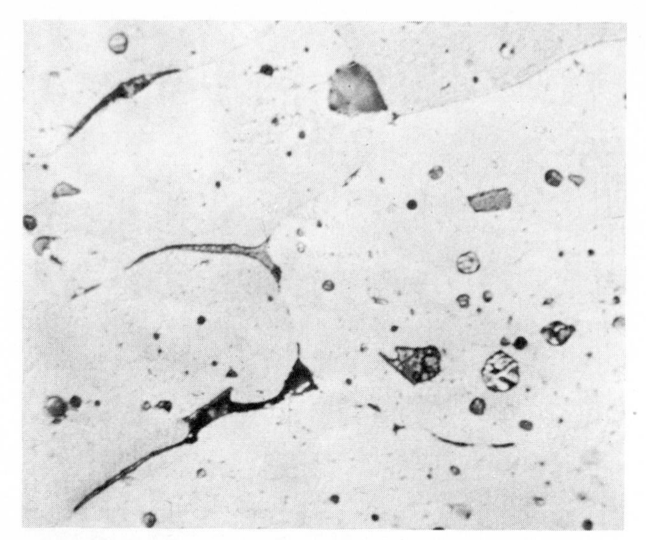

Fig. 6·13 Photomicrograph of an aluminum alloy in which some melting has occurred at the grain boundaries during heating. After cooling, these portions of the grain boundaries appear as dark broad lines; magnification, 1,000X. (Alcoa Research Laboratories, Aluminum Company of America.)

added, and since either metal can be considered as the solvent, the maximum electrical resistivity should occur in the center of the composition range. This is verified by the values given in the table. As copper is added to nickel, the strength of the alloy increases, and as nickel is added to copper, the strength of that alloy also increases. Therefore, between pure copper and pure nickel there must be an alloy which shows the maximum strength. It turns out to be at approximately two-thirds nickel and one-third copper. This is a very useful commercial alloy known as *Monel*. It shows good strength and good ductility along with high corrosion resistance. The same behavior is also true of hardness in that there will be an alloy that shows the maximum hardness, although the maximum tensile strength and hardness do not necessarily come at the same composition.

6·12 Variations of Type I Every alloy in the Type I system covered has a melting point between the melting points of *A* and *B*. It is possible to have a system in which the liquidus and solidus lines go through a minimum or a maximum (Fig. 6·14*a*, *b*). The alloy composition *x* in Fig. 6·14*a* behaves just like a pure metal. There is no difference in the liquid and solid composition. It begins and ends solidification at a constant temperature with no change in composition, and its cooling curve will show a horizontal line. Such alloys are known as *congruent-melting alloys*. Because alloy *x* has the lowest melting point in the series, and the equilibrium diagram resembles

TABLE 6·1 Properties of Annealed Copper-Nickel Alloys*

COMPOSITION % NICKEL	TENSILE STRENGTH, PSI	ELONGATION, % IN 2 IN.	BHN, 10 MM, 500 KG	LATTICE PARAMETER, 10^{-8} CM	ELECTRICAL RESISTIVITY, MICROHMS PER CU CM
0	30,000	53	36	3.6073	1.7
10	35,000	47	51	3.5975	14
20	39,000	43	58	3.5871	27
30	44,000	40	67	3.5770	38
40	48,000	39	70	3.5679	46
50	50,000	41	73	3.5593	51
60	53,000	41	74	3.5510	50
70	53,000	42	73	3.5432	40
80	50,000	43	68	3.5350	30
90	48,000	45	61	3.5265	19
100	43,000	48	54	3.5170	6.8

*By permission from R. M. Brick and A. Phillips, "Structure and Properties of Alloys," 2d ed., McGraw-Hill Book Company, New York, 1949.

the eutectic type to be discussed shortly, it is sometimes known as a *pseudoeutectic alloy.* Examples of alloy systems that show a minimum are Cu-Au and Ni-Pd. Those showing a maximum are rare, and there are no known metallic systems of this type.

6·13 Type II—Two Metals Completely Soluble in the Liquid State and Completely Insoluble in the Solid State Technically, no two metals are completely insoluble in each other. However, in some cases the solubility is so restricted that for practical purposes they may be considered insoluble.

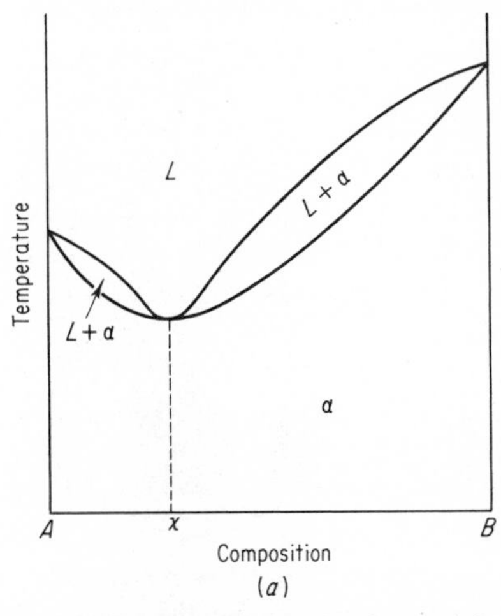

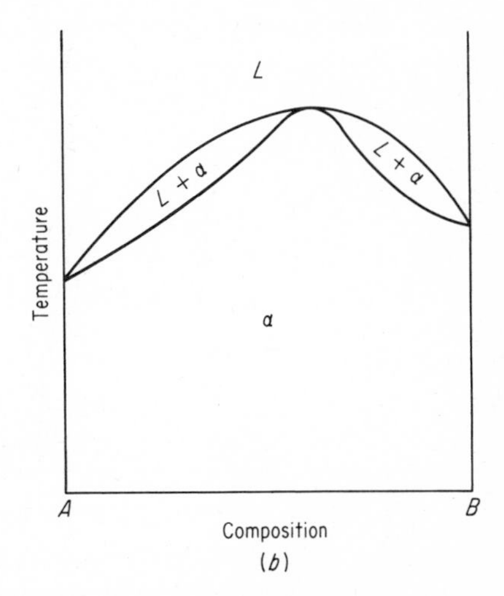

Fig. 6·14 (*a*) Solid-solution system showing a minimum. (*b*) Solid-solution system showing a maximum.

Raoult's law states that the freezing point of a pure substance will be lowered by the addition of a second substance provided the latter is soluble in the pure substance when liquid and insoluble when solidified. The amount of lowering of the freezing point is proportional to the molecular weight of the solute.

This phase diagram can be developed from a series of cooling curves in a manner analogous to that used for the solid solution diagram described previously, but in this case, the experimental curves show a different kind of behavior. The series of cooling curves for the pure metals and various alloys, and the room-temperature microstructures, are shown in Fig. 6·15. The cooling curves for the pure metals A and B show a single horizontal line at their freezing points, as expected. As B is added to A, the temperature for the beginning of solidification is lowered. As A is added to B, the temperature for the beginning of solidification for those alloys is also

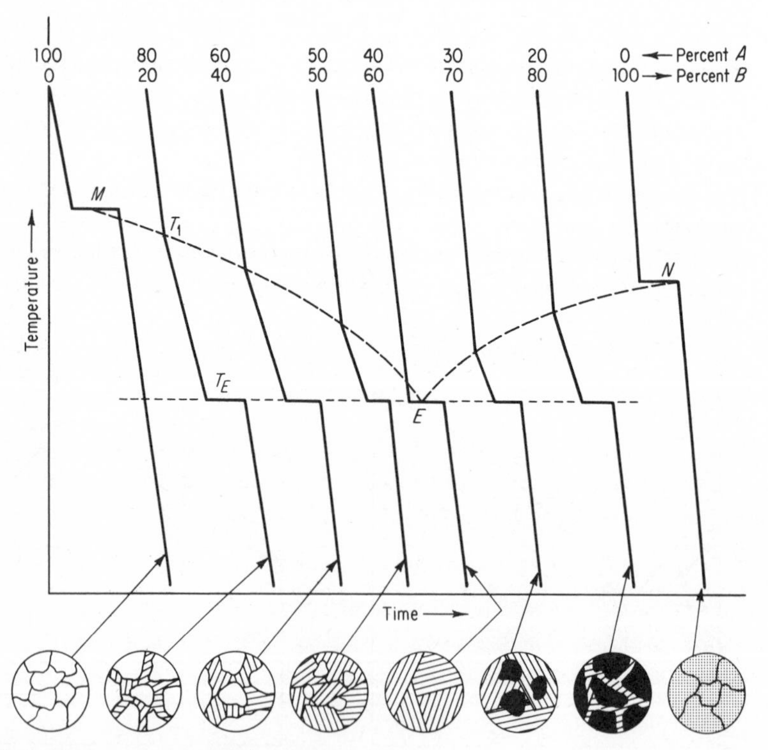

Fig. 6·15 Cooling curves and room-temperature microstructures for a series of alloys of two metals that are insoluble in the solid state. The upper dotted line indicates the form of the liquidus and the lower dotted line the form of the solidus line.

lowered. Therefore, since each metal lowers the freezing point of the other, the line connecting the points showing the beginning of solidification, the liquidus line, must show a minimum. This is illustrated by the upper dotted line in Fig. 6·15, showing a minimum at point E, known as the eutectic point, for a composition of 40A-60B. Notice that over a wide range of compositions, a portion of the cooling curve that shows the end of solidification occurs at a fixed temperature. This lower horizontal line at T_E, shown dotted in Fig. 6·15, is known as the eutectic temperature. In one alloy, the eutectic composition 40A-60B, complete solidification occurs at a single temperature, the eutectic temperature. Although the freezing of the eutectic composition thus resembles that of a pure metal, it is not a congruent melting alloy since we will shortly see that the resulting solid is composed of two phases.

The actual phase diagram may now be constructed by transferring the breaks on the cooling curves to a plot of temperature vs. composition, as shown in Fig. 6·16. The melting points of the two pure metals, points M and N, are plotted on the vertical lines that represent the pure metals. For an alloy containing 80A-20B the beginning of solidification T_1 and the end of solidification T_E are plotted as shown. The same procedure is followed for the remaining alloys. The upper line on the phase diagram connecting the two melting points, MEN, is the liquidus line and shows the beginning of solidification. The point at which the liquidus lines intersect, the minimum point E, is known as the *eutectic point.* T_E is called the *eutectic temperature* and 40A-60B the *eutectic composition.* The solidus line is always a continuous line connecting the melting points of the pure metals, so that the complete solidus line is $MFGN$.

This phase diagram consists of four areas. The area above the liquidus line is a single-phase homogeneous liquid solution, since the two metals are soluble in the liquid state. The remaining three areas are two-phase areas. Every two-phase area on a phase diagram must be bounded along a horizontal line by single phases. If the single-phase areas are labeled first, then the two-phase areas may be easily determined. For example, in Fig. 6·16, to determine the phases that exist in the two-phase area MFE, a horizontal tie line OL is drawn. This line intersects the liquidus at L, which means that the liquid is one of the phases existing in the two-phase area and intersects the left axis at point O. The left axis represents a single phase, the pure metal A, which below its melting point is solid. Therefore, the two phases existing in the area MFE are liquid and solid A. The same reasoning is applied to determine the two phases that exist in area NEG. These are liquid and solid B. The above ideas may be applied to any phase diagram and will be useful to the student for the labeling of more complex diagrams.

Since the two metals are assumed to be completely insoluble in the solid

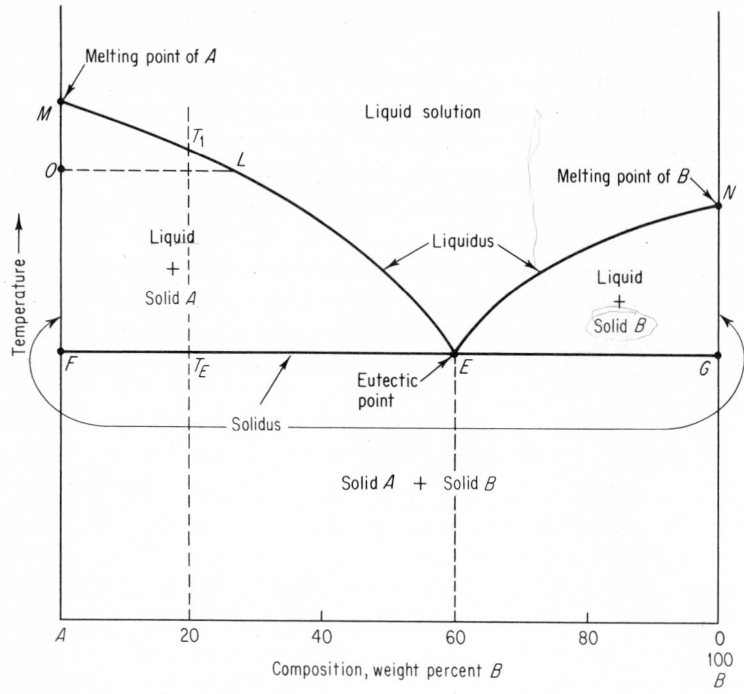

Fig. 6·16 Eutectic-type phase diagram.

state, it should be apparent that when freezing starts the only solid that can form is a pure metal. Also, every alloy when completely solidified must be a mixture of the two pure metals. It is common practice to consider alloys to the left of the eutectic composition as *hypoeutectic alloys* and those to the right as *hypereutectic alloys*. The way in which solidification takes place is of interest and will now be studied by following the slow cooling of several alloys.

Alloy 1 in Fig. 6·17 is the eutectic composition 40*A*-60*B*. As it is cooled from temperature T_0, it remains a uniform liquid solution until point *E*, the eutectic-temperature line, is reached. Since this is the intersection of the liquidus and solidus lines, the liquid must now start to solidify, and the temperature cannot drop until the alloy is completely solid. The liquid will solidify into a mixture of two phases. These phases are always the ones that appear at either end of the horizontal eutectic-temperature line, in this case point *F*, which is the pure metal *A*, and point *G*, the pure metal *B*. Let us assume that a small amount of pure metal *A* is solidified. This leaves the remaining liquid richer in *B*; the liquid composition has shifted slightly to the right. To restore the liquid composition to its equilibrium value, *B* will solidify. If slightly too much B is solidified, the liquid composition will have shifted to the left, requiring A to solidify to restore equilibrium. Therefore, at constant temperature, the liquid solidifies alternately pure A and pure B, resulting in an extremely fine mixture usually visible only under the microscope. This is known as the *eutectic mixture* (Fig. 6·18). The change of this liquid of composition E into two solids at constant temperature is known as the *eutectic reaction* and may be written as

$$\text{Liquid} \underset{\text{heating}}{\overset{\text{cooling}}{\rightleftharpoons}} \underbrace{\text{solid } A + \text{solid } B}_{\text{eutectic mixture}}$$

Since solidification of the eutectic alloy occurs at constant temperature, its cooling curve would be the same as that for a pure metal or any congruent-melting alloy. The eutectic solidification, however, is incongruent since there is a difference in composition between the liquid and the individual solid phases.

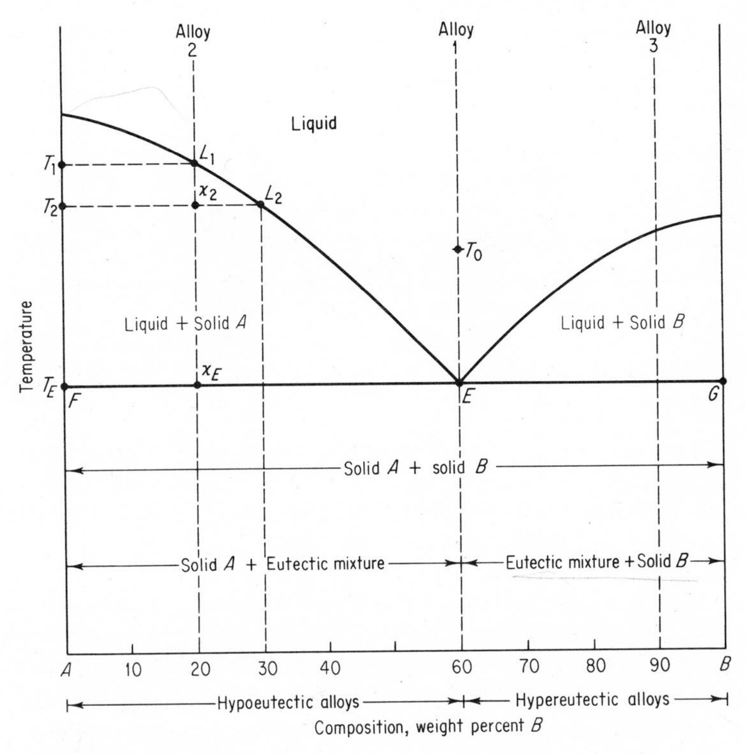

Fig. 6·17 Eutectic-type phase diagram.

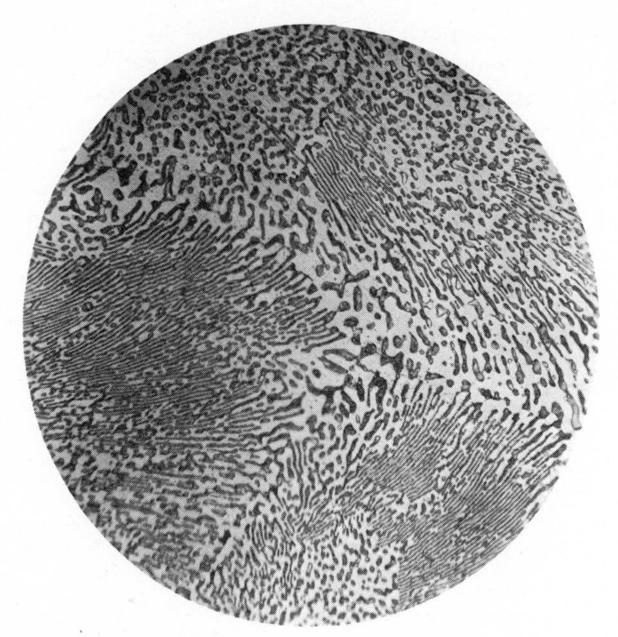

Fig. 6·18 Lead-bismuth eutectic mixture, 100X. (Research Laboratories, National Lead Company.)

Alloy 2, a hypoeutectic alloy composed of 80*A*-20*B*, remains a uniform liquid solution until the liquidus line, temperature T_1, is reached. At this point the liquid L_1 is saturated in *A*, and as the temperature is dropped slightly, the excess *A* must solidify. The liquid, by depositing crystals of pure *A*, must become richer in *B*. Applying Rule I at temperature T_2 shows the solid phase to be pure *A* and the liquid composition L_2 as 70*A*-30*B*. The amount which has solidified up to this temperature would be found by applying Rule II:

$$A \text{ (percent)} = \frac{x_2L_2}{T_2L_2} \times 100 = \frac{10}{30} \times 100 = 33 \text{ percent}$$

$$L_2 \text{ (percent)} = \frac{T_2x_2}{T_2L_2} \times 100 = \frac{20}{30} \times 100 = 67 \text{ percent}$$

The microstructure would appear as in Fig. 6·19*a*. As solidification continues, the amount of pure solid *A* increases gradually by continued precipitation from the liquid. The liquid composition, becoming richer in *B*, is slowly traveling downward and to the right along the liquidus curve, while the amount of liquid is gradually decreasing. When the alloy reaches x_E, the eutectic line, the liquid is at point *E*. The conditions existing just a fraction of a degree above T_E are:

Phases	Liquid	Solid A
Composition	40*A*-60*B*	100*A*
Relative amount	$\frac{T_E x_E}{T_E E} \times 100 = 33\%$	$\frac{x_E E}{T_E E} \times 100 = 67\%$

The microstructure would appear as in Fig. 6·19*b*. The remaining liquid (33 percent), having reached the eutectic point, now solidifies into the fine intimate mixture of *A* and *B* as described under alloy 1. When solidified, the alloy will consist of 67 percent of grains of primary *A* or proeutectic *A* (which formed between T_1 and T_E or before the eutectic reaction) and 33 percent eutectic ($A + B$) mixture (Fig. 6·19*c*). Every alloy to the left of the eutectic point *E*, when solidified, will consist of grains of proeutectic *A* and the eutectic mixture. The closer the alloy composition is to the eutectic composition, the more eutectic mixture will be present in the solidified alloy (see microstructures in Fig. 6·15).

Alloy 3, a hypereutectic alloy composed of 10*A*-90*B*, undergoes the same cooling process as alloy 2 except that when the liquidus line is reached the liquid deposits crystals of pure *B* instead of *A*. As the temperature is decreased, more and more *B* will solidify, leaving the liquid richer in *A*. The amount of liquid gradually decreases, and its composition gradually moves down and to the left along the liquidus line until point *E* is reached at the eutectic temperature. The remaining liquid now solidifies into the eutectic ($A + B$) mixture. After solidification, the alloy will consist of 75 percent grains of primary *B* or proeutectic *B* and 25 percent eutectic ($A + B$) mixture. The student should verify these figures and sketch the microstructure at room temperature. Every alloy to the right of the eutectic point, when solidified, will consist of grains of proeutectic *B* and the eutectic mixture. The only difference will be in the relative amounts (see microstructures in Fig. 6·15). The relationship between alloy composition and

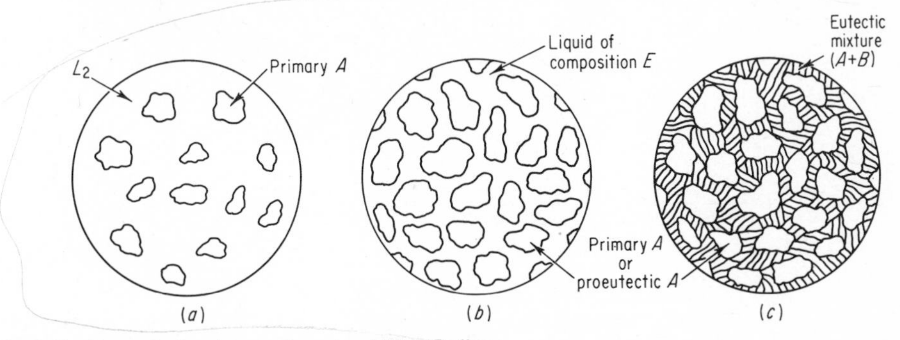

Fig. 6·19 Stages in the slow cooling of an 80*A*-20*B* alloy.

microstructure may be shown by using the eutectic composition as an imaginary boundary line. The area below the solidus line and to the left of the eutectic composition is labeled solid A + eutectic mixture, and that to the right, solid B + eutectic mixture (Fig. 6·17). Figure 6·20 shows the relation between alloy composition and relative amounts.

From the previous discussion it is apparent that, regardless of alloy composition, the same reaction takes place whenever the eutectic-temperature line is reached, namely,

$$\text{Liquid} \underset{\text{heating}}{\overset{\text{cooling}}{\rightleftharpoons}} \underbrace{\text{solid } A + \text{solid } B}_{\text{eutectic mixture}}$$

The above reaction applies specifically to this diagram; however, the eutectic reaction may be written in general as

$$\text{Liquid} \underset{\text{heating}}{\overset{\text{cooling}}{\rightleftharpoons}} \underbrace{\text{solid}_1 + \text{solid}_2}_{\text{eutectic mixture}}$$

the only requirement being that the eutectic mixture consist of two different solid phases. This mixture may be two pure metals, two solid solutions, two intermediate phases, or any combination of the above.

The simplified aluminum-silicon phase diagram is shown in Fig. 6·21, neglecting the slight solubility of silicon in aluminum. The numbers at the bottom of this diagram refer to the photomicrographs in Fig. 6·22. Beginning with alloy 1 at the left of Fig. 6·21, the microstructure of pure aluminum is shown in Fig. 6·22*a*. Alloy 2 (Fig. 6·22*b*), containing 8 percent silicon, consists of dendrites of primary or proeutectic aluminum surrounded by the eutectic mixture of aluminum and silicon. Notice the fine alternate light and dark structure of the eutectic. Since the eutectic is formed from

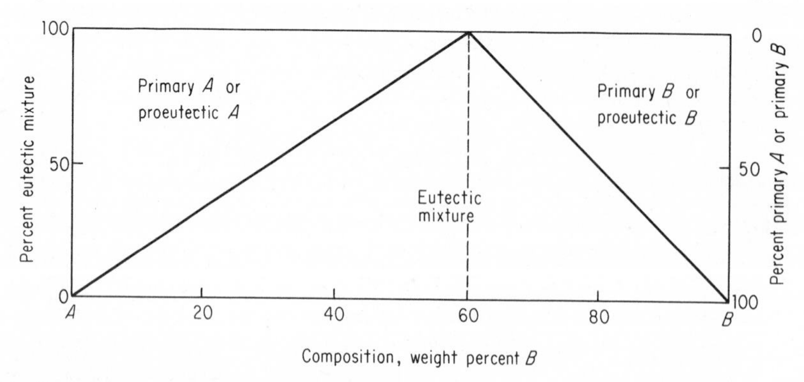

Fig. 6·20 Diagram showing the linear relationship between the parts of the microstructure and alloy composition for the eutectic system of Fig. 6·17.

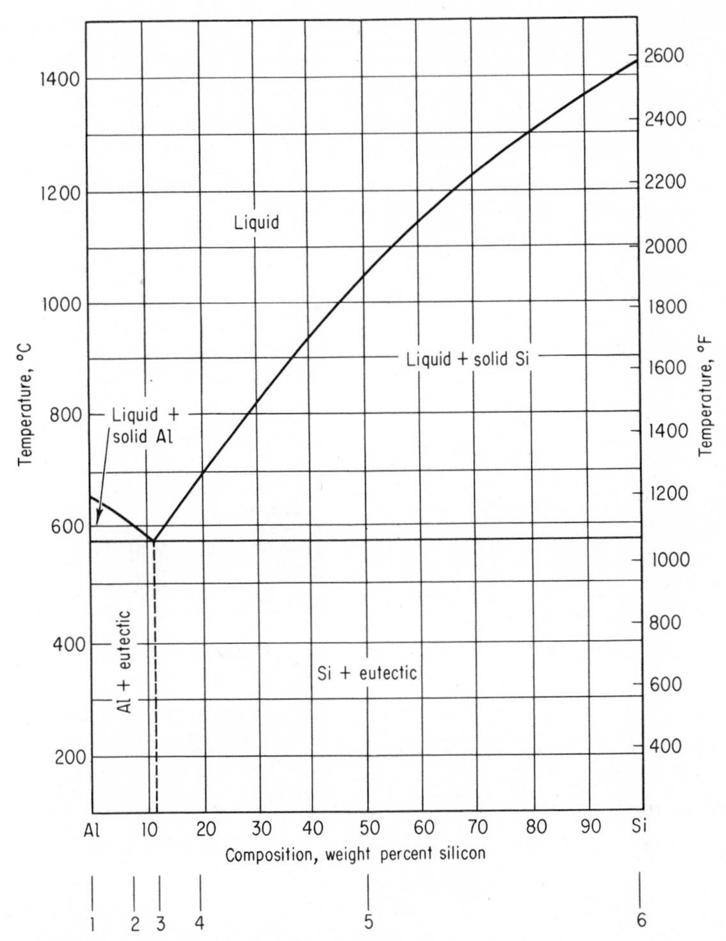

Fig. 6·21 Simplified aluminum-silicon phase diagram. The numbers on the bottom correspond to the photomicrographs in Fig. 6·22.

the last liquid to solidify, it fills the spaces between the arms of the dendrites. Alloy 3 (Fig. 6·22*c*) is the eutectic composition of 12 percent silicon and consists entirely of the eutectic mixture. As we move to the right, the microstructure will consist of primary silicon (black) and the eutectic mixture, the amount of primary silicon increasing with increasing silicon content as shown in Fig. 6·22*d* and *e*. Finally, Fig. 6·22*f* shows the microstructure of pure silicon. It is therefore possible to predict from an equilibrium diagram, with reasonable accuracy, the proportions of each phase which will exist in an alloy after slow cooling to room temperature.

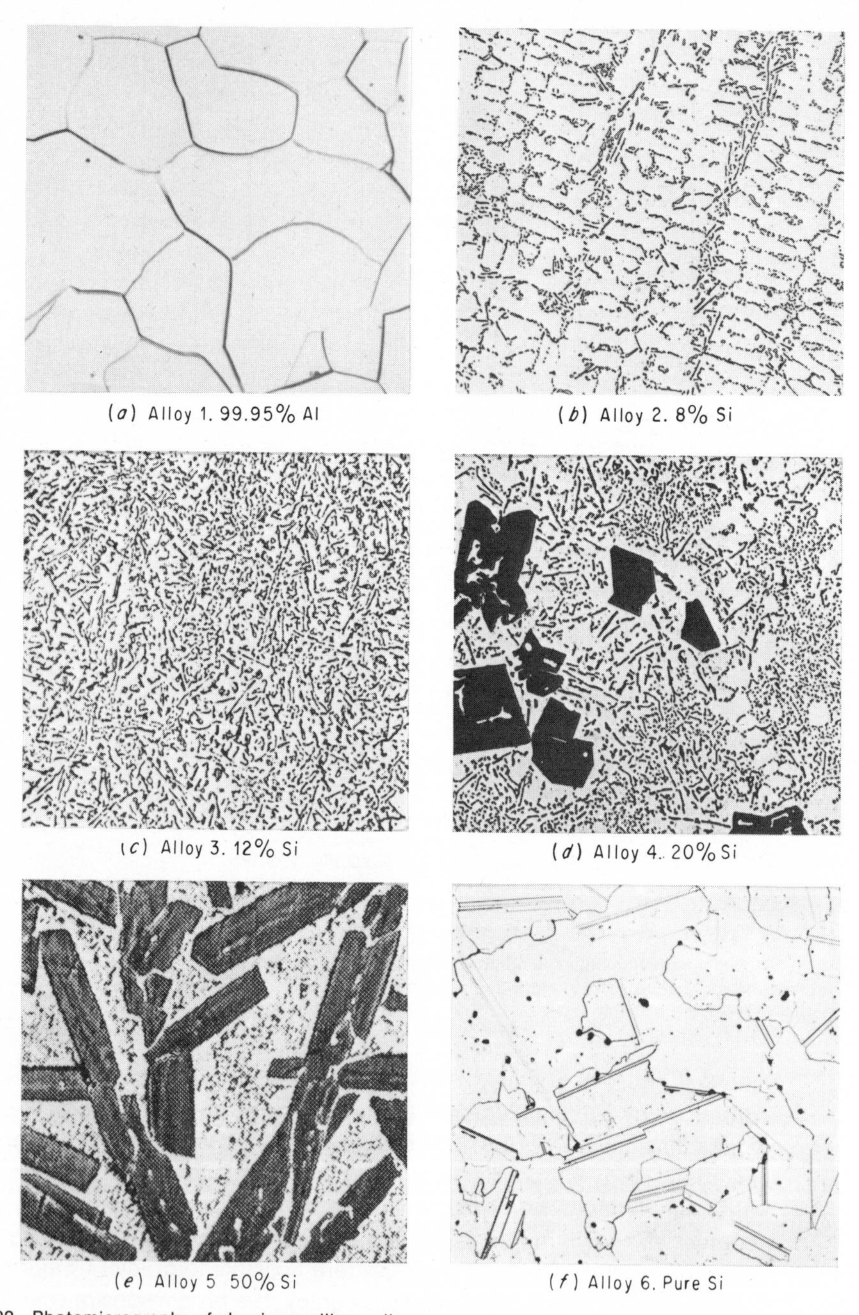

Fig. 6·22 Photomicrographs of aluminum-silicon alloys as numbered at the bottom of Fig. 6·21. (Research Laboratories, Aluminum Company of America.)

6·14 Type III—Two Metals Completely Soluble in the Liquid State but Only Partly soluble in the Solid State Since most metals show some solubility for each other in the solid state, this type is the most common and, therefore, the most important alloy system.

From the discussion of the previous two types, it is assumed that the student is familiar with the method of determining a phase diagram from a series of cooling curves. The remaining types will be drawn and studied directly.

The phase diagram of this type is shown in Fig. 6·23. The melting points of the two pure metals are indicated at points T_A and T_B, respectively. The liquidus line is T_AET_B, and the solidus line is T_AFEGT_B. The single-phase areas should be labeled first. Above the liquidus line, there is only a single-phase liquid solution. At the melting points, where the liquidus and solidus lines meet, the diagram resembles the cigar-shaped diagram of Type I (complete solid solubility), and since these metals are partly soluble in the solid state, a solid solution must be formed. Alloys in this system never solidify crystals of pure *A* or pure *B* but always a solid solution or mixture of solid solutions. The single-phase α (alpha) and β (beta) solid-solution

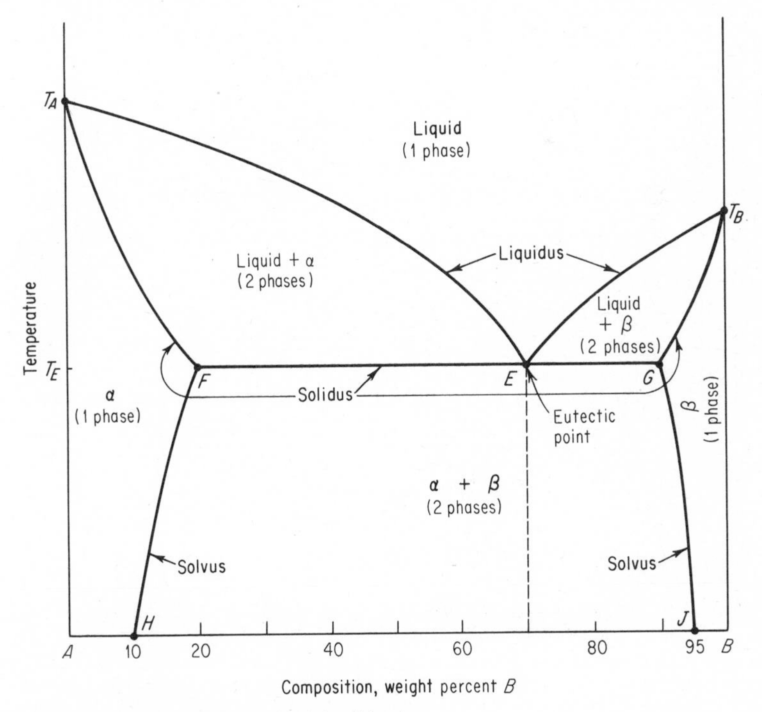

Fig. 6·23 Phase diagram illustrating partial solid solubility.

areas are now labeled. Since these solid solutions are next to the axes, they are known as *terminal solid solutions*. The remaining three two-phase areas may now be labeled as liquid + α, liquid + β, and $\alpha + \beta$. At T_E, the α solid solution dissolves a maximum of 20 percent B as shown by point F and the β solid solution a maximum of 10 percent A as shown by point G. With decreasing temperature, the maximum amount of solute that can be dissolved decreases, as indicated by lines FH and GJ. These lines are called *solvus* lines and indicate the maximum solubility (saturated solution) of B in A (α solution) or A in B (β solution) as a function of temperature. Point E, where the liquidus lines meet at a minimum, as in Type II, is known as the *eutectic point*. The slow cooling of several alloys will now be studied.

Alloy 1 (Fig. 6·24), composed of 95A-5B, when slow-cooled will follow a process exactly the same as any alloy in Type I. When the liquidus line is crossed at T_1, it will begin to solidify by forming crystals of α solid solution extremely rich in A. This process continues, with the liquid getting richer in B and gradually moving down along the liquidus line. The α

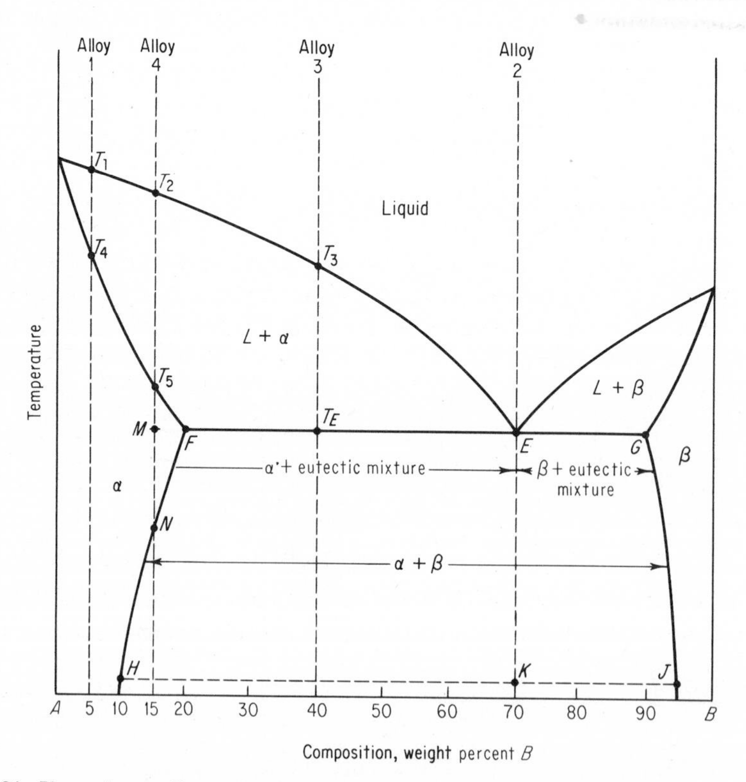

Fig. 6·24 Phase diagram illustrating partial solid solubility.

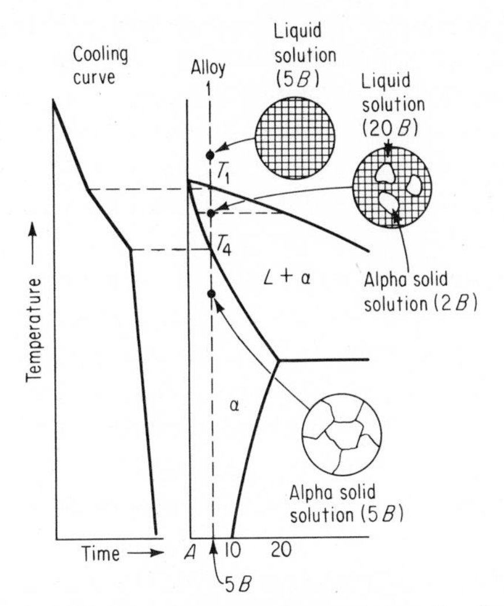

Fig. 6·25 The cooling curve and microstructure at various temperatures during solidification of a 95*A*-5*B* alloy.

solid solution, also getting richer in *B*, is moving down along the solidus line. When the solidus line is finally crossed at T_4 and with diffusion keeping pace with crystal growth, the entire solid will be a homogeneous α solid solution and will remain that way down to room temperature. The process of solidification and the cooling curve for this alloy are shown in Fig. 6·25.

Alloy 2, 30*A*-70*B*, is the eutectic composition and remains liquid until the eutectic temperature is reached at point *E*. Since this is also the solidus line, the liquid now undergoes the eutectic reaction, at constant temperature, forming a very fine mixture of two solids. The two solids that make up the eutectic mixture are given by the extremities of the eutectic-temperature line, α of composition *F* and β of composition *G*. The eutectic reaction may be written as

$$\text{Liquid} \underset{\text{heating}}{\overset{\text{cooling}}{\rightleftharpoons}} \underbrace{\alpha + \beta}_{\text{eutectic mixture}}$$

This reaction is the same as the one which occurred in the Type II diagram, except for the substitution of solid solutions for pure metals. The relative amounts of α and β in the eutectic mixture may be determined by applying Rule II (lever rule):

$$\alpha\ (\text{percent}) = \frac{EG}{FG} \times 100 = \frac{20}{70} \times 100 = 28.6\ \text{percent}$$

$$\beta\ (\text{percent}) = \frac{EF}{FG} \times 100 = \frac{50}{70} \times 100 = 71.4\ \text{percent}$$

Because of the change in solubility of B in A, line FH, and of A in B, line GJ, there will be a slight change in the relative amounts of α and β as the alloy is cooled to room temperature. The relative amounts of α and β at room temperature are

$$\alpha\text{ (percent)} = \frac{KJ}{HJ} \times 100 = \frac{25}{85} \times 100 = 29.4 \text{ percent}$$

$$\beta\text{ (percent)} = \frac{HK}{HJ} \times 100 = \frac{60}{85} \times 100 = 70.6 \text{ percent}$$

The eutectic mixture is shown in Fig. 6·29*c*. Notice the similarity between this picture and the eutectic mixture formed in Type II (Fig. 6·18). It is not possible to tell microscopically whether the eutectic mixture is made up of two solid solutions or two pure metals.

Alloy 3, 60*A*-40*B*, remains liquid until the liquidus line is reached at T_3. The liquid starts to solidify crystals of primary or proeutectic α solid solution very rich in A. As the temperature decreases the liquid becomes richer and richer in B, gradually moving down and to the right along the liquidus line until it reaches point E. Examining the conditions which exist just above the eutectic temperature T_E, there are two phases present:

Phases	Liquid	Primary α
Chemical composition	30*A*-70*B*	80*A*-20*B*
Relative amounts	40%	60%

The student should verify the above numbers by applying Rules I and II at the eutectic temperature. Since the remaining liquid (40 percent) is at point E, the right temperature and composition to form the eutectic mixture, it now solidifies by forming alternately crystals of α and β of the composition appearing at the ends of the eutectic temperature line (points F and G). The temperature does not drop until solidification is complete, and when complete, the microstructure appears as shown in Fig. 6·26.

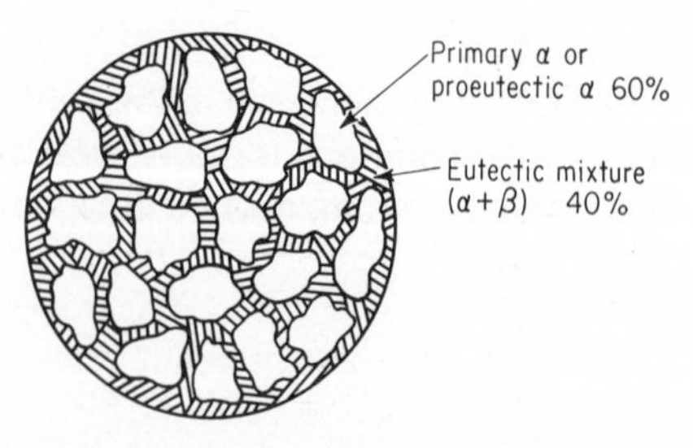

Fig. 6·26 Schematic picture of the microstructure, after solidification, of alloy 3 in Fig. 6·24.

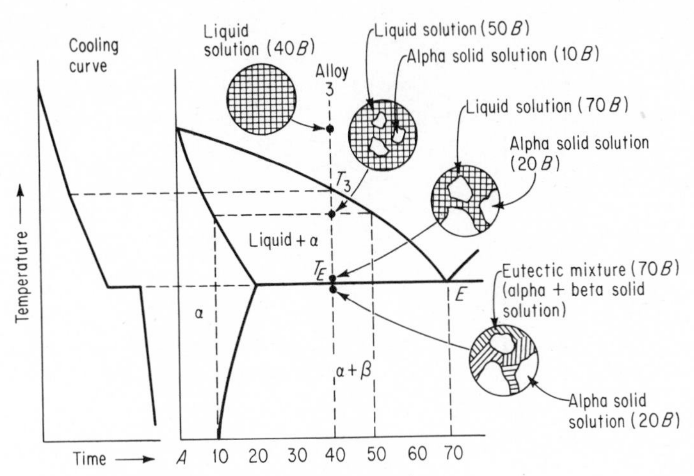

Fig. 6·27 The cooling curve and microstructure at various temperatures during solidification of a 60*A*-40*B* alloy.

Notice the similarity in microstructure between this alloy and Fig. 6·19*c*. As the alloy cools to room temperature because of the change in solubility indicated by the solvus line *FH*, some excess β is precipitated from the solution. The process of solidification and the cooling curve for this alloy are shown in Fig. 6·27.

Alloy 4, 85*A*-15*B*, follows the same process as described for alloy 1. The microstructure at various temperatures and the cooling curve for this alloy are shown in Fig. 6·28. Solidification starts at T_2 and is complete at T_5, the resultant solid being a homogeneous single phase, the α solid solution. At point *M* the solution is unsaturated. The solvus line *FH*, as explained previously, shows the decrease in solubility of *B* in *A* with decreasing temperature. As the alloy cools, the solvus line is reached at point *N*. The α solution is now saturated in *B*. Below this temperature, under conditions of slow cooling, the excess *B* must come out of solution. Since *A* is soluble in *B*, the precipitate does not come out as the pure metal *B*, but rather the β solid solution. At room temperature, the alloy will consist largely of α with a small amount of excess β, primarily along the grain boundaries (Fig. 6·28). The student should determine the amount of excess β by applying the lever rule at the line *HJ* (Fig. 6·24).

If the β phase is relatively brittle, the alloy will not be very strong or ductile. The strength of an alloy to a large extent is determined by the phase that is continuous through the alloy. In this case, although the β solution constitutes only about 5 percent of the alloy, it exists as a continuous network along the grain boundaries. Therefore, the alloy will tend to rupture along these boundaries. This alloy, however, may be made to undergo a

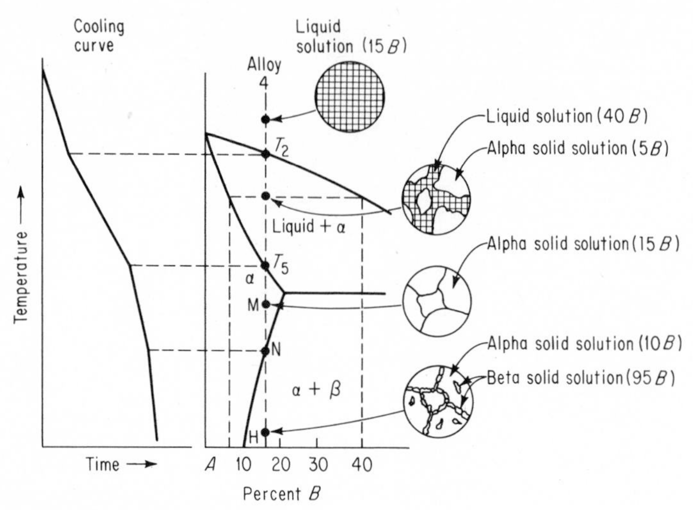

Fig. 6·28 The cooling curve and microstructure at various temperatures for an 85*A*-15*B* alloy.

significant change in strength and hardness after being properly heat-treated.

The lead-antimony equilibrium diagram and photomicrographs of various alloys in this system are shown in Fig. 6·29. Alloy 1 (Fig. 6·29*b*), containing 6.5 percent antimony, illustrates a typical hypoeutectic structure of primary α dendrites (black) and the eutectic mixture filling the spaces between the dendrites. Alloy 2 (Fig. 6·29*c*), containing 11.5 percent antimony, consists entirely of the eutectic mixture of α and β solid solutions. To the right of the eutectic composition, the alloys consist of primary β (white) surrounded by the eutectic mixture (Fig. 6·29*d*) and differ only in the relative amounts of the phases present. The amount of the eutectic mixture decreases as the alloy composition moves away from the eutectic composition.

The lead-tin equilibrium diagram and photomicrographs of various alloys in this system are shown in Fig. 6·30. Alloy 1 (Fig. 6·30*b*), containing 70 percent tin, is to the right of the eutectic composition. The microstructure consists of primary β dendrites (white) surrounded by the eutectic mixture. Alloy 2 (Fig. 6·30*c*) is the eutectic composition and consists entirely of a very fine mixture of α and β solid solutions. Alloys 3 and 4 (Fig. 6·30*d* and *e*), containing 60 and 50 percent tin, respectively, consist of dendrites of the lead-rich primary α solid solution (black) surrounded by the eutectic mixture, the amount of α increasing as the alloy composition moves to the left. Notice the similarity of the photomicrographs shown in Figs. 6·22, 6·29, and 6·30.

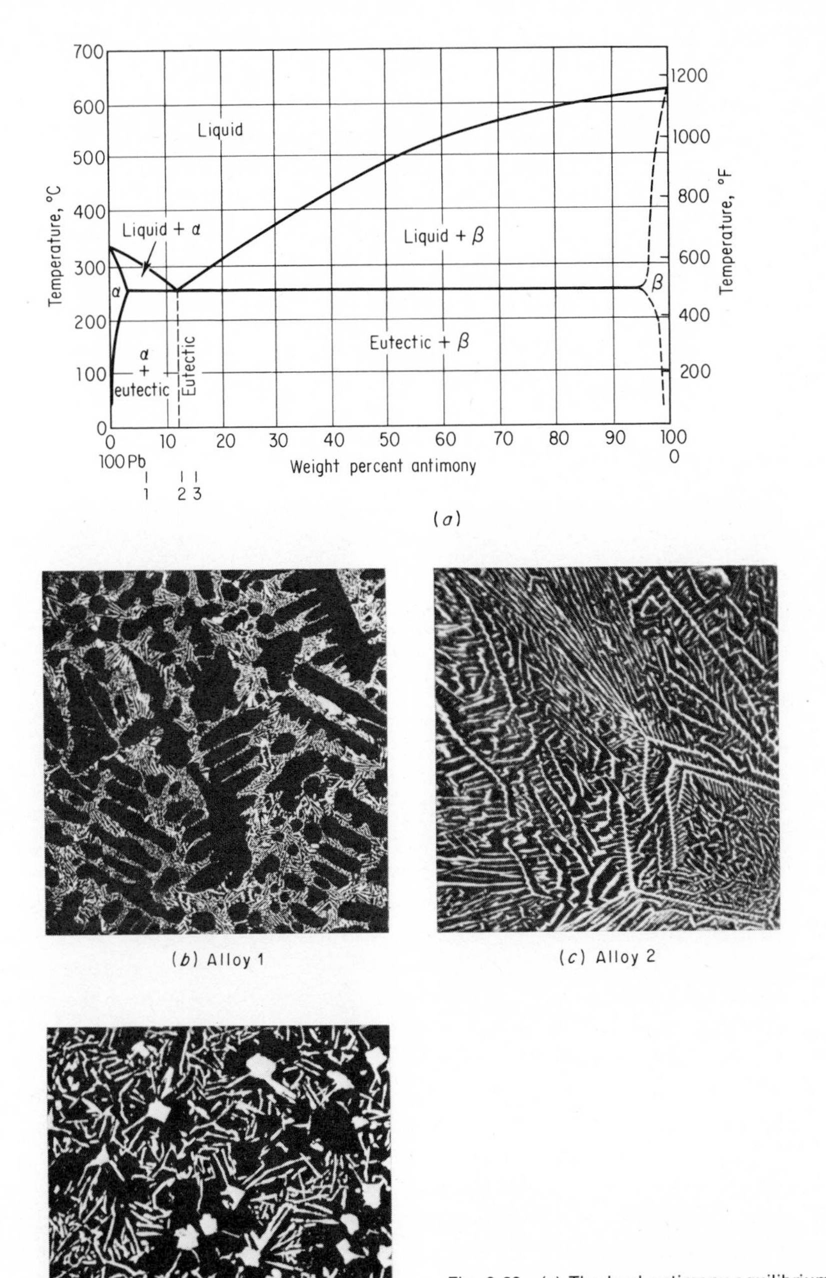

Fig. 6·29 (*a*) The lead-antimony equilibrium diagram. (*b*) 6.5 percent antimony alloy, 75X. (*c*) Eutectic alloy, 11.5 percent antimony, 250X. (*d*) 12.25 percent antimony alloy, 250X. (American Smelting and Refining Company.)

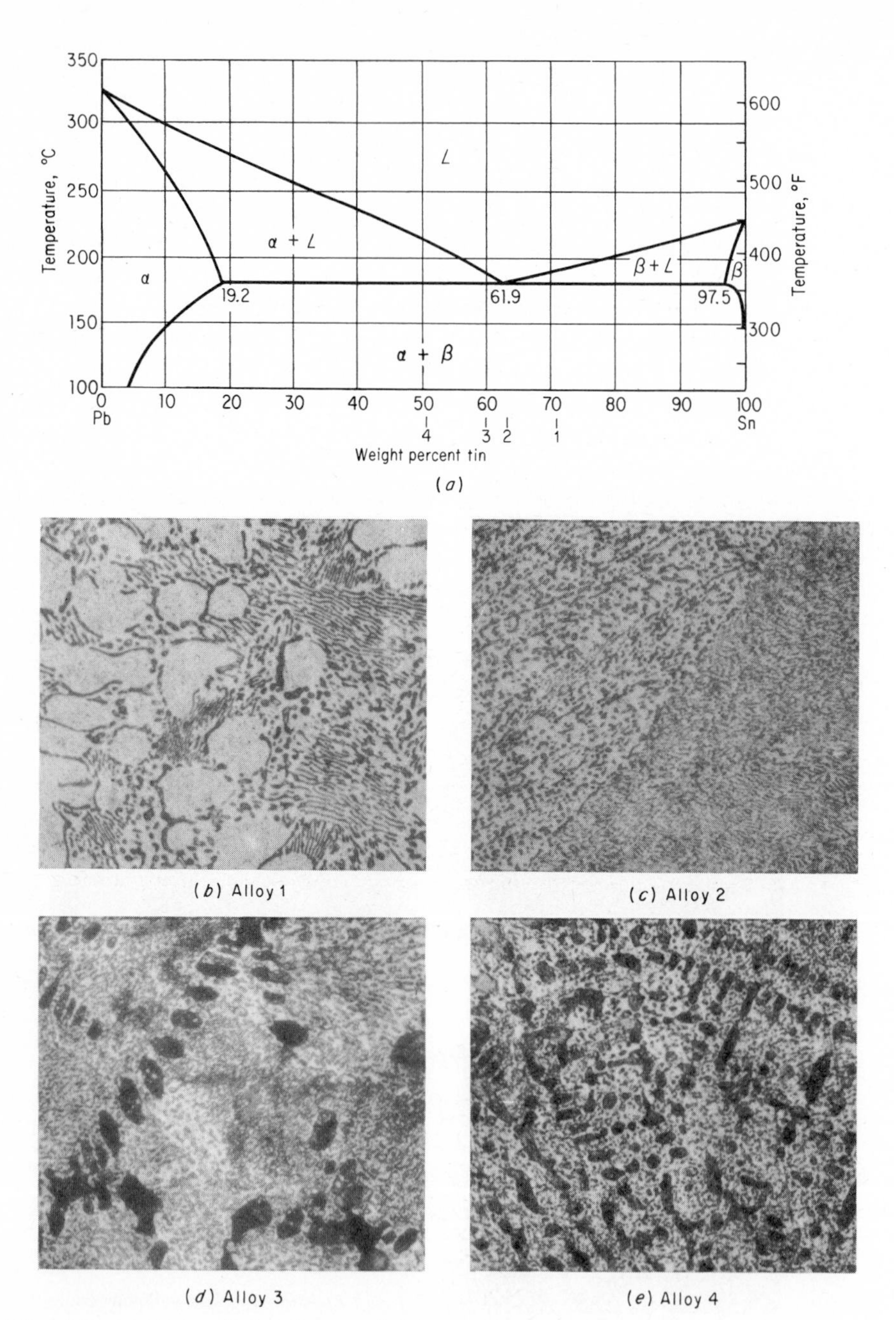

Fig. 6·30 (a) The lead-tin equilibrium diagram. (b) 70 percent tin alloy. (c) Eutectic alloy. (d) 60 percent tin alloy. (e) 50 percent tin alloy. All photomicrographs at 200X. (From H. Manko, "Solders and Soldering," McGraw-Hill Book Company, New York, 1964.)

6·15 Properties in Eutectic Alloy Systems It was shown in Fig. 6·20 that there is a linear relationship between the constituents appearing in the microstructure and the alloy composition for a eutectic system. This would seem to indicate that the physical and mechanical properties of a eutectic system should also show a linear variation. In actual practice, however, this ideal behavior is rarely found. The properties of any multiphase alloy depend upon the individual characteristics of the phases and how these phases are distributed in the microstructure. This is particularly true for eutectic alloy systems. Strength, hardness, and ductility are related to the size, number, distribution, and properties of the crystals of both phases. In many commercially important eutectic alloy systems, one phase is relatively weak and plastic while the other phase is relatively hard and brittle. As the eutectic composition is approached from the plastic-phase side, there will be an increase in the strength of the alloy. There will be a decrease in strength beyond the eutectic composition due to the decrease in the amount of the small eutectic particles and the increase in size and amount of the proeutectic brittle phase. Therefore, in this kind of system the eutectic composition will generally show maximum strength. This is illustrated in Fig. 6·31, which shows the variation in tensile strength and elongation for cast aluminum-silicon alloys containing up to 14 percent silicon. The tensile strength shows a maximum at very near the eutectic composition. The aluminum-silicon phase diagram is given in Fig. 6·21.

Another important point is that the resulting properties of a mixture most nearly resemble those of the phase which is continuous—that is, the phase that forms the background or matrix in which particles of the other phase are imbedded. The eutectic mixture is always the microconstituent which is continuous, since it is the last liquid to solidify and surrounds the primary grains. It is generally true that the phase which makes up the greater proportion in the eutectic mixture will be the continuous phase. If this phase

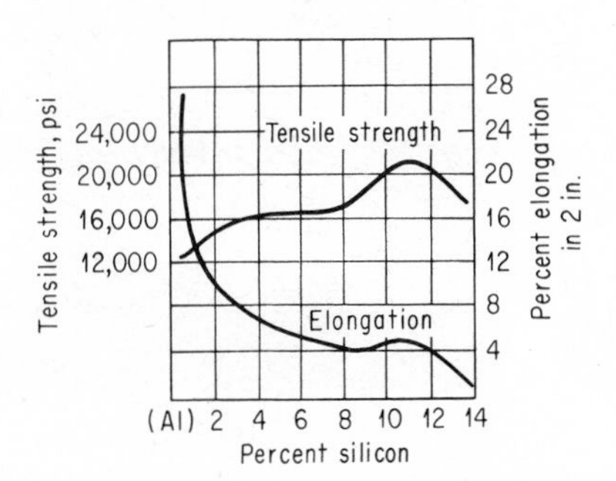

Fig. 6·31 Variation of typical properties of cast aluminum-silicon alloys up to 14 percent silicon. (From Guy, "Elements of Physical Metallurgy," Addison-Wesley Publishing Co., Reading, Mass., 1959.)

is plastic, the entire series of alloys will show some plasticity. If this phase is brittle, the entire series of alloys will be relatively brittle.

In addition to the above factors, an increase in cooling rate during freezing may result in a finer eutectic mixture, a greater amount of eutectic mixture, and smaller primary grains, which in turn will influence the mechanical properties considerably.

6·16 Age Hardening There are only two principal methods for increasing the strength and hardness of a given alloy: cold working or heat treatment. The most important heat-treating process for nonferrous alloys is age hardening, or precipitation hardening. In order to apply this heat treatment, the equilibrium diagram must show partial solid solubility, and the slope of the solvus line must be such that there is greater solubility at a higher temperature than at a lower temperature. These conditions are satisfied by Fig. 6·24. Let us consider the left side of the diagram involving the α solid solution. Alloy compositions that can be age-hardened are usually chosen between point *F* containing 20 percent *B* and point *H* containing 10 percent *B*. Commercially, most age-hardenable alloys are chosen of compositions slightly to the left of point *F*, although the maximum hardening effect would be obtained by an alloy containing 20 percent *B*. The phase which is dissolved may be a terminal solid solution, as in this case, or an intermediate alloy phase.

Two stages are generally required in heat treatment to produce age hardening: solution treatment and aging.

Solution Treatment If alloy 4 (Fig. 6·24, microstructure shown in Fig. 6·32*a*) is reheated to point *M*, all the excess β will be dissolved and the structure will be a homogeneous α solid solution. The alloy is then cooled rapidly (quenched) to room temperature. A supersaturated solution results, with the excess β trapped in solution. The quench is usually carried out in a cold-water bath or by a water spray. Drastic quenching tends to set

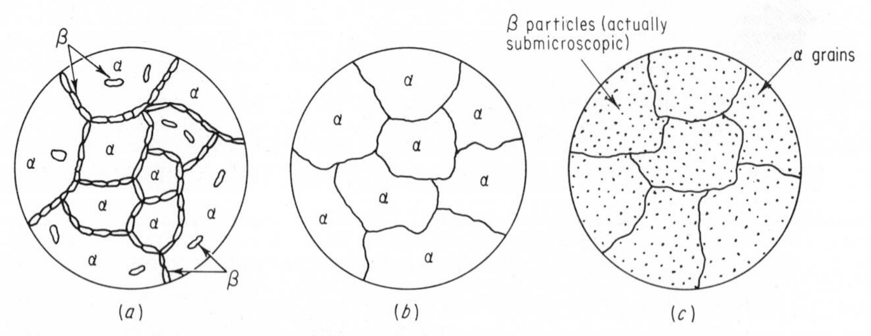

Fig. 6·32 Microstructure of an 85*A*-15*B* alloy. (*a*) After slow cooling; (*b*) after reheating and rapid cooling to room temperature; (*c*) after aging.

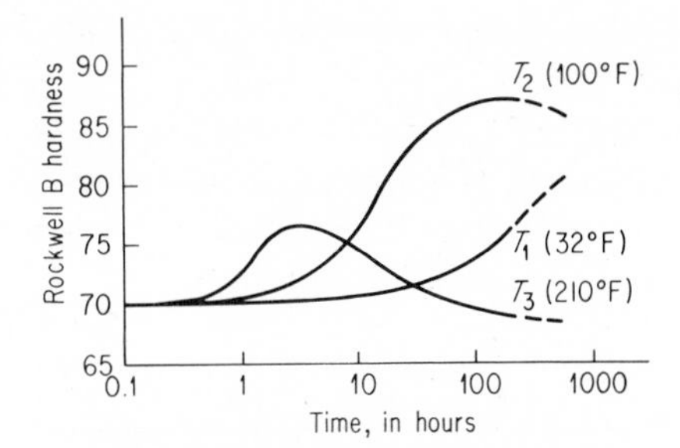

Fig. 6·33 Effect of temperature on the aging curves during precipitation hardening. Curves are for a 0.06 percent carbon steel. (After Davenport and Bain.)

up stresses which often result in distortion, especially if the parts are intricately designed. In such instances, boiling water may be used as a quench medium to minimize distortion. If α is a ductile phase, the alloy will be ductile immediately after quenching. This allows warped or distorted parts to be straightened easily. The straightening operations should be carried out as soon as possible after quenching. The microstructure is shown schematically in Fig. 6·32*b*.

Aging Process The alloy, as quenched, is a supersaturated solid solution and is in an unstable state. The excess solute will tend to come out of solution. The speed at which precipitation occurs varies with temperature. Figure 6·33 shows the effect of three temperatures on the aging curves of an iron alloy. At the low temperature the diffusion rate is so slow that no appreciable precipitation occurs. At T_3, hardening occurs quickly, due to rapid diffusion, but softening effects are also accelerated, resulting in a lower maximum hardness. The optimum temperature seems to be T_2, at which maximum hardening occurs within a reasonable length of time. Those alloys in which precipitation takes place at room temperature—so that they obtain their full strength after 4 or 5 days at room temperature—are known as *natural-aging* alloys. Those alloys which require reheating to elevated temperatures to develop their full strength are *artificial-aging* alloys. However, these alloys also age a limited amount at room temperature, the rate and extent of the strengthening depending upon the alloys (see Table 6·2). Refrigeration retards the rate of natural aging. At 32°F, the beginning of the aging process is delayed for several hours, while dry ice (−50 to −100°F) retards aging for an extended period. Use is made of this fact in the aircraft industry when aluminum-alloy rivets, which normally age at room temperature, are kept in deep-freeze refrigerators until they are driven. The rivets have previously been solution-treated, and as a single phase they are very ductile. After being driven, aging will take place at room temperature, with a resulting increase in strength and hardness. In the early

TABLE 6·2 Effect of Aging on Properties of Aluminum Alloy 2014 (3.5 to 4.5 Percent Copper)

ALLOY AND CONDITION	ULTIMATE STRENGTH, PSI	YIELD STRENGTH, PSI	ELONGATION, % IN 2 IN.	BHN, 500 KG, 10 MM	SHEAR STRENGTH, PSI
Annealed	27,000	14,000	18	45	18,000
Solution-treated, naturally aged	62,000	42,000	20	105	38,000
Solution-treated, artificially aged	70,000	60,000	13	135	42,000

theory of the aging process, it was thought that the excess phase comes out of solution as fine submicroscopic particles, many of which fall on the slip planes (Fig. 6·32*c*). These particles were considered to have a keying action, thereby interfering with movement along planes of ready slip, thus increasing strength and hardness.

Subsequent studies have led to a more complete understanding of the age-hardening process. The strengthening of a heat-treatable alloy by aging is not due merely to the presence of a precipitate. It is due to both the uniform distribution of a finely dispersed submicroscopic precipitate and the distortion of the lattice structure by those particles before they reach a visible size.

It is not possible to state definitely in what manner precipitate particles harden the matrix or solvent lattice. While there are several theories of precipitation hardening, the most useful is the *coherent lattice theory.* After solution treatment and quenching, the alloy is in a supersaturated condition, with the solute atoms distributed at random in the lattice structure, Fig. 6·34*a*. During an incubation period, the excess solute atoms tend to migrate to certain crystallographic planes, forming *clusters* or embryos of the precipitate. During aging, these clusters form an intermediate crystal structure, or *transitional lattice*, maintaining registry (coherency) with the lattice structure of the matrix. The excess phase will have different lattice parameters from those of the solvent, and as a result of the atom matching (coherency), there will be considerable distortion of the matrix, Fig. 6·34*b*. The distortion of the matrix extends over a larger volume than would be the case if the excess phase were a discrete particle. It is this distortion that interferes with the movement of dislocations and accounts for the rapid increase in hardness and strength during aging (see Fig. 6·35). Eventually, the equilibrium excess phase is formed with its own lattice structure (Fig. 6·34*c*). This causes a loss of coherency with the matrix and less distortion. Hardness and strength will decrease, and the alloy is "over-aged." There will now be a boundary between the excess phase and the matrix so that the precipitated particle will be visible under the microscope. Electrical conductivity decreases during aging because of lattice distortion;

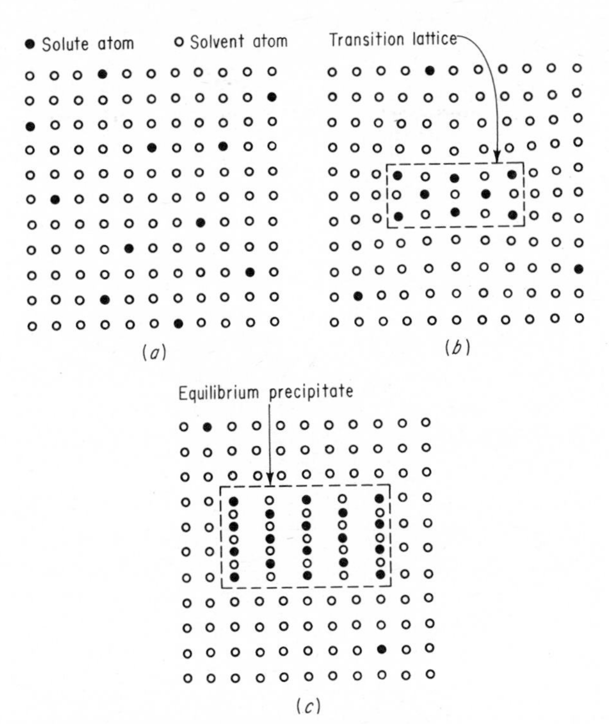

Fig. 6·34 The stages in the formation of an equilibrium precipitate. (*a*) Supersaturated solid solution. (*b*) Transition lattice coherent with the solid solution. (*c*) Equilibrium precipitate essentially independent of the solid solution. (From Guy, "Elements of Physical Metallurgy," Addison-Wesley Publishing Company, Inc., Reading, Mass., 1959.)

then it increases when precipitation takes place. The effect of aging time on some properties is shown in Fig. 6·35.

Aging does not have the same effect on the properties of all alloys. In some alloys the change in hardness and strength may be small; in others the changes may be large. This is not due to the amount dissolved in excess of the saturation limit, but to the effect of the precipitate on lattice distortion. For example, magnesium can dissolve 46 percent lead at the eutectic temperature but only 2 percent lead at room temperature. With propert heat treatment, precipitation takes place but no age hardening occurs. This is due to the absence of a transition lattice and very little localized distortion during precipitation. On the other hand, if the precipitation involves extensive changes in the lattice, a large amount of distortion occurs with wide changes in properties. In this case, the greater the amount of metal dissolved in excess of the saturation limit, the greater the distortion produced and the greater will be the effect on hardness and

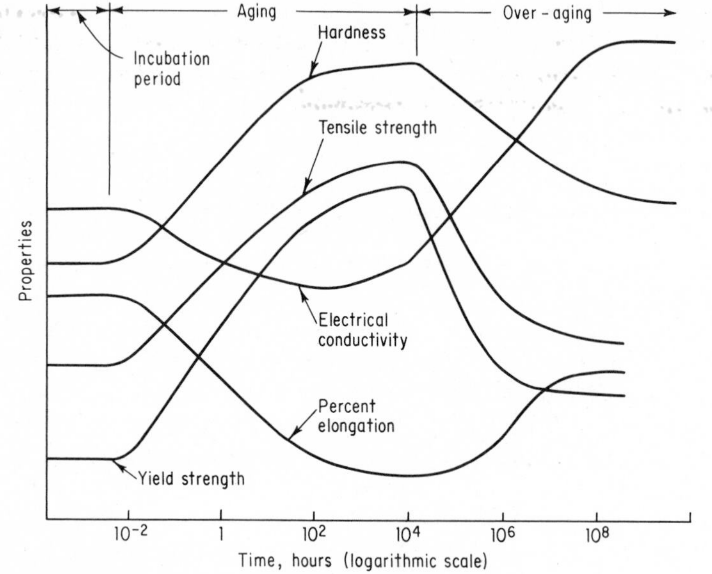

Fig. 6·35 Effect of aging time on properties. (By permission from L. F. Mondolfo and O. Zmeskal, "Engineering Metallurgy," McGraw-Hill Book Company, New York, 1955.)

strength. This is illustrated for some copper-beryllium alloys in Table 6·3. Figure 12·16 shows the copper-rich portion of the copper-beryllium alloy system. Notice that copper dissolves up to a maximum of 2.2 percent beryllium at 1600°F. At room temperature the solubility of beryllium in copper is approximately 0.2 percent. Referring to Table 6·3, while there is no difference in strength and hardness of both alloys after solution annealing (solution treatment), there is a difference after aging. The alloy containing 1.90–2.15 percent beryllium is closer to the maximum solubility of beryllium in copper, and upon aging more of the γ (gamma) phase will be precipitated, resulting in a greater hardness and strength as shown.

TABLE 6·3 Effect of Aging and Composition on the Mechanical Properties of Some Copper-Beryllium Alloys

ALLOY AND CONDITION	TENSILE STRENGTH, PSI	ELONGATION, % IN 2 IN.	ROCKWELL HARDNESS
1.90–2.15% Be:			
Solution-annealed	60,000–80,000	35–50	B 45–65
After aging	165,000–180,000	5–8	C 36–40
1.60–1.80% Be:			
Solution-annealed	60,000–80,000	35–50	B 45–65
After aging	150,000–165,000	5–8	C 33–37

6·17 Type IV—The Congruent-melting Intermediate Phase When one phase changes into another phase isothermally (at constant temperature) and without any change in chemical composition, it is said to be *congruent phase change* or *congruent transformation.* All pure metals solidify congruently. We have previously seen an example of a congruent-melting alloy as a variation of a Type I phase diagram. The alloy *x* in Fig. 6·14*a* goes from a liquid phase to a single solid phase at constant temperature without a change in composition, and it is therefore a congruent-melting alloy. Intermediate phases are so named because they are single phases that occur between the terminal phases on a phase diagram. Type IV will consider the formation of an intermediate phase by congruent melting, while Type V will cover the incongruent melting intermediate phase. Any intermediate phase may be treated as another component on a phase diagram. If the intermediate phase has a narrow range of composition, as do intermetallic compounds and interstitial compounds, it is then represented on the diagram as a vertical line and labeled with the chemical formula of the compound. If the intermediate phase exists over a range of composition, it is usually an electron compound and is labeled with a Greek letter. In recent years some authors tend to use Greek letters for all intermediate phases.

In Fig. 6·36, the intermediate alloy phase is shown as a vertical line. Since it is a compound, it is indicated as A_mB_n, where m and n are subscripts which indicate the number of atoms combined in the compound. For example, magnesium and tin form an intermediate phase which has the chemical formula Mg_2Sn. In this case, Mg is equivalent to A; 2 is equivalent to m; tin, Sn, is equivalent to B; and n is equal to 1.

It is apparent from Fig. 6·36 that the *A-B* system may be separated into two independent parts, one to show all the alloys between *A* and the com-

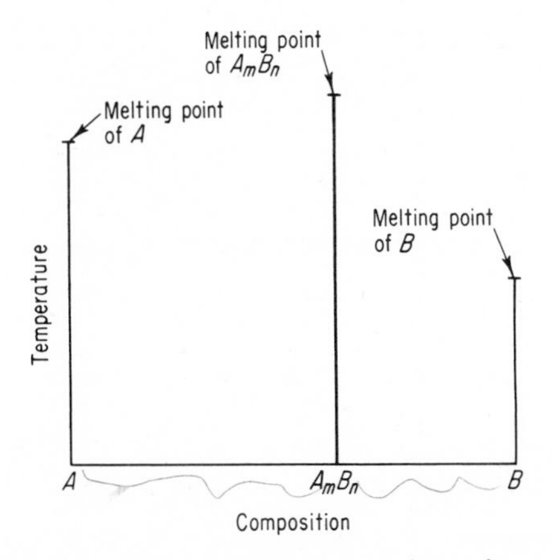

Fig. 6·36 Composition and melting point of pure *A*, pure *B*, and a compound *AmBn*.

pound A_mB_n and the other to show those between A_mB_n and B. The portion of the diagram between A and A_mB_n may be any of the types studied in this chapter; similarly for the portion between A_mB_n and B. If the compound shows no solubility for either pure metal and the pure metals show some solubility for each other, the equilibrium diagram will be as shown in Fig. 6·37. This diagram shows two different eutectic mixtures. The eutectic equations may be written as follows:

At T_1: $$\text{Liquid} \underset{\text{heating}}{\overset{\text{cooling}}{\rightleftharpoons}} \alpha + A_mB_n$$

At T_2: $$\text{Liquid} \underset{\text{heating}}{\overset{\text{cooling}}{\rightleftharpoons}} A_mB_n + \beta$$

The study of many actual systems that show the formation of several congruent-melting intermediate phases may be simplified by the above approach.

6·18 Type V—The Peritectic Reaction In the peritectic reaction a liquid and a solid react isothermally to form a new solid on cooling. The reaction is expressed in general as

$$\text{Liquid} + \text{solid}_1 \underset{\text{heating}}{\overset{\text{cooling}}{\rightleftharpoons}} \text{new solid}_2$$

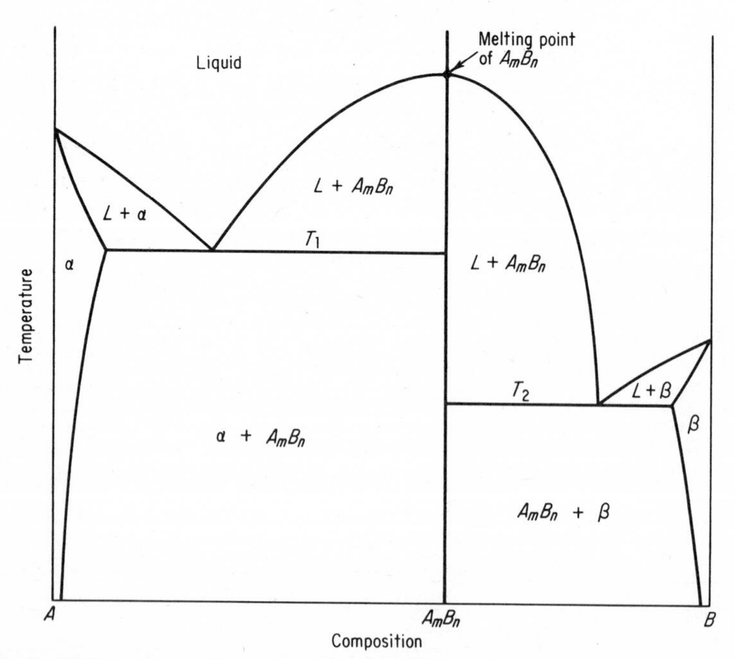

Fig. 6·37 Equilibrium diagram illustrating an intermediate alloy which is an intermetallic compound.

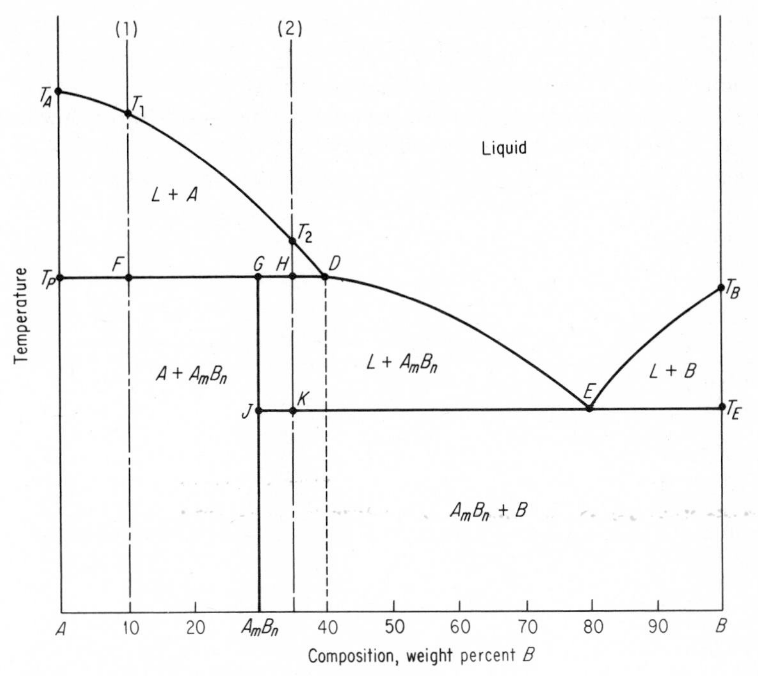

Fig. 6·38 Phase diagram showing the formation of an incongruent melting intermediate phase by a peritectic reaction.

The new solid formed is usually an intermediate phase (Fig. 6·38), but in some cases it may be a terminal solid solution (Fig. 6·39).

Consideration of Fig. 6·38 shows that the compound A_mB_n, 70*A*-30*B*, when heated to the peritectic temperature, point *G*, decomposes into two phases, liquid and solid *A*. Therefore, this is an example of an incon-

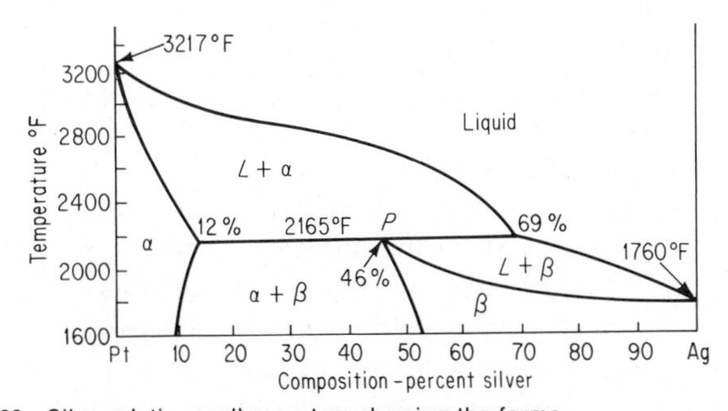

Fig. 6·39 Silver-platinum alloy system showing the formation of a terminal solid solution by a peritectic reaction.

gruent-melting intermediate alloy. The student should realize that the peritectic reaction is just the reverse of the eutectic reaction, where a single phase formed two new phases *on cooling*. The liquidus line is T_ADET_B and the solidus line is $T_AT_PGJT_ET_B$. The peritectic-reaction line is T_PD. Notice that only part of this line, the length T_PG, coincides with the solidus line. The slow cooling of several alloys will now be studied.

Alloy 1, 90*A*-10*B*, remains liquid until the liquidus line is reached at T_1. Solidification now takes place by forming crystals of the pure metal *A*. As the temperature falls, the liquid is decreasing in amount, and its composition is moving down along the liquidus line. Let us examine the conditions that exist just above the peritectic temperature T_P:

Phases	Liquid	Solid *A*
Composition	60*A*-40*B*	100 *A*
Relative amount	$\frac{T_PF}{T_PD} \times 100 = 25\%$	$\frac{FD}{T_PD} \times 100 = 75\%$

The conditions that exist just below the peritectic temperature are:

Phases	A_mB_n	Solid *A*
Composition	70*A*-30*B*	100 *A*
Relative amount	$\frac{T_PF}{T_PG} \times 100 = 33\%$	$\frac{FG}{T_PG} \times 100 = 67\%$

A first glance at these two areas seems to indicate that the liquid has disappeared at the horizontal line and in its place is the compound A_mB_n. Consideration of the chemical compositions shows that this is not possible. The liquid contains 60*A*, while A_mB_n contains 70*A*. The liquid is not rich enough in *A* to form the compound by itself. The liquid must therefore react with just the right amount of solid *A*, in this case 8 percent, to bring its composition to that of the compound A_mB_n. The following reaction must have taken place at the peritectic temperature:

Composition:	60*A*	100*A*		70*A*
Equation:	Liquid	+ solid *A*	$\xrightarrow{\text{cooling}}$	solid A_mB_n
Relative amount:	25%	8%		33%

The reaction takes place all around the surface of each grain of solid *A* where the liquid touches it. When the correct composition is reached, the layer solidifies into A_mB_n material surrounding every grain of *A*. Further reaction is slow since it must wait for the diffusion of atoms through the peritectic wall of A_mB_n in order to continue (see Fig. 6·40). When diffusion is completed, all the liquid will have been consumed, and since only 8 percent of pure *A* was required for the reaction, there will be 67 percent of *A* left. The final microstructure will show grains of primary *A* surrounded by

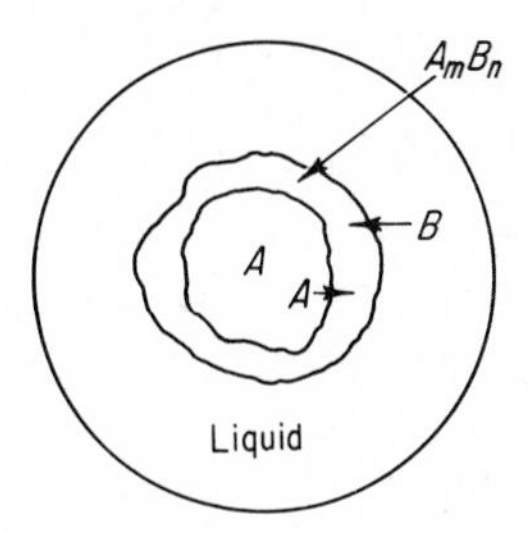

Fig. 6·40 Schematic picture of the peritectic reaction. Envelope of A_mB_n increases in thickness by diffusion of A atoms outward and B atoms inward.

the compound A_mB_n. Figure 6·41 shows the microstructure at various temperatures in the slow cooling of this alloy. The story will be the same for any alloy to the left of point G. The only difference will be in the amount of excess A remaining after the peritectic reaction is complete. The closer the alloy composition is to the composition of the compound, the less primary A will remain.

Alloy 2, 65A-35B, solidifies pure A when the liquidus line is crossed at T_2, and as solidification continues, the liquid becomes richer in B. When point H is reached, the liquid composition is 60A-40B. Applying the lever rule for this alloy, there is $^{35}/_{40} \times 100$ or 87.5 percent liquid and 12.5 percent solid A. Since the line GD is not part of the solidus line, some liquid must remain after the reaction takes place. It is therefore the solid A which must disappear in reacting with some of the liquid to form the compound A_mB_n.

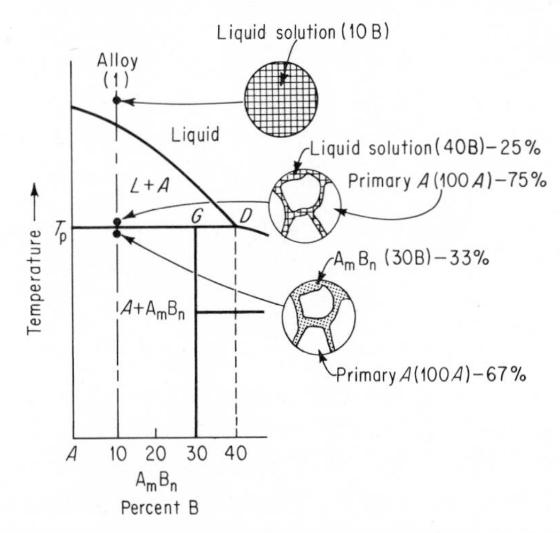

Fig. 6·41 Slow cooling of a 90A-10B alloy showing the microstructure at various temperatures.

The same reaction takes place again:

Composition:	$60A$	$100A$	$70A$
Reaction:	Liquid + solid A	$\xrightarrow{\text{cooling}}$	A_mB_n

The amount of liquid entering into the above reaction may be determined by applying the lever rule below the reaction temperature.

$$\text{Liquid (percent)} = \frac{GH}{DG} \times 100 = \frac{10}{20} \times 100 = 50 \text{ percent}$$

Since there was 87.5 percent liquid before the reaction and 50 percent liquid after the reaction, it is apparent that 37.5 percent of the liquid reacted with 12.5 percent of solid A to give 50 percent of the compound A_mB_n at the peritectic temperature. As cooling continues, the liquid now separates crystals of A_mB_n. The liquid becomes richer in B, and its composition gradually moves down and to the right along the liquidus line until it reaches point E, the eutectic temperature. At this temperature, there is only $5/50 \times 100$ or 10 percent liquid left. Since the liquid has reached the eutectic point, it now solidifies into the eutectic mixture of $A_mB_n + B$. This alloy, at room temperature, will consist of 90 percent primary or proeutectic A_mB_n surrounded by 10 percent of the eutectic ($A_mB_n + B$) mixture. Figure 6·42 shows the cooling curve and the changes in microstructure at various points in the slow cooling of this alloy.

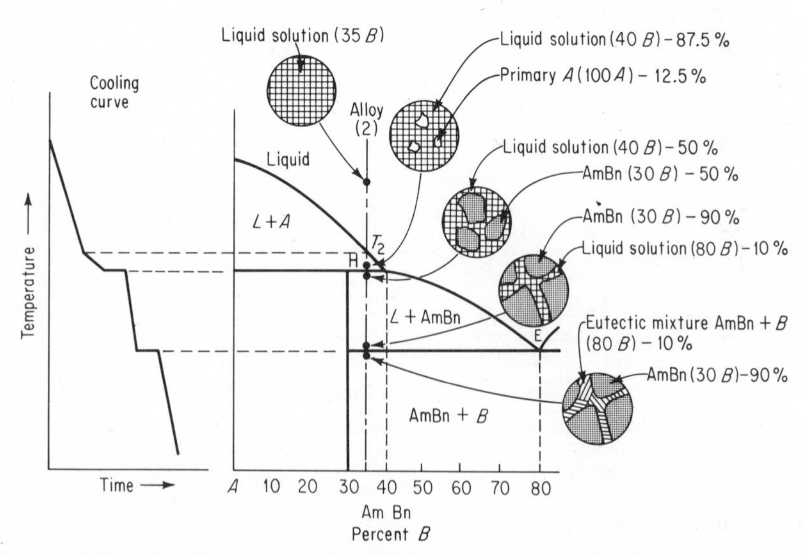

Fig. 6·42 The cooling curve and the microstructure at various temperatures during the slow cooling of a 65A-35B alloy.

The student should study the slow cooling of alloys on either side of the peritectic point P in the equilibrium diagram that illustrates the formation of a terminal solid solution by a peritectic reaction (Fig. 6·39).

The peritectic reaction was described under equilibrium conditions. In actual practice, however, this condition is rarely attained. Since the new phase forms an envelope around the primary phase, it will act as a hindrance to the diffusion which is essential to continue the reaction (see Fig. 6·40). As the layer of the new phase becomes thicker, the diffusion distance increases, so that the reaction is frequently incomplete. For example, according to the phase diagram of the silver-platinum alloy system, Fig. 6·39, a 60 percent Ag alloy should be a single phase β at room temperature. The actual cast structure, Fig. 6·43, however, is not a single phase. The light areas are primary α grains surrounded by the dark two-

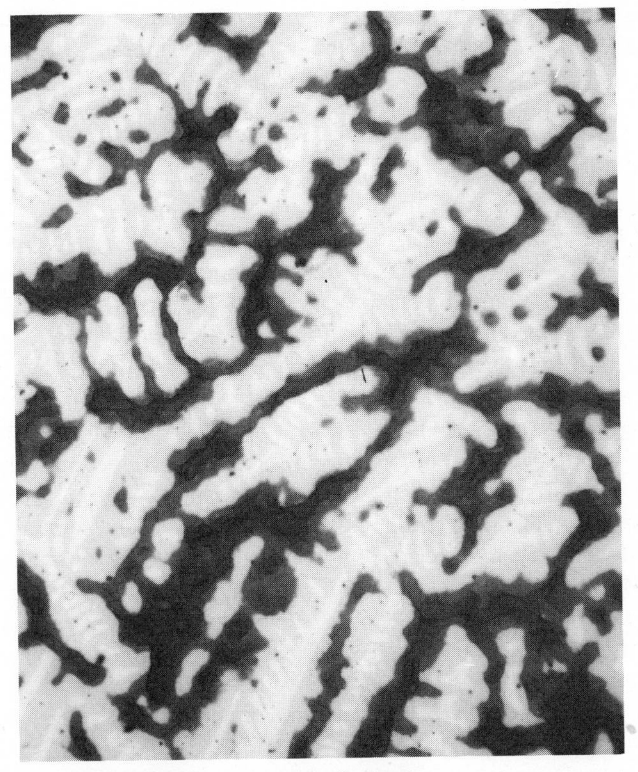

Fig. 6·43 40 percent Platinum + 60 percent silver cast alloy. Light areas are primary α; dark two-toned areas are β. Magnification 100X. (By permission from F. N. Rhines, "Phase Diagrams in Metallurgy," McGraw-Hill Book Company, New York, 1956.)

toned areas of β, indicating that the peritectic reaction was not complete.

Since peritectic alloys are usually two-phase mixtures, their mechanical properties follow the same principles stated for eutectic alloys with two differences: (1) the individual phases are more likely to be different than predicted for equilibrium conditions, and (2) cast grain size is usually coarse.

6·19 Type VI—Two Liquids Partly Soluble in the Liquid State: The Monotectic Reaction In all the types discussed previously, it was assumed that there was complete solubility in the liquid state. It is quite possible, however, that over a certain composition range two liquid solutions are formed that are not soluble in each other. Another term for solubility is miscibility. Substances that are not soluble in each other, such as oil and water, are said to be immiscible. Substances that are partly soluble in each other are said to show a miscibility gap, and this is Type VI.

The equilibrium diagram for this type is shown in Fig. 6·44. The liquidus line is T_ACFET_B, and the solidus line is $T_AT_EJT_B$. Alloys having compositions between point C and point F at a temperature just above T_M will consist of two liquid solutions, L_1 and L_2. The lines CD and FG show the composition of the liquid phases in equilibrium with each other at higher temperatures. In most cases, these lines are shown dotted because experimental difficulties at high temperatures usually prevent an accurate determination of their position. Since these lines tend to approach each other, it is possible that at higher temperatures the area will be closed and a single homogeneous liquid solution will exist. This area should be treated like any other two-phase area, and the same rules may be applied to determine the chemical composition of L_1 and L_2 and their relative amounts at any temperature. L_1 is a liquid solution of B dissolved in A, whereas L_2 is a liquid solution of A dissolved in B.

Let us study the slow cooling of several alloys. Alloy x containing 10 percent B is a single-phase liquid solution L_1, and it remains that way until the liquidus line is crossed at x_1. Solidification starts by forming crystals of the pure metal A. The liquid becomes richer in B, gradually moving down and to the right along the liquidus line. When the alloy has reached the monotectic temperature line T_M at point x_2, the liquid composition is given by point C, which is 80A and 20B. The horizontal line on any phase diagram indicates that a reaction must take place. What is the reaction in this case? Below the line, the two phases present are solid A and L_2. Offhand, it seems that L_1 has disappeared and that in its place we have L_2, but a more careful study of the diagram indicates that this could not have happened. The composition of L_2 is given by point F, which is 40A and 60B, whereas L_1 has a composition of 80A and 20B; so although L_1 has disappeared, by itself it could not have formed L_2. Remember that L_1 is a liquid solution rich in A, whereas L_2 is a liquid solution rich in B. The problem is that L_1 has too much A. Therefore, what must occur at the horizontal line is that enough solid A is precipitated from L_1 to bring its composition to the right one to form L_2.

To prove this, we can apply Rule II both above and below the horizontal line T_M. Above the line we have 50 percent solid A ($10/20 \times 100$) and 50 percent L_1. Below the line, we have 17 percent L_2 ($10/60 \times 100$) and 83 percent solid A. Therefore, at the horizontal line, the 50 percent of L_1 must have formed 17 percent L_2 and 33 percent solid A. The 33 percent, plus the 50 percent already existing, gives a total of 83 percent solid A as determined by the calculations. The reaction is summarized as follows:

Composition:	80A	40A	100A
Equation:	$L_1 \xrightarrow{\text{cooling}}$	L_2 +	solid A
Relative amount:	50%	17%	33%

When one liquid forms another liquid, plus a solid, on cooling, it is known as a monotectic reaction; the general equation for the monotectic reaction

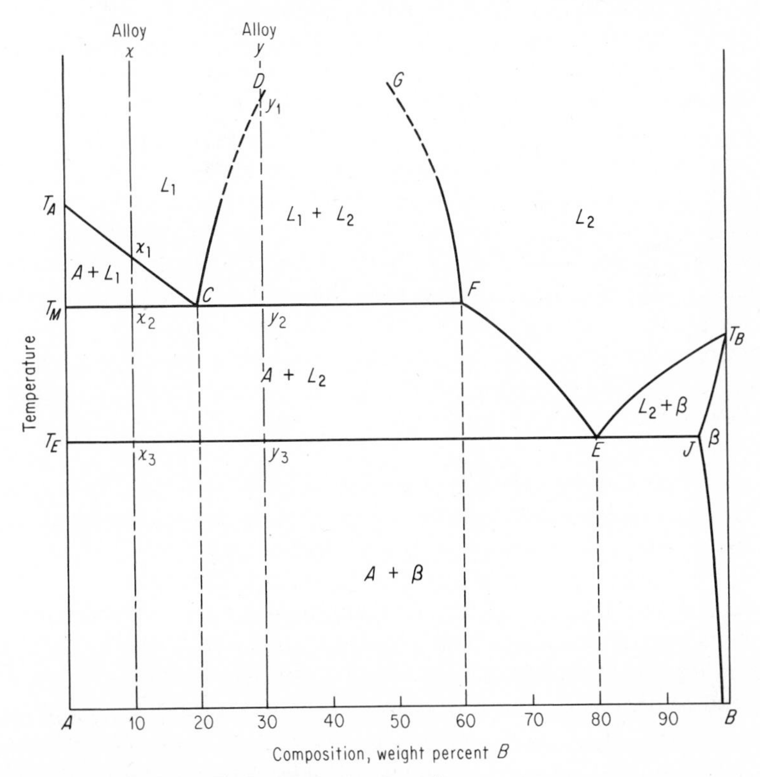

Fig. 6·44 Hypothetical equilibrium diagram of two metals partly soluble in the liquid state: the monotectic reaction.

may be written as:

$$L_1 \underset{\text{heating}}{\overset{\text{cooling}}{\rightleftharpoons}} L_2 + \text{solid}$$

Point C is known as the monotectic point. It should be apparent that the monotectic reaction resembles the eutectic reaction, the only difference being that one of the products is a liquid phase instead of a solid phase. It turns out that all known binary monotectic points in metal systems are located nearer the composition of the solid phase, so that the solid phase predominates in the reaction. In this case, 33 percent solid A was formed compared with 17 percent L_2. As in the case of the eutectic, alloys to the left of point C, such as alloy x, are known as hypomonotectic alloys, whereas alloys to the right of point C up to point F are known as hypermonotectic alloys.

We now continue with the discussion of the slow cooling of alloy x. After the monotectic reaction is complete and as the temperature is dropped, more solid A will be formed from L_2. When the eutectic temperature is reached at T_E, the alloy is at x_3 and L_2 will be at point E. This is the right temperature and right composition to form the eutectic mixture. The eutectic reaction now takes place, with L_2 forming a very fine mixture of solid A, plus solid β. The final microstructure will consist of 87.5 percent grains of primary A surrounded by 12.5 percent of the eutectic $(A + \beta)$ mixture. The student should verify these percentages.

The occurrence of two liquids in the hypermonotectic alloys, that is, alloys of compositions between C and F, above the monotectic temperature, introduces structural considerations which have not been discussed up to this point. Given enough time, the two liquids will separate into two layers according to density, with the lighter layer on top. It is quite possible, however, to have two liquids existing as an emulsion wherein tiny droplets of one liquid remain suspended in the other liquid. Unfortunately, knowledge of this behavior with respect to metals is very limited at the present time.

Consider the slow cooling of a hypermonotectic alloy y containing 70A-30B. At the elevated temperature, this alloy will be composed of a single homogeneous liquid phase L_1. Upon cooling, the limit of liquid immiscibility is crossed at y_1, and the second liquid L_2 will make its appearance, probably at the surface of the confining vessel, and possibly also at various points through the liquid bath. The composition of L_2 may be obtained by drawing a tie line in the two-phase region and applying Rule I. As the temperature decreases, the quantity of L_2 increases, so that just above the monotectic temperature, at point y_2, the amount of L_2 present would be equal to $10/40 \times 100$, or 25 percent. Conditions being favorable, this liquid will exist as a separate layer in the crucible or mold. That portion of the mixture which is composed of L_1 now reacts according to the monotectic equation to form more of L_2 + solid A. With continued cooling, more solid A is formed from L_2, its composition becoming richer in B, until the eutectic temperature is approached at point y_3. At that temperature, the remaining L_2 (37.5 percent) undergoes the eutectic reaction and solidifies into a very fine mixture of $A + \beta$.

An example of an alloy system showing a monotectic reaction is that between copper and lead given in Fig. 6·45. Notice that in this case the $L_1 + L_2$ region is closed. Also, although the terminal solids are indicated as α and β, the solubility is actually so small that they are practically the pure metals, copper and lead.

6·20 Type VII—Two Metals Insoluble in the Liquid and Solid States This will complete the study of basic phase or equilibrium diagrams that involve the liquid and solid states. If points C and F in Fig. 6·44 are moved in opposite directions, they will eventually hit the axes to give the diagram shown in Fig. 6·46. There are many combinations of metals which are practically insoluble in each other. When cooled, the two metals appear to solidify at their individual freezing points into two distinct layers with a sharp line of contact and almost no diffusion.

An alloy system which comes very close to this type is that between aluminum and lead shown in Fig. 6·47. Notice that the two-phase liquid region extends almost entirely across the diagram. This condition corresponds

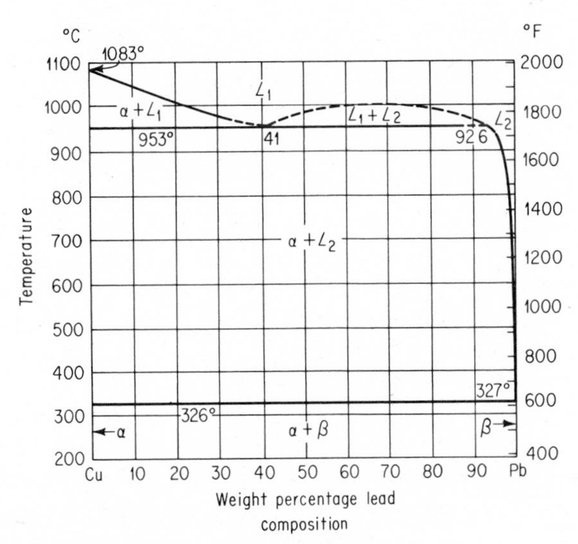

Fig. 6·45 The copper-lead equilibrium diagram. (From Metals Handbook, 1948 ed., p. 1200, American Society for Metals, Metals Park, Ohio.)

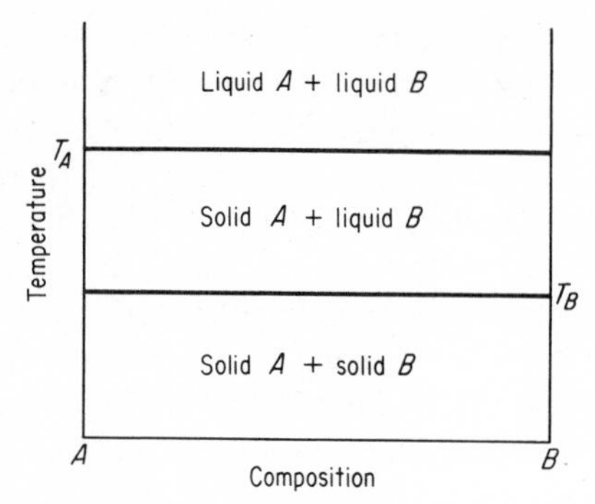

Fig. 6·46 Hypothetical equilibrium diagram for two metals insoluble in the liquid and solid states.

to a limiting case of the monotectic reaction and the eutectic reaction. The upper of the two horizontal lines represents a monotectic reaction in which the monotectic point is very close to the composition and melting point of pure aluminum. The lower horizontal line represents a eutectic reaction in which the eutectic point is practically coincident with the composition and melting point of pure lead.

6·21 Interrelation of Basic Types The various types of equilibrium diagrams that have been discussed may be combined in many ways to make up actual diagrams. It is important for the student to understand the interrelation

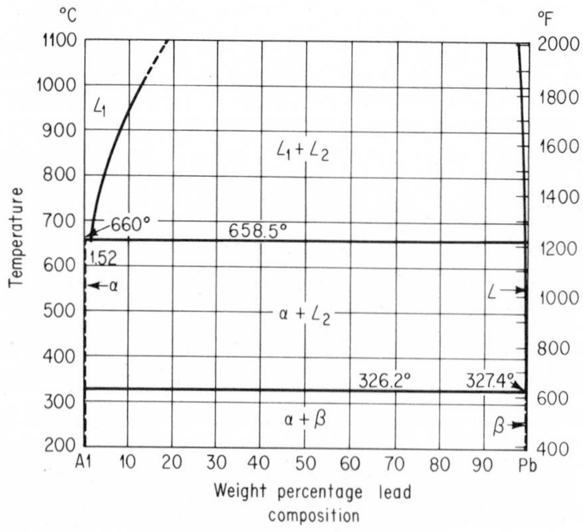

Fig. 6·47 The aluminum-lead alloy system. (From Metals Handbook, 1948 ed., p. 1165, American Society for Metals, Metals Park, Ohio.)

between the basic types to make the study of complex diagrams much simpler. The first three types differ only by the solubility in the solid state. Starting with a completely insoluble system of Type II (Fig. 6·48*a*), if the points at either end of the eutectic line (*F* and *G*) are moved toward each other, that is, toward greater solubility in the solid state, this will result in a diagram of Type III, partly soluble in the solid state (Fig. 6·48*b*). If they are moved until they coincide with the eutectic composition at *E*, a completely soluble system results (Fig. 6·48*c*). Types IV and V are determined by the intermediate phase. If this phase decomposes on heating (incongruent melting), the diagram will show a peritectic reaction. If the intermediate

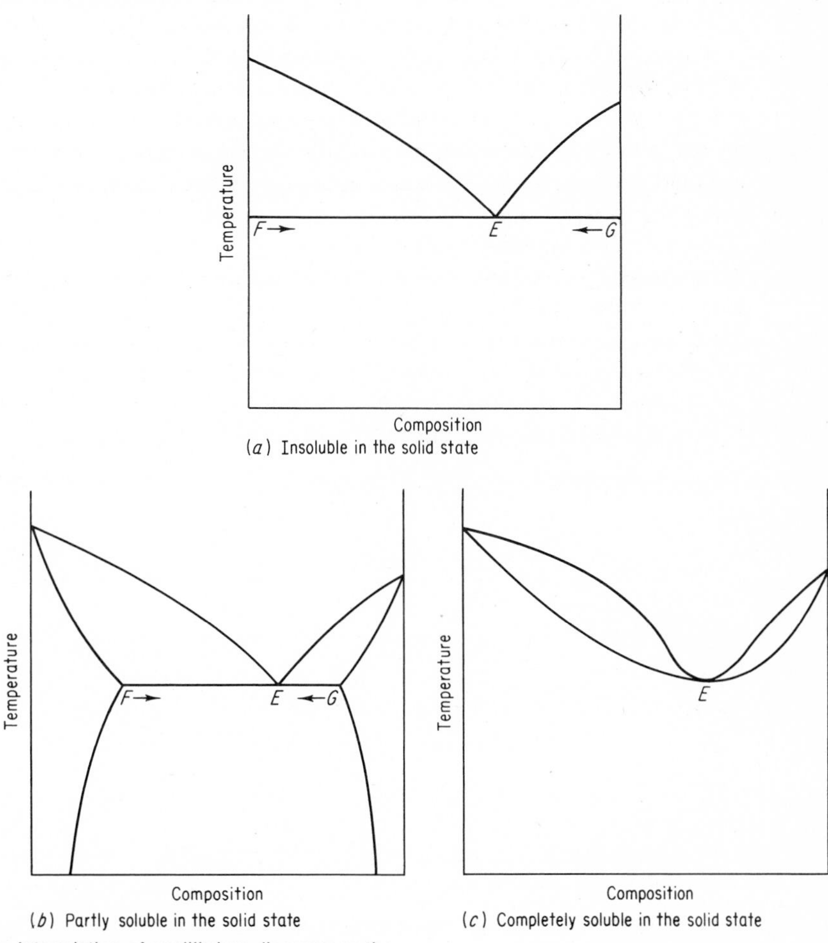

Fig. 6·48 Interrelation of equilibrium diagrams as the solubility in the solid state is varied.

phase shows a true melting point (congruent melting), the diagram may show a eutectic reaction.

TRANSFORMATIONS IN THE SOLID STATE

There are several equilibrium changes and reactions which take place entirely in the solid state.

6·22 Allotropy During the discussion of metals and crystal structure in Chap. 2, it was pointed out that several metals may exist in more than one type of crystal structure depending upon temperature. Iron, tin, manganese, and cobalt are examples of metals which exhibit this property, known as *allotropy.* On an equilibrium diagram, this allotropic change is indicated by a point or points on the vertical line which represents the pure metal. This is illustrated in Fig. 6·49. In this diagram, the gamma solid-solution field is "looped." The pure metal *A* and alloys rich in *A* undergo two transformations. Many of the equilibrium diagrams involving iron such as Fe-Si, Fe-Mo, and Fe-Cr show this looped solid-solution field. Since the type of iron that exists in this temperature range is gamma iron, the field is usually called the *gamma loop.*

In some alloy systems involving iron, the gamma loop is not closed. This is illustrated by the iron-nickel equlibrium diagram shown in Fig. 6·50. This diagram shows the freezing point of pure iron at 1539°C(2795°F), forming the δ solid solution, which is body-centered cubic. The γ solid solution is formed by a peritectic reaction at 1512°C(2757°F). Notice that for pure iron the allotropic change from δ (b.c.c.) crystal structure to the γ (f.c.c.) form occurs at 1400°C(2554°F), but for the alloy this change begins at a higher

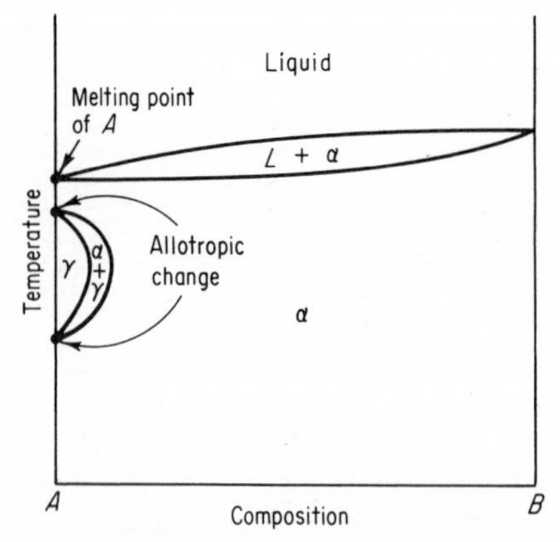

Fig. 6·49 Hypothetical equilibrium diagram showing metal *A* undergoing two allotropic changes.

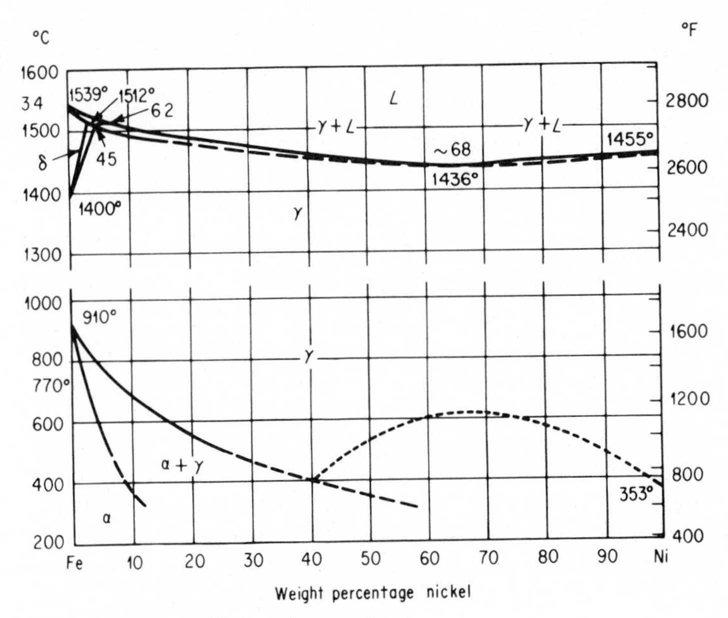

Fig. 6·50 The iron-nickel equilibrium diagram. (From Metals Handbook, 1948 ed., p. 1211, American Society for Metals, Metals Park, Ohio.)

temperature. The last allotropic change takes place at 910°C(1666°F), forming the α (b.c.c.) crystal structure.

6·23 Order-disorder Transformation Ordinarily in the formation of a substitutional type of solid solution the solute atoms do not occupy any specific position but are distributed at random in the lattice structure of the solvent. The alloy is said to be in a "disordered" condition. Some of these random solid solutions, if cooled slowly, undergo a rearrangement of the atoms where the solute atoms move into definite positions in the lattice. This structure is now known as an *ordered solid solution* or *superlattice* (Fig. 6·51). Ordering is most common in metals that are completely soluble in the solid state, and usually the maximum amount of ordering occurs at a simple atomic ratio of the two elements. For this reason, the ordered phase is sometimes given a chemical formula, such as AuCu and $AuCu_3$ in the gold-copper alloy system. On the equilibrium diagram, the ordered solutions are frequently designated as α', β', etc. or α', α'', etc., and the area in which they are found is usually bounded by a dot-dash line. The actual equilibrium diagram for the Au-Cu system is shown in Fig. 6·52.

When the ordered phase has the same lattice structure as the disordered phase, the effect of ordering on mechanical properties is negligible. Hard-

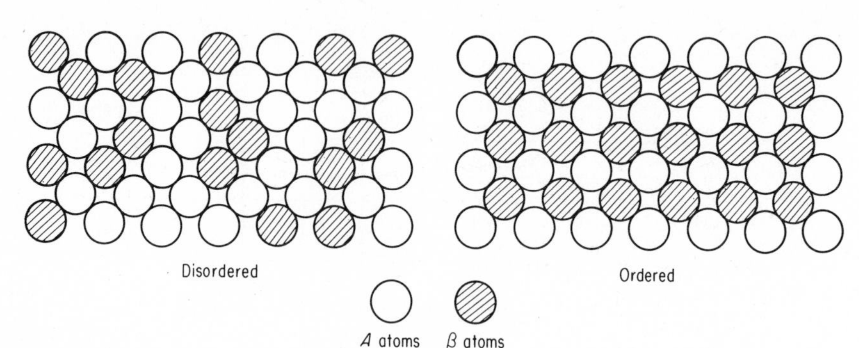

Fig. 6·51 Atomic arrangements in a disordered and ordered solid solution.

ening associated with the ordering process is most pronounced in those systems where the shape of the unit cell is changed by ordering. Regardless of the structure formed as a result of ordering, an important property change produced, even in the absence of hardening, is a significant reduction in electrical resistance (Fig. 6·53). Notice the sharp decrease in

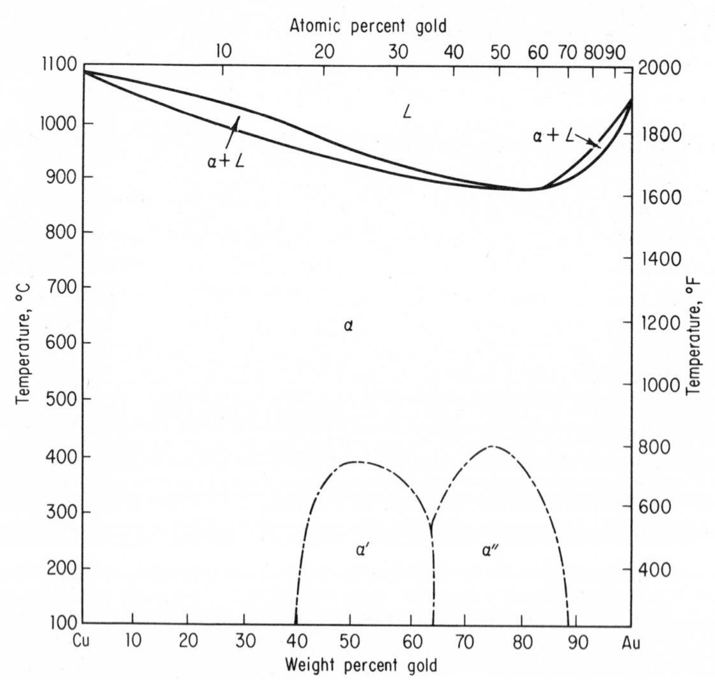

Fig. 6·52 The gold-copper equilibrium diagram. (From Metals Handbook, 1948., p. 1171, American Society for Metals, Metals Park, Ohio.)

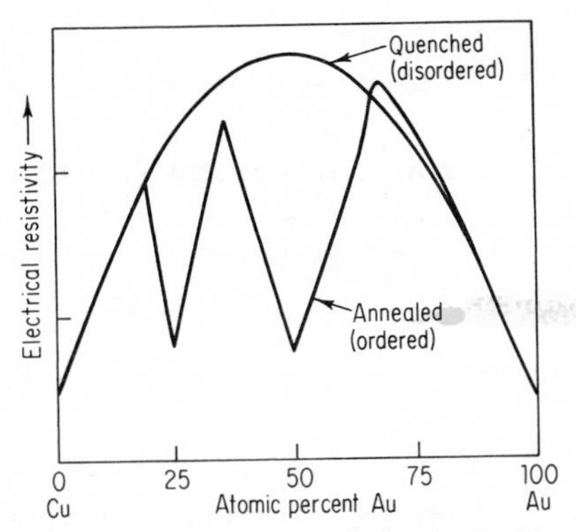

Fig. 6·53 Electrical resistivity vs. composition for the gold-copper system. (By permission from C. S. Barrett, "Structure of Metals," 2d ed., McGraw-Hill Book Company, New York, 1952.)

electrical resistivity at the compositions which correspond to the ordered phases $AuCu_3$ and AuCu. In the gold-copper equilibrium diagram there was no two-phase region between the disordered and ordered solid solutions. In some cases there will be a two-phase region between the ordered and disordered solid solutions. Most frequently this is associated with the formation of a crystal structure which is different from that of the disordered phase from which it forms. This is illustrated by the copper-palladium alloy system shown in Fig. 6·54. This diagram shows three ordered phases: α', α'', and β. Copper and palladium are both face-cen-

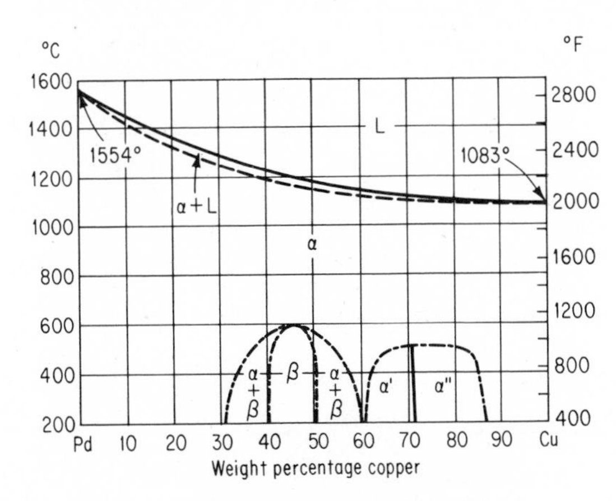

Fig. 6·54 The copper-palladium alloy system. (From Metals Handbook, 1948 ed., p. 1201, American Society for Metals, Metals Park, Ohio.)

tered cubic, and the ordered solutions α' and α'' are face-centered cubic; however, β is body-centered cubic and shows a two-phase region on each side.

6·24 The Eutectoid Reaction This is a common reaction in the solid state. It is very similar to the eutectic reaction but does not involve the liquid. In this case, a solid phase transforms on cooling into two new solid phases. The general equation may be written as

$$\text{Solid}_1 \underset{\text{heating}}{\overset{\text{cooling}}{\rightleftharpoons}} \underbrace{\text{solid}_2 + \text{solid}_3}_{\text{eutectoid mixture}}$$

The resultant eutectoid mixture is extremely fine, just like the eutectic mixture. Under the microscope both mixtures generally appear the same, and it is not possible to determine microscopically whether the mixture resulted from a eutectic reaction or a eutectoid reaction. An equilibrium diagram illustrating the eutectoid reaction is shown in Fig. 6·55.

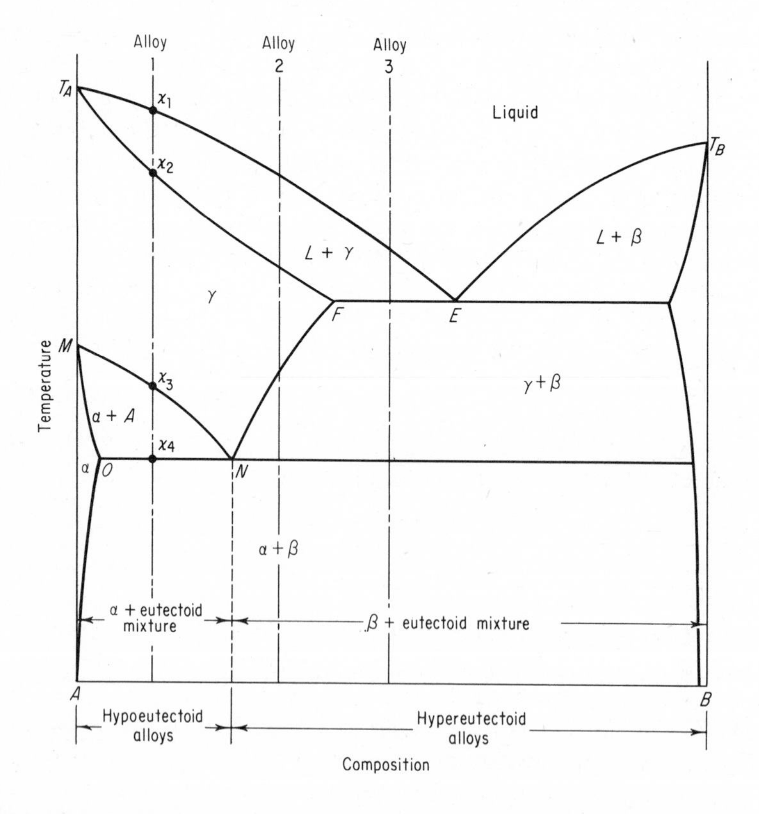

Fig. 6·55 Phase diagram illustrating the eutectoid reaction.

The liquidus line is T_AET_B and the solidus line is T_AFGT_B. The eutectic mixture is composed of the phases that occur at both ends of the eutectic temperature line, namely, γ solid solution (point F) and β solid solution (point G). Point M indicates an allotropic change for the pure metal A. The significance of the solvus line MN is that, as the alloy composition is increased in B, the temperature at which the allotropic change takes place is decreased, reaching a minimum at point N. The solvus line FN shows the decrease in solubility of B in γ as the temperature is decreased. Point N is known as the *eutectoid point.* Its composition is the eutectoid composition, and the line OP is the eutectoid temperature line. Like the eutectic diagram, it is common practice to call all alloys to the left of the eutectoid composition *hypoeutectoid alloys* and those to the right of point N *hypereutectoid alloys.*

When the hypoeutectoid alloy 1 is slow-cooled, γ solid solution is formed when the liquidus line is crossed at x_1. More and more γ is formed until the solidus line is crossed at x_2. It remains a uniform solid solution until the solvus line is crossed at x_3. The pure metal A must now start to undergo an allotropic change, forming the α solid solution. Notice that the α solid solution dissolves much less of B than does the γ solid solution. Some of the B atoms that are dissolved in the area that will undergo the allotropic change must now diffuse out of that area. When sufficient diffusion of B atoms has taken place, the remaining A atoms rearrange themselves into the new crystal structure, forming the α solid solution. The excess B atoms dissolve in the remaining γ solution, which becomes richer in B as the temperature falls. The composition of the remaining γ is gradually moving down and to the right along the solvus line MN. When the alloy reaches the eutectoid temperature x_4, the remaining γ has now reached the eutectoid point N. The significance of the eutectoid line is that this temperature is the end of the crystal structure change that started at x_3, and the remaining γ must now transform by the eutectoid reaction, forming alternate layers of α and β in an extremely fine mixture. The reaction may be written as

$$\gamma \underset{\text{heating}}{\overset{\text{cooling}}{\rightleftharpoons}} \underbrace{\alpha + \beta}_{\text{eutectoid mixture}}$$

The microstructure at room temperature consists of primary α or proeutectoid α which was formed between x_3 and x_4, surrounded by the eutectoid mixture of $\alpha + \beta$. This is shown in Fig. 6·56. In drawing the cooling curve for a given alloy from the phase diagram, it is important to remember that whenever a line is crossed on the phase diagram, there must be a corresponding break in the cooling curve. Also, when a horizontal line is crossed on the phase diagram, indicating a reaction, this will show on the cooling curve as a horizontal line. The cooling curve for alloy 1 is shown in Fig.

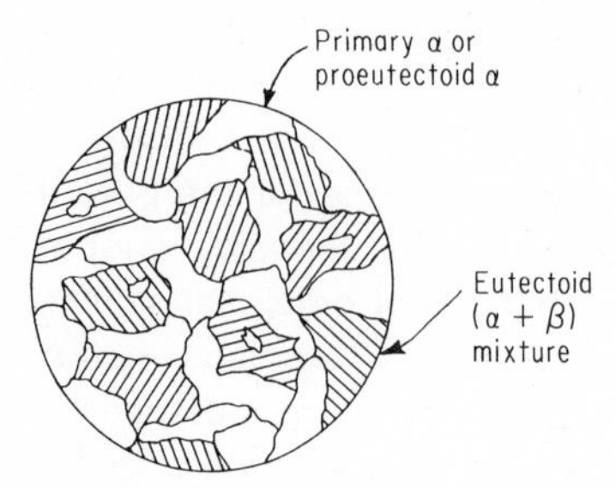

Fig. 6·56 Microstructure of a slow-cooled hypoeutectoid alloy, alloy 1 of Fig. 6·55.

6·57. The description of the slow cooling of the hypereutectoid alloys 2 and 3 is left as an exercise for the student.

6·25 The Peritectoid Reaction This is a fairly common reaction in the solid state and appears in many alloy systems. The peritectoid reaction may be written in general as

$$\text{Solid}_1 + \text{solid}_2 \underset{\text{heating}}{\overset{\text{cooling}}{\rightleftharpoons}} \text{new solid}_3$$

The new solid phase is usually an intermediate alloy, but it may also be a solid solution. The peritectoid reaction has the same relationship to the peritectic reaction as the eutectoid has to the eutectic. Essentially, it is the replacement of a liquid by a solid. Two hypothetical phase diagrams to illustrate the peritectoid reaction are shown in Figs. 6·58 and 6·59.

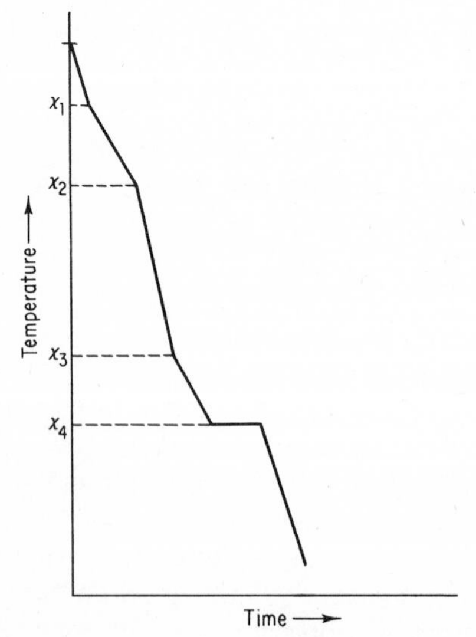

Fig. 6·57 Cooling curve for alloy 1 of Fig. 6·55.

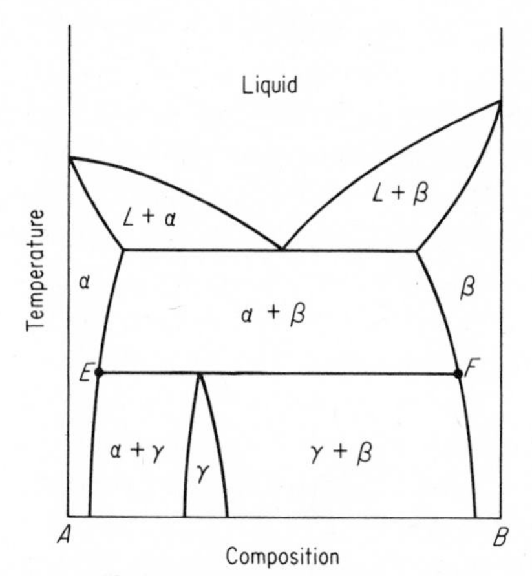

Fig. 6·58 Phase diagram showing the formation of an intermediate phase γ by a peritectoid reaction.

In Fig. 6·58, two solid phases α and β react at the peritectoid-temperature line *EF* to form an intermediate phase γ. The equation may be written as

$$\alpha + \beta \underset{\text{heating}}{\overset{\text{cooling}}{\rightleftharpoons}} \gamma$$

In Fig. 6·59, two solid phases, the pure metal *A* and β solid solution, react at the peritectoid-temperature line *CD* to form a new solid phase, the terminal solid solution γ. The equation may be written as

$$A + \beta \underset{\text{heating}}{\overset{\text{cooling}}{\rightleftharpoons}} \gamma$$

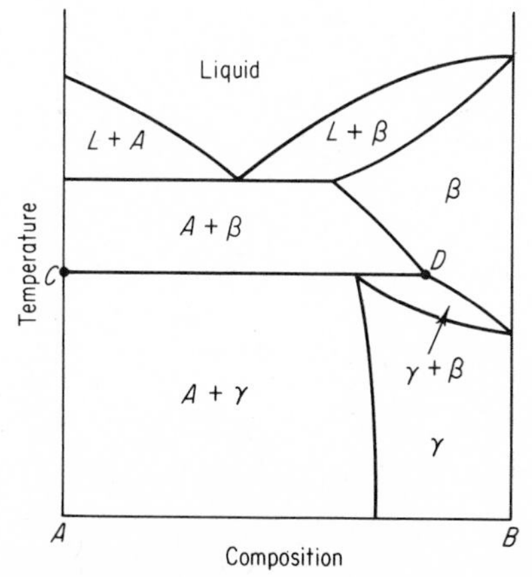

Fig. 6·59 Phase diagram showing the formation of the terminal solid solution γ by a peritectoid reaction.

It was pointed out in the discussion of the peritectic reaction that the microstructure of a peritectic alloy rarely shows complete transformation (Fig. 6·43). This is because diffusion through the new phase is required in order to reach equilibrium. Since the peritectoid reaction occurs entirely in the solid state and usually at lower temperatures than the peritectic reaction, the diffusion rate will be slower and there is less likelihood that equilibrium structures will be reached. Figure 6·60 shows a portion of the silver-aluminum phase diagram containing a peritectoid reaction. If a 7 percent aluminum alloy is rapidly cooled from the two-phase area just above the peritectoid temperature, the two phases will be retained, and the microstructure will show a matrix of γ with just a few particles of α (Fig. 6·61*a*). If the same alloy is slow-cooled to just below the peritectoid temperature and held there for 20 min before rapid cooling, some transformation will take place. The microstructure, Fig. 6·61*b*, shows that some γ has transformed to the new phase β, while much of the original α remains. The phase diagram indicates that there should be only a single-phase β, so obviously equilibrium has not been achieved. Even after holding for 2 hr at just below the peritectoid temperature, a single-phase structure is still not produced (Fig. 6·61*c*).

The similarity, in both the general equation and the appearance on an equilibrium diagram, for the monotectic, eutectic, and eutectoid reactions and the peritectic and peritectoid reactions is apparent from a study of Table 6·4. These reactions are by no means the only ones that may occur on equilibrium diagrams. However, they are by far the most common ones, and the student should become familiar with them.

6·26 Complex Diagrams Some of the equilibrium diagrams discussed under the simple types are the same as actual ones. Many alloy systems have diagrams which show more than one type of reaction and are generally more complex than the simple types. However, even the most complex diagrams show mainly the reactions that have been covered. The student

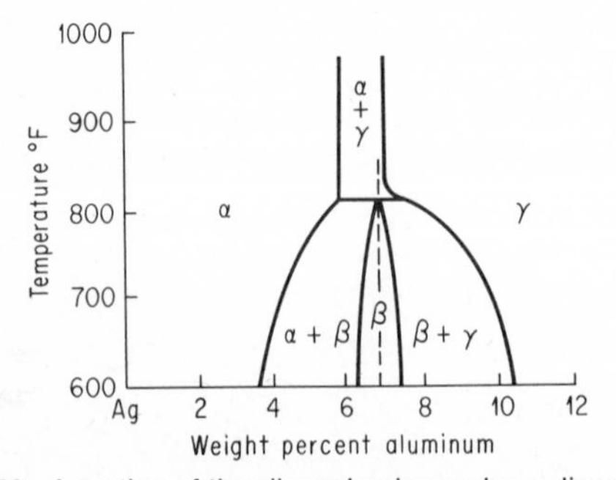

Fig. 6·60 A portion of the silver-aluminum phase diagram.

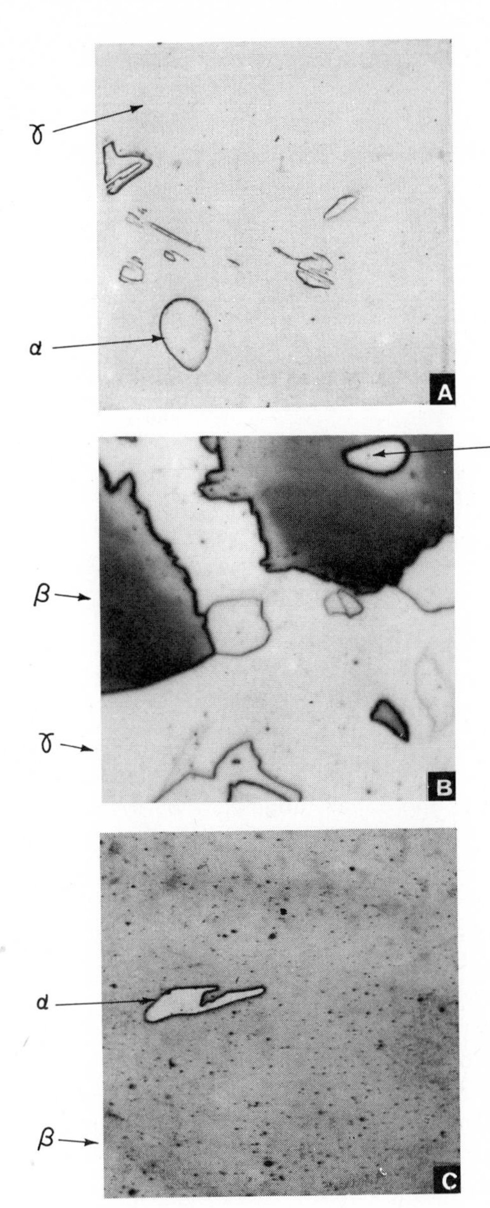

Fig. 6·61 (a) A peritectoid alloy, Ag + 7 percent Al, stabilized by long heat treatment above the peritectoid temperature and then quenched. "Islands" are the γ phase embedded in a matrix of γ. Magnification 150X. (b) Same as (a), cooled to a temperature slightly below that of the peritectoid and held 20 min before quenching. Much of the light-colored γ has transformed to the dark-colored β without greatly affecting the islands of γ. Magnification 150X. (c) Same as (a), cooled to a temperature slightly below the peritectoid and held for 2 hr before quenching. Dark matrix is β; light area is residual α that has not yet been dissolved by the β. Magnification 150X. (By permission from F. N. Rhines, "Phase Diagrams in Metallurgy," McGraw-Hill Book Company, New York, 1956.)

TABLE 6·4 Equilibrium-diagram Reactions

NAME OF REACTION	GENERAL EQUATION	APPEARANCE ON DIAGRAM
Monotectic	$L_1 \underset{\text{heating}}{\overset{\text{cooling}}{\rightleftharpoons}} L_2 + \text{solid}$	L_1 over L_2 + solid
Eutectic	$\text{Liquid} \underset{\text{heating}}{\overset{\text{cooling}}{\rightleftharpoons}} \text{solid}_1 + \text{solid}_2$	L over Solid$_1$ + solid$_2$
Eutectoid	$\text{Solid}_1 \underset{\text{heating}}{\overset{\text{cooling}}{\rightleftharpoons}} \text{solid}_2 + \text{solid}_3$	Solid$_1$ over Solid$_2$ + solid$_3$
Peritectic	$\text{Liquid} + \text{solid}_1 \underset{\text{heating}}{\overset{\text{cooling}}{\rightleftharpoons}} \text{new solid}_2$	Liquid + solid$_1$ over New solid$_2$
Peritectoid	$\text{Solid}_1 + \text{solid}_2 \underset{\text{heating}}{\overset{\text{cooling}}{\rightleftharpoons}} \text{new solid}_3$	Solid$_1$ + solid$_2$ over New solid$_3$

should be able to label a phase diagram completely; understand the significance of every point, line, and area; determine the various reactions that occur at the horizontal lines; and describe the slow cooling and microstructure of any alloy on a binary equilibrium diagram. The application of some of the principles in this chapter will now be illustrated for a complex equilibrium diagram such as the cobalt-tungsten alloy system shown in Fig. 6·62.

The freezing point of cobalt is shown on the left axis as 1495°C. The freezing point of tungsten, 3410°C, is where the two dotted lines on the right would meet and is above the range of temperatures shown. Since two-phase areas must be bounded by single phases on either side of a horizontal line, it is necessary to first label the single-phase areas. The upper line on the diagram is the liquidus line, so above the liquidus line there is a single homogeneous liquid solution indicated by L. On the left, from the freezing point of cobalt, there is a very thin, cigar-shaped area which resembles Type I. Therefore, there must be a solid solution formed below this area. The solid solution is labeled β. Once the β area has been labeled, the portion between the very thin lines is labeled as $\beta + L$. Next to this area is a portion of the diagram which looks like a variation of Type I and shows a maximum at 1500°C. The area between the points 35, 45, and 1500°C would also be $\beta + L$. Along the right axis, the dotted line indicates a small solid solution area which is labeled as epsilon, ϵ. Now the area above 1690°C may be labeled as $L + \epsilon$. The first horizontal line is at 1690°C. Above the line there is $L + \epsilon$. Below the line a new solid solution δ appears. The student should recognize that this is a peritectic reaction. After the δ region has been labeled, the area to the right may be labeled as $\delta + \epsilon$. The area to the left of the δ region above 1465°C may now be labeled as $L + \delta$. The area between 1100° and 1465° may also be labeled as $\beta + \delta$. The students should recognize the eutectic point at 45 percent *W* and 1465°C. Above the line is a liquid solution and below the line two solids β and δ. At 1100°C is another horizontal line and therefore, another reaction. Above the line are two solids δ and β. Below the line a new phase appears, an intermediate alloy phase, which is labeled γ. This is a peritectoid reaction. Once the γ region has been labeled, the area to the right becomes $\gamma + \delta$ and the area to the left $\beta + \gamma$. The point *A* on the vertical line representing cobalt must be an allotropic change. Below point *A* there is a very tiny amount of solubility of tungsten in cobalt. The solid solution area is too small to be shown on the diagram, but it is labeled α as indicated. The triangle *AB*3 may now be labeled as $\alpha + \beta$, although it is not labeled in the diagram. The dotted horizontal line at 350°C indicates another reaction. This is the eutectoid reaction, with the eutectoid point at 3 percent *W* and 350°C. Above the line is β solid solution, while below the line are two solid solutions $\alpha + \gamma$. The diagram is now completely labeled. Two intermediate alloy phases are shown on the diagram, γ and δ. Since their range of composition is not very great, they are most likely intermetallic compounds. The γ phase appears to correspond to the formula Co_7W_2 and the δ phase to the formula CoW. Another significant fact that may be determined from the diagram is the very large difference in solubility of tungsten in cobalt depending upon the type of

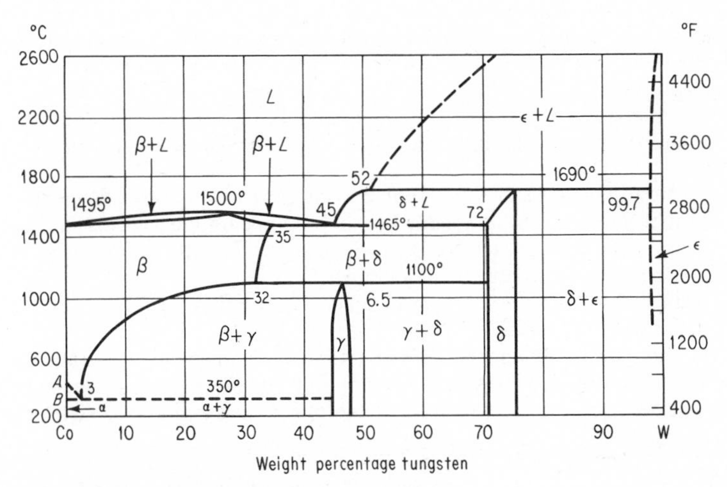

Fig. 6·62 The cobalt-tungsten alloy system. (From Metals Handbook, 1948 ed., p. 1193, American Society for Metals, Metals Park, Ohio.)

crystal structure. In the β solid solution, where cobalt is face-centered cubic, the maximum solubility of tungsten in cobalt is given as 35 percent at 1465°C. However, once the allotropic change has occurred and cobalt becomes close-packed hexagonal below point A, the solubility of tungsten in cobalt is almost negligible. The reactions and specific equations that occur at each horizontal line on this diagram are given below.

TEMPERATURE	REACTION	EQUATION
1690°C	Peritectic	$L + \epsilon \xrightleftharpoons{\text{cooling}} \delta$
1465°C	Eutectic	$L \xrightleftharpoons{\text{cooling}} \beta + \delta$
1100°C	Peritectoid	$\beta + \delta \xrightleftharpoons{\text{cooling}} \gamma$
350°C	Eutectoid	$\beta \xrightleftharpoons{\text{cooling}} \alpha + \gamma$

QUESTIONS

6·1 What information may be obtained from an equilibrium diagram?

6·2 Explain the importance of equilibrium diagrams in the development of new alloys.

6·3 Platinum and gold are completely soluble in both the liquid and solid states. The melting point of platinum is 3225°F and that of gold is 1945°F. An alloy containing 40 percent gold starts to solidify at 2910°F by separating crystals of 15 percent gold. An alloy containing 70 percent gold starts to solidify at 2550°F by separating crystals of 37 percent gold.

a Draw the equilibrium diagram to scale on a piece of graph paper and label all points, lines, and areas.

b For an alloy containing 70 percent gold (1) give the temperature of initial solidification; (2) give the temperature of final solidification; (3) give the chemical composition and relative amounts of the phases present at 2440°F; (4) draw the cooling curve.

6·4 Bismuth and antimony are completely soluble in both the liquid and solid states.

a Check the crystal-structure factor and calculate the relative-size factor for these metals.

b Bismuth melts at 520°F and antimony melts at 1170°F. An alloy containing 50 percent bismuth starts to solidify at 940°F by separating crystals of 90 percent antimony. An alloy containing 80 percent bismuth starts to solidify at 750°F by separating crystals of 75 percent antimony.

1 Draw the equilibrium diagram to scale on a piece of graph paper labeling all lines, points, and areas.

2 For an alloy containing 40 percent antimony, (*a*) give the temperature of initial solidification; (*b*) give the temperature of final solidification; (*c*) give the chemical composition and relative amounts of the phases present at 800°F; (*d*) draw the cooling curve.

6·5 Bismuth (melting point 520°F) and cadmium (melting point 610°F) are assumed to be completely soluble in the liquid state and completely insoluble in the solid state. They form a eutectic at 290°F containing 40 percent cadmium.

1 Draw the equilibrium diagram to scale on a piece of graph paper labeling all points, lines, and areas.

2 For an alloy containing 70 percent cadmium (*a*) give the temperature of initial solidification; (*b*) give the temperature of final solidification; (*c*) give the chemical composition and relative amounts of the phases present at a temperature of 100°F below (*a*); (*d*) sketch the microstructure at room temperature; (*e*) draw the cooling curve.

3 Same as part 2 but for an alloy containing 10 percent cadmium.

6·6 Lead melts at 620°F and tin melts at 450°F. They form a eutectic containing 62 percent tin at 360°F. The maximum solid solubility of tin in lead at this temperature is 19 percent; of lead in tin, 3 percent. Assume the solubility of each at room temperature is 1 percent.

1 Draw the equilibrium diagram to scale on a piece of graph paper labeling all points, lines, and areas.

2 Describe the solidification of a 40 percent tin alloy. Sketch its microstructure at room temperature, giving the chemical composition and relative amounts of the phases present.

3 Draw the cooling curve for the above alloy.

4 Repeat 2 and 3 for an alloy containing 90 percent tin.

6·7 Calcium (melting point 1560°F) and magnesium (melting point 1200°F) form a compound $CaMg_2$ which contains 45 percent calcium and melts at 1320°F. This compound forms a eutectic with pure magnesium at 960°F and contains 16 percent calcium. The solubility of the compound in magnesium is about 2 percent at the eutectic temperature and decreases to almost zero at room temperature. Magnesium is not soluble in the compound. A second eutectic is formed between the compound and calcium at 830°F containing 78 percent calcium, and there is no solid solubility between the compound and pure calcium.

1 Draw the equilibrium diagram to scale on a piece of graph paper labeling all points, lines, and areas.

2 Describe the slow cooling of an alloy containing 30 percent calcium. Sketch the microstructures at room temperature and give the relative amounts of the phases present.

3 Draw the cooling curve.

4 Write the specific equation of the reaction that takes place at each eutectic temperature.

6·8 **1** Label Fig. 6·63 completely.

2 Write the specific equation of the reaction that takes place at each horizontal line.

3 Sketch the microstructure of alloy *x* when slow-cooled to room temperature.

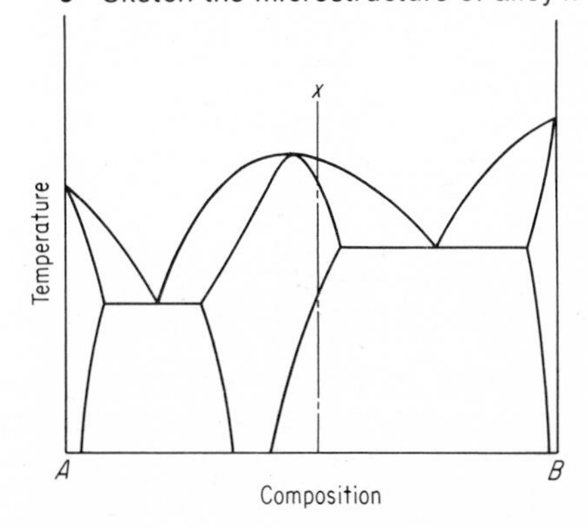

Fig. 6·63

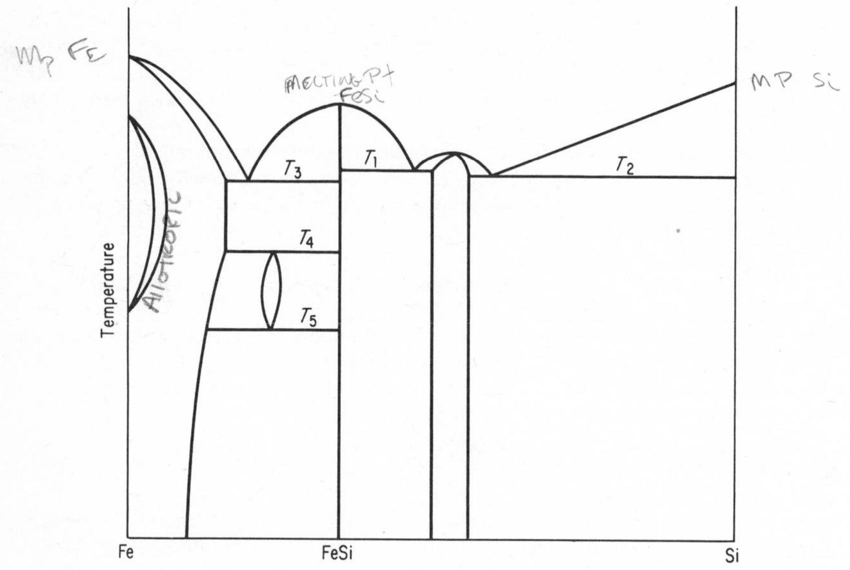

Fig. 6·64 The modified iron-silicon system.

6·9 **1** Describe the magnesium-nickel system so that it could be plotted from the description (see Metals Handbook, 1948 ed., p. 1225).

2 Plot this diagram to scale on a piece of graph paper and label all the areas.

3 Write the reaction that takes place at each horizontal line.

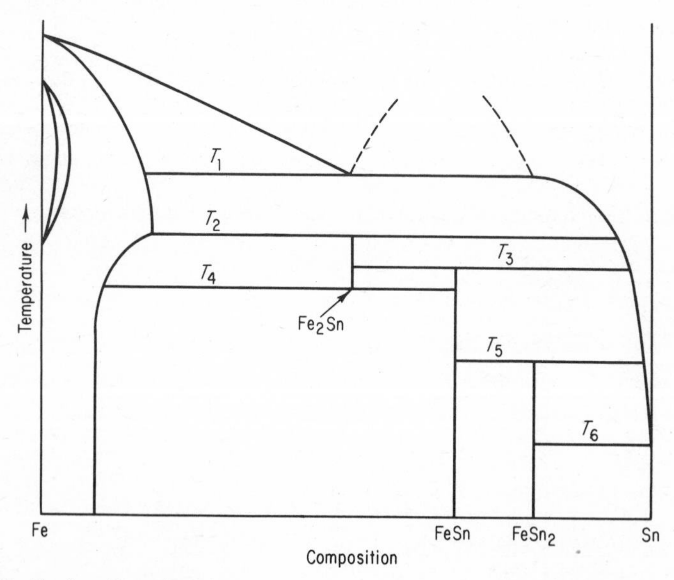

Fig. 6·65 The iron-tin system.

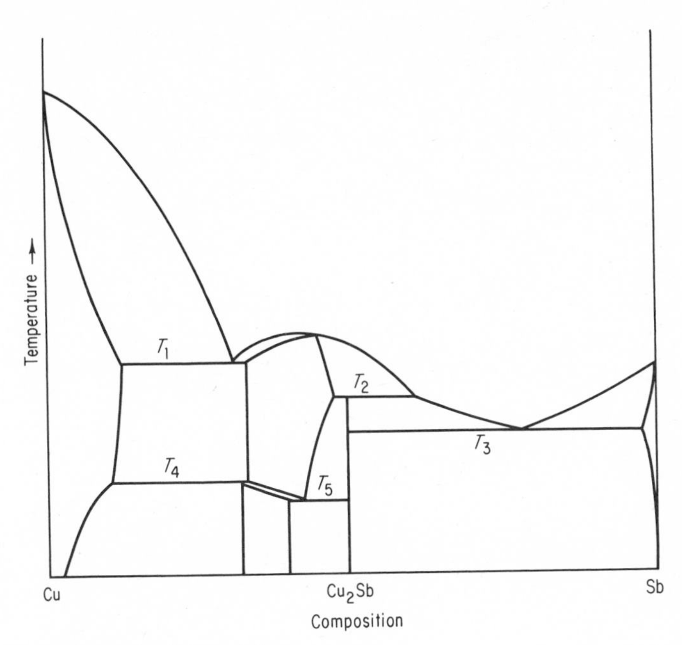

Fig. 6·66 The copper-antimony system.

4 Draw the cooling curve of a 40 percent magnesium alloy.

5 Describe the slow cooling of this alloy, sketch its microstructure at room temperature, and give the relative amounts of the phases present.

6·10 **1** Label completely the phase diagrams given in Figs. 6·64 to 6·66.

2 Give the name and write the specific equation of the reaction that takes place at each horizontal line.

3 Discuss the significance of each line on the diagrams.

REFERENCES

American Society for Metals: "Metals Handbook," 1948 ed., Metals Park, Ohio.

———: "Metals Handbook," 8th ed., 1973, Metals Park, Ohio.

Gordon, P.: "Principles of Phase Diagrams in Materials Systems," McGraw-Hill Book Company, New York, 1968.

Guy, A. G.: "Elements of Physical Metallurgy," Addison-Wesley Publishing Company, Inc., Reading, Mass., 1959.

Hansen, M., and K. Anderko: "Constitution of Binary Alloys," 2d ed., McGraw-Hill Book Company, New York, 1958.

Hume-Rothery, W., J. W. Christian, and W. B. Pearson: "Metallurgical Equilibrium Diagrams," The Institute of Physics, London, 1952.

———, R. E. Smallman, and C. W. Haworth: "The Structure of Metals and Alloys," Institute of Metals, London, 1969.

Marsh, J. S.: "Principles of Phase Diagrams," McGraw-Hill Book Company, New York, 1935.

Mondolfo, L. F., and O. Zmeskal: "Engineering Metallurgy," McGraw-Hill Book Company, New York, 1955.

Rhines, F. H.: "Phase Diagrams in Metallurgy." McGraw-Hill Book Company, New York, 1956.

Van Vlack, L. H.: "Elements of Materials Science," Addison-Wesley Publishing Company, Inc., Reading, Mass., 1959.

THE IRON–IRON CARBIDE EQUILIBRIUM DIAGRAM

7·1 Introduction The metal iron is a primary constituent of some of the most important engineering alloys. In an almost pure form, known as *ingot iron*, it is used for drainage culverts, roofing, and ducts, and as a base for porcelain enamel in refrigerator cabinets, stoves, washing machines, etc. A typical analysis for ingot iron is:

carbon	0.012 percent
manganese	0.017
phosphorus	0.005
sulfur	0.025
silicon	trace

Typical mechanical properties of ingot iron are as follows:

Tensile strength	40,000 psi
Elongation in 2 in.	40 percent
Rockwell B hardness	30

Iron is an allotropic metal, which means that it can exist in more than one type of lattice structure depending upon temperature. A cooling curve for pure iron is shown in Fig. 7·1.

When iron first solidifies at 2800°F, it is in the b.c.c. (body-centered cubic) δ (delta) form. Upon further cooling, at 2554°F, a phase change occurs and the atoms rearrange themselves into the γ (gamma) form, which is f.c.c. (face-centered cubic) and nonmagnetic. When the temperature reaches 1666°F, another phase change occurs from f.c.c. nonmagnetic γ iron to b.c.c. nonmagnetic α (alpha) iron. Finally, at 1414°F, the α iron becomes magnetic without a change in lattice structure. Originally, nonmagnetic α iron was called β iron until subsequent x-ray studies showed no change in lattice structure at 1414°F. Since this magnetic transformation does not affect the heat treatment of iron-carbon alloys, it will be disre-

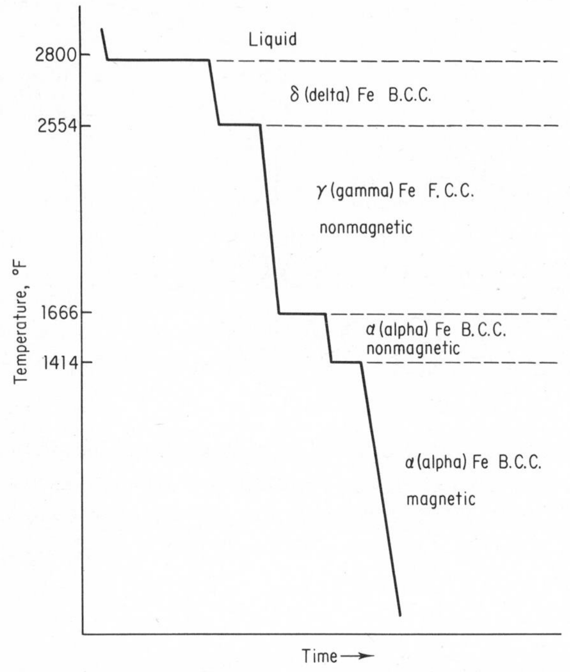

Fig. 7·1 Cooling curve for pure iron.

garded in our discussion. All the allotropic changes give off heat (exothermic) when iron is cooled and absorb heat (endothermic) when iron is heated.

7·2 Manufacture of Wrought Iron Wrought iron is essentially a two-component metal consisting of high-purity iron and slag. The slag is composed mainly of iron silicate. The small and uniformly distributed particles of slag exist physically separate in the iron. There is no fusion or chemical relationship between the slag and the iron.

Wrought iron was originally produced by the hand-puddling process, later by mechanical puddling, and since 1930 by the Byers or Aston process. Regardless of the process, there are three essential steps in the manufacture of wrought iron: first, to melt and refine the base metal; second, to produce and keep molten a proper slag; and third, to granulate, or disintegrate, the base metal and mechanically incorporate with it the desired amount of slag.

In the Byers process each step is separated and carried out in individual pieces of equipment. The raw materials of pig iron, iron oxide, and silica are melted in cupolas. The pig iron is purified to a highly refined state in a bessemer converter and then transferred to the ladle of the processing machine. An exact iron silicate slag made by melting together iron oxide and certain siliceous materials in an open-hearth furnace is poured into a ladle and moved directly below the processing machine.

The next step is the key operation of the process—that of base-metal disintegration and slag incorporation. The liquid refined iron at a temperature of about 2800°F is poured at a predetermined rate into the ladle containing the molten slag, which is at about 2300°F (Fig. 7·2). To ensure a uniform distribution of the refined metal into the slag, the processing machine automatically oscillates as well as moves forward and backward. Since the slag is maintained at a temperature considerably lower than the freezing point of the iron, the iron is continuously and rapidly solidified. The liquid iron contains large quantities of gases in solution, but when the metal solidifies, the gases are no longer soluble in it. This rapid solidification liberates the gases in the form of many small explosions of sufficient force to shatter the metal into small fragments which settle to the bottom of the slag ladle. Because of the noise of the explosions, this operation is called *shotting.* Since the iron is at a welding temperature, and because of the fluxing action of the siliceous slag, these fragments stick

Fig. 7·2 The key operation in the manufacture of wrought iron. (A. M. Byers Company.)

together to form a spongelike ball of iron globules coated with silicate slag.

The excess slag is poured off and the sponge ball, weighing between 6,000 and 8,000 lb, is placed in a press. The press squeezes out the surplus slag and welds the cellular mass of slag-coated particles of plastic iron into a bloom. The bloom is reduced in cross section to a billet, which is reheated and rolled into plate, bars, rods, tubing, etc.

7·3 Properties and Applications of Wrought Iron Quality wrought iron is distinguished by its low carbon and manganese contents. The carbon content is generally below 0.08 percent and the manganese content below 0.06 percent. The phosphorous content is usually higher than that of steel and ordinarily ranges from 0.05 to 0.160 percent. The sulfur content is kept low, and the silicon content of between 0.10 and 0.20 percent is concentrated almost entirely in the slag. The slag content usually varies from about 1 to 3 percent by weight.

A typical chemical analysis of wrought iron is as follows:

	PERCENT
Carbon	0.06
Manganese	0.045
Silicon	0.101
Phosphorus	0.068
Sulfur	0.009
Slag, by weight	1.97

Since wrought iron is a composite material, there are many methods of distinguishing between wrought iron and steel. Figure 7·3 shows the typical fibrous fracture of wrought iron; steel, on the other hand, shows a crystalline or granular break.

The uniform distribution of the slag throughout the ferrite matrix is clearly shown by microscopic examination of a transverse section (Fig.

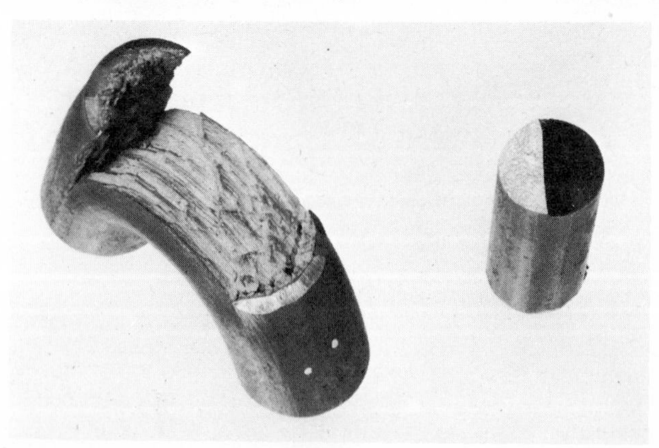

Fig. 7·3 (Left) Fibrous fracture of wrought iron; (right) crystalline fracture of steel.

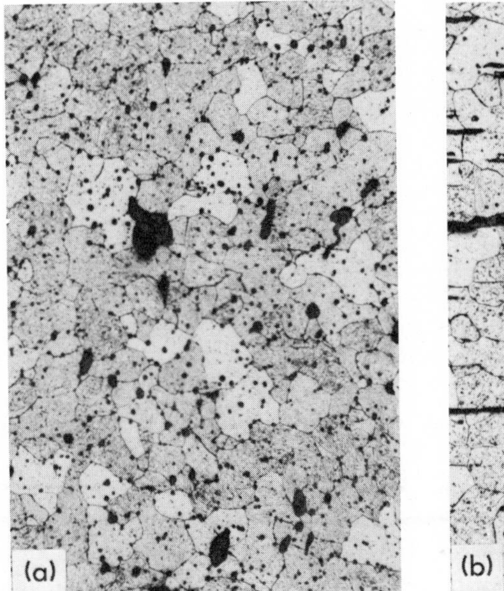

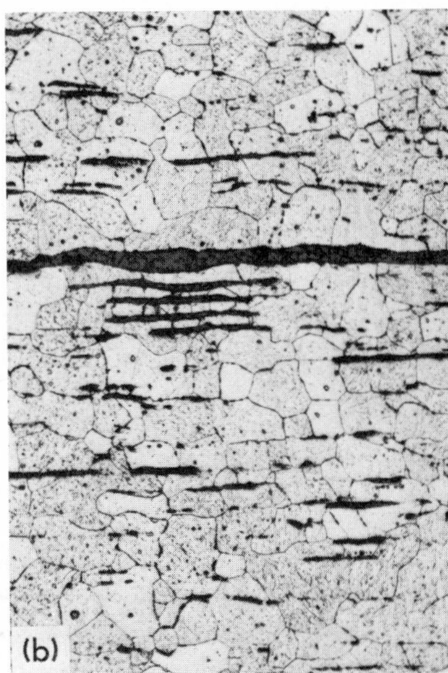

Fig. 7·4 The microstructure of wrought iron. Slag in a ferrite matrix. (*a*) Transverse section; (*b*) longitudinal section. Etched in 2 percent nital; 100X.

7·4*a*). The threadlike appearance of the slag is evident from microscopic examination of a longitudinal section, that is, a section parallel to the direction of rolling (Fig. 7·4*b*).

The mechanical properties of wrought iron are largely those of pure iron. Because of the nature of the slag distribution, however, the tensile strength and ductility are greater in the longitudinal or rolling direction than in the direction transverse to rolling. Typical mechanical properties of wrought iron in the longitudinal and transverse directions are given in Table 7·1. Improvement of rolling procedure has made possible the equalization of the tensile strength and ductility in both directions.

It is possible to improve the strength of wrought iron by alloying. The most popular alloy wrought irons are those containing between 1.5 and 3.5

TABLE 7·1 Tensile Properties of Wrought Iron

PROPERTY	LONGITUDINAL	TRANSVERSE
Tensile strength, psi	48,000–50,000	36,000–38,000
Yield point, psi	27,000–30,000	27,000–30,000
Elongation, % in 8 in.	18–25	2–5
Reduction in area, %	35–45	3–6

percent nickel. The comparative mechanical properties of unalloyed and nickel wrought iron are given in Table 7·2.

Charpy impact tests reveal that nickel-alloy wrought iron retains its impact strength to a high degree at subzero temperatures.

One of the principal virtues of wrought iron is its ability to resist corrosion. When exposed to corrosive media, it is quickly coated with an oxide film. As corrosion continues, the slag fibers begin to function as rust resistors. The dense, uniform, initial oxide film is securely fastened to the surface of the metal by the pinning effect of the slag fibers and protects the surfaces from further oxidation.

Wrought iron is used for standard pipe, nails, barbed wire, rivets, and welding fittings. It is available in plates, sheets, tubular forms, and structural shapes. Wrought iron has many applications in the railroad, shipbuilding, and oil industries, as well as for architectural purposes and for farm implements.

7·4 The Iron–Iron Carbide Diagram The temperature at which the allotropic changes take place in iron is influenced by alloying elements, the most important of which is carbon. The portion of the iron-carbon alloy system which is of interest is shown in Fig. 7·5. This is the part between pure iron and an interstitial compound, iron carbide, Fe_3C, containing 6.67 percent carbon by weight. Therefore, we will call this portion the iron–iron carbide equilibrium diagram. Before going into a study of this diagram, it is important for the student to understand that this is not a true equilibrium diagram, since equilibrium implies no change of phase with time. It is a fact, however, that the compound iron carbide will decompose into iron and carbon (graphite). This decomposition will take a very long time at room temperature, and even at 1300°F it takes several years to form graphite. Iron carbide is called a *metastable* phase. Therefore, the iron–iron carbide diagram, even though it technically represents metastable conditions, can be considered as representing equilibrium changes, under conditions of relatively slow heating and cooling.

The diagram shows three horizontal lines which indicate isothermal reactions. Figure 7·5 has been labeled in general terms with Greek letters to represent the solid solutions. However, it is common practice to give special names to most of the structures that appear on the diagram. The

TABLE 7·2 Tensile Properties of Unalloyed and Nickel Wrought Iron

PROPERTY	UNALLOYED WROUGHT IRON	NICKEL WROUGHT IRON
Tensile strength, psi	48,000	60,000
Yield point, psi	30,000	45,000
Elongation, % in 8 in.	25	22
Reduction in area, %	45	40

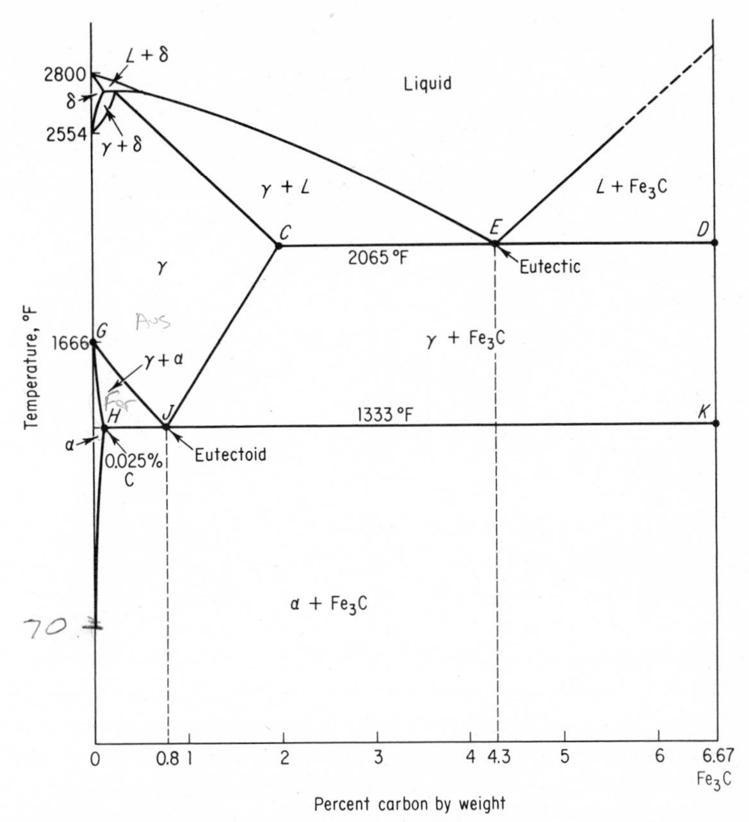

Fig. 7·5 The iron–iron carbide equilibrium diagram labeled in general terms.

γ solid solution is called *austenite.* The portion of the diagram in the upper left-hand corner is expanded in Fig. 7·6. This is known as the *delta region* because of the δ solid solution. The student should recognize the horizontal line at 2720°F as being a peritectic reaction. The equation of the peritectic reaction may be written as

$$\text{Liquid} + \delta \underset{\text{heating}}{\overset{\text{cooling}}{\rightleftharpoons}} \text{austenite}$$

The maximum solubility of carbon in b.c.c. δ Fe is 0.10 percent (point *M*), while in f.c.c. γ Fe the solubility is much greater. The presence of carbon influences the $\delta \rightleftharpoons \gamma$ allotropic change. As carbon is added to iron, the temperature of the allotropic change increases from 2554 to 2720°F at 0.10 percent C. Consider the significance of the line *NMPB*. On cooling, the portion

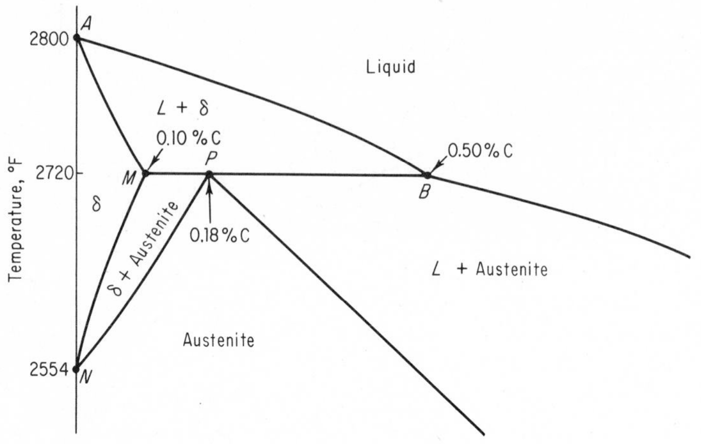

Fig. 7·6 The delta region of the iron–iron carbide diagram.

NM represents the beginning of the crystal structure change from b.c.c. δ Fe to f.c.c. γ Fe for alloys containing less than 0.10 percent C. The portion *MP* represents the beginning of this crystal structure change by means of a peritectic reaction for alloys between 0.10 and 0.18 percent C. For alloys containing less than 0.18 percent C, on cooling, the end of the crystal structure change is given by the line *NP*. The portion *PB* represents the beginning and the end of the crystal structure change by means of the peritectic reaction. In other words, for alloys between 0.18 and 0.50 percent C, the allotropic change begins and ends at a constant temperature. Notice that any alloy containing more than 0.50 percent C will cut the diagram to the right of point *B* and will solidify austenite directly. The delta solid solution and the allotropic change will be completely bypassed. Since no commercial heat treatment is done in the delta region, there will be no reason to refer to this portion of the diagram again.

The diagram in Fig. 7·7, which has the common names inserted, shows a eutectic reaction at 2065°F. The eutectic point, *E*, is at 4.3 percent C and 2065°F. Since the horizontal line *CED* represents the eutectic reaction, whenever an alloy crosses this line the reaction must take place. Any liquid that is present when this line is reached must now solidify into the very fine intimate mixture of the two phases that are at either end of the horizontal line, namely austenite and iron carbide (called *cementite*). This eutectic mixture has been given the name *ledeburite*, and the equation may be written as

$$\text{Liquid} \underset{\text{heating}}{\overset{\text{cooling}}{\rightleftharpoons}} \underbrace{\text{austenite} + \text{cementite}}_{\text{eutectic mixture—ledeburite}}$$

The eutectic mixture is not usually seen in the microstructure, since austenite is not stable at room temperature and must undergo another reaction during cooling.

There is a small solid solution area to the left of line *GH*. We know that 1666°F represents the change in crystal structure of pure iron from f.c.c. γ to b.c.c. α. That area is a solid solution of a small amount of carbon dissolved in b.c.c. α Fe and is called *ferrite*. The diagram shows a third horizontal line *HJK*, which represents a eutectoid reaction. The eutectoid point, *J*, is at 0.80 percent C and 1333°F. Any austenite present must now transform into the very fine eutectoid mixture of ferrite and cementite, called pearlite. The equation may be written as

$$\text{Liquid} \underset{\text{heating}}{\overset{\text{cooling}}{\rightleftharpoons}} \underbrace{\text{ferrite} + \text{cementite}}_{\text{eutectoid mixture—pearlite}}$$

Below the eutectoid temperature line every alloy will consist of a mixture of ferrite and cementite as indicated.

On the basis of carbon content it is common practice to divide the iron–iron carbide diagram into two parts. Those alloys containing less than 2 percent carbon are known as *steels*, and those containing more than 2 percent carbon are known as *cast irons*. The steel range is further sub-

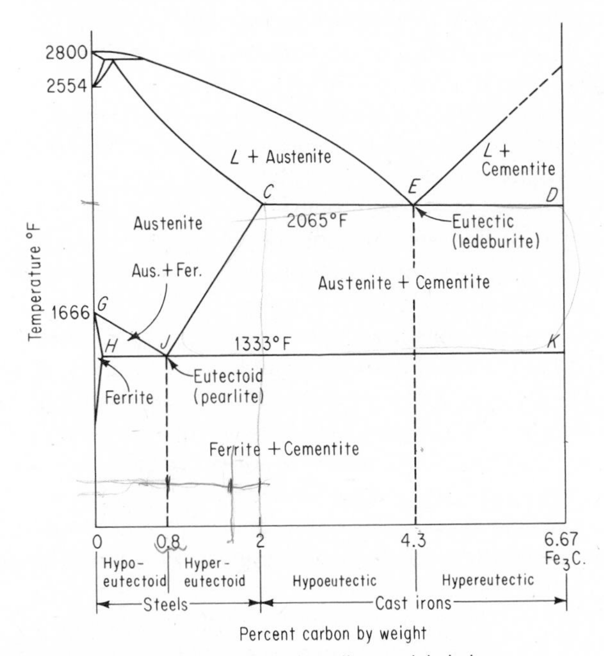

Fig. 7·7 The iron–iron carbide equilibrium diagram labeled with the common names for the structures.

divided by the eutectoid carbon content (0.8 percent C). Steels containing less than 0.8 percent C are called *hypoeutectoid steels*, while those containing between 0.8 and 2.0 percent C are called *hypereutectoid steels*. The cast iron range may also be subdivided by eutectic carbon content (4.3 percent C). Cast irons that contain less than 4.3 percent C are known as *hypoeutectic cast irons*, whereas those that contain more than 4.3 percent C are called *hypereutectic cast irons*.

7·5 Definition of Structures The names which, for descriptive or commemorative reasons, have been assigned to the structures appearing on this diagram will now be defined.

Cementite or iron carbide, chemical formula Fe_3C, contains 6.67 percent C by weight. It is a typical hard and brittle interstitial compound of low tensile strength (approx. 5,000 psi) but high compressive strength. It is the hardest structure that appears on the diagram. Its crystal structure is orthorhombic.

Austenite is the name given to the γ solid solution. It is an interstitial solid solution of carbon dissolved in γ (f.c.c.) iron. Maximum solubility is 2 percent C at 2065°F (point *C*). Average properties are: tensile strength, 150,000 psi; elongation, 10 percent in 2 in.; hardness, Rockwell *C* 40, approx.; and toughness, high. It is normally not stable at room temperature. Under certain conditions it is possible to obtain austenite at room temperature, and its microstructure is shown in Fig. 7·8*a*.

Ledeburite is the eutectic mixture of austenite and cementite. It contains 4.3 percent C and is formed at 2065°F.

Ferrite is the name given to the α solid solution. It is an interstitial solid solution of a small amount of carbon dissolved in α (b.c.c.) iron (Fig. 7·8*b*). The maximum solubility is 0.025 percent C at 1333°F (point *H*), and it dissolves only 0.008 percent C at room temperature. It is the softest structure that appears on the diagram. Average properties are: tensile strength, 40,000 psi; elongation, 40 percent in 2 in.; hardness, less than Rockwell C 0 or less than Rockwell B 90.

Pearlite (point *J*) is the eutectoid mixture containing 0.80 percent C and is formed at 1333°F on very slow cooling. It is a very fine platelike or lamellar mixture of ferrite and cementite. The fine fingerprint mixture called pearlite is shown in Fig. 7·8*c*. The white ferritic background or matrix which makes up most of the eutectoid mixture contains thin plates of cementite. The same structure, magnified 17,000 times with the electron microscope, is shown in Fig. 7·8*d*. Average properties are: tensile strength, 120,000 psi; elongation, 20 percent in 2 in.; hardness, Rockwell C 20, Rockwell B 95–100, or BHN 250–300.

7·6 Carbon Solubility in Iron Austenite, being f.c.c. with four atoms per unit cell, represents a much denser packing of atoms than ferrite, which is b.c.c. with two atoms per unit cell. This is shown by the expansion that takes

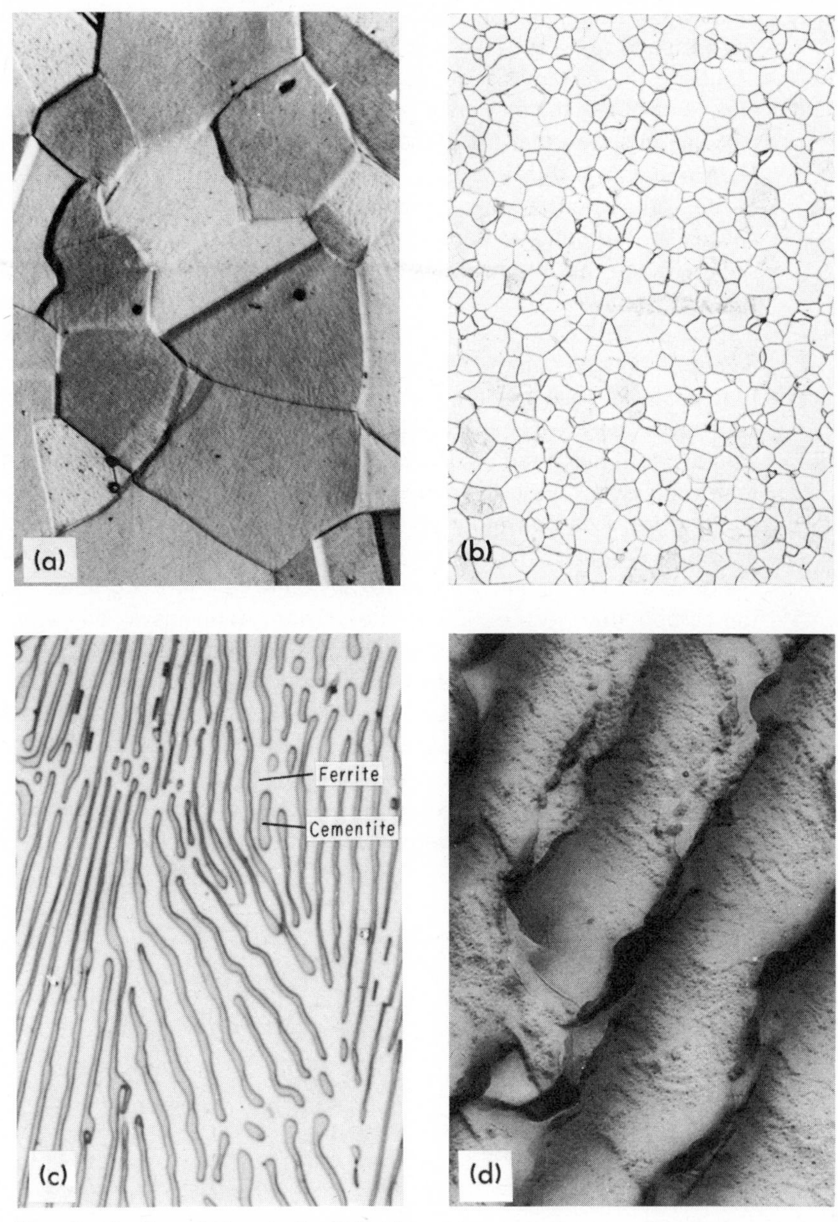

Fig. 7·8 The microstructure of (*a*) austenite, 500X; (*b*) ferrite, 100X; (*c*) pearlite, 2,500X; (*d*) pearlite, electron micrograph, 17,000X; enlarged 3X in printing. (*a*, *b*, and *c*, Research Laboratory, U.S. Steel Corporation.)

place when austenite changes to ferrite on slow cooling. If the iron atoms are assumed to be spheres, it is possible, from the lattice dimensions and assuming the distance of closest approach to be equal to the atom diameter, to calculate the amount of empty space in both crystal structures. The calculation shows that the percentage of unfilled space in the f.c.c. lattice is 25 percent, and in the b.c.c. lattice 32 percent. In both austenite and ferrite, the carbon atoms are dissolved interstitially, that is, in the unfilled spaces of the lattice structure. In view of the above calculations, it may seem strange that the solubility of carbon in austenite is so much greater than it is in ferrite. This seemingly unusual behavior may be explained by a study of Fig. 7·9. The largest hole in b.c.c. ferrite is halfway between the center of the face and the space between the two corner atoms. Two of the four possible positions for a carbon atom on the front face of a body-centered cube are shown in Fig. 7·9. The largest interstitial sphere that would just fit has a radius of $0.36(10^{-8})$ cm. The largest hole in f.c.c. austenite is midway along the edge between two corner atoms. One possible position for a carbon atom on the front face of a face-centered cube is shown in Fig. 7·9. The largest interstitial sphere that would just fit has a radius of $0.52(10^{-8})$ cm. Therefore, austenite will have a greater solubility for carbon than ferrite. Since the carbon atom has a radius of about $0.70(10^{-8})$ cm, the iron atoms in austenite are spread apart by the solution of carbon so that, at the maximum solubility of 2 percent, only about 10 percent of the holes are filled. The distortion of the ferrite lattice by the carbon atom is much greater than in the case of austenite; therefore, the carbon solubility is much more restricted.

7·7 Slow Cooling of Steel The steel portion of the iron–iron carbide diagram is of greatest interest, and the various changes that take place during the very

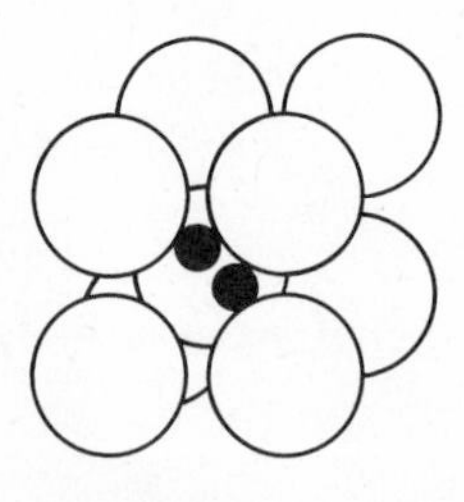

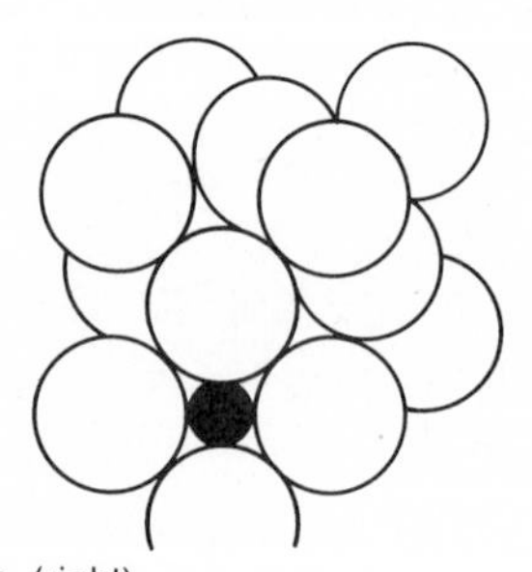

Fig. 7·9 Interstices of the b.c.c. (left) and f.c.c. (right). The maximum-diameter foreign sphere (black) that can enter the b.c.c lattice is indicated by the black atom with two of the four possible positions on one face shown here as filled. The f.c.c. lattice has far fewer holes, but, as shown by the black sphere, the hole is much larger. (By permission from Brick, Gordon, and Phillips, "Structure and Properties of Alloys," 3d ed., McGraw-Hill Book Company, 1965.)

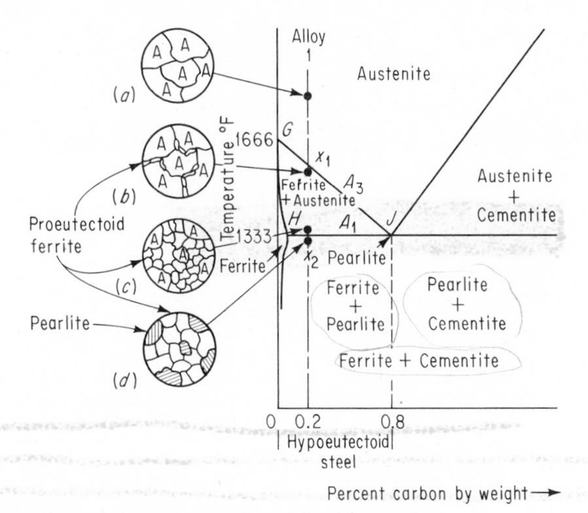

Fig. 7·10 Schematic representation of the changes in microstructure during the slow cooling of 0.20 percent carbon steel. (*a*) Austenite; (*b*) formation of ferrite grains at austenite grain boundaries; (*c*) growth of ferrite grains—composition of austenite is now 0.8 percent carbon; (*d*) austenite transforms to pearlite at 1333°F.

slow cooling, from the austenite range, of several steels will be discussed. Alloys containing more than 2 percent carbon will be discussed in Chap. 11.

Alloy 1 (Fig. 7·10) is a hypoeutectoid steel containing 0.20 percent carbon. In the austenite range, this alloy consists of a uniform interstitial solid solution. Each grain contains 0.20 percent carbon dissolved in the spaces of the f.c.c. iron lattice structure (Fig. 7·10*a*). Upon slow cooling nothing happens until the line *GJ* is crossed at point x_1. This line is known as the *upper-critical-temperature line* on the hypoeutectoid side and is labeled A_3. The allotropic change from f.c.c. to b.c.c. iron takes place at 1666°F for pure iron and decreases in temperature with increasing carbon content, as shown by the A_3 line. Therefore, at x_1, ferrite must begin to form at the austenite grain boundaries (Fig. 7·10*b*). Since ferrite can dissolve very little carbon, in those areas that are changing to ferrite the carbon must come out of solution before the atoms rearrange themselves to b.c.c. The carbon which comes out of solution is dissolved in the remaining austenite, so that, as cooling progresses and the amount of ferrite increases, the remaining austenite becomes richer in carbon. Its carbon content is gradually moving down and to the right along the A_3 line. Finally, the line *HJ* is reached at point x_2. This line is known as the *lower-critical-temperature line* on the hypoeutectoid side and is labeled A_1. The A_1 line is the eutectoid-temperature line and is the lowest temperature at which f.c.c.

iron can exist under equilibrium conditions. Just above the A_1 line, the microstructure consists of approximately 25 percent austenite and 75 percent ferrite (Fig. 7·10*c*). The remaining austenite, about 25 percent of the total material (Rule II) and containing 0.8 percent carbon, now experiences the eutectoid reaction

$$\text{Austenite} \underset{\text{heating}}{\overset{\text{cooling}}{\rightleftharpoons}} \underbrace{\text{ferrite} + \text{cementite}}_{\text{pearlite}}$$

Note that it is only austenite which is changing at the A_1 line. Therefore, when the reaction is complete the microstructure will show approximately 25 percent pearlite and 75 percent ferrite (Fig. 7·10*d*).

Let us consider the eutectoid reaction in a little more detail. Austenite is to change to ferrite. Austenite is an interstitial solid solution with each remaining grain dissolving 0.8 percent C in f.c.c. Fe. Ferrite, however, is b.c.c. Fe and dissolves very little carbon, so the change in crystal structure cannot occur until the carbon atoms come out of solution. Therefore, the first step is the precipitation of the carbon atoms to form plates of cementite (iron carbide). In the area immediately adjacent to the cementite plate the iron is depleted of carbon, and the atoms may now rearrange themselves to form b.c.c. ferrite. Thin layers of ferrite are formed on each side of the cementite plate. The process continues by the formation of alternate layers of cementite and ferrite to give the fine fingerprint mixture known as *pearlite*. The reaction usually starts at the austenite grain boundary, with the pearlite growing along the boundary and into the grain, see Fig. 7·11.

Since ferrite and pearlite are stable structures, the microstructure remains substantially the same down to room temperature and consists of approximately 75 percent proeutectoid ferrite (formed between the A_3 and A_1 lines) and 25 percent pearlite (formed from austenite at the A_1 line). Figure 7·12*a* shows the microstructure of a 0.2 percent C steel slow-cooled. As predicted, it consists of 75 percent proeutectoid ferrite (light areas) and 25 percent pearlite (dark areas). The dark areas in this micro certainly do not look like a mixture, which pearlite is supposed to

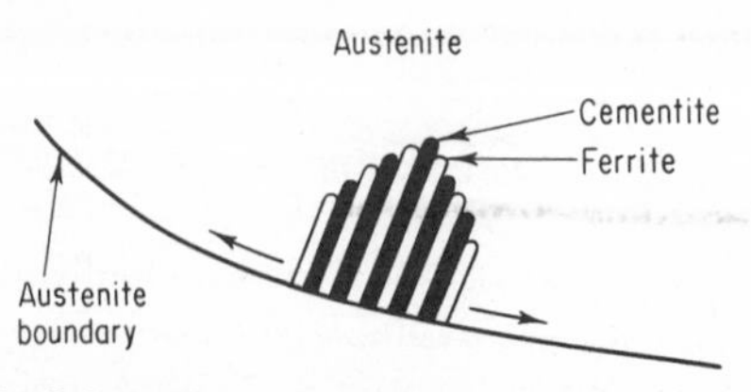

Fig. 7·11 Schematic picture of the formation and growth of pearlite.

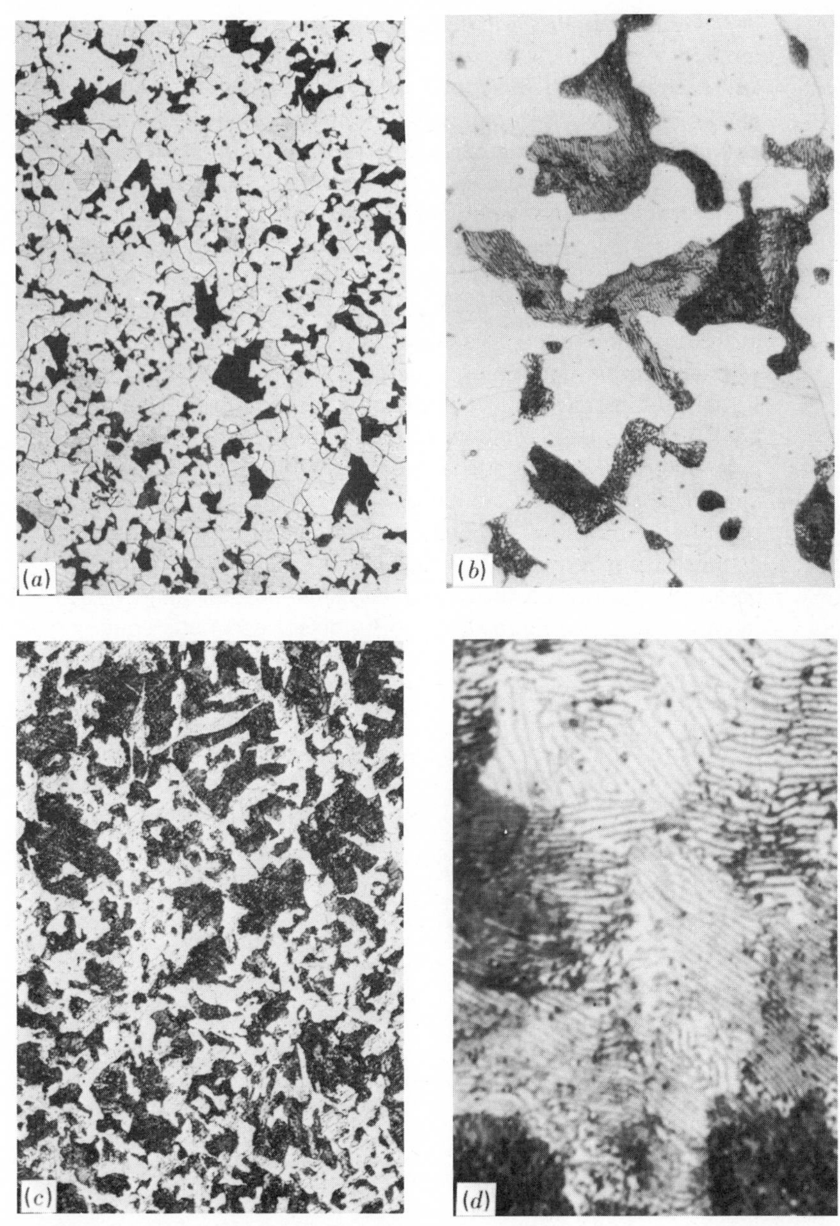

Fig. 7·12 Photomicrographs of (*a*) 0.20 percent carbon steel, slow-cooled, 100X; (*b*) same as (*a*), but at 500X; (*c*) 0.40 percent carbon steel, slow-cooled, 100X; (*d*) eutectoid (0.80 percent carbon) steel, slow-cooled, 500X. All samples etched with 2 percent nital. Dark areas are pearlite; light areas are ferrite.

be. Higher magnification (Fig. 7·12*b*), however, reveals the fine fingerprint mixture of pearlite.

The changes just described would be the same for any hypoeutectoid steel. The only difference would be in the relative amount of ferrite and pearlite. The closer the carbon content to the eutectoid composition (0.8 percent C), the more pearlite will be present in the microstructure. The microstructure of a 0.4 percent C steel slow-cooled (Fig. 7·12*c*) shows approximately 50 percent pearlite, while the eutectoid composition (0.8 percent C) shows 100 percent pearlite (Fig. 7·12*d*).

Alloy 2 (Fig. 7·13) is a hypereutectoid steel containing 1 percent carbon. In the austenite range, this alloy consists of a uniform f.c.c. solid solution with each grain containing 1 percent carbon dissolved interstitially (Fig. 7·13*a*). Upon slow cooling nothing happens until the line *CJ* is crossed at point x_3. This line is known as the *upper-critical-temperature line* on the hypereutectoid side and is labeled A_{cm}. The A_{cm} line shows the maximum amount of carbon that can be dissolved in austenite as a function of temperature. Above the A_{cm} line, austenite is an unsaturated solid solution. At the A_{cm} line, point x_3, the austenite is saturated in carbon. As the temperature is decreased, the carbon content of the austenite, that is, the maximum amount of carbon that can be dissolved in austenite, moves down

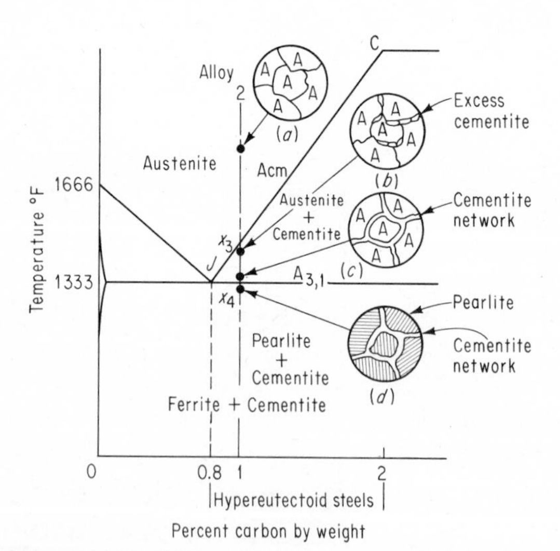

Fig. 7·13 Schematic representation of the changes in microstructure during the slow cooling of a 1.0 percent carbon steel. (*a*) Austenite; (*b*) formation of excess cementite at austenite grain boundaries; (*c*) growth of excess cementite to form a network—austenite composition is now 0.8 percent carbon; (*d*) Austenite transforms to pearlite at 1333°F.

along the A_{cm} line toward point J. Therefore, as the temperature decreases from x_3 to x_4, the excess carbon above the amount required to saturate austenite is precipitated as cementite primarily along the grain boundaries (Fig. 7·13*b,c*). Finally, the eutectoid-temperature line is reached at x_4. This line is called the *lower-critical-temperature line* on the hypereutectoid side and is labeled $A_{3,1}$. Just above the $A_{3,1}$ line the microstructure consists largely of austenite, with the excess proeutectoid cementite as a network surrounding the austenite grains (Fig. 7·13*c*). Applying Rule II with cementite on the right side of the line, the amount of cementite would be

$$\% \text{ cementite} = \frac{1.0 - 0.8}{6.67 - 0.8} \times 100 = 3.4\%$$

and the amount of austenite would be

$$\% \text{ austenite} = \frac{6.67 - 1.0}{6.67 - 0.8} \times 100 = 96.6\%$$

The $A_{3,1}$ line for hypereutectoid steels represents the beginning and the end of the allotropic change from f.c.c. austenite to b.c.c. ferrite. By the same process described earlier, the remaining austenite (containing 0.8 percent carbon) transforms to the eutectoid mixture, pearlite (Fig. 7·10*d*). At room temperature the microstructure consists of 96.6 percent pearlite (formed from austenite at the $A_{3,1}$ line) and a network of 3.4 percent proeutectoid cementite (formed between the A_{cm} and $A_{3,1}$ lines). Look closely at Fig. 7·14*a*, particularly where the pearlite areas meet, to see the thin, white proeutectoid cementite network. The story just described would be the same for any hypereutectoid steel, slow-cooled. As the carbon content of the alloy increases, the thickness of the proeutectoid cementite network generally increases. Figure 7·14*b* shows the microstructure of a 1.2 percent carbon steel. Both photomicrographs show very clearly the lamellar (platelike) structure of pearlite.

Note the difference in significance of the upper-critical-temperature lines, the A_3 and the A_{cm}. The former line involves an allotropic change, whereas the latter involves only a change in carbon solubility.

The mechanical properties of an alloy depend upon the properties of the phases and the way in which these phases are arranged to make up the structure. As was pointed out in Sec. 7·5, ferrite is relatively soft with low tensile strength, while cementite is hard with very low tensile strength. The combination of these two phases in the form of the eutectoid (pearlite) produces an alloy of much greater tensile strength than that of either phase. Since the amount of pearlite increases with an increase in carbon content for hypoeutectoid steels, the strength and Brinell hardness will also increase up to the eutectoid composition of 0.80 percent carbon. The ductility, as expressed by percent elongation and reduction in area, and

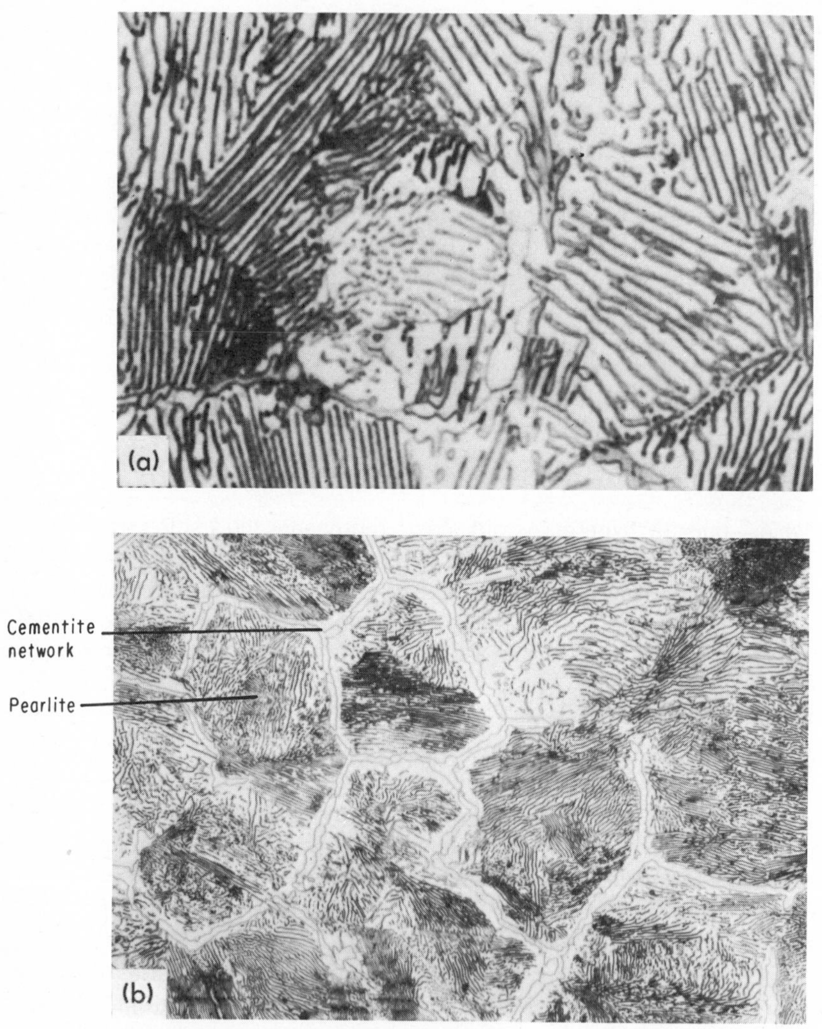

Fig. 7·14 Photomicrographs of (*a*) 1 percent carbon steel, slow-cooled, 500X; (*b*) 1.2 percent carbon steel, slow-cooled, 300X. Pearlite areas surrounded by a white proeutectoid cementite network. Note the increase in thickness of the cementite network with the increase in carbon content.

impact strength decrease with increasing carbon content. Beyond the eutectoid composition, the strength levels off and may even show a decrease due to the brittle cementite network. The Brinell hardness, however, continues to increase due to the greater proportion of hard cementite. The

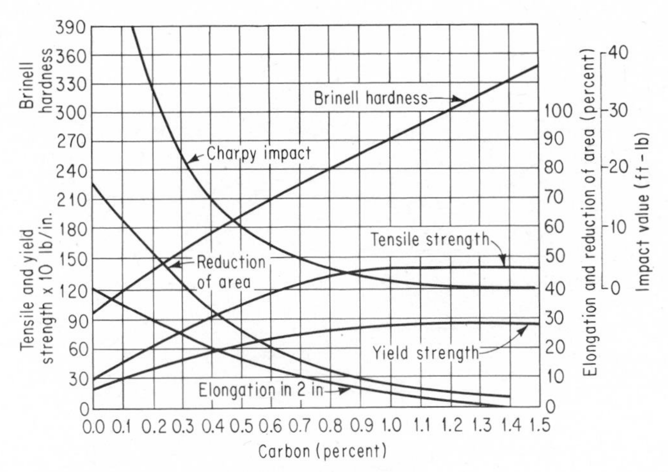

Fig. 7·15 Effect of carbon on mechanical properties of hot-worked steel. (By permission from *Sisco,* "Alloys of Iron and Carbon," McGraw-Hill Book Company, New York, 1937.)

variation in properties is shown in Fig. 7·15. The heat treatment of steel to produce significant changes in the mechanical properties will be discussed in detail in Chap. 8.

7·8 The Critical-temperature Lines In Fig. 7·10, the upper- and lower-critical-temperature lines are shown as single lines under equilibrium conditions and are sometimes indicated as Ae_3, Ae_1, etc. When the critical lines are actually determined, it is found that they do not occur at the same temperature. The critical line on heating is always higher than the critical line on cooling. To distinguish critical lines on heating from those occurring on cooling, the former are called *Ac* (*c* from the French word *chauffage*, which means heating) and the latter *Ar* (*r* from the French word *refroidissement*, which means cooling). Therefore, the upper critical line of a hypoeutectoid steel on heating would be labeled Ac_3 and the same line on cooling Ar_3.

The rate of heating and cooling has a definite effect on the temperature gap between these lines. The slower the rate of heating and cooling the nearer will the two lines approach each other, so that with infinitely slow heating and cooling they would probably occur at exactly the same temperature.

The results of thermal analysis of a series of carbon steels with an average heating and cooling rate of 11°F/min are shown in Fig. 7·16. The diagram shows the effect of this rate of heating and cooling on the position of the critical lines. Also shown are the Ac_2 and Ar_2 lines, which are due to the magnetic change in iron at 1414°F.

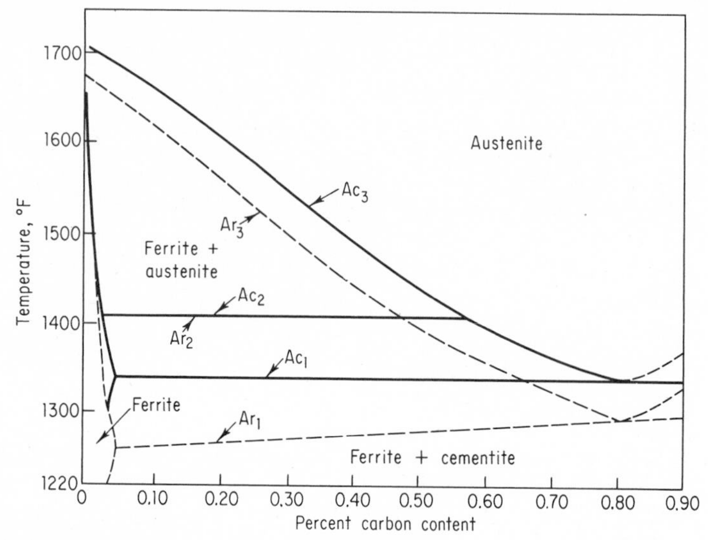

Fig. 7·16 Hypoeutectoid portion of the iron–iron carbide diagram. (By permission from Dowdell, "General Metallography," John Wiley & Sons, Inc., New York, 1943.)

7·9 Classification of Steels Several methods may be used to classify steels:

Method of Manufacture This gives rise to bessemer steel, open-hearth steel, electric-furnace steel, crucible steel, etc.

Use This is generally the final use for the steel, such as machine steel, spring steel, boiler steel, structural steel, or tool steel.

Chemical Composition This method indicates, by means of a numbering system, the approximate content of the important elements in the steel. This is the most popular method of classification and will be discussed in greater detail.

The steel specifications represent the results of the cooperative effort of the American Iron and Steel Institute (AISI) and the Society of Automotive Engineers (SAE) in a simplification program aimed at greater efficiency in meeting the steel needs of American industry.

The first digit of the four- or five-numeral designation indicates the type to which the steel belongs. Thus 1 indicates a carbon steel, 2 a nickel steel, 3 a nickel-chromium steel, etc. In the case of simple alloy steels, the second digit indicates the approximate percentage of the predominant alloying element. The last two or three digits usually indicate the mean carbon content divided by 100. Thus the symbol 2520 indicates a nickel steel of approximately 5 percent nickel and 0.20 percent carbon.

In addition to the numerals, AISI specifications may include a letter prefix to indicate the manufacturing process employed in producing the steel. SAE specifications now employ the same four-digit numerical designations as the AISI specifications, with the elimination of all letter prefixes.

The basic numbers for the four-digit series of the various grades of carbon and alloy steels with approximate percentages of identifying elements are:

10xx	Basic open-hearth and acid bessemer carbon steels
11xx	Basic open-hearth and acid bessemer carbon steels, high sulfur, low phosphorus
12xx	Basic open-hearth carbon steels, high sulfur, high phosphorus
13xx	Manganese 1.75
23xx	Nickel 3.50 (series deleted in 1959)
25xx	Nickel 5.00 (series deleted in 1959)
31xx	Nickel 1.25, chromium 0.60 (series deleted in 1964)
33xx	Nickel 3.50, chromium 1.50 (series deleted in 1964)
40xx	Molybdenum 0.20 or 0.25
41xx	Chromium 0.50, 0.80, or 0.95, molybdenum 0.12, 0.20, or 0.30
43xx	Nickel 1.83, chromium 0.50 or 0.80, molybdenum 0.25
44xx	Molybdenum 0.53
46xx	Nickel 0.85 or 1.83, molybdenum 0.20 or 0.25
47xx	Nickel 1.05, chromium 0.45, molybdenum 0.20 or 0.35
48xx	Nickel 3.50, molybdenum 0.25
50xx	Chromium 0.40
51xx	Chromium 0.80, 0.88, 0.93, 0.95, or 1.00
5xxxx	Carbon 1.04, chromium 1.03 or 1.45
61xx	Chromium 0.60 or 0.95, vanadium 0.13 or 0.15 min.
86xx	Nickel 0.55, chromium 0.50, molybdenum 0.20
87xx	Nickel 0.55, chromium 0.50, molybdenum 0.25
88xx	Nickel 0.55, chromium 0.50, molybdenum 0.35
92xx	Silicon 2.00
93xx	Nickel 3.25, chromium 1.20, molybdenum 0.12 (series deleted in 1959)
98xx	Nickel 1.00, chromium 0.80, molybdenum 0.25 (series deleted in 1964)
94Bxx	Nickel 0.45, chromium 0.40, molybdenum 0.12, boron 0.0005 min.

"Series deleted" does not mean that these steels are no longer made. It simply means that the tonnage produced is below a certain minimum to be included in the listing of standard grades. This listing is revised periodically by AISI.

Some representative standard-steel specifications are given for plain-carbon and free-machining steels in Table 7·3 and for alloy steels in Table 7·4.

TABLE 7·3 Some Representative Standard-steel Specifications

AISI NO.*	% C	% Mn	% P max	% S max	SAE NO.
		PLAIN-CARBON STEELS			
C1010	0.08–0.13	0.30–0.60	0.04	0.05	1010
C1015	0.13–0.18	0.30–0.60	0.04	0.05	1015
C1020	0.18–0.23	0.30–0.60	0.04	0.05	1020
C1025	0.22–0.28	0.30–0.60	0.04	0.05	1025
C1030	0.28–0.34	0.60–0.90	0.04	0.05	1030
C1035	0.32–0.38	0.60–0.90	0.04	0.05	1035
C1040	0.37–0.44	0.60–0.90	0.04	0.05	1040
C1045	0.43–0.50	0.60–0.90	0.04	0.05	1045
C1050	0.48–0.55	0.60–0.90	0.04	0.05	1050
C1055	0.50–0.60	0.60–0.90	0.04	0.05	1055
C1060	0.55–0.65	0.60–0.90	0.04	0.05	1060
C1065	0.60–0.70	0.60–0.90	0.04	0.05	1065
C1070	0.65–0.75	0.60–0.90	0.04	0.05	1070
C1074	0.70–0.80	0.50–0.80	0.04	0.05	1074
C1080	0.75–0.88	0.60–0.90	0.04	0.05	1080
C1085	0.80–0.93	0.70–1.00	0.04	0.05	1085
C1090	0.85–0.98	0.60–0.90	0.04	0.05	1090
C1095	0.90–1.03	0.30–0.50	0.04	0.05	1095
		FREE-MACHINING CARBON STEELS			
B1112	0.13 max	0.70–1.00	0.07–0.12	0.16–0.23	1112
B1113	0.13 max	0.70–1.00	0.07–0.12	0.24–0.33	1113
C1110	0.08–0.13	0.30–0.60	0.04	0.08–0.13	
C1113	0.10–0.16	1.00–1.30	0.04	0.24–0.33	
C1115	0.13–0.18	0.60–0.90	0.04	0.08–0.13	1115
C1120	0.18–0.23	0.70–1.00	0.04	0.08–0.13	1120
C1137	0.32–0.39	1.35–1.65	0.04	0.08–0.13	1137
C1141	0.37–0.45	1.35–1.65	0.04	0.08–0.13	1141
C1212	0.13 max	0.70–1.00	0.07–0.12	0.16–0.23	1112
C1213	0.13 max	0.70–1.00	0.07–0.12	0.24–0.33	1113
C12L14†	0.15 max	0.80–1.20	0.04–0.09	0.25–0.35	12L14

* Prefix AISI letters: B = acid bessemer carbon steel; C = basic open-hearth carbon steel.
† Lead, 0.15 to 0.35 percent.

Steels are sometimes classified by the broad range of carbon content such as:

Low-carbon steel: up to 0.25 percent carbon

Medium-carbon steel: 0.25 to 0.55 percent carbon

High-carbon steel: above 0.55 percent carbon

TABLE 7·4 Some Representative Alloy-steel Specifications

AISI NO.	% C	% Mn	% Ni	% Cr	% Mo	% V	SAE NO.	TYPE
1330	0.28–0.33	1.60–1.90	...	...	...	...	1330	Mn steels
1340	0.38–0.43	1.60–1.90	...	...	...	...	1340	
2317	0.15–0.20	0.40–0.60	3.25–3.75	...	...	...	2315	3% Ni steels
2330	0.28–0.33	0.60–0.80	3.25–3.75	...	...	...	2330	
E2512*	0.09–0.14	0.45–0.60	4.75–5.25	...	...	...	...	5% Ni steels
2515	0.12–0.17	0.40–0.60	4.75–5.25	...	...	...	2515	
3115	0.13–0.18	0.40–0.60	1.10–1.40	0.55–0.75	...	...	3115	Ni-Cr steels
3130	0.28–0.33	0.60–0.80	1.10–1.40	0.55–0.75	...	...	3130	
3140	0.38–0.43	0.70–0.90	1.10–1.40	0.55–0.75	...	...	3140	
E3310	0.08–0.13	0.45–0.60	3.65–3.75	1.40–1.75	...	...	3310	
4023	0.20–0.25	0.70–0.90	...	...	0.20–0.30	...	4023	Mo Steels
4037	0.35–0.40	0.70–0.90	...	...	0.20–0.30	...	4037	
4419	0.18–0.23	0.45–0.65	...	...	0.45–0.60	...	4419	
4118	0.18–0.23	0.70–0.90	...	0.40–0.60	0.08–0.15	...	4118	Cr-Mo steels
4130	0.28–0.33	0.40–0.60	...	0.80–1.10	0.15–0.25	...	4130	
4140	0.38–0.43	0.75–1.00	...	0.80–1.10	0.15–0.25	...	4140	
4150	0.48–0.53	0.75–1.00	...	0.80–1.10	0.15–0.25	...	4150	
4320	0.17–0.22	0.45–0.60	1.65–2.00	0.40–0.60	0.20–0.30	...	4320	Ni-Cr-Mo steels
4340	0.38–0.43	0.60–0.80	1.65–2.00	0.70–0.90	0.20–0.30	...	4340	
4720	0.17–0.22	0.50–0.70	0.90–1.20	0.35–0.55	0.15–0.25	...	4720	
4620	0.17–0.22	0.45–0.60	1.65–2.00	...	0.20–0.30	...	4620	Ni-Mo steels
4626	0.24–0.29	0.45–0.65	0.70–1.00	...	0.15–0.25	...	4626	
4820	0.18–0.23	0.50–0.70	3.25–3.75	...	0.20–0.30	...	4820	
5120	0.17–0.22	0.70–0.90	...	0.70–0.90	...	...	5120	Cr steels
5130	0.28–0.33	0.70–0.90	...	0.80–1.10	...	...	5130	
5140	0.38–0.43	0.70–0.90	...	0.70–0.90	...	...	5140	
5150	0.48–0.53	0.70–0.90	...	0.70–0.90	...	...	5150	
E52100*	0.95–1.10	0.25–0.45	...	1.30–1.60	...	...	52100	
6118	0.16–0.21	0.50–0.70	...	0.50–0.70	...	0.12	6118	Cr-V steels
6150	0.48–0.53	0.70–0.90	...	0.80–0.10	...	0.15	6150	
8620	0.18–0.23	0.70–0.90	0.40–0.70	0.40–0.60	0.15–0.25	...	8620	Low Ni-Cr-Mo steels
8630	0.28–0.33	0.70–0.90	0.40–0.70	0.40–0.60	0.15–0.25	...	8630	
8640	0.38–0.43	0.75–1.00	0.40–0.70	0.40–0.60	0.15–0.25	...	8640	
8720	0.18–0.23	0.70–0.90	0.40–0.70	0.40–0.60	0.20–0.30	...	8720	
8740	0.38–0.43	0.75–1.00	0.40–0.70	0.40–0.60	0.20–0.30	...	8740	
8822	0.20–0.25	0.75–1.00	0.40–0.70	0.40–0.60	0.20–0.40	...	8822	
			Si					
9260	0.56–0.64	0.75–1.00	1.80–2.20	...	...	...	9260	Si steel
			Ni					
E9310*	0.08–0.13	0.45–0.65	3.00–3.50	1.00–1.40	0.08–0.15	...	9310	Higher Ni-Cr-Mo steels
9840	0.38–0.43	0.70–0.90	0.85–1.15	0.70–0.90	0.20–0.30	...	9840	
9850	0.48–0.53	0.70–0.90	0.85–1.15	0.70–0.90	0.20–0.30	...	9850	
94B30	0.48–0.53	0.70–0.90	0.85–1.15	0.70–0.90	0.20–0.30	...	94B30	Boron steel

* E = basic electric-furnace process. All others are normally manufactured by the basic open-hearth process.

7·10 Effect of Small Quantities of Other Elements The previous discussion of the iron–iron carbide equilibrium diagram assumed that only iron and iron carbide were present. However, examination of Table 7·3 indicates that commercial plain carbon steels contain small quantities of other elements besides iron and carbon as part of the normal composition.

Sulfur Sulfur in commercial steels is generally kept below 0.05 percent. Sulfur combines with iron to form iron sulfide (FeS). Iron sulfide forms a low-melting-point eutectic alloy with iron which tends to concentrate at the grain boundaries. When the steel is forged or rolled at elevated temperatures, the steel becomes brittle, or *hot-short*, due to the melting of the iron sulfide eutectic which destroys the cohesion between grains, allowing cracks to develop. In the presence of manganese, sulfur tends to form manganese sulfide (MnS) rather than iron sulfide. The manganese sulfide may pass out in the slag or remain as well-distributed inclusions throughout the structure. It is recommended that the amount of manganese be 2 to 8 times the amount of sulfur. In the free-machining steels, the sulfur content is increased to between 0.08 and 0.35 percent. The improvement in machinability is due to the presence of more numerous sulfide inclusions which break up the chips, thus reducing tool wear.

Manganese Manganese is present in all commercial plain carbon steels, in the range of 0.03 to 1.00 percent. The function of manganese in counteracting the ill effects of sulfur was just pointed out. When there is more manganese present than the amount required to form MnS, the excess combines with some of the carbon to form the compound Mn_3C which is found associated with the iron carbide, Fe_3C, in cementite. Manganese also promotes the soundness of steel casting through its deoxidizing action on liquid steel. The effect of larger quantities of manganese will be discussed in Chap. 9 on alloy steels.

Phosphorus The phosphorus content is generally kept below 0.04 percent. This small quantity tends to dissolve in ferrite, increasing the strength and hardness slightly. In some steels 0.07 to 0.12 percent phosphorus seems to improve cutting properties. In larger quantities, phosphorus reduces ductility, thereby increasing the tendency of the steel to crack when cold worked, making it *cold-short*.

Silicon Most commercial steels contain between 0.05 and 0.3 percent silicon. Silicon dissolves in ferrite, increasing the strength of the steel without greatly decreasing the ductility. It promotes the deoxidation of molten steel through the formation of silicon dioxide, SiO_2, thus tending to make for greater soundness in casting. Silicon is an important element in cast iron, and its effect will be discussed in Chap. 11.

8 THE HEAT TREATMENT OF STEEL

8·1 Introduction The definition of heat treatment given in the *Metals Handbook* is: "A combination of heating and cooling operations, timed and applied to a metal or alloy in the solid state in a way that will produce desired properties." All basic heat-treating processes for steel involve the transformation or decomposition of austenite. The nature and appearance of these transformation products determine the physical and mechanical properties of any given steel.

The first step in the heat treatment of steel is to heat the material to some temperature in or above the critical range in order to form austenite. In most cases, the rate of heating to the desired temperature is less important than other factors in the heat-treating cycle. Highly stressed materials produced by cold work should be heated more slowly than stress-free materials to avoid distortion. The difference in temperature rise within thick and thin sections of articles of variable cross section should be considered, and whenever possible, provision should be made for slowing the heating of the thinner sections to minimize thermal stress and distortion. Usually less overall damage will be done to the steel by utilizing as slow a heating rate as is practical.

8·2 Full Annealing This process consists in heating the steel to the proper temperature and then cooling slowly through the transformation range, preferably in the furnace or in any good heat-insulating material. The slow cooling is generally continued to low temperatures.

The purpose of annealing may be to refine the grain, induce softness, improve electrical and magnetic properties, and, in some cases, to improve machinability.

Since the entire mass of the furnace must be cooled down along with the material, annealing is a very slow cooling process and therefore comes closest to following the iron–iron carbide equilibrium diagram.

Assume that we have a coarse-grained 0.20 percent carbon (hypoeutectoid) steel and that it is desired to refine the grain size by annealing. The microstructure is shown in Fig. 8·1*a*. When this steel is heated, no change will occur until the A_1 (lower-critical) line is crossed. At that temperature the pearlite areas will transform to small grains of austenite by means of the eutectoid reaction, but the original large ferrite grains will remain unchanged (Fig. 8·1*b*). Cooling from this temperature will not refine the grain. Continued heating between the A_1 and A_3 lines will allow the large ferrite grains to transform to small grains of austenite, so that above the A_3 (upper-critical) line the entire microstructure will show only small grains of austenite, Fig. 8·1*c*. Subsequent furnace cooling will result in small grains of proeutectoid ferrite and small areas of coarse lamellar pearlite (Fig. 8·1*d*). Therefore, the proper annealing temperature for hypoeutectoid steels is approximately 50°F above the A_3 line.

Refinement of the grain size of hypereutectoid steel will occur about

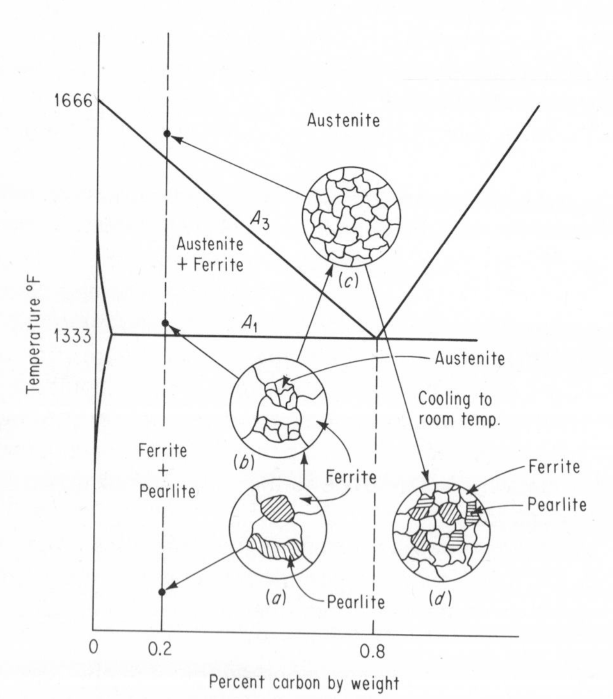

Fig. 8·1 Schematic representation of the changes in microstructure during the annealing of a 0.20 percent carbon steel. (*a*) Original structure, coarse-grained ferrite and pearlite. (*b*) Just above the A_1 line; pearlite has transformed to small grains of austenite, ferrite unchanged. (*c*) Above the A_3 line; only fine-grained austenite. (*d*) After cooling to room temperature; fine-grained ferrite and small pearlite areas.

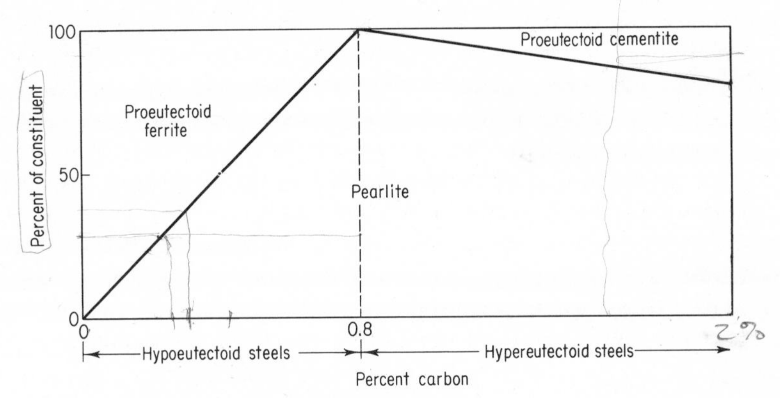

Fig. 8·2 Proportion of the constituents present in the microstructure of annealed steels as a function of carbon content.

50°F above the lower-critical-temperature ($A_{3,1}$) line. Heating above this temperature will coarsen the austenitic grains, which, on cooling, will transform to large pearlitic areas. The microstructure of annealed hypereutectoid steel will consist of coarse lamellar pearlite areas surrounded by a network of proeutectoid cementite (see Fig. 7·14). Because this excess cementite network is brittle and tends to be a plane of weakness, annealing should never be a final heat treatment for hypereutectoid steels. The presence of a thick, hard, grain boundary will also result in poor machinability.

A careful microscopic study of the proportions of ferrite and pearlite or pearlite and cementite present in an annealed steel can enable one to determine the approximate carbon content of the steel (Fig. 8·2).

The approximate tensile strength of annealed hypoeutectoid steels may be determined by the proportion of ferrite and pearlite present:

$$\text{Approx tensile strength} = \frac{40{,}000(\text{percent ferrite}) + 120{,}000(\text{percent pearlite})}{100}$$

For example, an annealed 0.20 percent carbon steel contains approximately 25 percent pearlite and 75 percent ferrite. Applying the above formula,

$$\text{Approx tensile strength} = 40{,}000(0.75) + 120{,}000(0.25) = 60{,}000 \text{ psi}$$

This same idea cannot be applied to hypereutectoid steels, since their strength is determined by the cementite network which forms the continuous phase. The presence of the brittle network results in a drop in tensile strength above 0.8 percent carbon (see Table 8·1). The proper annealing range for hypoeutectoid and hypereutectoid steels is shown in Fig. 8·3.

8·3 Spheroidizing As was pointed out earlier, an annealed hypereutectoid steel with a microstructure of pearlite and a cementite network will generally give poor machinability. Since cementite is hard and brittle, the cutting tool cannot cut through these plates. Instead, the plates have to be broken. Therefore, the tool is subjected to continual shock load by the cementite plates, and a ragged surface finish results. A heat-treating process which will improve the machinability is known as *spheroidize annealing.* This process will produce a spheroidal or globular form of carbide in a ferritic matrix, as shown in Fig. 8·4. One of the following methods may be used:

1 Prolonged holding at a temperature just below the lower critical line.
2 Heating and cooling alternately between temperatures that are just above and just below the lower critical line.
3 Heating to a temperature above the lower critical line and then either cooling very slowly in the furnace or holding at a temperature just below the lower critical line.

Prolonged time at the elevated temperature will completely break up the

Table 8·1 Mechanical Properties of Normalized and Annealed Steels*

CARBON, %	YIELD POINT, 1,000 PSI	TENSILE STRENGTH, 1,000 PSI	ELONGATION, % IN 2 IN.	REDUCTION IN AREA, %	BHN
Normalized (hot-rolled steel):					
0.01	26	45	45	71	90
0.20	45	64	35	60	120
0.40	51	85	27	43	165
0.60	60	109	19	28	220
0.80	70	134	13	18	260
1.00	100	152	7	11	295
1.20	100	153	3	6	315
1.40	96	148	1	3	300
Annealed:					
0.01	18	41	47	71	90
0.20	36	59	37	64	115
0.40	44	75	30	48	145
0.60	49	96	23	33	190
0.80	52	115	15	22	220
1.00	52	108	22	26	195
1.20	51	102	24	39	200
1.40	50	99	19	25	215

* By permission from Brick, Gordon, and Phillips, "The Structure and Properties of Alloys," 3d ed., McGraw-Hill Book Company, New York, 1965.

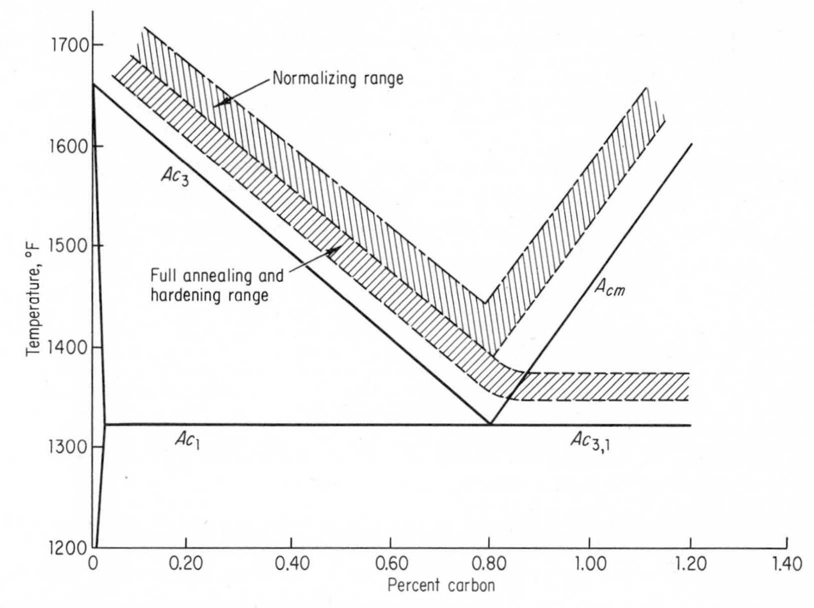

Fig. 8·3 Annealing, normalizing, and hardening range for plain-carbon steels.

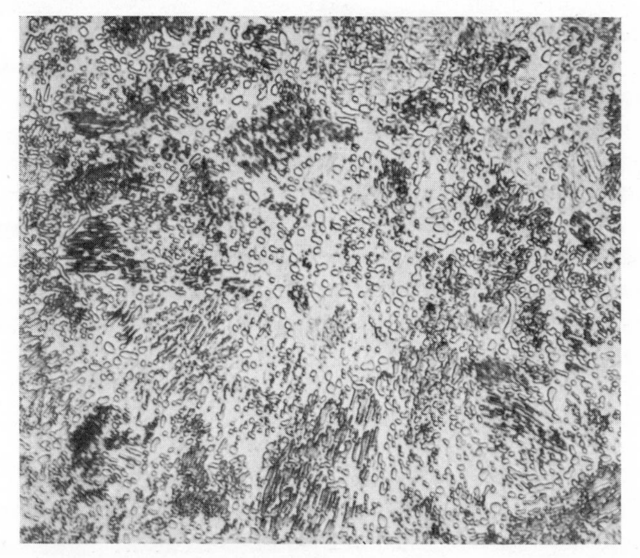

Fig. 8·4 A 1 percent carbon steel spheroidize-annealed, showing spheroidized cementite in a ferrite matrix. Etched in 2 percent nital, 750X.

pearlitic structure and cementite network. The cementite will become spheres, which is the geometric shape in greatest equilibrium with its surroundings. The cementite particles and the entire structure may be called *spheroidite* (see Fig. 8·4). Contrast this microstructure with the one shown in Fig. 7·14. Notice that in both cases the microstructure is made up of ferrite and cementite. The difference is in the form of the cementite, and this greatly affects the properties of the materials. The spheroidized structure is desirable when minimum hardness, maximum ductility, or (in high-carbon steels) maximum machinability is important. Low-carbon steels are seldom spheroidized for machining, because in the spheroidized condition they are excessively soft and "gummy." The cutting tool will tend to push the material rather than cut it, causing excessive heat and wear on the cutting tip. Medium-carbon steels are sometimes spheroidize-annealed to obtain maximum ductility for certain working operations. If the steel is kept too long at the spheroidize-annealing temperature, the cementite particles will coalesce and become elongated, thus reducing machinability.

8·4 Stress-relief Annealing This process, sometimes called *subcritical* annealing, is useful in removing residual stresses due to heavy machining or other cold-working processes. It is usually carried out at temperatures below the lower critical line (1000 to 1200°F).

8·5 Process Annealing This heat treatment is used in the sheet and wire industries and is carried out by heating the steel to a temperature below the lower critical line (1000 to 1250°F). It is applied after cold working and softens the steel, by recrystallization, for further working. It is very similar to stress-relief annealing.

8·6 Normalizing The normalizing of steel is carried out by heating approximately 100°F above the upper-critical-temperature (A_3 or A_{cm}) line followed by cooling in still air to room temperature. The temperature range for normalizing is shown in Fig. 8·3. The purpose of normalizing is to produce a harder and stronger steel than full annealing, so that for some applications normalizing may be a final heat treatment. Therefore, for hypereutectoid steels, it is necessary to heat above the A_{cm} line in order to dissolve the cementite network. Normalizing may also be used to improve machinability, modify and refine cast dendritic structures, and refine the grain and homogenize the microstructure in order to improve the response in hardening operations. The increase in cooling rate due to air cooling as compared with furnace cooling affects the transformation of austenite and the resultant microstructure in several ways. Since we are no longer cooling under equilibrium conditions, the iron–iron carbide diagram cannot be used to predict the proportions of proeutectoid ferrite and pearlite or proeutectoid cementite and pearlite that will exist at room temperature. There is less time for the formation of the proeutectoid constituent; consequently there will be less proeutectoid ferrite in normalized hypoeutectoid steels and less proeutectoid cementite in hypereutectoid steels as compared with annealed ones. Figure 8·5 shows the microstructure of a normalized 0.50 percent carbon steel. In the annealed condition this steel would have approximately 62 percent pearlite and 38 percent proeutectoid ferrite. Due to air cooling, this sample has only about 10 percent proeutectoid ferrite, which is the white network surrounding the dark pearlite areas. For hypereutectoid steels, normalizing will reduce the continuity of the proeutectoid cementite network, and in some cases it may be suppressed entirely. Since it was the presence of the cementite network which reduced the strength of annealed hypereutectoid steels, normalized steels should show an increase in strength. This is illustrated by the strength values given in Table 8·1, particularly for steels containing more than 0.8 percent carbon.

Aside from influencing the amount of proeutectoid constituent that will form, the faster cooling rate in normalizing will also affect the temperature of austenite transformation and the fineness of the pearlite. In general, the faster the cooling rate, the lower the temperature of austenite transformation and the finer the pearlite. The difference in spacing of the cementite plates in the pearlite between annealing and normalizing is shown schematically in Fig. 8·6. Ferrite is very soft, while cementite is very hard. With the cementite plates closer together in the case of normalized medium

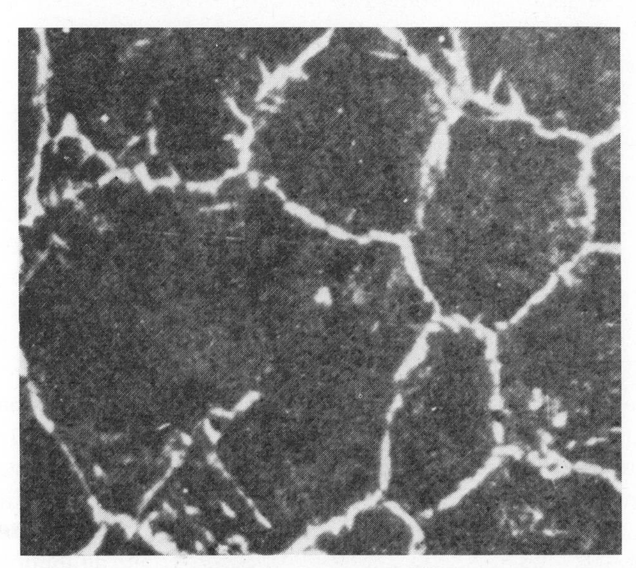

Fig. 8·5 Normalized 0.50 percent carbon steel, heated to 1800°F and air-cooled; 100X. Proeutectoid ferrite surrounding pearlite areas.

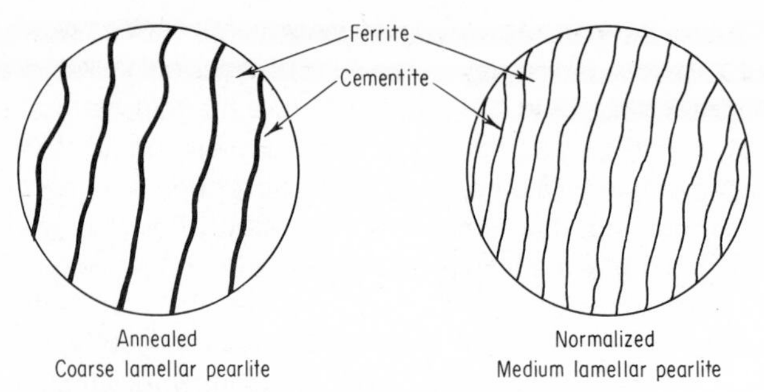

Fig. 8·6 Schematic picture of the difference in pearlitic structure due to annealing and normalizing.

pearlite, they tend to stiffen the ferrite so it will not yield as easily, thus increasing hardness. If the annealed coarse pearlite has a hardness of about Rockwell C 10, then the normalized medium pearlite will be about Rockwell C 20. Nonequilibrium cooling also shifts the eutectoid point toward lower carbon content in hypoeutectoid steels and toward higher carbon content in hypereutectoid steels. The net effect is that normalizing produces a finer and more abundant pearlite structure than is obtained by annealing, which results in a harder and stronger steel.

While annealing, spheroidizing, and normalizing may be employed to improve machinability, the process to be used will depend upon carbon content. Based on many studies, the optimum microstructures for machining steels of different carbon contents are usually as follows:

CARBON, %	OPTIMUM MICROSTRUCTURE
0.06 to 0.20	As cold-rolled
0.20 to 0.30	Under 3-in. dia., normalized; Over 3-in. dia., as cold-rolled
0.30 to 0.40	Annealed, to give coarse pearlite
0.40 to 0.60	Annealed, to give coarse pearlite or coarse spheroidite
0.60 to 1.00	100% spheroidite, coarse to fine

8·7 Hardening Under slow or moderate cooling rates, the carbon atoms are able to diffuse out of the austenite structure. The iron atoms then move slightly to become b.c.c. (body-centered cubic). This gamma-to-alpha transformation takes place by a process of nucleation and growth and is time-dependent. With a still further increase in cooling rate, insufficient time is allowed for the carbon to diffuse out of solution, and although some movement of the iron atoms takes place, the structure cannot become b.c.c. while the carbon is trapped in solution. The resultant structure, called *martensite*, is a supersaturated solid solution of carbon trapped in a body-centered tetragonal structure. Two dimensions of the unit cell are equal, but the third is slightly expanded because of the trapped carbon. The axial ratio c/a increases with carbon content to a maximum of 1.08 (see Fig. 8·7). This highly distorted lattice structure is the prime reason for

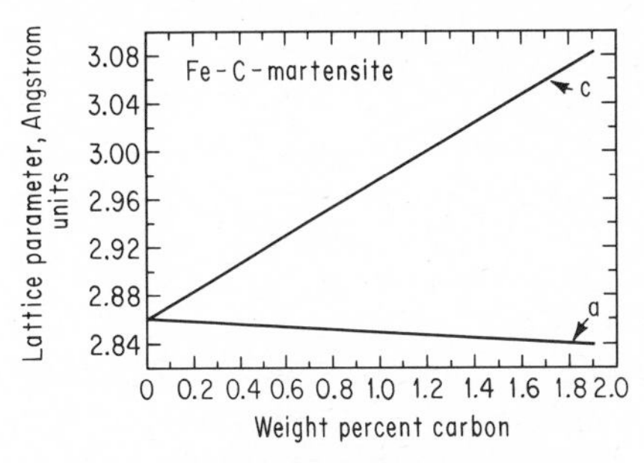

Fig. 8·7 Variation of lattice constants *a* and *c* of martensite with carbon content in plain-carbon steels. (By permission from C. Barrett and T. B. Massalski, "Structure of Metals," 3d ed., McGraw-Hill Book Company, New York, 1966.)

the high hardness of martensite. Since the atoms of martensite are less densely packed than in austenite, an expansion occurs during the transformation. This expansion during the formation of martensite produces high localized stresses which result in plastic deformation of the matrix. After drastic cooling (quenching), martensite appears microscopically as a white needlelike or acicular structure sometimes described as a pile of straw. In most steels, the martensitic structure appears vague and unresolvable (Fig. 8·8). In high-carbon alloys where the background is retained austenite, the acicular structure of martensite is more clearly defined (Fig. 8·9).

There are several important characteristics of the martensite transformation:

1 The transformation is diffusionless, and there is no change in chemical composition. Small volumes of austenite suddenly change crystal structure by a combination of two shearing actions.

2 The transformation proceeds only during cooling and ceases if cooling is interrupted. Therefore, the transformation depends only upon the decrease in temperature and is independent of time. A transformation of this type is said to be *athermal*, in contrast to one that will occur at constant temperature (*isothermal* transformation). The amount of martensite formed with decreasing temperature is not linear. The number of martensite needles produced at first is small; then the number increases, and finally, near the end, it decreases again (Fig. 8·10). The temperature of the start of martensite formation is known as the M_s *temperature* and that of the end of martensite formation as the M_f *temperature*. If the steel is held at any temperature below

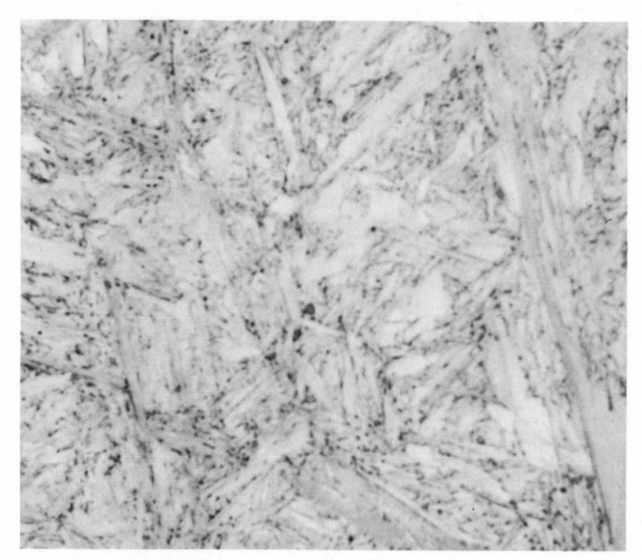

Fig. 8·8 The structure called martensite, 2500X. (Courtesy of Research Laboratory, U.S. Steel Corporation.)

the M_s, the transformation to martensite will stop and will not proceed again unless the temperature is dropped.

3 The martensite transformation of a given alloy cannot be suppressed, nor can the M_s temperature be changed by changing the cooling rate. The temperature range

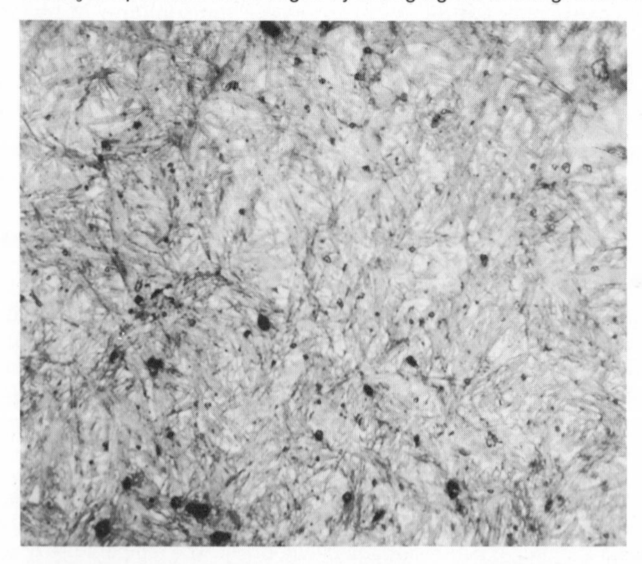

Fig. 8·9 A 1 percent carbon steel water-quenched. Etched in 2 percent nital, 750X. Martensitic needles in a retained austenite matrix.

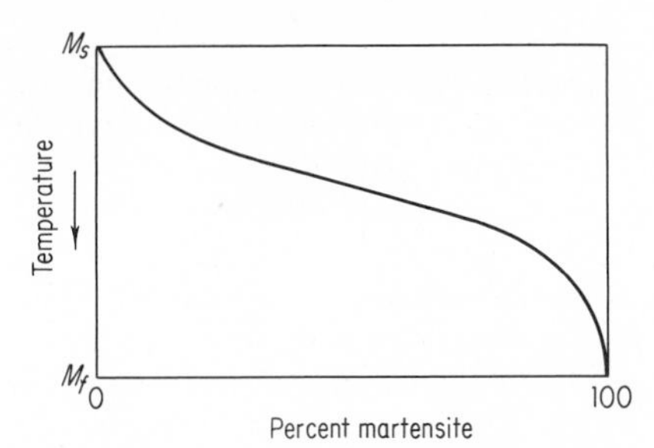

Fig. 8·10 Schematic representation of percentage of martensite formed as a function of temperature. (From "Metals Handbook," 1948 ed., American Society for Metals, Metals Park, Ohio.)

of the formation of martensite is characteristic of a given alloy and cannot be lowered by increasing the cooling rate. The M_s temperature seems to be a function of chemical composition only, and several formulas have been developed by which it may be calculated. One such formula (R. A. Grange and H. M. Stewart, *Metals Technology*, June, 1946) is

$$M_s\ (^\circ\mathrm{F}) = 1{,}000 - (650 \times \%\ \mathrm{C}) - (70 \times \%\ \mathrm{Mn}) - (35 \times \%\ \mathrm{Ni}) - (70 \times \%\ \mathrm{Cr}) - (50 \times \%\ \mathrm{Mo})$$

The influence of carbon on the M_s and M_f temperatures is shown in Fig. 8·11. The M_f temperature line is shown dotted because it is usually not clearly defined. Theoretically, the austenite to martensite transformation is never complete, and small amounts of retained austenite will remain even at low temperatures. The transformation of the last traces of austenite becomes more and more difficult as the amount of austenite decreases. This is apparent from the slope of the curve near 100 percent martensite in Fig. 8·10. Curves such as shown in these figures are based on visual estimations of the structures of metallographic samples, and small amounts of retained austenite are very difficult to measure when there are many overlapping martensite needles. Therefore, the M_f temperature is usually taken as the temperature at which the transformation is complete as far as one is able to determine by

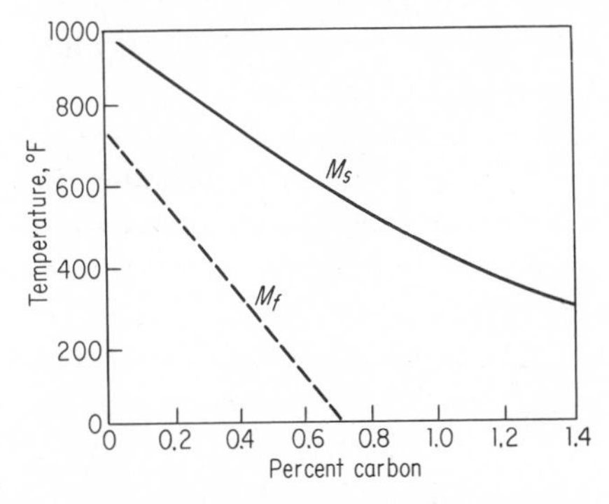

Fig. 8·11 Influence of carbon on the martensite range. (From "Metals Handbook," 1948 ed., American Society for Metals, Metals Park, Ohio.)

visual means.

4 Martensite is probably never in a condition of real equilibrium, although it may persist indefinitely at or near room temperature. The structure can be considered as a transition between the unstable austenite phase and the final equilibrium condition of a mixture of ferrite and cementite.

5 The most significant property of martensite is its potential of very great hardness. Although martensite is always harder than the austenite from which it forms (Fig. 8·12), extreme hardnesses are possible only in steels that contain sufficient carbon. The hardness of martensite increases rapidly at first with increase in carbon content, reaching about 60 Rockwell C at 0.40 percent carbon. Beyond that point the curve levels off, and at the eutectoid composition (0.80 percent carbon), the hardness is about Rockwell C 65. The leveling off is due to the greater tendency to retain austenite in higher-carbon steels. The high hardness of martensite is believed to be a result of the severe lattice distortions produced by its formation, since the amount of carbon present is many times more than can be held in solid solution. *The maximum hardness obtainable from a steel in the martensitic condition seems to be a function of carbon content only.*

The martensite transformation, for many years, was believed to be unique for steel. However, in recent years, this martensite type of transformation has been found in a number of other alloy systems, such as iron-nickel, copper-zinc, and copper-aluminum. The transformation is therefore recognized as a basic type of reaction in the solid state, and the term *martensite* is no longer confined only to the metallurgy of steel.

The basic purpose of hardening is to produce a fully martensitic structure, and the minimum cooling rate (°F per second) that will avoid the formation of any of the softer products of transformation is known as the *critical cooling rate*. The critical cooling rate, determined by chemical composition and austenitic grain size, is an important property of a steel since it indicates how fast a steel must be cooled in order to form only martensite.

8·8 The Isothermal-transformation Diagram It is apparent from the previous discussion that the iron–iron carbide equilibrium diagram is of little value in the study of steels cooled under nonequilibrium conditions. Many metallurgists realized that time and temperature of austenite transformation had a profound influence on the transformation products and the subsequent properties of the steel. However, this was not given scientific basis until E. S. Davenport and E. C. Bain published their classic paper (*Trans. AIME*, vol. 90, p. 117, 1930) on the study of the transformation of austenite at constant subcritical temperature. Since austenite is unstable below the lower critical temperature Ae_1, it is necessary to know at a particular subcritical temperature how long it will take for the austenite to start to transform, how long it will take to be completely transformed, and what will be the nature of the transformation product.

The best way to understand the isothermal-transformation diagrams

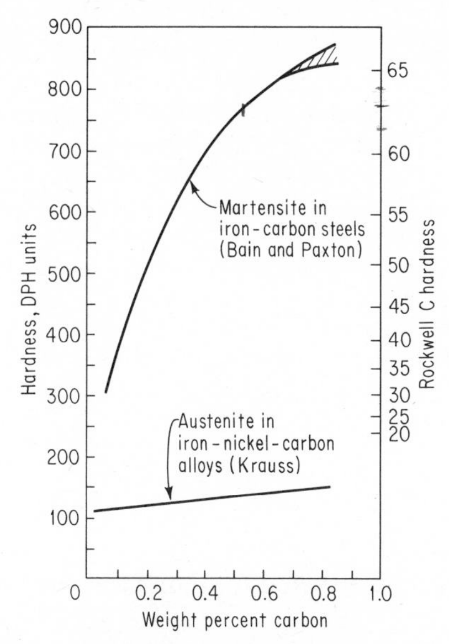

Fig. 8·12 The effect of carbon on the hardness of austenite and on the hardness of martensite. The shaded area of the upper curve represents the effect of retained austenite. (By permission from Brick, Gordon, and Phillips, "Structure and Properties of Alloys," 3d ed., McGraw-Hill Book Company, New York, 1965.)

is to study their derivation. The eutectoid composition of 0.8 percent carbon is the simplest one to study since there is no proeutectoid constituent present in the microstructure.

The steps usually followed to determine an isothermal-transformation diagram are:

Step 1 Prepare a large number of samples cut from the same bar. One method of handling the small samples during heat treatment is by means of a wire threaded through a hole in the sample, as shown in Fig. 8·13. The cross section has to be small in order to react quickly to changes in temperature.

Step 2 Place the samples in a furnace or molten salt bath at the proper austenitizing temperature. For a 1080 (eutectoid) steel, this temperature is approximately 1425°F. They should be left at the given temperature long enough to become completely austenite.

Step 3 Place the samples in a molten salt bath which is held at a constant subcritical temperature (a temperature below the Ae_1 line), for example, 1300°F.

Fig. 8·13 A typical sample which is used to determine an I-T diagram.

Step 4 After varying time intervals in the salt bath, each sample is quenched in cold water or iced brine.

Step 5 After cooling, each sample is checked for hardness and studied microscopically.

Step 6 The above steps are repeated at different subcritical temperatures until sufficient points are determined to plot the curves on the diagram.

We are really interested in knowing what is happening to the austenite at 1300°F, but the samples cannot be studied at that temperature. Therefore, we must somehow be able to relate the room-temperature microscopic examination to what is occurring at the elevated temperature. Two facts should be kept in mind:

1 Martensite is formed only from austenite almost instantaneously at low temperatures.

2 If austenite transforms at a higher temperature to a structure which is stable at room temperature, rapid cooling will not change the transformation product. In other words, if pearlite is formed at 1300°F, the pearlite will be exactly the same at room temperature no matter how drastically it is quenched, since there is no reason for the pearlite to change.

Steps 3, 4, and 5 are shown schematically in Fig. 8·14. Sample 1, after 30 s at 1300°F and quenched, showed only martensite at room temperature. Since martensite is formed only from austenite at low temperatures, it means that at the end of 30 s at 1300°F there was only austenite present and the transformation had not yet started. Sample 2, after 6 h at 1300°F and quenched, showed about 95 percent martensite and 5 percent coarse pearlite at room temperature (see Fig. 8·15). It means that at the end of 6 h at 1300°F there was 95 percent austenite and 5 percent coarse pearlite. The transformation of austenite at 1300°F has already started, and the transformation product is coarse pearlite. Using the same reasoning as above, the student should be able to follow the progress of austenite transformation by studying samples 3, 4, 5, and 6. The typical isothermal-

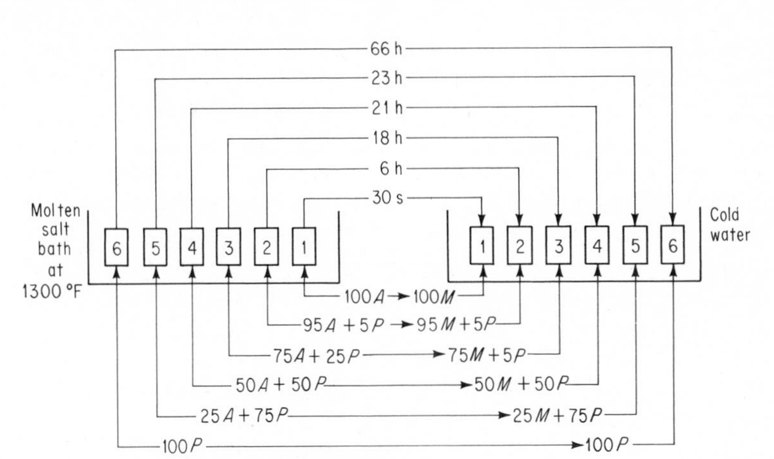

Fig. 8·14 The progress of austenite transformation to coarse pearlite at 1300°F as related to the structure at room temperature; *A* is austenite, *M* is martensite, *P* is pearlite.

transformation curve at 1300°F and several of the room-temperature microstructures are shown in Fig. 8·15. The light areas are martensite. Notice that the transformation from austenite to pearlite is not linear. Initially, the rate of transformation is very slow, then it increases rapidly, and finally it slows down toward the end.

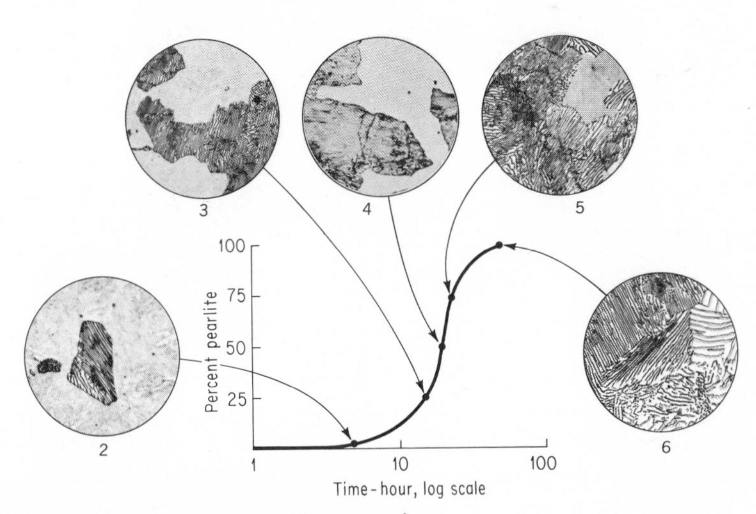

Fig. 8·15 Typical isothermal-transformation curve of austenite to pearlite for a 1080(eutectoid) steel at 1300°F; martensite is the light area. Micros correspond to the sample numbers given in Fig. 8·14. Magnification 500X. (Micros courtesy of Research Laboratory, U.S. Steel Corporation.)

As a result of this experiment, two points may be plotted at 1300°F, namely, the time for the beginning and the time for the end of transformation. It is also common practice to plot the time for 50 percent transformed. The entire experiment is repeated at different subcritical temperatures until sufficient points are determined to draw one curve showing the beginning of transformation, another curve showing the end of transformation, and a dotted curve in between showing 50 percent transformed (Fig. 8·16). The principal curves on the I-T diagram are drawn as broad lines to emphasize that their exact location on the time scale is not highly precise. Portions of these lines are often shown as dashed lines to indicate a much higher degree of uncertainty. Time is plotted on a logarithmic scale so that times of 1 min or less as well as times of 1 day or week, can be fitted into a reasonable space and yet permit an open scale in the region of short times. The diagram is known as an *I-T* (*isothermal-transformation*) *diagram.* Other names for the same curves are TTT (transformation, temperature, time) curves or S curves. Construction of a reasonably accurate diagram requires

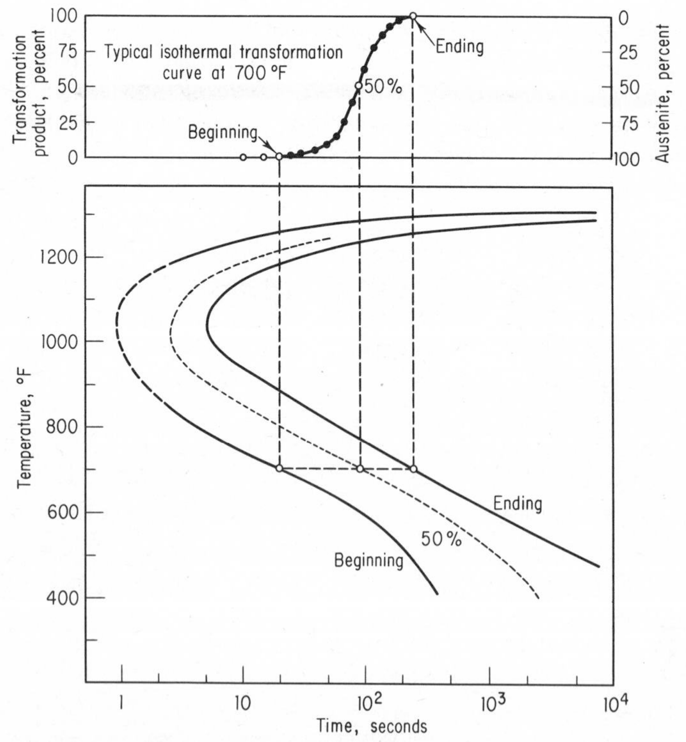

Fig. 8·16 Diagram showing how measurements of isothermal transformation are summarized by the I-T diagram. (From "Atlas of Isothermal Transformation Diagrams," U.S. Steel Corporation.)

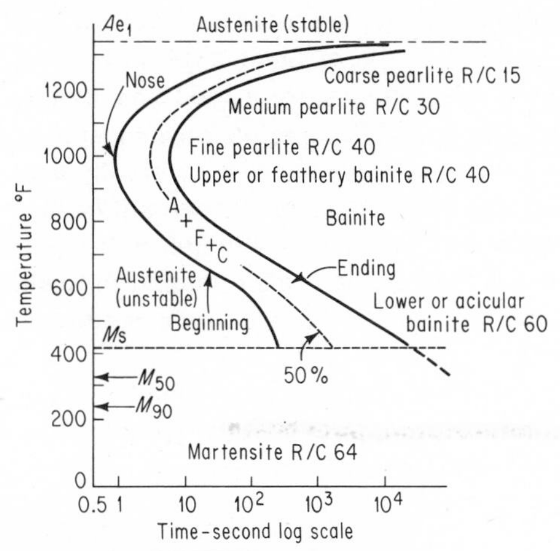

Fig. 8·17 Isothermal-transformation diagram for a 1080(eutectoid) steel.

the heat treatment and metallographic study of more than one hundred individual samples.

The I-T diagram for a 1080 eutectoid steel is shown in Fig. 8·17. Above the Ae_1 austenite is stable. The area to the left of the beginning of transformation consists of unstable austenite. The area to the right of the end-of-transformation line is the product to which austenite will transform at constant temperature. The area between the beginning and the end of transformation labeled $A + F + C$ consists of three phases, austenite, ferrite, and carbide, or austenite plus the product to which it is transforming. The point on the beginning of the transformation line farthest to the left is known as the *nose* of the diagram. In all diagrams, except the one for eutectoid steel, there is an additional line above the nose region. The first line to the left indicates the beginning of austenite transformation to proeutectoid ferrite in hypoeutectoid steels (Fig. 8·29) or proeutectoid cementite in hypereutectoid steels. The second line indicates the beginning of austenite transformation to pearlite. The area between the two lines is labeled $A + F$ (austenite plus proeutectoid ferrite), or $A + C$ (austenite plus proeutectoid cementite). The two lines generally merge at the nose region.

The M_s temperature is indicated as a horizontal line. Arrows pointing to the temperature scale indicate the temperature at which 50 and 90 percent of the total austenite will, on quenching, have transformed to martensite. In some diagrams, the data on martensite formation were obtained by direct measurement using a metallographic technique. In other diagrams, the temperatures were calculated by an empirical formula. To determine the progress of martensite formation metallographically, let us drastically quench a sample to a temperature below the M_s line, say 350°F. Approximately 20 percent of the austenite will have transformed to martensite. If this sample is now reheated for a short time to a temperature below the lower critical line, say 800°F, the martensite just formed will become dark. The austenite will remain unchanged. Upon quenching, this sample will show 20 percent dark martensite and 80 percent fresh or white martensite, and the amount formed at 350°F may therefore be determined microscopically. A typical example of the transformation of austenite to martensite determined by the heat-treating procedure described above is shown in Fig. 8·18.

8·9 Transformation to Pearlite and Bainite Returning to Fig. 8·17, the transformation product above the nose region is pearlite. The pearlite microstructure is the characteristic lamellar structure of alternate layers of fer-

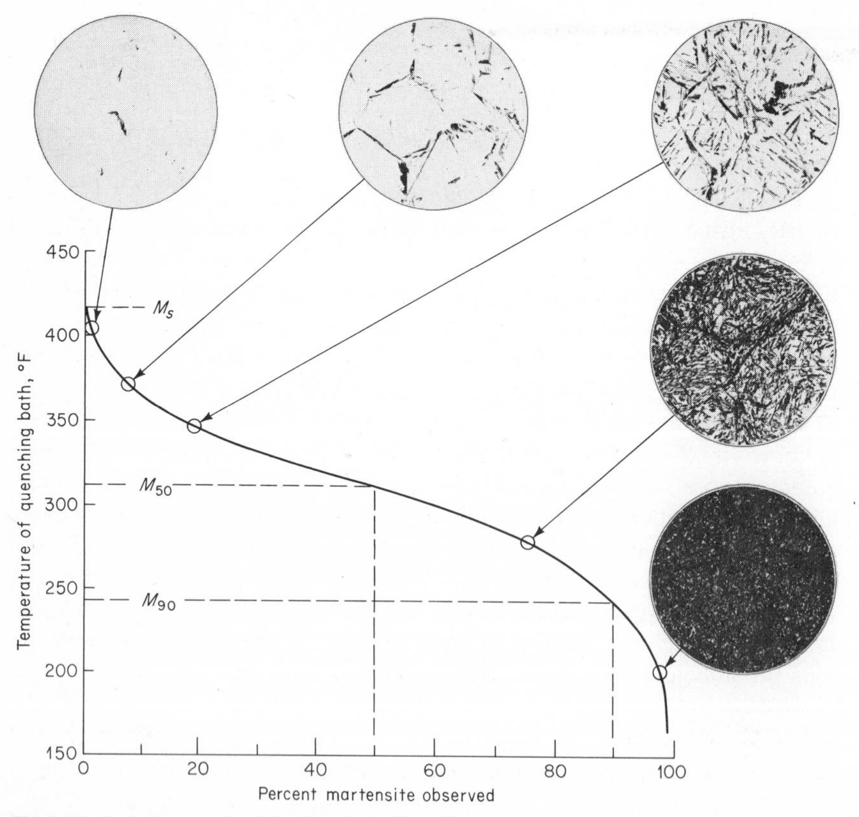

Fig. 8·18 Typical example of the transformation of austenite to martensite. Microstructures at 500X. (From "Atlas of Isothermal Transformation Diagrams," U.S. Steel Corporation.)

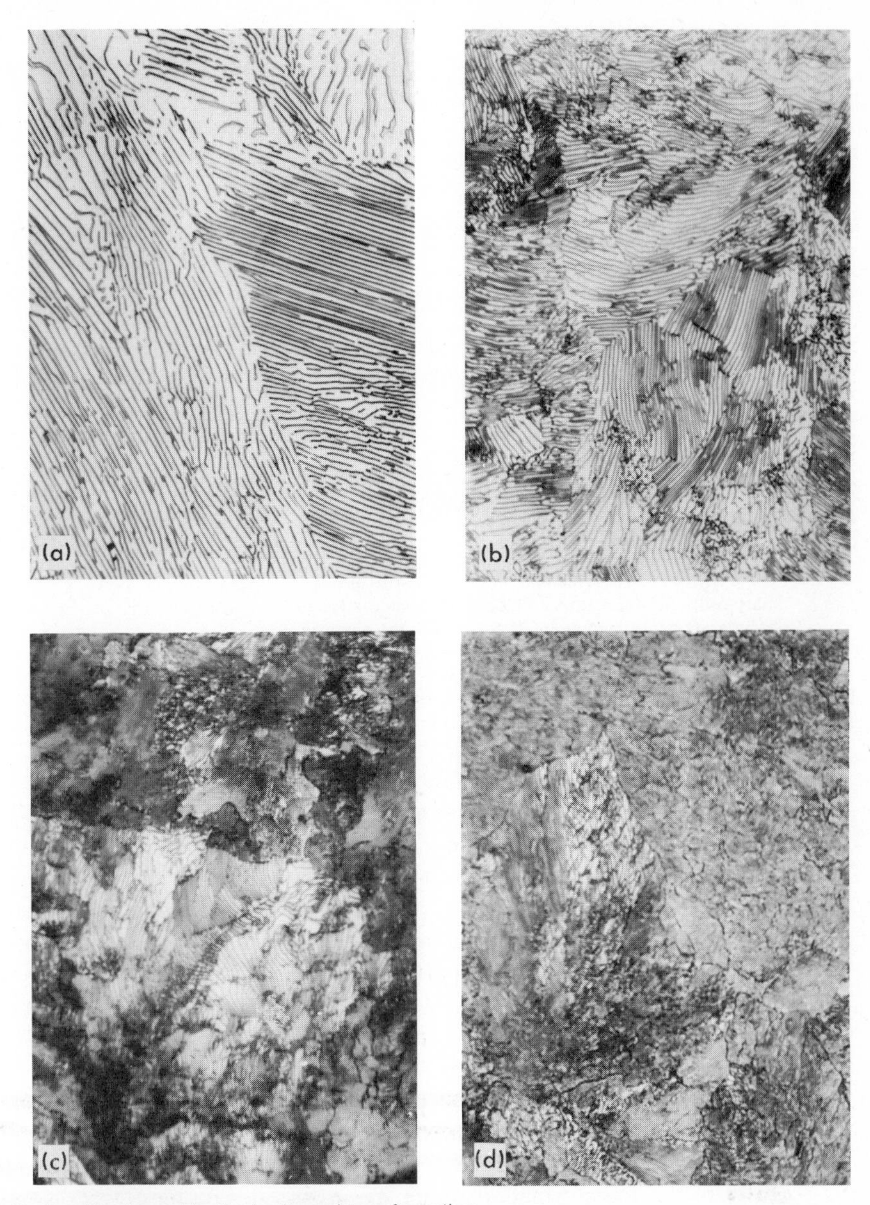

Fig. 8·19 Pearlites formed by the isothermal transformation of austenite at various subcritical temperatures; (*a*) 1300°F, (*b*) 1225°F, (*c*) 1150°F, (*d*) 1075°F. Magnified 1,500X. Note the increase in the fineness of pearlite with decreasing transformation temperature. (Courtesy of Research Laboratory, U.S. Steel Corporation.)

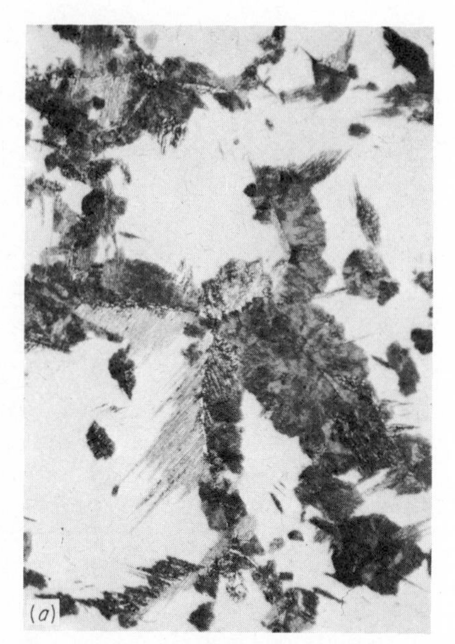

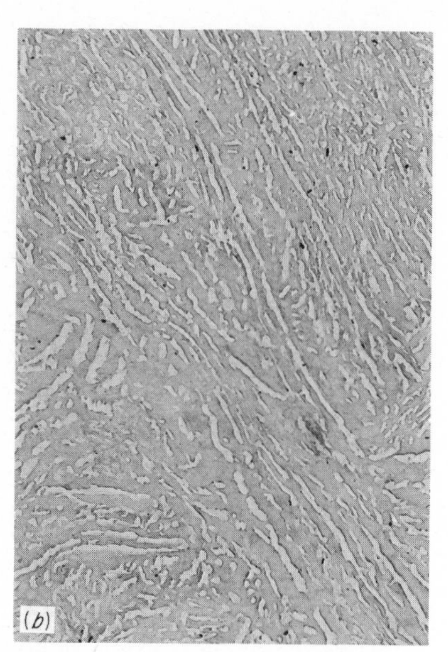

Fig. 8·20 (*a*) Feathery bainite and fine pearlite in a martensitic (white) matrix, 1,000X. (*b*) The microstructure of bainite transformed at 850°F, taken with the electron microscope, 15,000X. (Courtesy of Research Laboratory, U.S. Steel Corporation.)

rite and cementite described in Chap. 7. Just below the Ae_1 line, coarse lamellar pearlite is formed with a hardness of about Rockwell C 15. As the transformation temperature decreases, the characteristic lamellar structure is maintained, but the spacing between the ferrite and carbide layers becomes increasingly smaller until the separate layers cannot be resolved with the light microscope (see Fig. 8·19). As the temperature of transformation and the fineness of the pearlite decreases, it is apparent that the hardness will increase. As explained under normalizing, this hardness increase is the result of decreasing the spacing between plates of the hard constituent, cementite, within the soft ferrite matrix.

Between the nose region of approximately 950°F and the M_s temperature, a new, dark-etching aggregate of ferrite and cementite appears. This structure, named after E. C. Bain, is called *bainite*. At upper temperatures of the transformation range, it resembles pearlite and is known as upper or feathery bainite (Fig. 8·20*a*). At low temperatures it appears as a black needlelike structure (Fig. 8·21*a*) resembling martensite and is known as lower or acicular bainite. These photomicrographs show the gross structure of bainite, and the electron microscope is required to resolve

(a)

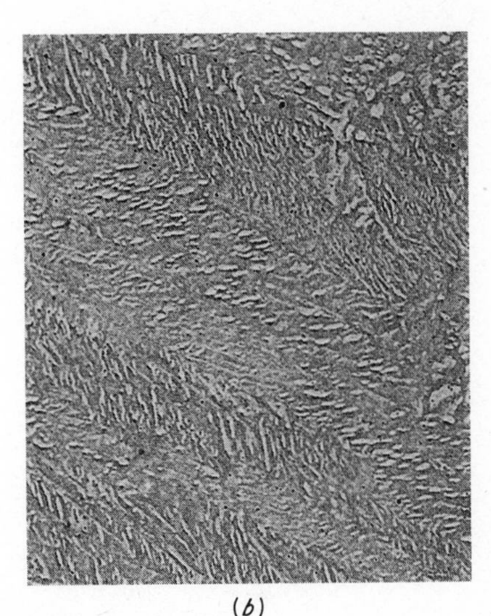

(b)

Fig. 8·21 (a) Acicular or lower bainite, black needles in a martensitic (white) matrix, 2,500X. (b) The microstructure of bainite transformed at 500°F, taken with the electron microscope, 15,000X. (Courtesy of Research Laboratory, U.S. Steel Corporation.)

the details. Figure 8·20b shows the structure of bainite transformed at 850°F and magnified 15,000 times. It consists of tiny carbide platelets generally oriented parallel with the long direction of the ferrite needles which make up the matrix. As the transformation temperature decreases, the ferrite needles become thinner and the carbide platelets become smaller and more closely spaced. The carbide platelets are usually oriented at an angle of about 60° to the long axis of the ferrite needles, rather than parallel to this direction (see Fig. 8·21b, which shows the structure of bainite transformed at 500°F and magnified 15,000 times). Whereas pearlite is nucleated by a carbide crystal, bainite is nucleated by a ferrite crystal, and this results in a different growth pattern, as illustrated in Fig. 8·22.

The hardness of bainite varies from about Rockwell C 40 for upper bainite to about Rockwell C 60 for lower bainite. This hardness increase, as with pearlite, is a reflection of the decrease in size and spacing of the carbide platelets as the transformation temperature decreases. The mechanical properties of a completely bainitic structure will be covered later in this chapter under austempering.

8·10 Cooling Curves and the I-T Diagram A cooling curve is determined experimentally by placing a thermocouple at a definite location in a steel sample and then measuring the variation of temperature with time. Since the coordinates of the I-T diagram are the same as those for a cooling curve, it is possible to superimpose various cooling curves on the I-T diagram. This was done in Fig. 8·23.

Cooling curve 1 shows a very slow cooling rate typical of conventional annealing. The diagram indicates that the material will remain austenitic for a relatively long period of time. Transformation will start when the cooling curve crosses the beginning of transformation at point x_1. The transformation product at that temperature will be very coarse pearlite. Transformation will continue until point x'_1. Since there is a slight difference in temperature between the beginning and end of transformation, there will be a slight difference in the fineness of pearlite formed at the beginning and at the end. The overall product will be coarse pearlite with low hardness. Below the temperature of x'_1 the rate of cooling will have no effect on the microstructure or properties. The material may now be cooled rapidly without any change occurring. This is of great value to companies doing commercial annealing, since the diagram indicates that it is not necessary to cool in the furnace to room temperature but that the material may be removed at a relatively high temperature after transformation and cooled in air.

Cooling curve 2 illustrates "isothermal" or "cycle annealing" and was developed directly from the I-T diagram. The process is carried out by cooling the material rapidly from above the critical range to a predetermined temperature in the upper portion of the I-T diagram and holding for the time indicated to produce complete transformation. In contrast to conventional annealing, this treatment produces a more uniform microstructure and hardness, in many cases with a shorter time cycle.

Cooling curve 3 is a faster cooling rate than annealing and may be considered typical of normalizing. The diagram indicates that transformation will start at x_3, with the formation of coarse pearlite, in a much shorter time than annealing. Transformation will be complete at x'_3 with the formation of medium pearlite. Since there is a greater temperature difference between x_3 and x'_3 than there is between x_1 and x'_1, the normalized micro-

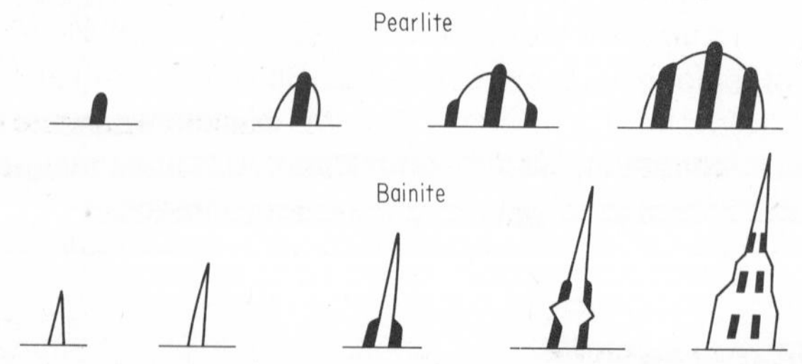

Fig. 8·22 Growth of pearlite, nucleated by a carbide crystal, and of bainite, nucleated by a ferrite crystal with carbide rejected as discontinuous small platelets. (After Hultgren.)

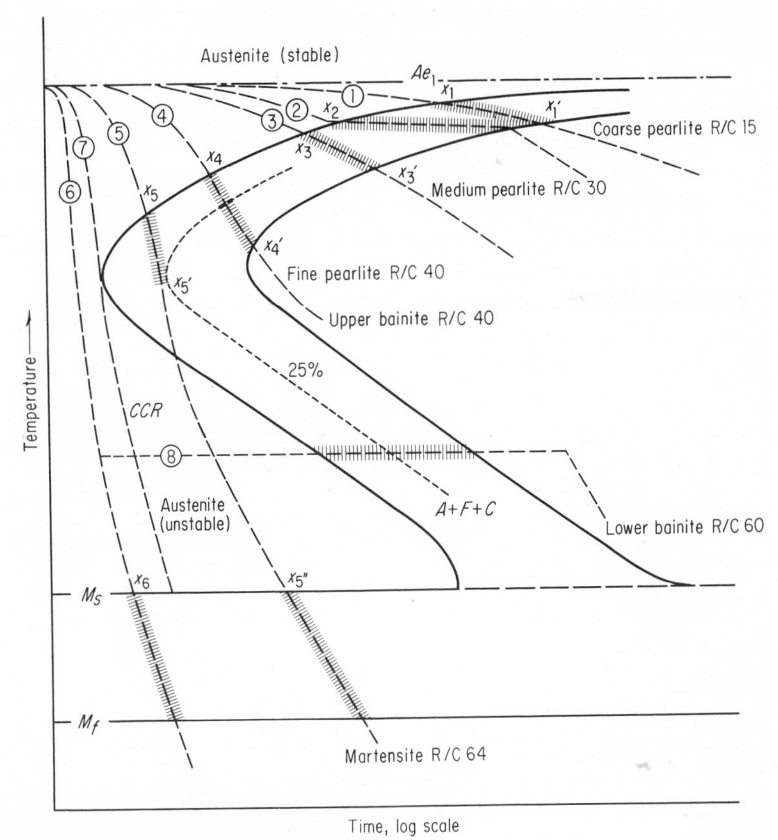

Fig. 8·23 Cooling curves superimposed on a hypothetical I-T diagram for a eutectoid steel. Cross-hatched portion of the cooling curve indicates transformation.

structure will show a greater variation in the fineness of pearlite and a smaller proportion of coarse pearlite than the annealed microstructure.

Cooling curve 4, typical of a slow oil quench, is similar to the one just described, and the microstructure will be a mixture of medium and fine pearlite.

Cooling curve 5, typical of an intermediate cooling rate, will start to transform (at x_5) to fine pearlite in a relatively short time. The transformation to fine pearlite will continue until the curve becomes tangent to some percentage transformed, say 25 percent, at x'_5. Below this temperature, the cooling curve is going in a direction of decreasing percent transformed. Since pearlite cannot form austenite on cooling, the transformation must stop at x'_5. The microstructure at this point will consist 25 percent of fine, nodular pearlite largely surrounding the existing austenitic grains. It will remain in this condition until the M_s line is crossed at x''_5. The remaining austenite now transforms to martensite. The final microstructure at room temperature will consist of 75 percent martensite (white areas) and 25 percent fine nodular pearlite largely concentrated along the original austenite grain boundaries (Fig. 8·24). If only a small amount of pearlite is present, the black etching nodules of pearlite in the white martensitic matrix nicely reveal the former austenitic grain size when the transformation to pearlite took place.

Cooling curve 6, typical of a drastic quench, is rapid enough to avoid

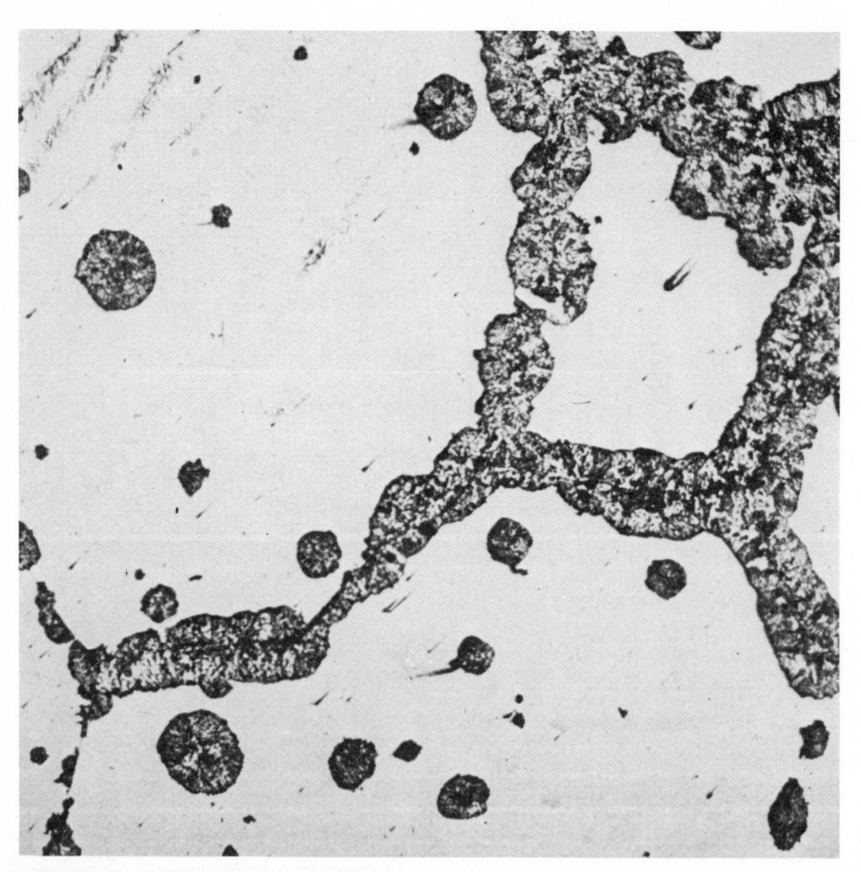

Fig. 8·24 Microstructure illustrating an intermediate cooling rate. Fine modular pearlite largely at the original austenite grain boundaries in a martensitic matrix. Etched in 2 percent nital, 150X. (From A. Sauveur, "The Metallography and Heat Treatment of Iron and Steel," McGraw-Hill Book Company, New York, 1935.)

transformation in the nose region. It remains austenitic until the M_s line is reached at x_6. Transformation to martensite will take place between the M_s and M_f lines. The final microstructure will be entirely martensite of high hardness.

It is apparent that to obtain a fully martensitic structure it is necessary to avoid transformation in the nose region. Therefore, cooling rate 7, which is tangent to the nose, would be the approximate critical cooling rate (CCR) for this steel. Any cooling rate slower than the one indicated will cut the curve above the nose and form some softer transformation product. Any cooling rate faster than the one illustrated will form only martensite. Different steels may be compared on the basis of their critical cooling rates.

Notice that it is possible to form 100 percent pearlite or 100 percent martensite by continuous cooling, but it is not possible to form 100 percent bainite. A complete bainitic structure may be formed only by cooling rapidly enough to miss the nose of the curve and then holding in the temperature range at which bainite is formed until transformation is complete. This is illustrated by cooling rate 6 then 8 in Fig. 8·23. It is apparent that continuously cooled steel samples will contain only small amounts of bainite, and this is probably the reason why this structure was not recognized until the isothermal study.

8·11 Transformation on Continuous Cooling Theoretically, cooling-rate curves should not be superimposed on the I-T diagram as was done in the previous section. The I-T diagram shows the time-temperature relationship for austenite transformation only as it occurs at constant temperature, but most heat treatments involve transformation on continuous cooling. It is possible to derive from the I-T diagram another diagram which will show the transformation under continuous cooling.* This is referred to as the C-T diagram (cooling-transformation diagram). Figure 8·25 shows the C-T diagram for a eutectoid steel superimposed on the I-T diagram from which it was derived. Consideration of the I-T diagram in relation to the location of lines of the C-T diagram shows that the "nose" has been moved downward and to the right by continuous cooling. The critical cooling rate tangent to the nose of the C-T diagram is shown as 250°F/s. This is somewhat slower than the rate indicated by the I-T diagram. Therefore, the use of isothermal "nose" times to determine required cooling rates will lead to some error; however, the error will be on the safe side in indicating a slightly faster cooling rate than is actually necessary to form only martensite. Notice the absence of an austenite-to-bainite region in the C-T diagram. In this steel the bainite range is "sheltered" by the overhanging pearlite nose, and bainite is not formed in any appreciable quantity on ordi-

* R. A. Grange and J. M. Kiefer, Transformation of Austenite on Continuous Cooling and Its Relation to transformation at Constant Temperature, *Trans. A.S.M.*, vol. 29, no. 85, 1941.

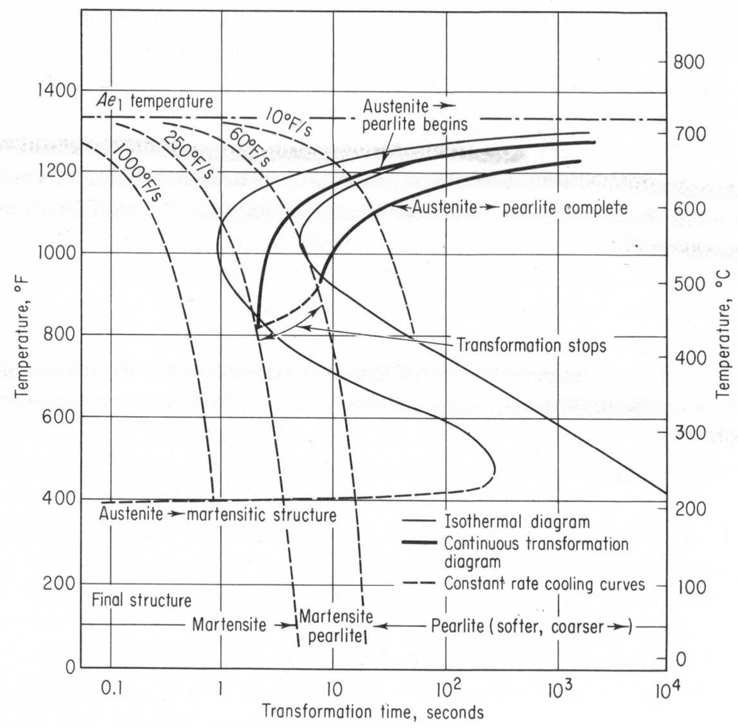

Fig. 8·25 Continuous cooling-transformation (C-T) diagram derived from the isothermal-transformation diagram for a plain-carbon eutectoid steel. (From "Atlas of Isothermal Transformation Diagrams," U.S. Steel Corporation.)

nary continuous cooling. This situation is generally different for alloy steels. Figure 8·26 shows the C-T diagram for a triple-alloy steel. This is a hypoeutectoid steel, so there is an additional area, austenite-to-ferrite, which was not present in the eutectoid steel. In this alloy steel, the pearlite zone lies relatively far to the right and does not shelter the bainite region. Therefore, with cooling rates between 2,100 and 54,000°F/h it is possible to obtain appreciable amounts of bainite in the microstructure. Notice that the cooling rate tangent to the "upper nose" (2,100°F/h) is not the critical cooling rate. The cooling rate tangent to the "lower nose" or "knee" of the diagram (54,000°F/h) would have to be exceeded to form only martensite.

The derivation of a C-T diagram is a tedious task and, for many purposes, not essential provided that the fundamental relationship of the C-T diagram to the corresponding I-T diagram is understood. Isothermal studies have greatly aided the classification of the microstructure of steel transformed during continuous cooling, and with the I-T diagram it is possible to visualize at what stage of the cooling cycle different structures are formed. The I-T diagram is useful in planning heat treatments and in understanding why steel responds as it does to a particular heat treatment, but it cannot be used directly to predict accurately the course of transformation under continuous cooling.

It should be noted that a particular I-T diagram exactly represents only one group of samples; samples from other heats, or even from other locations in the same heat, are likely to have slightly different I-T diagrams. When used with its limitations in mind, the I-T diagram is useful in inter-

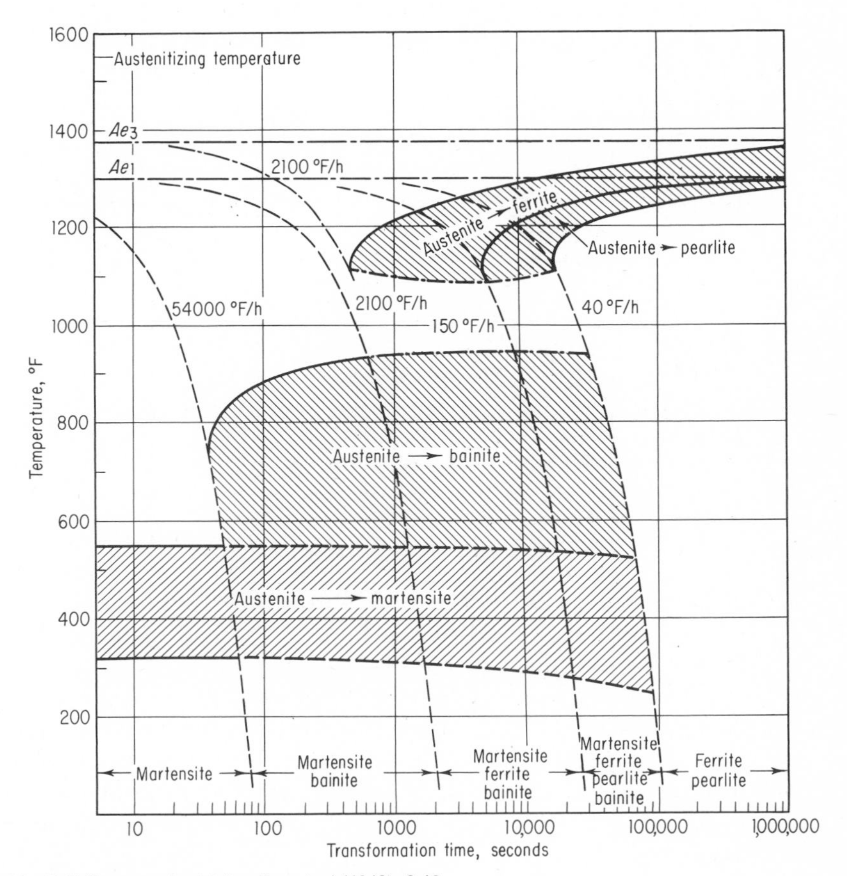

Fig. 8·26 C-T diagram of a triple-alloy steel (4340): 0.42 percent carbon, 0.78 percent manganese, 1.79 percent nickel, 0.80 percent chromium, 0.33 percent molybdenum. (From "U.S.S. Carilloy Steels," U.S. Steel Corporation.)

preting and correlating observed transformation phenomena on a rational basis, even though austenite transforms during continuous cooling rather than at constant temperature.

8·12 Position of the I-T Curves There are only two factors that will change the position of the curves of the I-T diagram, namely, chemical composition and austenitic grain size. With few exceptions, an increase in carbon or alloy content or in grain size of the austenite always retards transformation (moves the curves to the right), at least at temperatures at or above the nose region. This in turn slows up the critical cooling rate, making it easier to form martensite. This retardation is also reflected in the greater hardenability, or depth of penetration of hardness, of steel with higher alloy content or larger austenitic grain size.

The effect of increasing carbon content may be seen by referring to Figs. 8·27 and 8·29. Figure 8·27 shows the I-T diagram of a 1035 steel. The M_s

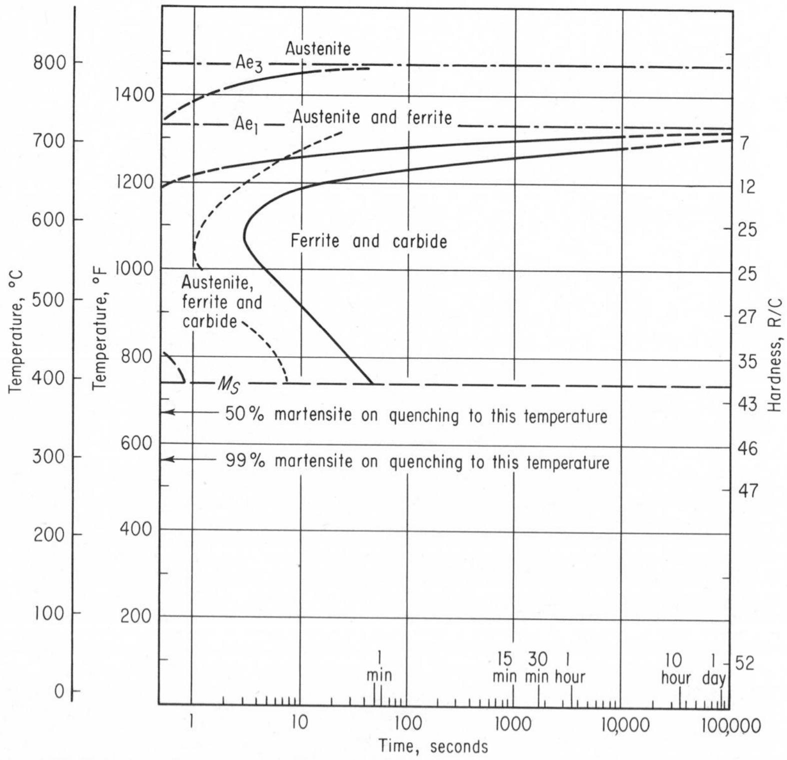

Fig. 8·27 I-T diagram of a 1035 steel: 0.35 percent carbon, 0.37 percent manganese. Grain size: 75 percent, 2 to 3; 25 percent, 7 to 8. Austenitized at 1550°F. (From "Atlas of Isothermal Transformation Diagrams," U.S. Steel Corporation.)

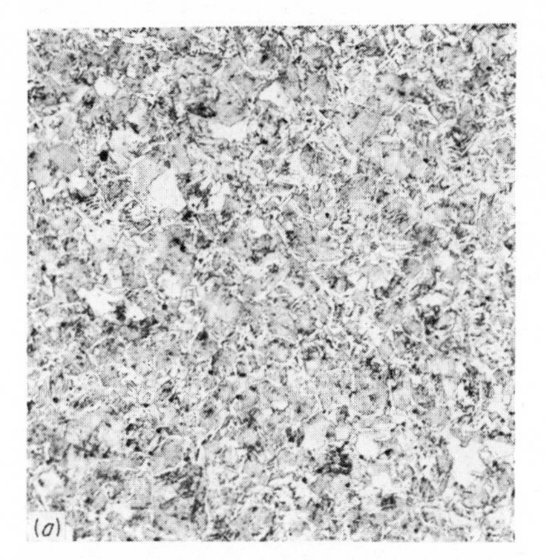

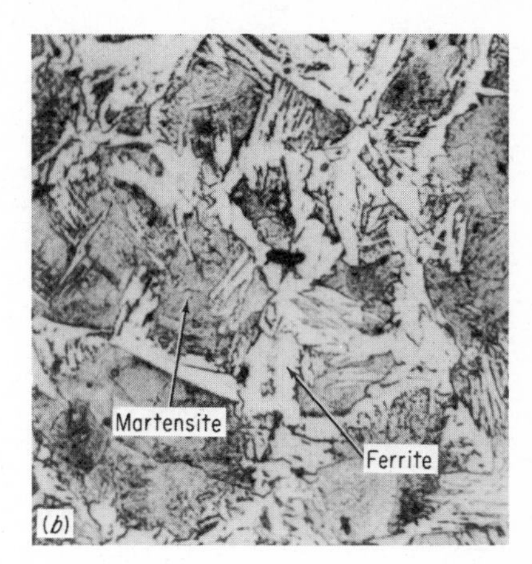

Fig. 8·28 Microstructure of a low-carbon steel, water-quenched, showing a white ferrite network surrounding the gray low-carbon martensite areas. (*a*) 100x; (*b*) 500x. Etched in 2 percent nital.

temperature is approximately 750°F. Since this is a hypoeutectoid steel, notice the presence of the austenite-to-ferrite region. The nose of the curve is not visible, indicating that it is very difficult to cool this steel fast enough to obtain only martensite. The microstructure of a low-carbon steel water-quenched, Fig. 8·28, shows a white ferrite network surrounding the gray low-carbon martensite areas. Figure 8·29 shows the I-T diagram for a 1050 steel. The increase in carbon content has shifted the curve far enough to the right to make the nose just visible, and the M_s temperature has been reduced to 620°F. Theoretically, in order to form only martensite, it is necessary to cool fast enough to get by 1000°F in approximately 0.7 s. Notice that the $A + F$ region has become much narrower and disappears in the vicinity of the nose. The microstructure of a medium-carbon steel water-quenched, Fig. 8·30, shows dark areas of fine pearlite that seem to outline some of the previous austenite grain boundaries, some dark feathery bainite, and substantially more martensite as the matrix than appeared in the low-carbon steel.

Although alloy additions tend in general to delay the start of transformation and to increase the time for its completion, they differ greatly in both magnitude and nature of their effects. Figure 8·31 shows the I-T diagram of a 1335 manganese steel. Comparison of this diagram to Fig. 8·27 shows that the addition of 1.50 percent manganese has shifted the entire curve to the right. The nose of the curve is now visible, and it should be possible to completely harden this steel much more easily than the plain carbon steel. Notice that the end-of-transformation line now shows an S shape. Where there is a pronounced minimum time in the ending line at relatively

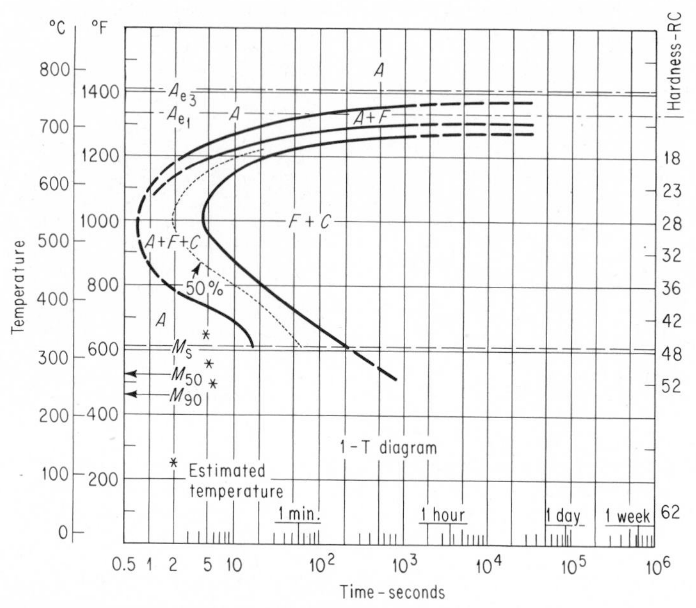

Fig. 8·29 I-T diagram of a 1050 steel: 0.50 percent carbon, 0.91 percent manganese. Grain size: 7 to 8; austenitized at 1670°F. (From "Atlas of Isothermal Transformation Diagrams," U.S. Steel Corporation.)

high temperature, it is possible to take advantage of this minimum to design a short annealing cycle. In this diagram the upper minimum (point x) is approximately 5 min at 1100°F for the end of transformation, whereas at 1200°F the end of transformation is approximately 1 h. Therefore, isother-

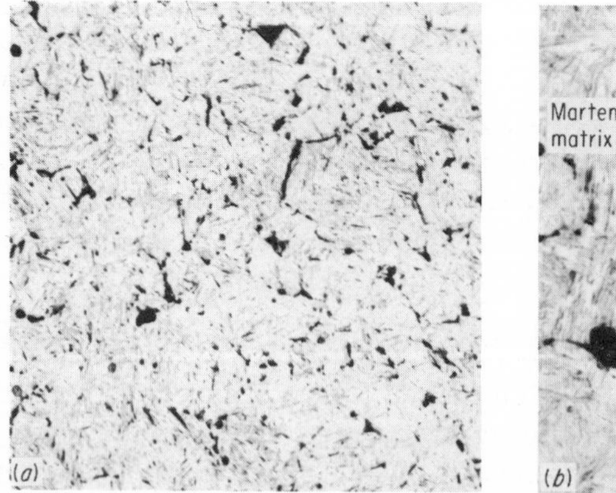

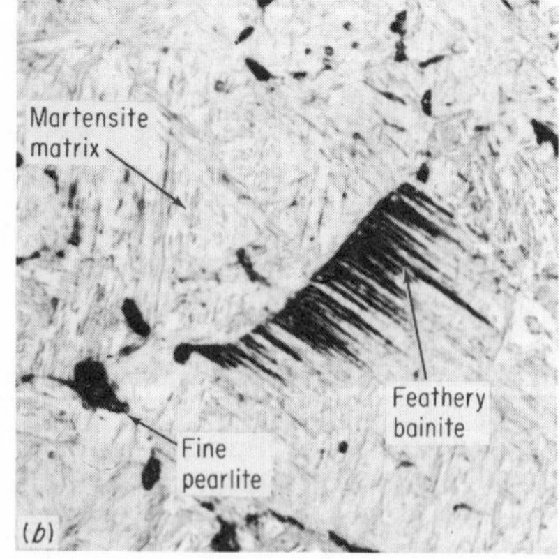

Fig. 8·30 Microstructure of a medium-carbon steel, water-quenched, showing dark areas of fine pearlite that seem to outline some of the previous austenite grain boundaries, some dark feathery bainite, and a matrix of martensite. (*a*) 100x; (*b*) 750x. Etched in 2 percent nital.

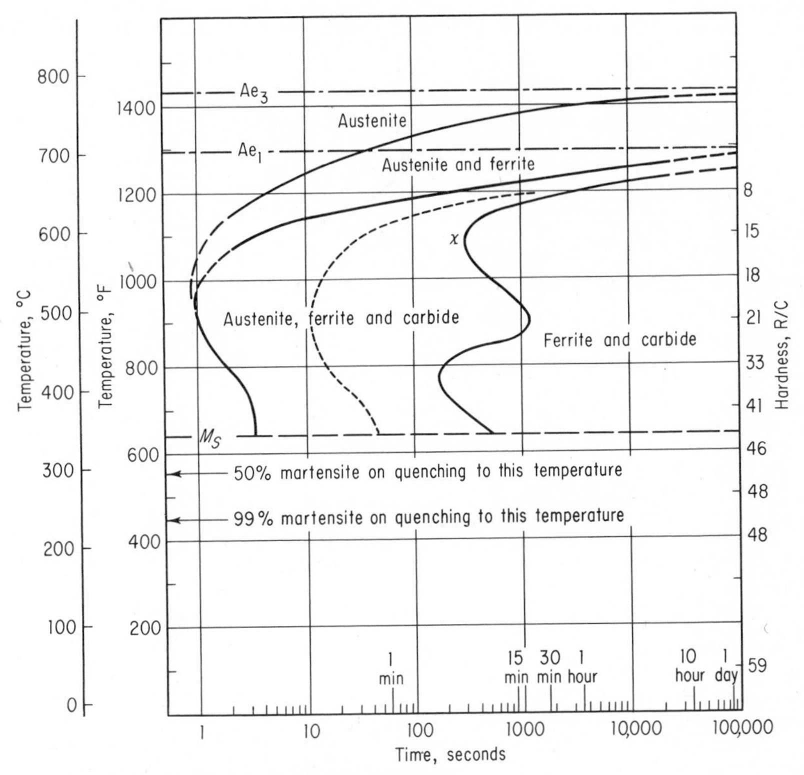

Fig. 8·31 I-T diagram of a 1335 steel. 0.35 percent carbon, 1.85 percent manganese. Grain size: 70 percent, 7; 30 percent, 2. Austenitized at 1550°F. (From "Atlas of Isothermal Transformation Diagrams," U.S. Steel Corporation.)

mal annealing at 1100°F would be complete in a relatively short time. Figure 8·32 shows the effect of chromium. The addition of 2 percent chromium has not only shifted the curve to the right but has changed its shape, particularly in the region of 900 to 1200°F. The effect of alloy additions is cumulative. This is illustrated in the C-T diagram for a triple-alloy steel (Fig. 8·26). The critical cooling rate for this steel is shown to be approximately 54,000°F/h, or about 15°F/s as compared with the critical cooling rate of 250°F/s for the eutectoid steel shown in Fig. 8·25. It is apparent that this eutectoid steel has a very fast critical cooling rate and low hardenability. The triple-alloy steel has a slow critical cooling rate, is deep-hardening, and illustrates how much easier it is to form martensite by the addition of alloying elements. This is one of the principal reasons for alloying steel. In general, the relative effect of the common alloying elements on the movement of the nose of the I-T diagram to the right is: vanadium (strongest), tungsten, molybdenum, chromium, manganese, silicon, and nickel (weakest).

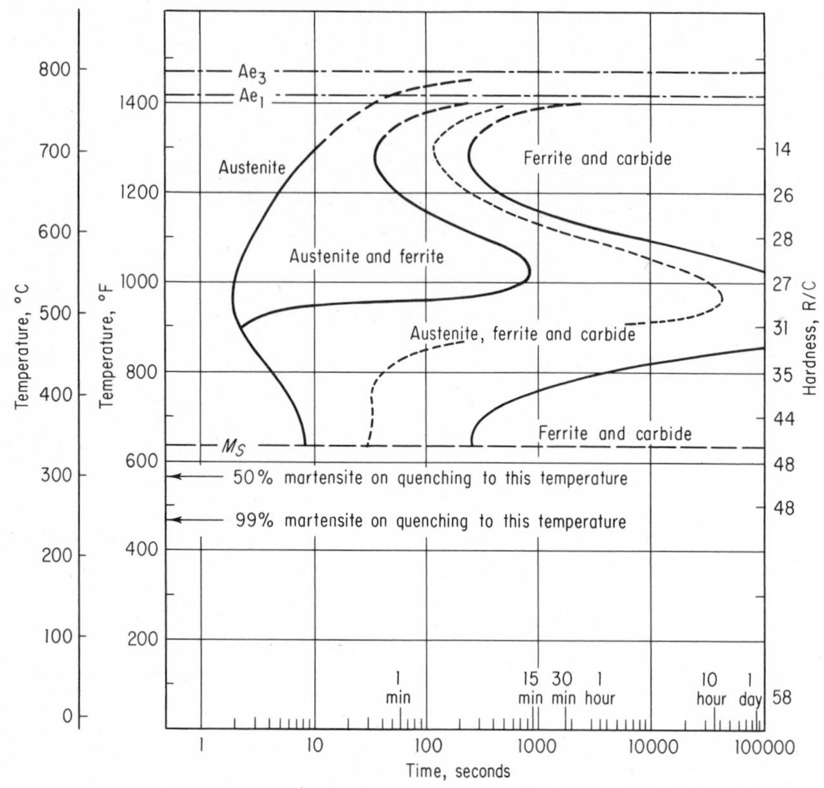

Fig. 8·32 I-T diagram of a 0.33 percent carbon, 0.45 percent manganese, 1.97 percent chromium steel. Grain size: 6 to 7. Austenitized at 1600°F. (From "Atlas of Isothermal Transformation Diagrams." U.S. Steel Corporation.)

The effect on the retardation of the critical cooling rate by coarsening the grain size of the austenite is shown in Fig. 8·33. The numbers that indicate grain size are related to the number of grains in a unit area, so that the higher the number the finer the grain size. Notice the considerable shift of the curves to the right by the higher austenitizing temperature and the resultant coarse austenitic grain size.

To summarize, there are only two factors that will decrease the critical cooling rate or move the I-T diagram to the right. One is increasing the amount of carbon and alloying elements added, and the other is coarsening the austenitic grain size. While the addition of alloying elements does not affect the maximum hardness obtainable from the steel, that property being controlled by carbon content only, they make it much easier to completely harden the steel. A plain carbon steel may have to be water-quenched to obtain full hardness, while the same steel when alloyed may attain the same hardness and may be hardened to a greater depth even when cooled more slowly by oil quenching. The use of a slower cooling rate reduces the danger of distortion and cracking during heat treatment. While coarsening the austenitic grain size has an effect similar to that of adding alloying elements, the coarser grain will tend to reduce the toughness of the steel. Therefore, if it is desired to reduce the critical cooling rate, this may be done best by changing the chemical composition rather than by coarsening the austenitic grain.

8·13 Hardening or Austenitizing Temperature The recommended austenitizing temperature for hypoeutectoid steels is about 50°F above the upper-critical-temperature (A_3) line. This is the same as the recommended annealing temperature. At any temperature below the A_3 line there will be some proeutectoid ferrite present which will remain after quenching, giving rise to soft spots and lower hardness.

For plain-carbon hypereutectoid steels the recommended austenitizing

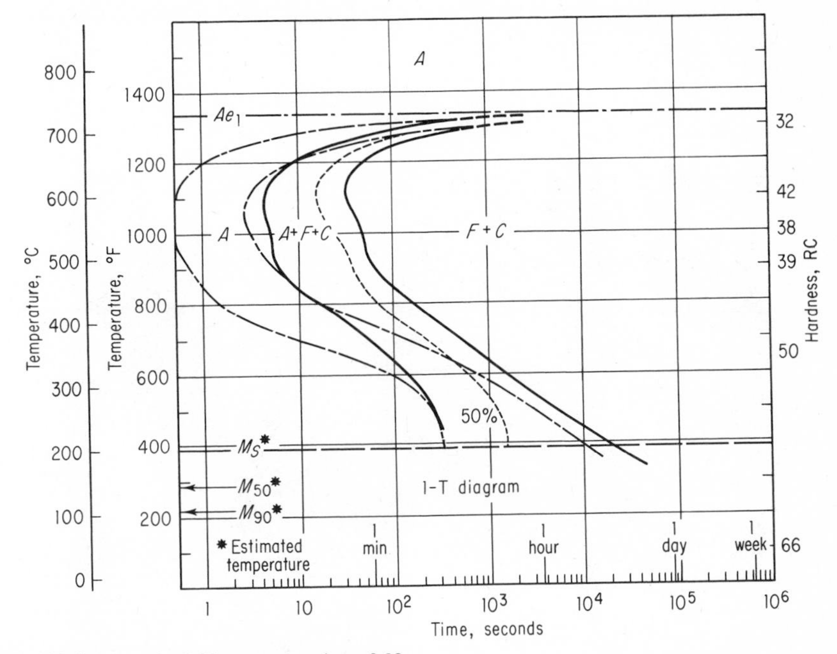

Fig. 8·33 I-T diagram of a 0.87 percent carbon, 0.30 percent manganese, 0.27 percent vanadium steels.
——— Grain size: 2 to 3, austenitized at 1925°F.
-------- Grain size: 11, austenitized at 1500°F, shows the shift of the I-T curve to the right by coarsening the austenitic grain. (From "Atlas of Isothermal Transformation Diagrams," U.S. Steel Corporation.)

temperature is usually between the A_{cm} and $A_{3,1}$ lines (Fig. 8·3); therefore, undissolved carbides would tend to be present in the microstructure at room temperature. Figure 8·34 shows the microstructure of a hypereutectoid steel, oil-quenched from between the A_{cm} and $A_{3,1}$ lines. Notice the small white undissolved carbide spheroids clearly visible in the dark fine pearlite areas. They are much more difficult to see in the light martensitic matrix.

The A_{cm} line rises so steeply that an excessively high temperature may be required to dissolve all the proeutectoid cementite in the austenite. This tends to develop undesirable coarse austenitic grain size, with danger of cracking on cooling.

8·14 Homogeneity of Austenite This refers to the uniformity in carbon content of the austenite grains. If a hypoeutectoid steel is heated for hardening, when the A_1 line is crossed, the austenite grains formed from pearlite will contain 0.8 percent carbon. With continued heating, the austenite grains formed from proeutectoid ferrite will contain very little carbon, so that when the A_3 line is crossed, the austenite grains will not be uniform in carbon content. Upon quenching, the austenite grains leaner in carbon, having a

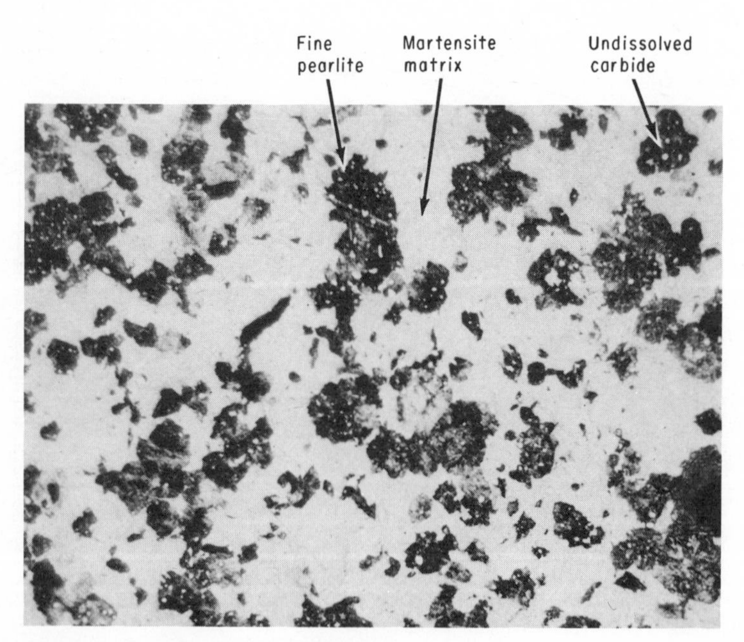

Fig. 8·34 Microstructure of a high-carbon steel, oil-quenched from between the A_{cm} and $A_{3,1}$ lines. Fine pearlite (dark) in a martensitic (light) matrix. Etched in 2 percent nital, 500X. Notice the small white undissolved carbide spheroids in the dark pearlitic areas.

fast critical cooling rate, tend to transform to nonmartensitic structures, while those richer in carbon, having a slower critical cooling rate, tend to form martensite. This results in a nonuniform microstructure with variable hardness. This condition may be avoided by very slow heating so that uniformity is established by carbon diffusion during heating. However, the excessive time required by this method does not make it commercially practical. A more suitable method is to soak the material at the austenitizing temperature. At this temperature diffusion of carbon is rapid, and uniformity will be established in a short time. To be on the safe side, it is recommended that the material be held at the austenitizing temperature 1 h for each inch of thickness or diameter.

8·15 Mechanism of Heat Removal During Quenching The structure, hardness, and strength resulting from a heat-treating operation are determined by the actual cooling rating obtained by the quenching process. If the actual cooling rate exceeds the critical cooling rate, only martensite will result. If the actual cooling rate is less than the critical cooling rate, the part will not completely harden. The greater the difference between the two cooling rates the softer will be the transformation products and the lower the hardness. At this point, it is necessary to understand the mechanism of heat removal during quenching.

To illustrate, a typical cooling curve for a small steel cylinder quenched in warm water is shown in Fig. 8·35. Instead of showing a constant cooling rate throughout the quench, the cooling curve shows three stages. Keep in mind the difference between a cooling curve and a cooling rate. A cooling curve shows the variation of temperature with time during quenching. A cooling rate, however, shows the rate of change of temperature with time. The cooling rate at any temperature may be obtained from the cooling curve by drawing a tangent to the curve at that temperature and determining the slope of the tangent. The more nearly horizontal the tangent, the slower is the cooling rate. Visualizing tangents drawn at various points on the cooling curve in Fig. 8·35, it is apparent that the cooling rate is constantly changing during cooling.

Stage A—Vapor-blanket Cooling State In this first stage, the temperature of the metal is so high that the quenching medium is vaporized at the surface of the metal and a thin stable film of vapor surrounds the hot metal. Cooling is by conduction and radiation through the gaseous film, and since vapor films are poor heat conductors, the cooling rate is relatively slow through this stage.

Stage B—Vapor-transport Cooling Stage This stage starts when the metal has cooled to a temperature at which the vapor film is no longer stable. Wetting of the metal surface by the quenching medium and violent boiling occur. Heat is removed from the metal very rapidly as the latent heat of vaporization. This is the fastest stage of cooling.

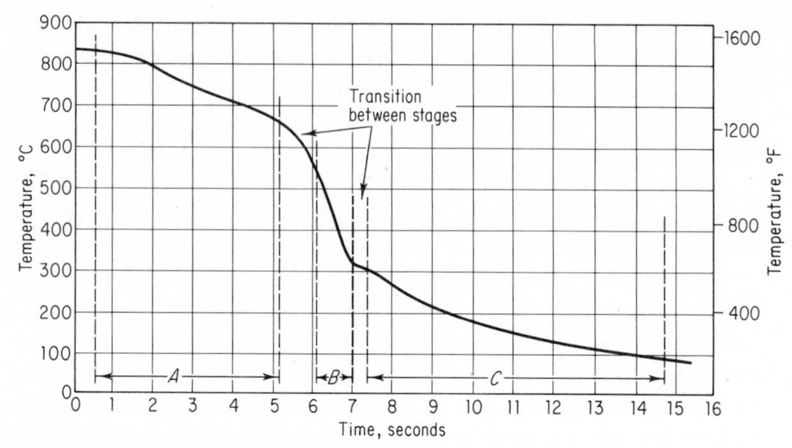

Fig. 8·35 Typical cooling curve for a small cylinder quenched in warm water. (Gulf Oil Corporation.)

Stage C—Liquid Cooling Stage This stage starts when the surface temperature of the metal reaches the boiling point of the quenching liquid. Vapor no longer forms, so cooling is by conduction and convection through the liquid. The rate of cooling is slowest in this stage.

Many factors determine the actual cooling rate. The most important are the type of quenching medium, the temperature of the quenching medium, the surface condition of the part, and the size and mass of the part.

8·16 Quenching Medium In view of the mechanism of heat removal, the ideal quenching medium would show a high initial cooling rate to avoid transformation in the nose region of the I-T diagram and then a slow cooling rate throughout the lower-temperature range to minimize distortion. Unfortunately, there is no quenching medium that exhibits these ideal properties. Water and water solutions of inorganic salts have high initial cooling rates through the A and B stages, but these high cooling rates persist to low temperatures where distortion and cracking tend to occur. Conventional quenching oils have a longer A, or vapor-blanket, stage and a shorter B stage with a slower rate of cooling.

The following industrial quenching media are listed in order of decreasing quenching severity:

1 Water solution of 10 percent sodium chloride (brine)
2 Tap water
3 Fused or liquid salts
4 Soluble oil and water solutions
5 Oil
6 Air

The cooling curves obtained by different media in the center of a 1/2-in.-

diameter stainless-steel bar are shown in Fig. 8·36. The curve furthest to the left is a 10 percent brine solution at 75°F. Notice that this quenching medium has a very short vapor stage lasting about 1 s and then drops quickly into the boiling stage, where the cooling rate is very rapid. It finally goes into the third stage at about 10 s. Looking at the cooling curve for tap water at 75°F, notice that the vapor stage is slightly longer than for brine. It drops into the boiling stage after about 3 s. The cooling rate during this stage, while very rapid, is not quite so fast as that for brine. The third stage is reached after about 15 s. Now examine the cooling curve for fused salt. This is usually an inorganic low-melting-point salt which is heated until it is liquid; the liquid is then used as a quenching medium. In this case, the fused salt is at 400°F. Notice that the fused salt has a very short vapor stage, approximately equal to that of brine. However, the cooling rate during the boiling stage is not so rapid as that for brine or tap water, and it reaches the third stage at about 10 s. The next two curves deal with oil, the dotted line being Gulf Super-Quench oil at 125°F and the solid line slow oil at 125°F. They both show a relatively long vapor stage, the difference being that Gulf Super-Quench enters the boiling stage after about 7 s, whereas the slow oil enters the boiling stage after about 13 s. The third stage is reached by Gulf Super-Quench after about 15 s and about

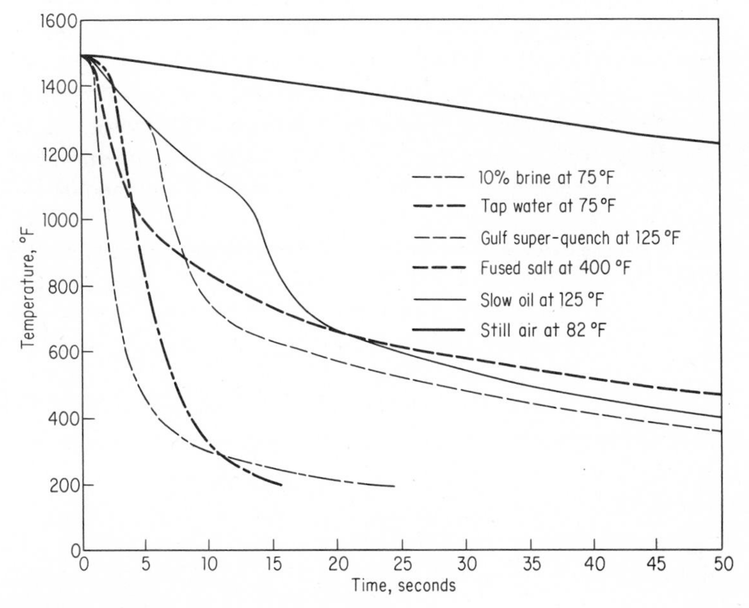

Fig. 8·36 Center-cooling curves for stainless-steel specimens, 1/2 in. diameter by 2 1/2 in. long. No agitation. (Gulf Oil Corporation.)

22 s for the slow oil. The final cooling curve for still air at 82°F never gets out of the vapor stage and therefore shows a very slow cooling rate over the entire range.

The usual methods of comparing the quenching speed of different media are by determining the rate of cooling at some fixed temperature or the average rate between two temperatures. Cooling rates of several media are given in Table 8·2.

8·17 Temperature of Quenching Medium Generally, as the temperature of the medium rises, the cooling rate decreases (Fig. 8·37; Table 8·2). This is due to the increase in persistence of the vapor-blanket stage. Since the medium is closer to its boiling point, less heat is required to form the vapor film. This is particularly true of water and brine. The figures for Gulf Super-Quench in Table 8·2, however, show an increase in cooling rate with a rise in temperature of the medium, which seems contrary to the previous statement. In the case of oil, there are two opposing factors to be considered. As the temperature of the oil rises there is a tendency for the cooling rate to decrease due to the persistence of the vapor film. However, as the temperature of the oil rises it also becomes more fluid, which increases the rate of heat conduction through the liquid. What happens to the cooling rate is determined by which factor has the greatest influence. If the increase in the rate of heat conduction is greater than the decrease due to the persistence of the vapor film, the net result will be an increase in the actual cooling rate, as in the case of Gulf Super-Quench. However, if the reverse is true, then the net result will be a decrease in actual cooling rate, as shown by the figures for slow oil. The optimum rates of cooling are obtained with conventional quenching oils at bath temperatures between 120 and 150°F. To prevent a temperature rise in the medium during quenching, it is always necessary to provide sufficient volume of medium. In some

TABLE 8·2 Cooling Rates at Center of 1/2-in.-diameter by 2 1/2-in.-long Stainless-steel Specimen When Quenched from 1500°F in Various Media*

BATH	RATE AT 1300°F, °F/s		RATE AT 1200°F, °F/s		AVERAGE RATE 1250–900°F, °F/s	
	75°	125°	75°	125°	75°	125°
Brine (10%)	382	296	382	325	383	287
Tap water	211	46	223	117	220	176
Gulf Super-Quench	80	85	170	180	135	137
Slow oil	36	32	30	26	39	44
10% soluble oil, 90% water	36	30	36	30	34	28
Still air	5	...	4	...	3	
Fused salt (at 400°F)	162		130		66	

* Courtesy of Gulf Oil Corp.

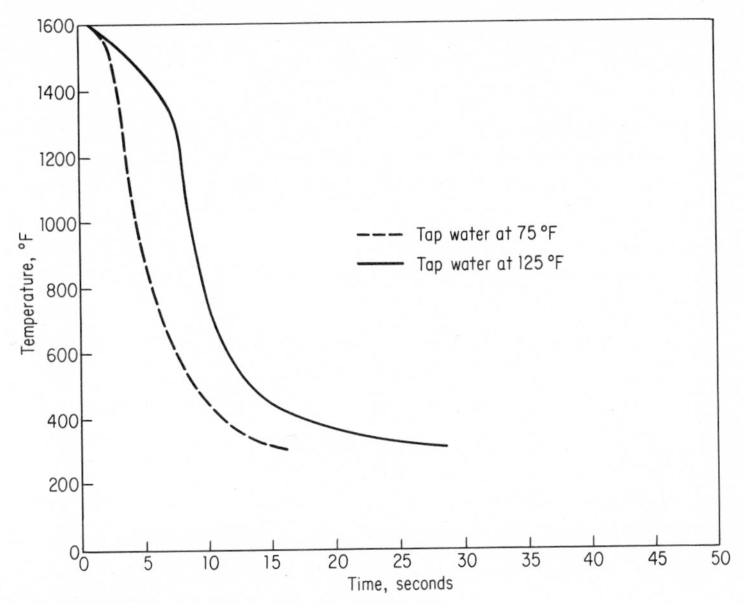

Fig. 8·37 Center-cooling curves for specimens quenched in tap water at bath temperatures of 75 and 125°F. No agitation. (Gulf Oil Corporation.)

cases, cooling coils are inserted in the quenching tank to control the temperature of the medium.

The cooling rate may be improved and the temperature of the medium kept constant by circulation of the medium and agitation of the piece. This effectively wipes off the vapor film as quickly as it forms, reducing the length of the vapor-blanket stage, and results in faster cooling (Fig. 8·38). The quenching severity, relative to still water as 1.0, is shown in Table 8·3 for various conditions of quench. Circulation is a factor which is sometimes overlooked in the quenching process. It is possible, by proper choice of circulation, to obtain a wide variety of cooling rates with an oil quench.

TABLE 8·3 Quenching Severity Relative to Still Water As 1.0 for Various Conditions of Quench*

METHOD OF COOLING	OIL	WATER	BRINE
No circulation of liquid or agitation of piece	0.25–0.30	0.9–1.0	2
Mild circulation or agitation	0.30–0.35	1.0–1.1	2–2.2
Moderate circulation	0.35–0.40	1.2–1.3	
Good circulation	0.40–0.50	1.4–1.5	
Strong circulation	0.50–0.80	1.6–2.0	
Violent circulation	0.80–1.10	4	5

* From M. A. Grossmann, "Principles of Heat Treatment," American Society for Metals, Metals Park, Ohio, 1953.

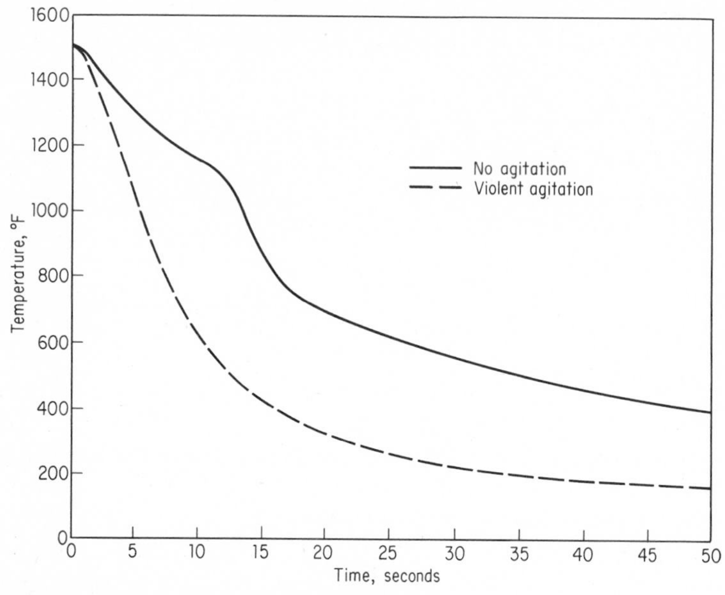

Fig. 8·38 Effect of agitation on center-cooling curves of a stainless-steel specimen quenched in conventional quenching oil. Oil temperature 125°F. (Gulf Oil Corporation.)

8·18 Surface Condition When steel is exposed to an oxidizing atmosphere, because of the presence of water vapor or oxygen in the furnace a layer of iron oxide called *scale* is formed. Experiments have shown that a thin layer of scale has very little effect on the actual cooling rate, but that a thick layer of scale (0.005 in. deep) retards the actual cooling rate (see Fig. 8·39). There is also the tendency for parts of the scale to peel off the surface when the piece is transferred from the furnace to the quench tank, thus giving rise to a variation in cooling rate at different points on the surface. The presence of scale need be considered only if the actual cooling rate is very close to the critical cooling rate. More important, since scale is softer than the hardened steel, there will be a tendency for the scale to clog up the grinding wheel during the finishing operations. In any case, the formation of scale is usually avoided in commercial heat treatment.

Many methods are used industrially to minimize the formation of scale. The choice of method depends upon the part being heat-treated, type of furnace used, availability of equipment, and cost.

Copper Plating A flash coating of only a few ten-thousandths of an inch of copper will protect the surface of a steel against the formation of scale. This method is economical when copper-plating tanks are available in the plant.

Protective Atmosphere An atmosphere that is inert with respect to steel may be introduced under pressure into the furnace. Gases used for this purpose are hydrogen, dissociated ammonia, and combusted gas resulting from the partial or complete combustion of hydrocarbon-fuel gases such as methane and propane in special generators.

Liquid-salt Pots The part to be heat-treated may be immersed in a liquid-salt furnace that is neutral with respect to the steel. The piece, being completely surrounded by the neutral liquid salt, cannot be oxidized to form scale.

Cast-iron Chips The part is buried in a container having cast-iron chips. Any oxygen entering the furnace reacts with the cast iron before it reaches the steel.

8·19 Size and Mass Since it is only the surface of a part which is in contact with the quenching medium, the ratio of surface area to mass is an important factor in determining the actual cooling rate. This ratio is a function of the geometric shape of the part and is largest for a spherical part. Thin plates and small-diameter wires have a large ratio of surface area to mass and therefore rapid cooling rates. Consider a long cylinder so that the surface area of the ends is negligible. The surface area is equal to the circumference times the length of the cylinder, and the mass is equal to the cross-sectional area times the length times the density of the material. The ratio is

$$\frac{\text{Surface area}}{\text{Mass}} = \frac{\pi DL}{(\pi/4)D^2L\rho} = \frac{4}{D\rho}$$

where ρ = density.

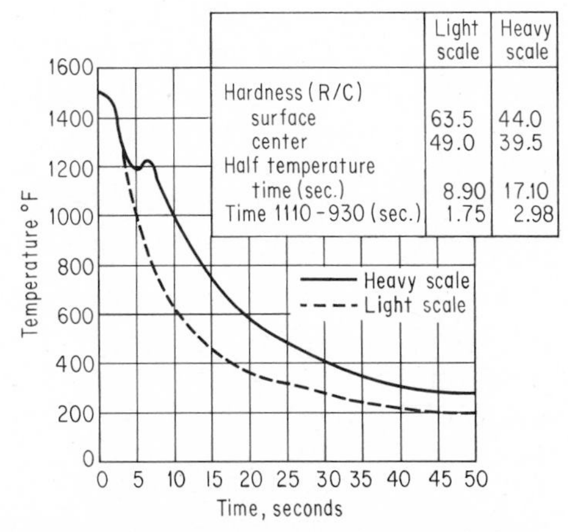

Fig. 8·39 Effect of scale on center-cooling curves of SAE 1095 steel specimens quenched in Gulf Super-Quench. Oil temperature 125°F, violent agitation. (Gulf Oil Corporation.)

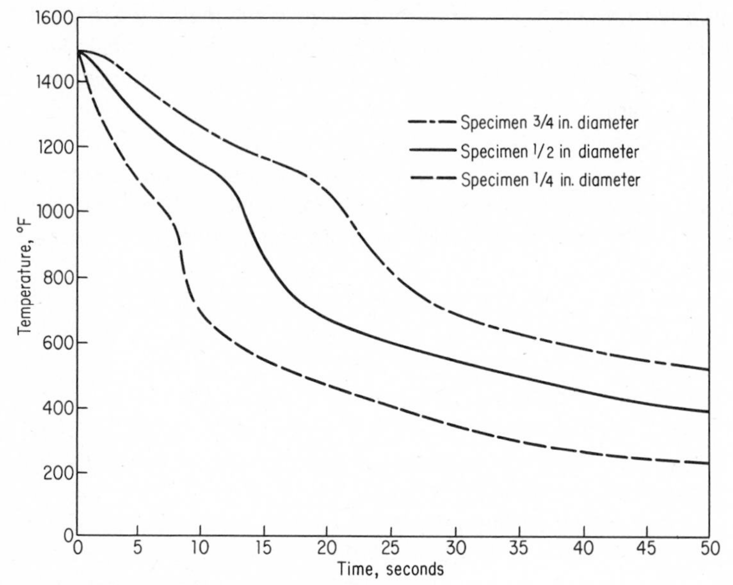

Fig. 8·40 Effect of mass on center-cooling curves of stainless-steel specimens quenched in conventional quenching oil. Oil temperature 125°F. (Gulf Oil Corporation.)

The calculation shows that the ratio is inversely proportional to diameter. If the diameter is increased, the ratio of surface area to mass decreases, and the cooling rate decreases. In other words, with a fixed quenching medium, a large piece will be cooled more slowly than a small piece (Fig. 8·40). As the diameter increases, the duration of the vapor-blanket stage increases. The vapor-transport stage is less distinct, and the transition from the vapor-transport stage to the last stage becomes more gradual. The rate of cooling in all three stages decreases sharply (Fig. 8·41).

Let us now perform an experiment on a medium-carbon steel of about 0.45 percent carbon. A series of pieces ranging from 1/2 to 5 in. diameter are heated to the proper austenitizing temperature and quenched in water. When the surface hardness was determined on these pieces, the following results were obtained:

DIAMETER OF PIECE WATER-QUENCHED, IN.	SURFACE HARDNESS, ROCKWELL C
0.5	59
1	58
2	41
3	35
4	30
5	24

From the above data we can conclude that the actual cooling rate at the surface of the 1/2- and 1-in. pieces exceeded the critical cooling rate for this steel, so that a fully martensitic structure was obtained with maximum hardness. The surface of the 2- and 3-in. pieces received intermediate cooling, and the structure is probably a mixture of martensite, fine pearlite and a small amount of ferrite. The surface of the 4- and 5-in. pieces received slow cooling, with a resulting structure of pearlite and ferrite. As a

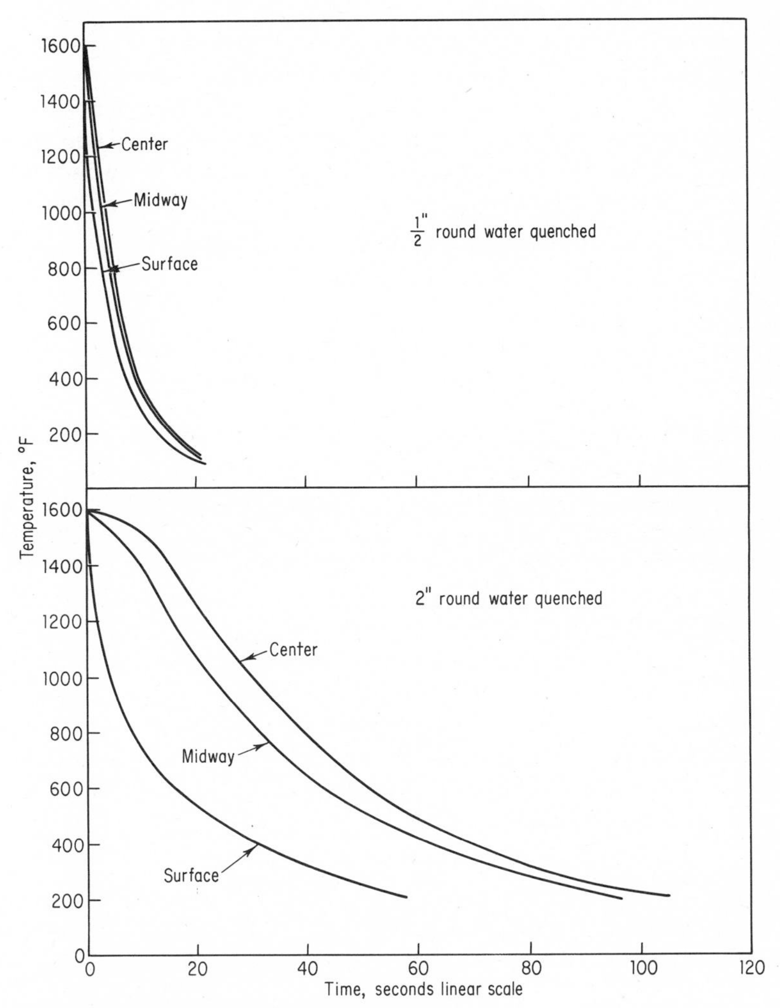

Fig. 8·41 Cooling curves at the surface, midway on the radius, and at the center of two different-sized bars when water-quenched. (From "Suiting the Heat Treatment to the Job," U.S. Steel Corporation.)

matter of fact, for the 5-in. piece, the amount of heat to be removed is so large compared with the surface area available that the water quench is ineffective, and very nearly the same hardness would have been obtained if that piece had been cooled in the furnace. The approximate relation of the actual cooling curves of the surface to the I-T diagram is shown in Fig. 8·42.

Up to this point, the discussion has concerned itself only with the surface hardness. The surface, being in actual contact with the quenching medium, was cooled most rapidly in quenching. The heat in the interior of the piece must be removed by conduction, through the body of the piece, eventually reaching the surface and the quenching medium. Therefore, the cooling rate in the interior is less than that at the surface. Figure 8·43 shows the time-temperature cooling curves at different positions in a 1-in.-diameter bar during a drastic quench. If such a variation in cooling rates exists across the radius of a bar during cooling, it is to be anticipated that variations in hardness would be evident when the bars are cut and

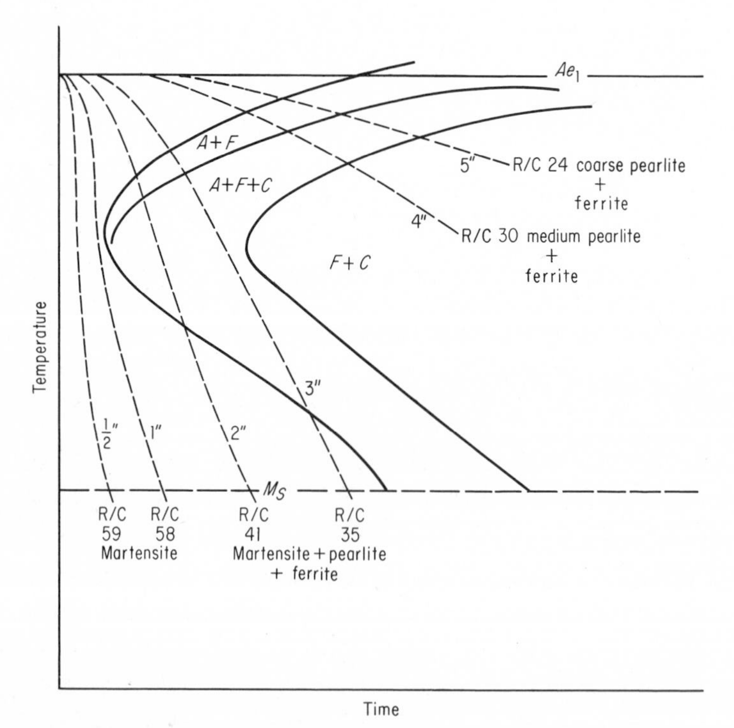

Fig. 8·42 Surface-cooling curves, the final structure and hardness of different-sized rounds related to the I-T diagram of a 0.45 percent carbon steel.

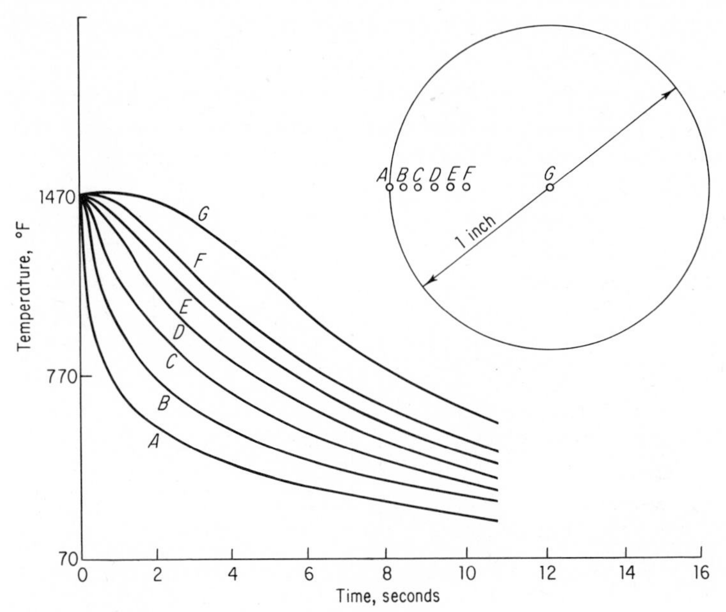

Fig. 8·43 Time-temperature cooling curves at different positions in a 1-in.-diameter bar quenched drastically in water. (From M. A. Grossmann, "Principles of Heat Treatment," American Society for Metals, Metals Park, Ohio, 1955.)

hardness surveys made on the cross section. Notice that it is possible to have a considerable temperature difference between the surface and the center during quenching. If a vertical line is drawn at 2 s intersecting the curves, the surface will have come down to approximately 600°F while the center is still at 1470°F. Therefore there is a temperature difference of about 870° between the surface and the center at the end of 2 s. This temperature difference will give rise to stresses during heat treatment which may result in distortion and cracking of the piece. These stresses will be discussed in greater detail in the section on residual stresses later in this chapter.

The results of a hardness survey on the different diameters, whose surface hardness was previously mentioned, are shown in Fig. 8·44*a*. This may be called a *hardness-penetration* or *hardness-traverse diagram*, since it shows at a glance to what extent the steel has hardened during quenching. Hardenability is related to the depth of penetration of the hardness, and in no case has the hardness penetrated deeply, so that this steel is said to have *low hardenability*. As anticipated, it is seen that the

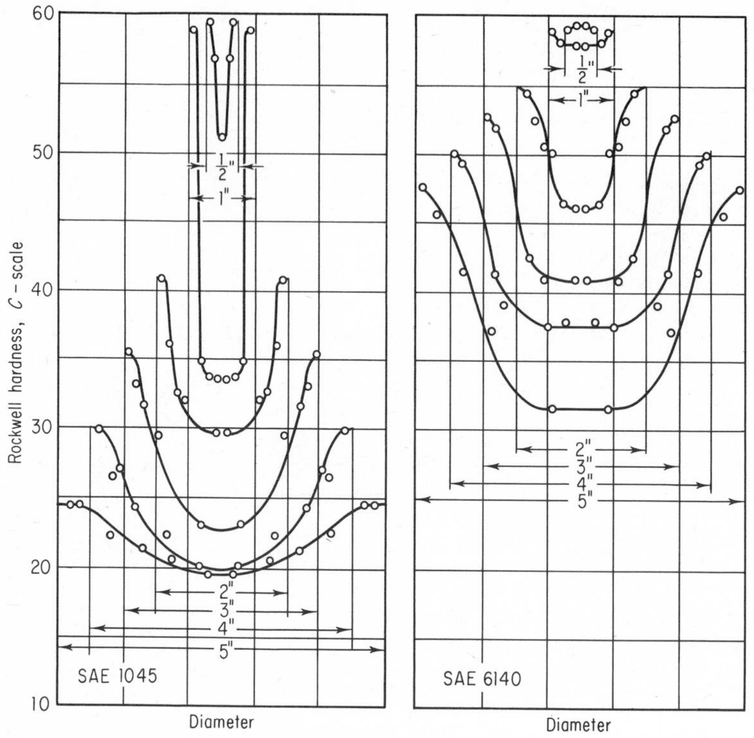

Fig. 8·44 Hardness-penetration or hardness-traverse curves for various sizes quenched in water. (*a*) SAE 1045 steel; (*b*) SAE 6140 chrome-vanadium steel. (From M. A. Grossmann, "Principles of Heat Treatment," American Society for Metals, Metals Park, Ohio, 1955.)

hardness of the quenched piece is less as its size increases, and also that each piece is lower in hardness at the center than at the surface.

A study of the curves of Fig. 8·44*a* shows an interesting situation. The hardness of Rockwell C 30 was obtained (1) at the surface of a 4-in. round, (2) at about 1/2 in. under the surface of a 3-in. round, and (3) at almost the center of a 2-in. round. These three points are equivalent and have reached the same hardness because the actual cooling rate was the same at each location. This leads to a very important conclusion: *that for a steel of fixed composition and austenitic grain size, regardless of the shape or size of the piece and the quenching conditions, wherever the actual cooling rate is the same, the hardness must be the same.* The student should realize that the converse of this statement is not necessarily true. Wherever the hardness is the same in a steel of fixed composition and austenitic grain size, the actual cooling rate may or may not have been the same. If the actual rate at the center of a piece exceeds the critical cooling rate for the steel, the hardness at the center will be the same as that at the surface, but the actual cooling rate at both locations will be different. This is shown in Fig. 8·44*b* for a ½-in. bar.

Increase in the hardenability or depth of penetration of the hardness may be accomplished by either of two methods:

1 With the actual cooling rates fixed, slow up the critical cooling rate (shift the I-T curve to the right) by adding alloying elements or coarsening the austenitic grain size.

2 With the I-T curve fixed, increase the actual cooling rates by using a faster quenching medium or increasing circulation.

Since increasing cooling rates increase the danger of distortion or cracking, the addition of alloying elements is the more popular method of increasing hardenability. Figure 8·44*b* shows the hardness-penetration diagrams, after water quenching, for different-size rounds of a chromium-vanadium steel of about the same carbon content as Fig. 8·44*a*. The hard-

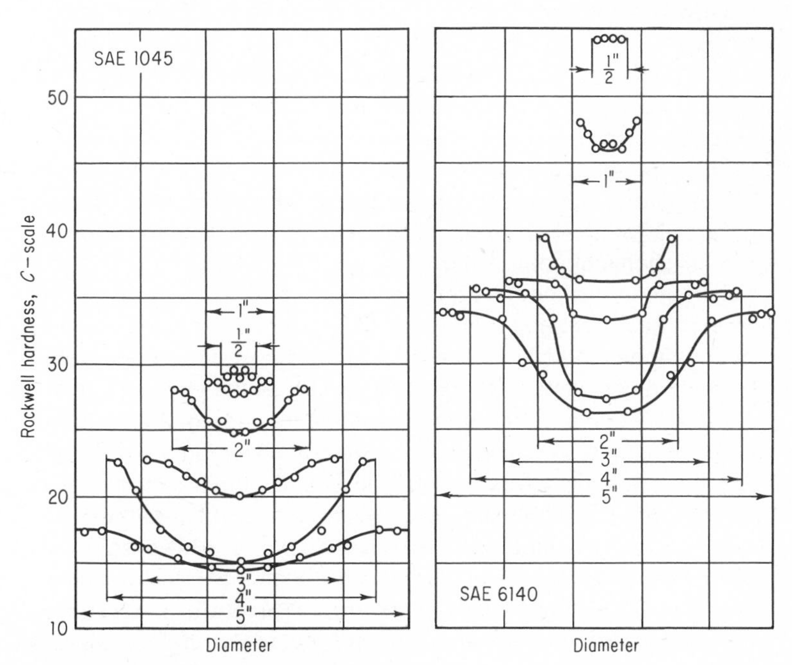

Fig. 8·45 Hardness-penetration or hardness-traverse curves for various sizes quenched in oil. (*a*) SAE 1045 steel; (*b*) SAE 6140 chrome-vanadium steel. (From M. A. Grossmann, "Principles of Heat Treatment," American Society for Metals, Metals Park, Ohio, 1955.)

ness level of all sizes has been raised appreciably, and notice that the 1-in. size has almost achieved a fully martensitic structure across its diameter. If the actual cooling rate is reduced by using an oil quench, the hardness level of all pieces made of the plain-carbon steel will drop, as shown by Fig. 8·45*a*. Even the chromium-vanadium steel (Fig. 8·45*b*) shows a drop in hardness level in an oil quench as compared with the water quench, but note that the ½-in. round still attained full hardness.

During the entire discussion of cooling rates, it was assumed that the thermal conductivity of all steels is the same. This is not technically true, but the variation in thermal conductivity between different steels is so small compared with the other variables in the quenching process that it may be considered constant with little error.

8·20 Hardenability The usual method of purchasing steel is on the basis of chemical composition. This allows a considerable variation in the carbon and alloy content of the steel. For example, AISI 4340 steel has the following composition range: 0.38 to 0.43 percent C, 0.60 to 0.80 percent Mn, 0.20 to 0.35 percent Si, 1.65 to 2.00 percent Ni, 0.70 to 0.90 percent Cr, and 0.20 to 0.30 percent Mo. Let us determine the percent variation of each element within the stated limits. For example, in the case of carbon, the difference between 0.43 and 0.38 is 0.05. If we divide 0.05 by the average between the limits or 0.40 and express this as a percentage, it turns out to be 12.5 percent. Following the same procedure for the other elements, the percent variation turns out to be even greater: 28.7 percent for Mn, 53.8 percent for Si, 19.1 percent for Ni, 25 percent for Cr, and 40 percent for Mo. These figures illustrate that it is possible when expressed on a percentage basis to have a considerable variation in chemical composition. This variation in chemical composition within a particular grade will cause a variation in the critical cooling rate and in turn a variation in the response of the steel to heat treatment. Figure 8·46 is a schematic representation of what might happen in an actual case. Assume the I-T curve on the left shows the beginning of transformation of a steel with all the elements on the low side and the I-T curve on the right shows the beginning of transformation with all the elements on the high side. Superimposed on the diagram is a cooling curve for a steel part quenched under certain conditions. If the elements are on the high side, this cooling curve will miss the nose of the I-T diagram, and the steel will attain full hardness. However, if the elements are on the low side, the cooling curve will hit the I-T curve above the nose, and the steel will not attain full hardness. Therefore, buying a steel according to chemical composition is no assurance that full hardness will be attained under certain quenching conditions. Since strength is the prime factor in design, unless special properties are desired, it would seem more economical to base the material specification on the response to heat treatment (hardenability) rather than chemical composition. It is therefore necessary to have a test that will predict the hardenability of the steel. The most widely used method of determining hardenability is the *end-quench hardenability test*, or the *Jominy test.*

The test has been standardized by the ASTM, SAE, and AISI. In conducting this test, a 1-in.-round specimen 4 in. long is heated uniformly to the proper austenitizing temperature. It is then removed from the furnace and placed on a fixture where a jet of water impinges on the bottom face of the specimen (Fig. 8·47). The size of the orifice, the distance from the orifice to the bottom of the specimen, and the temperature and circulation of the water are all standardized, so that every specimen quenched in this fixture receives the same rate of cooling. After 10 min on the fixture, the specimen is removed, and two parallel flat surfaces are ground longitudinally to a depth of 0.015 in. Rockwell C scale hardness readings are taken at 1/16-in. intervals from the quenched end. The results are expressed as a curve of hardness values vs. distance from the quenched end. A typical hardenability curve is shown in Fig. 8·48. Details pertaining to the testing procedure may be obtained by referring to ASTM Designation A255-48T End Quench Test for Hardenability of Steel.

Each location on the Jominy test piece, quenched in a standard manner, represents a certain cooling rate, and since the thermal conductivity of all steels is assumed to be the same, this cooling rate is the same for a given position on the test piece regardless of the composition of the steel from which the test piece is made. Each specimen is thus subjected to a series of cooling rates varying continuously from very rapid at the quenched

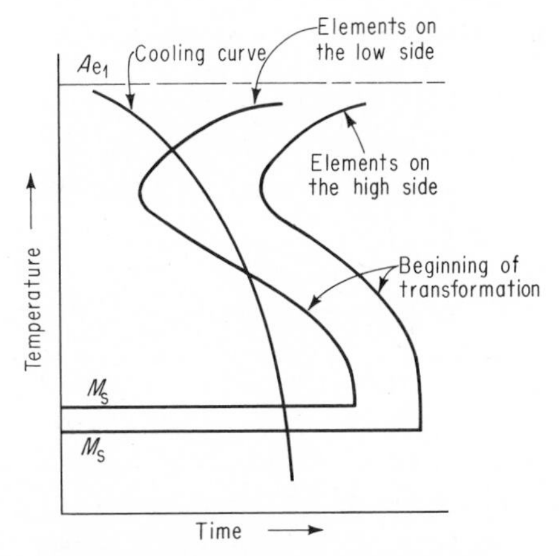

Fig. 8·46 Schematic representation of the effect of variation in chemical composition for a given grade of steel. Cooling curve superimposed on the beginning of transformation curve.

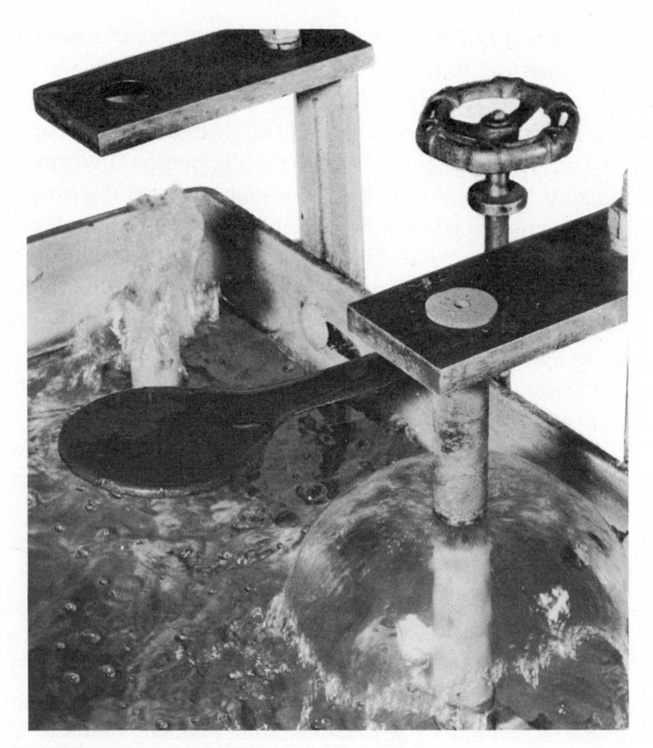

Fig. 8·47 End-quench hardenability specimen being quenched. (Bethlehem Steel Company.)

end to very slow at the air-cooled end (Table 8·4). Since each distance along the quenched bar is equivalent to a certain actual cooling rate, you could just as well plot Rockwell C hardness vs. actual cooling rate as Rockwell C hardness vs. distance. This is exactly what is done on the ASTM

TABLE 8·4 Cooling Rates at Distances from the Water-cooled End of the Standard End-quench Hardenability Test Bar

DISTANCE FROM QUENCHED END, IN.	COOLING RATE, °F/s AT 1300°F	DISTANCE FROM QUENCHED END, IN.	COOLING RATE, °F/s AT 1300°F
1/16	490	11/16	19.5
1/8	305	3/4	16.3
3/16	195	13/16	14.0
1/4	125	7/8	12.4
5/16	77.0	15/16	11.0
3/8	56.0	1	10.0
7/16	42	1 1/4	7.0
1/2	33	1 1/2	5.1
9/16	26	1 3/4	4.0
5/8	21.4	2	3.5

graph paper (Fig. 8·48). Notice that the upper part of the graph contains a scale parallel to the distance scale which has readings of the approximate actual cooling rate at 1300°F that correspond to the values given in Table 8·4. Therefore, the end-quench hardenability curve really shows how hardness varies with different actual cooling rates for a particular steel.

Figure 8·49 shows the continuous cooling–transformation diagram for an alloy steel of the 8630 type on which cooling curves, representing those at selected locations along the end-quench test bar, have been superimposed. This serves to clarify the relationship between the end-quench

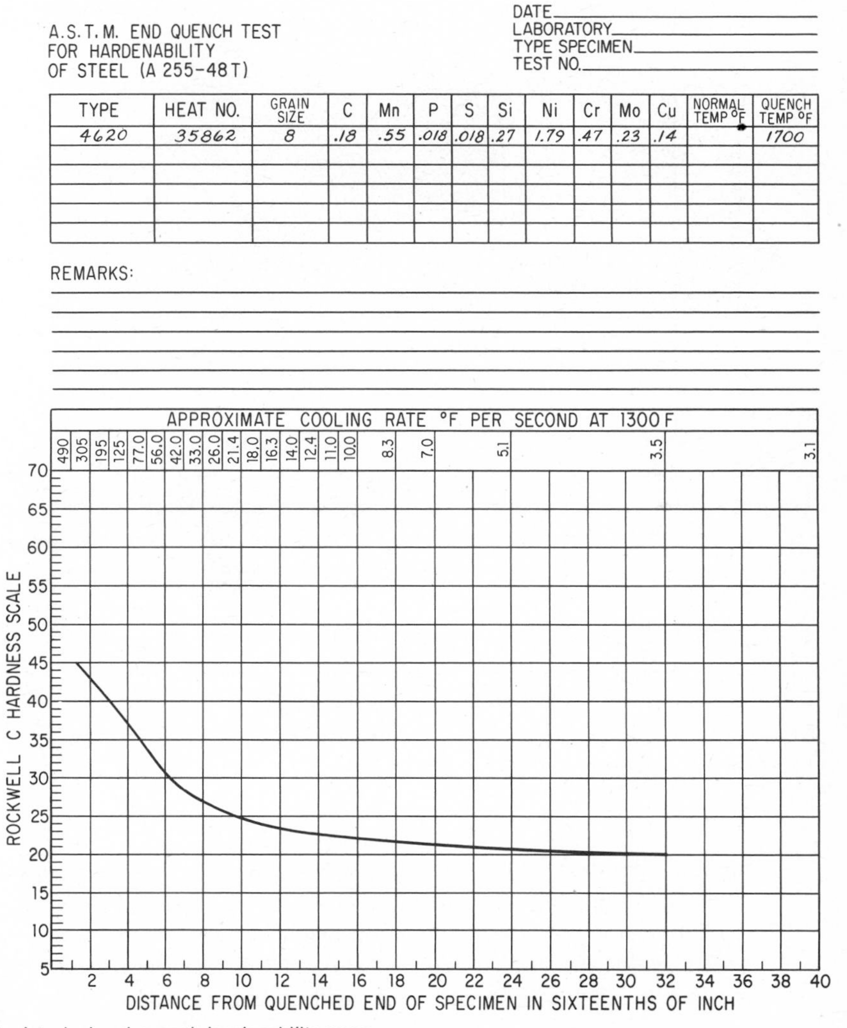

A.S.T.M. END QUENCH TEST
FOR HARDENABILITY
OF STEEL (A 255-48 T)

DATE_____
LABORATORY_____
TYPE SPECIMEN_____
TEST NO._____

TYPE	HEAT NO.	GRAIN SIZE	C	Mn	P	S	Si	Ni	Cr	Mo	Cu	NORMAL TEMP °F	QUENCH TEMP °F
4620	35862	8	.18	.55	.018	.018	.27	1.79	.47	.23	.14		1700

REMARKS:

Fig. 8·48 A typical end-quench hardenability curve.

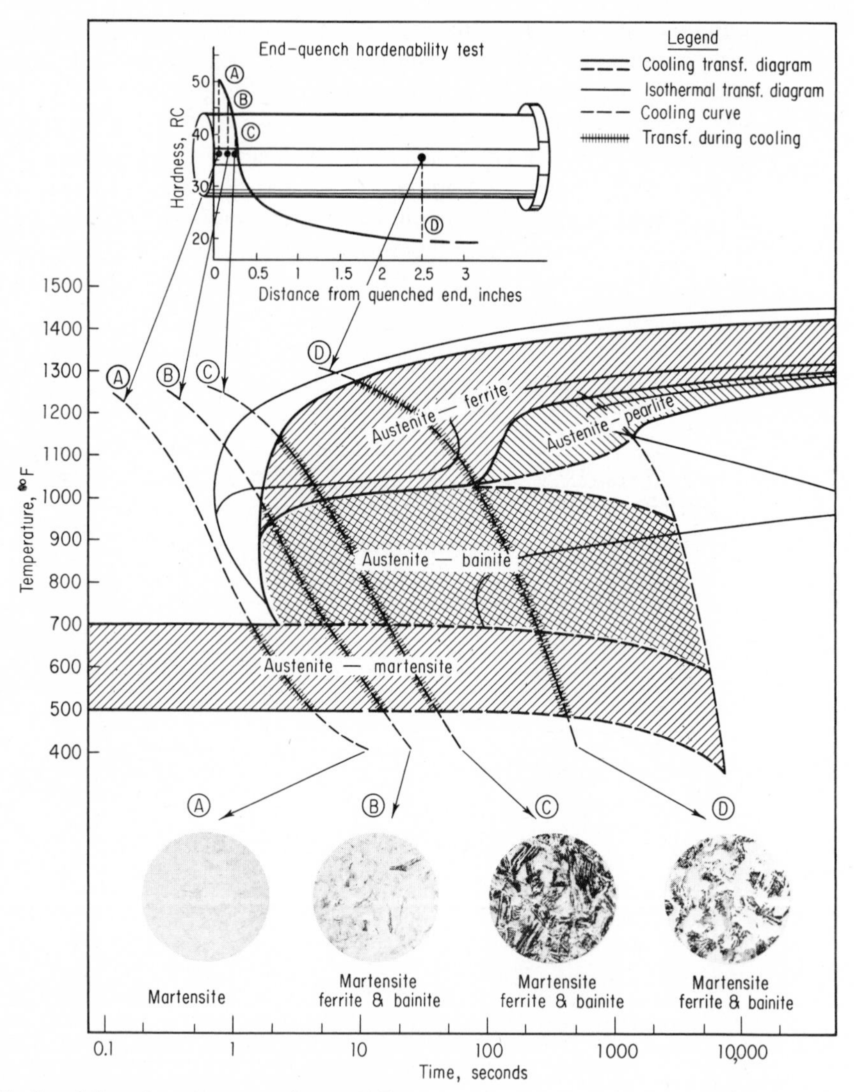

Fig. 8·49 Correlation of continuous-cooling and I-T diagram with the end-quench hardenability test data for an 8630-type steel. (From "U.S.S. Carilloy Steels," U.S. Steel Corporation.)

hardenability test and the transformation behavior previously discussed.

Analyzing the microstructural changes on cooling at these various rates, it will be seen that rate *A*, which represents the rate nearest the quenched end, exceeds the critical cooling rate and will result in transformation to martensite. Rates *B*, *C*, and *D* all result in transformation to various mixtures of ferrite, bainite, and martensite, the amount of martensite decreasing with the decrease in cooling rate. It should be noted that a very slow cooling rate is necessary to obtain pearlite in this steel.

Although hardenability is usually expressed in terms of hardness changes, it is the changes in microstructure reflected by those hardness values which are of importance in the properties of steel. Alloying elements, in general, increase hardenability by delaying transformation in the pearlite and bainite regions, thus permitting the formation of martensite with slower rates of cooling.

The end-quench curves of three alloy steels, each containing 0.40 percent carbon but of different hardenability, are shown in Fig. 8·50. All three steels develop the same maximum hardness of Rockwell C 52.5 at the water-quenched end, since this is primarily a function of carbon content only. However, in the high-hardenability steel 4340, this hardness is maintained for a considerable distance, whereas in the lower-hardenability steels 4140 and 5140 the hardness drops off almost immediately.

By the analysis of data collected from hundreds of heats of each grade of steel, the AISI has established minimum and maximum hardenability curves known as *hardenability bands*. A typical hardenability band is shown in Fig. 8·51. The suffix H denotes steels that may be bought on the basis of a hardenability specification, with chemical composition, grain size, etc., being of secondary importance.

Two points are usually designated in specifying hardenability by one of the following methods:

A The minimum and maximum hardness values at any desired distance. The distance selected should be that distance on the end-quench test bar which corre-

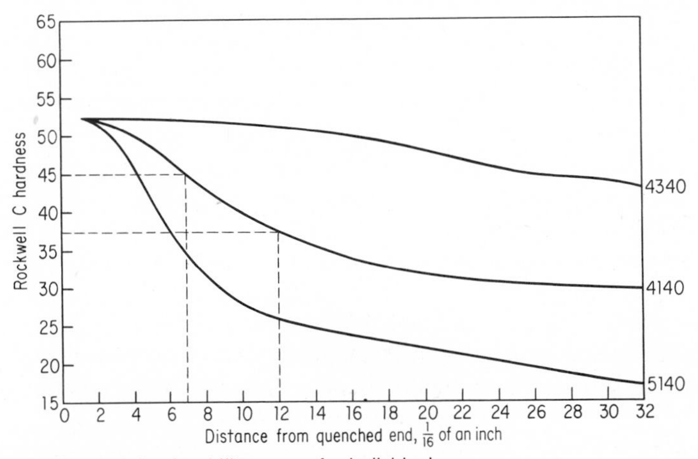

Fig. 8·50 End-quench hardenability curves for individual samples of 4340, 4140, and 5140 alloy steels.

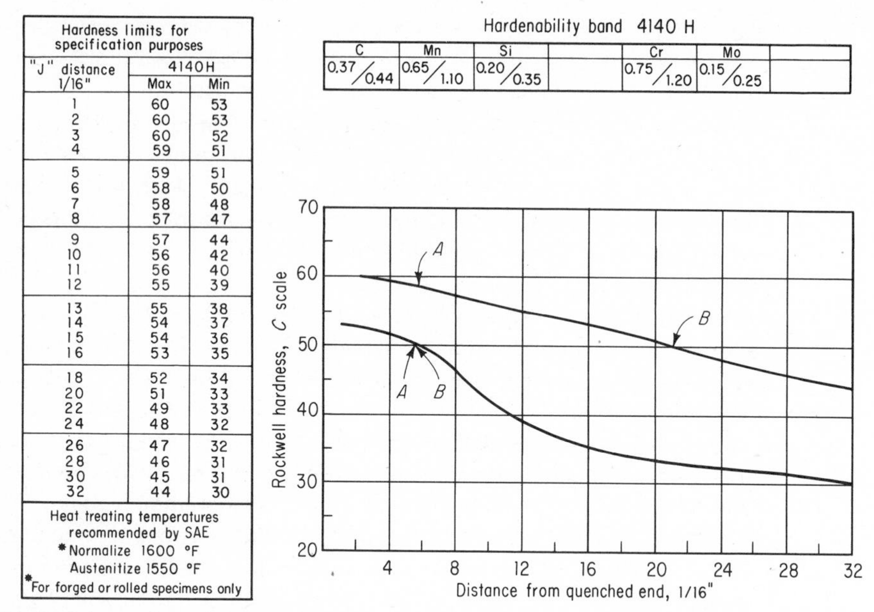

Hardness limits for specification purposes		
"J" distance 1/16"	4140H	
	Max	Min
1	60	53
2	60	53
3	60	52
4	59	51
5	59	51
6	58	50
7	58	48
8	57	47
9	57	44
10	56	42
11	56	40
12	55	39
13	55	38
14	54	37
15	54	36
16	53	35
18	52	34
20	51	33
22	49	33
24	48	32
26	47	32
28	46	31
30	45	31
32	44	30

Heat treating temperatures recommended by SAE
*Normalize 1600 °F
Austenitize 1550 °F
*For forged or rolled specimens only

Fig. 8·51 Test data and standard hardenability band for 4140H steel. (American Iron and Steel Institute.)

sponds to the section used by the purchaser. For example, in Fig. 8·51, the specification could be J50/58 = 6/16 in.

B The minimum and maximum distances at which any desired hardness value occurs. This method of specification, in Fig. 8·51, could be J50 = 6/16 to 21/16 in.

8·21 Use of Hardenability Data To select a steel to meet a minimum hardness at a given location in a part quenched under given conditions, the cooling rate at the given location must first be known and the reference point on the end-quench test bar having the same cooling rate must be determined. If the part is simple in cross section, such as round, flat, or square, numerous charts are available in the literature which give the cooling rate at different locations under various quenching conditions or the equivalent point on the end-quench test bar that has the same cooling rate. The relation between the end-quench test bar and the center and mid-radius locations of various sizes of rounds quenched under different conditions is shown in Figs. 8·52 and 8·53. The severity of quench is designated by Table 8·5. Let us consider a practical application of the end-quench hardenability test. Assume that a company is required to manufacture a steel shaft 2 in. in diameter to a specified minimum hardness at the center after hardening of Rockwell C 42. They plan to use a good oil quench and moderate agitation ($H = 0.35$, Table 8·5). They would like to use a bar of 4140 steel whose hardenability curve is Fig. 8·50. The problem is to determine whether that steel will meet the above specifications. In order to solve the problem it is first necessary to know what the actual cooling rate is at the center of a 2-in.-diameter round when it is quenched under the given conditions, or the distance along the end-quench hardenability test bar that has the same cooling rate. Referring to Fig. 8·52, for a 2-in. bar diameter and $H = 0.35$, point X is located. Therefore, 3/4 or 12/16 from the water-cooled end of the end-quench hardenability test bar has the same cooling rate as the center of a 2-in. round quenched under the given conditions. Referring to Fig. 8·50, a vertical line is drawn at 12/16 intersecting the curve for 4140 steel as shown. The hardness, read to the left, is Rockwell C 37. Since the required hardness was Rockwell 42, this steel will not satisfy that requirement under these quenching conditions. Suppose the quenching medium were changed to water with no agitation ($H = 1.0$, Table 8.5). Returning to Fig. 8·52, point Y is located, which gives a distance of 7/16 from the water-cooled end. Drawing a vertical line at 7/16 in Fig. 8·50 intersecting the 4140 curve shows that the hardness now will be Rockwell C 45. Therefore, going to a water quench will meet the hardness require-

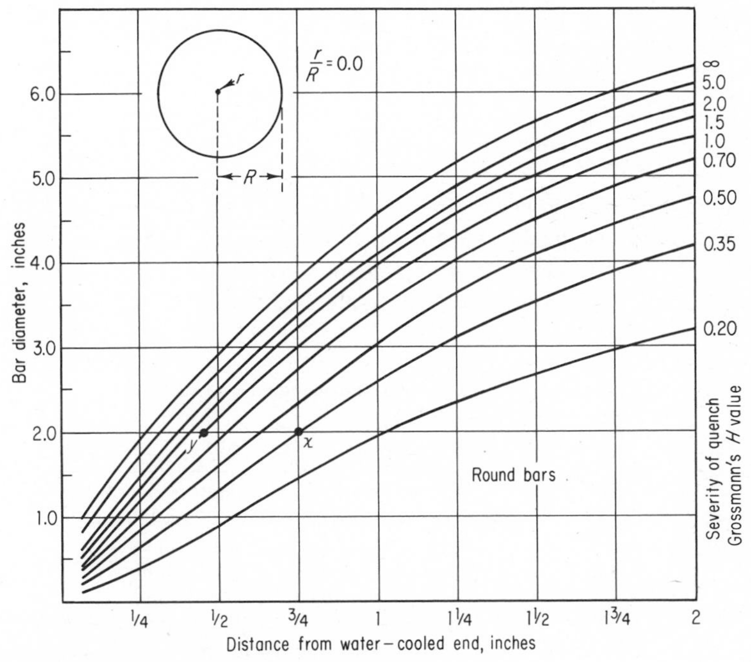

Fig. 8·52 Location on the end-quench hardenability test bar corresponding to the center of round bars under various quenching conditions. (From "U.S.S. Carilloy Steels," U.S. Steel Corporation.)

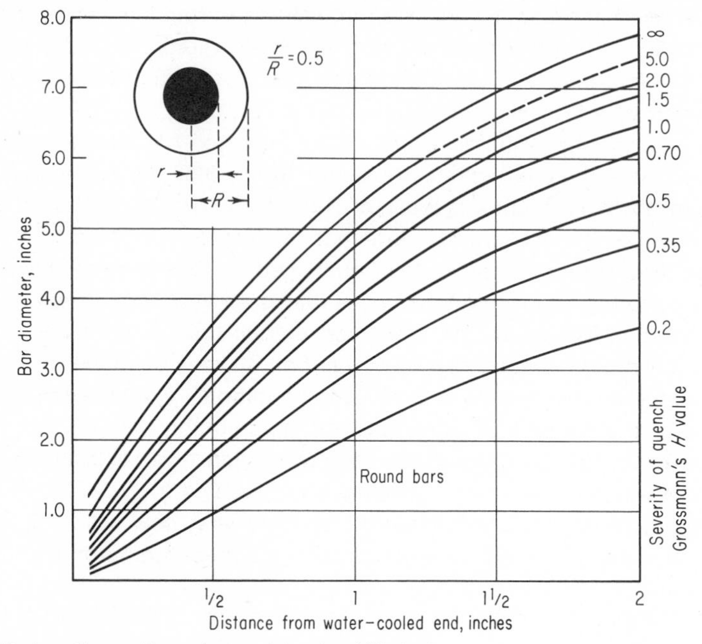

Fig. 8·53 Location on the end-quench hardenability test bar corresponding to the mid-radius position of round bars under various quenching conditions. (From "U.S.S. Carilloy Steels," U.S. Steel Corporation.)

ment. Suppose, however, that a water quench cannot be used. Then there is no alternative except to go to a steel of higher hardenability. Reference to Fig. 8·50 shows that 4340 steel will certainly meet the original requirements, but the hardness of Rockwell C 52 is probably too high.

TABLE 8·5 H or Severity of Quench Values for Various Quenching Conditions*

H VALUE	QUENCHING CONDITION
0.20	Poor oil quench—no agitation
0.35	Good oil quench—moderate agitation
0.50	Very good oil quench—good agitation
0.70	Strong oil quench—violent agitation
1.00	Poor water quench—no agitation
1.50	Very good water quench—strong agitation
2.00	Brine quench—no agitation
5.00	Brine quench—violent agitation
∞	Ideal quench

* From "U.S.S. Carilloy Steels," U.S. Steel Corporation.

Examination of published minimum hardenability limits will result in the selection of several steels to meet the original requirements. The final selection will be based on other manufacturing requirements and will show the greatest overall economy.

The typical U curves (Fig. 8·44) or hardness-penetration curves of rounds quenched under given conditions may be calculated from the standard hardenability band (Fig. 8·51) and a series of curves such as those in Figs. 8·52 and 8·53. The calculated results will show the maximum and minimum hardness variation across the cross section of various sizes of rounds.

It is possible to show the relation between the minimum as-quenched hardness at the center location and the diameter of different-sized rounds quenched under the same conditions by means of the standard hardenability band (Fig. 8·51) and the curves of Fig. 8·52. The same may be done at other locations by curves similar to Fig. 8·53.

The approximate cooling rate, under fixed quenching conditions, at any location in an irregularly shaped part may be determined if the hardenability curve for that steel is available. Assume that a hardness of Rockwell C 40 was obtained at a particular location in a part made of 4140 steel whose hardenability curve is shown in Fig. 8·50. Rockwell C 40 is obtained at a distance of $^{10}/_{16}$ on the end-quench test bar, and Table 8·4 gives the cooling rate at that location as 21.4°F/s.

When steel is purchased on the basis of a hardenability specification, the purchaser is certain that he will obtain the desired mechanical properties after heat treatment. This results in fewer rejections or retreatments and greater economy.

8·22 Tempering In the as-quenched martensitic condition, the steel is too brittle for most applications. The formation of martensite also leaves high residual stresses in the steel. Therefore, hardening is almost always followed by tempering or drawing, which consists in heating the steel to some temperature below the lower critical temperature. The purpose of tempering is to relieve residual stresses and to improve the ductility and toughness of the steel. This increase in ductility is usually attained at the sacrifice of the hardness or strength.

In general, over the broad range of tempering temperatures, hardness decreases and toughness increases as the tempering temperature is increased. This is true if toughness is measured by reduction of area in a tensile test. However, this is not entirely true if the notched bar such as Izod or Charpy is used as a measure of toughness. Most steels actually show a decrease in notched-bar toughness when tempered between 400 and 800°F, even though the piece at the same time loses hardness and strength. The reason for this decrease in toughness is not fully understood. The variation of hardness and notched-bar toughness with tempering temperature shown in Fig. 8·54 is typical of plain-carbon and low-alloy steels.

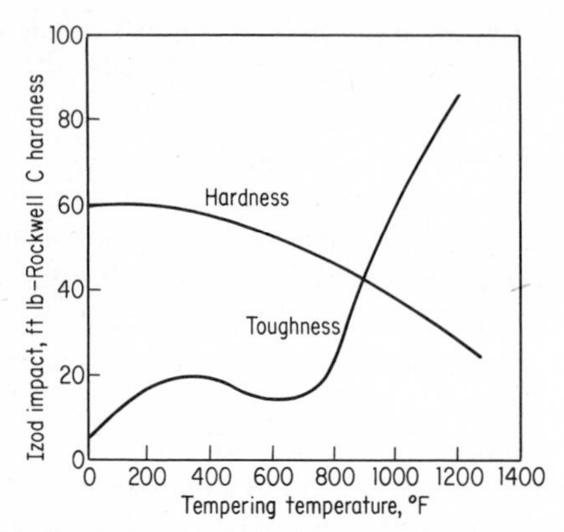

Fig. 8·54 Hardness and notched-bar toughness of 4140 steel after tempering 1 h at various temperatures. (From "Suiting the Heat Treatment to the Job," U.S. Steel Corporation.)

The tempering range of 400 to 800°F is a dividing line between applications that require high hardness and those requiring high toughness. If the principal desired property is hardness or wear resistance, the part is tempered below 400°F; if the primary requirement is toughness, the part is tempered above 800°F. If the part does not have any "stress raisers" or notches, the change in ductility may be a better indication of toughness than the notched-bar test, and tempering in the range of 400 to 800°F may not be detrimental. The effect of tempering temperature on the mechanical properties of a low-alloy steel 4140 is shown in Fig. 8·55.

Residual stresses are relieved to a large extent when the tempering temperature reaches 400°F, and by 900°F they are almost completely gone.

Certain alloy steels exhibit a phenomenon known as *temper brittleness*, which is a loss of notched-bar toughness when tempered in the range of 1000 to 1250°F followed by relatively slow cooling. Toughness is maintained, however, if the part is quenched in water from the tempering temperature. The precise mechanism which causes temper brittleness has not been established, although the behavior suggests some phase which precipitates along the grain boundaries during slow cooling. High manganese, phosphorus, and chromium appear to promote susceptibility, while molybdenum seems to have a definite retarding effect.

Martensite, as defined previously, is a supersaturated solid solution of carbon trapped in a body-centered tetragonal structure. This is a metastable condition, and as energy is applied by tempering, the carbon will be precipitated as carbide and the iron will become b.c.c. There will be diffusion and coalescence of the carbide as the tempering temperature is raised.

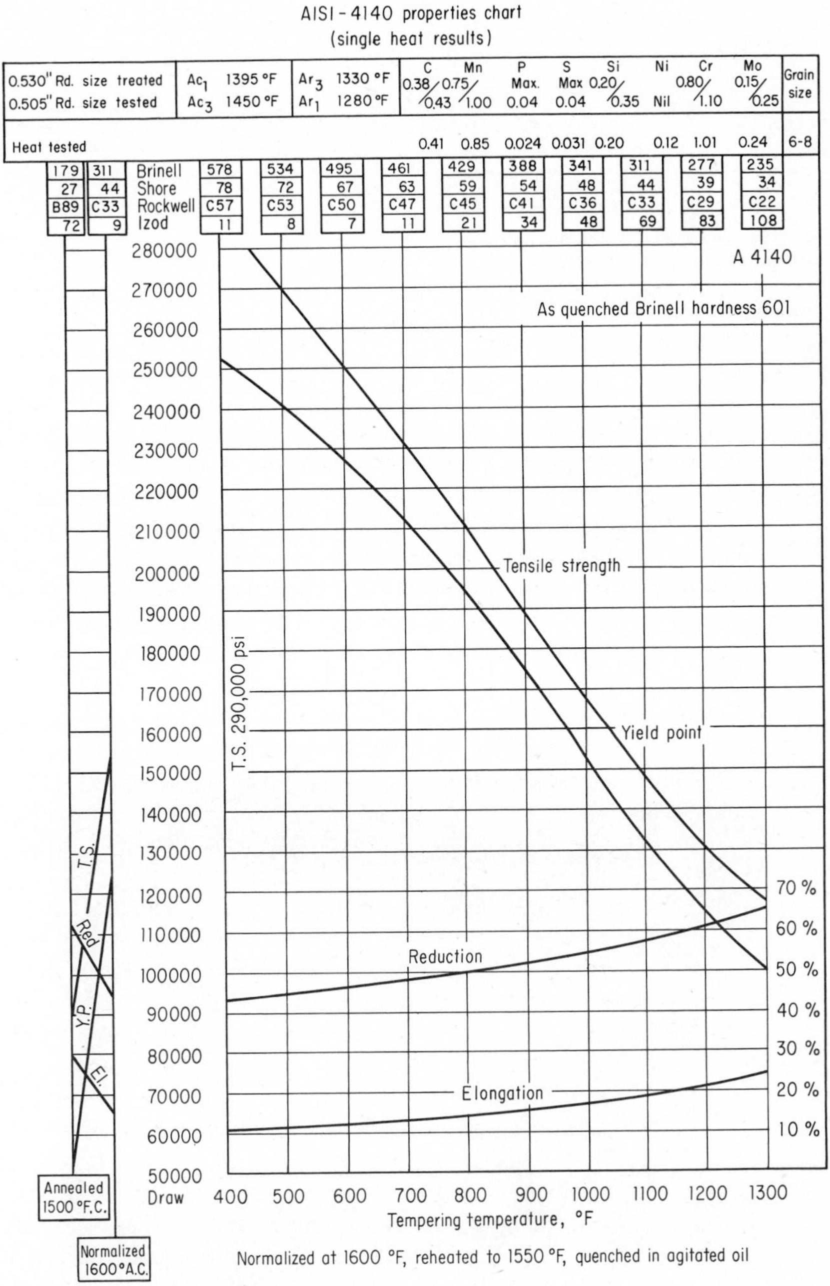

AISI - 4140 properties chart
(single heat results)

0.530" Rd. size treated 0.505" Rd. size tested	Ac_1 1395 °F Ac_3 1450 °F	Ar_3 1330 °F Ar_1 1280 °F	C 0.38/0.43	Mn 0.75/1.00	P Max. 0.04	S Max 0.04	Si 0.20/0.35	Ni Nil	Cr 0.80/1.10	Mo 0.15/0.25	Grain size
Heat tested			0.41	0.85	0.024	0.031	0.20	0.12	1.01	0.24	6-8

Annealed 1500 °F.C.	Normalized 1600 °A.C.		400	500	600	700	800	900	1000	1100	1200	1300
179	311	Brinell	578	534	495	461	429	388	341	311	277	235
27	44	Shore	78	72	67	63	59	54	48	44	39	34
B89	C33	Rockwell	C57	C53	C50	C47	C45	C41	C36	C33	C29	C22
72	9	Izod	11	8	7	11	21	34	48	69	83	108

A 4140

As quenched Brinell hardness 601

280000
270000
260000
250000
240000
230000
220000
210000
200000
190000
180000
170000
160000
150000
140000
130000
120000
110000
100000
90000
80000
70000
60000
50000

T.S. 290,000 psi

Tensile strength

Yield point

Reduction

Elongation

T.S.
Red.
Y.P.
El.

70 %
60 %
50 %
40 %
30 %
20 %
10 %

Draw 400 500 600 700 800 900 1000 1100 1200 1300

Tempering temperature, °F

Annealed 1500 °F.C.

Normalized 1600 °A.C.

Normalized at 1600 °F, reheated to 1550 °F, quenched in agitated oil

Fig. 8·55 Mechanical properties of 4140 steel after oil quenching and tempering at various temperatures. (Bethlehem Steel Company.)

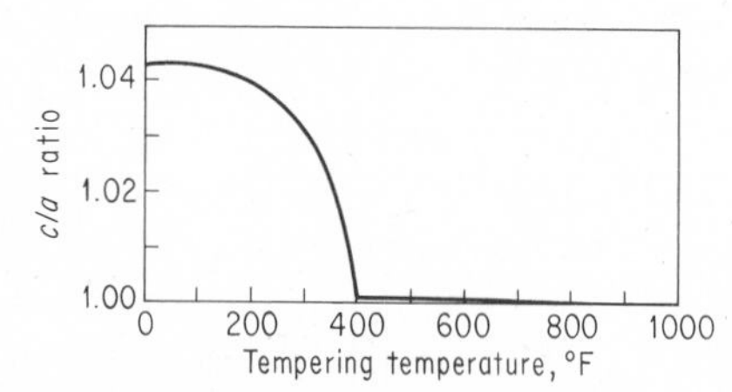

Fig. 8·56 The axial ratio c/a of martensite as a function of tempering temperature. When $c/a = 1.00$, the martensite has decomposed into ferrite and a carbide phase. (From Brick, Gordon, and Phillips, "Structure and Properties of Alloys," 3d ed., McGraw-Hill Book Company, New York, 1965.)

When plain carbon steel is heated in the range of 100 to 400°F, the structure etches dark and is sometimes known as *black martensite*. The original as-quenched martensite is beginning to lose its tetragonal crystal structure by the formation of a hexagonal close-packed *transition carbide* (epsilon carbide) and low-carbon martensite. X-ray studies, Fig. 8·56, show the decrease in c/a ratio as carbon is precipitated from martensite forming epsilon carbide. The precipitation of the transition carbide may cause a slight increase in hardness, particularly in high-carbon steels. The steel

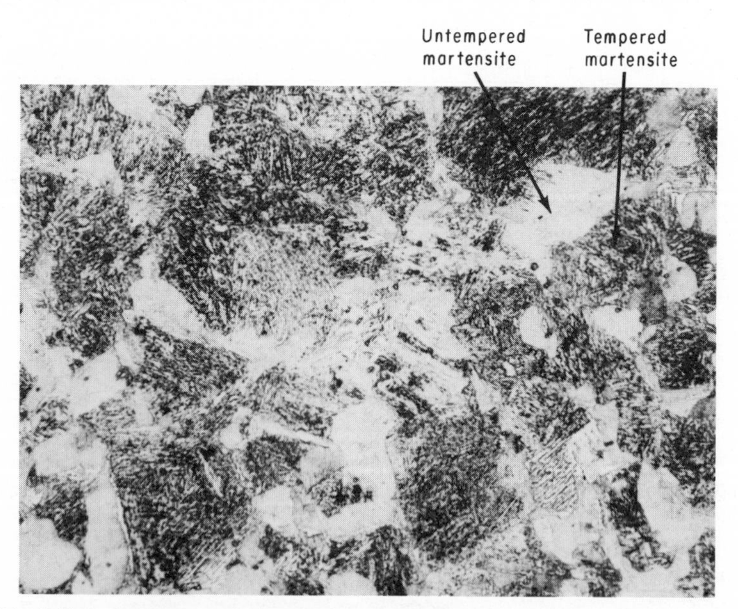

Fig. 8·57 1045 steel water-quenched and tempered at 600°F for 1 h. Tempered martensite (dark) and untempered martensite (light). Etched in 2 percent nital, 500X.

has high strength, high hardness, low ductility, and low toughness, and many of the residual stresses are relieved.

Heating in the range from 450 to 750°F changes the epsilon carbide to orthorhombic cementite (Fe_3C), the low-carbon martensite becomes b.c.c. ferrite, and any retained austenite is transformed to lower bainite. The carbides are too small to be resolved by the optical microscope, and the entire structure etches rapidly to a black mass formerly called *troostite* (Fig. 8·57). If the sample is magnified 9,000 times using the electron microscope, the carbide precipitate is clearly seen. Some of the carbide has come out along the original martensitic plate directions (Fig. 8·58). While the tensile strength has dropped, it is still very high, over 200,000 psi. The ductility has increased slightly, but the toughness is still low. Hardness has decreased to between Rockwell C40 and 60 depending upon the tempering temperature.

Tempering in the range of 750 to 1200°F continues the growth of the cementite particles. This coalescence of the carbide particles allows more of the ferrite matrix to be seen, causing the sample to etch lighter than the lower-temperature product. In this structure, formerly known as *sorbite*, the carbide is just about resolvable at 500x (Fig. 8·59) and is clearly seen in the electron micrograph (Fig. 8·60). Mechanical properties in this range are: tensile strength 125,000–200,000 psi, elongation 10–20 percent in 2 in., hardness Rockwell C 20–40. Most significant is the rapid increase in toughness, as shown by Fig. 8·54.

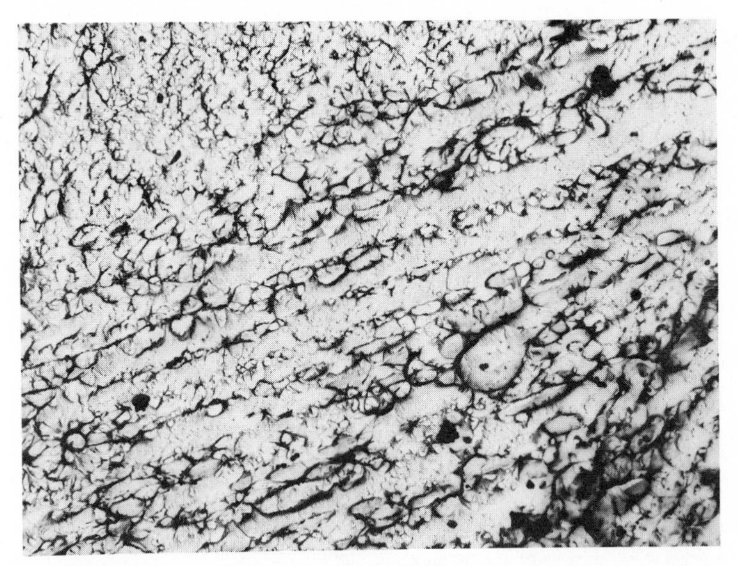

Fig. 8·58 Same sample as Fig. 8·57, taken with the electron microscope, 9,000X.

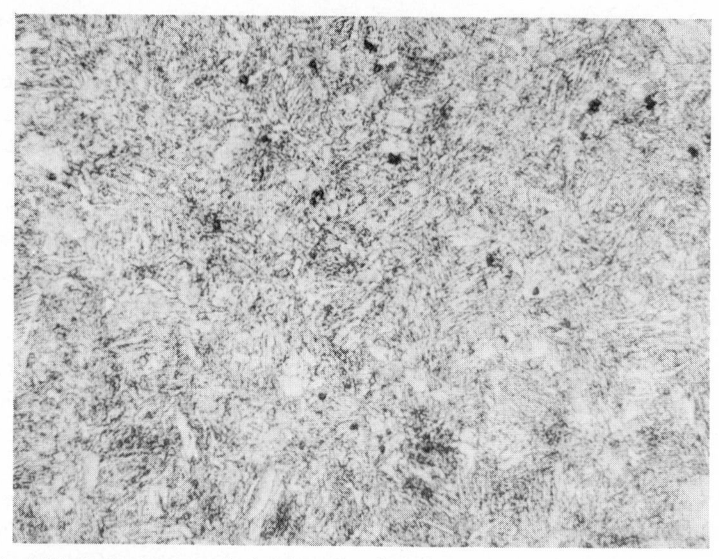

Fig. 8·59 1045 steel water-quenched and tempered at 1150°F for 1 h. Precipitated carbide particles in a ferrite matrix. Etched in 2 percent nital, 500X.

Heating in the range from 1200 to 1333°F produces large, globular cementite particles. This structure is very soft and tough and is similar to the spheroidized cementite structure obtained directly from austenite by a spheroidized anneal (see Fig. 8·4).

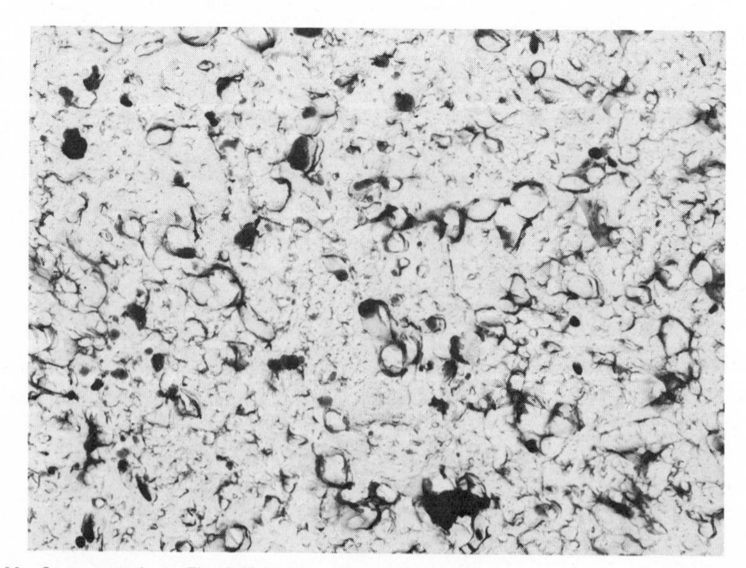

Fig. 8·60 Same sample as Fig. 8·59, taken with the electron microscope, 9,000X.

For many years, metallurgists divided the tempering process into definite stages. The microstructure appearing in these stages was given names like troostite and sorbite. However, the changes in microstructure are so gradual that it is more realistic to call the product of tempering at any temperature simply *tempered martensite*. The transformation products of austenite and martensite are summarized in Fig. 8·61.

In the above discussion, time of tempering has been assumed to be constant. Since tempering is a process that involves energy, both time and temperature are factors. The same effect can be achieved by using a shorter time at a higher temperature as by using a longer time at a lower temperature. Figure 8·62 shows the effect of time at four tempering temperatures for a eutectoid steel. Note that most of the softening action occurs in the first few minutes and that little further reduction in hardness results from increasing the time of tempering from, say, 1 to 5 h.

It is important to realize that, when toughness measurements are made in order to compare different steels, the comparisons must be made at the same hardness or strength levels and at the same temperature of testing.

If a medium tensile strength is desired, one may ask why it is necessary first to form a fully martensitic structure and then to reduce the strength substantially in tempering, when the same tensile strength may be obtained, with less difficulty in quenching, from mixtures of martensite and

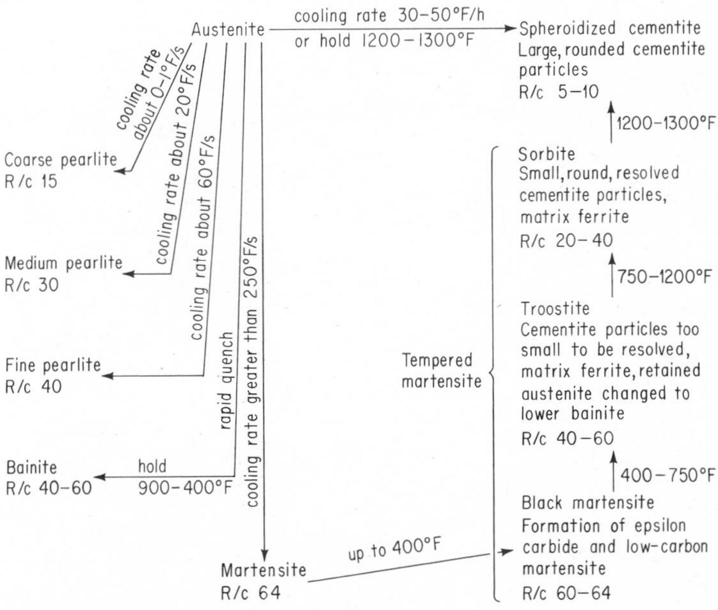

Fig. 8·61 Transformation products of austenite and martensite for a eutectoid steel.

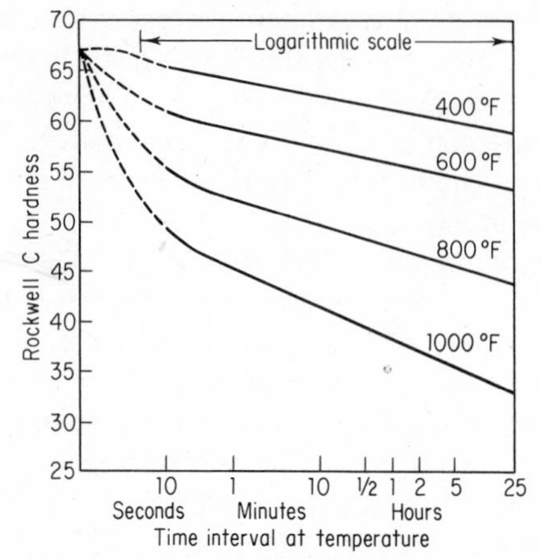

Fig. 8·62 Effect of time interval at four tempering temperatures upon the softening of a quenched 0.82 percent carbon steel. (From Bain and Paxton, "Alloying Elements in Steel," American Society for Metals, Metals Park, Ohio, 1961.)

bainite or martensite and pearlite. Samples of a medium-carbon alloy steel were heat-treated in three different ways: (1) quenched to martensite, (2) partially transformed isothermally to bainite and quenched to form a mixture of bainite and martensite, (3) partially transformed isothermally

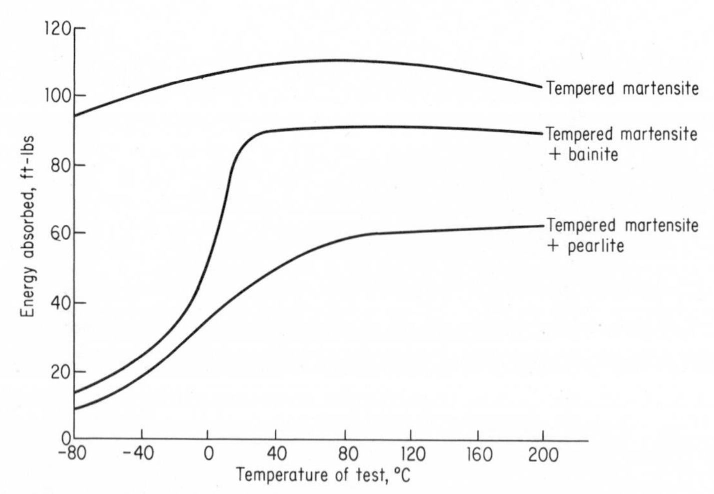

Fig. 8·63 Variation of notched-bar toughness with testing temperature for three structures tempered to the same tensile strength of 125,000 psi. (From Brick and Phillips, "Structure and Properties of Alloys," 2d ed., McGraw-Hill Book Company, New York, 1949.)

to ferrite and pearlite and then quenched, resulting in a mixture of largely pearlite and martensite. The three samples were then tempered to the same tensile strength of 125,000 psi and tested. The sample that was fully martensitic before tempering had the highest yield strength, the highest ductility, the highest fatigue strength, and the greatest toughness. Figure 8·63 shows the notched-bar toughness of the three structures at different testing temperatures. The curves also indicate that the presence of bainite in the tempered martensite is less detrimental at room temperature and above than pearlite.

As a further aid in the selection of a steel for a given application, it is possible to extend the usefulness of the end-quench hardenability test by subjecting additional end-quench samples to various tempering temperatures (Fig. 8·64).

8·23 Austempering This is a heat-treating process developed from the I-T diagram to obtain a structure which is 100 percent bainite. It is accomplished by first heating the part of the proper austenitizing temperature followed by cooling rapidly in a salt bath held in the bainite range (usually between 400 and 800°F). The piece is left in the bath until the transformation to bainite is complete. The steel is caused to go directly from austenite to bainite, and at no time is it in the fully hardened martensitic state. Actually, austempering is a complete heat treatment, and no reheating is involved as in tempering. Figure 8·65 illustrates austempering schematically, showing

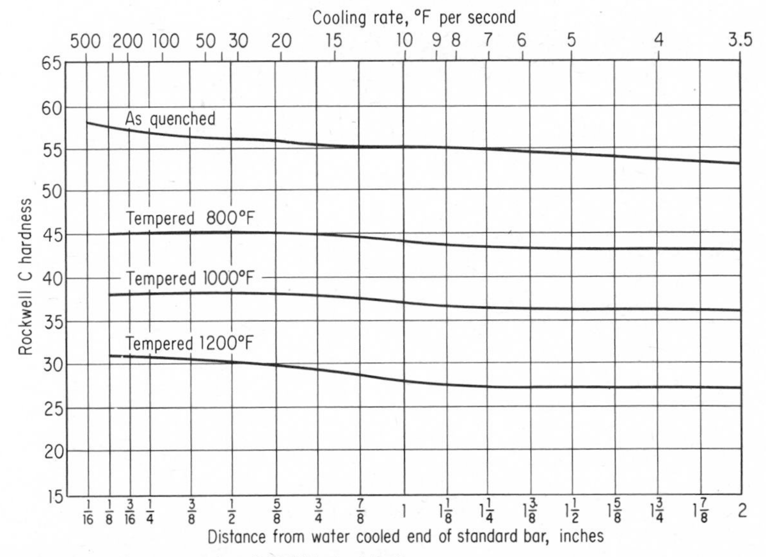

Fig. 8·64 End-quench test results of 4340H steel in the as-quenched condition and after tempering at the indicated temperatures. (Joseph T. Ryerson & Son, Inc.)

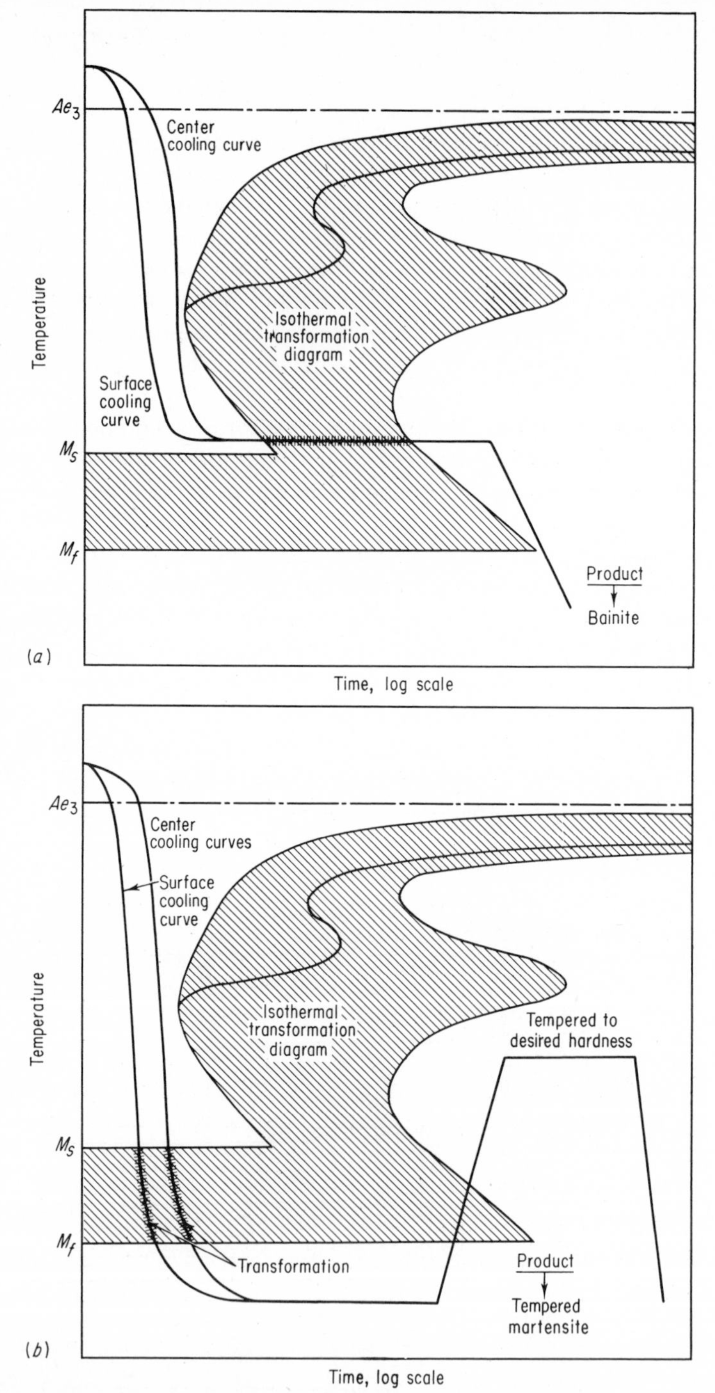

Fig. 8·65 (*a*) Schematic transformation diagram for austempering; (*b*) schematic transformation diagram for conventional quench and temper method. (From "U.S.S. Carilloy Steels," U.S. Steel Corporation.)

TABLE 8·6 Summary of Tensile and Impact Properties of 0.180-in. Round Rods Heat-treated by Quench and Temper Method and by Austempering*

PROPERTY MEASURED	QUENCH AND TEMPER METHOD	AUSTEMPERING
Rockwell C hardness	49.8	50.0
Ultimate tensile strength, psi	259,000	259,000
Elongation, % in 2 in.	3.75	5.0
Reduction in area, %	26.1	46.4
Impact, ft-lb (unnotched round specimen)	14.0	36.6
Free-bend test	Ruptured at 45°	Greater than 150° without rupture

Analysis: 0.78% C; 0.58% Mn; 0.146% Si; 0.042% P; 0.040% S
Heat treatment resulting in grain size (1450°F) 5 to 6 with 6 predominating:

Quench and Temper	*Austempering*
Pb bath 1450°F, 5 min	Pb bath 1450°F, 5 min
Oil quench	Transformed in Pb-Bi bath at 600°F, 20 min
Tempered 650°F, 30 min	

* Research Laboratory, U.S. Steel Corporation.

the difference between austempering and the conventional quench and temper method.

The comparison of mechanical properties developed by austempering and the quench and temper method is usually made at the same hardness or strength (Table 8·6). The superiority of austempering shows up in such properties as reduction of area in tension, resistance to impact, and the slow-bend test (Fig. 8·66). A striking demonstration of the resiliency of an austempered shovel is shown in Fig. 8·67. The marked improvement in the impact strength of austempered parts is most pronounced in the hardness range of Rockwell C 45 to 55 (Fig. 8·68).

Aside from the advantage of greater ductility and toughness along with high hardness as a result of austempering, there is also less distortion and danger of quenching cracks because the quench is not so drastic as that of the conventional method.

The primary limitation of austempering is the effect of mass of the part being heat-treated. Only sections which can be cooled fast enough to avoid transformation to pearlite, in the termperature range of 900 to 1200°F, are suitable. Therefore, most industrial applications have been in sections less than 1/2 in. thick (Fig. 8·69). This thickness may be increased somewhat by the use of alloy steels, but then the time for completion of transformation to bainite may become excessive.

8·24 Surface Heat Treatment or Case Hardening Numerous industrial applications require a hard wear-resistant surface called the *case*, and a relatively

Fig. 8·66 Improved toughness and ductility of austempered rods compared with quenched and tempered rods of the same hardness. (Courtesy of Research Laboratory, U.S. Steel Corporation.)

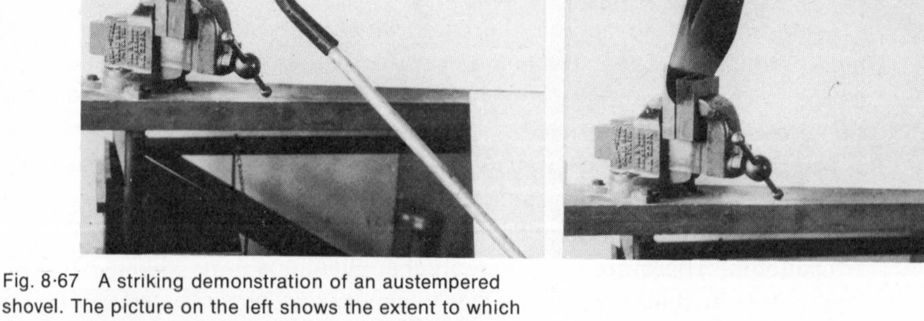

Fig. 8·67 A striking demonstration of an austempered shovel. The picture on the left shows the extent to which the shovel can be bent without failure, and that on the right shows how the bent shovel, after removal of the bending force, returns to its original position without permanent deformation. (Courtesy of Research Laboratory, U.S. Steel Corporation.)

soft, tough inside called the *core*. There are five principal methods of case hardening:

1. Carburizing
2. Nitriding
3. Cyaniding or carbonitriding
4. Flame hardening
5. Induction hardening

The first three methods change the chemical composition, carburizing by the addition of carbon, nitriding by the addition of nitrogen, and cyaniding by the addition of both carbon and nitrogen. The last two methods do not change the chemical composition of the steel and are essentially shallow-hardening methods. In flame and induction hardening the steel must be capable of being hardened; therefore, the carbon content must be about 0.30 percent or higher.

8·25 Carburizing This is the oldest and one of the cheapest methods of case hardening. A low-carbon steel, usually about 0.20 percent carbon or lower, is placed in an atmosphere that contains substantial amounts of carbon monoxide. The usual carburizing temperature is 1700°F. At this tempera-

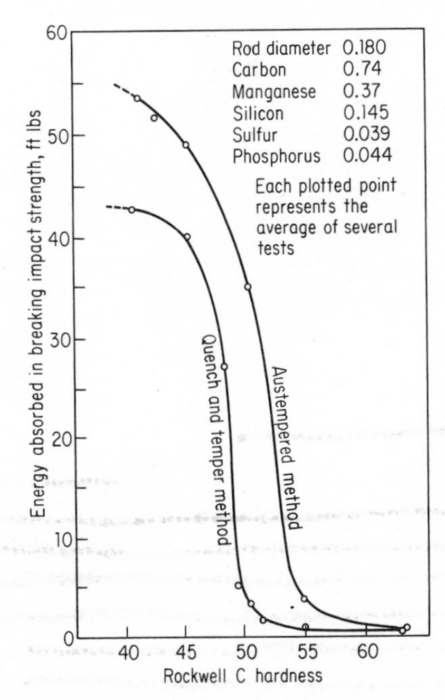

Fig. 8·68 Comparison between austempering and quenched and tempered steel. (From "Suiting the Heat Treatment to the Job," U.S. Steel Corporation.)

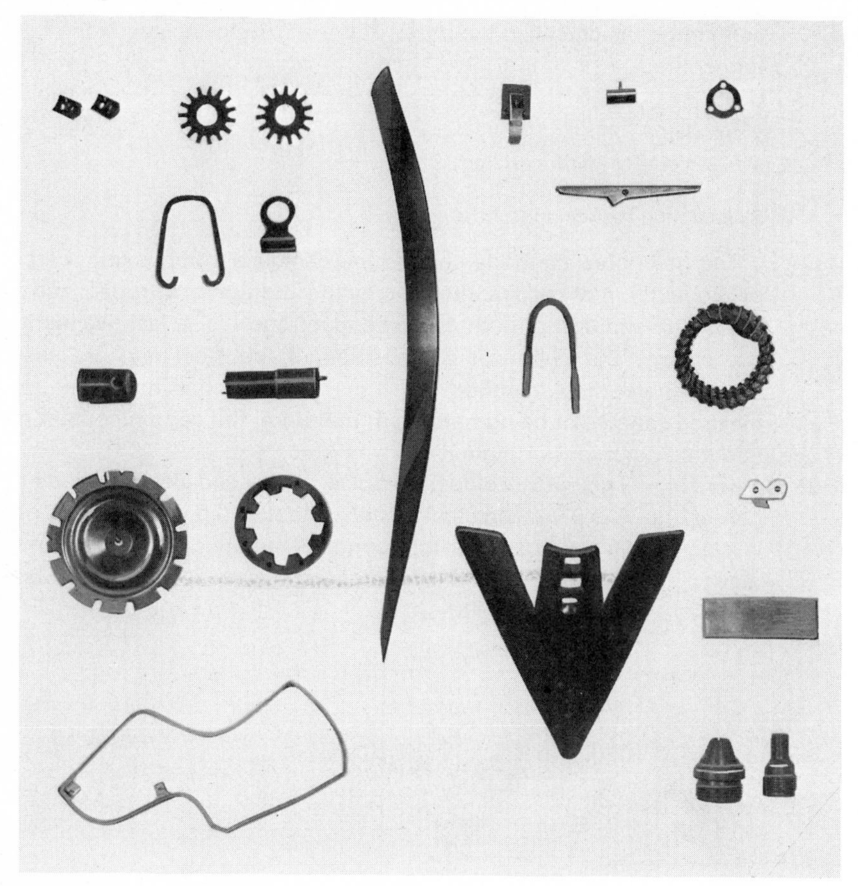

Fig. 8·69 Variety of industrial articles that are austempered. (Ajax Electric Company.)

ture, the following reaction takes place:

$$Fe + 2CO \rightarrow Fe_{(C)} + CO_2$$

where $Fe_{(C)}$ represents carbon dissolved in austenite. The maximum amount of carbon that can be dissolved in austenite at 1700°F is indicated on the iron–iron carbide equilibrium diagram at the A_{cm} line. Therefore, very quickly, a surface layer of high carbon (about 1.2 percent) is built up. Since the core is of low carbon content, the carbon atoms trying to reach equilibrium will begin to diffuse inward. The rate of diffusion of carbon in austenite, at a given temperature, is dependent upon the diffusion coefficient and the carbon-concentration gradient. Under known and standard operating conditions, with the surface at a fixed carbon concentration, the form of the carbon gradient may be predicted, with reasonable accuracy, as a function of elapsed time. After diffusion has taken place for the required amount of time depending upon the case depth desired, the part is removed from the furnace and cooled. If the part is furnace-cooled and examined microscopically, the carbon gradient will be visible in the gradual change of the structure. At the surface is the hypereutectoid zone consisting of pearlite with a white cementite network, followed by the eutectoid zone of only pearlite and finally the hypoeutectoid zone of pearlite and ferrite, with the amount of ferrite increasing until the core is reached. This is illustrated in Fig. 8·70. The case depth may be measured microscopically with a micrometer eyepiece. The carbon gradient and the case depth may be determined experimentally by placing the part in a lathe and machining samples for chemical analysis at increments of 0.005 in. until the core is reached. Analysis to determine carbon content is made, and the results can be plotted graphically, as in Fig. 8·71. The relation of time and temperature to case depth is shown in Fig. 8·72 and Table 8·7.

The carburizing equation given previously, $Fe + 2CO \rightarrow Fe_{(c)} + CO_2$, is reversible and may proceed to the left, removing carbon from the surface layer if the steel is heated in an atmosphere containing carbon dioxide (CO_2). This is called *decarburization.* Other possible decarburizing reactions are

$$Fe_{(C)} + H_2O \rightarrow Fe + CO + H_2$$
$$Fe_{(C)} + O_2 \rightarrow Fe + CO_2$$

Decarburization is a problem primarily with high-carbon steels and tool steels. The surface, depleted of carbon, will not transform to martensite on subsequent hardening, and the steel will be left with a soft skin. For many tool applications, the stresses to which the part is subjected in service are maximum at or near the surface, so that decarburization is harmful. Figure 8·73 shows decarburization on the surface of a high-carbon steel. Decarburization may be prevented by using an endothermic gas atmosphere in the furnace to protect the surface of the steel from oxygen, carbon dioxide, and water vapor. An endothermic gas atmosphere is prepared by reacting relatively rich mixtures of air and hydrocarbon gas (usually natural gas) in an externally heated generator in the presence of a nickel catalyst. The gas produced consists of 40 percent nitrogen, 40 percent hydrogen, and 20 percent carbon monoxide.

Commercial carburizing may be accomplished by means of pack carburizing, gas carburizing, and liquid carburizing. In *pack carburizing*, the work is surrounded by a carburizing compound in a closed container. The

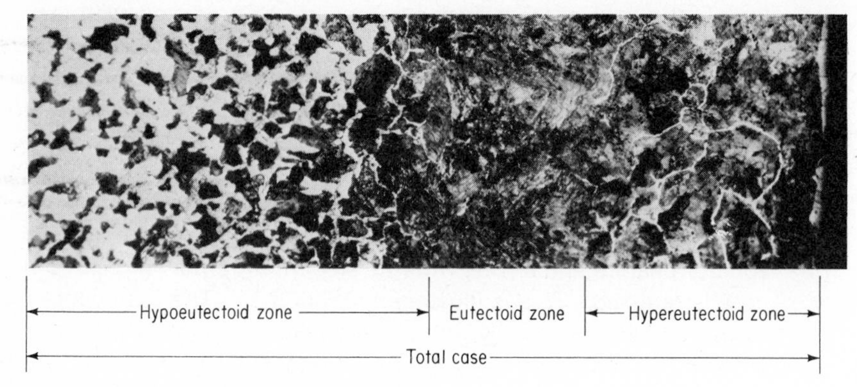

Fig. 8·70 0.20 percent carbon steel, pack-carburized at 1700°F for 6 h and furnace-cooled. Etched in 2 percent nital, 30X.

container is heated to the proper temperature for the required amount of time and then slow-cooled. This is essentially a batch method and does not lend itself to high production. Commercial carburizing compounds usually consist of hardwood charcoal, coke, and about 20 percent of barium car-

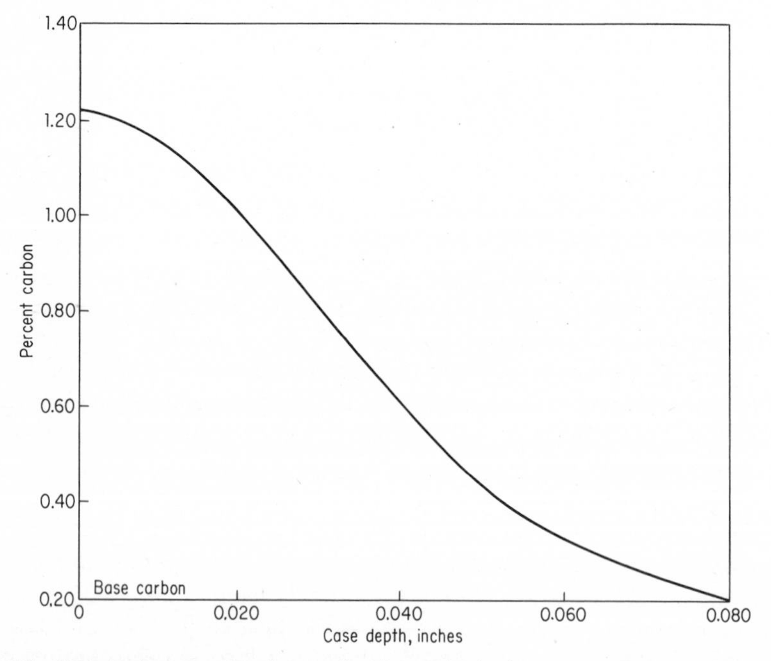

Fig. 8·71 Carbon-concentration gradient in a carburized steel with 0.080 in. total case.

TABLE 8·7 Case Depth in Inches by Carburizing*

TIME, H	TEMP, °F					
	1500	1550	1600	1650	1700	1750
1	0.012	0.015	0.018	0.021	0.025	0.029
2	0.017	0.021	0.025	0.030	0.035	0.041
3	0.021	0.025	0.031	0.037	0.043	0.051
4	0.024	0.029	0.035	0.042	0.050	0.059
5	0.027	0.033	0.040	0.047	0.056	0.066
6	0.030	0.036	0.043	0.052	0.061	0.072
7	0.032	0.039	0.047	0.056	0.066	0.078
8	0.034	0.041	0.050	0.060	0.071	0.083
9	0.036	0.044	0.053	0.063	0.075	0.088
10	0.038	0.046	0.056	0.067	0.079	0.093
11	0.040	0.048	0.059	0.070	0.083	0.097
12	0.042	0.051	0.061	0.073	0.087	0.102
13	0.043	0.053	0.064	0.076	0.090	0.106
14	0.045	0.055	0.066	0.079	0.094	0.110
15	0.047	0.057	0.068	0.082	0.097	0.114
16	0.048	0.059	0.071	0.084	0.100	0.117
17	0.050	0.060	0.073	0.087	0.103	0.121
18	0.051	0.062	0.075	0.090	0.106	0.125
19	0.053	0.064	0.077	0.092	0.109	0.128
20	0.054	0.066	0.079	0.094	0.112	0.131
21	0.055	0.067	0.081	0.097	0.114	0.134
22	0.056	0.069	0.083	0.099	0.117	0.138
23	0.058	0.070	0.085	0.101	0.120	0.141
24	0.059	0.072	0.086	0.103	0.122	0.144

* Courtesy of Republic Steel Corp.

bonate as an energizer. The carburizing compound is in the form of coarse particles or lumps, so that, when the cover is sealed on the container, sufficient air will be trapped inside to form carbon monoxide. The principal advantages of pack carburizing are that it does not require the use of a prepared atmosphere and that it is efficient and economical for individual processing of small lots of parts or of large, massive parts. The disadvantages are that it is not well suited to the production of thin carburized cases that must be controlled to close tolerances; it cannot provide the close control of surface carbon that can be obtained by gas carburizing; parts cannot be direct-quenched from the carburizing temperature; and excessive time is consumed in heating and cooling the charge. Because of the inherent variation in case depth and the cost of packing materials, pack carburizing is not used on work requiring a case depth of less than 0.030 in., and tolerances are at least 0.010 in.

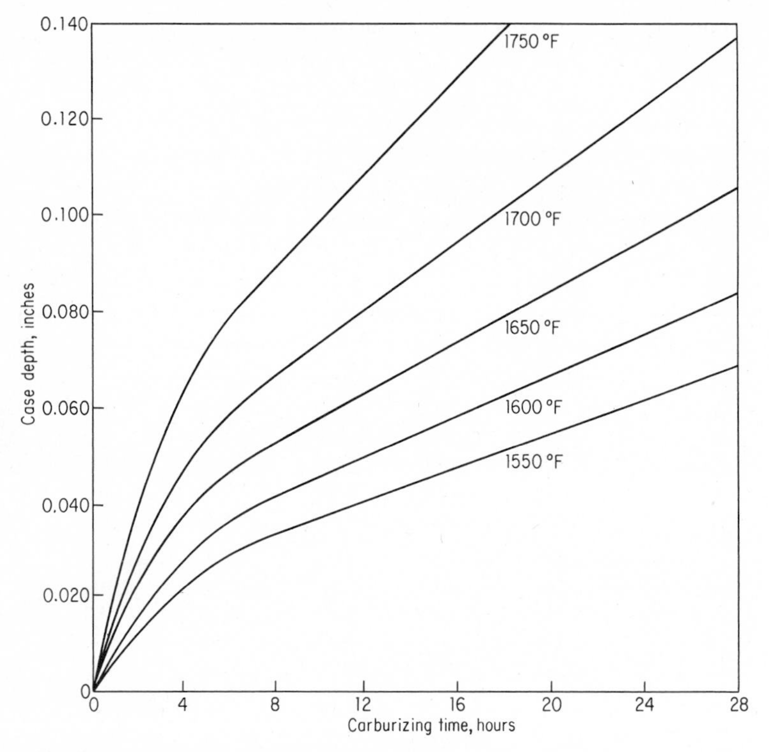

Fig. 8·72 Relation of time and temperature to case depth.

Gas carburizing may be either batch or continuous and lends itself better to production heat treatment. The steel is heated in contact with carbon monoxide and/or a hydrocarbon which is readily decomposed at the carburizing temperature. The hydrocarbon may be methane, propane, natural gas, or vaporized fluid hydrocarbon. Commercial practice is to use a carrier gas, such as obtained from an endothermic generator, and enrich it with one of the hydrocarbon gases.

It was mentioned previously that carburized parts will usually have a thin outer layer of high carbon. There are two reasons why it may be desirable to avoid this hypereutectoid layer. First, if the piece is cooled slowly from the carburizing temperature, a proeutectoid cementite network will form at the grain boundaries. On subsequent hardening, particularly if the steel is heated to below the A_{cm} line, some grain-boundary cementite will remain in the finished piece and is a frequent cause for failure. Second, the hyper-eutectoid surface-carbon content will increase the amount of retained austenite. Therefore, if the steel is highly alloyed, the carbon content of the case should be no greater than the eutectoid content of 0.80 percent carbon. By using a *diffusion period*, during which the gas is turned off but the temperature maintained, gas carburizing allows the surface carbon to be reduced to any desired value. Use of the diffusion period also produces much cleaner work by dissipation of carbon deposit (soot) during the time when no gas is flowing. Gas carburizing allows quicker handling by direct quenching, lower cost, cleaner surroundings, closer quality control, and greater flexibility of operation in comparison with pack carburizing.

Liquid carburizing is a method of case-hardening steel by placing it in a bath of molten cyanide so that carbon will diffuse from the bath into the metal and produce a case comparable to one resulting from pack or gas carburizing. Liquid carburizing may be distinguished from cyaniding by

Fig. 8·73 Decarburized layer of ferrite on the surface of a high-carbon annealed steel. Etched in 2 percent nital, 200X.

the character and composition of the case produced. The cyanide case is higher in nitrogen and lower in carbon; the reverse is true of liquid-carburized cases. Cyanide cases are seldom to a depth greater than 0.010 in.; liquid-carburizing baths permit cases as deep as 0.250 in. Low-temperature salt baths (light-case) usually contain a cyanide content of 20 percent and operate between 1550 and 1650°F. These are best suited for case depths up to 0.030 in. High-temperature salt baths (deep-case) usually have a cyanide content of 10 percent and operate between 1650 and 1750°F. High-temperature salt baths are used for producing case depths of 0.030

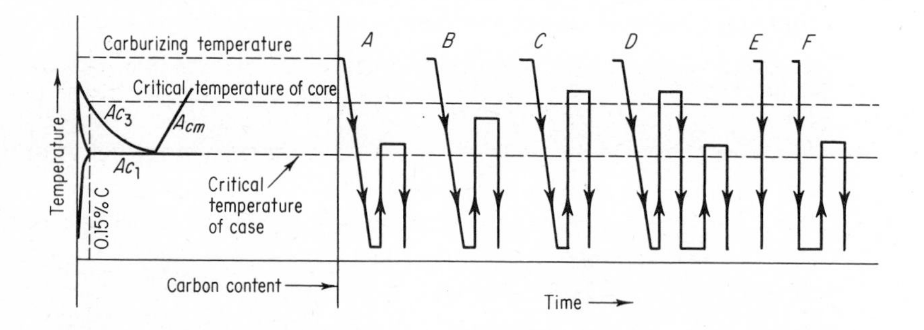

TREATMENT	CASE	CORE
A—best adapted to fine-grained steels	Refined; excess carbide not dissolved	Unrefined; soft and machinable
B—best adapted to fine-grained steels	Slightly coarsened; some solution of excess carbide	Partially refined; stronger and tougher than *A*
C—best adapted to fine-grained steels	Somewhat coarsened; solution of excess carbide favored; austenite retention promoted in highly alloyed steels	Refined; maximum core strength and hardness; better combination of strength and ductility than *B*
D—best treatment for coarse-grained steels	Refined; solution of excess carbide favored; austenite retention minimized	Refined; soft and machinable; maximum toughness and resistance to impact
E—adapted to fine-grained steels only	Unrefined with excess carbide dissolved; austenite retained; distortion minimized	Unrefined but hardened
F—adapted to fine-grained steels only	Refined; solution of excess carbide favored; austenite retention minimized	Unrefined; fair toughness

Fig. 8·74 Various heat treatments for carburized steels. (From "Metals Handbook," 1948 ed., American Society for Metals, Metals Park, Ohio.)

to 0.120 in., although it is possible to go as high as 0.250 in. In general, liquid carburizing is best suited to small and medium-size parts, since large parts are difficult to process in salt baths. The advantages of liquid carburizing are: (1) freedom from oxidation and sooting problems, (2) uniform case depth and carbon content, (3) a rapid rate of penetration, and (4) the fact that the bath provides high thermal conductivity, thereby reducing the time required for the steel to reach the carburizing temperature. Disadvantages include: (1) parts must be thoroughly washed after treatment to prevent rusting; (2) regular checking and adjustment of the bath composition is necessary to obtain uniform case depth; (3) some shapes cannot be handled because they either float or will cause excessive dragout of salt; and (4) cyanide salts are poisonous and require careful attention to safety.

8·26 Heat Treatment after Carburizing Since steel is carburized in the austenite region, direct quenching from the carburizing temperature will harden both the case and core if the cooling rate is greater than the critical cooling rate. Direct quenching of coarse-grained steels often leads to brittleness and distortion, so that this treatment should be applied only to fine-grained steels. Alloy steels are rarely used in the direct-quenched condition because of the large amount of retained austenite in the hardened case. Figure 8·74 shows a diagrammatic representation of various hardening treatments for carburized steels together with case and core properties.

When a carburized part is hardened, the case will appear as a light martensite zone followed by a darker transition zone (Fig. 8·75). The *hard*

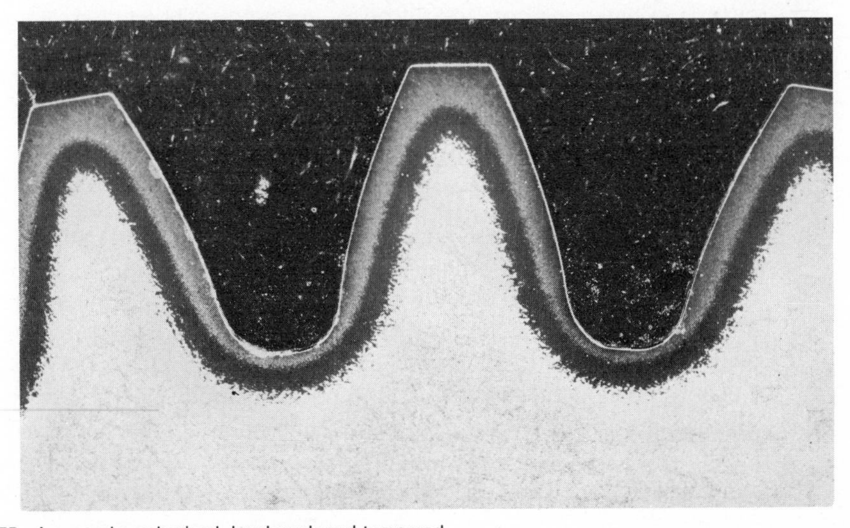

Fig. 8·75 A properly carburized, hardened, and tempered gear. Etched in 2 percent nital, 7X.

case or *effective case* is measured from the outer edge to the middle of the dark zone. From the nature of the carbon gradient, the hard case contains the portion of the case above 0.40 percent carbon and is approximately equal to two-thirds of the total case. Hardness-traverse measurements may also be used to determine the depth of the effective case, since the middle of the transition zone is at approximately Rockwell C 50.

8·27 Cyaniding and Carbonitriding Cases that contain both carbon and nitrogen are produced in liquid salt baths (cyaniding) or by use of gas atmospheres (carbonitriding). The temperatures used are generally lower than those used in carburizing, being between 1400 and 1600°F. Exposure is for a shorter time, and thinner cases are produced, up to 0.010 in. for cyaniding and up to 0.030 in. for carbonitriding.

In *cyaniding*, the proportion of nitrogen and carbon in the case produced by a cyanide bath depends on both composition and temperature of the bath, the latter being the most important. Nitrogen content is higher in baths operating at the lower end of the temperature range than in those operating at the upper end of the range. Generally, carbon content of the case is lower than that produced by carburizing, ranging from about 0.5 to 0.8 percent. The case also contains up to about 0.5 percent nitrogen; therefore, file-hard cases can be obtained on quenching in spite of the relatively low carbon content. Several mixtures of cyanides are available for the bath. Although baths of higher sodium cyanide concentrations are employed, the most commonly used mixture is made up of 30 percent sodium cyanide, 40 percent sodium carbonate, and 30 percent sodium chloride. This mixture has a melting point of 1140°F and remains quite stable under continuous operating conditions. The active hardening agents of cyaniding baths, carbon and nitrogen, are not produced directly from sodium cyanide (NaCN). Molten cyanide decomposes in the presence of air at the surface of the bath to produce sodium cyanate (NaNCO), which in turn decomposes as follows:

$$2NaCN + O_2 \rightarrow 2NaNCO$$

$$4NaNCO \rightarrow Na_2CO_3 + 2NaCN + CO + 2N$$

The carbon content of the case developed in the cyanide bath increases with an increase in the cyanide concentration of the bath, thus providing considerable flexibility. A bath operating at 1550°F and containing about 3 percent cyanide may be used to restore carbon to decarburized steels, while a 30 percent cyanide bath at the same temperature will develop a 0.005-in. case on the surface of a 0.65 percent carbon steel in 45 min. This process is particularly useful for parts requiring a very thin hard case, such as screws, small gears, nuts, and bolts. The principal disadvantages of cyaniding are the same as those mentioned under liquid carburizing.

Carbonitriding is a case-hardening process in which a steel is heated in a gaseous atmosphere of such composition that carbon and nitrogen are absorbed simultaneously. The term carbonitriding is misleading because it implies a modified nitriding process. Actually carbonitriding is a modification of carburizing, and the name "nitrocarburizing" would be more descriptive. The process is also known as *dry cyaniding*, *gas cyaniding*, and *nicarbing*. The atmospheres used in carbonitriding generally comprise a mixture of carrier gas, enriching gas, and ammonia. The carrier gas is usually a mixture of nitrogen, hydrogen, and carbon monoxide produced in an endothermic generator, as in gas carburizing. The carrier gas is supplied to the furnace under positive pressure to prevent air infiltration and acts as a diluent for the active gases (hydrocarbons and ammonia), thus making the process easier to control. The enriching gas is usually propane or natural gas and is the primary source for the carbon added to the surface. At the furnace temperature, the added ammonia (NH_3) breaks up or dissociates to provide the nitrogen to the surface of the steel. Figure 8·76 shows a carbonitrided case obtained by heating C1213

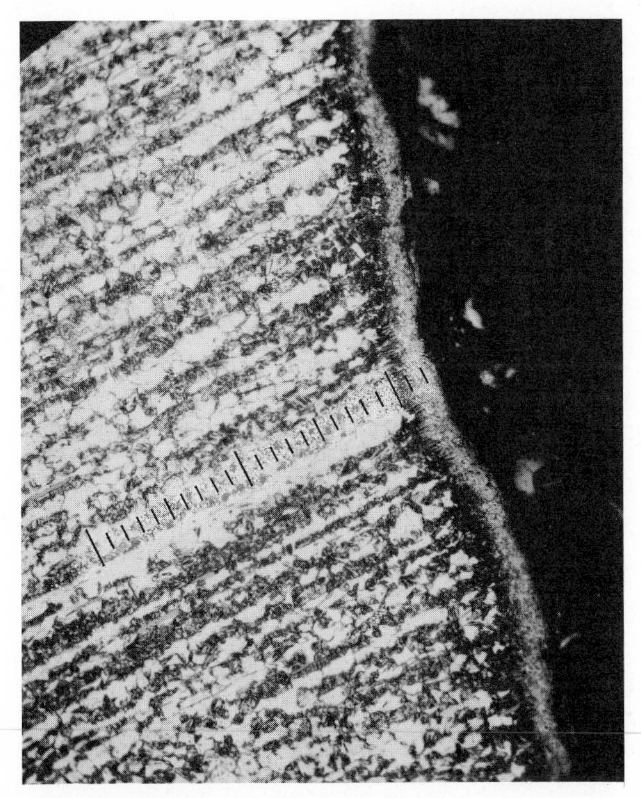

Fig. 8·76 Carbonitrided case on AISI C1213 steel. Heated at 1550°F for 20 min in an ammonia-propane atmosphere, then oil-quenched. Etched in 2 percent nital, 125X. Case depth approximately 0.0025 in.

steel in an ammonia-propane atmosphere at 1550°F for 20 min followed by oil quenching. Each division of the micrometer eyepiece is 0.001 in., and the effective case depth measured to the middle of the dark transition zone is approximately 0.0025 in.

The presence of nitrogen in the austenite accounts for the major differences between carbonitriding and carburizing. Carbon-nitrogen austenite is stable at lower temperatures than plain-carbon austenite and transforms more slowly on cooling. Carbonitriding therefore can be carried out at lower temperatures and permits slower cooling rates than carburizing in the hardening operation. Because of the lower temperature treatment and oil quenching rather than water quenching, distortion is reduced, and there is less danger of cracking. Since nitrogen increases the hardenability, carbonitriding the less expensive carbon steels for many applications will provide properties equivalent to those obtained in gas-carburized alloy steels. It has also been found that the resistance of a carbonitrided surface to softening during tempering is markedly superior to that of a carburized surface.

8·28 Nitriding This is a process for case hardening of alloy steel in an atmosphere consisting of a mixture in suitable proportions of ammonia gas and dissociated ammonia. The effectiveness of the process depends on the formation of nitrides in the steel by reaction of nitrogen with certain alloying elements. Although at suitable temperatures and with the proper atmosphere all steels are capable of forming iron nitrides, the best results are obtained in those steels that contain one or more of the major nitride-forming alloying elements. These are aluminum, chromium, and molybdenum. The nitrogen must be supplied in the atomic or nascent form; molecular nitrogen will not react.

The parts to be nitrided are placed in an airtight container through which the nitriding atmosphere is supplied continuously while the temperature is raised and held between 925 and 1050°F. The nitriding cycle is quite long, depending upon the case depth desired. As shown in Fig. 8·77, a 60-h cycle will give a case depth of approximately 0.024 in. at 975°F.

A nitrided case consists of two distinct zones. In the outer zone the

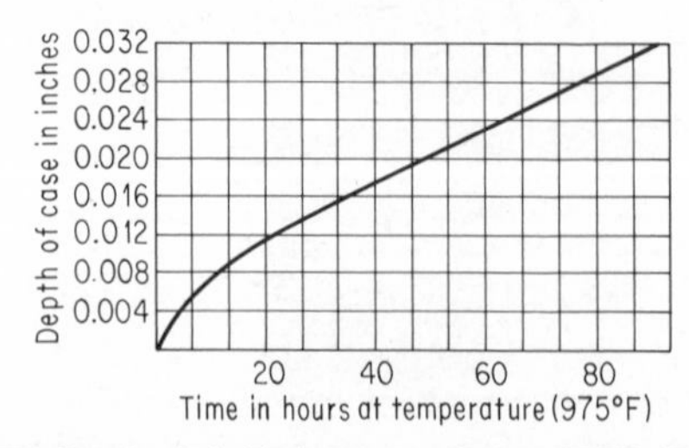

Fig. 8·77 Depth of nitrided case vs. time at 975°F. (From "Heat Treatment of Steels," Republic Steel Corporation.)

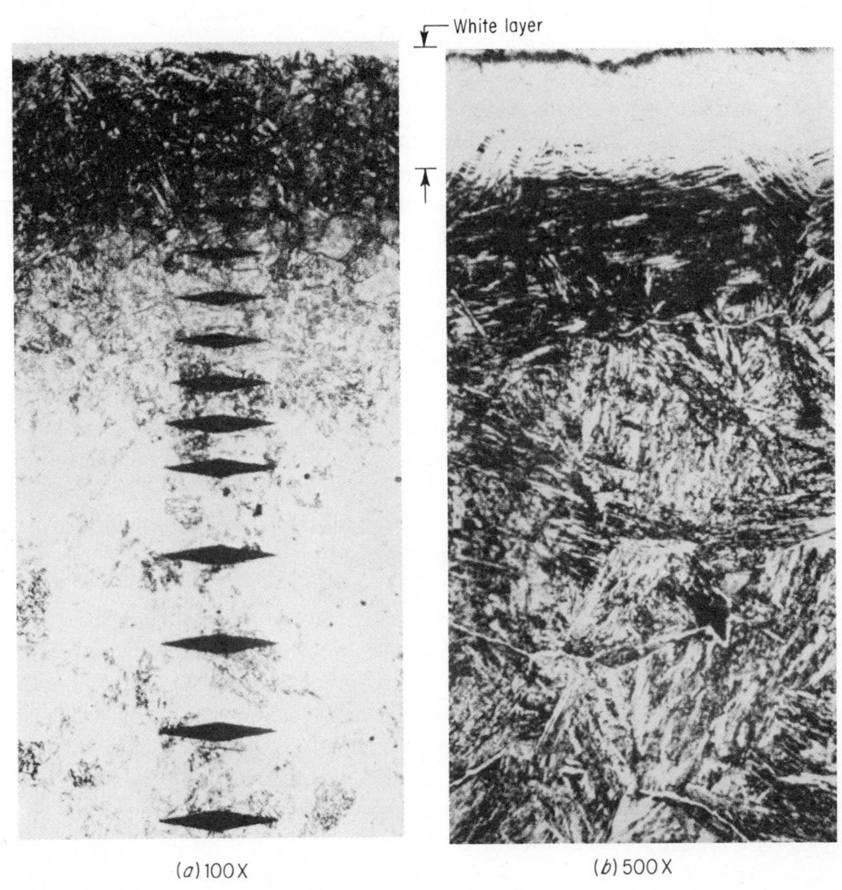

Fig. 8·78 Microstructure illustrating a nitrided case produced by a single-stage process. Nitrided for 48 h at 975°F and 30 percent ammonia dissociation. Diamond shapes are Knoop hardness impressions. (Courtesy of The Nitralloy Corporation.)

nitride-forming elements, including iron, have been converted to nitrides. This region, which varies in thickness up to a maximum of about 0.002 in., is commonly known as the "white layer" because of its appearance after a nital etch. In the zone beneath this white layer, alloy nitrides only have been precipitated. A typical microstructure, illustrated in Fig. 8·78*b*, shows the white layer and the underlying nitride case. At the lower magnification, illustrated in Fig. 8·78*a*, the lighter core structure can be seen beneath the nitride case. The depth of nitride case is determined by the rate of diffusion of nitrogen from the white layer to the region beneath. The nitriding medium, therefore, needs to contain only sufficient active nitrogen to maintain the white layer. Any increase beyond this point serves to increase the depth of white layer and does not affect the thickness of the inner layer. The concentration of active nitrogen on the surface of the steel, which determines the depth of the white layer, is fixed by the degree of dissociation of the ammonia. In the single-stage nitriding process this dissociation is held between 15 and 30 percent by adjusting the rate of flow. A temperature in the 925 to 975°F range is employed. The double-stage process, also known as the Floe process, has the advantage of reducing the thickness of the white nitride layer. In the first stage of the double-stage process, the ammonia dissociation is held at 20 percent for a period of 5 to 10 h at 975°F. During this period the white layer is established, and the useful nitride starts to form by diffusion of nitrogen out of it. In the second stage, the ammonia dissociation is increased to 83 to 86 percent, and the temperature is usually raised to 1025 to 1050°F. During this second stage the gas composition is such that it maintains only a thin white layer on the finished part. A typical structure of the case produced by this method is shown in Fig. 8·79.

The white layer is brittle and tends to chip or spall from the surface if it has a thickness in excess of 0.0005 in. Thicker white layers produced by the single-stage process must be removed by grinding or lapping after nitriding. Ordinarily an allowance of at least 0.002 in. on a side is made in the finish machining dimensions if grinding is necessary after nitriding. If the double-stage process is used, however, grinding or other finishing operations may be omitted except insofar as they are required in order to meet dimensional tolerances. The very thin white layer obtained by this method, usually from 0.0002 to 0.0004 in. in depth, does not chip or pit, and the frictional characteristics of the surface are excellent. This layer also has good wear-in properties and may be expected to improve corrosion resistance.

Hardest cases, approximately R/C 70, are obtained with aluminum alloy steels known as *Nitralloys*. These are medium-carbon steels containing also chromium and molybdenum. For some applications where lower hardness is acceptable, medium-carbon standard steels containing chromium and molybdenum (AISI 4100, 4300 series) are used. Nitriding has also been applied to stainless steels and tool steels for certain applications.

The steel is usually hardened and tempered between 1100 and 1300°F to produce a sorbitic structure of maximum core toughness and then nitrided. Since nitriding is performed at relatively low temperatures and no quenching is required, distortion is reduced to a minimum, although some growth does occur due to the increase in volume of the case. However, this growth is constant and predictable for a given part and cycle, so that in most cases parts may be machined very close to final dimensions before nitriding. This

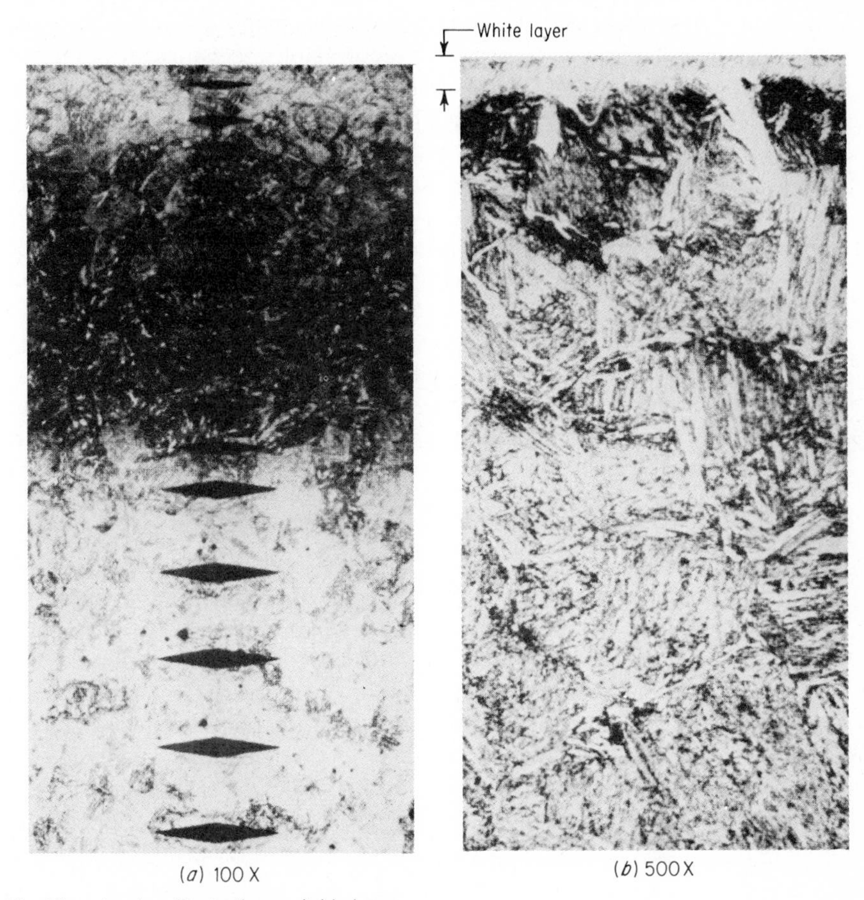

Fig. 8·79 Microstructure illustrating a nitrided case produced by the double-stage Floe process. Nitrided for 8 h at 975°F and 20 percent ammonia dissociation followed by 42 h at 1050°F and 83 to 86 percent ammonia dissociation. Diamond shapes are Knoop hardness impressions. Notice the much smaller white layer compared with that in Fig. 8·78. (Courtesy of The Nitralloy Corporation.)

is an advantage of nitriding over carburizing. Some complex parts which cannot be case-hardened satisfactorily by carburizing have been nitrided without difficulty. Wear resistance is an outstanding characteristic of the nitrided case and is responsible for its selection in most applications. The hardness of a nitrided case is unaffected by heating to temperatures below the original nitriding temperature. Substantial hardness is retained to at least 1150°F, in marked contrast with a carburized case, which begins to lose its hardness at relatively low temperatures. Fatigue resistance is also an important advantage. Tool marks and surface scratches have little effect on the fatigue properties of nitrided steels. Although it is sometimes indicated that nitriding improves the corrosion resistance of a steel, this is true only if the white layer is not removed. Corrosion resistance of stainless steels is reduced considerably by nitriding, a factor which must be taken into account when nitrided stainless steels are used in corrosive atmospheres. Disadvantages of nitriding include the long cycles usually required, the brittle case, use of special alloy steels if maximum hardness is to be obtained, cost of ammonia atmosphere, and the technical control required. Nitriding is used extensively for aircraft engine parts such as cams, cylinder liners, valve stems, shafts, and piston rods.

8·29 Flame Hardening The remaining two methods, flame hardening and induction hardening, do not change the chemical composition of the steel. They are essentially shallow hardening methods. Selected areas of the surface of a steel are heated into the austenite range and then quenched to form martensite. Therefore, it is necessary to start with a steel which is capable of being hardened. Generally, this is in the range of 0.30 to 0.60 percent carbon.

In flame hardening, heat may be applied by a single oxyacetylene torch, as shown in Fig. 8·80, or it may be part of an elaborate apparatus which automatically heats, quenches, and indexes parts. Depth of the hardened zone may be controlled by an adjustment of the flame intensity, heating time, or speed of travel. Skill is required in adjusting and handling manually operated equipment to avoid overheating the work because of high flame temperature. Overheating can result in cracking after quenching and excessive grain growth in the region just below the hardened zone. Four methods are in general use for flame hardening: (1) stationary; (2) pro-

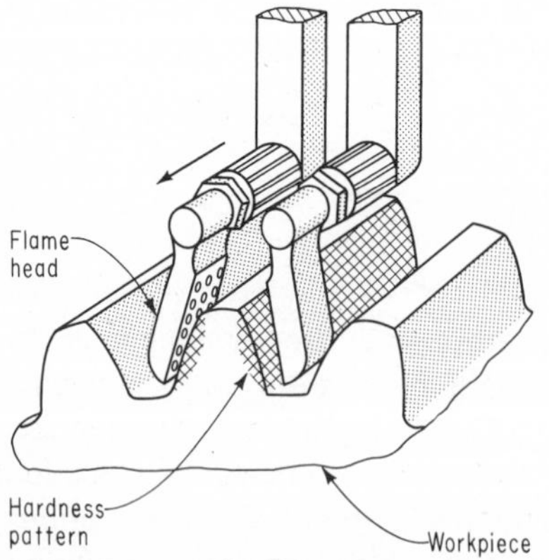

Fig. 8·80 Progressive method of flame hardening, showing the hardness pattern developed. (From "Metals Handbook." vol. 2, American Society for Metals, Metals Park, Ohio, 1964.)

gressive; (3) spinning; (4) progressive-spinning. In the first, both torch and work are stationary. This method is used for the spot hardening of small parts such as valve stems and open-end wrenches. In the progressive method, the torch moves over a stationary work piece; this is used for hardening of large parts, such as the ways of a lathe, but is also adaptable to the treatment of teeth of large gears (Fig. 8·80). In the spinning method, the torch is stationary while the work rotates. This method is used to harden parts of circular cross section, such as precision gears, pulleys, and similar components. The progressive-spinning method, in which the torch moves over a rotating workpiece, is used to surface-harden long circular parts such as shafts and rolls.

In all procedures, provision must be made for rapid quenching after the surface has been heated to the required temperature. This may be accomplished by the use of water sprays, by quenching the entire piece in water or oil, or even by air-cooling for some steels. After quenching, the part should be stress-relieved by heating in the range of 350 to 400°F and air-cooled. Such a treatment does not appreciably reduce surface hardness. The hardened zone is generally much deeper than that obtained by carburizing, ranging from 1/8 to 1/4 in. in depth. Thinner cases of the order of 1/16 in. can be obtained by increasing the speed of heating and quenching.

Among the advantages of flame hardening are adaptability and portability. The equipment can be taken to the job and adjusted to treat only the area which requires hardening. Parts too large to be placed in a furnace can be handled easily and quickly with the torch. Another advantage is the ability to treat components after surface finishing, since there is little scaling, decarburization, or distortion. Disadvantages include (1) the possibility of overheating and thus damaging the part and (2) difficulty in producing hardened zones less than 1/16 in. in depth.

8·30 Induction Hardening Induction hardening depends for its operation on localized heating produced by currents induced in a metal placed in a rapidly changing magnetic field. The operation resembles a transformer in which the primary or work coil is composed of several turns of copper tubing that are water-cooled, and the part to be hardened is made the secondary of a high-frequency induction apparatus. Five basic designs of work coils for use with high-frequency units and the heat patterns developed by each are shown in Fig. 8·81*a* through *e*. These basic shapes are: (*a*) a simple solenoid for external heating, (*b*) a coil to be used internally for heating bores, (*c*) a "pie-plate" type of coil designed to provide high current densities in a narrow band for scanning applications, (*d*) a single-turn-coil for scanning a rotating surface, provided with a contoured half-turn that will aid in heating the fillet, and (*e*) a "pancake" coil for spot heating.

When high-frequency alternating current passes through the work coil,

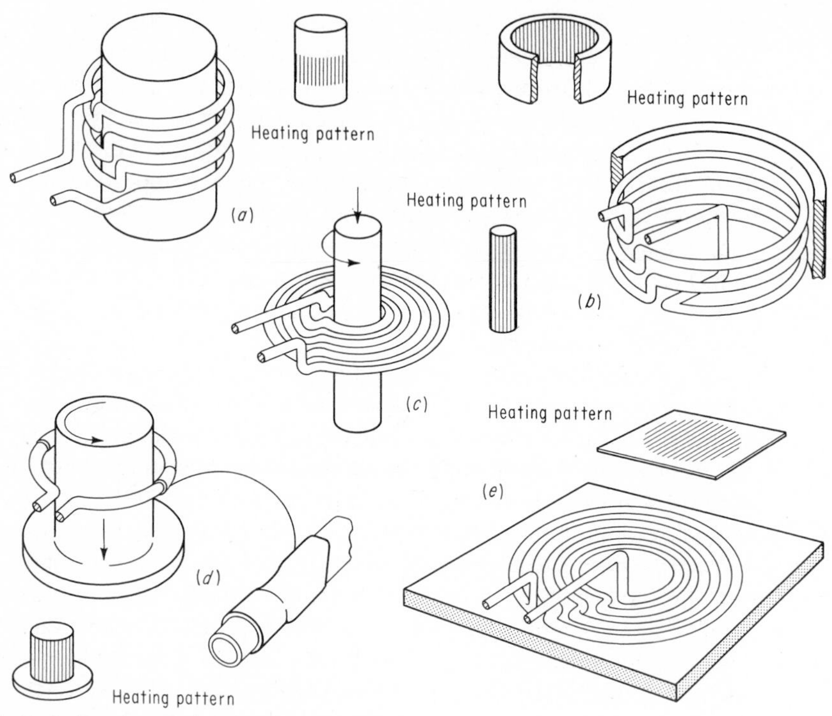

Fig. 8·81 Typical works coils for high-frequency units and the heat patterns developed by each. (From "Metals Handbook," vol. 2, American Society for Metals, Metals Park, Ohio, 1964.)

a high-frequency magnetic field is set up. This magnetic field induces high-frequency eddy currents and hysteresis currents in the metal. Heating results from the resistance of the metal to passage of these currents. The high-frequency induced currents tend to travel at the surface of the metal. This is known as *skin effect*. Therefore, it is possible to heat a shallow layer of the steel without heating the interior. However, heat applied to the surface tends to flow toward the center by conduction, and so time of heating is an important factor in controlling the depth of the hardened zone. The surface layer is heated practically instantaneously to a depth which is inversely proportional to the square root of the frequency. The range of frequencies commonly used is between 10,000 and 500,000 Hz. Table 8·8 shows the effect of frequency on depth of case hardness. Greater case depths may be obtained at each frequency by increasing the time of heating.

In batch processes, temperature is generally controlled by automatically timing the cycle. In continuous processes, the speed of passage of the

TABLE 8·8 Effect of Frequency on Depth of Case Hardness*

FREQUENCY, HZ	THEORETICAL DEPTH OF PENETRATION OF ELECTRICAL ENERGY, IN.	PRACTICAL DEPTH OF CASE HARDNESS, IN.
1,000	0.059	0.180 to 0.350
3,000	0.035	0.150 to 0.200
10,000	0.020	0.100 to 0.150
120,000	0.006	0.060 to 0.100
500,000	0.003	0.040 to 0.080
1,000,000	0.002	0.010 to 0.030

* From Metals Handbook, vol. 2, p. 180, American Society for Metals, Metals Park, Ohio, 1964.

work through the coils is adjusted to obtain the required temperature. Because these methods of temperature control are indirect, conditions producing the required case depth are generally determined by experiment. A radiation pyrometer may be used to measure and control the actual temperature of the work and improve uniformity of hardening. As in flame hardening, provision must be made for rapid quenching of the part after it has reached the desired temperature.

The case obtained by induction hardening is similar to that obtained by flame hardening, and thinner cases are possible. The steels used are similar to those used for flame hardening. Plain-carbon steels of medium carbon content are used for most applications, particularly for the production of thin cases. The carbon dissolves completely even in the short time required to heat the steel to the quenching temperature. Alloy steels can also be induction-hardened and are needed particularly for deep cases. Low-alloy steels are readily surface-hardened by this method, but highly alloyed steels are more sluggish and may require an increase in temperature to achieve the desired structure for satisfactory hardening. However, because of the rapid heating, the alloy steels can be heated to temperatures from 100 to 200°F higher by induction hardening than by conventional methods without danger of excessive grain growth. Steel parts that have been surface-hardened by induction generally exhibit less total distortion than the same parts quenched from a furnace. The microstructure of the steel before induction hardening is important in selecting the heating cycle to be used. Structures obtained after quenching and tempering so that the carbides are small and uniformly dispersed are most readily austenitized; accordingly, minimum case depths can be developed with maximum surface hardness while using very rapid rates of heating. Pearlite-ferrite structures typical of normalized, hot-rolled, and annealed steels containing 0.40 to 0.50 percent carbon also respond successfully to induction hardening. Another advantage of induction hardening is the ability to fit the equipment directly into the production line and use relatively unskilled labor since the operation is practically automatic. Among disadvantages are the cost of the equipment, the fact that small quantities or irregular-shaped parts cannot be handled economically, and high maintenance costs. Typical parts that have been induction-hardened are piston rods, pump shafts, spur gears, and cams.

8·31 Residual Stresses These are stresses that remain in the part after the force has disappeared. Residual stresses always arise from a nonuniform plastic deformation. In the case of heat treatment, this nonuniform plastic deformation may be caused by the temperature gradient or the phase change or usually a combination of both factors during cooling. Residual stresses are a very serious problem in heat treatment, since they often result in distortion or cracking and in some cases in premature failure of the part in service. Actually the problem of residual stresses is quite complex, but it is hoped that the following discussion, although simplified, will give the student some insight into and appreciation of the factors that give rise to these stresses.

Consider first the effect of temperature gradient alone. It was shown earlier, under the effect of size and mass, that during quenching the surface is cooled more rapidly than the inside. This results in a temperature gradient across the cross section of the piece or a temperature difference between the surface and the center. For example, let us examine the cooling curves of a 2-in. round water-quenched (Fig. 8·41). At the end of 10 s the surface has cooled to about 700°F, while the center is at about 1500°F. Almost all solids expand as they are heated and contract as they are cooled. This means that at the end of 10 s the surface, since it is at a much lower temperature, should have contracted much more than the inside. However, since the outside and inside are attached to each other, the inside, being longer, will prevent the outside from contracting as much as it should. It will therefore elongate the outside layers, putting them in tension while the inside in turn will be in compression. The approximate magnitude of this thermal stress may be calculated from the following formula:

$$s = \alpha E \, \Delta T$$

where s = thermal stress, psi

α = coefficient of linear expansion, in./(in.) (°F)

E = modulus of elasticity, psi

ΔT = difference in temperature, °F

Assuming an average value for the coefficient of expansion for steel as 6.5×10^{-6} in./(in.)(°F) and $E = 30 \times 10^{6}$ psi, insertion of these values in the above equation with $\Delta T = 800$ (1500 to 700°F) gives

$$s = 6.5 \times 10^{-6} \times 30 \times 10^{6} \times 800 = 156{,}000 \text{ psi}$$

This is the approximate value of the thermal stress existing between the outside and inside layers because of the temperature difference of 800°F.

This total stress of 156,000 psi must now be distributed between the inside and outside layers, and the average stress is inversely proportional to the area available to support this stress. Assuming that the outside layers constitute one-fourth of the cross-sectional area, the average tensile stress on the outside would be equal to $\frac{3}{4} \times 156{,}000$, or 117,000 psi, while the average compressive stress on the inside would be $\frac{1}{4} \times 156{,}000$, or 39,000 psi. This stress distribution is plotted schematically in Fig. 8·82. The area in tension must balance the area in compression in order for the stresses to be in equilibrium across the cross section. The plot in Fig. 8·82 shows a sharp drop in stress at the junction of the inside and outside layers due to a sharp drop in temperature from 1500 to 700 F. Actually the temperature does not drop sharply but changes gradually across the cross section, as shown by the curves of Fig. 8·43. A truer representation of the stress distribution is shown by the dotted curve in Fig. 8·82. The above discussion shows that the tensile stress on the surface may reach a very high value. If this stress exceeds the ultimate strength of the material, cracking will occur. This is what usually happens when glass is subjected to a large temperature difference. In the case of steel, however, thermal stresses alone very rarely lead to cracking. If the stress is below the yield strength of the steel, the stress will be borne elastically. When the entire piece has reached room temperature, $\Delta T = 0$, and therefore, since the thermal stress will be zero, there will be no distortion. If the stress exceeds the yield strength, the surface layer will be plastically deformed or permanently elongated. At room temperature the surface will have residual compressive

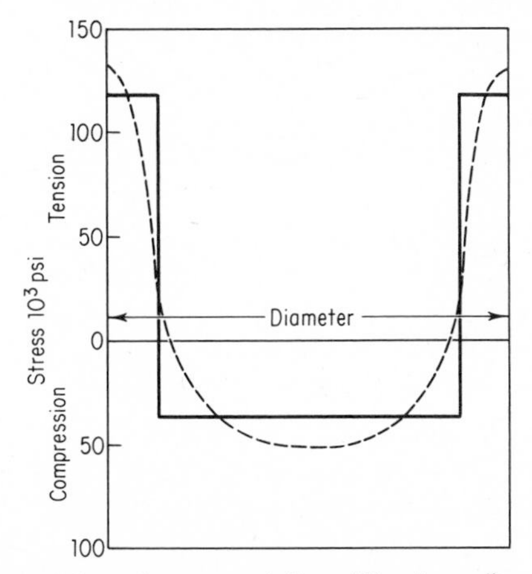

Fig. 8·82 Schematic representation of the stress distribution across the diameter due to a temperature gradient. Dotted curve indicates a truer representation of the stress distribution.

stress and the inside, residual tensile stress. If the piece was originally cylindrical, it will now be barrel-shaped.

Austenite, being f.c.c. (face-centered cubic), is a denser structure than any of its transformation products. Therefore, when austenite changes to ferrite, pearlite, bainite, or martensite, an expansion occurs. The austenite-to-martensite expansion is the largest and amounts to a volume increase of about 4.6 percent. The martensite expansion will be greater the lower the M_s temperature. Figure 8·83 shows the changes in length, during cooling, of a small-diameter cylinder as measured in a dilatometer. The piece is austenitic at the elevated temperature, and normal contraction of the austenite takes place until the M_s temperature is reached. Between the M_s and the M_f the transformation of austenite to martensite causes an expansion in length. After the M_f temperature, the martensite undergoes normal contraction.

Let us now consider the combined effect of temperature gradient and phase change for two possibilities: (1) through-hardened steel and (2) shallow-hardened steel.

Figure 8·84 shows the surface- and center-cooling curves superimposed on the I-T diagram for the through-hardened steel. Since the center-cooling rate exceeds the critical cooling rate, the part will be fully martensitic across its diameter. During the first stage, to time t_1, the stresses present are due to the temperature gradient. The surface, prevented from contracting as much as it should by the center, will be in tension while the center will be in compression. During the second stage, between times t_1 and t_2, the surface, having reached the M_s temperature, transforms to mar-

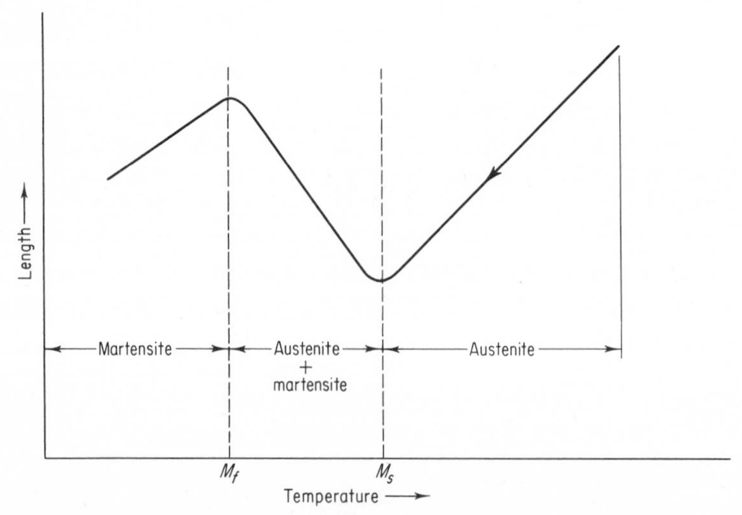

Fig. 8·83 Schematic dilation curve for martensite formation.

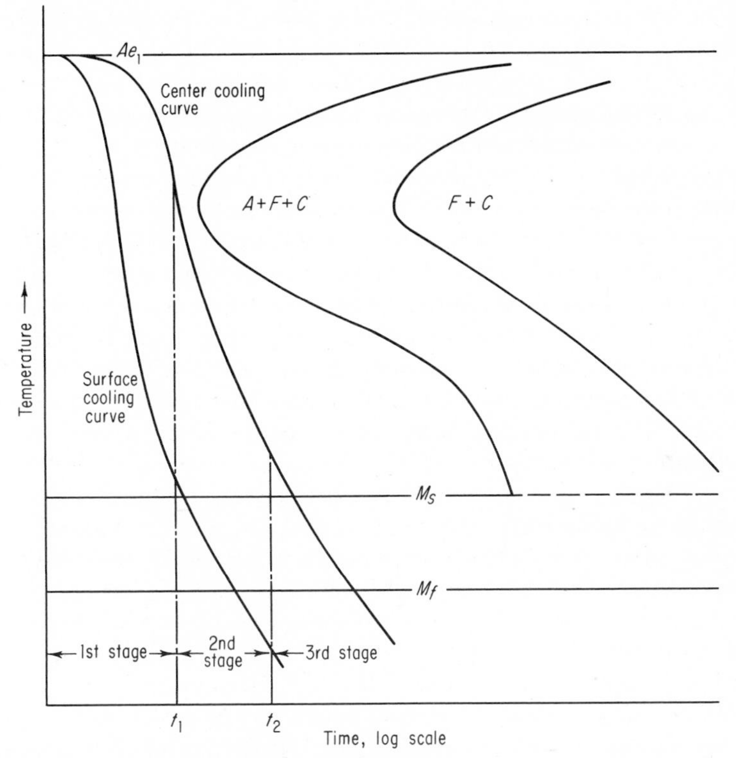

Fig. 8·84 Center- and surface-cooling curves superimposed on the I-T diagram to illustrate the through-hardened condition.

tensite and expands. The center, however, is undergoing normal contraction due to cooling. The center contracting will prevent the surface from expanding as much as it should, and the surface will tend to be in compression while the center will tend to be in tension. After t_2, the surface has reached room temperature and will be a hard, brittle, martensitic structure. During the third stage, the center finally reaches the M_s temperature and begins to expand, forming martensite. The center, as it expands, will try to pull the surface along with it, putting the surface in tension. The stress condition in the three stages is summarized below.

STAGE	STRESS CONDITION	
	SURFACE	CENTER
First (temperature gradient)	Tension	Compression
Second ($A \rightarrow M$ of surface)	Compression	Tension
Third ($A \rightarrow M$ of center)	Tension	Compression

To initiate and propagate a crack it is necessary for tensile stress to be present. Let us examine the three stages with regard to the danger of cracking. In the first stage, the surface is in tension; however, it is austenitic, and if the stress is high enough, rather than cracking, it will deform plastically, relieving the stress. In the second stage, the center is in tension and is austenitic, so that the tendency is to produce plastic deformation rather than cracking. In the last stage, the surface is again in tension. Now, however, the surface is hard, unyielding martensite. As the center expands, there is little likelihood of plastic deformation. It is during this stage that the greatest danger of cracking exists. Depending upon the difference in time between the transformation of the surface and center, the cracking may occur soon after the quench or sometimes many hours later. Figure 8·85 shows schematically the type of failure that may occur. The crack will take place in the tension layers and will be widest at the surface. It will gradually disappear when it gets to the compression layers on the inside. Very rarely does one end up with two pieces, because the crack cannot be propagated through the compression layers. By a microscopic study of the crack in the cross section, it is possible to determine how much of the cross section was in tension and how much was in compression. One heat-treating rule which minimizes the danger of cracking is that parts should be tempered immediately after hardening. Tempering will give the surface martensite some ductility before the center transforms.

Another very effective method of minimizing distortion and cracking is by *martempering* or *marquenching*, illustrated in Fig. 8·86. It is carried out by heating to the proper austenitizing temperature, quenching rapidly in a liquid-salt bath held just above the M_s temperature, and holding for a period of time. This allows the surface and the center to reach the same temperature; air cooling to room temperature then follows. Since air cooling from just above the martensite-formation range introduces very little temperature gradient, the martensite will be formed at nearly the same time throughout the piece. Thus martempering minimizes residual stresses and greatly reduces the danger of distortion and cracking. The heat treatment is completed by tempering the martensite to the desired hardness.

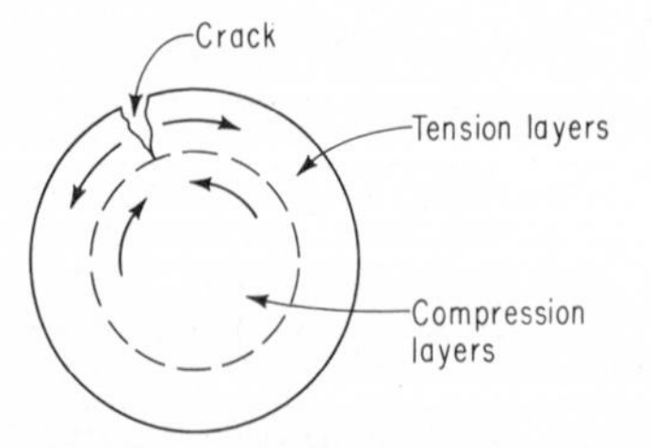

Fig. 8·85 Sketch of possible fracture in a through-hardened steel.

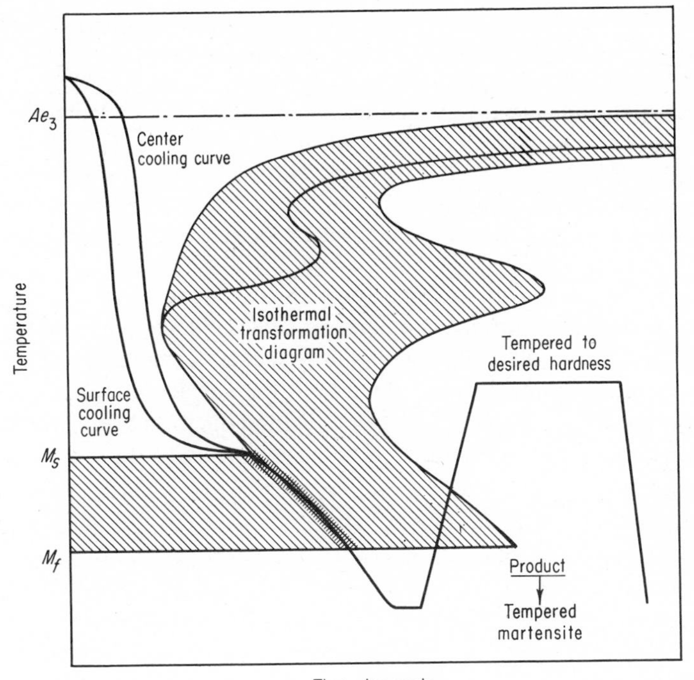

Fig. 8·86 Schematic transformation diagram illustrating martempering or marquenching. (From "U.S.S. Carilloy Steels," U.S. Steel Corporation.)

Figure 8·87 shows the surface- and center-cooling curves superimposed on the I-T diagram for the shallow-hardened steel. During the first stage, up to time t_1, the stresses present are due only to the temperature gradient, and as in the through-hardened condition, the surface will be in tension while the center will be in compression. During the second stage, between times t_1 and t_2, both the surface and center will transform. The surface will transform to martensite while the center will transform to a softer product, like pearlite. The entire piece is expanding, but since the expansion resulting from the formation of martensite is greater than that resulting from the formation of pearlite, the surface tends to expand more than the center. This tends to put the center in tension while the surface is in compression. After t_2, the center will contract on cooling from the transformation temperature to room temperature. The surface, being martensitic and having reached room temperature much earlier, will prevent the center from contracting as much as it should. This will result in higher tensile stresses in the center. The stress condition in the three stages is summarized on page 342.

STAGE	STRESS CONDITION	
	SURFACE	CENTER
First (temperature gradient)	Tension	Compression
Second ($A \rightarrow M$ of surface, $A \rightarrow P$ of center)	Compression	Tension
Third (cooling of center to room temperature)	Greater compression	Greater tension

Let us examine the three stages with regard to the danger of cracking. In the first stage, the surface is in tension, but being austenitic, if the stress is high enough, it will yield rather than crack. In the second stage, the center is in tension. However, since both the surface and center are expanding, the tensile stress will be small. During the third stage, the surface, as a hard, rigid shell of martensite, will prevent the center from contract-

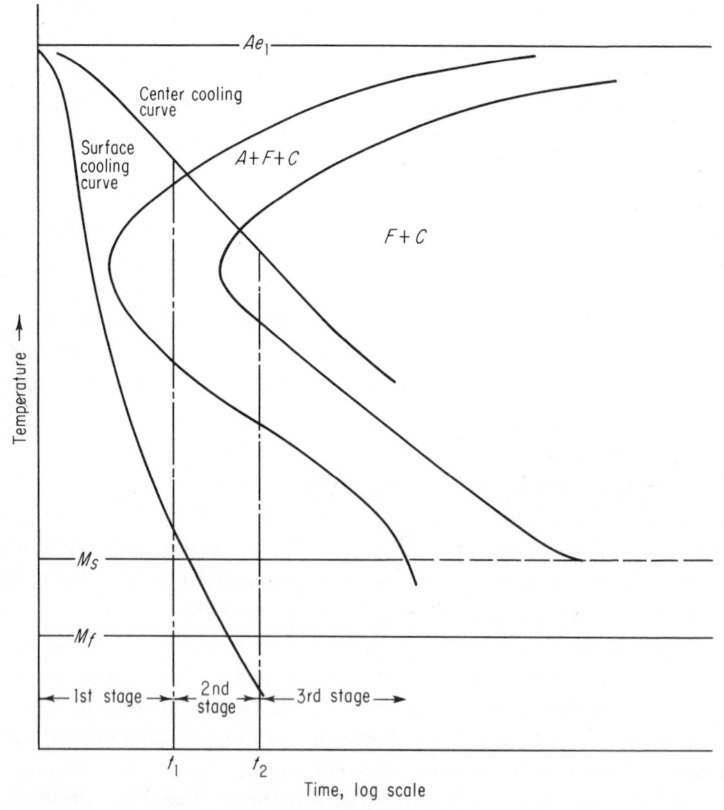

Fig. 8·87 Center- and surface-cooling curves superimposed on the I-T diagram to illustrate the shallow-hardened condition.

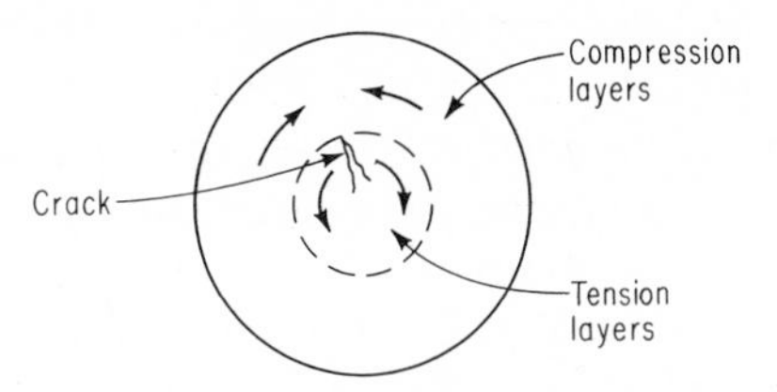

Fig. 8·88 Sketch of possible fracture in a shallow-hardened steel.

ing as it cools to room temperature. The tensile stresses in the center may reach a high value, and since the center is pearlite of relatively low tensile strength, it is during this stage that the greatest danger of cracking exists. Figure 8·88 shows schematically the type of failure that may occur in a shallow-hardened steel. The tensile crack is internal and cannot come to the surface because of the compressive stress in the surface layers. Since they are internal, these cracks are difficult to detect. X-ray testing or in some cases Magnaflux inspection may show the presence of internal fissures. Very often these parts are placed in service without knowledge of the internal quenching cracks. As soon as there is the slightest bit of tensile stress in the surface due to the external load, the crack will come through and the part will fail.

In many applications, the tensile stress developed by the external force is maximum at or near the surface. For these applications, shallow-hardened or case-hardened parts are preferred, since the surface residual stresses are usually compressive. In order for the surface to be in tension, the residual compressive stresses must first be brought to zero. This effectively increases the strength of the surface. The same beneficial effect and greatly increased life have been found for leaf springs where residual surface compressive stresses were induced by shot peening before the springs were placed in service.

8·32 Hardenable Carbon Steels Carbon steels are produced in greater tonnage and have wider use than any other metal because of their versatility and low cost. The low-carbon steels (0.10 to 0.25 percent carbon) are usually subjected to process annealing or case-hardening treatments. Since low-carbon steels have low hardenability and form little or no martensite on quenching, the improvement in mechanical properties is so small that it is hardly worth the cost, and this heat treatment is rarely applied. Process annealing of these steels is applied principally to sheet and strip at the mills to prepare it for forming or drawing, or between operations to relieve the work strains and permit further working. This operation is carried out at temperatures between the recrystallization temperature and the lower transformation temperature (A_1 line). The effect is to soften the steel by recrystallization and some grain growth of the ferrite. A stress-relieving treatment at 1000°F is applied to low-carbon cold-headed bolts. The lower temperature relieves much of the stress induced by cold working yet retains most of the strength and provides ample toughness. Case hardening of low-carbon steels was described in detail earlier in this chapter.

The medium-carbon steels (0.25 to 0.55 percent carbon), because of their higher carbon content, are generally used in the hardened and tempered condition. By varying the quenching medium and tempering temperature a wide range of mechanical properties can be produced. They are the most versatile of the three groups of carbon steels and are most commonly used for crankshafts, couplings, tie rods, and many other machinery parts where the required hardness values are in the range of R/C 20 to 48. The medium-carbon steels are usually either normalized or annealed before hardening in order to obtain the best mechanical properties after hardening and tempering. Cold-headed products are commonly made from these steels, especially the ones containing less than 0.40 percent carbon. Process treatment before cold working is usually necessary because the higher carbon decreases the workability. Frequently, a spheroidizing treatment is used depending upon the application. Water is the quenching medium most commonly used because it is the cheapest and easiest to install. In some cases, where a faster quench is desired, brine or caustic soda solution may be used. When the section is thin or the properties required after heat treatment are not high, oil quenching is used. This nearly always solves the breakage problem and is very effective in reducing distortion. Many of the common hand tools, such as pliers, open-end wrenches, and screwdrivers are made from medium-carbon steels. They are usually quenched in water, either locally or completely, and then suitably tempered.

High-carbon steels (above 0.55 percent carbon) are more restricted in their application since they are more costly to fabricate and have decreased machinability, formability, and weldability compared with medium-carbon steels. They are also more brittle in the heat-treated condition. Higher-carbon steels such as 1070 to 1095 are especially suitable for springs, where resistance to fatigue and permanent set are required. Most of the parts made from steels in this group are hardened by conventional quenching. Water quenching is used for heavy sections of the lower-carbon steels and for cutting edges, while oil quenching is for general use. Austempering and martempering are often successfully applied to take advantage of the considerably reduced distortion and in some cases the greater toughness at high hardness. It is important to remember that, even with the use of a drastic quench, these steels are essentially shallow-hardening compared with alloy steels and that there is therefore a definite limitation to the size of section that can be hardened. Screwdrivers, pliers, wrenches (except the Stillson type), and similar hand tools are usually hardened by oil quenching followed by tempering to the required hardness range. Even when no reduction in the as-quenched hardness is desired, stress relieving at 300 to 375°F is employed to help prevent sudden service failures. In the Stillson-type wrenches, the jaw teeth are really cutting edges and are usually quenched in water or brine to produce a hardness of Rockwell C 50 to 60. Either the jaws may be locally heated and quenched or the parts may be heated all over and the jaws locally time-quenched in water or brine. The entire part is then quenched in oil for partial hardening of the remainder. Hammers require high hardness on the striking face and somewhat lower hardness on the claws. They are usually locally hardened and tempered on each end. Final hardness on the striking face is usually Rockwell C 50 to 58; on the claws, 40 to 47. Cutting tools such as axes, hatchets, and mower blades must have high hardness and high relative toughness in their cutting edge, as well as the ability to hold a sharp edge. For hardening, the cutting edges of such tools are usually heated in liquid baths to the lowest temperature at which the piece can be hardened, and then quenched in brine. The final hardness at the cutting edge is Rockwell C 55 to 60.

QUESTIONS

8·1 Describe completely the changes that take place during the slow cooling of a 0.5 percent carbon steel from the austenite range.

8·2 Calculate the relative amounts of the structural constituents present in furnace-cooled steels containing (*a*) 0.30 percent carbon, (*b*) 0.60 percent carbon, (*c*) 0.80 percent carbon, (*d*) 1.2 percent carbon, (*e*) 1.7 percent carbon.

8·3 What are the limitations on the use of the iron–iron carbide diagram?

8·4 What is the effect of increasing cooling rate on (*a*) temperature of austenite transformation, (*b*) fineness of pearlite, (*c*) amount of proeutectoid constituent?

8·5 Is it possible to determine the approximate carbon content of a normalized steel from microscopic study? Explain.

8·6 In Table 8·1, why do annealed steels show a decrease in tensile strength above 0.80 percent carbon?

8·7 In Table 8·1, why do normalized steels show an increase in tensile strength up to 1.2 percent and then a decrease?

8·8 Define *critical cooling rate.*

8·9 What factors influence the critical cooling rate? Explain.

8·10 Define *actual cooling rate.*

8·11 What factors influence the actual cooling rate? Explain.

8·12 How is the actual cooling rate determined?

8·13 Calculate the surface-area-to-mass ratio of a 2-in.-diameter 10-ft-long cylinder, and compare it with a sphere of the same mass.

8·14 Explain why the cooling rate of oil may be increased by increasing the oil temperature.

8·15 Explain two ways in which the hardness-traverse curve of a given steel may show a straight horizontal line.

8·16 Explain why the surface hardness of quenched high-carbon steel may be less than the hardness under the surface.
8·17 What are the advantages of specifying steel on the basis of hardenability?
8·18 How will the microstructure differ for three samples of a 0.20 percent C steel after the following heat treatments? (*a*) Heated to 1700°F and furnace-cooled; (*b*) heated to 1800°F and furnace-cooled; (*c*) heated to 1700°F and air-cooled.
8·19 How will the microstructure differ in four samples of a 0.40 percent carbon steel after the following heat treatments? (*a*) Heated to 1500°F and air-cooled; (*b*) heated to 1500°F and oil-quenched; (*c*) heated to 1500°F and water-quenched; (*d*) heated to 1350°F and water-quenched.
8·20 If the samples in Question 8·19 are 2 in. in diameter, sketch the approximate hardness-traverse curves after the given heat treatments.
8·21 Give two different methods of obtaining a spheroidized cementite structure.
8·22 Sketch the I-T diagram of a 1080 steel (Fig. 8·17) and (*a*) show a cooling curve that will result in a structure of 50 percent martensite and 50 percent pearlite; (*b*) show a cooling curve that will result in a uniform pearlitic structure of Rockwell C 40.
8·23 A steel showed a hardness of Rockwell C 40 at the quenched end of a hardenability test. What was the approximate carbon content?
8·24 What will be the approximate hardness at the quenched end of (*a*) a 1050 hardenability test specimen; (*b*) a 6150 hardenability test specimen; (*c*) a 4150 hardenability test specimen?
8·25 Describe how an I-T diagram is determined experimentally.
8·26 What are the limitations on the use of the I-T diagram?
8·27 What will be the hardness at the center and mid-radius position of (*a*) 2-in.-diameter 4140 steel with a poor oil quench, strong oil quench, brine quench—no agitation; (*b*) same as (*a*) for 5140 steel; (*c*) same as (*a*) and (*b*) for a $2\frac{1}{2}$-in.-diameter bar; (*d*) same as (*a*) and (*b*) for a 3-in.-diameter bar?
8·28 Plot the change in hardness and Izod impact strength as a function of tempering from the data in Fig. 8·55.
8·29 What are the principal advantages of austempering compared with the conventional quench and temper method?
8·30 What are the limitations of austempering?
8·31 From the data in Table 8·7, plot case depth vs. time at 1500, 1600, and 1700°F. What conclusions may be drawn from the shape of these curves?
8·32 From the data in Table 8·7, plot case depth vs. temperature at 4, 10, and 20 h. What conclusions may be drawn from these curves?
8·33 What are the limitations on the use of high carburizing temperatures such as 1900 and 2000°F?
8·34 What are the advantages of gas carburizing compared with pack carburizing?
8·35 Define *hard case* or *effective case.*
8·36 Describe an application and the heat treatment used so that advantage may be taken of the residual stresses resulting from heat treatment.
8·37 What will be the nature of the residual stresses after carburizing? Explain.
8·38 What are the limitations of martempering?
8·39 You have been given a gear with a broken tooth that has failed prematurely in service. The normal heat treatment is carburize, harden, and temper. Describe completely how this gear would be studied to determine possible metallurgical cause for failure.
8·40 Assume that a cold chisel is to be made of plain-carbon steel. Analyze the application for properties required, select the hardness range desired, select the carbon content, and specify the heat treatment.

REFERENCES

American Iron and Steel Institute: "Steel Products Manual, Alloy Steel-Semifinished; Hot Rolled and Cold Finished Bars," New York, 1970.
American Society for Metals: "Metals Handbook," 1948 ed., Metals Park, Ohio.
———: "Metals Handbook," vol. 2, 1964, and vol. 7, 1972, Metals Park, Ohio.
Bethlehem Steel Corporation: "Modern Steels and Their Properties," Bethlehem, Pa., 1959.
Brick, R. M., R. B. Gordon, and A. Phillips: "Structure and Properties of Alloys," McGraw-Hill Book Company, New York, 1965.
Bullens, D. K.: "Steel and Its Heat Treatment," vols. 1 to 3, John Wiley & Sons, Inc., New York, 1948–1949.
Clark, D. S., and W. R. Varney: "Physical Metallurgy for Engineers," Van Nostrand Reinhold Company, New York, 1962.
Crafts, W., and J. L. Lamont: "Hardenability and Steel Selection," Pitman Publishing Corporation, New York, 1949.
DuMond, T. C.: "Quenching of Steels," American Society for Metals, Metals Park, Ohio, 1959.
Felbeck, D. K.: "Introduction to Strengthening Mechanisms," Prentice-Hall, Inc., Englewood Cliffs, N.J., 1968.
Grossmann, M. A.: "Elements of Hardenability," American Society for Metals, Metals Park, Ohio, 1952.
——— and E. C. Bain: "Principles of Heat Treatment," American Society for Metals, Metals Park, Ohio, 1964.
Guy, A. G.: "Elements of Physical Metallurgy," 2d ed., Addison-Wesley Publishing Company, Inc., Reading, Mass., 1959.
———: "Physical Metallurgy for Engineers," Addison-Wesley Publishing Company, Inc., Reading, Mass., 1962.
Hultgren, R.: "Fundamentals of Physical Metallurgy," Prentice-Hall, Inc., Englewood Cliffs, N.J., 1952.
Hume-Rothery, W.: "Structure of Alloys and Iron," Pergamon Press Inc., New York, 1966.
Peckner, D.: "The Strengthening of Metals," Van Nostrand Reinhold Company, New York, 1964.
Reed-Hill, R. E.: "Physical Metallurgy Principles," Van Nostrand Reinhold Company, New York, 1964.
Republic Steel Corporation: "Heat Treatment of Steels," Cleveland, Ohio.
Rogers, B. A.: "The Nature of Metals," American Society for Metals, Metals Park, Ohio, 1951.
Smith, C. O.: "The Science of Engineering Materials," Prentice-Hall, Inc., Englewood Cliffs, N.J., 1969.
Smith, M. C.: "Alloy Series in Physical Metallurgy," Harper & Row, Publishers, Incorporated, New York, 1956.
U.S. Steel Corporation: "Atlas of Isothermal Transformation Diagrams," Pittsburgh, 1963.
———: "Suiting the Heat Treatment to the Job," Pittsburgh, Pa., 1967.
Williams, R. S., and V. O. Homerberg: "Principles of Metallurgy," 5th ed., McGraw-Hill Book Company, New York, 1948.

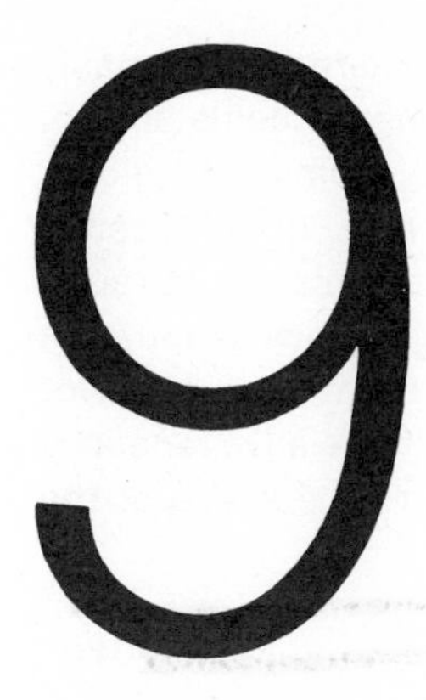

ALLOY STEELS

9·1 Introduction Plain-carbon steels are very satisfactory where strength and other requirements are not too severe. They are also used successfully at ordinary temperatures and in atmospheres that are not highly corrosive, but their relatively low hardenability limits the strength that can be obtained except in fairly thin sections. Almost all hardened steels are tempered to reduce internal stresses. It was pointed out in the previous chapter that plain-carbon steels show a marked softening with increasing tempering temperature. This behavior will lessen their applicability for parts that require hardness above room temperature. Most of the limitations of plain-carbon steels may be overcome by the use of alloying elements.

An *alloy steel* may be defined as one whose characteristic properties are due to some element other than carbon. Although all plain-carbon steels contain moderate amounts of manganese (up to about 0.90 percent) and silicon (up to about 0.30 percent), they are not considered alloy steels because the principal function of the manganese and silicon is to act as deoxidizers. They combine with oxygen and sulfur to reduce the harmful effect of those elements.

9·2 Purpose of Alloying Alloying elements are added to steels for many purposes. Some of the most important are:

1. Increase hardenability
2. Improve strength at ordinary temperatures
3. Improve mechanical properties at either high or low temperatures
4. Improve toughness at any minimum hardness or strength
5. Increase wear resistance
6. Increase corrosion resistance
7. Improve magnetic properties

Alloying elements may be classified according to the way they may be distributed in the two main constituents of annealed steel.

Group 1 Elements which dissolve in ferrite
Group 2 Elements which combine with carbon to form simple or complex carbides

9·3 Effect of Alloying Elements upon Ferrite Technically, there is probably some solubility of all the elements in ferrite, but some elements are not found extensively in the carbide phase. Thus nickel, aluminum, silicon, copper, and cobalt are all found largely dissolved in ferrite. In the absence of carbon, considerable proportions of the group 2 elements will be found dissolved in ferrite. Therefore, the carbide-forming tendency is apparent only when there is a significant amount of carbon present. The behavior of the individual elements is shown in Table 9·1, and the relative tendency of certain elements to exist in both groups is shown by the size of the arrowhead.

Any element dissolved in ferrite increases its hardness and strength in accordance with the general principles of solid solution hardening. The order of increasing effectiveness in strengthening iron, based upon equal additions by weight, appears to be about as follows: chromium, tungsten, vanadium, molybdenum, nickel, manganese, and silicon (Fig. 9·1). The hardening effect of the dissolved elements is actually small and illustrates how relatively little is the contribution of the strengthening of the ferrite to the overall strength of the steel. This is shown in Fig. 9·2 for low-carbon chromium alloys. The upper curve indicates the influence of chromium to change the tensile strength by changing the structure, while the lower curve indicates the minor influence of chromium in essentially constant structures.

9·4 Effects of Alloying Elements upon Carbide The influence of the amount of carbide and the form and dispersion of the carbide on the properties of steel has been discussed in Chap. 8. Since all carbides found in steel are hard and brittle, their effect on the room-temperature tensile properties is similar regardless of the specific composition.

TABLE 9·1 Behavior of the Individual Elements in Annealed Steel*

ALLOYING ELEMENT	GROUP 1 DISSOLVED IN FERRITE		GROUP 2 COMBINED IN CARBIDE
Nickel	Ni		
Silicon	Si		
Aluminum	Al		
Copper	Cu		
Manganese	Mn	⟵⟶	Mn
Chromium	Cr	⟵⟶	Cr
Tungsten	W	⟵⟶	W
Molybdenum	Mo	⟵⟶	Mo
Vanadium	V	⟵⟶	V
Titanium	Ti	⟵⟶	Ti

* Adapted from E. C. Bain and H. W. Paxton, "Alloying Elements in Steel," 2d ed., American Society for Metals, Metals Park, Ohio, 1961.

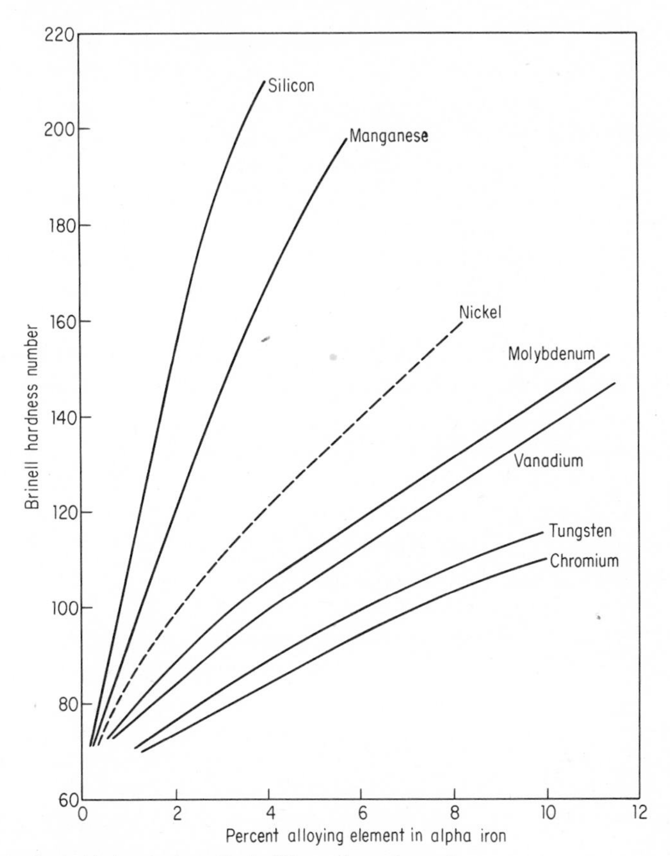

Fig. 9·1 Probable hardening effect of the various elements as dissolved in alpha iron. (From E. C. Bain and H. W. Paxton, "Alloying Elements in Steel," American Society for Metals, Metals Park, Ohio, 1961.)

The presence of elements that form carbides influences the hardening temperature and soaking time. Complex carbides are sluggish to dissolve and tend to remain out of solution in austenite. This serves to lower the carbon and alloy content of austenite below that of the steel as a whole. Undissolved carbides also act to reduce grain growth. Both effects tend to reduce hardenability. When dissolved in austenite, the carbide-forming elements are very powerful deep-hardening elements.

While all the carbides found in steel are hard, brittle compounds, chromium and vanadium carbides are outstanding in hardness and wear resistance. The hardness and wear resistance of alloy steels rich in carbides are in a large measure determined by the amount, size, and distribution of these hard particles. These factors, in turn, are controlled by chemical composition, method of manufacture, and heat treatment.

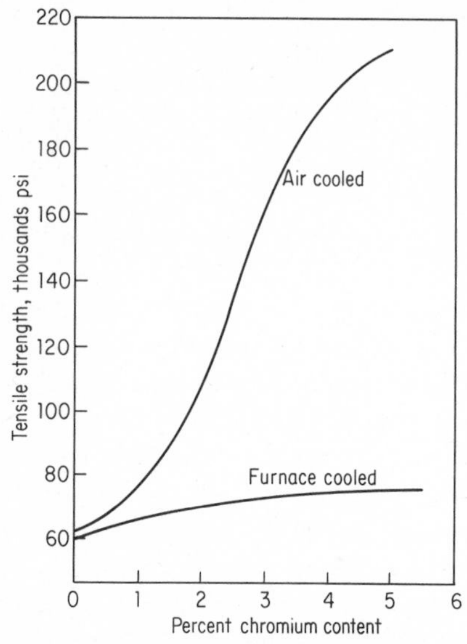

Fig. 9·2 The minor effect of chromium in annealed steels compared with its powerful effect as a strengthener through its influence on structure in air-cooled steels. (From E. C. Bain and H. W. Paxton, "Alloying Elements in Steel," American Society for Metals, Metals Park, Ohio, 1961.)

9·5 Influence of Alloying Elements on the Iron–Iron Carbide Diagram Although no mention was made in Chap. 7 of the possible modification of the iron–iron carbide diagram by the presence of elements other than carbon, in determining the effects of alloying elements this possibility must be given due consideration. When a third element is added to steel, the binary iron–iron carbide diagram no longer represents equilibrium conditions. Although the construction and interpretation of ternary equilibrium diagrams are outside the scope of this book, the presence of alloying elements will change the critical range, the position of the eutectoid point, and the location of the alpha and gamma fields indicated by the binary iron–iron carbide diagram.

Nickel and manganese tend to lower the critical temperature on heating, while molybdenum, aluminum, silicon, tungsten, and vanadium tend to raise it. The change in critical temperature produced by the presence of alloying elements is important in the heat treatment of alloy steels since it will either raise or lower the proper hardening temperature as compared with the corresponding plain-carbon steel.

The eutectoid point is shifted from the position it normally has in the iron–iron carbide diagram. All the alloying elements tend to reduce the carbon content of the eutectoid, but only nickel and manganese reduce the eutectoid temperature (Fig. 9·3). Increasing amounts of nickel and manganese may lower the critical temperature sufficiently to prevent the transformation of austenite on slow cooling; they are known as austenite-stabilizing elements. Therefore, austenite will be retained at room temperature. This situation occurs in the austenitic stainless steels.

Certain alloying elements, notably molybdenum, chromium, silicon, and titanium, in increasing amounts, tend to contract the pure austenitic region and enlarge the field in which alpha (α) or delta (δ) iron is found. This change is shown in Fig. 9·4, where the solid lines represent the contraction of the austenitic field with increasing amounts of the alloying element. Alloy compositions to the right of the "triangles" will be largely austenite with increasing amounts of carbide, while to the left of the austenite areas, austenite with more or less ferrite (solutions in α or δ iron) will be found.

9·6 Effect of Alloying Elements in Tempering In the discussion of tempering of plain-carbon steels, it was shown that hardened steels are softened by reheating. As the tempering temperature is increased, the hardness drops continuously. The general effect of alloying elements is to retard the softening rate, so that alloy steels will require a higher tempering temperature

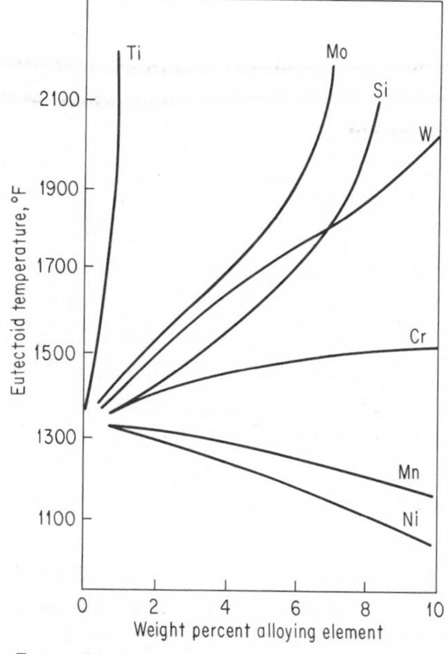

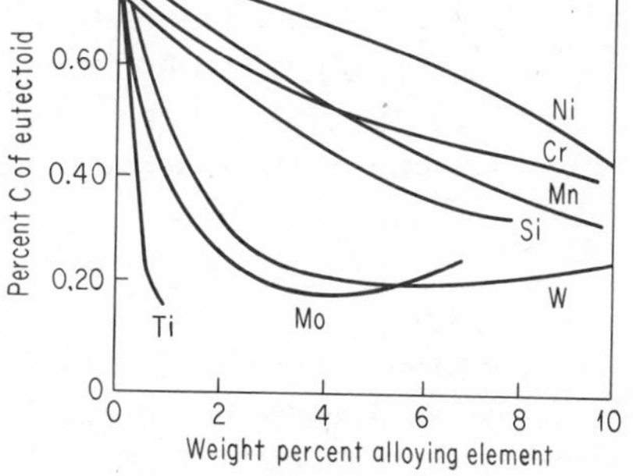

Fig. 9·3 Eutectoid composition and eutectoid temperature as influenced by several alloying elements. (From E. C. Bain and H. W. Paxton, "Alloying Elements in Steel," American Society for Metals, Metals Park, Ohio, 1961.)

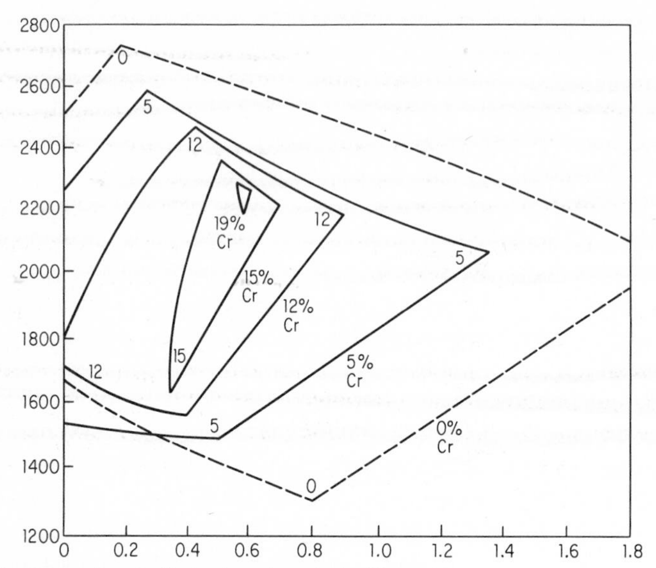

Fig. 9·4 Range of austenite in chromium steels. (From "Metals Handbook," 1948 ed., American Society for Metals, Metals Park, Ohio.)

to obtain a given hardness. The elements that remain dissolved in ferrite, such as nickel, silicon, and to some extent manganese, have very little effect on the hardness of tempered steel.

The complex carbide-forming elements such as chromium, tungsten, molybdenum, and vanadium, however, have a very noticeable effect on the retardation of softening. Not only do they raise the tempering temperature, but when they are present in higher percentages, the softening curves for these steels will show a range in which the hardness may actually increase with increase in tempering temperature. This characteristic behavior of alloy steels containing carbide-forming elements is known as *secondary hardness* and is believed to be due to delayed precipitation of fine alloy carbides. The effect of increasing chromium content is illustrated in Fig. 9·5.

The specific effects of the alloying elements in steel are summarized in Table 9·2.

Classification of alloy steels according to chemical composition as in the AISI series has been covered in Chap. 7, and compositions of some representative alloy steels are given in Table 8·2.

The preceding sections were devoted to a general discussion of the alloying elements in steel. Since a large number of alloy steels are manufactured, it is not feasible to discuss the individual alloy steels; however, a brief consideration of the specific effects of the common alloy elements and their application will be given.

9·7 Nickel Steels (2xxx Series) Nickel is one of the oldest, most fundamental steel-alloying elements. It has unlimited solubility in gamma (γ) iron and is highly soluble in ferrite, contributing to the strength and toughness of this phase. Nickel lowers the critical temperatures of steel, widens the temperature range for successful heat treatment, retards the decomposition of austenite, and does not form any carbides which might be difficult to dissolve during austenitizing. Nickel also reduces the carbon content of the eutectoid; therefore, the structure of unhardened nickel steels contains a higher percentage of pearlite than similarly treated plain-carbon steels. Since the pearlite forms at a lower temperature, it is finer and tougher than the pearlite in unalloyed steels. These factors permit the attainment of given strength levels at lower carbon contents, thus increasing toughness, plasticity, and fatigue resistance. Nickel steels are highly suited for high-strength structural steels which are used in the as-rolled

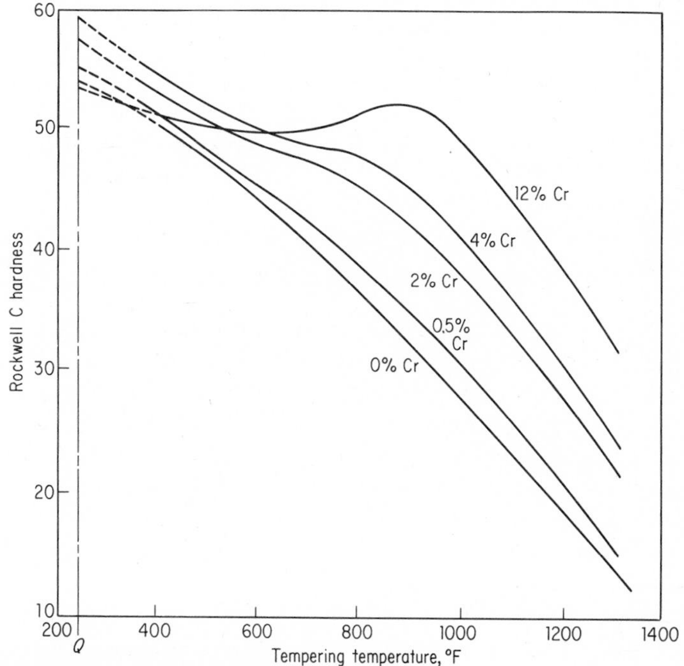

Fig. 9·5 The softening, with increasing tempering temperature, of quenched 0.35 percent carbon steels as influenced by chromium content. (From E. C. Bain and H. W. Paxton, "Alloying Elements in Steel," American Society for Metals, Metals Park, Ohio, 1961.)

TABLE 9·2 Specific Effects of Alloying Elements in Steel*

ELEMENT	SOLID SOLUBILITY		INFLUENCE ON FERRITE	INFLUENCE ON AUSTENITE (HARDENABILITY)	INFLUENCE EXERTED THROUGH CARBIDE		PRINCIPAL FUNCTIONS
	IN GAMMA IRON	IN ALPHA IRON			CARBIDE-FORMING TENDENCY	ACTION DURING TEMPERING	
Aluminum	1.1% (increased by C)	36%	Hardens considerably by solid solution	Increases hardenability mildly, if dissolved in austenite	Negative (graphitizes)	. . .	1. Deoxides efficiently 2. Restricts grain growth (by forming dispersed oxides or nitrides) 3. Alloying element in nitriding steel
Chromium	12.8% (20% with 0.5% C)	Unlimited	Hardens slightly; increases corrosion resistance	Increases hardenability moderately	Greater than Mn; less than W	Mildly resists softening	1. Increases resistance to corrosion and oxidation 2. Increases hardenability 3. Adds some strength at high temperatures 4. Resists abrasion and wear (with high carbon)
Cobalt	Unlimited	75%	Hardens considerably by solid solution	Decreases hardenability as dissolved	Similar to Fe	Sustains hardness by solid solution	1. Contributes to red-hardness by hardening ferrite
Manganese	Unlimited	3%	Hardens markedly; reduces plasticity somewhat	Increases hardenability moderately	Greater than Fe; less than Cr	Very little, in usual percentages	1. Counteracts brittleness from the sulfur 2. Increases hardenability inexpensively
Molybdenum	3%± (8% with 0.3% C)	37.5% (less with lowered temperature)	Provides age-hardening system in high Mo-Fe alloys	Increases hardenability strongly (Mo > Cr)	Strong; greater than Cr	Opposes softening, by secondary hardening	1. Raises grain-coarsening temperature of austenite 2. Deepens hardening 3. Counteracts tendency toward temper brittleness 4. Raises hot and creep strength, red-hardness 5. Enhances corrosion resistance in stainless steel 6. Forms abrasion-resisting particles
Nickel	Unlimited	10% (irrespective of carbon content)	Strengthens and toughens by solid solution	Increases hardenability mildly, but tends to retain austenite with higher carbon	Negative (graphitizes)	Very little in small percentages	1. Strengthens unquenched or annealed steels 2. Toughens pearlitic-ferritic steels (especially at low temperature) 3. Renders high-chromium iron alloys austenitic

TABLE 9·2 (Continued)

ELEMENT	SOLID SOLUBILITY		INFLUENCE ON FERRITE	INFLUENCE ON AUSTENITE (HARDENABILITY)	INFLUENCE EXERTED THROUGH CARBIDE		PRINCIPAL FUNCTIONS
	IN GAMMA IRON	IN ALPHA IRON			CARBIDE-FORMING TENDENCY	ACTION DURING TEMPERING	
Phosphorus	0.5%	2.8% (irrespective of carbon content)	Hardens strongly by solid solution	Increases hardenability	Nil	. . .	1. Strengthens low-carbon steel 2. Increases resistance to corrosion 3. Improves machinability in free-cutting steels
Silicon	2%± (9% with 0.35% C)	18.5% (not much changed by carbon)	Hardens with loss in plasticity ($Mn < Si < P$)	Increases hardenability moderately	Negative (graphitizes)	Sustains hardness by solid solution	1. Used as general-purpose deoxidizer 2. Alloying element for electrical and magnetic sheet 3. Improve oxidation resistance 4. Increases hardenability of steels carrying nongraphitizing elements 5. Strengthens low-alloy steels
Titanium	0.75% (1% ± with 0.20% C)	6% ± (less with lowered temperature)	Provides age-hardening system in high Ti-Fe alloys	Probably increases hardenability very strongly as dissolved. The carbide effects reduce hardenability	Greatest known (2% Ti renders 0.50% carbon steel unhardenable)	Persistent carbides probably unaffected. Some secondary hardening	1. Fixes carbon in inert particles *a.* Reduces martensitic hardness and hardenability in medium-chromium steels *b.* Prevents formation of austenite in high-chromium steels *c.* Prevents localized depletion of chromium in stainless steel during long heating
Tungsten	6% (11% with 0.25% C)	33% (less with lowered temperature)	Provides age-hardening system in high W-Fe alloys	Increases hardenability strongly in small amounts	Strong	Opposes softening by secondary hardening	1. Forms hard, abrasion-resistant particles in tool steels 2. Promotes hardness and strength at elevated temperature
Vanadium	1% (4% with 0.20% C)	Unlimited	Hardens moderately by solid solution	Increases hardenability very strongly, as dissolved	Very strong ($V < Ti$ or Cb)	Maximum for secondary hardening	1. Elevates coarsening temperature of austenite (promotes fine grain) 2. Increases hardenability (when dissolved) 3. Resists tempering and causes marked secondary hardening

* "Metals Handbook," 1948 ed., American Society for Metals, Metals Park, Ohio.

condition or for large forgings which are not adapted to quenching. The 3.5 percent nickel steels (23xx series) with low carbon are used extensively for carburizing of drive gears, connecting-rod bolts, studs, and kingpins. The 5 percent nickel steels (25xx series) provide increased toughness and are used for heavy-duty applications such as bus and truck gears, cams, and crankshafts. Nickel has only a mild effect on hardenability but is outstanding in its ability to improve toughness, particularly at low temperatures.

While nickel steels of the 2xxx series have been deleted from the AISI-SAE classification of standard alloy steels, it does not mean that they are not manufactured. Removal from the classification simply means that the tonnage produced is below a certain minimum. The steels in this series have been largely replaced in many applications by the lower-cost triple-alloy steels of the 86xx series.

9·8 Chromium Steels (5xxx Series) Chromium is a less expensive alloying element than nickel and forms simple carbides (Cr_7C_3, Cr_4C) or complex carbides [$(FeCr)_3C$]. These carbides have high hardness and good wear resistance. Chromium is soluble up to about 13 percent in γ iron and has unlimited solubility in α ferrite. In low-carbon steels, chromium tends to go into solution, thus increasing the strength and toughness of the ferrite. When chromium is present in amounts in excess of 5 percent, the high-temperature properties and corrosion resistance of the steel are greatly improved.

The plain-chromium steels of the 51xx series contain between 0.15 and 0.64 percent carbon and between 0.70 and 1.15 percent chromium. The low-carbon alloy steels in this series are usually carburized. The presence of chromium increases the wear resistance of the case, but the toughness in the core is not so high as the nickel steels. With medium carbon, these steels are oil-hardening and are used for springs, engine bolts, studs, axles, etc. A high-carbon (1 percent) high-chromium (1.5 percent) alloy steel (52100) is characterized by high hardness and wear resistance. This steel is used extensively for ball and roller bearings and for crushing machinery. A special type of chromium steel containing 1 percent carbon and 2 to 4 percent chromium has excellent magnetic properties and is used for permanent magnets.

The high-chromium steels containing over 10 percent chromium are noted for their high resistance to corrosion and will be discussed in Sec. 9·15.

9·9 Nickel-chromium Steels (3xxx Series) In these steels the ratio of nickel to chromium is approximately 2½ parts nickel to 1 part chromium. A combination of alloying elements usually imparts some of the characteristic properties of each one. The effect of nickel in increasing toughness and ductility is combined with the effect of chromium in improving hardenability and wear resistance. It is important to remember that the combined effect of two or more alloying elements on hardenability is usually greater than the sum of the effects of the same alloying elements used separately.

The low-carbon nickel-chromium alloy steels are carburized. The chromium supplies the wear resistance to the case, while both alloying elements improve the toughness of the core. With 1.5 percent nickel and 0.60 percent chromium (31xx series) they are used for worm gears, piston pins, etc. For heavy-duty applications, such as aircraft gears, shafts, and cams, the nickel content is increased to 3.5 percent and the chromium content to 1.5 percent (33xx series). The medium-carbon nickel-chromium steels are used in the manufacture of automotive connecting rods and drive shafts.

As in the case of the nickel steels, the steels in this series have also been deleted from the classification. In many cases, these steels have been replaced by the triple-alloy steels of the 87xx and 88xx series because of lower cost.

The very high nickel-chromium alloy steels will be discussed in Sec. 9·15.

9·10 Manganese Steels (31xx Series) Manganese is one of the least expensive alloying elements and is present in all steels as a deoxidizer. Manganese also reduces the tendency toward *hot-shortness* (red-shortness) resulting from the presence of sulfur, thereby enabling the metal to be hot-worked. When manganese is absent or very low, the predominant sulfide is FeS, which forms a eutectic with iron and has a tendency to form thin continuous films around the primary crystals during solidification of the steel. These films are liquid at the rolling temperature of steel and produce a condition of hot-shortness which is a tendency to crack through the grain boundaries during working. Manganese is outstanding in its power to combine with sulfur, and manganese sulfide has a much higher melting point than the iron sulfide eutectic. The manganese sulfide remains solid at the rolling temperature and has a less adverse effect on the hot-working properties of steel.

It is only when the manganese content exceeds about 0.80 percent that the steel may be classed as an alloy steel. Manganese contributes markedly to strength and hardness, but to a lesser degree than carbon, and is most effective in the higher-carbon steels. This element is a weak carbide former and has a moderate effect on hardenability. Like nickel, manganese lowers the critical range and decreases the carbon content of the eutectoid.

Fine-grained manganese steels attain unusual toughness and strength. These steels are often used for gears, spline shafts, axles, and rifle barrels. With a moderate amount of vanadium added, manganese steels are used for large forgings that must be air-cooled. After normalizing, this steel will yield properties equivalent to those obtained in a plain-carbon steel after a full hardening and tempering operation.

When the manganese content exceeds about 10 percent, the steel will be austenitic after slow cooling. A special steel, known as *Hadfield manganese steel*, usually contains 12 percent manganese. After a properly controlled heat treatment, this steel is characterized by high strength, high ductility, and excellent resistance to wear. It is an outstanding material for resisting severe service that combines abrasion and wear as found in power-shovel buckets and teeth, grinding and crushing machinery, and railway-track work. If this alloy is slow-cooled from 1750°F, the structure will consist of large brittle carbides surrounding austenite grains. This structure has low strength and ductility. In this condition the tensile strength is about 70,000 psi, with elongation values down to 1 percent. If the same alloy, after allowing the carbides to dissolve, is quenched from 1850°F, the structure will be fully austenitic with a tensile strength of about 120,000 psi, elongation of 45 percent, and a BHN of 180. The alloy now has much greater strength and ductility as compared with the annealed condition. The steel is usually reheated below 500°F to reduce quenching stresses. In the austenitic condition following rapid cooling, the steel is not very hard; however, when it is placed in service and subjected to repeated impact, the hardness increases to about 500 BHN. This increase in hardness is due to the ability of manganese steels to work-harden rapidly and to the conversion of some austenite to martensite.

9·11 Molybdenum Steels (4xxx Series) Molybdenum is a relatively expensive alloying element, has a limited solubility in γ and α iron, and is a strong carbide former. Molybdenum has a strong effect on hardenability and, like chromium, increases the high-temperature hardness and strength of steels. Steels containing molybdenum are less susceptible to temper brittleness than other alloy steels. This element is most often used in combination with nickel or chromium or both nickel and chromium. For carburizing applications it improves the wear resistance of the case and the toughness of the core.

The plain-molybdenum steels (40xx and 44xx series) with low carbon content are generally carburized and are used for spline shafts, transmission gears, and similar applications where service conditions are not too severe. With higher carbon they have been used for automotive coil and leaf springs. The chromium-molybdenum steels (41xx series) are relatively cheap and possess good deep-hardening characteristics, ductility, and weldability. They have been used extensively for pressure vessels, aircraft structural parts, automobile axles, and similar applications. The nickel-molybdenum steels (46xx and 48xx series) have the advantage of the high strength and ductility from nickel, combined with deep-hardening and improved machinability imparted by molybdenum. They have good toughness combined with high fatigue strength and wear resistance. They are used for transmission gears, chain pins, shafts, and bearings. The triple-alloy nickel-chromium-molybdenum steels (43xx and 47xx series) have the advantages of the nickel-chromium steels along with the high hardenability imparted by molybdenum. They are used extensively in the aircraft industry for the structural parts of the wing assembly, fuselage, and landing gear.

9·12 Tungsten Steel Tungsten has a marked effect on hardenability, is a strong carbide former, and retards the softening of martensite on tempering. In general, the effect of tungsten in steel is similar to that of molybdenum, although larger quantities are required. Approximately 2 to 3 percent tungsten is equivalent to 1 percent molybdenum. Since tungsten is relatively expensive and large quantities are necessary to obtain an appreciable effect, it is not used in general engineering steels. Tungsten is used primarily in tool steels.

9·13 Vanadium Steels Vanadium is the most expensive of the common alloying elements. It is a powerful deoxidizer and a strong carbide former which inhibits grain growth. Vanadium additions of about 0.05 percent produce a sound, uniform, fine-grain casting. When dissolved, vanadium has a marked effect on hardenability, yielding high mechanical properties on air cooling. Therefore, carbon-vanadium steels are used for heavy locomotive and machinery forgings that are normalized.

The low-carbon chromium-vanadium steels (61xx series) are used in the case-hardened condition in the manufacture of pins and crankshafts. The medium-carbon chromium-vanadium steels have high toughness and strength and are used for axles and springs. The high-carbon grade with high hardness and wear resistance is used for bearings and tools.

9·14 Silicon Steels (92xx Series) Silicon, like manganese, is present in all steels as a cheap deoxidizer. When a steel contains more than 0.60 percent silicon, it is classed as a silicon steel. Like nickel, silicon is not a carbide former but rather dissolves in ferrite, increasing strength and toughness. A steel containing 1 to 2 percent silicon known as *navy steel* is used for structural applications requiring a high yield point. *Hadfield silicon steel* with less than 0.01 percent carbon and about 3 percent silicon has excellent magnetic properties for use in the cores and poles of electrical machinery.

A properly balanced combination of manganese and silicon produces a steel with unusually high strength and with good ductility and toughness. This silicon-manganese steel (9260) is widely used for coil and leaf springs and also for chisels and punches.

9·15 Stainless Steels Stainless steels are used for both corrosion- and heat-resisting applications. A three-numeral numbering system is used to identify stainless steels. The last two numerals have no particular significance, but the first numeral indicates the group as follows:

SERIES DESIGNATION	GROUPS
2xx	Chromium-nickel-manganese; nonhardenable, austenitic, nonmagnetic
3xx	Chromium-nickel; nonhardenable, austenitic, nonmagnetic
4xx	Chromium; hardenable, martensitic, magnetic
4xx	Chromium; nonhardenable, ferritic, magnetic
5xx	Chromium; low chromium, heat-resisting

The corrosion-resisting property is due to a thin, adherent, stable chromium oxide or nickel oxide film that effectively protects the steel against many corroding media. This property is not evident in the low-chromium structural steels previously discussed and is apparent only when the chromium content exceeds about 10 percent.

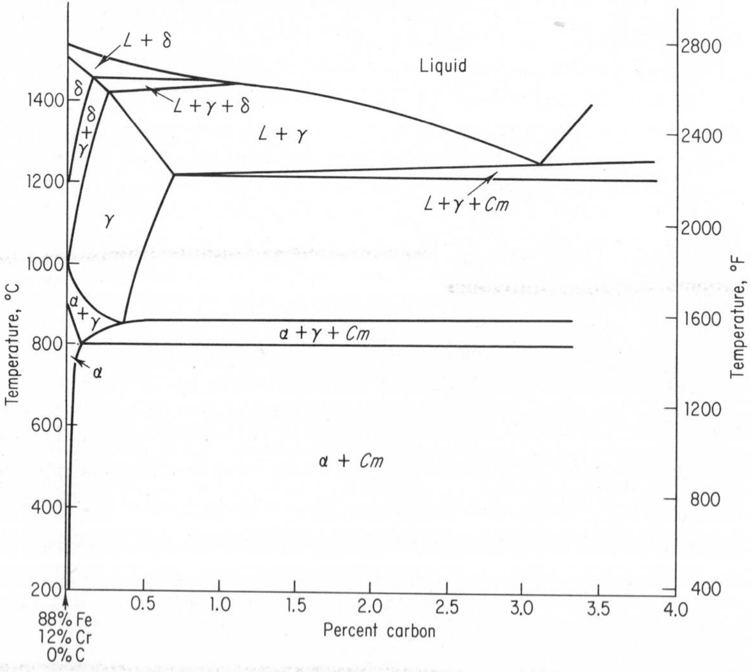

Fig. 9·6 Cross-section diagram for steels containing 12 percent chromium. (From E. E. Thum, "Book of Stainless Steels," 2d ed., American Society for Metals, Metals Park, Ohio, 1935.)

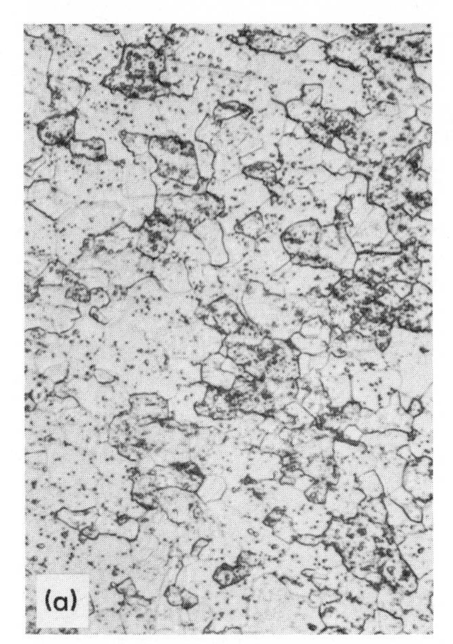

Fig. 9·7 Microstructure of a 12 percent chromium steel. (*a*) Annealed; small carbide particles in a ferrite matrix; (*b*) Quenched from 1850°F, tempered at 600°F; structure is tempered martensite. Etched in picric-hydrochloric acid, 500x. (Research Laboratory, Universal-Cyclops Steel Corporation.)

Since stainless steels contain relatively large amounts of chromium, the iron-chromium-carbon alloys belong to a ternary system. Figures 9·6 and 9·8 represent plane sections through such a ternary system. While these plane figures are not true equilibrium diagrams, they are useful in a study of phase changes and in interpreting structures.

Figure 9·6 shows a diagram for steels with 12 percent chromium and varying carbon. In comparison with the iron–iron carbide diagram, the presence of this amount of chromium has raised the critical temperatures and reduced the austenite area. However, with the proper amount of carbon, these steels may be heat-treated to a martensitic structure, as were the plain-carbon steels.

The microstructure of a 12 percent chromium steel in the annealed condition is shown in Fig. 9·7*a*. It consists of ferrite and small carbide particles. This same steel after quenching from 1850°F and tempering at 600°F consists of tempered martensite and bainite (Fig. 9·7*b*).

Figure 9·8 shows a diagram with 18 percent chromium and varying carbon. Consideration of this diagram indicates that, if the carbon content of the steel is low, austenite will not be formed on heating. These steels are

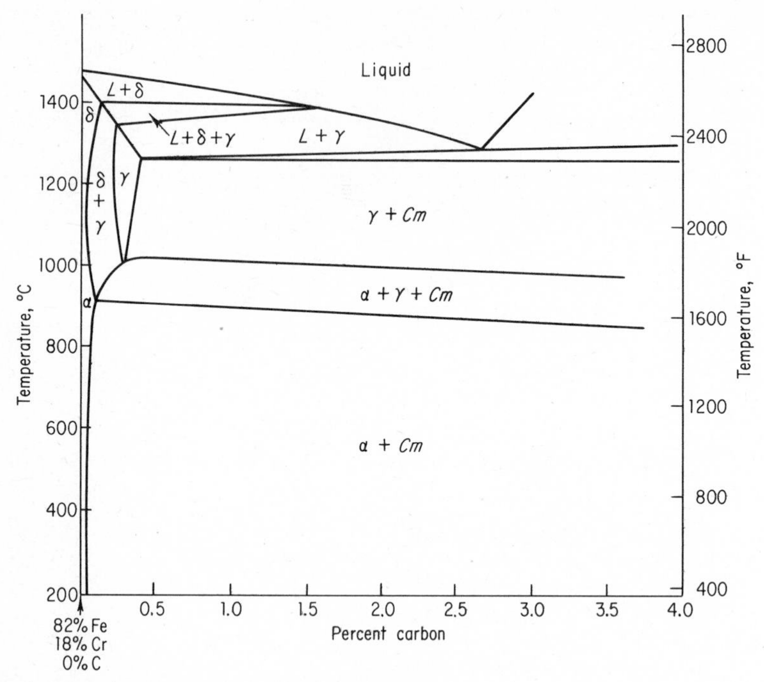

Fig. 9·8 Cross-section diagram for iron-carbon alloys containing 18 percent chromium. (From E. E. Thum, "Book of Stainless Steels," 2d ed., American Society for Metals, Metals Park, Ohio, 1935.)

nonhardenable, since subsequent quenching will only form ferrite of low hardness. Figure 9·9*a* illustrates the ferritic microstructure obtained by quenching a 0.03 percent carbon, 18 percent chromium steel from the delta region. If the carbon content is increased so that on heating the steel will be in the δ plus γ field, some hardness will result on quenching because of the transformation of γ iron. This is illustrated in Fig. 9·9*b* for a 0.075 percent carbon, 18 percent chromium steel water-quenched from 1850°F. The microstructure consists of ferrite (light area) and transformation product (dark area). If the carbon is still further increased so that the steel is in the austenite $\gamma + Cm$ field on heating, subsequent quenching will produce full hardness. This is shown in Fig. 9·9*c* for a high-carbon, 18 percent chromium steel water-quenched from 1850°F and tempered at 600°F. The microstructure consists of tempered martensite plus some undissolved carbides.

The addition of nickel to the chromium steel will produce further modifications in the diagram. Figure 9·10 indicates the trend in the changes of steel with 18 percent chromium, 8 percent nickel, and varying carbon content. The austenite formed at the elevated temperature is a particularly stable phase reluctant to transform and tends to be retained after annealing. Figure 9·11*a* shows the fully austenitic microstructure of an 18 percent chromium, 8 percent nickel steel after annealing. The microstructure of this steel after cold working is shown in Fig. 9·11*b*.

The response of stainless and heat-resisting steels to heat treatment depends upon their composition. They are divided into three general groups.

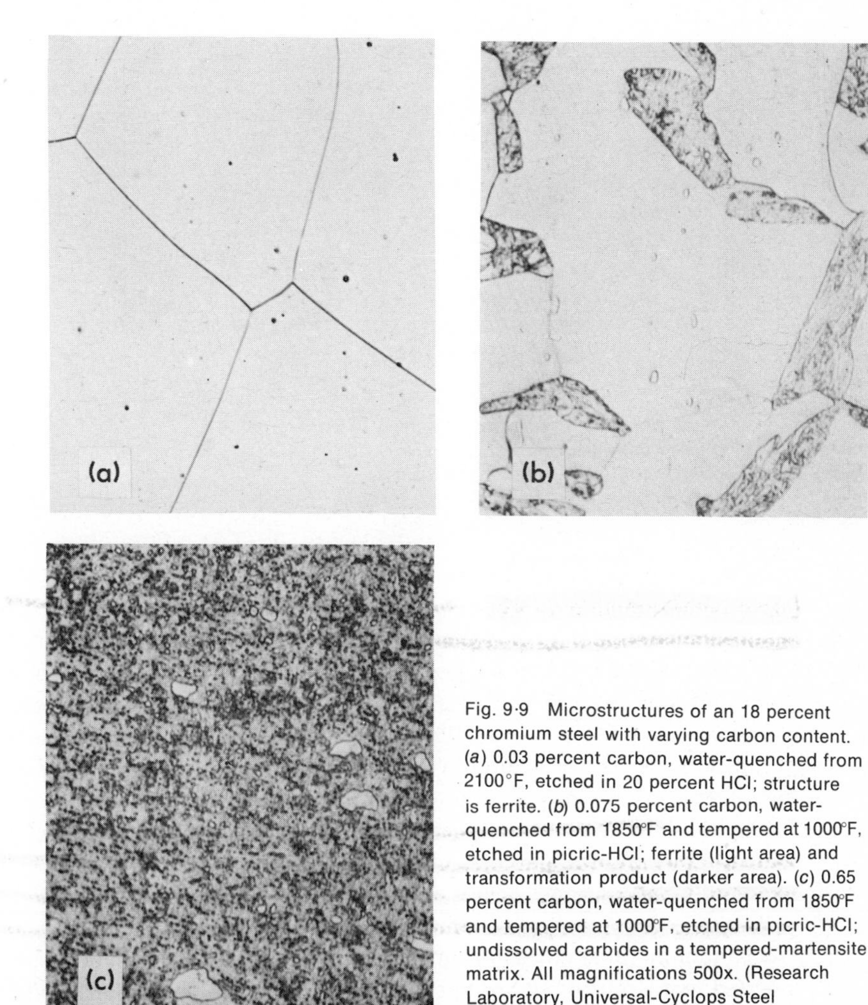

Fig. 9·9 Microstructures of an 18 percent chromium steel with varying carbon content. (*a*) 0.03 percent carbon, water-quenched from 2100°F, etched in 20 percent HCl; structure is ferrite. (*b*) 0.075 percent carbon, water-quenched from 1850°F and tempered at 1000°F, etched in picric-HCl; ferrite (light area) and transformation product (darker area). (*c*) 0.65 percent carbon, water-quenched from 1850°F and tempered at 1000°F, etched in picric-HCl; undissolved carbides in a tempered-martensite matrix. All magnifications 500x. (Research Laboratory, Universal-Cyclops Steel Corporation.)

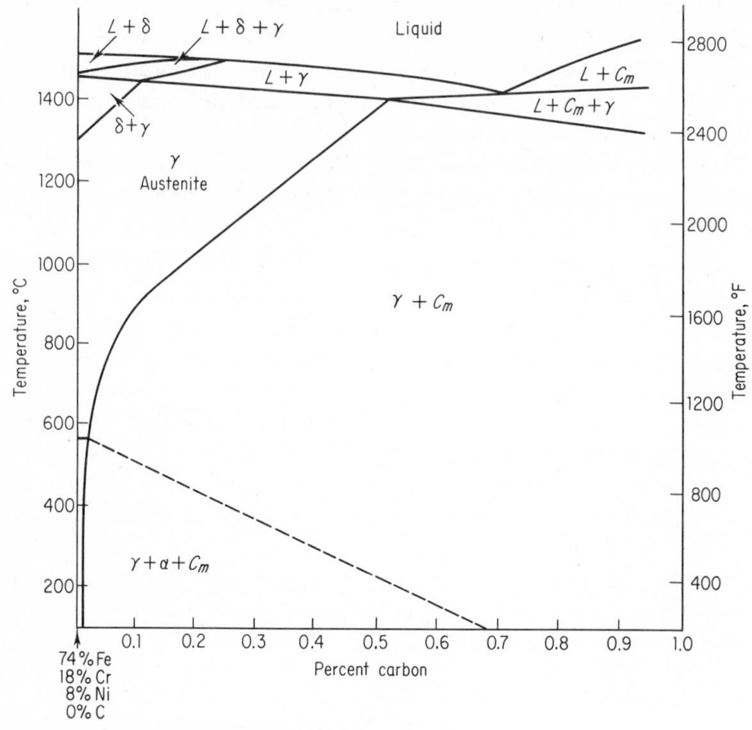

Fig. 9·10 Tentative cross-section diagram showing trend of reactions in steels alloyed with 18 percent chromium and 8 percent nickel. (From E. E. Thum, "Book of Stainless Steels," 2d ed., American Society for Metals, Metals Park, Ohio, 1935.)

Martensitic Stainless Steels These steels are primarily straight chromium steels containing between 11.5 and 18 percent chromium. Some examples of this group are types 403, 410, 416, 420, 440A, 501, and 502. Some of the properties and applications of the martensitic stainless steels are given in Fig. 9·12. Types 410 and 416 are the most popular alloys in this group and are used for turbine blades and corrosion-resistant castings. The chemical composition and typical mechanical properties are given in Table 9·3. The martensitic types of stainless steels are magnetic, can be cold-worked without difficulty, especially with low carbon content, can be machined satisfactorily, have good toughness, show good corrosion resistance to weather and to some chemicals, and are easily hot-worked. They attain the best corrosion resistance when hardened from the recommended temperature but are not as good as the austenitic or ferritic stainless steels.

The heat treatment process for martensitic steels is essentially the same as for plain-carbon or low-alloy steels, where maximum strength and hardness depend chiefly on carbon content. The principal difference is that the high alloy content of the stainless grades causes the transformation to be so sluggish, and the hardenability so high, that maximum hardness is produced by air cooling. These steels are normally hardened by heating them above the transformation range to temperatures near 1850°F, then cooling in air or oil. Time at temperature is held to a minimum to prevent decarburization or excessive grain growth. Tempering of steels in this group should not be done in the range of 750 to 950°F because of a drop in impact properties. Tempering is usually done above 1100°F. The higher tempering temperatures will cause some precipitation of carbides with a subsequent reduction in corrosion resistance. However, with the carbon content on the low side of the range, the lowering of corrosion resistance is not too severe.

Stainless steels as a group are much more difficult to machine than plain-carbon steels. The use of a small amount of sulfur in types 416 and selenium in type 416Se improves the machinability considerably. The use of selenium has less effect in reducing the corrosion resistance than sulfur. Stainless steels of type 440, with carbon content between 0.60 and 1.20 percent and 16 to 18 percent chromium, will have high corrosion resistance,

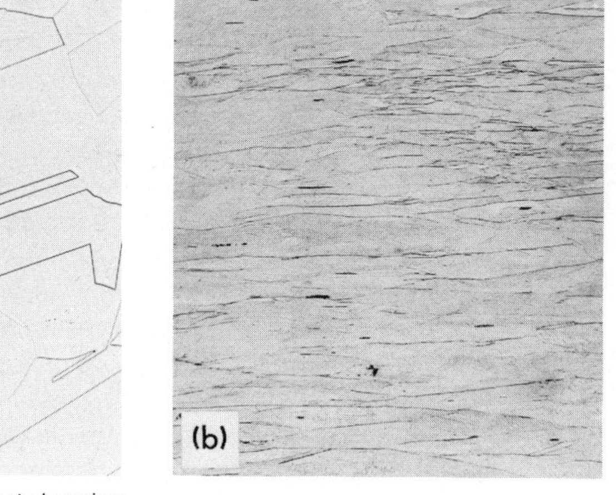

Fig. 9·11 Microstructures of an 18 percent chromium, 8 percent nickel steel. (*a*) After annealing, all austenite; (*b*) after cold working. Etched in glyceregia, 100x. (The International Nickel Company.)

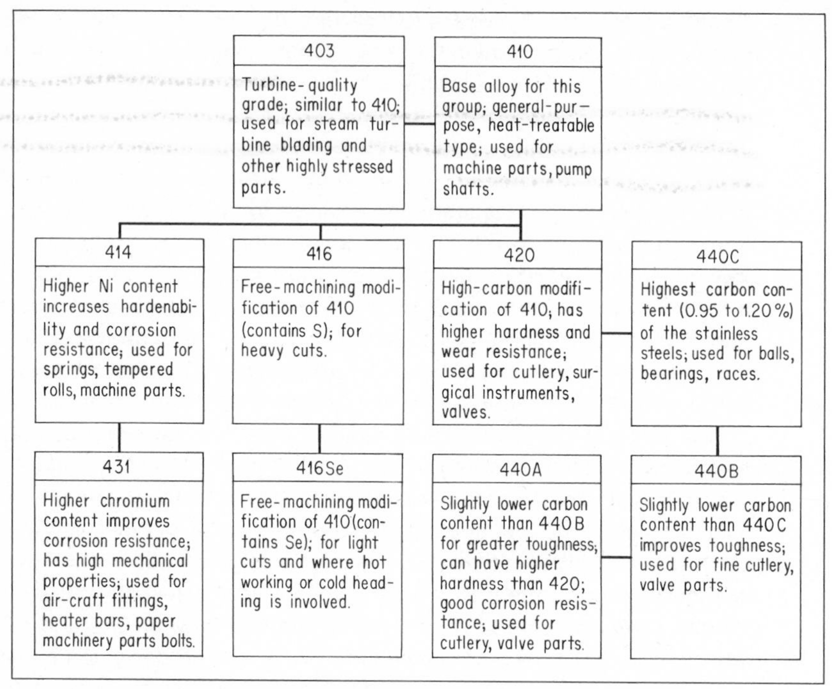

Fig. 9·12 The martensitic stainless steels. (From Machine Design, Metals Reference Issue, The Penton Publishing Co., Cleveland, Ohio, 1967.)

strength, and wear resistance. These alloys are used for cutlery, valve parts, and bearings.

The addition of about 2 percent nickel to the 16 to 18 percent chromium, low-carbon alloys (type 431) extends the austenite region and thus renders them heat-treatable. They are usually air-cooled, and the heat treatment requires careful control of composition and quenching temperature because of the possible presence of delta ferrite at the austenitizing temperature. Type 431 has been used for aircraft fittings, paper machinery parts, pumps, and bolts.

The relatively low-chromium alloy steels containing 4 to 6 percent chromium (types 501 and 502) have excellent resistance to oxidation and much better corrosion resistance than ordinary steel. These steels may be hardened by oil quenching or in some cases air cooling. The properties attained are really intermediate between the series 5xxx alloy steels and the type 400 martensitic stainless steels. Therefore, they are suitable for mild corrosion conditions or at temperatures below 1000°F. They have been used extensively for petroleum refining equipment such as heat exchangers, valve bodies, pump rings, and other fittings.

Ferritic Stainless Steels This group of straight-chromium stainless steels contain approximately 14 to 27 percent chromium and include types 405, 430, and 446 (see Fig. 9·13). Low in carbon content, but generally higher in chromium than the martensitic grades, these steels are not hardened by heat treatment and are only moderately hardened by cold working. They are magnetic and can be cold-worked or hot-worked, but they develop their maximum softness, ductility, and corrosion resistance in the annealed condition. In the annealed condition, the strength of these steels is approximately 50 percent higher than that of carbon steels, and they are superior to the martensitic stainless steels in corrosion resistance and machinability. Since the ferritic steels may be cold-formed easily, they are used extensively for deep-drawn parts such as vessels for chemical and food industries and for architectural and automotive trim.

The only heat treatment applied to truly ferritic steels is annealing. This treatment serves primarily to relieve welding or cold-working stresses. An important form of brittleness peculiar to the ferritic grades can develop from prolonged exposure to, or slow cooling within, the temperature range from about 750 to 950°F. Notch-impact strength is most adversely affected. Although the precise cause of the brittleness has not been determined, its effects increase rapidly with chromium content, reaching a maximum in type 446. Certain heat treatments, such as furnace cooling for maximum ductility, must be controlled to avoid embrittlement. Ferritic steels are usually annealed at temperatures above the range for 850°F embrittlement

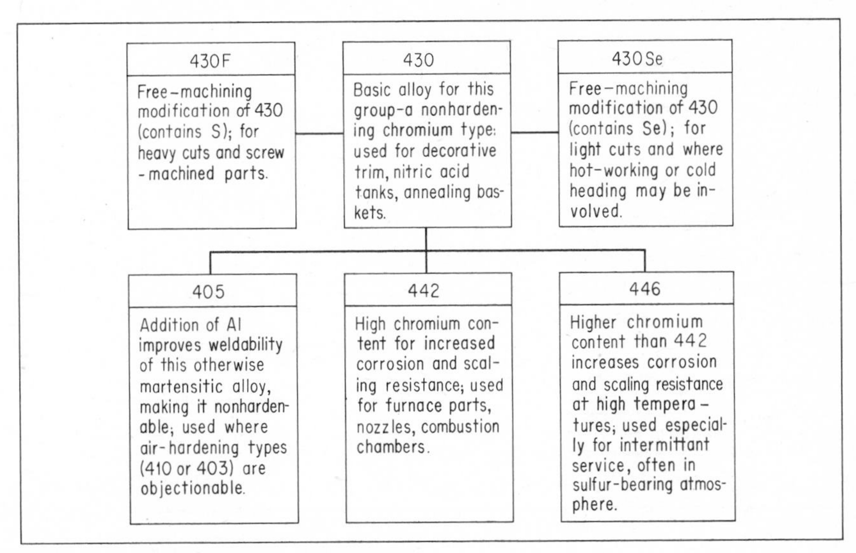

Fig. 9·13 The ferritic stainless steels. (From Machine Design, Metals Reference Issue, The Penton Publishing Co., Cleveland, Ohio, 1967.)

TABLE 9·3 Chemical Composition and Typical Mechanical Properties of Some Stainless Steels*

GROUP	AUSTENITIC GROUP					
TYPE NUMBER	201	202	301	302	309	316
Analysis, %:						
Chromium	16.0-18.0	17.0-19.0	16.0-18.0	17.0-19.0	22.0-24.0	16.0-18.0
Nickel	3.5-5.5	4.0-6.0	6.0-8.0	8.0-10.0	12.0-15.0	10.0-14.0
Other elements	N_2 0.25 max	N_2 0.25 max	...	...	...	Mo 2.0-3.0
Carbon	0.15 max	0.15 max	0.15 max	0.15 max	0.20 max	0.08 max
Manganese	5.5-7.5	7.5-10.0	2.0 max	2.0 max	2.0 max	2.0 max
Silicon	1.00 max	1.0 max	1.0 max	1.0 max	1.0 max	1.0 max
Temperature, °F:						
Forging—start	2300	2300	2200	2200	2150	2200
Annealing—ranges	1850—200	1850—2000	1950-2050	1850-2050	2050-2150	1975-2150
Annealing—cooling†	WQ (AC)	WQ (AC)	WQ (AC)	WQ (AC)	WQ (AC)	WQ (AC)
Hardening—ranges	‡	‡	‡	‡	‡	‡
Quenching	...	...	...	...	...	...
Tempering—for intermediate hardness	...	...	...	...	...	...
Drawing—for relieving stresses	...	...	...	...	...	...
Mechanical properties —annealed:						
Structure annealed	A	A	A	A	A	A
Yield strength, 1,000 psi min	40	40	35	30	30	30
Ultimate strength, 1,000 psi min	115	100	100	80	75	75
Elongation, % in 2 in. min	40.0	40.0	50.0	50.0	40.0	40.0
Reduction in area, % min	...	...	60.0	60.0	50.0	50.0
Modulus of elasticity in tension, 10^6 psi	29.0	29.0	29.0	29.0	29.0	29.0
Hardness, Brinell	210 max	210 max	180 max	180 max	200 max	200 max
Hardness, Rockwell	B 95 max	B 95 max	B 90 max	B 90 max	B 95 max	B 95 max
Impact values, Izod, ft-lb	85 min	85 min	85 min	85 min	80 min	70 min
Mechanical properties —heat-treated:						
Yield strength, 1,000 psi	...	...	...	...	...	...
Ultimate strength, 1,000 psi	§	§	§	...	...	...
Elongation, % in 2 in.	...	...	...	...	...	...
Hardness, Brinell	...	...	...	...	...	...
Hardness, Rockwell	...	...	...	...	...	...

Abbreviations: AC = air cool, FC = furnace cool, SFC = slow furnace cool, WQ = water quench, O = oil quench, AC = air cool, F = ferrite, C = carbide, A = austenite.
* From "Stainless Steel Handbook," Allegheny Ludlum Steel Corp.
† Thin sections of 300 series, marked WQ (AC) are usually air-cooled, heavy sections water-quenched.
‡ Hardenable only by cold working.
§ Ultimate strengths up to 350,000 psi for wire and 250,000 psi for strip can be obtained by cold working.
¶ Generally used in the annealed condition only.

and below temperatures at which austenite might form. When heat treated above the A_1 line to obtain maximum ductility, these steels are cooled slowly. They are not tempered, since the amount of martensite formed is negligible, and because of possible embrittlement in the 850°F range.

	MARTENSITIC GROUP					FERRITIC GROUP		
410	416	420	440A	501	502	405	430	446
11.5-13.5	12.0-14.0	12.0-14.0	16.0-18.0	4.0-6.0	4.0-6.0	11.5-14.5	14.0-18.0	23.0-27.0
0.50 max	...	...	0.50 max	...	...	0.50 max	0.50 max	0.50 max
...	...	...	Mo 0.75 max	Mo 0.4-0.65	Mo 0.4-0.65	Al 0.10-0.30	...	N_2 0.25 max
0.15 max	0.15 max	0.15 max	0.60-0.75	0.10 min	0.10 max	0.08 max	0.12 max	0.20 max
1.0 max	1.25 max	1.0 max	1.0 max	1.0 max	1.0 max	1.0 max	1.0 max	1.5 max
1.0 max	1.0 max	1.0 max	1.0 max	1.0 max	1.0 max	1.0 max	1.0 max	1.0 max
2100	2150	2000	2100	2150	2150	2100	2100	2150
1500-1650	1500-1650	1550-1650	1550-1650	1525-1600	1525-1600	1350-1500	1400-1500	1450-1600
SFC	FC	FC	FC	FC	FC	AC	FC	WQ
1700-1850	1700-1850	1800-1900	1850-1900	1600-1700	¶		Nonhardenable	
O or A	O or A	O or A	O or A	O				
Over 1100	Up to 1200	Below 700	Over 1100	Over 1000				
Under 700	Below 700	Below 700	Under 700	Under 700				
F-C	F-C	F-C	F-C	F-C	F-C	F-C	F-C	F-C
32	40-50	50-60	55	30	25	32	35	45
60	60-80	90-100	95	70	65	60	60	75
20.0	30-20	25-20	20.0	28.0	30.0	20.0	20.0	20.0
50.0	60-50	50-40	40.0	65.0	75.0	50.0	40.0	40.0
29.0	29.0	29.0	30.0	29.0	29.0	·29.0	29.0	29.0
200 max	145-185	200-230	240 max	160	150	180 max	200 max	200 max
B 95 max	B 79-90	B 93-98	B 100 max	...	B 75	B 90 max	B 95 max	B 95 max
85 min	50-30	...	Low	...	85 min	25 min	3-85	Low
35-180	60-130	120-220	55-240	90-135				
60-200	90-160	150-250	95-275	115-175				
25-2	20-10	12-2	20-2	20-15				
120-400	180-300	275-500	200-555	240-370				
B 70-C 45	B 88-107	C 30-52	B 95-C 55					

Austenitic Stainless Steels These are the chromium-nickel (type 3xx) and chromium-nickel-manganese stainless steels (type 2xx). These types are austenitic, are essentially nonmagnetic in the annealed condition, and do not harden by heat treatment. The total content of nickel and chromium is at least 23 percent. They can be hot-worked readily and can be cold-worked when proper allowance is made for their rapid work hardening. Cold working develops a wide range of mechanical properties, and the steel in this condition may become slightly magnetic. They are extremely shock-resistant and difficult to machine unless they contain sulfur and selenium (types 303 and 303Se). These steels have the best high-temperature strength and resistance to scaling of the stainless steels. The corrosion resistance of the austenitic stainless steels is usually better than that of the martensitic or ferritic steels.

Type 302, the basic alloy of the austenitic stainless steels, has been modified into a family of 22 related alloys (see Fig. 9·14). For example, lowering the carbon content to 0.08 percent maximum led to type 304 with improved weldability and decreased tendency toward carbide precipitation. To avoid carbide precipitation during welding, a lower-carbon version, type 304L, was developed which contains only 0.03 percent carbon maximum. Although type 304L suppresses carbide precipitation during cooling through the range of 1500 to 800°F after welding, potentially more serious precipitation problems are encountered in multiple-pass welding or service in the 800 to 1500°F range. To meet these requirements, the stabilized grades, type 321 with Ti added, and type 347 with Cb or Ta added, are recommended. In both alloys, a carbide other than chromium carbide precipitates, thus chromium is retained in solution and the alloy maintains its corrosion resistance. A stabilizing heat treatment consists of holding either annealed or welded types at 1600 to 1650°F for 2 to 4 h, followed by rapid cooling in air or water. The purpose is to precipitate all carbon as a carbide of titanium or columbium in order to prevent subsequent precipitation of chromium carbide.

Although all stainless steels can be hardened somewhat by cold work, the response becomes pronounced in the austenitic alloys, reaching a maximum in types 201 and 301. Table 9·4 compares the work-hardening behavior of the 17-7 Cr-Ni type 301 with the more stable 18-8 type 302.

Figure 9·14 shows the interrelation and wide variety of applications for which the austenitic stainless steels have been used, and Table 9·3 gives

TABLE 9·4 Effect of Cold Rolling in the Longitudinal Direction on Types 301 and 302 Austenitic Stainless Steels(*)

COLD REDUCTION, %	CONDITION OF METAL	TENSION				COMPRESSION	
		YIELD STRENGTH,† PSI	TENSILE STRENGTH, PSI	ELONGATION IN 2 IN., %	ROCKWELL HARDNESS NUMBER	YIELD STRENGTH,† PSI	BUCKLING STRENGTH PSI
		TYPE 301 STAINLESS STEEL					
0	Annealed	33,000	117,800	68	B 85	40,000	57,800
10	Cold-rolled	67,000	147,600	47	C 32	54,000	89,400
25	Cold-rolled	127,000	165,200	24	C 38	96,000	151,400
35	Cold-rolled	164,000	196,000	15	C 43	139,000	184,500
45	Cold-rolled	200,000	225,000	7	C 46	163,000	218,000
		TYPE 302 STAINLESS STEEL					
0	Annealed	36,000	94,000	61	B 80	36,000	50,250
20	Cold-rolled	121,000	139,300	22	C 29	74,000	120,400
35	Cold-rolled	131,000	155,300	15	C 36	95,000	151,800
50	Cold-rolled	151,000	177,400	6	C 38	99,000	155,200

* Adapted from "Metals Handbook," vol. 1, American Society for Metals, Metals Park, Ohio, 1961.
† 0.2 percent offset.

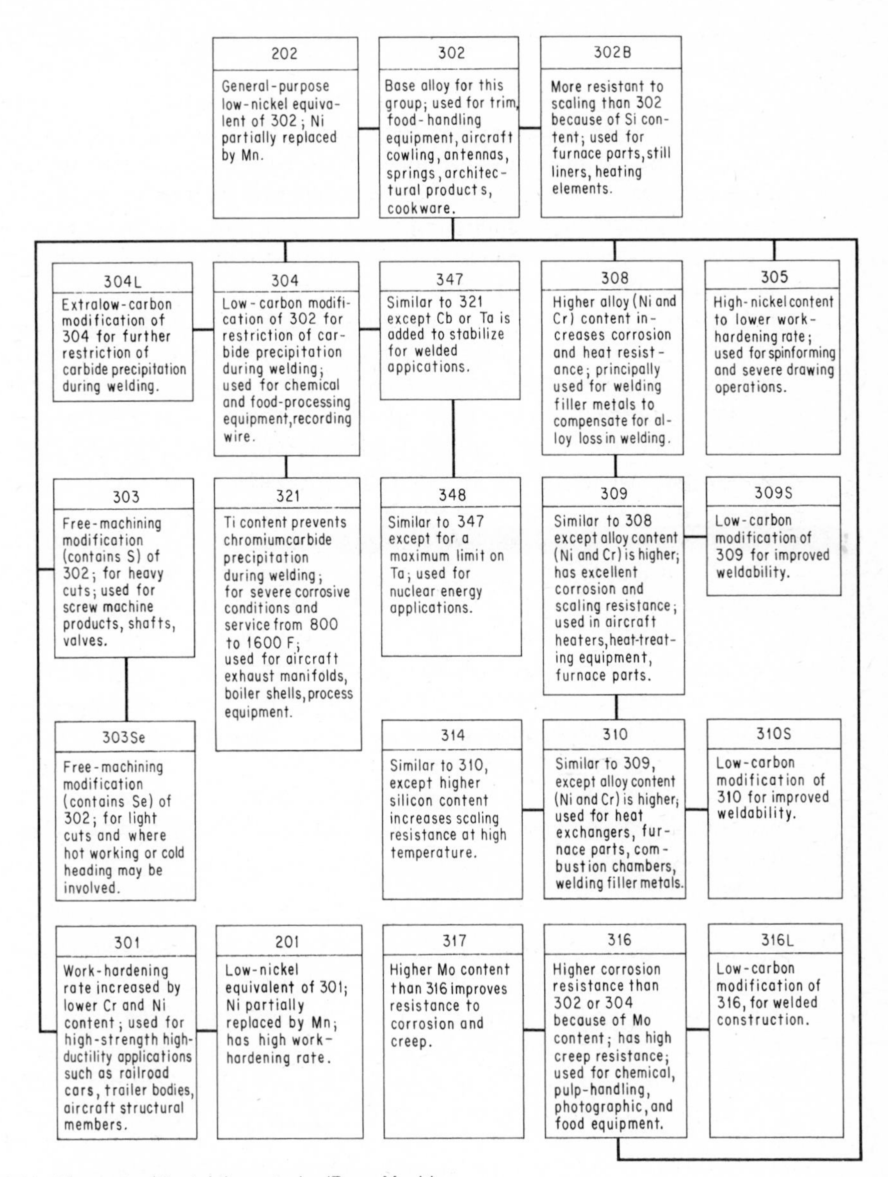

Fig. 9·14 The austenitic stainless steels. (From Machine Design Metals Reference Issue, The Penton Publishing Co., Cleveland, Ohio, 1967.)

the chemical composition and typical mechanical properties of some stainless steels.

The shortage of nickel in times of national emergency has presented a a serious problem to stainless-steel producers and consumers. Started during World War II and continued through the Korean emergency, development work covering the substitution of manganese for nickel in stainless steel led to the production of types 201 and 202, the chromium-nickel-manganese stainless steels. Type 201 with a nominal composition of 17 percent Cr, 4.5 percent Ni, and 6.5 percent Mn is a satisfactory substitute for type 301 (17 percent Cr, 7 percent Ni) where machinability and severe forming characteristics are not essential. Where those characteristics are essential, type 202 with a nominal composition of 18 percent Cr, 5 percent Ni, and 8 percent Mn is more desirable because the higher manganese reduces the rate of work hardening. Although types 201 and 202 have somewhat less resistance to chemical corrosion than 301 and 302, their resistance to atmospheric corrosion is entirely comparable.

9·16 Precipitation-hardening Stainless Steels As a result of research during World War II a new group of stainless steels with precipitation-hardening characteristics were developed. The first of these nonstandard grades of stainless steels, 17-7PH, was made available in 1948. The nominal chemical composition of some representative precipitation-hardening stainless steels is given in Table 9·5. These steels are usually solution-annealed at the mill and supplied in that condition. After forming they are aged to attain the increase in hardness and strength. In general they have lower nickel content, thus reducing the stability of the austenite. These steels may also have elements such as copper and aluminum that tend to form coherent alloy precipitates.

The 17-4PH alloy is solution-treated at 1900°F followed by air cooling with the resultant transformation of austenite to martensite. Aging is carried out by reheating in the range from 900 to 1150°F to cause a precipitation effect. The lower temperature results in the highest strength and hardness. Figure 9·15 shows the microstructure of a 17-4PH alloy, solution-treated at 1900°F and air-cooled, then aged for 4 h at 925°F and air-cooled. Transformation to martensite is essentially complete, and the martensite has been tempered by the aging treatment. Figure 9·16 shows the micro-

TABLE 9·5 Nominal Composition of Precipitation-hardening Wrought Stainless Steels

GRADE	C %	Mn %	Si %	Cr %	Ni %	Mo %	OTHERS %
17-4 PH	0.04	0.40	0.50	16.50	4.25	. . .	0.25 Cb, 3.60 Cu
17-7 PH	0.07	0.70	0.40	17.00	7.00	. . .	1.15 Al
PH 15-7 Mo	0.07	0.70	0.40	15.00	7.00	2.25	1.15 Al
17-10 P	0.12	0.75	0.50	17.00	10.50	. . .	0.28 P

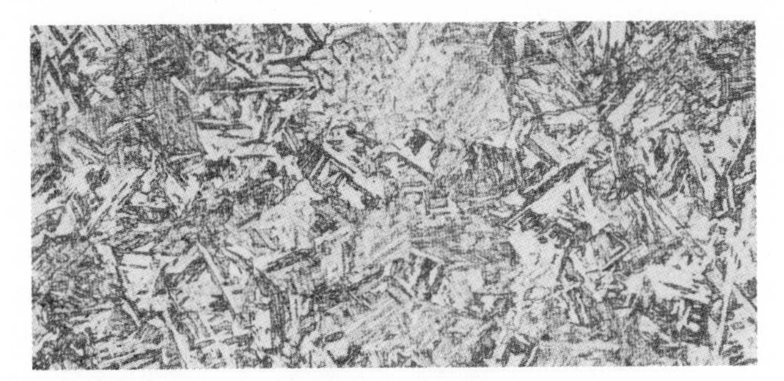

Fig. 9·15 17–4PH alloy solution treated at 1900°F and air-cooled, then aged 4 h at 925°F and air-cooled. Structure is tempered martensite. Etched in Fry's reagent, 100X. (From Metals Handbook, vol. 7, "Atlas of Microstructures," American Society for Metals, 1972.)

structure of the same alloy after the same treatment, except that it was aged at a higher temperature, 1100°F. The structure is tempered martensite that is more refined and has greater ductility but lower strength than the structure shown in Fig. 9·15. Table 9·6 gives the nominal mechanical properties of some precipitation-hardening stainless steels. The 17-4PH alloy should not be put into service in any application in the solution-treated condition because its ductility can be relatively low and its resistance to stress-corrosion cracking is poor. Aside from the increase in strength and ductility, aging also improves both toughness and resistance to stress-corrosion.

The 17-7PH and PH15-7Mo alloys are solution-annealed at 1950°F followed by air cooling. This treatment produces a structure of austenite with about 5 to 20 percent delta ferrite (see Fig. 9·17*a*). In this condition the alloy

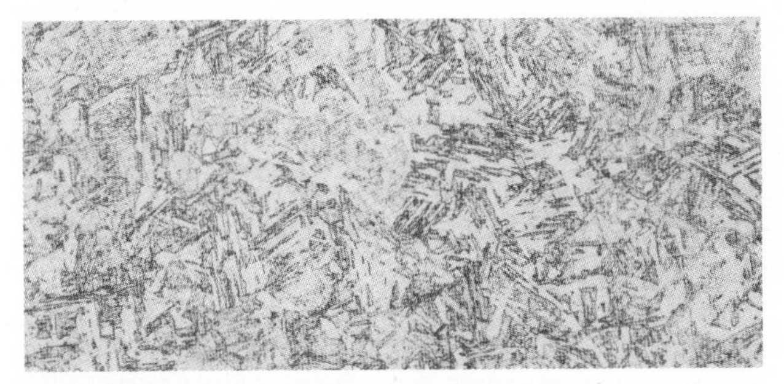

Fig. 9·16 Same alloy and treatment as Fig. 9·15, except aged at 1100°F. The tempered martensite is more refined and has greater ductility than the structure shown in Fig. 9·15. Etched in Fry's reagent, 100X. (From Metals Handbook, vol. 7, "Atlas of Microstructures," American Society for Metals, 1972.)

TABLE 9·6 Nominal Mechanical Properties of Precipitation-hardening Wrought Stainless Steels*

GRADE	CONDITION	TENSILE STRENGTH, PSI	0.2% YIELD STRENGTH, PSI	ELONGATION IN 2 IN., %	REDUCTION OF AREA, %	ROCKWELL HARDNESS
17-4 PH	Solution-annealed	150,000	110,000	10	45	C 33
	H 900°F†	200,000	178,000	12	48	C 44
	H 1025°F†	170,000	165,000	15	56	C 38
	H 1075°F†	165,000	150,000	16	58	C 36
	H 1150°F†	145,000	125,000	19	60	C 33
17-7 PH	Solution-annealed	130,000	40,000	35	...	B 85
	TH 1050°F‡	200,000	185,000	9	...	C 43
	RH 950°F§	235,000	220,000	6	...	C 48
	CH 950°F¶	265,000	260,000	2	...	...
PH 15-7 Mo	Solution-annealed	130,000	55,000	35	...	B 88
	TH 1050°F‡	210,000	200,000	7	...	C 44
	RH 950°F§	240,000	225,000	6	...	C 48
17-10 P	Solution-annealed	89,000	37,000	70	76	B 82
	Hardened, 1300°F, 24 h	143,000	98,000	20	32	C 32

* Adapted from "Metals Handbook," vol. 1, 1961, American Society for Metals, Metals Park, Ohio.
† Solution-annealed, then reheated at indicated temperature.
‡ Solution-annealed, reheated at 1400°F for 1½ h, air-cooled to room temperature, then reheated at 1050°F for 1½ h.
§ Solution-annealed, reheated at 1750°F for 10 min, air-cooled, then held at −100°F for 8 h, followed by reheating at 950°F for 1 h.
¶ Cold-rolled 60%, then reheated at 950°F for 1 h.

is soft and may be easily formed. Several hardening sequences are possible. In the TH (temper, hard) sequence, the austenite is conditioned by reheating to 1400°F. This will precipitate chromium carbides, thus reducing the carbon and chromium content of the austenite and allowing transformation on cooling. Cooling is continued to below 60°F but above 32°F in order to obtain the amount of martensite necessary for the strength levels desired. Aging is usually carried out at 1050°F for the best combination of strength and ductility (see Fig. 9·17*b*). The TH sequence gives better ductility but lower strength than other sequences (see Table 9·6). In the RH sequence, the austenite is conditioned at 1750°F. This higher temperature results in more carbon in solution in the austenite and therefore a lower M_s temperature. The transformation to martensite is obtained by a subzero treatment at −100°F with subsequent aging at 950°F (Fig. 9·17*c*). The 17-7PH alloy may also be supplied in the cold-rolled condition. Here, transformation is achieved by cold rolling, and heat treatment is reduced to a single aging at 950°F. Although strength is greatly increased, ductility is reduced and formability is limited. PH15-7Mo is a high-strength modification of 17-7PH and requires identical heat-treating procedures.

The 17-10P alloy is a nickel stainless with 0.28 percent phosphorus. This alloy is solution-annealed at 2100°F followed by rapid cooling to produce a supersaturated austenitic matrix with excellent ductility and forming characteristics. Subsequent aging in the range of 1200 to 1400°F will produce lower tensile strength than other precipitation-hardening alloys but much higher ductility (Table 9·6).

9·17 Maraging Steels A series of iron-base alloys capable of attaining yield strengths up to 300,000 psi in combination with excellent fracture toughness was made available early in 1960. These steels are low-carbon, containing 18 to 25 percent nickel together with other hardening elements, and are called *maraging* (martensitic plus aging). They are considered to be martensitic as annealed and attain ultrahigh strength on being aged in

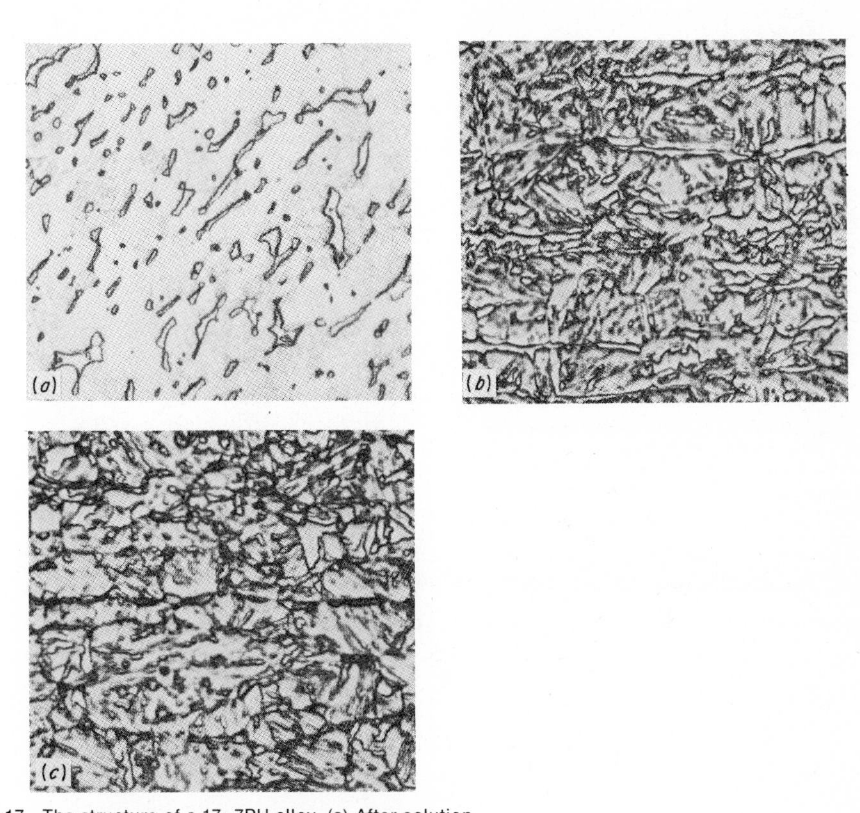

Fig. 9·17 The structure of a 17–7PH alloy. (*a*) After solution treatment 1 h at 1950°F, air-cooled. Ferrite islands in an austenite matrix, etched in Fry's reagent, 100X. (*b*) Solution-treated as in (*a*) and reheated to 1400°F for 1½ h, air-cooled to 60°F and held for ½ h, aged for 1½ h at 1050°F, air-cooled (condition TH1050). Ferrite islands and small chromium carbides in a martensite matrix; etched in Vilella's reagent, 1,000X. (*c*) Solution-treated as in (*a*) but reheated to 1750°F for 10 min and air-cooled, held 8 h at −100°F, aged for 1 h at 950°F, air-cooled (condition RH950). Ferrite stringers in a martensite matrix with less carbide out of solution compared with (*b*); etched in Vilella's reagent, 1,000X. (From Metals Hanbook, vol. 7, "Atlas of Microstructures," American Society for Metals, 1972.)

TABLE 9·7 Compositions of Nickel Maraging Steels*

DESIGNATION	C MAX	Mn MAX	Si MAX	S MAX	P MAX	Ni	Co	Mo	Ti	Al	Cb
25 Ni		0.10	0.10	0.01	0.01	25.0–26.0	. . .	. . .	1.3–1.6	0.15–0.30	0.30–0.50
20 Ni	0.03	0.10	0.10	0.01	0.01	19.0–20.0	. . .	. . .	1.3–1.6	0.15–0.30	0.30–0.50
18 Ni (300)	0.03	0.10	0.10	0.01	0.01	18.0–19.0	8.5–9.5	4.6–5.2	0.5–0.8	0.05–0.15	. . .
18 Ni (250)	0.03	0.10	0.10	0.01	0.01	17.0–19.0	7.0–8.5	4.6–5.2	0.3–0.5	0.05–0.15	. . .
18 Ni (200)	0.03	0.10	0.10	0.01	0.01	17.0–19.0	8.0–9.0	3.0–3.5	0.15–0.25	0.05–0.15	. . .

* From "Metals Handbook," vol. 2, American Society for Metals, Metals Park, Ohio, 1964.

the annealed, or martensitic, condition. The martensite formed is soft and tough rather than the hard, brittle martensite of conventional low-alloy steels. This ductile martensite has a low work-hardening rate and can be cold-worked to a high degree.

Thus far, the commercial steels developed fall into two distinct alloy groups which differ in the hardening elements used (see Table 9·7). The 18 percent nickel grades use mainly cobalt-molybdenum additions with small amounts of titanium and aluminum. The 20 percent nickel and 25 percent nickel grades use titanium-aluminum-columbium additions.

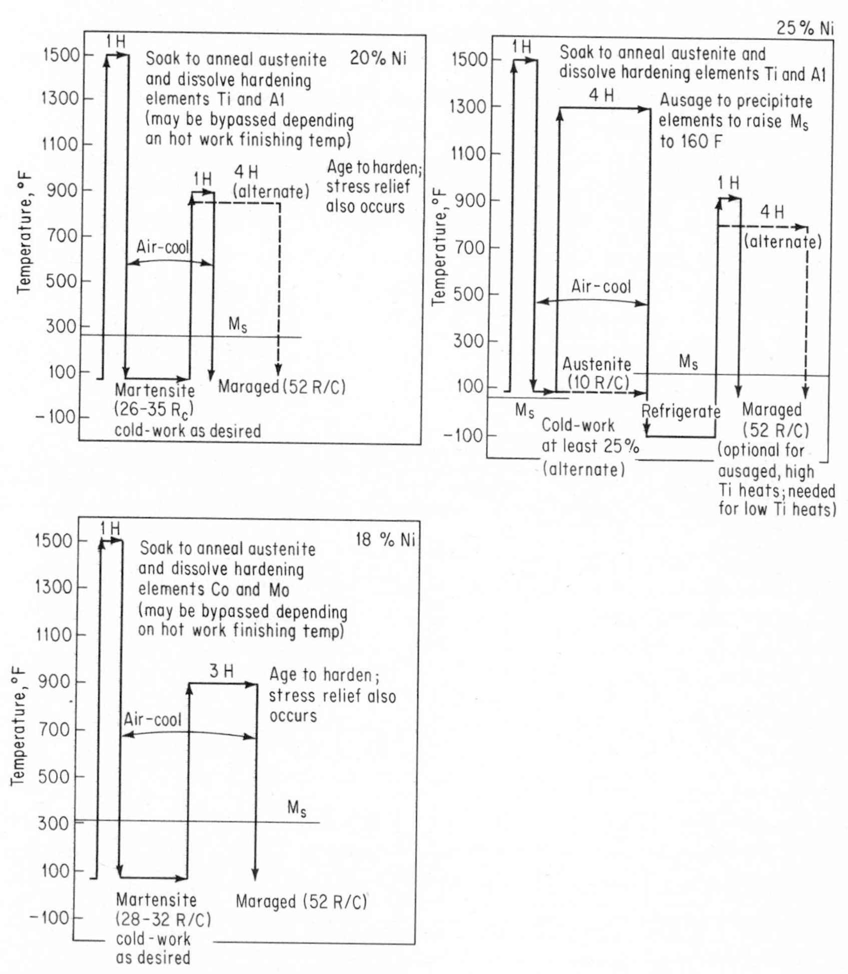

Fig. 9·18 Heat-treating cycles for the maraging steels. (The International Nickel Company.)

The greatest interest has been directed to the 18 percent nickel grades designed primarily for ultrahigh strength at room temperature. An important reason for the interest is their superior fracture toughness compared with quenched and tempered medium-carbon ultrahigh strength steels. All that is required is a simple heat treatment carried out at moderate temperature to develop full strength in these steels (see Fig. 9·18). Section size and heating and cooling rates are not important factors in the hardening process. Since these steels are extra low in carbon content, decarburization is not a problem, and a protective atmosphere is not required. The use of low aging temperatures reduces distortion to a minimum. This means that little machining or forming is required after hardening to produce parts with high dimensional accuracy. These steels are fully weldable and have good machinability in the annealed condition. The mechanical properties of five maraging steels are given in Table 9·8. Figure 9·19 shows the microstructure of an 18 percent nickel 250 grade maraging steel at 900°F for 3 h. The structure is made up of very fine precipitates in a martensitic matrix. The precise role of each of the elements in the hardening process is not completely understood, but some idea may be obtained by considering the effect of alloying elements on the hardness of the martensitic base. The hardness of the binary iron-nickel martensite is Rockwell C 25. A weak response to maraging is found after adding 7 percent cobalt to the alloy base. The addition of molybdenum alone gives a slight increase

TABLE 9·8 Mechanical Properties of Maraging Steels

	18%Ni			20%Ni	25%Ni
Annealed:					
Yield Strength, (0.2% offset), 1,000 psi	110			115	40
Tensile Strength, 1,000 psi	140			152	132
Elongation (in 1 in.), %	17			8	30
Reduction in area, %	75			...	72
Hardness, R/C	28–32			26–35	10–15
	18Ni(300)†	18Ni(250)†	18Ni(200)†	‡	§
After Maraging:					
Yield Strength, 1,000 psi	295–303	240–268	190–210	246	249–256
Tensile Strength, 1000 psi	297–306	250–275	200–220	256	265–270
Elongation, %	12	10–12	14–16	11	12
Reduction in area, %	60	45–58	65–70	45	53

* The International Nickel Company.
† Maraged 900°F, 3 h.
‡ Maraged 900°F, 1 h.
§ Conditioned 1300°F, 4 h, refrigerated, maraged 800–850°F, 1 h.

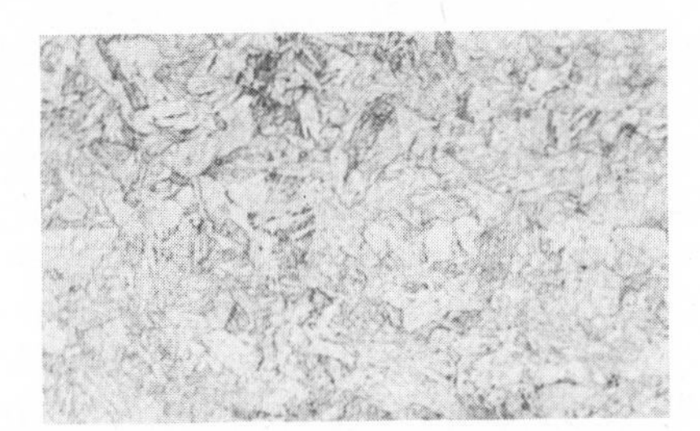

Fig. 9·19 The microstructure of an 18 percent nickel 250-grade maraging steel after aging at 900°F for 3 h. Structure is fine precipitates in a martensite matrix. Etched in modified Fry's reagent, 250X. (The International Nickel Company.)

in annealed hardness and considerable maraging response. However, as shown in Fig. 9·20, when molybdenum is added in the presence of 7 percent cobalt, an increase in hardness greater than the combined effect of both elements is obtained.

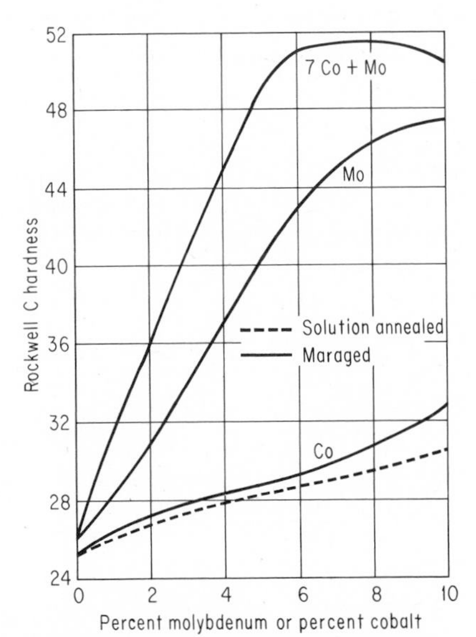

Fig. 9·20 Cobalt additions alone have little effect on maraged hardness. Molybdenum additions alone increase the hardness of the iron-nickel martensite after aging at 900°F. A strong additive effect of both cobalt and molybdenum can be seen. (The International Nickel Company.)

Compared with the 18 percent nickel grades, the 20 percent grade has the advantage of lower alloy content and freedom from cobalt and molybdenum, which may be desirable for some applications and environments. However, it is lower in toughness, in resistance to stress-corrosion cracking, and in dimensional stability during heat treatment. In comparison with the 25 percent nickel grade, this steel does not require a conditioning treatment at an intermediate temperature to become martensitic (see Fig. 9·18). Also, because of its lower nickel content, it has an M_s temperature above room temperature. However, to ensure complete transformation to martensite, refrigeration at –10°F is recommended before aging.

In contrast to the other grades of maraging steels, the 25 percent nickel grade is largely austenitic after annealing. To reach high strength levels, this steel after forming must be completely transformed to martensite befor maraging. This may be accomplished in either of two ways: (1) ausaging and (2) cold work and refrigeration (Fig. 9·18). In *ausaging* the steel is given a conditioning treatment at 1300°F for 4 h after forming. This treatment reduces the stability of the austenitic structure by causing nickel-titanium compounds to precipitate from the austenitic solid solution, and as a result it raises the M_s temperature so that austenite will transform to martensite on cooling to room temperature. In the second method the austenite is cold-worked at least 25 percent to start the transformation to martensite, which is completed by refrigeration at –100°F.

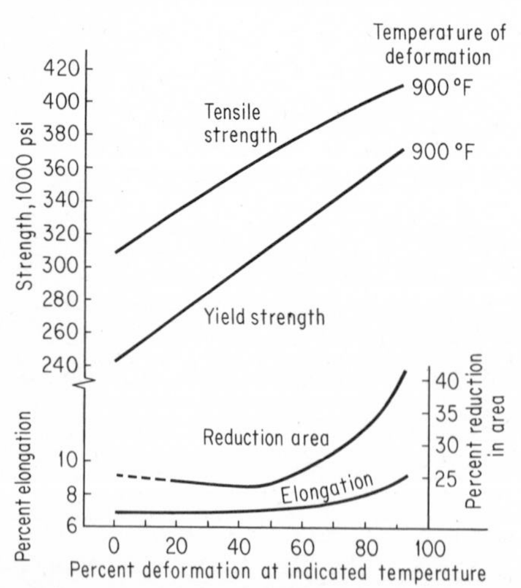

Fig. 9·21 Effect on mechanical properties of deforming austenite of H11 hot-work tool steel, austenitized at 1900°F, cooled to 900°F, deformed, cooled to room temperature, and tempered twice (double-tempered) at 950°F. (Data from Vanadium-Alloys Steel Company.)

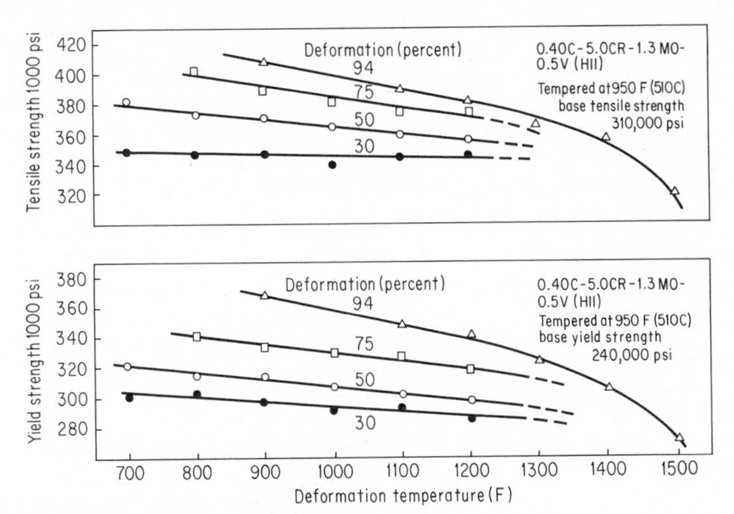

Fig. 9·22 Effect of temperature and deformation on yield and tensile strength of ausformed H11 hot-work tool steel. (Data from Vanadium-Alloys Steel Company.)

Applications for maraging steels are hulls for hydrospace vehicles, motor cases for missiles, and low-temperature structural parts. They have also been suggested for hot extrusion dies, cold-headed bolts, mortar and rifle tubing, and pressure vessels.

9·18 Ausforming A new type of extremely high-strength steels have recently been developed as a result of ausforming or austenitic forming. The technique consists of deforming unstable austenite of moderately alloyed steels at a temperature below the A_1 line in the "bay" that exists between the pearlite and bainite reactions, followed immediately by oil quenching to prevent the formation of nonmartensitic transformation products. The resultant microstructure consists of fine martensitic plates, the size and dispersion of which are determined by prior austenitic grain size and the amount of plastic deformation. While the technique is applicable to many steels, it has been used on AISI 4340 and H11 hot-work tool steel (listed in Table 10.1). The tensile strength of 4340 steel may be increased from 300,000 psi (by conventional heat treatment) to over 400,000 psi, with satisfactory ductility, by ausforming. The hot-working operation must be completed fairly quickly to avoid the transformation to softer products. Figure 9·21 shows that the tensile properties for H11 steel in sheet form may be increased to the vicinity of 400,000 psi, maintaining a useful level of ductility, by a 94 percent deformation at 900°F. The curves of Fig. 9·22 indicate that the temperature of deformation is not critical. Highly stressed structural parts for aircraft and automotive leaf springs are particularly attractive applications for ausforming.

QUESTIONS

9·1 Explain how alloying elements that dissolve in ferrite increase its strength.

9·2 What effect would the addition of 1 percent chromium have on the properties of a low-carbon steel? A high-carbon steel?

9·3 What factors determine the wear-resisting properties of a steel?

9·4 Devise and explain a practical test to compare the wear resistance of different steels.

9·5 If the primary consideration is hardenability, which alloy steel should be selected?

9·6 Look up the chemical composition of 4340 steel (Table 7·2). This steel is to be used as a structural member in an aircraft landing-gear assembly. What mechanical properties would be desirable for this application? Describe a heat treatment that would be applied to obtain these properties.

9·7 Same as Question 9·6 for 6150 steel to be used as a front coil spring in an automobile.

9·8 Look up the chemical composition of 4620 steel. On the basis of chemical composition, what mechanical properties would you expect this steel to have? Specify a heat treatment for this steel and give some possible applications.

9·9 Why was manganese chosen as a substitute for nickel in the development of the 2xx series of stainless steels?

9·10 Which stainless steel is best suited for surgical instruments? Explain.

9·11 Why do some pots and pans have a copper bottom and stainless steel inside? What type of stainless steel is best for this application? Why?

9·12 What is the difference in chemical composition between the standard stainless steels and the precipitation-hardenable stainless steels?

9·13 Describe three ways in which precipitation-hardenable stainless steels may be hardened.

9·14 Why does aging 17-4PH at 1150°F give lower mechanical properties than aging at 1050°F?

9·15 What effect does reduction of the carbon and chromium content of the austenite have on the M_s temperature?

9·16 Why is refrigeration or subzero treatment sometimes used for precipitation-hardening stainless steels?

9·17 Explain the term *maraging*.

9·18 What is the difference in heat treatment between the 18 percent Ni and the 25 percent Ni maraging steels?

9·19 Give at least four advantages of maraging steels as compared to regular stainless steels.

REFERENCES

Allegheny Ludlum Steel Corp.: "Stainless Steel Handbook," Pittsburgh, 1951.

American Iron and Steel Institute: "Stainless and Heat-Resisting Steels," Steel Products Manual, New York, 1963.

American Society for Metals: "Metals Handbook," 7th ed., 1948; 1954 Supplement; 8th ed., vol. 1, 1961, vol. 2, 1964 and vol. 7, 1972, Metals Park, Ohio.

Archer, R. S., J. F. Briggs, and C. M. Loeb: "Molybdenum; Steels, Irons, Alloys," Climax Molybdenum Co., New York, 1948.

Bain, E. C., and H. W. Paxton: "Alloying Elements in Steel," 2d ed., American Society for Metals, Metals Park, Ohio, 1961.

Bilby, Glover, and Wakeman: "Modern Theory in the Design of Alloys," The Institute of Metallurgists, London, 1967.
Burnham, T. H.: "Special Steels," 2d ed., Sir Isaac Pitman & Sons, Ltd., London, 1933.
Hall, A. M., and C. J. Slunder: "The Metallurgy, Behavior and Application of the 18 Per Cent Nickel Maraging Steels," NASA SP-5051, National Aeronautics and Space Administration, Washington, D.C., 1968.
Hull, A. M.: "Nickel in Iron and Steel," John Wiley & Sons, Inc., New York, 1954.
International Nickel Co.: "Nickel Alloy Steels," 2d ed., New York, 1949.
Monypenny, J. H. G.: "Stainless Iron and Steel," 2 vols., 3d ed., Chapman and Hall, London, 1951–1954.
Parr, J. G., and A. Hanson: "An Introduction to Stainless Steel," American Society for Metals, Metals Park, Ohio, 1965.
Thum, E. E.: "Book of Stainless Steels," 2d ed., American Society for Metals, Metals Park, Ohio, 1935.
U.S. Steel Corp.: "The Making, Shaping and Treating of Steel," 7th ed., Pittsburgh, 1957.
Zapffe, C. A.: "Stainless Steels," American Society for Metals, Metals Park, Ohio, 1949.

10 TOOL STEELS

10·1 Classification of Tool Steels Any steel used as a tool may be technically classed as a tool steel. However, the term is usually restricted to high-quality special steels used for cutting or forming purposes.

There are several methods of classifying tool steels. One method is according to the quenching media used, such as water-hardening steels, oil-hardening steels, and air-hardening steels. Alloy content is another means of classification, such as carbon tool steels, low-alloy tool steels, and medium-alloy tool steels. A final method of grouping is based on the application of the tool steel, such as hot-work steels, shock-resisting steels, high-speed steels, and cold-work steels.

The method of identification and type classification of tool steels adopted by the AISI (American Iron and Steel Institute) includes the method of quenching, applications, special characteristics, and steels for special industries. The commonly used tool steels have been grouped into seven major headings, and each group or subgroup has been assigned an alphabetical letter as follows:

GROUP	SYMBOL AND TYPE	
Water-hardening	W	
Shock-resisting	S	
Cold-work	O	Oil-hardening
	A	Medium-alloy air-hardening
	D	High-carbon high-chromium
Hot-work	H	(H1–H19, incl., chromium-base; H20–H39, incl., tungsten-base; H40–H59, incl., molybdenum-base)
High-speed	T	Tungsten-base
	M	Molybdenum-base
Mold	P	Mold steels (P1–P19, incl., low-carbon; P20–P39, incl., other types)
Special-purpose	L	Low-alloy
	F	Carbon-tungsten

TABLE 10·1 Identification and Type Classification of Tool Steels*

Type	IDENTIFYING ELEMENTS, %									
	C	Mn	Si	Cr	Ni	V	W	Mo	Co	Al
	WATER-HARDENING TOOL STEELS SYMBOL W									
W1	0.60/1.40†									
W2	0.60/1.40†	. . .	. . .	. . .	. . .	0.25				
W5	1.10	. . .	. . .	0.50						
	SHOCK-RESISTING TOOL STEELS SYMBOL S									
S1	0.50	. . .	. . .	1.50	. . .	. . .	2.50			
S2	0.50	. . .	1.00	. . .	. . .	. . .	. . .	0.50		
S5	0.55	0.80	2.00	. . .	. . .	. . .	. . .	0.40		
S7	0.50	. . .	. . .	3.25	. . .	. . .	. . .	1.40		
	COLD-WORK TOOL STEELS SYMBOL O, OIL-HARDENING TYPES									
O1	0.90	1.00	. . .	0.50	. . .	. . .	0.50	. . .		
O2	0.90	1.60								
O6‡	1.45	. . .	1.00	. . .	. . .	. . .	. . .	0.25		
O7	1.20			0.75			1.75			
	SYMBOL A, MEDIUM ALLOY AIR HARDENING TYPES									
A2	1.00	. . .	. . .	5.00	. . .	. . .	. . .	1.00		
A3	1.25	. . .	. . .	5.00	. . .	1.00	. . .	1.00		
A4	1.00	2.00	. . .	1.00	. . .	. . .	. . .	1.00		
A6	0.70	2.00	. . .	1.00	. . .	. . .	. . .	1.00		
A7	2.25	. . .	. . .	5.25	. . .	4.75	1.00	1.00		
A8	0.55	. . .	. . .	5.00	. . .	. . .	1.25	1.25		
A9	0.50	. . .	. . .	5.00	1.50	1.00	. . .	1.40		
A10‡	1.35	1.80	1.25	. . .	1.80	. . .	. . .	1.50		

Type	IDENTIFYING ELEMENTS, %								
	C	Mn	Si	Cr	Ni	V	W	Mo	Co	Al
SYMBOL D, HIGH-CARBON HIGH-CHROMIUM TYPES										
D2	1.50	...	...	12.00	...	...	...	1.00		
D3	2.25	...	...	12.00						
D4	2.25	...	...	12.00	...	...	...	1.00		
D5	1.50	...	...	12.00	...	...	...	1.00	3.00	
D7	2.35	...	...	12.00	...	4.00	...	1.00		
HOT-WORK TOOL STEELS SYMBOL H H1–H19, INCL., CHROMIUM-BASE TYPES										
H10	0.40	...	...	3.25	...	0.40	...	2.50		
H11	0.35	...	...	5.00	...	0.40	...	1.50		
H12	0.35	...	...	5.00	...	0.40	1.50	1.50		
H13	0.35	...	...	5.00	...	1.00	...	1.50		
H14	0.40	...	...	5.00	...	...	5.00			
H19	0.40	...	...	4.25	...	2.00	4.25	...	4.25	
H20–H39, INCL., TUNGSTEN-BASE TYPES (H27–H39 UNASSIGNED)										
H21	0.35	...	...	3.50	...	...	9.00			
H22	0.35	...	...	2.00	...	...	11.00			
H23	0.30	...	...	12.00	...	...	12.00			
H24	0.45	...	...	3.00	...	...	15.00			
H25	0.25	...	...	4.00	...	...	15.00			
H26	0.50	...	...	4.00	...	1.00	18.00			
H40–H59, INCL., MOLYBDENUM-BASE TYPES (H40,H44–H59 UNASSIGNED)										
H41	0.65	...	...	4.00	...	1.00	1.50	8.00		
H42	0.60	...	...	4.00	...	2.00	6.00	5.00		
H43	0.55	...	...	4.00	...	2.00	...	8.00		

TABLE 10·1 (Continued)

Type	IDENTIFYING ELEMENTS, %								
	C	Mn	Si	Cr	Ni	V	W	Mo	Co	Al
	HIGH-SPEED TOOL STEELS SYMBOL T, TUNGSTEN-BASE TYPES									
T1	0.70	...	...	4.00	...	1.00	18.00	...		
T2	0.80	...	...	4.00	...	2.00	18.00	...		
T4	0.75	...	...	4.00	...	1.00	18.00	...	5.00	
T5	0.80	...	...	4.00	...	2.00	18.00	...	8.00	
T6	0.80	...	...	4.50	...	1.50	20.00	...	12.00	
T8	0.75	...	...	4.00	...	2.00	14.00	...	5.00	
T15	1.50	...	...	4.00	...	5.00	12.00	...	5.00	
	SYMBOL M, MOLYBDENUM-BASE TYPES									
M1	0.80	...	...	4.00	...	1.00	1.50	8.00		
M2	0.85/1.00†	...	...	4.00	...	2.00	6.00	5.00		
M3	1.05	...	...	4.00	...	2.40	6.00	5.00		
M4	1.30	...	...	4.00	...	4.00	5.50	4.50		
M6	0.80	...	...	4.00	...	1.50	4.00	5.00	12.00	
M7	1.00	...	...	4.00	...	2.00	1.75	8.75		
M10	0.85	...	...	4.00	...	2.00	...	8.00		
M30	0.80	...	...	4.00	...	1.25	2.00	8.00	5.00	
M34	0.90	...	...	4.00	...	2.00	2.00	8.00	8.00	
M36	0.80	...	...	4.00	...	2.00	6.00	5.00	8.00	
M41	1.10	...	...	4.25	...	2.00	6.75	3.75	5.00	
M42	1.10	...	...	3.75	...	1.15	1.50	9.50	8.00	
M43	1.20	...	...	3.75	...	1.60	2.75	8.00	8.25	
M44	1.50	...	...	4.25	...	2.25	5.25	6.25	12.00	
M46	1.25	...	...	4.00	...	3.20	2.00	8.25	8.25	
M47	1.10	...	...	3.75	...	1.25	1.50	9.50	5.00	

Type	C	Mn	Si	Cr	Ni	V	W	Mo	Co	Al
	IDENTIFYING ELEMENTS, %									
SPECIAL-PURPOSE TOOL STEELS SYMBOL L. LOW-ALLOY TYPES										
L2	0.50/1.10†	...	...	1.00	...	0.20				
L3	1.00	...	...	1.50	...	0.20				
L6	0.70	...	...	0.75	1.50	...	...	0.25		
SYMBOL F, CARBON-TUNGSTEN TYPES										
F1	1.00	...	...	...	...	...	1.25			
F2	1.25	...	...	...	...	...	3.50			
MOLD STEELS, SYMBOL P P1–P19, INCL., LOW-CARBON TYPES (P7–P19 UNASSIGNED)										
P2	0.07	...	...	2.00	0.50	...	...	0.20		
P3	0.10	...	...	0.60	1.25					
P4	0.07	...	...	5.00						
P5	0.10	...	...	2.25						
P6	0.10	...	...	1.50	3.50					
P20–P39, INCL., OTHER TYPES (P22–P39 UNASSIGNED)										
P20	0.30	...	...	1.25	...	...	...	0.25		
P21	0.20	...	...	...	4.00	...	...	...	...	1.20

*From Steel Products Manual, "Tool Steels," American Iron and Steel Institute, January, 1970.
†Varying carbon contents may be available.
‡Contains free graphite in the microstructure to improve machinability.

The AISI identification and type classification of tool steels is given in Table 10·1.

10·2 Selection of Tool Steels The selection of a proper tool steel for a given application is a difficult task. The best approach is to correlate the metallurgical characteristics of tool steels with the requirements of the tool in operation.

In most cases, the choice of a tool steel is not limited to a single type or even to a particular family for a working solution to an individual tooling problem. Although many tool steels will perform on any given job, they will have to be judged on the basis of expected productivity, ease of fabrication, and cost. In the final analysis, it is the cost per unit part made by the tool that determines the proper selection.

Most tool-steel applications, with the exception of those to be made into machine parts, may be divided into types of operations: cutting, shearing, forming, drawing, extrusion, rolling, and battering. A cutting tool may have a single cutting edge which is in continuous contact with the work, such as a lathe or planer tool; or it may have two or more cutting edges which do continuous cutting, such as a drill or a tap; or it may have a number of cutting edges, with each edge taking short cuts and functioning only part of the time, such as a milling cutter or hob. When cutting is the chief function of the tool steel, it should have high hardness as well as good heat and wear resistance.

Shearing tools for use in shears, punches, and blanking dies require high wear resistance and fair toughness. These characteristics must be properly balanced, depending on the tool design, thickness of the stock being sheared, and temperature of the shearing operation.

Forming tools are characterized by imparting their form to the part being made. This may be done by forcing the solid metal into the tool impression either hot or cold by using a hot-forging or cold-heading die. This group also includes dies for die casting where the molten or semimolten metal is forced under pressure into the form of the die. Forming tools must have high toughness and high strength, and many require high red-hardness (resistance to heat softening).

Drawing and extrusion dies are characterized by substantial slippage between the metal being formed and the tool itself. Deep-drawing dies such as those used for the forming of cartridge cases generally require high strength and high wear resistance. Toughness to withstand outward pressures and wear resistance is most important for cold-extrusion dies, whereas dies for hot extrusion require, in addition, high red-hardness.

Thread-rolling dies must be hard enough to withstand the forces in forming the thread and must have sufficient wear resistance and toughness to adjust to the stresses developed. Battering tools include chisels and all forms of tools involving heavy shock loads. The most important characteristic for these tools is high toughness.

From the above discussion, it is apparent that, for most applications, hardness, toughness, wear resistance, and red-hardness are the most important selection factors in choosing tool steels. In individual applications, many other factors must be seriously considered. They include the amount of distortion which is permissible in the shape under consideration; the amount of surface decarburization which can be tolerated; hardenability or depth of hardness which can be obtained; resistance to heat checking; heat-treating requirements, including temperatures, atmospheres, and equipment; and finally, the machinability.

10·3 Comparative Properties The comparative properties of the most common tool steels are given in Table 10·2. Toughness, red-hardness, wear resistance, nondeforming properties, machinability, safety in hardening, and resistance to decarburization have been qualitatively rated good, fair, or poor. Depth of hardening is rated shallow, medium, or deep. The steels have been rated relative to each other rather than within any one particular class. The hot-work steels have been rated good or very good in red-hardness, but significant differences exist in "hot strength" of these steels, and a steel to be selected for hot die work requires careful study. Although the high speed steels are rated poor in toughness, there are differences in this property between the different high-speed steels which should be taken into account for a particular application.

10·4 Nondeforming Properties The tool steels in Table 10·2 have been rated on the basis of distortion obtained in hardening from the recommended hardening temperatures. Since steels expand and contract during heating and quenching, the extent to which dimensions change is most important for complex shapes. Intricately designed tools and dies must maintain their shape after hardening. Those steels rated good or best in nondeforming properties can be machined very close to size before heat treatment, so that little grinding will be required after hardening. Parts that involve rather drastic section changes should not be made of steels which are subject to excessive warpage in heating or quenching, as this will generally lead to cracking during hardening. In general, air-hardening steels exhibit the least distortion; those quenched in oil show moderate distortion; and water-hardening steels show the greatest distortion. The distortion is associated with the temperature gradient and the resulting dimensional changes during heating and cooling, which were discussed in detail under residual stresses in Chap. 8.

10·5 Depth of Hardening This is related to the hardenability of the individual tool steels. The hardenability ratings in Table 10·2 are based on the use of the recommended quenching medium. The shallow-hardening steels such as the carbon tool steels (group W), the tungsten finishing steels (group F), and several of the carburizing grades in group P are generally quenched in water. The hardenability increases with increasing alloy content. The only alloying element which decreases hardenability is cobalt. To develop high

TABLE 10·2 Comparative Properties of Some Tool Steels*

STEEL TYPE	HARDENING, °F	QUENCHING MEDIUM	TEMPERING RANGE, °F	APPROX HARDNESS ROCKWELL C†	DEPTH OF HARDENING
W1	1400–1550	Brine or water	300–650	65-50	Shallow
W2	1400–1550	Brine or water	300–650	65-50	Shallow
S1	1650–1800	Oil	400–1200	58–40	Medium
S5	1600–1700	Oil	350–800	60–50	Medium
O1	1450–1500	Oil	300–500	62–57	Medium
A2	1700–1800	Air	350–1000	62–57	Deep
A4	1500–1600	Air	350–800	62–54	Deep
D2	1800–1975	Air	400–1000	61–54	Deep
D3	1700–1800	Oil	400–1000	61–54	Deep
D4	1775–1850	Air	400–1000	61–54	Deep
H11	1825–1875	Air	1000–1200	54–38	Deep
H19	2000–2200	Air or oil	1000–1300	59–40	Deep
H21	2000–2200	Air or oil	1100–1250	54–36	Deep
H23	2200–2350	Air or oil	1200–1500	47–30	Deep
H26	2150–2300	Salt, oil, or air	1050–1250	58–43	Deep
H41	2000–2175	Salt, oil, or air	1050–1200	60–50	Deep
T1	2300–2375	Oil, air, or salt	1000–1100	65–60	Deep
T4	2300–2375	Oil, air, or salt	1000–1100	66–62	Deep
T6	2325–2400	Oil, air, or salt	1000–1100	65–60	Deep
M1	2150–2225	Oil, air, or salt	1000–1100	65–60	Deep
M2	2175–2250	Oil, air, or salt	1000–1100	65–60	Deep
M6	2150–2200	Oil, air, or salt	1000–1100	66–61	Deep
M41	2175–2220	Oil, air, or salt	1000–1100	70–65	Deep
L2	1450–1550 1550–1700	Water Oil	350–1000	63–45	Medium
L6	1475–1550	Oil	350–1000	62–45	Medium
F2	1450–1600	Water or brine	300–500	66–62	Shallow
P2	1525–1550‡	Oil	300–500	64–58§	Shallow
P20	1500–1600	Oil	900–1100	37–28	Shallow

*Adapted from tables in Steel Products Manual, "Tool Steels," American Iron and Steel Institute, 1970.
†After tempering.
‡After carburizing.
§Carburized case hardness.

strength throughout a large section, it is important to select a high-alloy steel.

10·6 Toughness The term *toughness* as applied to tool steels may be thought of as the ability to resist breaking rather than the ability to absorb energy during deformation, as defined in Chap. 1. Most tools must be rigid articles, and usually even slight plastic deformation makes the tool unfit for use. As might be expected, this property is best in the medium- and low-carbon tool steels of groups S and H, which form the basis of the shock-resisting tool steels. Shallow-hardening steels which end up with a relatively soft

NON-DEFORMING PROPERTIES	SAFETY IN HARDENING	TOUGH-NESS	RED-HARDNESS	WEAR RESISTANCE	MACHIN-ABILITY	RESISTANCE TO DECARBURI-ZATION
Poor	Fair	Good	Poor	Fair to good	Best	Best
Poor	Fair	Good	Poor	Fair to good	Best	Best
Fair	Good	Very good	Fair	Fair	Fair	Fair to good
Fair	Good	Best	Fair	Fair	Fair	Poor
Very good	Very good	Fair	Poor	Good	Good	Good
Best	Best	Fair	Fair	Very good	Fair	Fair
Best	Best	Fair	Fair	Good	Fair to poor	Good to fair
Best	Best	Poor	Good	Best	Poor	Fair
Very good	Good	Poor	Good	Best	Poor	Fair
Best	Best	Poor	Good	Best	Poor	Fair
Very good	Best	Good	Good	Fair	Fair	Fair
Good	Good	Good	Good	Fair	Fair	Fair
Air: good Oil: fair	Good	Good	Good	Fair to good	Fair	Fair
Air: good Oil: fair	Good	Fair	Very good	Fair to good	Fair	Fair
Salt, air: good Oil: fair	Good	Fair	Very good	Good	Fair	Fair
Salt, air: good Oil: fair	Fair	Poor	Very good	Good	Fair	Poor
Good	Good	Poor	Very good	Very good	Fair	Good
Good	Fair	Poor	Best	Very good	Fair	Fair
Good	Fair	Poor	Best	Very good	Fair	Poor
Good	Fair	Poor	Very good	Very good	Fair	Poor
Good	Fair	Poor	Very good	Very good	Fair	Fair
Good	Fair	Poor	Very good	Very good	Fair	Poor
Good	Fair	Poor	Very good	Very good	Fair	Poor
Water: poor Oil: fair	Water: poor Oil: fair	Very good	Poor	Good	Good	Good
Good	Good	Very good	Poor	Good	Fair	Good
Poor	Poor	Poor	Poor	Very good	Fair	Good
Good	Good	Good	Poor	Fair	Fair	Good
Good	Good	Good	Poor	Fair	Good	Good

tough core are also rated good in toughness. The cold-work tool steels which are high in carbon tend toward brittleness and low toughness.

10·7 Wear Resistance All the tool steels have relatively good wear resistance, but several are outstanding in this property. Wear resistance may be defined as the resistance to abrasion or resistance to the loss of dimensional tolerances. Wear resistance might be required on a single cutting edge or over the total surface of the part. In general, a correlation exists between the hard, undissolved carbide particles and wear resistance.

10·8 Red-hardness This property, also called *hot-hardness*, is related to the resistance of the steel to the softening effect of heat. It is reflected to some extent in the resistance of the material to tempering, which is an important selective factor for high-speed and hot-work tools. A tool steel with good red-hardness is essential when temperatures at which the tools must operate exceed 900°F. Alloying elements which form hard, stable carbides generally improve the resistance to softening at elevated temperature. Outstanding in this property are the tool steels that contain relatively large amounts of tungsten, chromium, and molybdenum.

10·9 Machinability This is the ability of the material to be cut freely and produce a good finish after being machined. The machinability ratings given in Table 10·2 merely show the relative difficulty which might be encountered in machining the steels in question during manufacture of tools and dies. The factors that affect machinability of tool steels are the hardness in the annealed condition, the microstructure of the steel, and the quantity of hard excess carbides.

When compared with the conventional alloy steels, tool steels are considerably more difficult to machine. The best machinable tool steel (W type) has a machinability rating of about 30 percent that of B1112 screw stock. It is therefore usual to compare the machinability of tool steels with W1 at an arbitrary rating of 100. On this basis, the machinability for each of the different types of tool steel is rated in Table 10·3.

The machinability and general workability of tool steels decrease with increasing carbon and alloy content. Low annealed hardnesses are usually more difficult to attain as the carbon and alloy content increases. The presence of carbon in combination with strong carbide-forming elements such as vanadium, chromium, and molybdenum reduces machinability by the formation of a large number of hard carbide particles which are out of solution after annealing.

TABLE 10·3 Machinability Ratings of Tool Steels

Water-hardening grades rated at 100

TOOL-STEEL GROUP	MACHINABILITY RATING
W	100
S	85
O	90
A	85
D	40–50
H (Cr)	75
H (W or Mo)	50–60
T	40–55
M	45
M	45–60
L	90
F	75
P	75–100

From "Metals Handbook," 8th ed., American Society for Metals, Metals Park, Ohio, 1961.

10·10 Resistance to Decarburization This is an important factor in the selection of tool steels since it influences the type of heat-treating equipment selected and the amount of material to be removed from the surface after hardening. Decarburization usually occurs when steels are heated above 1300°F, and unless some method is used to protect them (such as heating in a protective atmosphere), they are likely to lose some of their surface carbon. Decarburization will result in a soft rather than a hard surface after hardening. Tools that are intricately designed and cannot be ground after hardening must not show any decarburization.

The straight-carbon tool steels are least subject to decarburization. The shock-resisting tool steels are poor in this property; the hot-work tool steels are considered fair; and the majority of the other tool steels have good resistance to decarburization.

10·11 Brand Names For many years, a manufacturer requiring a tool steel for a particular application would indicate the nature of this application to a tool-steel producer and would receive a recommended tool steel. This practice has led to the use of brand names which the tool-steel producer gave to each of his different types of steel. Tool-steel producers made every effort to maintain the quality and thus the reputation of their brand names.

With the rapid growth of industry, many tool-steel consumers had more than one source of supply and had to issue specifications covering chemical composition and some physical properties of the steel. At the same time, many tool-steel manufacturers published information regarding the chemical composition and physical properties of their various brands of tool steel. Although the AISI has standardized the chemical composition of tool steels (Table 10·1), the use of trade names has persisted to the present day. Table 10·4 lists approximate comparable tool-steel brand names of various manufacturers.

10·12 Water-hardening Tool Steels (Group W) These are essentially plain-carbon tool steels, although some of the higher-carbon-content steels have small amounts of chromium and vanadium added to improve hardenability and wear resistance. The carbon content varies between 0.60 and 1.40 percent, and the steels may be roughly placed into three subdivisions according to carbon content.

0.60 to 0.75 percent carbon—for applications where toughness is the primary consideration, such as hammers, concrete breakers, rivet sets, and heading dies for short runs.

0.75 to 0.95 percent carbon—for applications where toughness and hardness are equally important, such as punches, chisels, dies, and shear blades.

0.95 to 1.40 percent carbon—for applications where increased wear resistance and retention of cutting edge are important. They are used for woodworking tools, drills, taps, reamers, and turning tools.

In general, the straight-carbon tool steels are less expensive than the

TABLE 10·4 Comparable Tool-steel Brand Names

GENERAL CLASSIFICATION	AISI NO.	ALLEGHENY-LUDLUM	BETHLEHEM	BRAEBURN	CARPENTER	COLUMBIA	CRUCIBLE	FIRTH-STERLING
Water-hardening tool steels:								
Standard quality	W1	Pompton	XCL	Standard	No. 11 Comet	Standard	Black Diamond	Sterling
Extra quality	W1	Pompton Extra	XX	Extra	No. 11 Extra	Extra	Labelle Extra	F-S Extra
Extra quality	W2	Python	Superior	. . .	No. 11 Extra Vanadium	Vanadium Extra	Alva Extra	Extra V
Shock-resisting tool steels	S1	Seminole	67 Chisel	Vibro	. . .	Buster	Atha Pneu	J-S Punch
	S5	AL 609	Omega	. . .	No. 481	CEC Smoother	Labelle Silicon No. 2	Chino
Cold-work tool steels:								
Oil-hardening types	O1	Saratoga	BTR	Kiski	. . .	EXL-Die	Ketos	Invaro No. 1
	O2	Deward	. . .	S.O.D.	Stentor	. . .	Paragon	Invaro No. 2
Medium-alloy air-hardening	A2	Sagamore	A-H5	Airque	No. 484	EZ-Die Smoothcut	Airkool	Airvan
High-carbon high-chromium	D2	Ontario	Lehigh H	Superior 3	No. 610	Atmodie	Airdi 150	Chromovan
	D3	Huron	Lehigh S	Superior 1	Hampden	. . .	HYCC	Triple Die
Hot-work tool steels, chromium base	H12	Potomac	Cromo-W Cromo-W^{V}	Pressurdie No. 2	No. 345	Alco Die	Chro-Mow	HWD-1
	H13	Potomac M	Cromo-High V	Pressurdie No. 3	No. 883	Vanadium Fire Die	Nudie V	HWD-3
Plastic-mold steels:								
Straight iron	P1	. . .	Duramold C	. . .	Mirromold	. . .	Crusca cold hubbing	. . .
5% chromium air-hardening	P4	. . .	Duramold A	. . .	Super Samson		Airmold	

TABLE 10·4 (Continued)

GENERAL CLASSIFICATION	AISI NO.	JESSOP	LATROBE	UDDEHOLM	UNIVERSAL CYCLOPS	VANADIUM-ALLOYS	VULCAN CRUCIBLE
Water-hardening tool steels:							
Standard quality	W1	Lion	Standard Carbon	UHB	Standard	Red Star Tool	Fort Pitt
Extra quality	W1	Lion Extra	Extra Carbon	UHB Extra	Extra	Extra L	Extra
Extra quality	W2	Lion extra Vanadium	. . .	UHB-VA	Extra Draco	Elvandi	
Shock-resisting tool steels	S1	Top Notch	XL Chisel	UHB-711	Alco M, Alco S	Par Exc	Q.A.
	S5	No. 259	. . .	UHB Resisto	Cyclops 67	Mosil	487D
Cold-work tool steels:							
Oil-hardening types	O1	Truform	Badger	UHB-46	Wando	Non-shrinkable Colonial No. 6	Oil hardening
	O2	Special oil hardening	Mangano	. . .	. . .	. . .	Non-shrinkable
Medium-alloy air-hardening	A2	Windsor	Select B	UHB-151	Sparta	Air Hard	Vuldie
High-carbon high-chromium	D2	CNS-1	Olympic	TRI-Mo	Ultradie No. 2 and 3	Ohio Die	Alidie
	D3	CNS-2	GSN	TRI-Van	Ultradie No. 1	. . .	Hi Pro
Hot-work tool steels, chromium base	H12	DICA B	LPD	UHB Special	Thermold B	Hot Form No. 1 and 2	TCM
	H13	DICA B Vanadium	V.D.C.	UHB Orvar	Thermold Av	Hot Form 5	Vulcast
Plastic-mold steels:							
Straight iron	P1	. . .	. . .	UHB Forma	. . .	. . .	Plastic Die
5% chromium air-hardening	P4	. . .	. . .	UHB Premo			

alloy tool steels, and with proper heat treatment they yield a hard martensitic surface with a tough core. These steels must be water-quenched for high hardness and are therefore subject to considerable distortion. They have the best machinability ratings of all the tool steels and are the best in respect to decarburization, but their resistance to heat is poor. Because of this low red-hardness, carbon steels cannot be used.as cutting tools under conditions where an appreciable amount of heat is generated at the cutting edge. Their use as cutting tools is limited to conditions involving low speeds and light cuts on relatively soft materials, such as wood, brass, aluminum, and unhardened low-carbon steels.

The typical microstructure of a W1 water-quenching tool steel, austenitized at 1450°F, brine-quenched, and tempered at 325°F, is shown in Fig. 10·1. The low tempering temperature has resulted in a matrix of dark-etching tempered martensite with some undissolved carbide particles (white dots), and a hardness of Rockwell C 64.

10·13 Shock-resisting Tool Steels (Group S) These steels were developed for those applications where toughness and the ability to withstand repeated shock are paramount. They are generally low in carbon content, the carbon varying between 0.45 and 0.65 percent. The principal alloying elements in these steels are silicon, chromium, tungsten, and sometimes molybdenum. Silicon strengthens the ferrite, while chromium increases hardenability and contributes slightly to wear resistance. Molybdenum aids in increasing hardenability, while tungsten imparts some red-hardness to

Fig. 10·1 W1 water-hardening tool steel austenitized at 1450°F, quenched in brine, and tempered at 325°F. Structure is tempered martensite with some undissolved spheroidal carbide (white dots). Etched in 3 percent nital, 1,000×. (From Metals Handbook, vol. 7, "Atlas of Microstructures," American Society for Metals, 1972.)

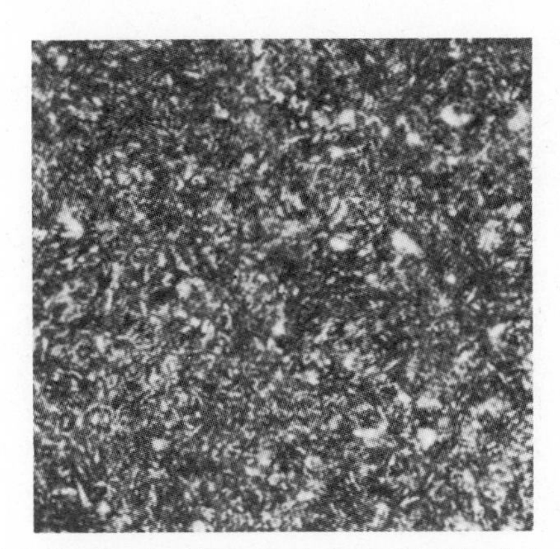

Fig. 10·2 S1 shock-resisting tool steel austenitized at 1750°F, oil-quenched, and tempered at 800°F. Structure is tempered martensite with some spheroidal-carbide particles (white dots). Etched in 3 percent nital, 1,000×. (From Metals Handbook, vol. 7, "Atlas of Microstructures," American Society for Metals, 1972.)

these steels. Most of these steels are oil-hardening, although some have to be water-quenched to develop full hardness.

The high silicon content tends to accelerate decarburization, and suitable precautions should be taken in heat treatment to minimize this. They are classed as fair in regard to red-hardness, wear resistance, and machinability, and hardness is usually kept below Rockwell C 60. The steels in this group are used in the manufacture of forming tools, punches, chisels, pneumatic tools, and shear blades.

Figure 10·2 shows spheroidal carbide particles in a tempered martensite matrix typical of an S1 shock-resisting tool steel austenitized at 1750°F, oil-quenched, and tempered at 800°F. The higher tempering temperature has caused the tempered martensite to etch lighter in comparison with Fig. 10·1.

10·14 Cold-work Tool Steels This is considered to be the most important group of tool steels, since the majority of tool applications can be served by one or more of the steels in this classification.

The oil-hardening low-alloy type (group O) contains manganese and smaller amounts of chromium and tungsten. They have very good nondeforming properties and are less likely to bend, sag, twist, distort, or crack during heat treatment than are the water-hardening steels.

Figure 10·3*a* shows the normal microstructure of spheroidal carbide particles in a tempered martensite matrix of an O1 oil-hardening tool steel austenitized at 1500°F, oil-quenched, and tempered at 425°F. Raising the

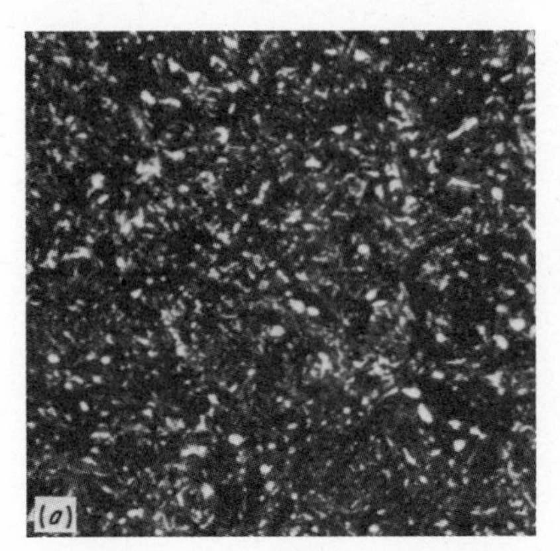

Fig. 10·3 O1 oil-hardening tool steel. (*a*) Austenitized at 1500°F, oil-quenched, and tempered at 425°F. The normal structure consists of spheroidal-carbide particles in a tempered-martensite matrix. (*b*) Austenitized at 1800°F, oil-quenched, and tempered at 425°F. Structure is coarse-martensite needles (black) in retained-austenite matrix (white), both resulting from overheating. Etched in 3 percent nital, 1,000×. (From Metals Handbook, vol. 7, "Atlas of Microstructures," American Society for Metals, 1972.)

austenitizing temperature to 1800°F results in coarse martensite needles (black) in a retained austenite matrix (white), both resulting from overheating (Fig. 10·3*b*). These steels are relatively inexpensive and their high carbon content produces adequate wear resistance for short-run applications at or near room temperature. The main function of the high silicon content in O6 steel is to induce graphitization of part of the carbide, thereby improving machinability in the annealed condition.

Figure 10·4 shows the structure of an O6 oil-hardening tool steel austenitized at 1500°F, oil-quenched, and tempered at 425°F. The large black spots are graphite, white spots are carbide particles, and the matrix is tempered martensite (gray). These steels have good machinability and good resistance to decarburization; toughness is only fair, and their red-hardness is as poor as the straight-carbon tool steels. They are used for taps, solid treading dies, form tools, and expansion reamers.

The medium-alloy type (group A), with 1 percent carbon, contains up to 3 percent manganese, up to 5 percent chromium, and 1 percent molybdenum. The increased alloy content, particularly manganese and molybdenum, confers marked air-hardening properties and increased hardenability. Figure 10·5 shows the structure of an A2 tool steel austenitized at 1800°F, air-cooled, and tempered at 350°F, Rockwell C 63. The large white spots are alloy carbide particles, the small white spots are spheroidal carbide particles, and the gray matrix is tempered martensite. The barely vis-

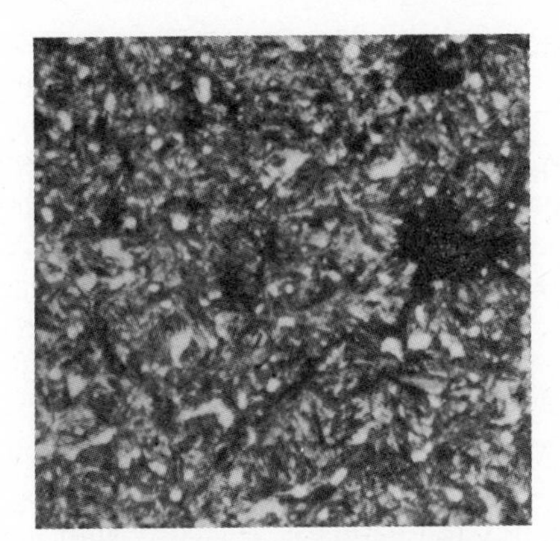

Fig. 10·4 O6 oil-hardening tool steel austenitized at 1500°F, oil-quenched, and tempered at 425°F. The structure consists of graphite (black) and carbide particles (white) in a matrix of tempered martensite (gray). Etched in 3 percent nital, 1,000×. (From Metals Handbook, vol. 7, "Atlas of Microstructures," American Society for Metals, 1972.)

ible black lines are grain boundaries. This group has excellent nondeforming properties, good wear resistance, and fair toughness, red-hardness, and resistance to decarburization, but only fair to poor machinability. These steels are used for blanking, forming, trimming, and thread-rolling dies.

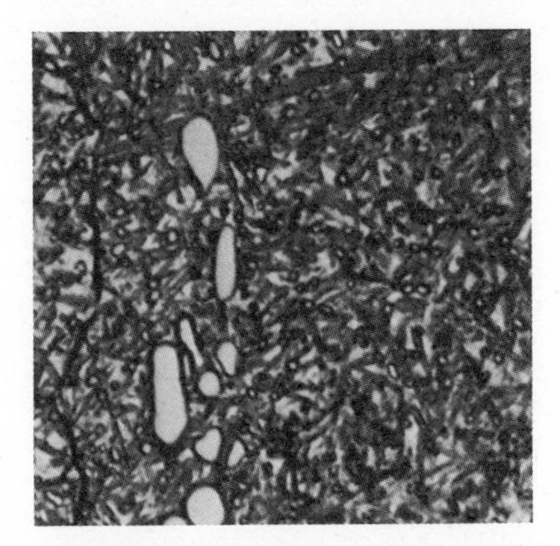

Fig. 10·5 A2 tool steel austenitized at 1800°F, air-cooled, and tempered at 350°F, Rockwell C 63. The structure consists of large alloy-carbide particles (white) and small spheroidal-carbide particles (white dots) in a matrix of tempered martensite. Grain boundaries (black lines) are barely visible. Etched in 3 percent nital, 1,000×. (From Metals Handbook, vol. 7, "Atlas of Microstructures," American Society for Metals, 1972.)

The high-carbon high-chromium types (group D) contain up to 2.25 percent carbon and 12 percent chromium. They may also contain molybdenum, vanadium, and cobalt. The combination of high carbon and high chromium gives excellent wear resistance and nondeforming properties. They have good abrasion resistance, and minimum dimensional change in hardening makes these steels popular for blanking and piercing dies; drawing dies for wire, bars, and tubes; thread-rolling dies; and master gauges.

10·15 Hot-work Tool Steels (Group H) In many applications the tool is subjected to excessive heat because the material is being worked, as in hot forging and extruding, die casting, and plastic molding. Tool steels developed for these applications are known as *hot-work tool steels* and have good red-hardness.

The alloying elements noted for red-hardness are chromium, molybdenum, and tungsten. However, the sum of these elements must be at least 5 percent before the property of red-hardness becomes appreciable.

The hot-work tool steels may be subdivided into three groups:

(1) *Hot-work chromium-base* (H11 to H19), containing a minimum of 3.25 percent chromium and smaller amounts of vanadium, tungsten, and molybdenum. They have good red-hardness because of their medium chromium content, supplemented by the addition of the carbide-forming elements of vanadium, tungsten, and molybdenum. The low carbon and relatively low total alloy content promote toughness at the normal working hardnesses of Rockwell C 40 to 55. Higher tungsten and molybdenum contents increase red-hardness but slightly reduce toughness. These steels are extremely deep-hardening and may be air-hardened to full working hardness in sections up to 12 in. The air-hardening qualities and balanced alloy content are responsible for low distortion in hardening. These steels are especially adapted to hot die work of all kinds, particularly extrusion dies, die-casting dies, forging dies, mandrels, and hot shears.

Figure 10·6 shows the structure of an H11 tool steel austenitized at 2050°F, oil-quenched, and double-tempered (2 h plus 2 h) at 1100°F, Rockwell C 46 to 48. The matrix is coarse tempered martensite with a few spheroidal carbide particles (white spots). The high austenitizing temperature has dissolved most of the carbide.

An interesting application is the use of the H11 steel for highly stressed structural parts, particularly for supersonic aircraft. The chief advantage of this steel over conventional high-strength steels is its ability to resist softening during continued exposure to temperatures up to 1000°F and at the same time to provide moderate toughness and ductility at tensile strengths of 250,000 to 300,000 psi.

Another important advantage of H11 for these applications is its exceptional ease of forming while in the austenitic condition by the process called *ausforming* described in Sec. 9.18. It also has good weldability, a relatively low coefficient of thermal expansion, and above average resistance to corrosion and oxidation.

(2) *Hot-work tungsten-base* (H21 to H26), containing at least 9 percent tungsten and 2 to 12 percent chromium. The higher alloy content increases resistance to high-temperature softening compared with the H11 to H19 steels, but it also makes them more susceptible to brittleness at the normal working hardnesses of Rockwell C 45 to 55. They can be air-hardened for low distortion, or they can be quenched in oil or hot salt to minimize scaling. Although they have much greater toughness, these

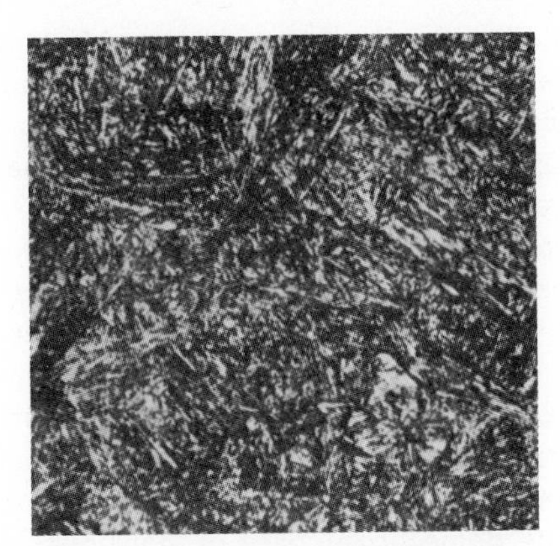

Fig. 10·6 H11 tool steel austenitized at 2050°F, oil-quenched, and double-tempered (2 h plus 2 h) at 1100°F, Rockwell C 46 to 48. Structure consists of a few spheroidal-carbide particles in a coarse tempered-martensite matrix. Etched in 2 percent nital, 500×. (From Metals Handbook, vol. 7, "Atlas of Microstructures," American Society for Metals, 1972.)

steels have many of the characteristics of the high-speed tool steels. In fact, H26 is a low-carbon version of T1 high-speed steel (see Table 10·1). They have been used for high-temperature applications such as mandrels and extrusion dies for brass, nickel alloys, and steel.

(3) *Hot-work molybdenum-base* (H41 to H43), containing 8 percent molybdenum, 4 percent chromium, and smaller amounts of tungsten and vanadium. These steels are similar to the tungsten hot-work steels, having almost identical characteristics and uses. In composition, they resemble the various types of molybdenum high-speed tool steels, although they have lower carbon content and greater toughness. Their principal advantage over the tungsten hot-work steels is lower initial cost. These steels are more resistant to heat cracking (checking) than the tungsten grades, but in common with all high-molybdenum steels they require greater care in heat treatment, particularly with regard to decarburization.

The hot-work tool steels as a group have good toughness because of low carbon content, good to excellent red-hardness, and fair wear resistance and machinability. They are only fair to poor in resistance to decarburization, but when air-hardened they show little or no distortion from heat treating.

10·16 High-speed Tool Steels These steels are among the most highly alloyed of the tool steels and usually contain large amounts of tungsten or molybdenum along with chromium, vanadium, and sometimes cobalt. The carbon content varies between 0.70 and 1 percent, although some types contain as much as 1.5 percent carbon.

The major application of high-speed steels is for cutting tools, but they are also used for making extrusion dies, burnishing tools, and blanking punches and dies.

Compositions of the high-speed steels are designed to provide excellent red-hardness and reasonably good shock resistance. They have good non-deforming properties and may be quenched in oil, air, or molten salts. They are rated as deep-hardening, have good wear resistance, fair machinability, and fair to poor resistance to decarburization.

The high-speed steels are subdivided into two groups: molybdenum-base (group M) and tungsten-base (group T). The most widely used tungsten-base type is known as 18-4-1 (T1), the numerals denoting the content, respectively, of tungsten, chromium, and vanadium in percentages. From the standpoint of fabrication and tool performance, there is little difference between the molybdenum and tungsten grades. The important properties of red-hardness, wear resistance, and toughness are about the same. Since there are adequate domestic supplies of molybdenum and since most of the tungsten must be imported, the molybdenum steels are lower in price, and over 80 percent of all the high-speed steel produced is of the molybdenum type.

When better than average red-hardness is required, steels containing cobalt are recommended. Higher vanadium content is desirable when the material being cut is highly abrasive. In T15 steel, a combination of cobalt plus high vanadium provides superiority in both red-hardness and abrasion resistance. The use of high-cobalt steels requires careful protection against decarburization during heat treatment, and since these steels are more brittle, they must be protected against excessive shock or vibration in service.

Figure 10·7 shows the structure of a T1 high-speed steel austenitized at 2335°F, salt-quenched to 1125°F, air-cooled, and double-tempered at 1000°F. White spots are undissolved carbides in a matrix of tempered martensite.

The presence of many wear-resistant carbides in a hard heat-resistant matrix makes these steels suitable for cutting tool applications, while their toughness allows them to outperform the sintered carbides in delicate tools and interrupted-cut applications. The tungsten and molybdenum high-speed steels are used in a wide variety of cutting tools, such as tool bits, drills, reamers, broaches, taps, milling cutters, hobs, saws, and woodworking tools.

10·17 Mold Steels (Group P) These steels contain chromium and nickel as the principal alloying elements, with molybdenum and aluminum as additives. Most of these steels are alloy carburizing steels produced to tool steel quality. They are generally characterized by very low hardness in the annealed condition and resistance to work hardening; both factors are favorable for hubbing operations. In hubbing, a master hub is forced into a soft blank. After the impression has been formed or cut, the steels are generally carburized and hardened to a surface hardness of Rockwell C 58 to 64 for

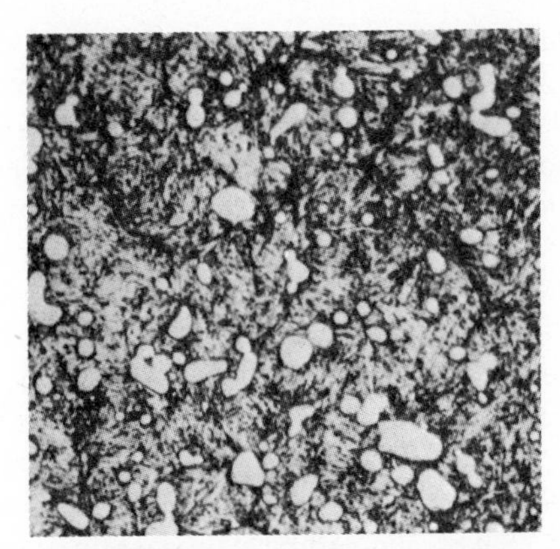

Fig. 10·7 T1 high-speed steel austenitized at 2335°F, salt-quenched to 1125°F, air-cooled, and double-tempered at 1000°F. Undissolved carbide particles in a matrix of tempered martensite. Etched in 4 percent nital, 1,000×. (From Metals Handbook, vol. 7, "Atlas of Microstructures," American Society for Metals, 1972.)

wear resistance. With the exception of P4, these steels have poor red-hardness and therefore are used almost entirely for low-temperature die-casting dies and for molds for injection or compression molding of plastics. Types P20 and P21 are normally supplied in a heat-treated condition to a hardness of Rockwell C 30 to 35, so that they can be readily machined into large intricate dies and molds. Since they are prehardened, no subsequent heat treatment is required, and distortion and size changes are avoided.

10·18 Special-purpose Tool Steels Many tool steels do not fall into the usual categories and are therefore designated as special-purpose tool steels. They have been developed to handle the peculiar requirements of one certain application and are more expensive for many applications than the more standard steels.

The low-alloy types (group L) contain chromium as the principal alloying element, with additions of vanadium, molybdenum, and nickel. The high chromium contents promote wear resistance by the formation of hard complex iron-chromium carbides, and together with molybdenum they increase hardenability. Nickel increases toughness, while vanadium serves to refine the grain. These steels are oil-hardening and thus only fair in resisting dimensional change. Typical uses are various machine-tool applications where high wear resistance with good toughness is required, as in bearings, rollers, clutch plates, cams, collets, and wrenches. The high-carbon types are used for arbors, dies, drills, taps, knurls, and gages.

The carbon-tungsten type (group F) are generally shallow-hardening, water-quenching steels with high carbon and tungsten contents to promote high wear resistance. Under some conditions of operation these steels have four to ten times the wear resistance of the plain-carbon group W tool steels. They are relatively brittle, so that in general they are used for high-wear, low-temperature, low-shock applications. Typical uses are paper-cutting knives, wire-drawing dies, plug gages, and forming and finishing tools.

10·19 Heat Treatment of Tool Steels Proper heat treatment of tool steels is one of the most important factors in determining how they will perform in service. Although the emphasis in heat treatment is usually on the cooling rate, it should be realized that as much damage may be done to the steel on heating as on cooling.

Tool steels should not be heated so rapidly as to introduce large temperature gradients in the piece. This may be avoided by slow heating or by preheating the steel at a lower temperature before placing it in the high-heat furnace. Some heat treaters prefer placing the tool-steel parts into a cold furnace and then bringing both the work and the furnace up to temperature together. It is also important that the tool steel be allowed to remain at the proper temperature for a sufficient time to make certain that the entire section has been heated uniformly.

To avoid overheating, tool steels should not be heated to too high a temperature or kept at heat too long. Quenching from excessive temperature may result in cracking. Overheating causes excessive grain growth and consequent loss in toughness.

It is essential that some means be used to protect the surface of the tool steel from excessive scaling or decarburization during heating. This has been discussed in Sec. 8·18. Any decarburized areas must be removed from tool-steel surfaces to provide satisfactory hardnesses.

The manner and media of quenching vary according to the steel being quenched and the speed required in quenching. The usual quenching media are water, brine, oil, and air. Carbon and low-alloy tool steels are quenched in brine or water; high-alloy tool steels are quenched in oil, air, or molten salts. While still air, fan cooling, and compressed-air blasts are used for cooling, still air is the preferred method, for it is more likely to provide uniform cooling. Interrupted quenching is also used for tool steels. By this method, the steel is quenched in a liquid bath of salt or lead between 900 and 1200°F, then cooled in air to about 150°F.

It is recommended procedure to temper tool steels immediately after quenching and before they have cooled to room temperature to minimize the danger of cracking due to strains introduced by quenching. The tempering, or drawing, operation relieves the stresses developed during hardening and provides more toughness. Preferred practice is to utilize long draws at comparatively low temperatures rather than short draws at high temperatures.

Carbon and low-alloy steels are generally tempered at temperatures between 300 and 500°F, while for high-alloy steels the range 300 to 500°F is used for hardness and 900 to 1200°F for toughness. The high-speed tool

Fig. 10·8 Die made of carbon tool steel. Cracking occurred in quench because of excessive stresses set up between the thin rim and the body. (At top) Longitudinal section through die. Numbers are Rockwell hardness values. (Bethlehem Steel Company.)

steels are tempered between 950 and 1100°F, and the use of a double draw which repeats the original cycle is common practice.

10·20 Tool Failures The analysis to determine the probable cause for premature tool failure is often complex. Thorough investigation, however, will generally reveal good reasons behind every tool failure. It is the purpose of this section to discuss briefly five fundamental factors that contribute to tool failure.

Faulty Tool Design This may lead to failure either in heat treatment or during service. When a tool is to be liquid-quenched, the use of heavy sections next to light sections should be avoided. During quenching, the light sections will cool rapidly and harden before the heavy sections. This will set up quenching stresses that often result in cracking. Figure 10·8 shows a failed die made of carbon tool steel, while Fig. 10·9 shows a cracked die made of manganese oil-hardening tool steel. In each case, cracking occurred during quenching because of excessive stresses set up by the drastic change in section. This type of failure may usually be avoided by making the tool as a two-piece assembly. The use of square holes is another prime source of tool failure due to faulty design. If it is essential to use adjacent heavy and light sections or sharp corners, the use of an air-hardening steel is recommended.

Faulty Steel Despite the careful control used in the manufacture and inspection of tool steels, occasionally there is some defect in the steel. There may be porous areas resulting from shrinkage during solidification of the ingot which are known as *voids* or *pipe*. There may be *streaks* or *laps* due to segregation or nonmetallic inclusions, which usually run longitudinally with respect to the original bar stock. Other defects are *tears,* which are transverse surface defects resulting from working the steel under conditions where it does not have sufficient ductility; internal-cooling cracks known as flakes; and surface-cooling cracks as a result of cooling too rapidly after the last forging or rolling operation.

Tools made from large bar stock (over 4 in. diameter) of high-chromium steels generally show a brittle carbide network due to insufficient hot work (Fig. 10·11*a*). The use of disks of small bar stock which are upset-forged provides additional hot work, which breaks up the brittle carbide network and ensures a more uniform carbide distribution. Figure 10·10 shows an 8-in.-diameter milling cutter which failed in service, and subsequent microstudy showed the presence of a brittle carbide network. This milling cutter was made of bar stock and did not undergo sufficient reduction by hot working to remove the remnants of the as-cast carbide network. The normal microstructure, with good carbide distribution resulting from a properly hot-worked high-speed steel after quenching and tempering, is illustrated in Fig. 10·11*b*.

Faulty Heat Treatment This factor is the cause of the large majority of tool failures. Tools should be properly handled during and after the quench. They should be removed from the quench while still warm and transferred immediately to a tempering furnace. As was pointed out in the previous section on heat treatment, tools should be quenched from the recommended hardening temperature. The use of excessively high hardening temperature causes grain coarsening, which is evident on a fractured surface. Evidence of overheating may usually be found by microexamination. Figure 10·12 shows a manganese oil-hardening tool-steel cam which cracked in hardening. The microstructure revealed coarse, acicular martensite typical of overheated steel, instead of the normal microstructure of fine tempered martensite and spheroidal carbides. It is estimated that this steel was heated to 1800°F instead of the proper temperature of 1475°F.

Improper Grinding Very high surface stresses may be set up in a hardened tool because of the grinding operation. These stresses may be high enough to cause cracks. Light grinding cracks tend to appear at 90° from the direction of grinding, while heavy grinding cracks present a characteristic

Fig. 10·9 Die made of manganese oil-hardening tool steel. Cracking occurred because of excessive quenching stress set up between the heavy body and the small protruding section. (Bethlehem Steel Company.)

Fig. 10·10 Failed milling cutters made of M2 high-speed steel. Note chipped teeth. Failure occurred because of poor carbide distribution resulting from insufficient reduction by hot working in the steel plant. (Bethlehem Steel Company.)

network pattern (see Fig. 1·33*a*). The presence of grinding cracks is best revealed by magnetic-particle testing.

Mechanical Overload and Operational Factors Mechanical factors that cause tool failures due to overload may be accidental or are the result of excessive stress concentration or improper clearances and alignment. This type of tool failure is often difficult to determine, since thorough investigation of the failed tool will not reveal any cause for the short life. Figure 10·13 shows a failed reamer made of M2 high-speed steel. The bushing on the left was drilled to an undersize hole, and when this reamer attempted to take an excessively heavy cut, breakage resulted.

A common method of failure due to operational factors occurs in tools used for hot-work operations. These tools are subjected to repeated thermal stresses because of alternate heating and cooling of the tools. This leads to a network of very fine hairline cracks known as *heat checks*. Figure 10·14 shows heat checks developed on the surface of a punch tip made of H12 hot-work tool steel. It should be realized that, under these severe operating conditions, eventual failure of the tool is to be expected and that there is no simple solution to the problem of avoiding failure due to heat checking.

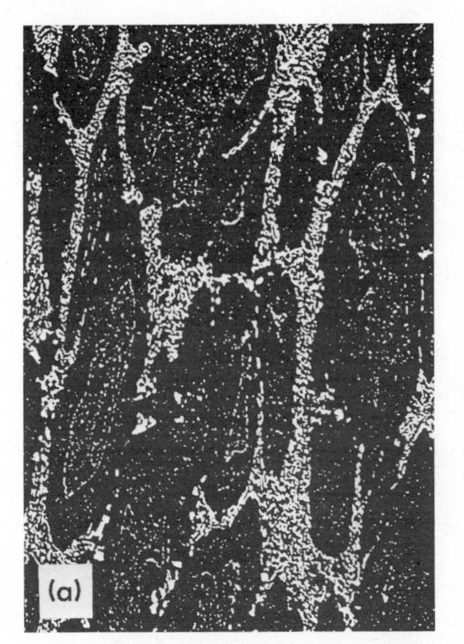

Fig. 10·11 M2 high-speed steel. (*a*) Longitudinal section of an 8-in. bar showing a carbide network due to insufficient hot working. (*b*) Longitudinal section of a 1-in. bar upset-forged showing normal spheroidal-carbide distribution after sufficient hot working. Etched in 4 percent nital, 100×. (Latrobe Steel Company.)

SPECIAL CUTTING MATERIALS

10·21 Stellites These are essentially cobalt-chromium-tungsten alloys. They contain from 25 to 35 percent chromium, 4 to 25 percent tungsten, 1 to 3 percent carbon, and the remainder cobalt. The hardness of stellite varies from Rockwell C 40 to 60, depending upon the tungsten and carbon content. Microscopically, the alloys consist mainly of tungstides and carbides. Their outstanding properties are high hardness, high resistance to wear and corrosion, and excellent red-hardness. This combination of properties makes them very suitable for cutting applications.

Stellite metal-cutting tools are widely used for machining steel, cast iron, cast steel, stainless steel, brass, and most machinable materials. They may be operated at higher speeds than those used with high-speed-steel tools. Stellite alloys are usually cast to the desired shape and size and are therefore not so tough as high-speed steels. They are also appreciably weaker and more brittle than high-speed steels, so that a careful analysis of specific machining operations is necessary in selecting the proper tool. Stellite cutting tools are used as single-point lathe tools, milling cutter blades for large inserted tooth cutters, spot facers, reamers, form tools, and burnish-

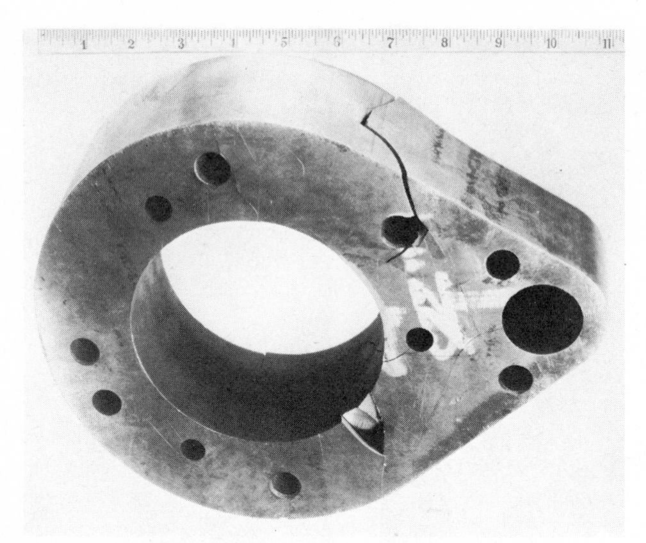

Fig. 10·12 Cam made of manganese oil-hardening tool steel which cracked in hardening because it was quenched from an excessively high temperature. (Bethlehem Steel Company.)

ing rollers. Stellite is also used as a hard-facing material for trimming dies and gauge blocks, on plowshares and cultivators for farm use, on the wearing parts of crushing and grinding machinery, and on bucket teeth for excavating and dredging equipment.

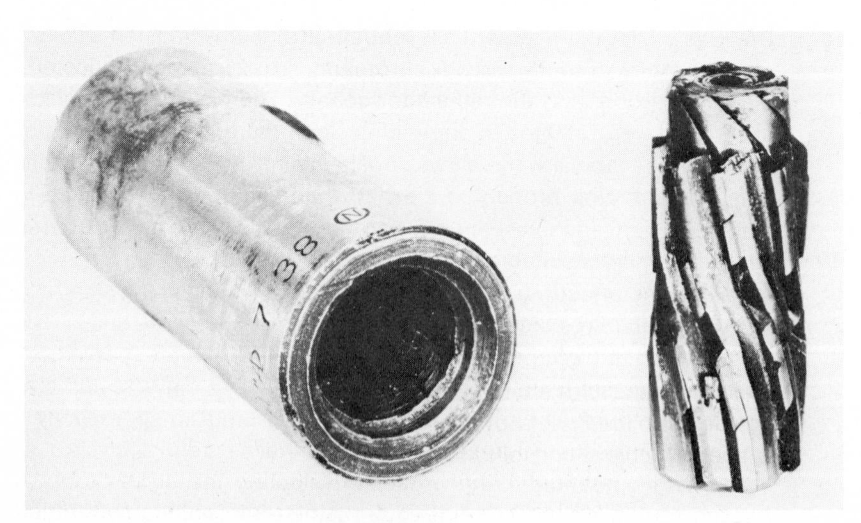

Fig. 10·13 (Right) Failed reamer made of M2 high-speed steel. (Left) Bushing on which failure occurred. (Bethlehem Steel Company.)

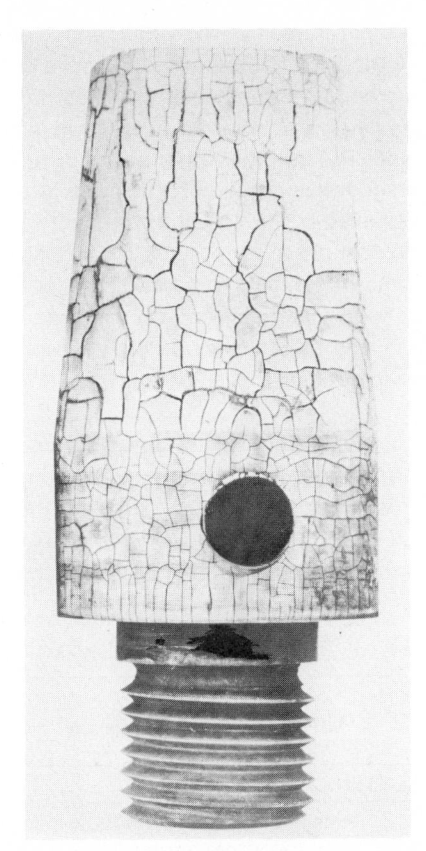

Fig. 10·14 Punch tip, $3\frac{3}{4}$ in. diameter by $8\frac{3}{4}$ in., made of HI2 hot-work steel. This tool became stuck in hot forgings several times and was heated to a much higher temperature than it was designed to withstand. Heat checks resulted. (Bethlehem Steel Company.)

10·22 Cemented Carbides These materials are made of very finely divided carbide particles of the refractory metals, cemented together by a metal or alloy of the iron group, forming a body of very high hardness and high compressive strength. Cemented carbides are manufactured by powder-metallurgy techniques. The process consists essentially in preparing the powder carbides of tungsten, titanium, or tantalum; mixing one or more of these powders with a binder, usually cobalt powder; pressing the blended powder into compacts of the desired shape; and sintering the pressed shapes to achieve consolidation.

The blended powders are formed into desired shapes by cold pressing, followed by sintering; or by hot pressing, during which pressing and sintering are done at the same time. Pressures used in cold pressing vary between 5 and 30 tons/sq in., depending upon the size and shape of the compact. Sintering is carried out at temperatures between 2500 and 2700°F for 30 to 60 min. At these elevated temperatures, the cobalt forms a eutectic with the carbides, and this eutectic becomes the cementing material. After cooling, the sintered compact has its final properties, since it does not respond to any known heat treatment. The carbides are present as individual grains, and also as a finely dispersed network resulting from the precipitation during cooling of carbide dissolved in the cobalt during sintering.

Carbides may be classified into two broad categories: (1) the straight-tungsten carbide grades, used primarily for machining cast iron, austenitic steel, and nonferrous and nonmetallic materials; and (2) the grades containing major amounts of titanium and tantalum carbides, used primarily for machining ferritic steel. A more detailed classification based on composition is shown in Table 10·5.

The exceptional tool performance of sintered carbide results from high hardness and high compressive strength combined with unusual red-hardness. The lowest hardness of sintered carbide is approximately the same as the highest hardness available in tool steel, Rockwell A 85 (Rockwell C 67). Typical compressive mechanical properties of sintered carbides are

TABLE 10·5 Classification of Cemented Carbides*

CARBIDE GROUP	COMPOSITION % REMAINDER WC Co	TaC + TiC	HARDNESS R/A	TYPICAL APPLICATIONS
		STRAIGHT TUNGSTEN CARBIDE		
1	2.5–6.5	0–3	93–91	Finishing to medium roughing cuts on cast iron, nonferrous alloys, and superalloys; low-impact dies
2	6.5–15	0–2	92–85	Rough cuts on cast iron; moderate-impact dies
3	15–30	0–5	88–85	High-impact dies
		ADDED CARBIDE PREDOMINANTLY TiC		
4	3–7	20–42	93.5–92	Light high-speed finishing cuts on steel
5	7–10	10–22	92.5–90	Medium arts and speeds on steel
6	10–12	8–15	92.0–89	Roughing cuts on steel
		ADDED CARBIDE PREDOMINANTLY TaC		
7	4.5–8	16–25	93-91	Light cuts on steel
8	8–10	12–20	92–90	General purpose and heavy cutting of steel
		ADDED CARBIDE EXCLUSIVELY TaC		
9	5.5–16	18–30	91.5–84.0	Wear-resistant applications, particularly involving heat

*From Metals Handbook, 8th ed., vol. 3, p. 316, American Society for Metals, Metals Park, Ohio, 1967.

TABLE 10·6 Typical Compressive Mechanical Properties of Sintered Carbides*

	MEASURED IN COMPRESSION						
CARBIDE GROUP	COMPRESSIVE STRENGTH PSI	ELASTIC LIMIT PSI	MODULUS OF ELASTICITY, MILLION PSI	POISSON'S RATIO	DUCTILITY, %	IMPACT STRENGTH, FT-LB†	FATIGUE LIMIT, 1000 PSI‡
1 (3% Co)	615,000	500,000	105	0.24	0.60	...	...
1 (6% Co)	614,000	286,000	105	0.28	0.85	0.73	95
2 (10% Co)	600,000	125,000	87	0.20	1.90	1.10	105
3 (16% Co)	545,000	95,000	76	0.22	2.70	1.75	...
5	625,000	230,000	78	0.22	1.00	0.60	90
6	533,000	97,000	80	0.22	2.00	0.40	90
7	635,000	173,000	82	0.21	0.90	...	...
8	631,000	250,000	81	0.22	1.00	0.60	85
9	705,000	240,000	86	0.22	1.70	0.60	85

*From Metals Handbook, 8th ed., vol. 3, p. 319, American Society for Metals, Metals Park, Ohio, 1967.
†Values for impact strength are from unnotched specimens in sections approximately ¼ in. square.
‡Values for fatigue limit are based on 20 million cycles of stress, for specimens of the R. R. Moore rotating-beam type.

shown in Table 10·6. The compressive strength for most grades is substantially more than 500,000 psi, along with very high modulus of elasticity. Both properties seem to decrease with increasing cobalt content.

Microstructure affects hardness and strength. The size of the carbide particles (grains), their distribution and porosity, and the quality of the bond between cobalt and carbide crystals are important factors. Increasing the tungsten carbide grain size lowers the hardness, because the softer cobalt "lakes," which are interspersed between grains, are also larger. This is illustrated in Fig. 10·15 for a 94 percent WC, 6 percent cobalt alloy. The hardness decreased from Rockwell A 93 to 92 to 91 as the WC grain size went from fine to intermediate to coarse.

Since cemented carbides have low toughness and tensile strength, the usual practice is to braze or mechanically fasten a small piece of carbide material (called an insert) to a steel shank, which provides rigid support under the cutting edge (Fig. 10·16). Where space permits, mechanical attachment is usually preferred. When a tip is brazed to the body or holder, the tool must be removed from the machine for resharpening by using a silicon carbide or diamond-impregnated grinding wheel. Under these conditions only one or two cutting edges of the tip can be used. In contrast, when the tip is mechanically secured in the holder, it can be loosened and turned to the next cutting edge without removing the tool from the machine. With this procedure, more cutting edges are available for use. After all the edges have been used, common practice is to discard the carbide tip because it is less expensive to replace it than to recondition it by grinding. Also, the hazard of damage to the insert by brazing is eliminated.

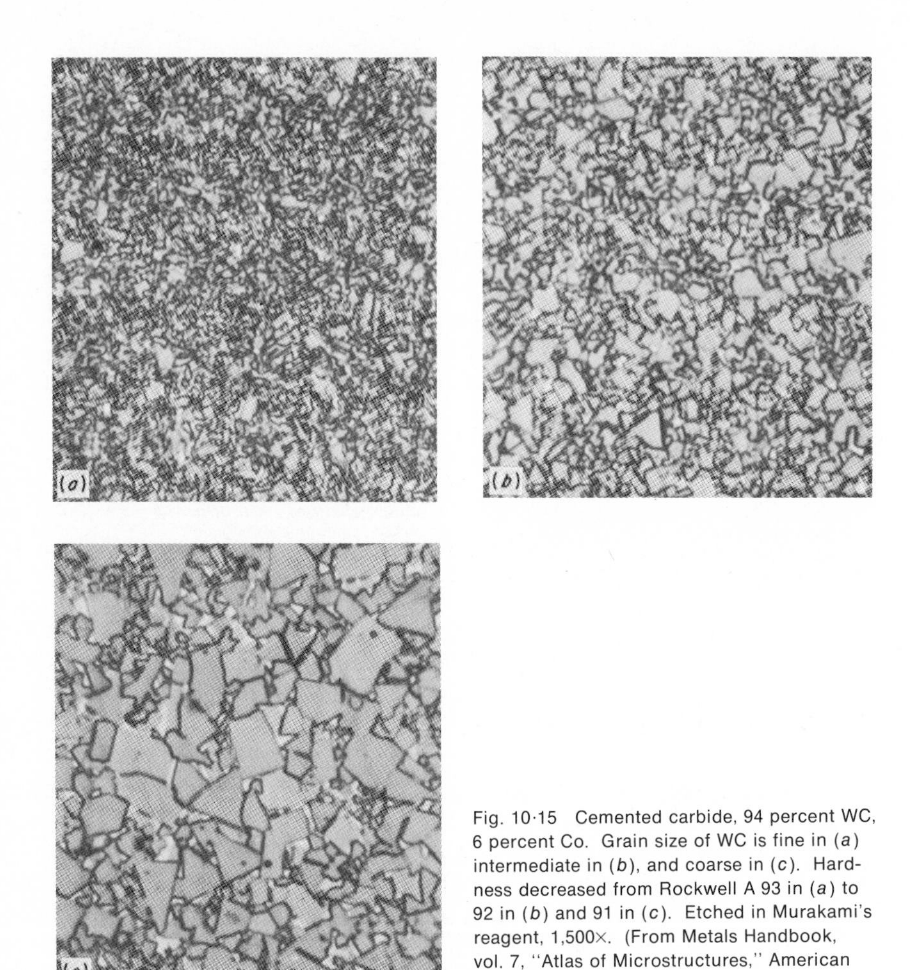

Fig. 10·15 Cemented carbide, 94 percent WC, 6 percent Co. Grain size of WC is fine in (*a*) intermediate in (*b*), and coarse in (*c*). Hardness decreased from Rockwell A 93 in (*a*) to 92 in (*b*) and 91 in (*c*). Etched in Murakami's reagent, 1,500×. (From Metals Handbook, vol. 7, "Atlas of Microstructures," American Society for Metals, 1972.)

More cemented carbides are consumed for metal cutting than for any other type of application. Because of their ability to retain a sharp cutting edge, the straight-tungsten carbide grades are virtually the only tool material used to cut abrasive materials such as Fiberglas and phenolic resins. Carbides which have the highest hardness are also being used for production cutting of white cast iron at Rockwell C 60. Cemented carbides are used for drills, reamers, boring and facing tools, and saws for the machining of both metals and nonmetals. Cutting speeds and feeds employed with carbide tools are generally higher than those used with high-speed steel or stellite.

The high hardness and wear resistance of cemented carbides make them well suited for earth drilling and mining applications. Various types of specialized rock bits for drilling in extremely hard and abrasive rock formations utilize carbide inserts instead of the conventional hard-faced steel teeth. They are also used for such applications as facings for hammers in hammer mills, facings for jaw crushers, sandblast nozzles, ring and plug gages, and gage blocks. Cemented carbide dies are used for the hot drawing of tungsten and molybdenum and for the cold drawing of wire, bar, and tubing made of steel, copper, aluminum, and other materials.

Cermets usually contain the carbides of titanium and chromium with nickel or a nickel-base alloy as the binder. In the most common grade, the hard phase is predominantly titanium carbide with chromium carbide additions and 30 to 70 percent nickel binder. This grade has high hardness, high resistance to oxidation, relatively high resistance to thermal shock, and relatively low density, but it also has low ductility and toughness. It is used where high-temperature abrasion resistance is the primary objective. When the binder content is less than 20 percent, the cermets have been

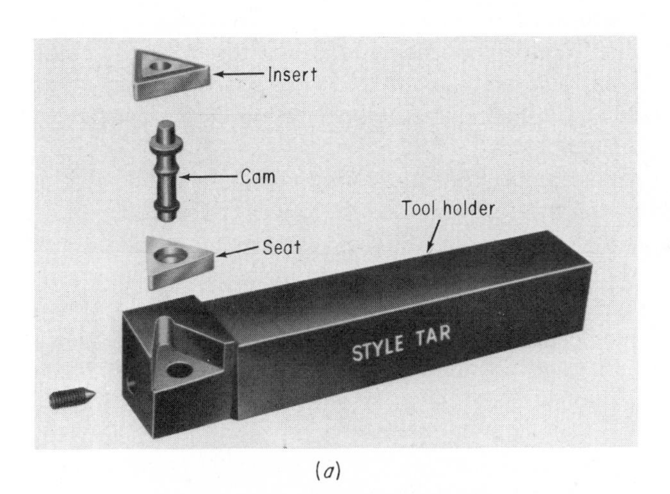

(a)

(b)

Fig. 10·16 Carbide tools: (a) mechanical holder and (b) brazed tool. (Courtesy of the DoALL Company.)

used for cutting both steel and cast iron at high speed with medium to light chip loads.

10·23 Ceramic Tools Most ceramic or cemented oxide cutting tools are manufactured primarily from aluminum oxide. Bauxite (a hydrated form of aluminum oxide) is chemically processed and converted into a denser, crystalline form called *alpha alumina*. Fine grains are obtained from the precipitation of the alumina or from the precipitation of the decomposed alumina compound.

Ceramic tool inserts are produced by either cold or hot pressing. In cold pressing, the fine alumina powder is compressed into the required form and then sintered in a furnace at 1600 to 1700°C. Hot pressing combines forming and sintering, pressure and heat being applied simultaneously. Small amounts of titanium oxide or magnesium oxide may be added for certain types of ceramics to aid in the sintering process and to retard growth. After the inserts have been formed, they are finished with diamond-impregnated grinding wheels.

Ceramic cutting tools are most commonly made as a disposable insert which is fastened in a mechanical holder. Disposable inserts are available in many styles, such as triangular, square, rectangular, and round. As with cemented carbide tools, when a cutting edge becomes dull, a sharp edge may be obtained by the insert being indexed (turned) in the holder. Ceramic inserts may also be fastened to a steel shank by epoxy glue. This method of holding the insert almost eliminates the strains caused by clamping in mechanical holders.

The principal elevated-temperature properties of alumina are high hardness, chemical inertness, and resistance to wear. The high hardness and wear resistance of alumina are the main reasons for its use in machining cast iron and hardened steel at high cutting speeds. The inertness of alumina to iron at high temperature prevents welding of the tool to steel or cast iron workpieces and contributes to the production of good surface finish. Table 10·7 lists some mechanical properties of ceramic and other cutting-tool materials.

TABLE 10·7 Mechanical Properties of Tool Materials*

PROPERTY	CERAMIC	HIGH-SPEED STEEL	C-2 CARBIDE
Transverse rupture strength, psi	90,000	500,000	230,000
Compressive strength, psi	500,000	600,000	650,000
Modulus of elasticity, psi	60×10^6	32×10^6	100×10^6
Hardness, Rockwell A	93	85	92
Microhardness, Knoop 100g	1780	740	1800

*From Metals Handbook, 8th ed., vol. 3, p. 323, American Society for Metals, Metals Park, Ohio, 1967.

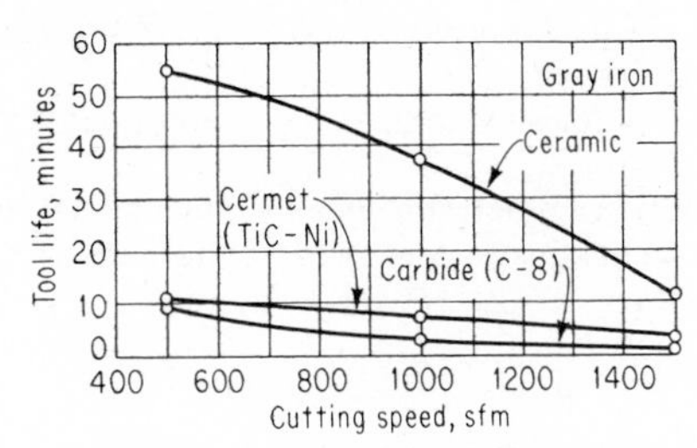

Fig. 10·17 Relation of tool life to cutting speed for ceramic, cermet, and carbide tools when cutting gray iron. (From Metals Handbook, 8th ed., vol. 3, p. 323, American Society for Metals, Metals Park, Ohio, 1967.)

When ceramic cutting tools are properly mounted in suitable holders and used on accurate rigid machines, they offer the following advantages:

1 Machining time is reduced because of the higher cutting speeds possible. Speeds ranging from 50 to 200 percent higher than those used for carbides are quite common (Fig. 10·17).
2 High stock-removal rates and increased productivity result because heavy depths of cut can be made at high surface speeds.
3 Under proper conditions a ceramic tool lasts from three to ten times longer than a carbide tool.
4 Ceramic tools retain their strength and hardness at high machining temperatures (in excess of 2000°F).
5 More accurate size control of the workpiece is possible because of the greater wear resistance of ceramic tools.
6 They withstand the abrasion of sand and of inclusions found in castings.
7 Heat treated materials as hard as Rockwell C 66 can be readily machined.
8 They produce a better surface finish than is possible with other cutting tools.

The disadvantages of ceramic tools are as follows:

1 They are brittle and tend to chip easily.
2 Initial cost is approximately twice that of carbide tools.
3 A more rigid machine is necessary than for other cutting tools.
4 Considerably more horsepower is required for ceramic tools to cut efficiently.

QUESTIONS

10·1 List the properties most important for tool steels and give one industrial application where each property would be required.
10·2 What would be the influence of each of the following alloying elements on the properties of a tool steel: chromium, tungsten, molybdenum, vanadium, silicon, manganese, and cobalt?
10·3 Describe the basis for selection of a tool steel to be used as a threadrolling die on AISI 1020 steel.

10·4 Describe the heat treatment you would apply to the tool steel selected in Question 10·3 and the reasons for this heat treatment.
10·5 Same as Questions 10·3 and 10·4, for a die to be used to cold-head AISI 1020 bearing rollers.
10·6 Using the equation given below, plot a graph showing the dimensional change, inches per inches, when spheroidite changes to austenite as the carbon content varies from 0 to 1.40 percent.

$$\text{Dimensional change, in./in.} = 0.1555 + 0.0075\ (\text{percent C})$$

10·7 Same as Question 10·6 for the change from austenite to martensite with varying carbon, using the following formula:

$$\text{Dimensional change, in./in.} = 0.0155 - 0.0018\ (\text{percent C})$$

10·8 Plot the net dimensional change for the reaction spheroidite → austenite → martensite as the carbon content varies from 0 to 1.40 percent. (Hint: Use the equations given in Questions 10·6 and 10·7 to determine an equation for the net dimensional change.)

REFERENCES

Allegheny Ludlum Steel Corporation: "Tool Steel Handbook," Pittsburgh, 1951.
American Iron and Steel Institute: Steel Products Manual, "Tool Steels," New York, 1970.
American Society for Metals: "Metals Handbook," 7th ed., 1948; 8th ed., 1961, 1964, and vol 7, 1972, Metals Park, Ohio.
American Society of Tool and Manufacturing Engineers: "Tool Engineers Handbook," 2d ed., McGraw-Hill Book Company, New York, 1959.
Bethlehem Steel Company: "The Tool Steel Trouble Shooter Handbook 322," Bethlehem, Pa., 1952.
Palmer, F. R., and G. V. Luerssen: "Tool Steel Simplified," 2d ed., Carpenter Steel Co., Reading, Pa., 1948.
Roberts, G. A., J. C. Hamaker, and A. R. Johnson: "Tool Steels," American Society for Metals, Metals Park, Ohio, 1962.
Seabright, L. H.: "The Selection and Hardening of Tool Steels," McGraw-Hill Book Company, New York, 1950.

11 CAST IRON

11·1 Introduction Cast irons, like steels, are basically alloys of iron and carbon. In relation to the iron–iron carbide diagram, cast irons contain a greater amount of carbon than that necessary to saturate austenite at the eutectic temperature. Therefore, cast irons contain between 2 and 6.67 percent carbon. Since high carbon content tends to make the cast iron very brittle, most commercially manufactured types are in the range of 2.5 to 4 percent carbon.

The ductility of cast iron is very low, and it cannot be rolled, drawn, or worked at room temperature. Most of the cast irons are not malleable at any temperature. However, they melt readily and can be cast into complicated shapes which are usually machined to final dimensions. Since casting is the only suitable process applied to these alloys, they are known as *cast irons*.

Although the common cast irons are brittle and have lower strength properties than most steels, they are cheap, can be cast more readily than steel, and have other useful properties. In addition, by proper alloying, good foundry control, and appropriate heat treatment, the properties of any type of cast iron may be varied over a wide range. Significant developments in foundry control have led to the production of large tonnages of cast irons whose properties are generally consistent.

11·2 Types of Cast Iron The best method of classifying cast iron is according to metallographic structure. There are four variables to be considered which lead to the different types of cast iron, namely the carbon content, the alloy and impurity content, the cooling rate during and after freezing, and the heat treatment after casting. These variables control the condition of the carbon and also its physical form. The carbon may be combined as iron carbide in cementite, or it may exist as free carbon in graphite. The shape and distribution of the free carbon particles will greatly influence the physical properties of the cast iron. The types of cast iron are as follows:

White cast irons, in which all the carbon is in the combined form as cementite.

Malleable cast irons, in which most or all of the carbon is uncombined in the form of irregular round particles known as *temper carbon.* This is obtained by heat treatment of white cast iron.

Gray cast irons, in which most or all of the carbon is uncombined in the form of graphite flakes.

Chilled cast irons, in which a white cast-iron layer at the surface is combined with a gray-iron interior.

Nodular cast irons, in which, by special alloy additions, the carbon is largely uncombined in the form of compact spheroids. This structure differs from malleable iron in that it is obtained directly from solidification and the round carbon particles are more regular in shape.

Alloy cast irons, in which the properties or the structure of any of the above types are modified by the addition of alloying elements.

11·3 White Cast Iron The changes that take place in white cast iron during solidification and subsequent cooling are determined by the iron–iron carbide diagram discussed in Chap. 7. All white cast irons are hypoeutectic alloys, and the cooling of a 2.50 percent carbon alloy will now be described.

The alloy, at x_1 in Fig. 11·1, exists as a uniform liquid solution of carbon dissolved in liquid iron. It remains in this condition as cooling takes place until the liquidus line is crossed at x_2. Solidification now begins by the formation of austenite crystals containing about 1 percent carbon. As the temperature falls, primary austenite continues to solidify, its composition moving down and to the right along the solidus line toward point *C.* The liquid in the meantime is becoming richer in carbon, its composition also moving down and to the right along the liquidus line toward point *E.* At the eutectic temperature, 2065°F, the alloy consists of austenite dendrites containing 2 percent carbon and a liquid solution, containing 4.3 percent carbon. The liquid accounts for $(2.5-2.0)/(4.3-2.0)$, or 22 percent of the alloy by weight. This liquid now undergoes the eutectic reaction isothermally to form the eutectic mixture of austenite and cementite known as *ledeburite.* Since the reaction takes place at a relatively high temperature, ledeburite tends to appear as a coarse mixture rather than the fine mixture typical of many eutectics. It is not unusual for ledeburite to be separated completely, with the eutectic austenite added to the primary austenite dendrites, leaving behind layers of massive, free cementite.

As the temperature falls, between x_3 and x_4, the solubility of carbon in austenite decreases, as indicated by the A_{cm} line *CJ.* This causes precipitation of proeutectoid cementite, most of which is deposited upon the cementite already present. At the eutectoid temperature, 1333°F, the remaining austenite containing 0.8 percent carbon and constituting

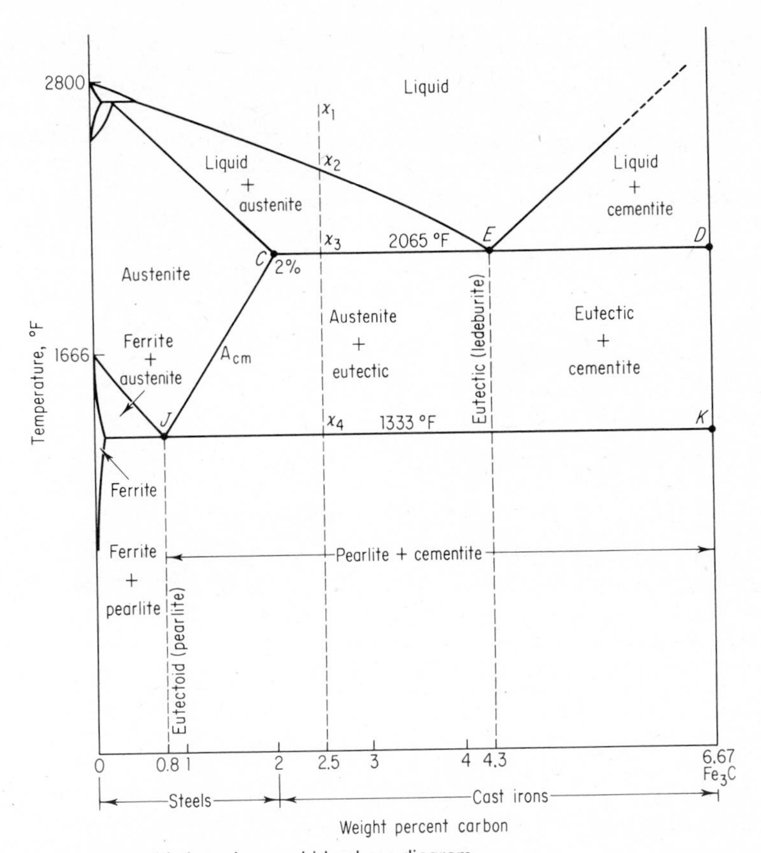

Fig. 11·1 The metastable iron–iron carbide phase diagram.

(6.67 − 2.5)/(6.67 − 0.8), or 70 percent of the alloy, undergoes the eutectoid reaction isothermally to form pearlite. During subsequent cooling to room temperature, the structure remains essentially unchanged.

The typical microstructure of white cast iron, consisting of dendrites of transformed austenite (pearlite) in a white interdendritic network of cementite, is illustrated in Fig. 11·2*a*. Higher magnification of the same sample (Fig. 11·2*b*) reveals that the dark areas are pearlite.

It was pointed out in Chap. 7 that cementite is a hard, brittle interstitial compound. Since white cast iron contains a relatively large amount of cementite as a continuous interdendritic network, it makes the cast iron hard and wear-resistant but extremely brittle and difficult to machine. "Completely white" cast irons are limited in engineering applications because of this brittleness and lack of machinability. They are used where

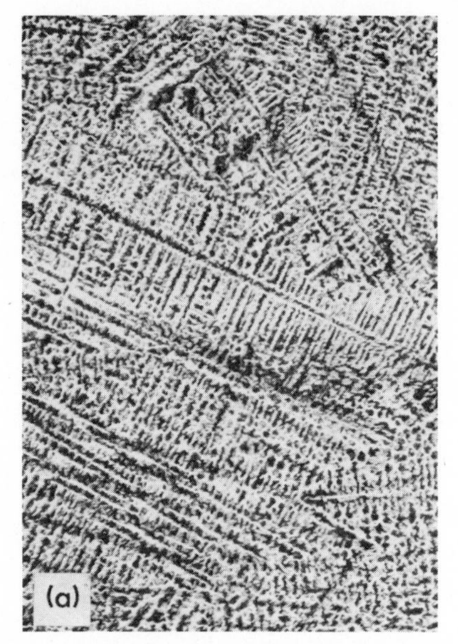

Fig. 11·2 The microstructure of white cast iron. (*a*) Dark areas are primary dendrites of transformed austenite (pearlite) in a white interdendritic network of cementite, 20×. (*b*) Same sample at 250×, showing pearlite (dark) and cementite (white). Etched in 2 percent nital.

resistance to wear is most important and the service does not require ductility, such as liners for cement mixers, ball mills, certain types of drawing dies, and extrusion nozzles. A large tonnage of white cast iron is used as a starting material for the manufacture of malleable cast iron. The range mechanical properties for unalloyed white irons is as follows: hardness Brinell 375 to 600, tensile strength 20,000 to 70,000 psi, compressive strength 200,000 to 250,000 psi, and modulus of elasticity 24 to 28 million psi.

11·4 Malleable Cast Iron It was pointed out in Sec. 7·4 that cementite (iron carbide) is actually a metastable phase. There is a tendency for cementite to decompose into iron and carbon, but under normal conditions it tends to persist indefinitely in its original form. Up to this point, cementite has been treated as a stable phase; however, this tendency to form free carbon is the basis for the manufacture of malleable cast iron.

The reaction $Fe_3C \rightleftharpoons 3Fe + C$ is favored by elevated temperatures, the existence of solid nonmetallic impurities, higher carbon contents, and the presence of elements that aid the decomposition of Fe_3C.

On the iron–iron carbide equilibrium diagram for the metastable system, shown in Fig. 11·3, are superimposed the phase boundaries of the stable iron-carbon (graphite) system as dotted lines.

The purpose of malleabilization is to convert all the combined carbon in white iron into irregular nodules of temper carbon (graphite) and ferrite. Commercially, this process is carried out in two steps known as the *first* and *second stages of the anneal.*

White irons suitable for conversion to malleable iron are of the following range of composition:

	Percent
Carbon	2.00–2.65
Silicon	0.90–1.40
Manganese	0.25–0.55
Phosphorus	Less than 0.18
Sulfur	0.05

In the first-stage annealing, the white-iron casting is slowly reheated to a temperature between 1650 and 1750°F. During heating, the pearlite is converted to austenite at the lower critical line. The austenite thus formed dis-

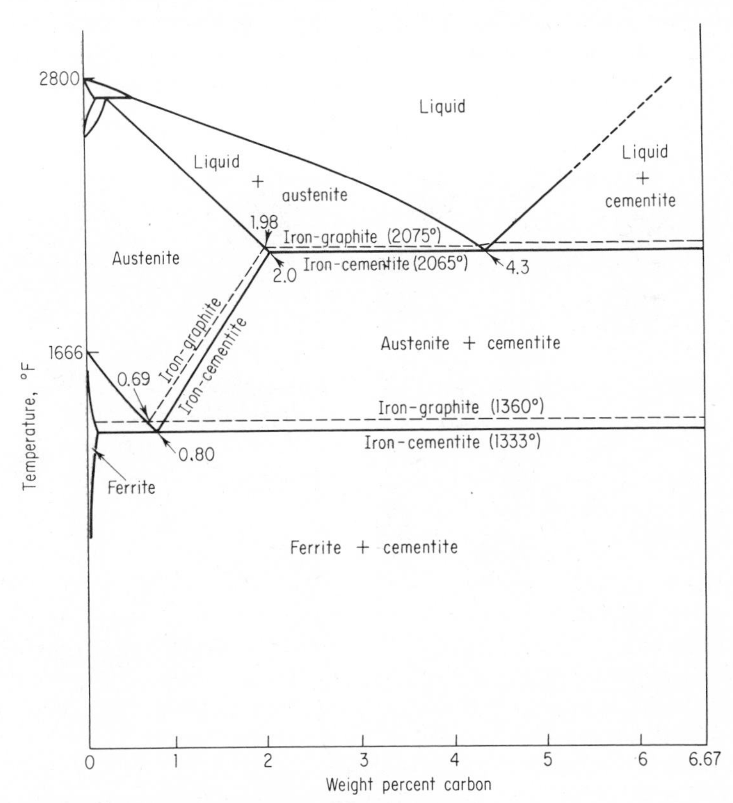

Fig. 11·3 The stable iron-graphite system (dotted lines) superimposed on the metastable iron–iron carbide system.

solves some additional cementite as it is heated to the annealing temperature.

Figure 11·3 shows that the austenite of the metastable system can dissolve more carbon than can austenite of the stable system. Therefore, a driving force exists for the carbon to precipitate out of the austenite as free graphite. This graphitization starts at the malleableizing temperature. The initial precipitation of a graphite nucleus depletes the austenite of carbon, and so more is dissolved from the adjacent cementite, leading to further carbon deposition on the original graphite nucleus. The graphite nuclei grow at approximately equal rates in all directions and ultimately appear as irregular nodules or spheroids usually called *temper carbon* (Fig. 11·4*a*). Temper carbon graphite is formed at the interface between primary carbide and saturated austenite at the first-stage annealing temperature, with growth around the nuclei by a reaction involving diffusion and carbide decomposition. Nucleation and graphitization are accelerated by the presence of submicroscopic particles that can be introduced into the iron by the proper melting practice. High silicon and carbon contents promote nucleation and graphitization, but these elements must be restricted to certain maximum levels since the iron must solidify as white iron. Therefore, graphitizing nuclei are best provided by proper annealing prac-

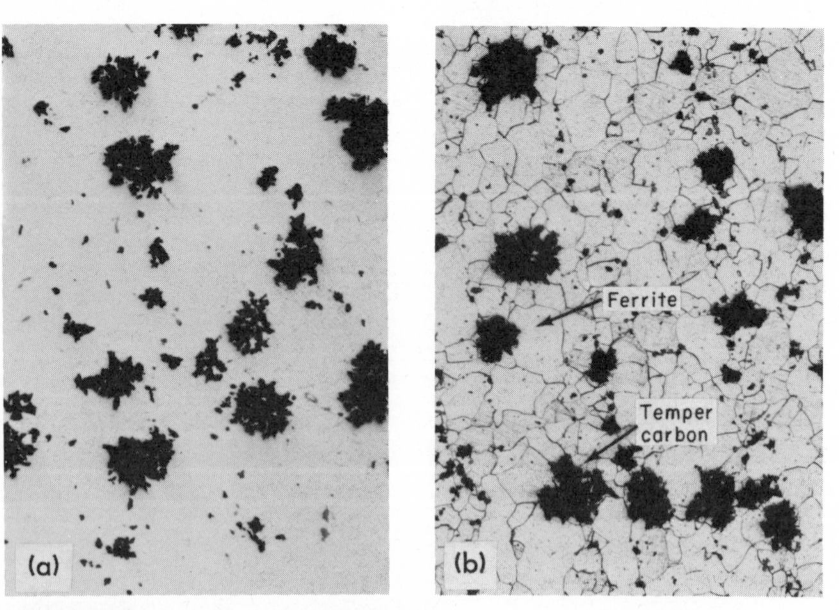

Fig. 11·4 (*a*) Malleable iron, unetched. Irregular nodules of graphite called *temper carbon*, 100×. (*b*) Ferritic malleable iron, temper carbon (black) in a ferrite matrix. Etched in 5 percent nital, 100×.

tice. The rate of annealing depends on chemical composition, nucleation tendency, and temperature of annealing. The temperature of first-stage annealing exerts considerable influence on the number of temper-carbon particles produced. Increasing annealing temperature accelerates the rate decomposition of primary carbide and produces more graphite particles per unit area. However, high first-stage annealing temperatures result in excessive distortion of castings during annealing and the need for straightening operations after heat treatment. Annealing temperatures are adjusted to provide maximum practical annealing rates and minimum distortion and are therefore controlled between 1650 and 1750°F. The white-iron casting is held at the first-stage annealing temperature until all massive carbides have been decomposed. Since graphitization is a relatively slow process, the casting must be soaked at temperature for at least 20 h, and large loads may require as much as 72 h. The structure at completion of first-stage graphitization consists of temper-carbon nodules distributed throughout the matrix of saturated austenite.

After first-stage annealing, the castings are cooled as rapidly as practical to about 1400°F in preparation for the second stage of the annealing treatment. The fast cooling cycle usually requires 2 to 6 h, depending on the equipment used.

In the second-stage annealing, the castings are cooled slowly at a rate of 5 to 15°F/h through the critical range at which the eutectoid reaction would take place. During the slow cooling, the carbon dissolved in the austenite is converted to graphite on the existing temper-carbon particles, and the remaining austenite transforms into ferrite. Once graphitization is complete, no further structural changes take place during cooling to room temperature, and the structure consists of temper-carbon nodules in a ferrite matrix (Fig. 11·4*b*). This type is known as *standard* or *ferritic malleable iron*. The changes in microstructure during the malleableizing cycle are shown schematically in Fig. 11·5.

In the form of compact nodules, the temper carbon does not break up the continuity of the tough ferritic matrix. This results in a higher strength and ductility than exhibited by gray cast iron. The graphite nodules also serve to lubricate cutting tools, which accounts for the very high machinability of malleable iron. Ferritic malleable iron has been widely used for automotive, agricultural, and railroad equipment; expansion joints and railing casting on bridges; chain-hoist assemblies, industrial casters; pipe fittings; and many applications in general hardware.

Alloyed malleable irons are those whose properties result from the addition of alloying elements not normally present in significant quantities in ferritic malleable iron. Since those alloyed malleable irons are completely malleableized, their influence is largely on the ferritic matrix. The two principal kinds are copper-alloyed malleable iron and copper-molybdenum-alloyed malleable iron. The effect of copper is to increase corrosion resist-

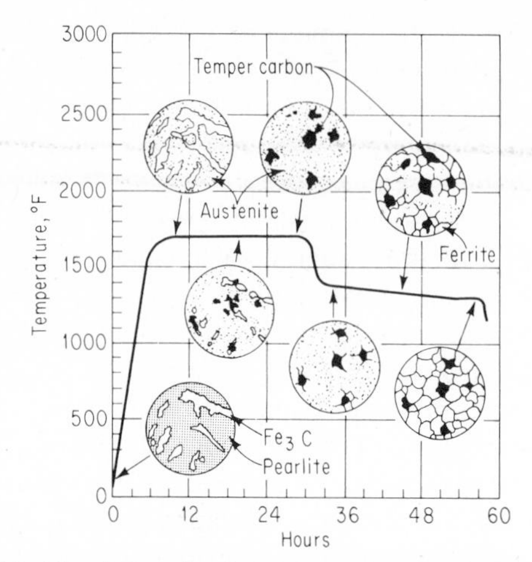

Fig. 11·5 The changes in microstructure as a function of the malleableizing cycle resulting in temper carbon in a ferrite matrix. (From "Malleable Iron Castings," Malleable Founders Society, Cleveland, Ohio, 1960.)

ance, tensile strength, and yield point at very slight reduction in ductility. Hardness is also increased, as shown in Fig. 11·6. The addition of copper and molybdenum in combination produces a malleable iron of superior corrosion resistance and mechanical properties. The mechanical properties of a copper-molybdenum-alloyed malleable iron are as follows:

Tensile strength, psi	58,000–65,000
Yield point, psi	40,000–45,000
Elongation, % in 2 in.	15–20
BHN	135–155

Compare these properties with those given for ferritic malleable iron in Table 11·1.

11·5 Pearlitic Malleable Iron If a controlled quantity of carbon, in the order of 0.3 to 0.9 percent, is retained as finely distributed iron carbide, an entirely different set of mechanical properties results. The strength and hardness of the castings will be increased over those of ferritic malleable iron by an amount which is roughly proportional to the quantity of combined carbon remaining in the finished product.

First-stage graphitization is a necessary prerequisite for all methods of manufacturing malleable-iron castings. If manganese is added, the regular cycle can be maintained to retain combined carbon throughout the matrix, or the second-stage annealing of the normal process may be replaced by a quench, usually air, which cools the castings through the eutectoid range fast enough to retain combined carbon throughout the matrix. The amount

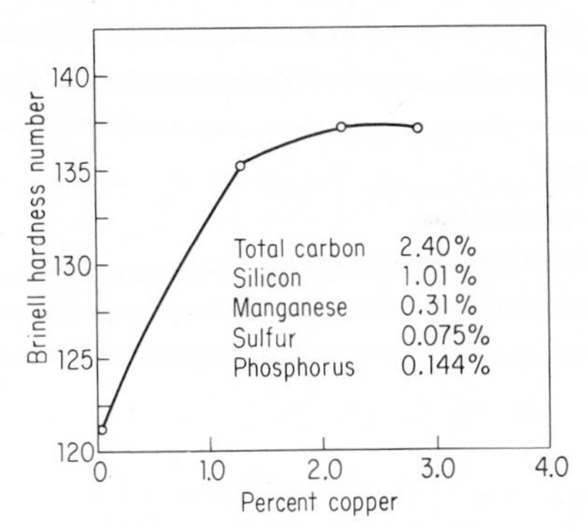

Fig. 11·6 Relationship between copper content and hardness for a malleable iron of the composition shown. (Malleable Founders Society.)

of pearlite formed depends upon the temperature at which the quench starts and the rate of cooling. High quench temperatures and fast cooling rates (air blast) result in greater amounts of retained carbon or pearlite. If the air quench produces a fast enough cooling rate through the eutectoid range, the matrix will be completely pearlitic (Fig. 11·7).

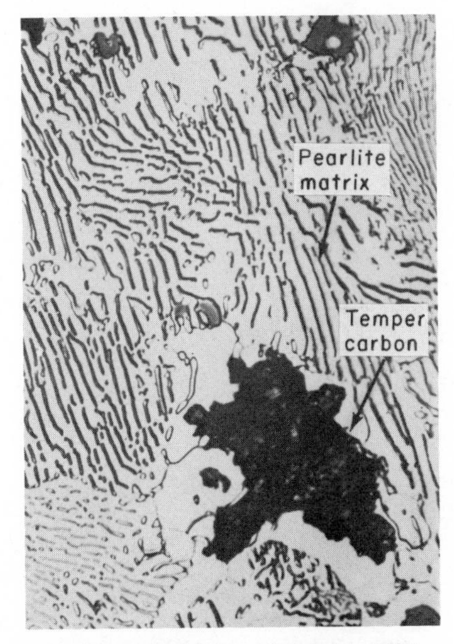

Fig. 11·7 Pearlitic malleable iron. Nital etch, 500×. (Malleable Founders Society.)

If the cooling rate through the critical range is not quite fast enough to retain all the combined carbon, the areas surrounding the temper-carbon nodules will be completely graphitized, while those at greater distance from the nodules will be pearlitic (Fig. 11·8). Because of its general appearance, this is referred to as a *bull's-eye* structure.

A fully ferritic malleable iron may be converted into pearlitic malleable iron by reheating above the lower critical temperature, followed by rapid cooling. The higher the temperature, the more carbon will be dissolved from the graphite nodules. Subsequent cooling will retain the combined carbon and develop the desired properties.

It is common practice to temper most pearlitic malleable irons after air cooling. Those having coarse pearlitic structures are tempered at relatively high temperatures (between 1200 and 1300°F) to spheroidize the pearlite (Fig. 11·9), improve machinability and toughness, and lower the hardness.

If it is desired to increase the mechanical properties of the matrix, it is necessary to reheat for 15 to 30 min at 1550 to 1600°F to re-austenitize and homogenize the matrix material. The castings are then quenched in heated and agitated oil, which develops a matrix of martensite and bainite with a hardness of Rockwell C 55 to 60. The amount of martensite formed will

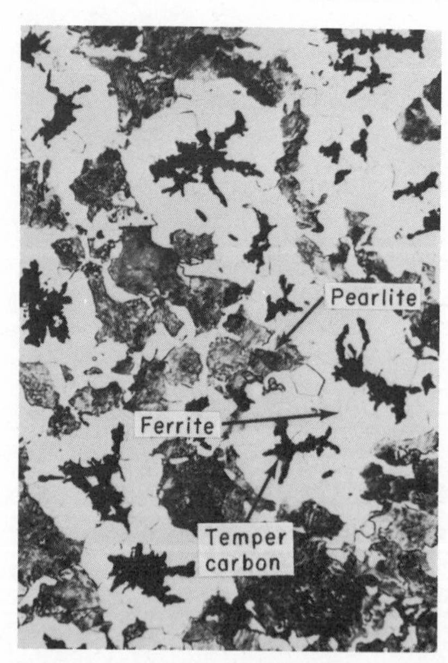

Fig. 11·8 Typical appearance of a "bull's-eye" structure. Temper-carbon nodules surrounded by ferritic areas (white), with lamellar pearlite (dark) located between the bull's eyes. Nital etch, 100×. (Malleable Founders Society.)

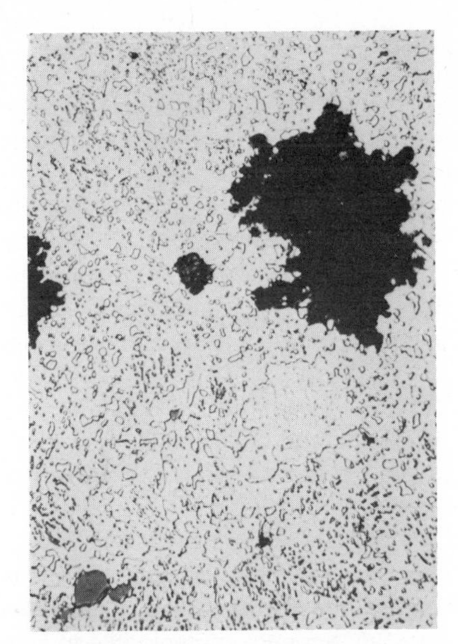

Fig. 11·9 Microstructure of a pearlitic malleable iron tempered to obtain a spheroidite matrix. Nital etch, 500×. (Malleable Founders Society.)

depend upon the quenching speed of the particular oil used, the temperature from which the work is quenched, the time at that temperature, the thickness of the casting, and the chemistry of the iron. The martensitic malleable iron is then tempered between 450 and 1320°F to develop the desired properties. The matrix microstructure consists of various types of tempered martensite, depending on the final hardness of the castings.

Welding of pearlitic malleable iron is seldom recommended because of the formation of a brittle, low-strength white-iron layer under the weld bead, caused by the melting and rapid freezing of the malleable iron. The pearlitic structure adjacent to the white iron in the welding zone also is altered through the redissolving of some temper carbon.

The tensile properties of ferritic and pearlitic malleable iron are given in Table 11·1.

TABLE 11·1 Tensile Properties of Malleable Cast Iron

TYPE	TENSILE STRENGTH, 1,000 PSI	YIELD STRENGTH, 1,000 PSI	ELONGATION, % IN 2 IN.	BHN
Ferritic	50–60	32–39	20–10	110–145
Pearlitic	65–120	45–100	16–2	163–269

Alloyed pearlitic malleable castings are made from white irons that contain one or more alloying elements so that the regular malleableizing anneal will not result in a ferritic matrix. The alloy additions usually do not affect first-stage graphitization but serve as carbide stabilizers during the eutectoid range or subeutectoid tempering treatments. Many of the alloying elements also increase hardenability and strengthen the matrix. Manganese and sulfur may be added in quantities not normally found in standard malleable iron. Copper may be added to improve strength, corrosion resistance, and graphite distribution. Suitably alloyed pearlitic malleable iron may be fully martensitic in sections as heavy as 2 in. after air quenching from 1600°F.

Some of the industrial applications of pearlitic malleable iron are for axle and differential housings, camshafts and crankshafts in automobiles; for gears, chain links, sprockets, and elevator brackets in conveyor equipment; for rolls, pumps, nozzles, cams, and rocker arms as machine parts; for gun mounts, tank parts, and pistol parts in ordinance; and finally for a variety of small tools such as wrenches, hammers, clamps, and shears.

11·6 Gray Cast Iron This group is one of the most widely used alloys of iron. In the manufacture of gray cast irons, the tendency of cementite to separate into graphite and austenite or ferrite is favored by controlling alloy composition and cooling rates. Most gray cast irons are hypoeutectic alloys containing between 2.5 and 4 percent carbon.

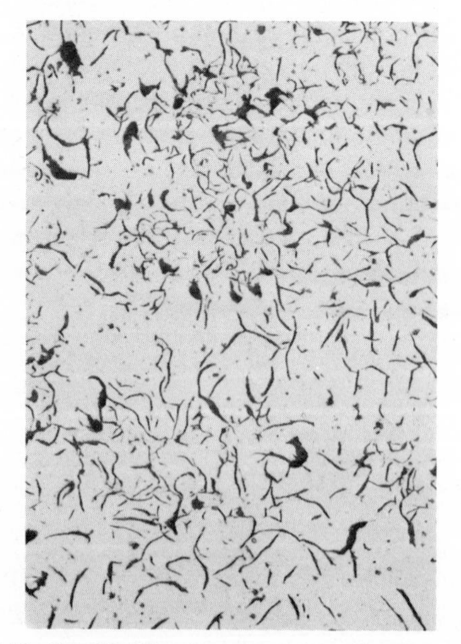

Fig. 11·10 Graphite flakes in gray cast iron. Unetched, 100×.

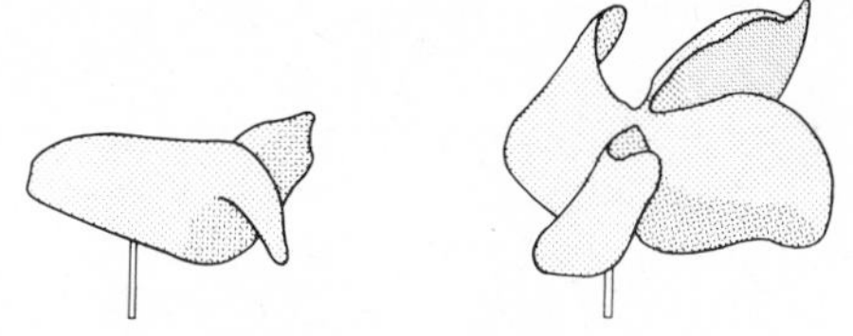

Fig. 11·11 Space models of flake graphite. (After MacKenzie.)

These alloys solidify by first forming primary austenite. The initial appearance of combined carbon is in the cementite resulting from the eutectic reaction at 2065°F. The graphitization process is aided by high carbon content, high temperature, and the proper amount of graphitizing elements, notably silicon.

There is experimental evidence that, with proper control of the above factors, the alloy will follow the stable iron-graphite equilibrium diagram (Fig. 11·3), forming austenite and graphite at the eutectic temperature of 2075°F. At any rate, any cementite which is formed will graphitize rapidly. The graphite appears as many irregular, generally elongated and curved plates which give gray cast iron its characteristic grayish or blackish fracture (Fig. 11·10). It should be emphasized that while the microstructure represents their appearance on a plane surface, the flakes are three-dimensional particles. They are, in effect, curved plates sometimes connected, and may be represented by the space models shown in Fig. 11·11.

During continued cooling, there is additional precipitation of carbon because of the decrease in solubility of carbon in austenite. This carbon is precipitated as graphite or as proeutectoid cementite which promptly graphitizes.

The strength of gray cast iron depends almost entirely on the matrix in which the graphite is embedded. This matrix is largely determined by the condition of the eutectoid cementite. If the composition and cooling rate are such that the eutectoid cementite also graphitizes, then the matrix will be entirely ferritic. On the other hand, if graphitization of the eutectoid cementite is prevented, the matrix will be entirely pearlitic. The constitution of the matrix may be varied from pearlite, through mixtures of pearlite and ferrite in different proportions, down to practically pure ferrite. The graphite-ferrite mixture is the softest and weakest gray iron; the strength and hardness increase with the increase in combined carbon, reaching a maximum with the pearlitic gray iron. Figure 11·12 shows the microstructure of gray cast iron with the matrix almost entirely pearlitic.

11·7 Silicon in Cast Iron Silicon is a very important element in the metallurgy of gray iron. It increases fluidity and influences the solidification of the molten alloy. The eutectic composition is shifted to the left approximately

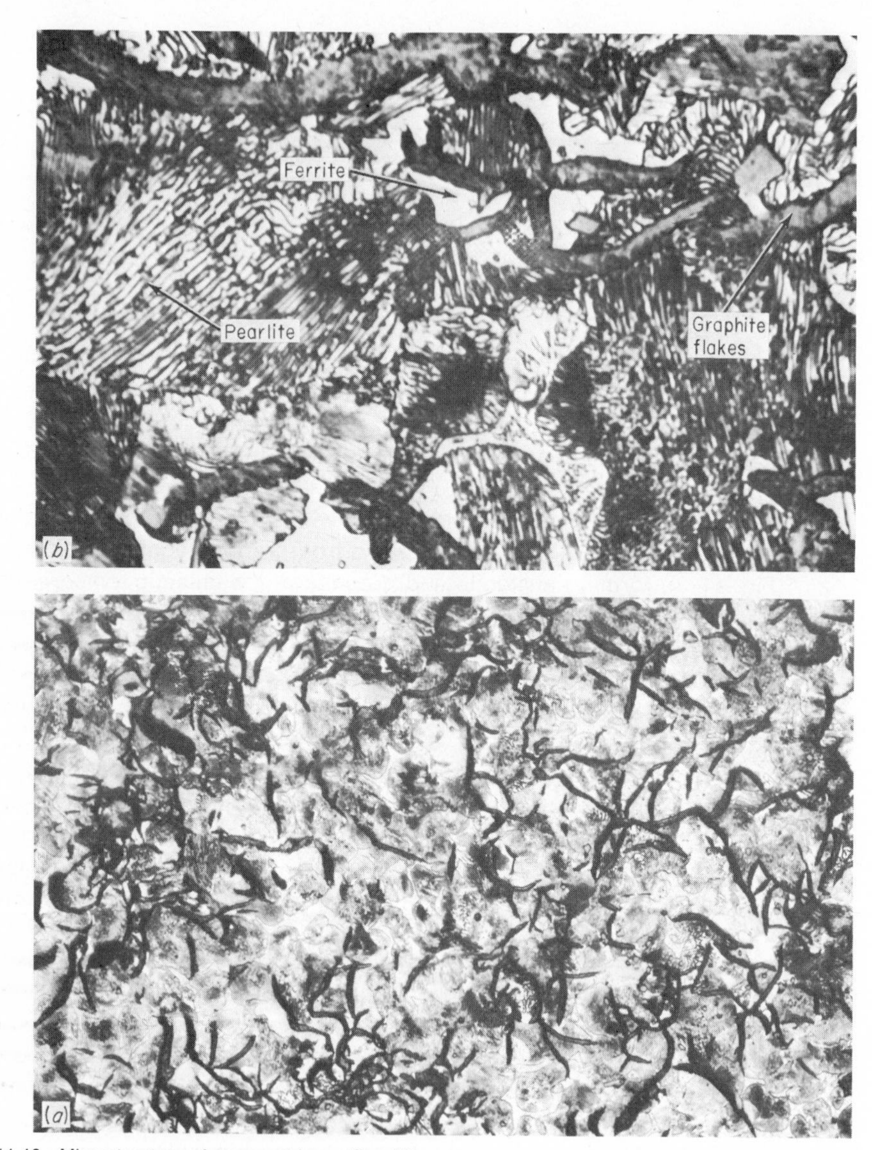

Fig. 11·12 Microstructure of gray cast iron. Graphite flakes in a pearlitic matrix with a small amount of ferrite (white areas). Etched in 2 percent nital. (*a*) 100×; (*b*) 500×.

0.30 percent carbon for each 1 percent silicon, which effectively depresses the temperature at which the alloy begins to solidify. As the silicon content is increased, the austenite field decreases in area, the eutectoid carbon content is lowered, and the eutectoid transformation occurs over a broadening range.

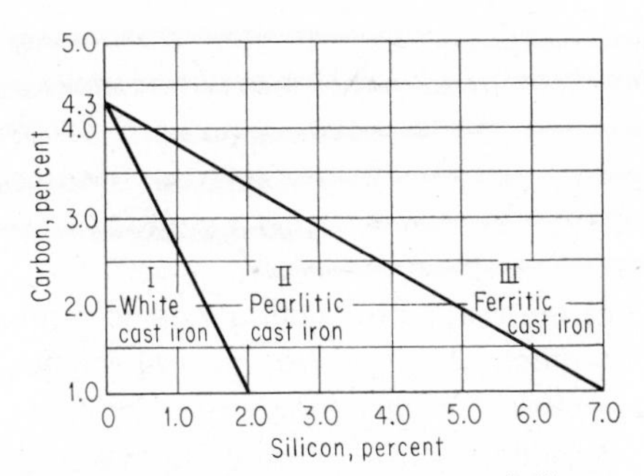

Fig. 11·13 Relation of structure to carbon and silicon content of cast iron. (After Maurer.)

Silicon is a graphitizer, and if not counterbalanced by carbide-promoting elements, it favors solidification according to the stable iron-graphite system. Therefore, during solidification in the presence of silicon, carbon is precipitated as primary graphite in the form of flakes. Once primary graphite has formed, its shape cannot be altered by any method. It is these weak graphite flakes that break up the continuity of the matrix and the notch effect at the end of these flakes that accounts for the low strength and low ductility of gray iron.

The relation of the carbon and silicon content to the structure of thin sections of cast iron is shown in Fig. 11·13. In region I, cementite is stable, so the structure will be white cast iron. In region II, there is enough silicon to cause graphitization of all the iron carbide except the eutectoid cementite. The microstructure will consist of graphite flakes in a matrix that is largely pearlite, as in Fig. 11·12. In region III, the large amount of silicon causes complete dissociation of the cementite to graphite and ferrite. This will result in a ferritic gray cast iron of very low strength.

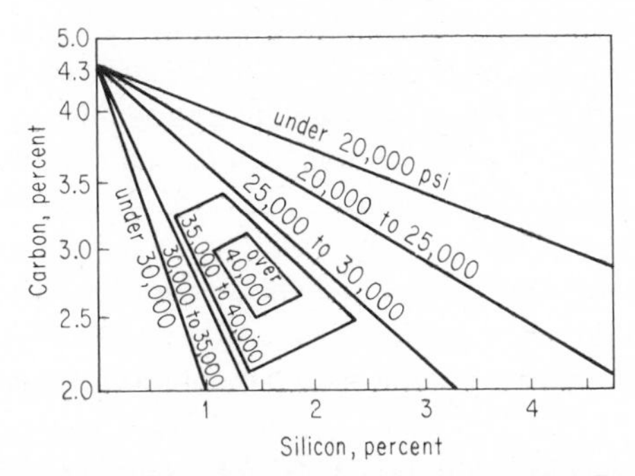

Fig. 11·14 Relation of tensile strength to carbon and silicon content of cast iron. (Coyle, *Trans. ASM*, vol. 12, p. 446, 1927.)

The influence of carbon and silicon content on the tensile strength is shown in Fig. 11·14. The highest tensile strength is obtained with a carbon content of 2.75 percent and a silicon content of 1.5 percent. These percentages in Fig. 11·13 will result in a pearlitic gray cast iron. If the percentages of carbon and silicon are such as to yield either a white cast iron or a ferritic gray cast iron, the tensile strength will be low.

Careful control of the silicon content and cooling rate is required to graphitize the eutectic and proeutectoid cementite, but not the eutectoid cementite, in order to end up with a pearlitic gray iron of high strength.

11·8 Sulfur in Cast Iron Most commercial gray irons contain between 0.06 and 0.12 percent sulfur. The effect of sulfur on the form of carbon is the reverse of silicon. The higher the sulfur content, the greater will be the amount of combined carbon, thus tending to produce a hard, brittle white iron.

Aside from producing combined carbon, sulfur tends to react with iron to form iron sulfide (FeS). This low-melting compound, present as thin interdendritic layers, increases the possibility of cracking at elevated temperatures (*red-short*). High sulfur tends to reduce fluidity and often is responsible for the presence of blowholes (trapped air) in castings.

Fortunately, manganese has a greater affinity for sulfur than iron, forming manganese sulfide (MnS). The manganese sulfide particles appear as small, widely dispersed inclusions which do not impair the properties of the casting. It is common commercial practice to use a manganese content of two to three times the sulfur content.

11·9 Manganese in Cast Iron Manganese is a carbide stabilizer, tending to increase the amount of combined carbon, but it is much less potent than sulfur. If manganese is present in the correct amount to form manganese sulfide, its effect is to reduce the proportion of combined carbon by removing the effect of sulfur. Excess manganese has little effect on solidification and only weakly retards primary graphitization. On eutectoid graphitization, however, manganese is strongly carbide-stabilizing.

11·10 Phosphorus in Cast Iron Most gray irons contain between 0.10 and 0.90 percent phosphorus originating from the iron ore. Most of the phosphorus combines with the iron to form iron phosphide (Fe_3P). This iron phosphide forms a ternary eutectic with cementite and austenite (pearlite at room temperature). The ternary eutectic is known as *steadite* and is a normal feature in the microstructure of cast irons (Fig. 11·15). Steadite is relatively brittle, and with high phosphorus content, the steadite areas tend to form a continuous network outlining the primary austenite dendrites. The condition reduces toughness and makes the cast iron brittle, so that the phosphorus content must be carefully controlled to obtain optimum mechanical properties.

Phosphorus increases fluidity and extends the range of eutectic freezing, thus increasing primary graphitization when the silicon content is high

Fig. 11·15 Gray iron showing steadite areas (arrow). Etched in 2 percent nital, 500×. (The International Nickel Company.)

and phosphorus content is low. It is therefore useful in very thin castings where a less fluid iron may not take a perfect impression of the mold.

If the silicon, sulfur, manganese, and phosphorus contents are controlled at proper levels, the only remaining variable affecting the strength of a pearlitic gray iron is the graphite flakes. Since graphite is extremely soft and weak, its size, shape, and distribution will determine the mechanical properties of the cast iron. It is the reduction of the size of the graphite flakes and the increase in their distribution that have accounted for the improvement in the quality of gray cast iron.

11·11 Heat Treatment of Gray Iron Stress relieving is probably the most frequently applied heat treatment for gray irons. Gray iron in the as-cast condition usually contains residual stresses because cooling proceeds at different rates throughout various sections of a casting. The resultant residual stresses may reduce strength, cause distortion, and in some extreme cases even result in cracking. The temperature of stress relieving is usually well below the transformation range of pearlite to austenite. For maximum relief of stress with minimum decomposition of carbide, a temperature range of 1000 to 1050°F is desirable. Figure 11·16 indicates that from 75 to 85

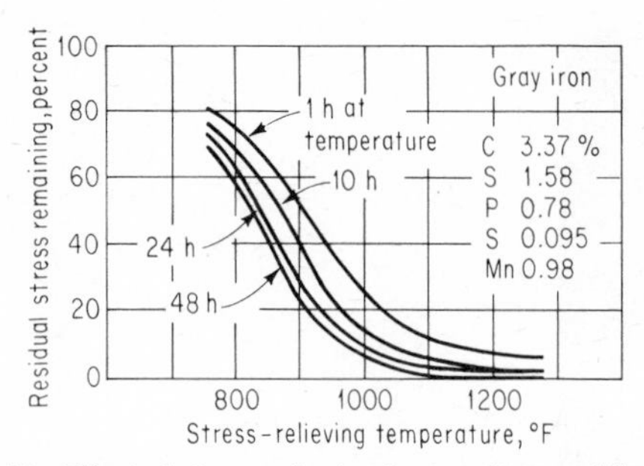

Fig. 11·16 Effect of stress-relieving temperature and time on residual stress in gray iron. (After G. N. J. Gilbert, from "Metals Handbook," vol. 2, American Society for Metals, Metals Park, Ohio, 1964.)

percent of the residual stress can be removed on holding for 1 h in this range. When almost complete stress relief (over 85 percent) is required, a minimum temperature of 1100°F can be employed.

Annealing of gray iron consists of heating it to a temperature that is high enough to soften it and thus improve the machinability. For most gray irons an annealing temperature between 1300 and 1400°F is recommended. Up to approximately 1100°F, the effect of temperature on the structure of gray iron is insignificant. As the temperature increases above 1100°F, the rate at which iron carbide decomposes to ferrite plus graphite increases markedly, reaching a maximum at about 1400°F for unalloyed or low-alloy iron. The casting must be held at temperature long enough to allow the graphitizing process to go to completion. At temperatures below 1300°F, an excessively long holding time is usually required.

Gray iron is normalized by being treated to a temperature above the transformation range, held at this temperature for a period of about 1 h/in. of maximum section thickness, and cooled in still air to room temperature. The temperature range for normalizing gray iron is approximately 1625 to 1700°F. Normalizing may be used to enhance mechanical properties, such as hardness and tensile strength, or to restore as-cast properties that have been modified by another heat treating process, such as graphitizing or the preheating and postheating associated with repair welding.

Gray iron, like steel, can be hardened when cooled rapidly or quenched from a suitable elevated temperature. The quenched iron may be tempered by reheating in the range from 300 to 1200°F to increase toughness and relieve stresses. Ordinarily gray iron is furnace-hardened from a temperature of 1575 to 1600°F. The quenching medium may be water, oil, hot salt, or air, depending on the composition and section size. Oil is the usual quenching medium for through-hardening. Quenching in water may be too drastic and may cause cracking and distortion unless the castings are massive and uniform in cross section. Water is often used for quenching with flame or induction hardening where only the outer surface is hardened. As-quenched gray iron is brittle. Tempering after quenching improves strength and toughness but decreases hardness. A temperature of about 700°F is required before the impact strength is restored to the as-cast level. After tempering at 700°F for maximum toughness, the hardness of the matrix is still about Rockwell C 50. Where toughness is not required and a tempering temperature of 300 to 500°F is acceptable, the matrix hardness is equivalent to Rockwell C 55 to 60. Heat treatment is not ordinarily used commercially to increase the strength of gray-iron castings, because the strength of the as-cast metal can be increased at less cost by reducing the silicon and total carbon contents or by adding alloying elements. Gray iron is usually quenched and tempered to increase the resistance to wear and abrasion by increasing the hardness. A structure consisting of graphite embedded in a hard martensitic matrix is produced by heat treatment. This process can replace the white-iron surface usually produced by chilling. It can be applied where chilling is not feasible, as with complicated shapes or large castings. The combination of high matrix hardness and graphite as a lubricant results in a surface with good wear resistance for some applications such as farm implement gears, sprockets, diesel cylinder liners, and automotive camshafts. Thus, heat treatment extends the field of application of gray iron as an engineering material.

11·12 Size and Distribution of Graphite Flakes Large graphite flakes seriously interrupt the continuity of the pearlitic matrix, thereby reducing the strength and ductility of the gray iron. Small graphite flakes are less damaging and are therefore generally preferred.

Graphite-flake sizes are usually determined by comparison with standard sizes prepared jointly by the AFS (American Foundrymen's Society) and the ASTM (American Society for Testing Materials). The procedure for preparation and measurement of flake size is given in ASTM Designation A247-67, 1971 Book of ASTM Standards, Part 31. The measurement is made of the lengths of the largest graphite flakes in a unetched section of the gray iron at 100×. Numbers are assigned as indicated in Table 11-2.

TABLE 11·2 Graphite Flake Sizes

AFS-ASTM FLAKE SIZE NUMBER	LENGTH OF LONGEST FLAKES AT 100× IN.	MM
1	4 or more	128
2	2–4	64
3	1–2	32
4	½–1	16
5	¼–½	8
6	⅛–¼	4
7	1/16–⅛	2
8	1/16 or less	1

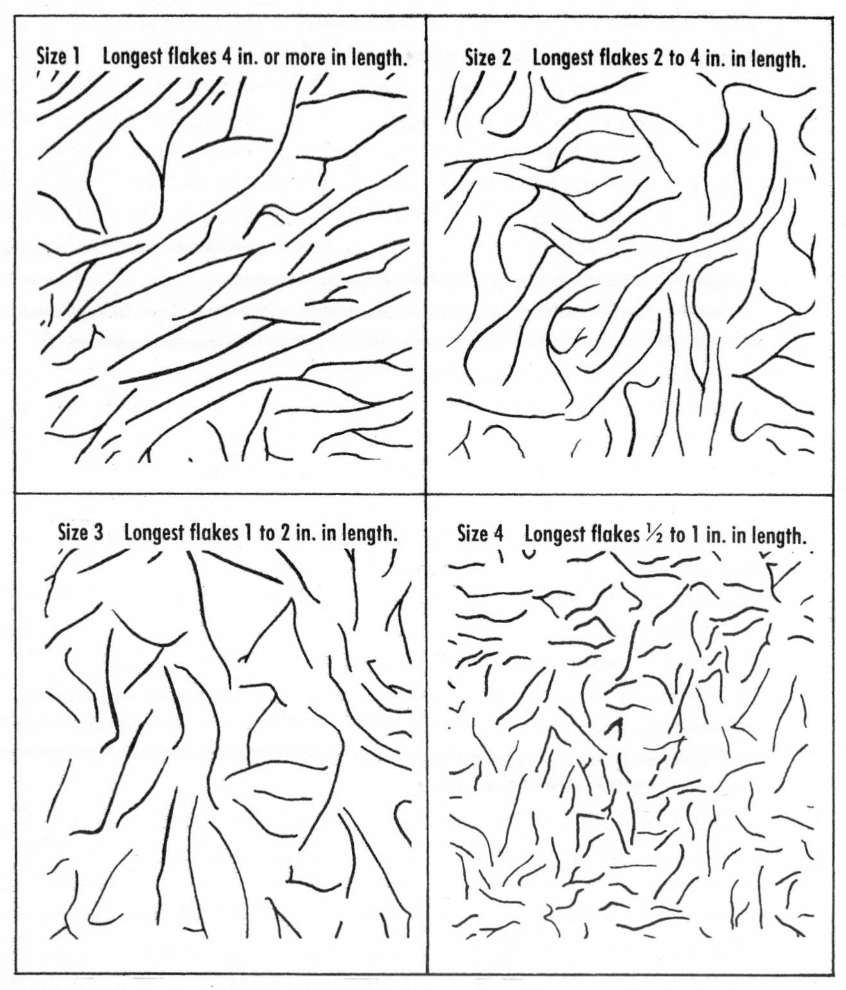

Fig. 11·17 Graphite-flake size chart illustrated by typical fields showing as nearly as possible the sizes represented. (Prepared jointly by ASTM and AFS.)

The flake lengths are illustrated in Fig. 11·17 by typical fields showing as nearly as possible the various sizes.

Slow cooling of hypoeutectic irons to favor graphitization also produces large crystals of primary austenite. This restricts the eutectic mixture or graphite to the grain boundaries and results in graphite flakes that are relatively few in number and coarse.

Increasing the carbon content to increase the amount of eutectic also increases the amount of graphite formed. This may weaken the cast iron more than a smaller flake size can strengthen it.

Increasing the silicon content increases the amount of eutectic formed,

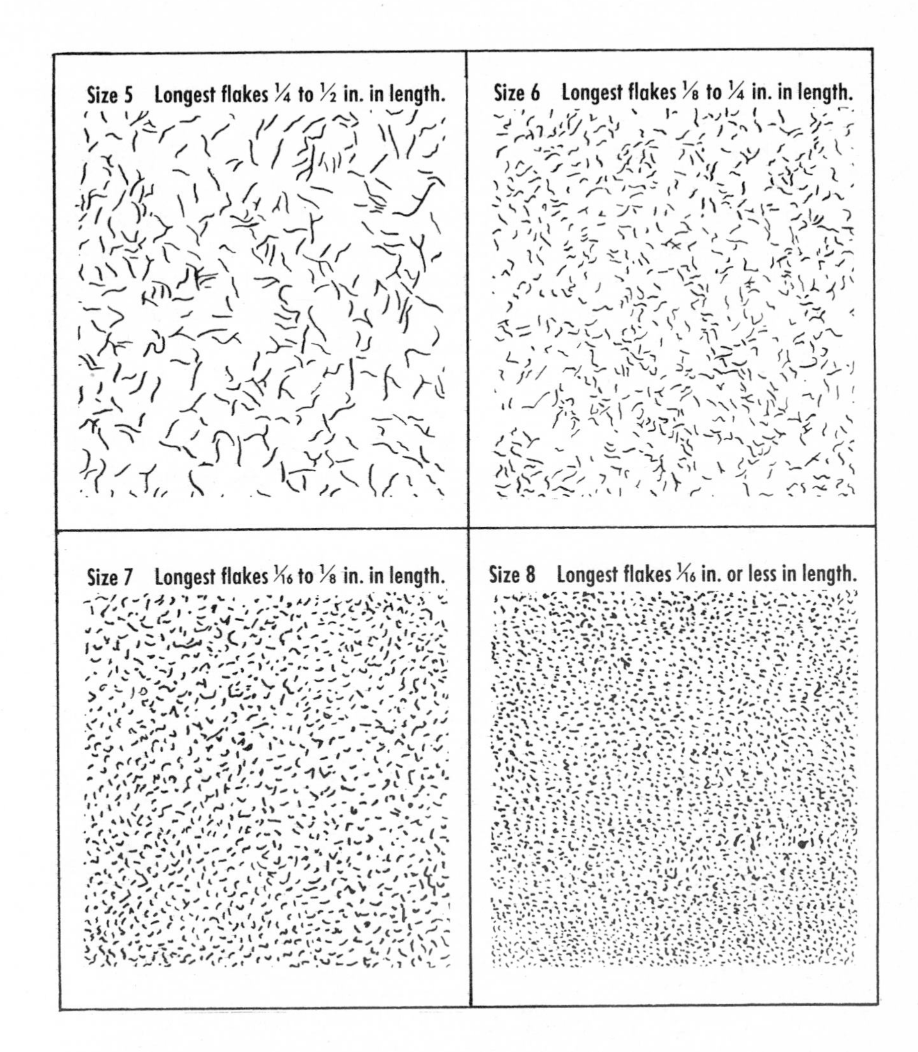

thus reducing flake size. However, since high silicon content has such a strong graphitizing influence, the matrix will probably be ferritic, resulting in a weak casting.

The best method of reducing the size and improving the distribution of the graphite flakes seems to be by the addition of a small amount of material known as an *inoculant*. Inoculating agents that have been used successfully are metallic calcium, aluminum, titanium, zirconium, silicon carbide, calcium silicide, or combinations of these. The exact mechanism by which they operate is not clearly understood. They probably promote the nucleation of primary austenite, resulting in small grains, which reduces the size and improves the distribution of the graphite flakes.

The way in which the graphite flakes are arranged in the microstructure of gray cast iron is usually indicated as one or more types that have been jointly prepared by the AFS and the ASTM. The five flake types are shown in Fig. 11·18.

Type D and type E flake patterns usually result from the graphitization of a normal eutectic structure. These types appear in irons of very high purity or in commercial irons that have been cooled rather rapidly during solidification. Although the graphitic flake size is small, the interdendritic pattern and high graphite content weaken the material. Therefore, types D and E flake patterns are undesirable in gray irons. When the cooling rate is slower, most commercial gray irons show complete divorcement of the eutectic, so that types D and E flake patterns do not occur.

The most desirable flake pattern in gray iron is represented by the uniform distribution and random orientation of type A. This results from a completely divorced eutectic structure. As was mentioned earlier, the size of the individual graphite flakes is determined by the size of the austenite crystals around which they form.

The rosette pattern of type B graphite flakes is common only in the intermediate region of a chilled cast iron. This region is known as the *mottled* region and consists of a mixture of gray and white cast iron (Fig.

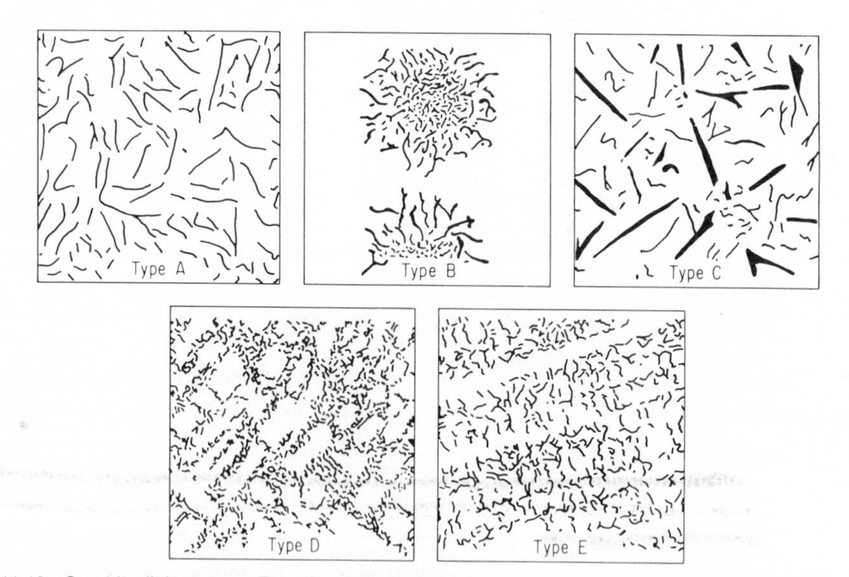

Fig. 11·18 Graphite-flake types. Type A—uniform distribution, random orientation; type B—rosette groupings, random orientation; type C—superimposed flake sizes, random orientation; type D—interdendritic segregation, random orientation; type E—interdendritic segregation, preferred orientation. (Prepared jointly by ASTM and AFS.)

11·20). The cooling rate in this region is the maximum that would permit graphitization.

The few large, straight graphite flakes present in type C always indicate that the iron is hypereutectic in carbon content. Silicon and several other alloying elements reduce the carbon content of the eutectic, and if they are present in sufficient amounts the eutectic composition may be reduced to below 3.5 percent carbon.

11·13 Mechanical Properties and Applications of Gray Cast Iron The most important classification of gray irons, from an engineering standpoint, is that employed in the ASTM Specification A48. The gray-iron castings are classed in seven classes (Nos. 20, 25, 30, 35, 40, 50, 60) which give the minimum tensile strength of test bars in thousands of pounds per square inch. For example, class 20 gray iron would have a minimum tensile strength of 20,000 psi; class 30, 30,000 psi; and so on. Table 11·3 gives typical mechanical properties of standard gray-iron test bars, as cast.

Tensile strength is important in selecting a gray iron for parts that are subjected to static loads in indirect tension or bending. Such parts include pressure vessels, housings, valves, fittings, and levers. Irons above 40,000 psi in tensile strength are usually considered *high-strength irons* and are somewhat more expensive to produce and more difficult to machine. Gray irons do not exhibit a well-defined yield point as do most mild steels. The stress-strain curve does not show a straight-line portion; thus a definite modulus of elasticity cannot be determined. Usual methods are to determine the "relative" modulus at 25 percent of the expected tensile strength, or the "tangent" modulus by drawing a tangent at some given stress value. The percent elongation is small for all cast irons, rarely exceeding 3 to 4 percent, and the reduction of area is too slight to be appreciable.

Compressive strength is important when the gray iron is used for machinery foundations or supports. Like all brittle materials, the compressive strength of gray iron is much greater than its tensile strength and is largely a function of the shearing strength. Failure in compression usually occurs along an oblique plane unless the specimen is long enough to allow failure by buckling.

Many grades of gray iron have higher torsional shear strength than some grades of steel. This characteristic, along with low notch sensitivity, makes gray iron a suitable material for various types of shafting.

The hardness of gray iron is an average result of the soft graphite in the iron and the metallic matrix. Variation in graphite size and distribution will cause wide variations in hardness (particularly Rockwell hardness). The Brinell tester, covering a larger area, tends to give a more accurate hardness value than the Rockwell tester.

Figure 11·19 shows the general correlation between Brinell hardness

TABLE 11·3 Typical Mechanical Properties of Standard Gray-iron Test Bars, As Cast*

ASTM CLASS	TENSILE STRENGTH, PSI	COMPRESSIVE STRENGTH, PSI	TORSIONAL SHEAR STRENGTH, PSI	MODULUS OF ELASTICITY, MILLION PSI		REVERSED BENDING FATIGUE LIMIT, PSI	TRANSVERSE STRENGTH OF 1.2-IN.-DIAM BAR 18-IN. SPAN, LB	BHN
				TENSION	TORSION			
20	22,000	83,000	26,000	9.6–14.0	3.9–5.6	10,000	1,850	156
25	26,000	97,000	32,000	11.5–14.8	4.6–6.0	11,500	2,175	174
30	31,000	109,000	40,000	13.0–16.4	5.2–6.6	14,000	2,525	201
35	36,500	124,000	48,500	14.5–17.2	5.8–6.9	16,000	2,850	212
40	42,500	140,000	57,000	16.0–20.0	6.4–7.8	18,500	3,175	235
50	52,500	164,000	73,000	18.8–22.8	7.2–8.0	21,500	3,600	262
60	62,500	187,500	88,500	20.4–23.5	7.8–8.5	24,500	3,700	302

*By permission from "Metals Handbook," 8th ed., American Society for Metals, Metals Park, Ohio, 1961.

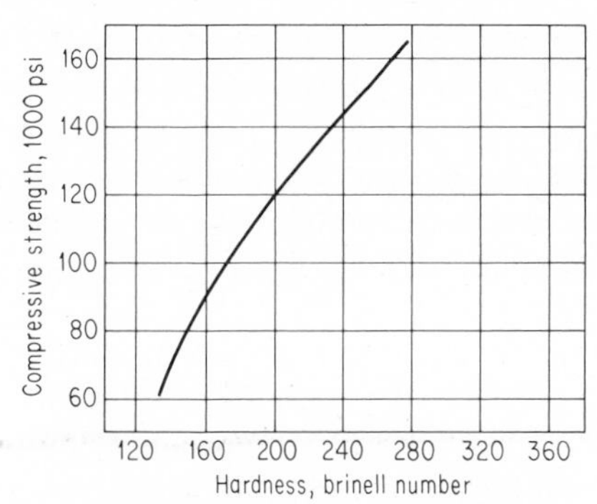

Fig. 11·19 General correlation between Brinell hardness and compressive strength. (After Donoho.)

and compressive strength. This correlation is much better than that between tensile strength and hardness, because the compressive strength steadily increases with increasing hardness and is not greatly influenced by microstructural variations as is the tensile strength.

The microstructure is the primary factor in determining the hardness of gray iron. Table 11·4 indicates the wide range of Brinell hardness numbers obtainable with various iron structures. The composition also exerts a considerable effect on the hardness. Increasing carbon and silicon contents will result in decreasing hardness, although the effect is not as marked on hardness as it is on tensile strength.

Because gray iron is the least expensive type of casting, it should always be considered first when a cast metal is being selected. Another metal should be chosen only when the mechanical and physical properties of gray iron are inadequate. Examples of applications requiring a bare mini-

TABLE 11·4 The Brinell Hardness of Iron Castings Classed by General Microstructure

TYPE	BHN
Ferritic (annealed) gray iron	110–140
Austenitic irons	140–160
Soft gray iron	140–180
Pearlitic irons	160–220
Pearlitic alloy iron of low alloy content	200–250
Tempered martensitic irons	260–350
Martensitic irons	350–450
White iron, unalloyed (according to carbon content)	280–500
Alloy white iron	450–550
Martensitic white iron	550–700
Nitrided iron (surface only)	900–1,000

mum of casting properties and lowest possible cost are counterweights for elevators and industrial furnace doors. Gray iron is widely used also for guards and frames around hazardous machinery. Many types of gear housings, enclosures for electrical equipment, pump housings, and steam turbine housings are cast in gray iron because of its low cost. Other similar gray-iron castings are used for motor frames, fire hydrants, and sewer covers.

11·14 Chilled Cast Iron Chilled-iron castings are made by casting the molten metal against a metal chiller, resulting in a surface of white cast iron. This hard, abrasion-resistant white-iron surface or case is backed up by a softer gray-iron core. This case-core structure is obtained by careful control of the overall alloy composition and adjustment of the cooling rate.

Freezing starts first, and the cooling rate is most rapid where the molten metal is in contact with the mold walls. The cooling rate decreases as the center of the casting is approached. A chilled-iron casting may be produced by adjusting the composition of the iron so that the normal cooling rate at the surface is just fast enough to produce white iron while the slower cooling rate below the surface will produce mottled or gray iron (Fig. 11·20).

If only selected surfaces are to be white iron, it is common practice to use a composition which would normally solidify as gray iron and employ metal liners (chills) to accelerate the cooling rate of the selected areas. The depth of the white-iron layer is controlled by using thin metal plates whenever a thin white-iron layer is desired and heavier metal plates where a deeper chill is necessary.

The depth of chill decreases and the hardness of the chilled zone increases with increasing carbon content. Since silicon is a graphitizer, the depth of chill is decreased with increasing silicon content.

The addition of manganese decreases the depth of chill until the sulfur has been neutralized by formation of manganese sulfide. Above this amount, manganese increases chill depth and hardness.

Phosphorus decreases the depth of chill. With carbon and silicon constant, an increase of 0.1 percent phosphorus will decrease the depth of chill about 0.1 in.

Nickel reduces the chill depth, and its influence is about one-fourth that of silicon. The reduction in chill depth is accompanied by a gradual increase in hardness until the nickel content reaches about 5 percent. Nickel also refines the carbide structure of the chill and the gray-iron structure below the chill.

Chromium is used in small amounts to control chill depth; 0.01 percent chromium will neutralize about 0.015 percent silicon. Because of the formation of chromium carbides, chromium is used in amounts of 1 to 4 percent in chilled irons to increase hardness and improve abrasion resistance.

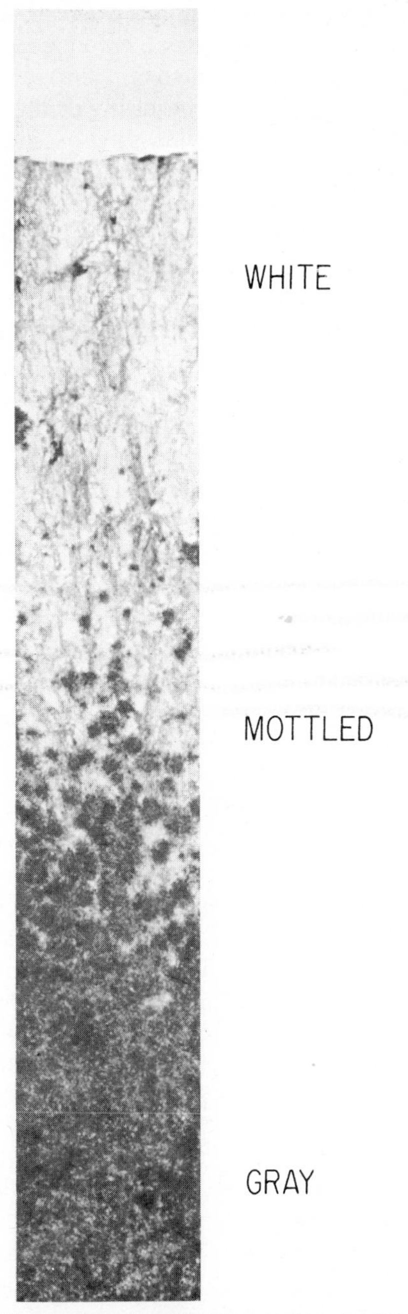

Fig. 11·20 Fracture of a chilled-iron casting showing the white, mottled, and gray portions, 3×.

It also stabilizes carbide and suppresses the formation of graphite in heavy sections. When added in amounts of 12 to 35 percent, chromium will impart resistance to corrosion and oxidation at elevated temperatures.

Copper, in additions of less than 4 percent, decreases the depth of chill, but in excess of this amount the chill depth and hardness increase. Copper also reduces the ratio of the mottled portion to the white-iron portion.

Molybdenum is only about one-third as effective as chromium in increasing the chill depth; however, it improves the resistance of the chilled face to spalling, pitting, chipping, and heat checking.

A constant chill depth may be obtained by using a combination of alloying elements that have opposite effects. Since nickel reduces chill depth, it is common practice to add chromium, which increases chill depth, to neutralize the nickel and result in a constant chill depth. The normal ratio employed for this purpose is 3 parts of nickel to 1 of chromium.

Chilled-iron casting is used for railway-car wheels, crushing rolls, stamp shoes and dies, sprockets, plowshares, and many other heavy-duty machinery parts. Table 11·5 gives the composition and hardness of typical chilled-iron castings.

11·15 Nodular Cast Iron Nodular cast iron, also known as *ductile iron, spheroidal graphite iron,* and *spherulitic iron,* is cast iron in which the graphite is present as tiny balls or spheroids. The compact spheroids interrupt the continuity of the matrix much less than graphite flakes; this results in higher strength and toughness compared with a similar structure of gray iron. Nodular cast iron differs from malleable iron in that it is usually obtained as a result of solidification and does not require heat treatment. The spheroids are more rounded than the irregular aggregates of temper carbon found in malleable iron (Fig. 11·21).

The total carbon content of nodular iron is the same as in gray cast iron. Spheroidal graphite particles form during solidification because of the presence of a small amount of certain alloying elements. The nodule-forming addition, usually magnesium or cerium, is made to the ladle just before casting. Since these elements have a strong affinity for sulfur, the base iron-alloy sulfur content must be below 0.015 percent for the treat-

TABLE 11·5 Composition and Hardness of Typical Chilled-iron Castings*

APPLICATION	C	Si	Mn	HARDNESS
Car wheels	3.35	0.50	0.55	62 scleroscope
Plowshares	3.40	1.35	0.60	514 Brinell
Mold boards	3.50	1.00	0.60	534 Brinell
Sprockets	3.30	1.80	0.65	477 Brinell

*By permission from "Metals Handbook," 8th ed., American Society for Metals, Metals Park, Ohio, 1961.

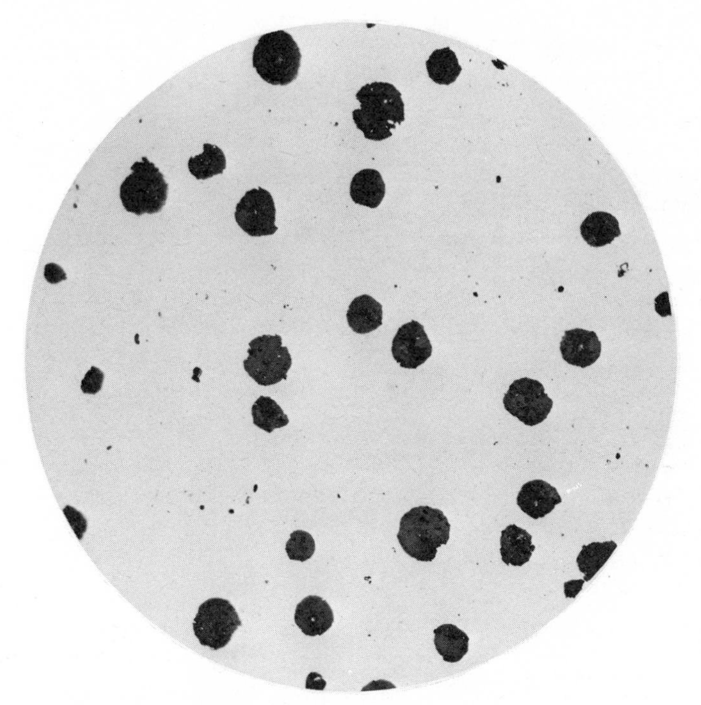

Fig. 11·21 Nodular iron, unetched, showing graphite spheroids, 125×. (The International Nickel Company.)

ment to be effective, and the alloys are described as "desulfurized."

The amount of ferrite in the as-cast matrix depends on composition and rate of cooling. Nodular irons with a matrix having a maximum of 10 percent pearlite are known as *ferritic irons* (Fig. 11·22). This structure gives maximum ductility, toughness, and machinability.

A matrix structure which is largely pearlite can be produced as cast or by normalizing. Normalizing is carried out by air cooling from a temperature of 1600 to 1650°F. Pearlitic ductile irons (Fig. 11·23) are stronger but less ductile than ferrite irons. A martensitic matrix may be obtained by quenching in oil or water from 1600 to 1700°F. The quenched structures are usually tempered, after hardening, to the desired strength and hardness levels.

Austenitic ductile irons are highly alloyed types which retain their austenitic structure down to at least –75°F. These irons are of interest because of their relatively high corrosion resistance and good creep properties at elevated temperatures.

The tensile mechanical properties of basic types of nodular iron are given in Table 11·6.

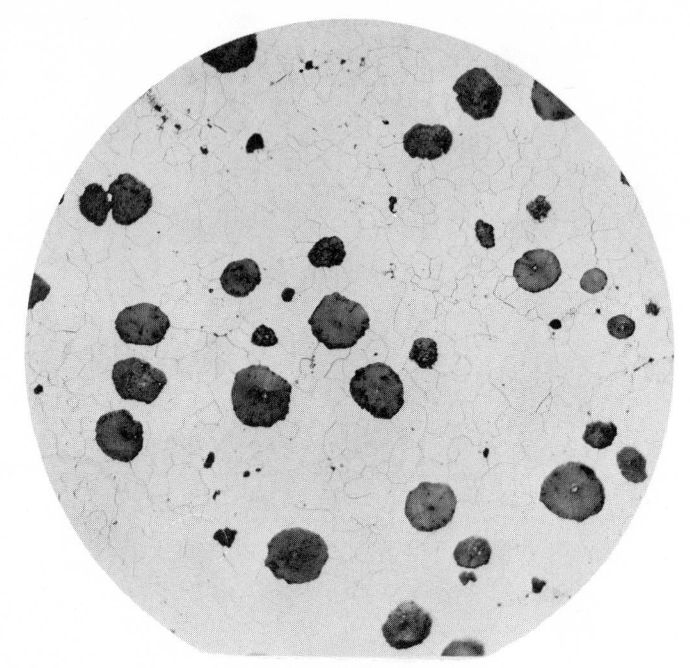

Fig. 11·22 Ferritic nodular iron showing graphite spheroids in a ferrite matrix. Etched in 2 percent nital, 125×. (The International Nickel Company.)

Some typical applications of nodular iron are agricultural—tractor and implement parts; automotive and diesel—crankshafts, pistons, and cylinder heads; electrical fittings, switch boxes, motor frames, and circuit-

TABLE 11·6 Mechanical properties of Basic Types of Nodular Iron*

TYPE	ALLOY CONTENT	TENSILE STRENGTH, PSI	YIELD STRENGTH, PSI	ELONGATION, % IN 2 IN.	BHN
Ferritic	Low	55,000	35,000	25	130
	High	90,000	70,000	12	210
Pearlitic	Low	80,000	60,000	10	200
	Low†	130,000	90,000	7	275
	High	130,000	110,000	2	275
Quenched	...	100,000	80,000	10	215
		150,000	130,000	2	320
Austenitic	‡	60,000	30,000	40	130
	§	60,000	40,000	10	160

*By permission from "Metals Handbook," 1954 Supplement, American Society for Metals, Metals Park, Ohio.
†Normalized.
‡3.00 percent C, 2.50 percent Si, 20.0 percent Ni, 2.0 percent Mn.
§3.00 percent C, 2.0 percent Si, 20.0 percent Ni, 1 percent Mn, 1.5 percent Cr.

breaker parts; mining—hoist drums, drive pulleys, flywheels, and elevator buckets; steel mill—work rolls, furnace doors, table rolls, and bearings; tool and die—wrenches, levers, handles, clamp frames, chuck bodies, and miscellaneous dies for shaping steel, aluminum, brass, bronze, and titanium.

A summary of the cast-iron microstructures and the phases coexisting at various temperatures is shown in Fig. 11·24.

11·16 Alloy Cast Irons An alloy cast iron is one which contains a specially added element or elements in sufficient amount to produce a measurable modification in the physical or mechanical properties. Elements normally obtained from raw materials, such as silicon, manganese, sulfur, and phosphorus, are not considered alloy additions.

Alloying elements are added to cast iron for special properties such as resistance to corrosion, heat, or wear, and to improve mechanical properties. Most alloying elements in cast iron will accelerate or retard graphitization, and this is one of the important reasons for alloying. The most common alloying elements are chromium, copper, molybdenum, nickel, and vanadium.

Chromium increases combined carbon by forming complex iron-chromium carbides that are more stable than iron carbide. Small amounts of chromium increase strength, hardness, depth of chill, and resistance to

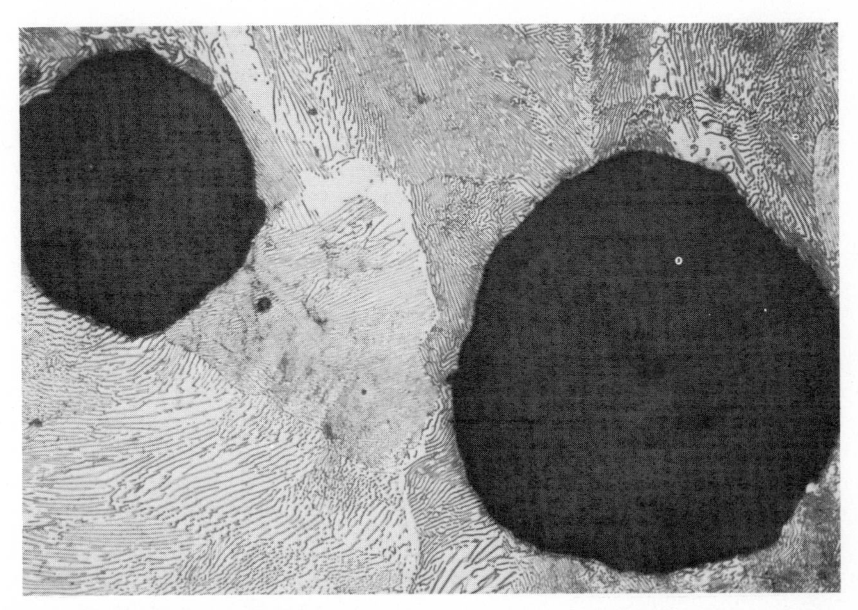

Fig. 11·23 Pearlitic nodular iron showing graphite spheroids in a pearlitic matrix. Etched in 2 percent nital, 500×. (The International Nickel Company.)

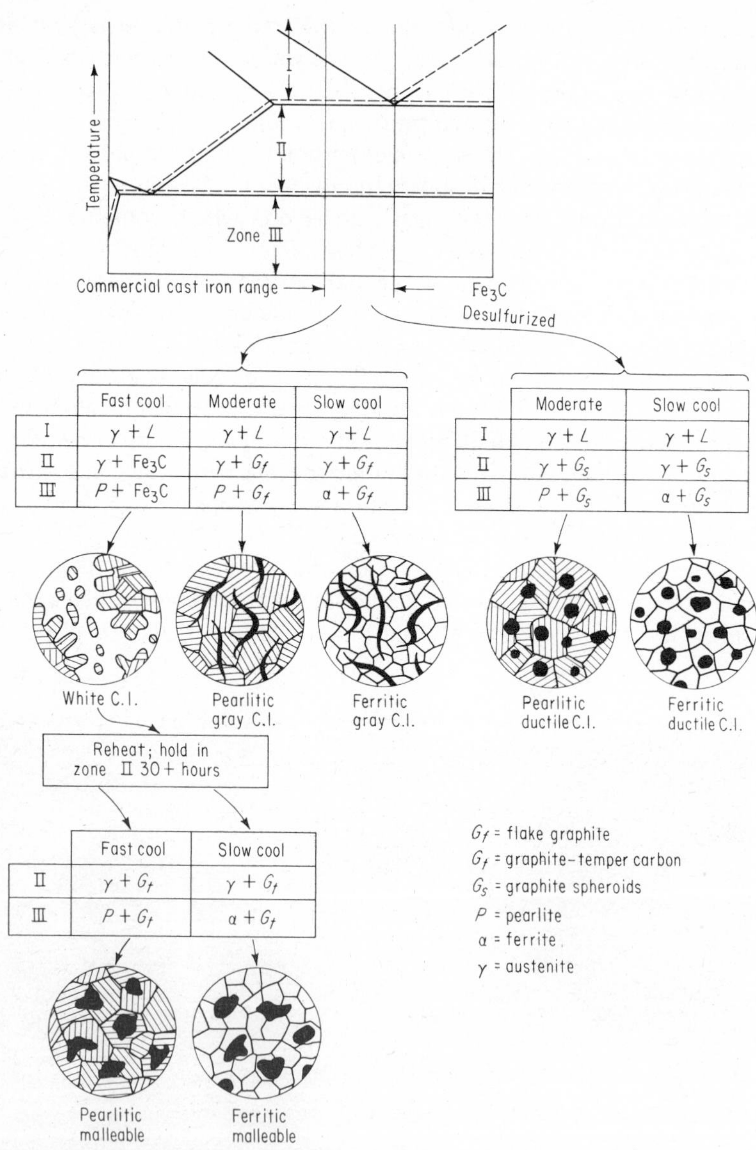

Fig. 11·24 Summary of cast-iron microstructures and the phases coexisting at various temperatures. (From Moffatt, Pearsall, and Wulff, "The Structure and Properties of Materials," John Wiley & Sons, Inc., New York, 1964.)

wear and heat but decrease machinability. The strong carbide-forming tendency of chromium is illustrated by the following effects on the microstructure of a soft gray iron as the chromium content is increased:

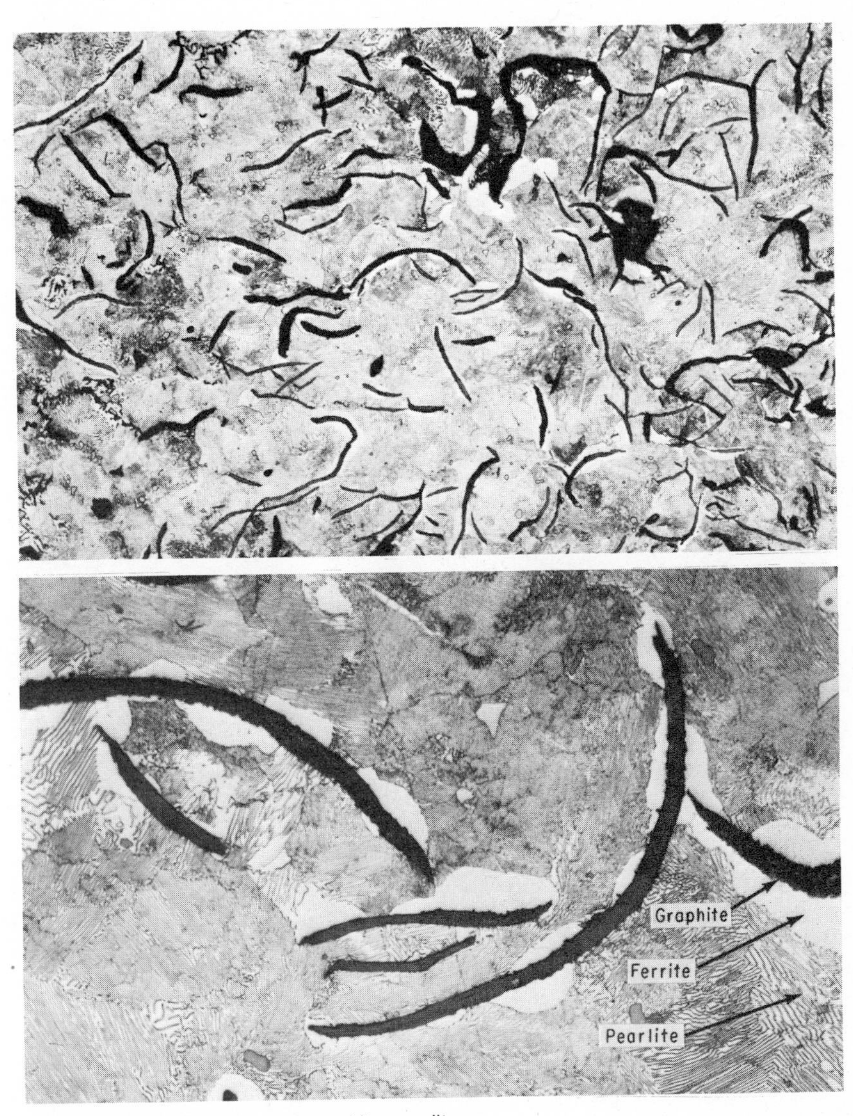

Fig. 11·25 Low-nickel cast iron. As-cast, graphite, pearlite, and a small amount of ferrite (white areas). Etched in 2 percent nital. Top 100×, bottom 500×. (The International Nickel Company.)

Percent	Structure
0	Ferrite and coarse graphite
0.3	Less ferrite, some pearlite, and finer graphite
0.6	Fine graphite and pearlite

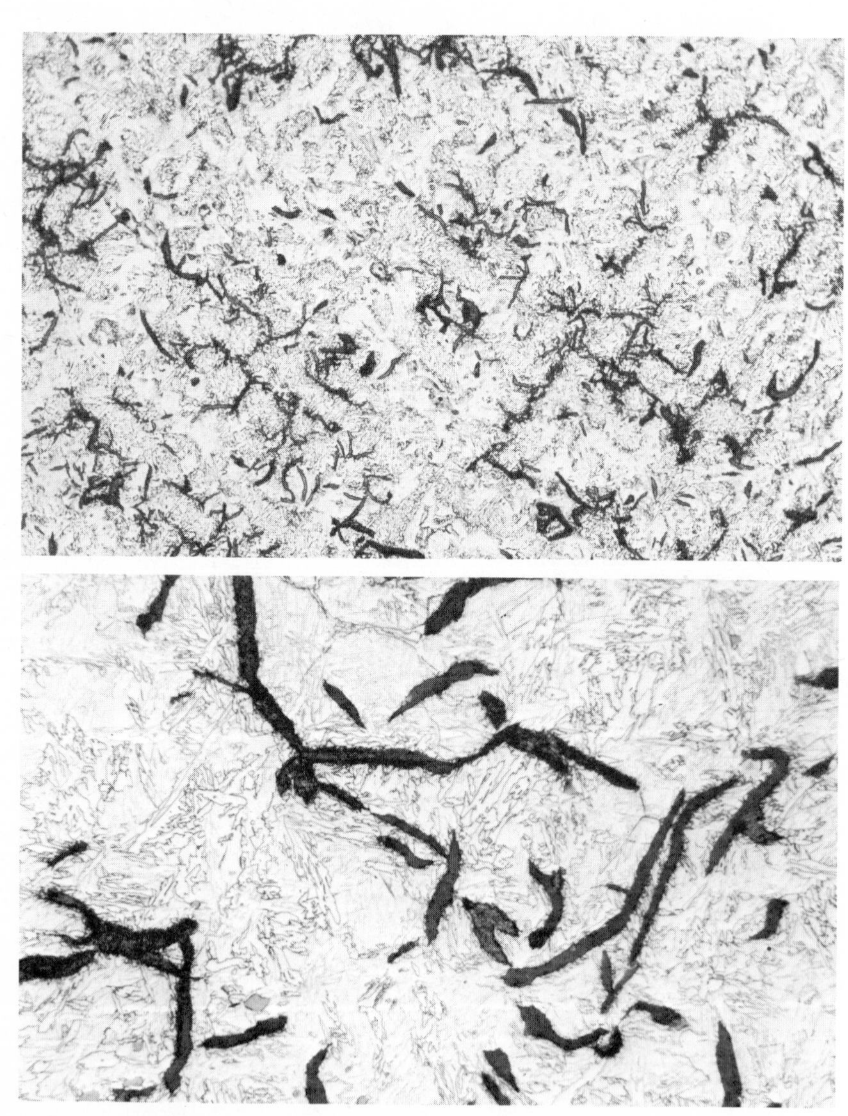

Fig. 11·26 Nickel-molybdenum cast iron (1.50 percent Ni, 1.62 percent Mo). As-cast, graphite flakes in a bainitic matrix. Etched in 2 percent nital. Top 100×, bottom 500×. (The International Nickel Company.)

1.0	Fine graphite, pearlite, and small carbide
3.0	Disappearance of graphite
5.0	Much massive carbide
10-30	Fine carbide

Chromium additions of less than 1 percent give a general improvement in mechanical properties. For resistance to corrosion or for use at high temperatures, as much as 35 percent chromium is used in combination with other alloying elements.

Copper is a graphitizer but is only about one-fifth as potent as silicon in this respect. For general engineering applications the copper content is between 0.25 and 2.5 percent. Copper tends to break up massive cementite and strengthen the matrix.

Molybdenum improves mechanical properties and is a mild stabilizer of carbides. Molybdenum is added in quantities from 0.25 to 1.25 percent, and its effect is similar to that in steel. Fatigue strength, tensile strength, transverse strength, heat resistance, and hardness of the cast iron are all improved. Molybdenum also retards the transformation of austenite, thus increasing hardenability and freedom from cracking and distortion. Molybdenum is always used in combination with other alloying elements.

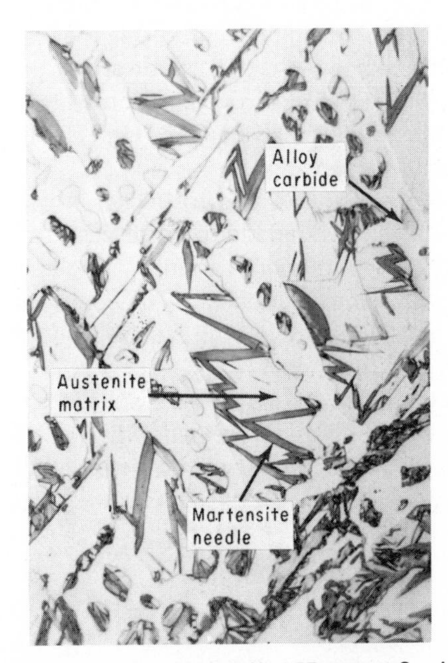

Fig. 11·27 Ni-Hard, 3.95 percent Ni, 1.57 percent Cr. As-cast, martensite needles in austenite plus alloy carbides. Etched in 2 percent nital, 250×. (The International Nickel Company.)

TABLE 11·7 Typical Low-alloy Cast Irons*

Silicon, %	2.25	2.09	2.23	2.12	2.11	2.57
Total carbon, %	3.53	3.33	3.38	3.44	3.41	2.81
Combined carbon, %	0.57	0.89	0.64	0.74	0.72	0.73
Sulfur, %	0.08	0.06	0.06	0.06	0.06	0.06
Phosphorus, %	0.14	0.16	0.14	0.16	0.16	0.16
Manganese, %	0.59	0.62	0.60	0.60	0.60	0.60
Nickel, %	0.71	1.77	1.46	1.00	2.00	2.00
Chromium, %	0.25	0.74	0.15	0.25	0.20	0.20
Molybdenum, %	...	...	...	0.60	0.90	0.90
Transverse strength, lb	2,400	2,625	26.5	3,000	3,300	3,700
Tensile strength, psi	28,000	36,000	30,000	39,000	44,000	66,000
BHN	180	214	187	212	250	300

*By permission from A. W. Grosvenor, "Basic Metallurgy," vol. 1, American Society for Metals, Metals Park, Ohio, 1954.

Vanadium is a very powerful carbide former, stabilizes cementite, and reduces graphitization. Vanadium additions, between 0.10 and 0.25 percent, increase tensile strength, transverse strength, and hardness.

Nickel is a graphitizer but only about one-half as effective as silicon in this respect. The purpose of nickel (0.5 to 6.0 percent) in the engineering gray irons is to control the structure by retarding austenite transformation, stabilizing pearlite, and maintaining combined carbon at the eutectoid quantity. Thus the microstructure of a low-nickel cast iron shows graphite, pearlite, and very little ferrite (Fig. 11·25). In combination with about 1 percent molybdenum, the matrix tends to be bainitic (Fig. 11·26). This structure has a hardness of about 385 BHN.

For excellent abrasion resistance, about 4 percent nickel in combination with about 1.50 percent chromium is added to white cast iron. The structure is shown in Fig. 11·27. The primary dendrites, originally austenite, have been partially transformed to martensite. The combination of iron carbides in a martensitic matrix results in high hardness (600 to 750 BHN) along with good strength and toughness.

The addition of 14 to 38 percent nickel to gray irons results in high heat resistance, high corrosion resistance, and low expansivity. Because of the large amount of nickel, the matrix will be austenitic (Fig. 11·28).

The composition and mechanical properties of some low-alloy cast irons are given in Table 11·7.

QUESTIONS

11·1 Explain the difference in microstructure and properties of white and gray cast iron.

11·2 Differentiate between free and combined carbon.

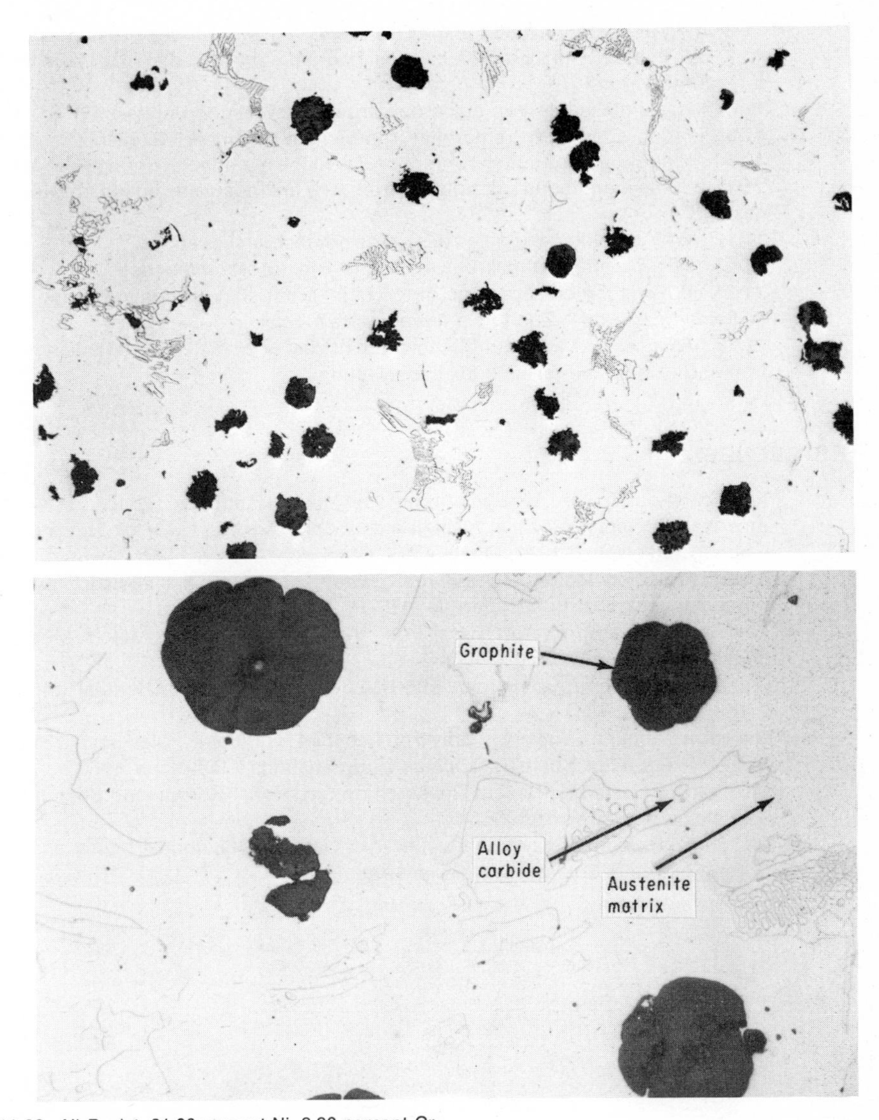

Fig. 11·28 Ni-Resist, 21.06 percent Ni, 2.20 percent Cr, 0.06 percent Mg. As-cast, nodular graphite and carbides in an austenitic matrix. Etched in 2 percent nital. Top 100×, bottom 500×. (The International Nickel Company.)

11·3 Discuss the effect of the amount of free carbon on the properties of gray cast iron.

11·4 How may the properties of gray cast iron be varied?

11·5 Differentiate, in microstructure, gray cast iron, malleable iron, and nodular iron.

11·6 Why are graphite flakes in gray iron very often surrounded by ferrite areas?
11·7 Why should the iron–iron carbide diagram not be used to determine the structures in gray iron?
11·8 Why is malleable iron made only from hypoeutectic white iron?
11·9 Is it possible to make nodular iron by heat treatment? Explain.
10·10 Why should the sulfur content be low in the manufacture of nodular iron?
11·11 In the manufacture of nodular iron, why are inoculants added only just before casting?
11·12 Why is welding of chilled cast irons not recommended?
11·13 Why is welding of pearlitic malleable iron not recommended?
11·14 What is the disadvantage of too high a first-stage annealing temperature for malleable cast iron? Too low a temperature? Explain.
11·15 Assume that a C clamp is to made of cast iron. Select a suitable type of cast iron and explain the reasons for the selection.

REFERENCES

American Foundrymen's Association: "Cast Metals Handbook," Chicago, 1944.
American Society for Metals: "Metals Handbook," 7th ed., 1948; 8th ed., vol. 1, 1961, vol. 2, 1964, vol. 7, 1972, Metals Park, Ohio.
American Society for Testing Materials: "Gray, Ductile and Malleable Iron Castings," Special Technical Publication no. 455, Philadelphia, 1969.
Boyles, A.: "The Structure of Cast Iron," American Society for Metals, Metals Park, Ohio, 1947.
Climax Molybdenum Company: "The Uses of Molybdenum in Nodular Irons," New York, 1964.
Gray Iron Founders' Society: "Gray Iron Castings Handbook," Cleveland, 1958.
Malleable Founders Society: "Malleable Iron Castings," Cleveland, 1960.
Merchant, H. D. (ed.): "Recent Research on Cast Iron," Gordon and Breach Science Publishers, New York, 1968.
Walton C. F. (ed.): "Gray and Ductile Iron Castings Handbook," Gray and Ductile Iron Founders' Society, Cleveland, 1971.

12 NONFERROUS METALS AND ALLOYS

12·1 Introduction Metallic materials, when considered in a broad sense, may be divided into two large groups, ferrous and nonferrous. The ferrous materials are iron-based, and the nonferrous materials have some element other than iron as the principal constituent. The bulk of the nonferrous materials is made up of the alloys of copper, aluminum, magnesium, nickel, tin, lead, and zinc. Other nonferrous metals and alloys that are used to a lesser extent include cadmium, molybdenum, cobalt, zirconium, beryllium, titanium, tantalum, and the precious metals gold, silver, and the platinum group.

This chapter will be concerned with the more important nonferrous metals and alloys.

COPPER AND COPPER ALLOYS

12·2 Copper The properties of copper that are most important are high electrical and thermal conductivity, good corrosion resistance, machinability, strength, and ease of fabrication. In addition, copper is nonmagnetic, has a pleasing color, can be welded, brazed, and soldered, and is easily finished by plating or lacquering. Certain of these basic properties may be improved by suitable alloying. Most of the copper that is used for electrical conductors contains over 99.9 percent copper and is identified as electrolytic tough-pitch copper (ETP) or oxygen-free high-conductivity copper (OFHC). Electrolytic tough-pitch copper is also used for roofing, gutters, downspouts, automobile radiators and gaskets, kettles, vats, pressure vessels, and distillery and other process equipment. Electrolytic tough-pitch copper contains from 0.02 to 0.05 percent oxygen, which is combined with copper as the compound cuprous oxide (Cu_2O). As cast, copper oxide and copper form an interdendritic eutectic mixture (Fig. 12·1). After working

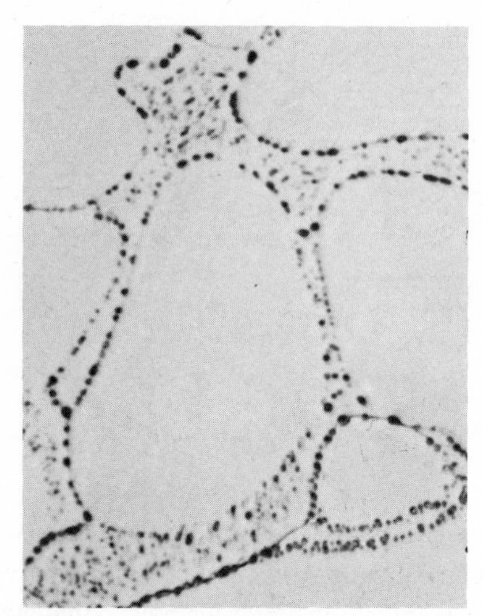

Fig. 12·1 Copper–copper oxide eutectic in cast tough-pitch copper. Lightly etched with sodium dichromate, 500×. (Revere Copper and Brass Company.)

and annealing, the interdendritic network is destroyed (Fig. 12·2), and the strength is improved. Oxygen-free copper is used in electronic tubes or similar applications because it makes a perfect seal to glass.

Arsenical copper containing about 0.3 percent arsenic has improved resistance to special corrosive conditions and is used for certain condenser and heat-exchanger applications.

Free-cutting copper with about 0.6 percent tellurium has excellent machining properties and is used for bolts, studs, welding tips, and electrical parts such as contact pins, switch gears, relays, and precision electrical equipment.

Silver-bearing copper has a silver content of 7 to 30 oz/ton. Silver raises the recrystallization temperature of copper, thus preventing softening during soldering of commutators. It is preferred in the manufacture of electric motors for railroad and aircraft use.

12·3 Temper Designation of Copper and Copper Alloys Since copper and most copper alloys are homogenous single phases, they are not susceptible to heat treatment, and their strength may be altered only by cold working. There are two general classes of temper for non-heat-treatable wrought-copper alloys: cold-worked and soft or annealed.

The different cold-worked tempers shown in Table 12·1 are obtained by cold-working the annealed material a definite amount. The percentage

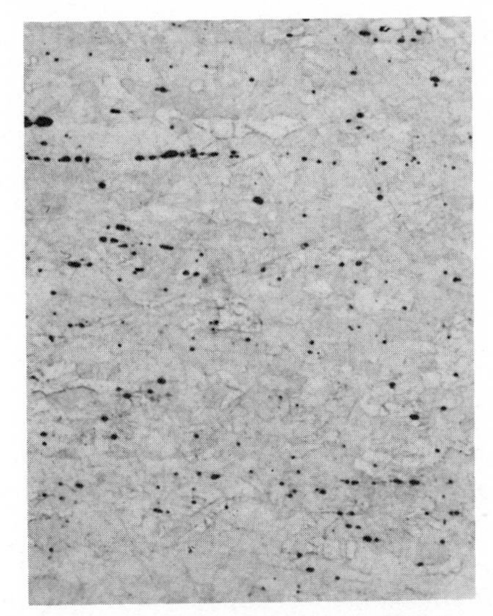

Fig. 12·2 Particles of copper oxide (black spots) in wrought tough-pitch copper. Lightly etched in ammonium hydroxide and hydrogen peroxide, 250×. (Revere Copper and Brass Company.)

reduction for strip is based on thickness difference and for wire on area difference.

Annealed tempers are used for forming at room temperature and are described by range of grain size or nominal grain size, expressed as average grain diameter in millimeters. Measurement of grain size is standardized by ASTM E79, Estimating the Average Grain Size of Wrought Copper

TABLE 12·1 Cold-worked Temper Designations*

DESCRIPTION	APPROXIMATE % REDUCTION BY COLD WORKING	
	STRIP	WIRE
Quarter hard	10.9	20.7
Half hard	20.7	37.1
Three-quarter hard	29.4	50.0
Hard	37.1	60.5
Extra hard	50.0	75.0
Spring	60.5	84.4
Extra spring	68.7	90.2

*From "Metals Handbook," 1961 ed., p. 1006, American Society for Metals, Metals Park, Ohio.

TABLE 12·2 Commonly Used Grain Sizes

GRAIN SIZE, MM	RECOMMENDED FOR
0.015	Slight forming operations; best polishing
0.025	Easy drawing, good polishing such as hubcaps
0.035	Good drawing and polishing as for headlight reflectors
0.050	Heavy drawing and spinning; more difficult to polish
0.100	Severe draws on heavy material

and Copper-base Alloys. The grain size best suited for a particular application depends upon the thickness of the metal, the depth of draw, and the type of surface required after the draw. Commonly used grain sizes are shown in Table 12·2.

12·4 Copper Alloys The most important commercial copper alloys may be classified as follows:

- **I** Brasses—alloys of copper and zinc
  - **A** Alpha brasses—alloys containing up to 36 percent zinc
    - **1** Yellow alpha brasses—20 to 36 percent zinc
    - **2** Red brasses—5 to 20 percent zinc
  - **B** Alpha plus beta brasses—54 to 62 percent copper
- **II** Bronzes—up to 12 percent of alloying element
  - **A** Tin bronzes
  - **B** Silicon bronzes
  - **C** Aluminum bronzes
  - **D** Beryllium bronzes
- **III** Cupronickels—alloys of copper and nickel
- **IV** Nickel silvers—alloys of copper, nickel, and zinc

12·5 Brasses—General Brasses are essentially alloys of copper and zinc. Some of these alloys have small amounts of other elements such as lead, tin, or aluminum. Variations in composition will result in desired color, strength, ductility, machinability, corrosion resistance, or a combination of such properties.

The portion of the binary copper-zinc phase diagram which is applicable to commercial alloys is shown in Fig. 12·3. The solubility of zinc in the alpha (α) solid solution increases from 32.5 percent at 1657°F to about 39 percent at 850°F. Since copper is f.c.c. (face-centered cubic), the α solid solution is f.c.c. The beta (β) phase is a b.c.c. (body-centered cubic) electron compound and undergoes ordering, indicated by a dot-dash line, in the region of 850 to 875°F. On cooling in this temperature range the b.c.c. β phase, with copper and zinc atoms randomly dispersed at lattice points, changes continuously to the ordered structure β', which is still b.c.c. but

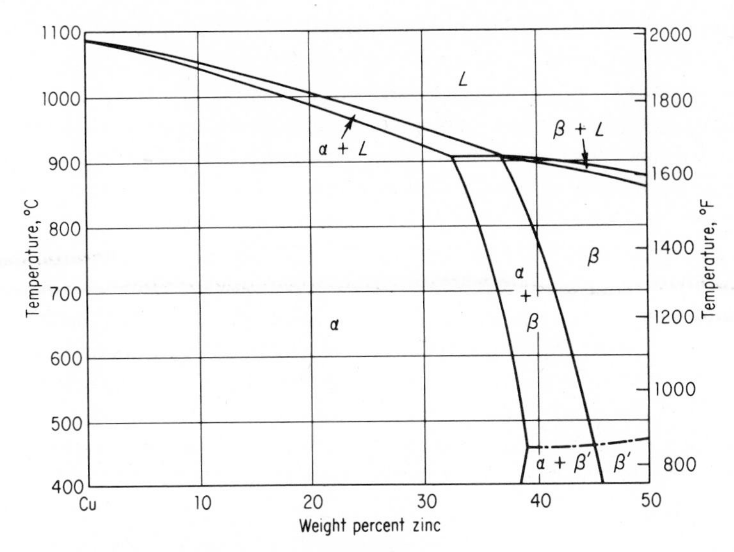

Fig. 12·3 Cu-rich portion of the Cu-Zn phase diagram. (From "Metals Handbook," 1948 ed., p. 1206, American Society for Metals, Metals Park, Ohio.)

with the copper atoms at the corners and zinc atoms at the centers of the unit cubes. The ordering reaction is so rapid that it cannot be retarded or prevented by quenching.

The effect of zinc on the tensile properties of annealed copper alloys is shown in Table 12·3.

In most cases, the addition of solid-solution elements tends to decrease ductility as strength increases. The addition of zinc, as shown in Table 12·3, increases ductility along with strength. The best combination of strength and ductility is obtained in 70Cu-30Zn brass.

TABLE 12·3 Effect of Zinc on Properties of Copper Alloys*

Zn, %	TENSILE STRENGTH, PSI	ELONGATION, % IN 2 IN.	BHN, 10 MM, 500 KG
0	32,000	46	38
5	36,000	49	49
10	41,000	52	54
15	42,000	56	58
20	43,000	59	56
25	45,000	62	54
30	46,000	65	55
35	46,000	60	55
40 (+β′)	54,000	45	75

*Data from Chase Brass & Copper Co. for commercial alloys of moderate grain size.

The commercial brasses may be divided into two groups, brasses for cold working (α brasses) and brasses for hot working (α plus β brasses).

12·6 Alpha Brasses Alpha brasses containing up to 36 percent zinc possess relatively good corrosion resistance and good working properties. The color of α brasses varies according to copper content from red for high-copper alloys to yellow at about 62 percent copper. The α brasses may be divided into two groups, yellow α brasses and red brasses.

Yellow α Brasses These contain 20 to 36 percent zinc, combine good strength with high ductility and are therefore suited for drastic cold-working operations. It is common practice to stress-relief anneal these brasses after severe cold working to prevent *season cracking.* Season cracking or stress-corrosion cracking is due to the high residual stresses left in the brass as a result of cold working. These stresses make the brass more susceptible to intergranular corrosion, particularly in ammonia atmospheres (Fig. 12·4). Stress relieving in the recovery range (up to about 500°F) or the

Fig. 12·4 Stress-corrosion cracking in α brass, 150×. (Revere Copper and Brass Company.)

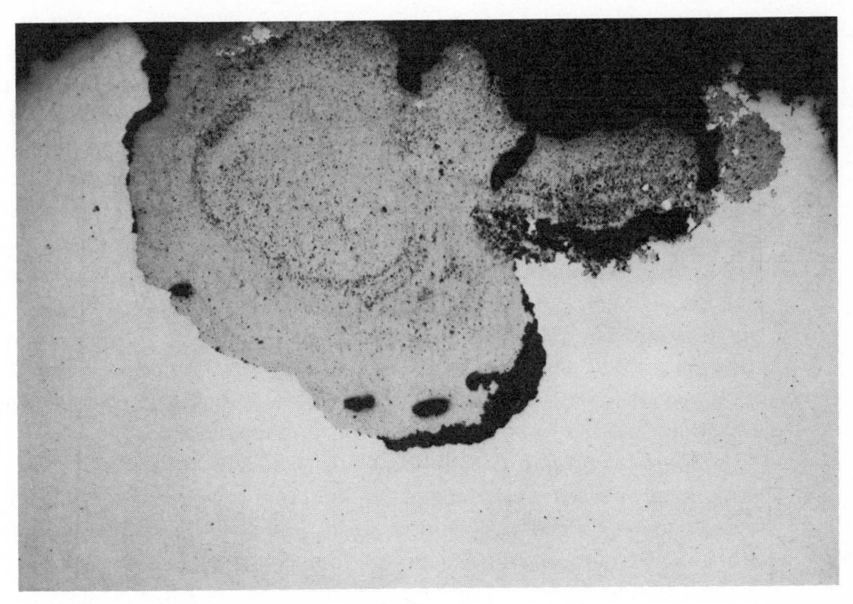

Fig. 12·5 Plug-type dezincification in brass, unetched, 50×. (Revere Copper and Brass Company.)

substitution of a less susceptible copper alloy will minimize the danger of stress-corrosion cracking.

Yellow α brasses are also subject to a pitting corrosion called *dezincification*. This type of corrosion usually occurs when brass is in contact with sea water or with fresh waters that have a high content of oxygen and carbon dioxide. Dezincification involves dissolution of the alloy and a subsequent deposition of porous nonadherent copper. Action of this kind, unless stopped, will eventually penetrate the cross section of the metal and lead to leakage through the porous copper layer. If it occurs in a localized area it is known as *plug-type dezincification* (Fig. 12·5). Small amounts of tin or antimony minimize dezincification in yellow brasses.

The most widely used yellow α brasses are *cartridge brass* (70Cu-30Zn) and *yellow brass* (65Cu-35Zn). Typical applications include automotive—radiator cores, tanks, headlight reflectors; electrical—flashlight shells, lamp fixtures, socket shells, screw shells; hardware—eyelets, fasteners, grommets, rivets, springs; plumbing accessories; and ammunition components.

The addition of 0.5 to 3 percent lead improves machinability so that leaded brass is used for screw-machine parts, engraving plates, keys, lock parts, tumblers, gears, and watch parts.

Two variations of yellow α brasses have been developed for special applications. *Admiralty metal* (71Cu-28Zn-1Sn), with the addition of 1 percent tin for improved strength and corrosion resistance, is used for condenser and heat-exchanger tubes in steam power plant equipment. Better corrosion resistance is obtained in *aluminum brass* (76Cu-22Zn-2Al). This alloy forms a tenacious, self-healing film which protects the tube against high cooling-water velocities in marine and land power stations.

Red Brasses These contain between 5 and 20 percent zinc. They generally have better corrosion resistance than yellow brasses and are not susceptible to season cracking or dezincification. The most common low-zinc brasses are gilding metal (95Cu-5Zn), commercial bronze (90Cu-10Zn), red brass (85Cu-15Zn), and low brass (80Cu-20Zn).

Gilding metal (95Cu-5Zn) has higher strength than copper and is used for coins, medals, tokens, fuse caps, primers, emblems, plaques, and as a base for articles to be gold-plated or highly polished.

Commercial bronze (90Cu-10Zn) has excellent cold-working and hot-working properties and is used for costume jewelry, compacts, lipstick cases, marine hardware, forgings, rivets, and screws. Leaded commercial bronze (1.75 percent Pb) is used for screws and other parts for automatic screw-machine work.

Fig. 12·6 Two-phase structure of annealed muntz metal. β' appears light surrounding the darker α grains. Etched in ammonium hydroxide and hydrogen peroxide, 150×. (Revere Copper and Brass Company.)

Fig. 12·7 Muntz metal water-quenched from 1515°F. β' phase is light, α phase is dark. Etched in ammonium hydroxide and hydrogen peroxide, 50×. (By permission from R. M. Brick, R. B. Gordon, and A. Phillips, "Structure and Properties of Alloys," 3d ed., McGraw-Hill Book Company, New York, 1965.)

Red brass (85Cu-15Zn) is used for electrical conduit, screw shells, sockets, hardware, condenser and heat exchanger tubes, plumbing pipe, lipstick cases, compacts, nameplates, tags, and radiator cores.

Low brass (80Cu-20Zn) is used for ornamental metalwork, medallions, thermostat bellows, musical instruments, flexible hose, and other deep-drawn articles.

12·7 Alpha Plus Beta Brasses These contain from 54 to 62 percent copper. Consideration of Fig. 12·3 shows that these alloys will consist of two phases, α and β'. The β' phase is harder and more brittle at room temperature than α; therefore, these alloys are more difficult to cold-work than the α brasses. At elevated temperatures the β phase becomes very plastic, and since most of these alloys may be heated into the single-phase β region, they have excellent hot-working properties.

The most widely used $\alpha + \beta'$ brass is *muntz metal* (60Cu-40Zn), which has high strength and excellent hot-working properties. The two-phase structure of annealed muntz metal is shown in Fig. 12·6. Rapid cooling from the β region may suppress the precipitation of most of the α phase. Figure 12·7 shows the microstructure of muntz metal after water quenching from 1515°F. The quench preserved most of the β, but some α has formed, particularly at the grain boundaries. Subsequent reheating to a low temperature will allow more of the α to come out of the supersaturated solution; thus it is possible to heat-treat this alloy. Muntz metal is used in sheet form for ship-sheathing, condenser heads, perforated metal, and architectural work. It is also used for valve stems, brazing rods, and condenser tubes. Leaded muntz metal containing 0.40 to 0.80 percent lead has improved machinability.

Free-cutting brass (61.5Cu-35.5Zn-3Pb) has the best machinability of any brass combined with good mechanical and corrosion-resistant properties. It is used for hardware, gears and automatic high-speed screw machine parts.

Forging brass (60Cu-38Zn-2Pb) has the best hot-working properties of any brass and is used for hot forgings, hardware, and plumbing parts.

Architectural bronze (57Cu-40Zn-3Pb) has excellent forging and free matching properties. Typical applications are handrails, decorative moldings, grilles, storefronts, hinges, lock bodies, and industrial forgings.

Naval brass (60Cu-39.25Zn-0.75Sn), also known as *tobin bronze,* has increased resistance to salt-water corrosion and is used for condenser plates, welding rod, propeller shafts, piston rods, and valve stems. Leaded naval brass with the addition of 1.75 Pb for improved machinability is used for marine hardware. Figure 12·8 shows the two-phase structure of hot-extruded naval brass. The β phase is darkened while the alpha phase is light.

Manganese bronze (58.5Cu-39Zn-1.4Fe-1Sn-0.1Mn), really a high-zinc brass, has high strength combined with excellent wear resistance and is used for clutch disks, extruded shapes, forgings, pump rods, shafting rod,

Fig. 12·8 Hot-extruded naval brass. β phase is dark, α phase is light. Etched in $FeCl_3$, 75×. (Revere Copper and Brass Company.)

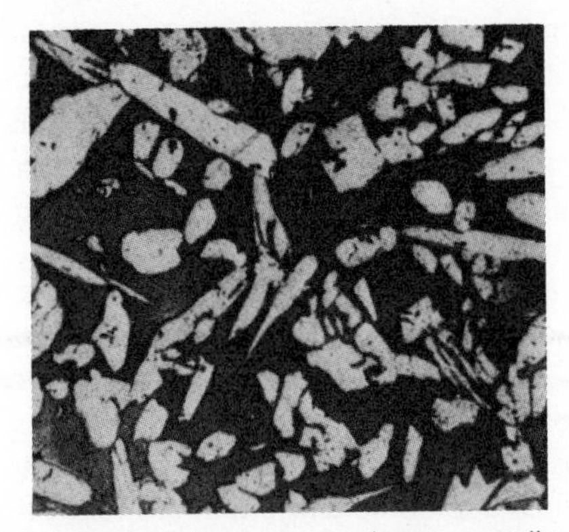

Fig. 12·9 Manganese bronze, as sand cast. α needles (light) in a β matrix (dark). Etched in $NH_4OH + H_2O_2$, then $FeCl_3$, 100×. (From Metals Handbook, vol. 7, "Atlas of Microstructures," American Society for Metals, 1972.)

valve stems, and welding rod. Figure 12·9 shows the structure of manganese bronze as sand cast. It consists of white needles of the α solid solution in a matrix of the β phase.

12·8 Cast Brasses The previous discussion was concerned primarily with wrought brasses, which are mainly binary alloys of copper and zinc. The cast brasses are similar in name to the wrought brasses but usually contain appreciable amounts of other alloying elements. Tin may be present from 1 to 6 percent and lead from 1 to 10 percent; some alloys may contain iron, manganese, nickel, and aluminum.

An example of a casting brass is *leaded red brass* (85Cu-5Sn-5Pb-5Zn),

Fig. 12·10 Leaded red brass as continuous cast. Cored α dendrites with uniformly distributed lead particles (black dots). Etched in $NH_4OH + H_2O_2$, 200×. (From Metals Handbook, vol. 7, "Atlas of Microstructures," American Society for Metals, 1972.)

which is used for general castings requiring fair strength, soundness, and good machining properties, such as low-pressure valves, pipe fittings, small gears, and small pump castings. The structure of leaded red brass, as continuous cast, is shown in Fig. 12·10. It consists of cored or segregated α dendrites, due to the relatively rapid cooling rate, and small, uniformly distributed lead particles (black dots).

12·9 Bronzes—General The term *bronze* was originally applied to the copper-tin alloys; however, the term is now used for any copper alloy, with the exception of copper-zinc alloys, that contains up to approximately 12 percent of the principal alloying element. Bronze, as a name, conveys the idea of a higher-class alloy than brass, and as indicated in the proceeding sections, it has been incorrectly applied to some alloys that are really special brasses. Commercial bronzes are primarily alloys of copper and tin, aluminum, silicon, or beryllium. In addition, they may contain phosphorus, lead, zinc, or nickel.

12·10 Tin Bronzes These are generally referred to as *phosphor bronzes* since phosphorus is always present as a deoxidizer in casting. The usual range

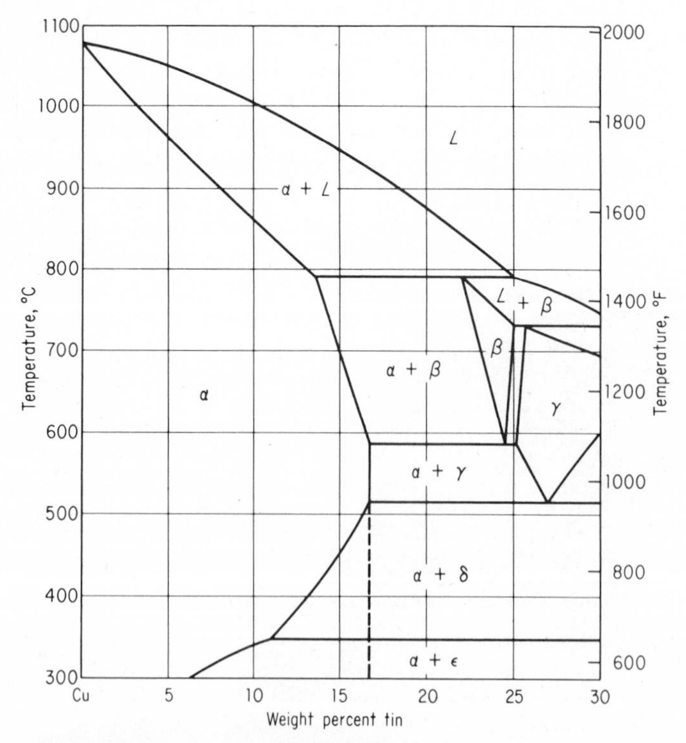

Fig. 12·11 Copper-rich portion of the copper tin phase diagram. (From "Metals Handbook," 1948 ed., p. 1204, American Society for Metals, Metals Park, Ohio.)

of phosphorus content is between 0.01 and 0.5 percent, and of tin between 1 and 11 percent.

The copper-rich portion of the copper-tin alloy system is shown in Fig. 12·11. The β phase forms as the result of a peritectic reaction at 1468°F.

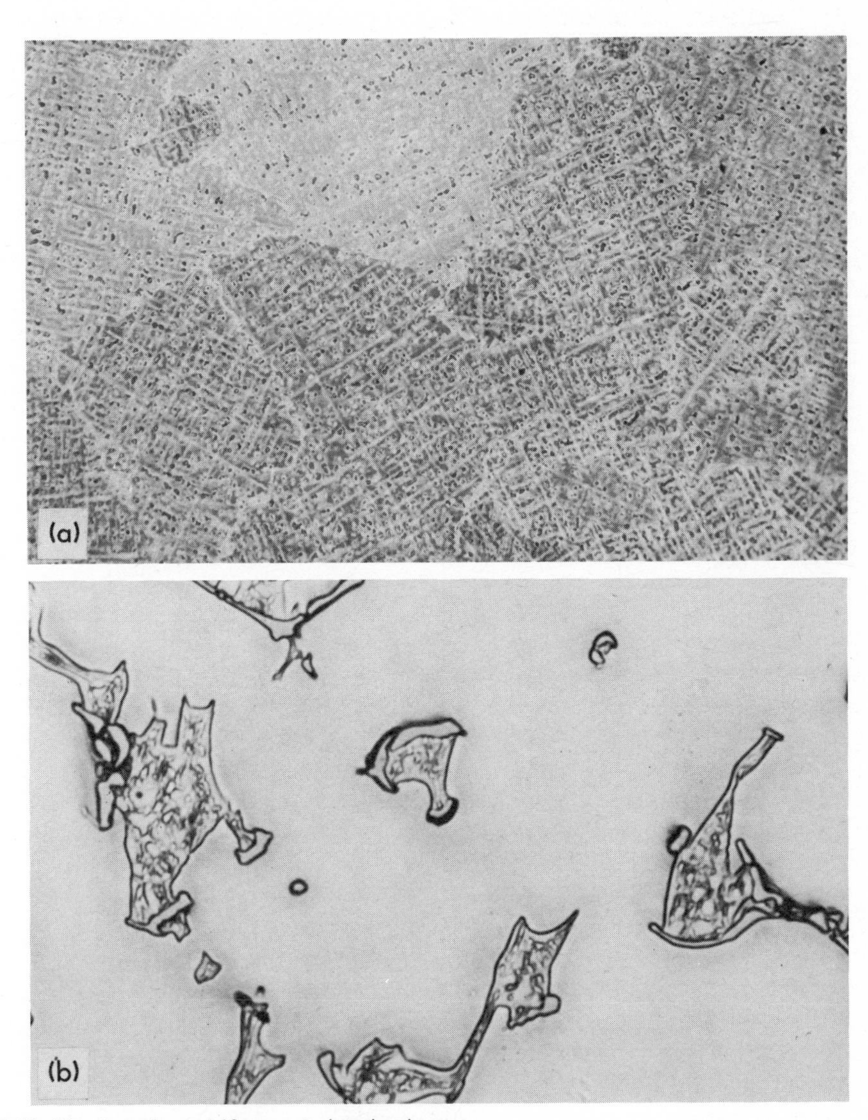

Fig. 12·12 Structure of a cast 10 percent phosphor-bronze alloy. (*a*) Rapidly cooled, particles of δ in a dendritic α matrix, 75×; (*b*) detail of the δ phase, 1,000×. (American Brass Company.)

At 1087°F, the β phase undergoes a eutectoid reaction to form the eutectoid mixture $(\alpha + \gamma)$. At 968°F, gamma (γ) also undergoes a eutectoid transformation to $(\alpha + \delta)$. The diagram also indicates the decomposition of the delta (δ) phase. This takes place by a eutectoid reaction at 662°F forming $(\alpha + \epsilon)$. This reaction is so sluggish that in commercial alloys, the epsilon (ϵ) phase is nonexistent. The slope of the solvus line below 968°F shows a considerable decrease in the solubility of tin in the α phase. The precipitation of the δ or ϵ phase due to this change in solubility is so slow that, for practical purposes, the solvus line is as indicated by the vertical dotted line below 968°F. For this reason, slow-cooled cast tin bronzes containing below 7 percent tin generally show only a single phase, the α solid solution. There is some of the δ phase in most castings containing over 7 percent tin. The structure of a rapidly cooled cast 10 percent phosphor bronze (Fig. 12·12*a*) shows small particles of the δ phase in a fine dendritic α matrix. Detail of the δ phase is shown in Fig. 12·12*b*.

The phosphor bronzes are characterized by high strength, toughness, high corrosion resistance, low coefficient of friction, and freedom from season cracking. They are used extensively for diaphragms, bellows, lock washers, cotter pins, bushings, clutch disks, and springs.

Zinc is sometimes used to replace part of the tin. The result is an im-

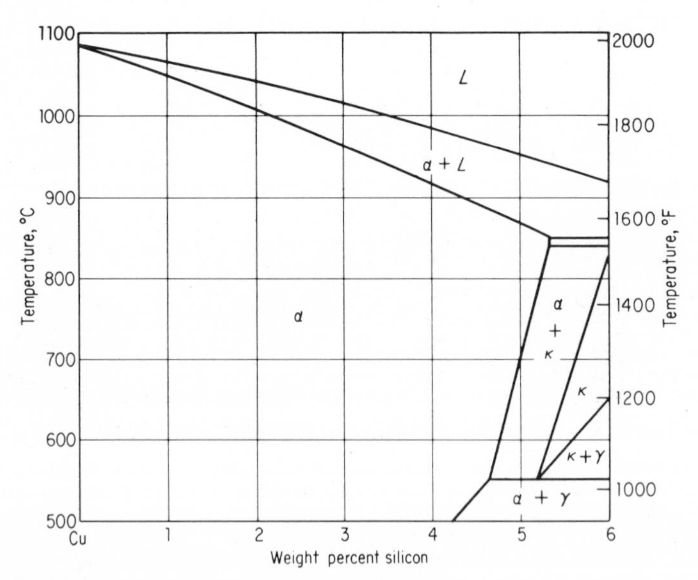

Fig. 12·13 Copper-rich portion of the copper-silicon alloy system. (From "Metals Handbook," 1948 ed., p. 1203, American Society for Metals, Metals Park, Ohio.)

provement in the casting properties and toughness with little effect on wear resistance.

Lead is often added to tin bronze to improve machinability and wear resistance. High-lead tin bronze may contain as much as 25 percent lead. The leaded alloys are used for bushing and bearings under moderate or light loads.

12·11 Silicon Bronzes The copper-rich portion of the copper-silicon alloy system is shown in Fig. 12·13. The solubility of silicon in the α phase is 5.3 percent at 1565°F and decreases with temperature. The eutectoid reaction at 1030°F is very sluggish, so that commercial silicon bronzes, which

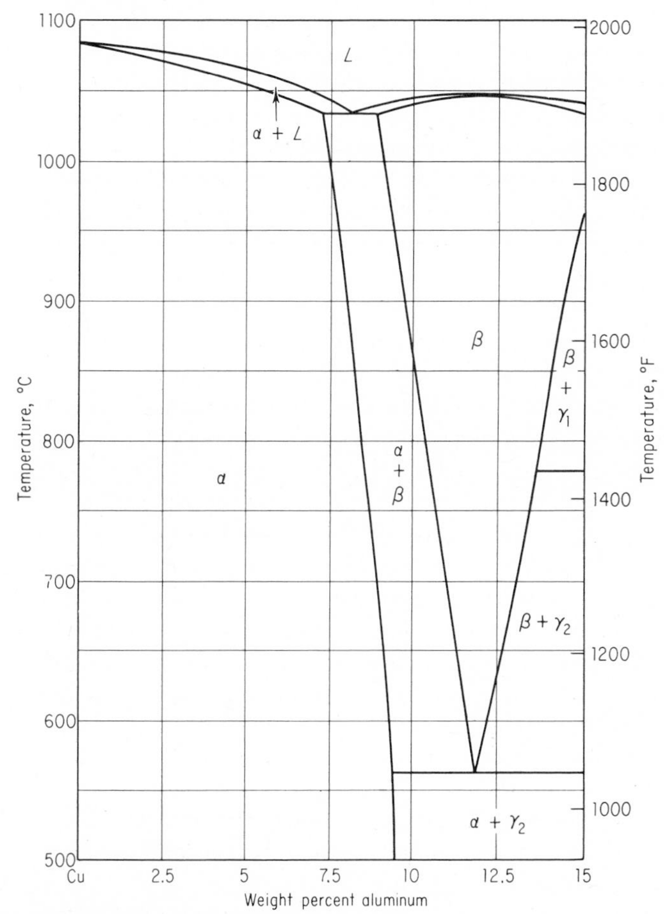

Fig. 12·14 Copper-rich portion of the copper-aluminum alloy system. (From "Metals Handbook," 1948 ed., p. 1160, American Society for Metals, Metals Park, Ohio.)

generally contain less than 5 percent silicon, are single-phase alloys.

Silicon bronzes are the strongest of the work-hardenable copper alloys. They have mechanical properties comparable to those of mild steel and corrosion resistance comparable to that of copper. They are used for tanks, pressure vessels, marine construction, and hydraulic pressure lines.

12·12 Aluminum Bronzes The copper-rich portion of the copper-aluminum alloy system is shown in Fig. 12·14. The maximum solubility of aluminum in the α solid solution is approximately 9.5 percent at 1050°F. The β phase undergoes a eutectoid reaction at 1050°F to form the $(\alpha + \gamma_2)$ mixture.

Most commercial aluminum bronzes contain between 4 and 11 percent aluminum. Those alloys containing up to 7.5 percent aluminum are generally single-phase alloys, while those containing between 7.5 and 11 percent aluminum are two-phase alloys. Other elements such as iron, nickel, manganese, and silicon are frequently added to aluminum bronzes. Iron (0.5 to 5.0 percent) increases strength and hardness and refines the grain; nickel (up to 5 percent) has the same effect as iron but is not so effective; silicon (up to 2 percent) improves machinability; manganese promotes soundness in castings by combining with gases and also improves strength.

The single-phase aluminum bronzes show good cold-working properties and good strength combined with corrosion resistance to atmospheric and water attack. They are used for condenser tubes, cold-work forms, corrosion-resistant vessels, nuts and bolts, and protective sheathing in marine applications.

The $\alpha + \beta$ aluminum bronzes are interesting because they can be heat-treated to obtain structures similar to those in steel. Figure 12·15*a* shows the structure of primary α and granular eutectoid $(\alpha + \gamma_2)$, representative of an as-cast 10 percent aluminum bronze. On furnace cooling from above the eutectoid temperatures, a lamellar structure resembling pearlite is formed (Fig. 12·15*b*). If the two-phase alloy is quenched from 1500 to 1600°F, a needlelike structure resembling martensite is formed (Fig. 12·15*c*). The quenched alloys are tempered between 700 and 1100°F to increase strength and hardness. Heat-treated aluminum bronzes are used for gears, propeller hubs, blades, pump parts, bearings, bushings, nonsparking tools, and drawing and forming dies.

12·13 Beryllium Bronzes The copper-rich portion of the copper-beryllium alloy system is shown in Fig. 12·16. The solubility of beryllium in the α solid solution decreases from 2.1 percent at 1590°F to less than 0.25 percent at room temperature. This change in solubility is always indicative of age-hardening possibilities.

The theory of age hardening is discussed in Sec. 6·16. The optimum mechanical properties are obtained in an alloy containing approximately

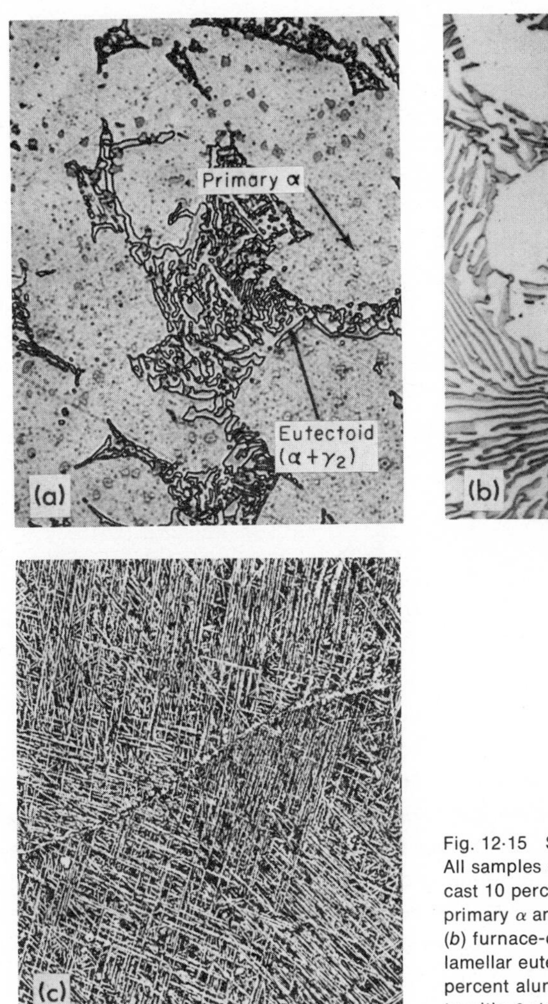

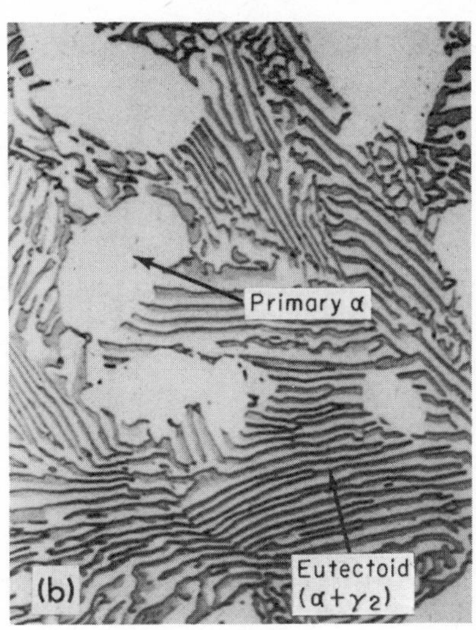

Fig. 12·15 Structures of aluminum bronze. All samples etched with ferric nitrate. (*a*) As-cast 10 percent aluminum bronze showing primary α and granular eutectoid, 750×; (*b*) furnace-cooled aluminum bronze showing lamellar eutectoid, 500×; (*c*) quenched 10.7 percent aluminum bronze showing a martensitic β structure, 100×. (Ampco Metal, Inc.)

2 percent beryllium. A typical heat-treating cycle for this alloy would be: solution-anneal at 1450°F, water-quench, cold-work, and finally age at 600°F.

Figure 12·17 shows the microstructures of a beryllium bronze containing a nominal 1.92 percent beryllium and 0.20 to 0.30 percent cobalt. Figure 12·17*a* shows the structure after a slow cool has allowed precipitation of the dark-etching γ phase primarily at the grain boundaries. Water quenching from the proper annealing temperature will result in a single-phase

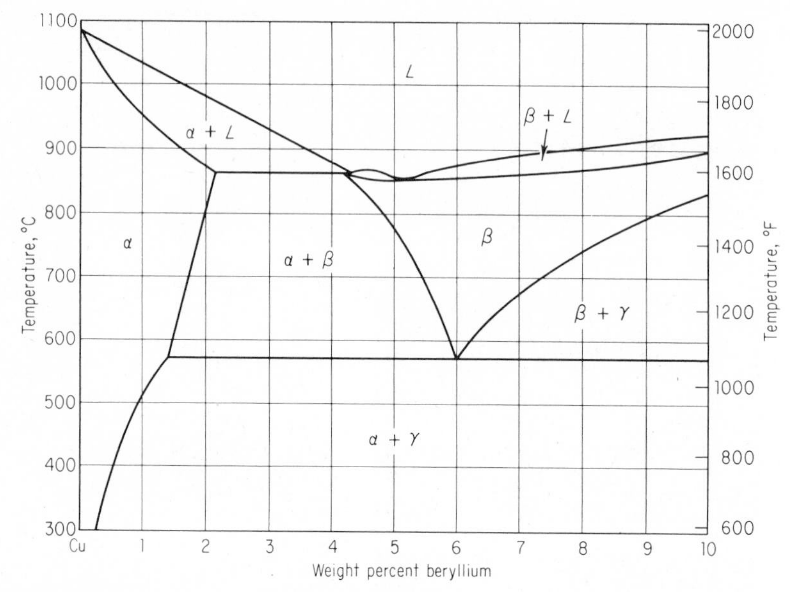

Fig. 12·16 Copper-rich portion of the copper-beryllium alloy system. (From "Metals Handbook," 1948 ed., p. 1176, American Society for Metals, Metals Park, Ohio.)

structure, and subsequent aging will allow precipitation of the γ phase as very fine particles throughout the α matrix (Fig. 12·17*b*). The use of too high an aging temperature causes grain-boundary coarsening typical of overaging (Fig. 12·17*c*). Under certain conditions, the β phase comes out of solution during solidification of the ingot. The primary β usually persists during subsequent processing and appears as densely populated bands known as *β stringers* (Fig. 12·17 *d*). The mechanical properties of two beryllium-bronze alloys are shown in Table 6·3.

Beryllium bronzes are used for parts requiring a combination of excellent formability in the soft condition with high yield strength, light fatigue strength, and creep resistance in the hardened condition (many springs); parts requiring corrosion resistance, high strength, and relatively high electrical conductivity (diaphragms, contact bridges, surgical instruments, bolts, and screws); hard parts that will wear well against hardened steel (firing pins, dies, nonsparking tools).

12·14 Cupronickels These are copper-nickel alloys that contain up to 30 percent nickel. The copper-nickel binary phase diagram (Fig. 6·11) shows complete solubility, so that all cupronickels are single-phase alloys. They are not susceptible to heat treatment and may have their properties altered only by cold working.

The cupronickel alloys have high resistance to corrosion fatigue and also high resistance to the corrosive and erosive action of rapidly moving sea water. They are widely used for condenser, distiller, evaporator, and heat-exchanger tubes for naval vessels and coastal power plants.

A copper–25 percent nickel alloy is used for cladding on both sides of electrolytic tough-pitch copper in the manufacture of the United States dime, as shown in Fig. 12·18.

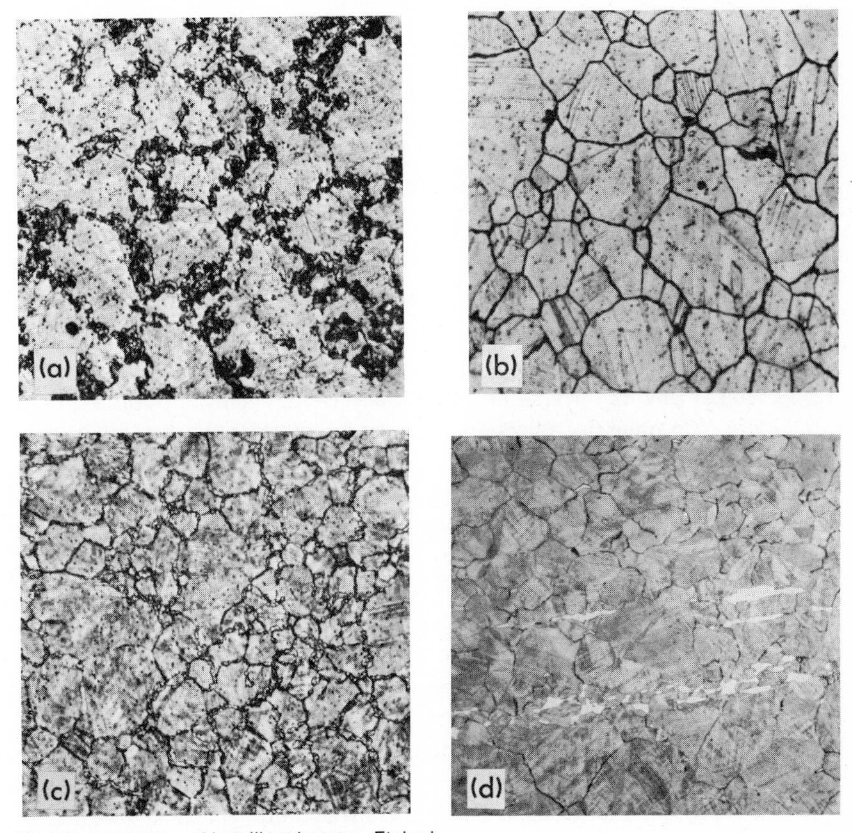

Fig. 12·17 Microstructures of beryllium bronze. Etched with aqueous solution of ammonium persulfate and ammonium hydroxide, 300×. (*a*) Slow-cooled and aged for 3 h at 600°F. γ phase (dark) mainly at α grain boundaries. (*b*) Solution-treated and aged for 3 h at 600°F. Fine particles of γ in an α matrix. (*c*) Solution-treated and aged for 2 h at 700°F. Grain-boundary coarsening typical of an overaged structure. (*d*) β "stringers" shown as densely populated bands. The β phase, originating in the cooling of the ingot, is not attacked by the etchant and appears white against a darkened α matrix. (The Beryllium Corporation.)

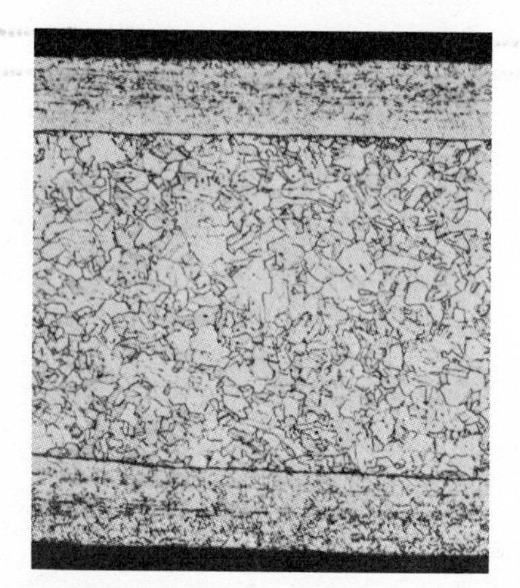

Fig. 12·18 Copper–25 percent nickel cladding on both sides of electrolytic tough-pitch copper in the United States dime. Etched in $K_2Cr_2O_7 + H_2SO_4 + HCl$, 50×. (From Metals Handbook, vol. 7, "Atlas of Microstructures," American Society for Metals, 1972.)

12·15 Nickel Silvers These are essentially ternary alloys of copper, nickel, and zinc. Commercial alloys are produced with the following range of composition: copper 50 to 70 percent, nickel 5 to 30 percent, zinc 5 to 40 percent.

The nickel silvers containing over 60 percent copper are single-phase alloys that show only fair hot-working properties but are ductile and easily worked at room temperature. The addition of nickel to the copper-zinc alloy gives it a pleasing silver-blue white color and good corrosion resistance to food chemicals, water, and atmosphere. These alloys make excellent base metals for plating with chromium, nickel, or silver. They are used for rivets, screws, table flatware, zippers, costume jewelry, nameplates, and radio dials.

The nickel silvers containing between 50 and 60 percent copper are two-phase $\alpha + \beta$ alloys. They have a relatively high modulus of elasticity and, like the $\alpha + \beta$ brasses, are readily hot-worked. Nickel silvers are less susceptible to stress corrosion than binary copper-zinc alloys of the same zinc content.

Typical applications of the $\alpha + \beta$ nickel silvers include springs and contacts in telephone equipment, resistance wire, hardware, and surgical and dental equipment.

The composition and typical mechanical properties of copper and some copper alloys are summarized in Table 12·4 on pages 482 and 483.

ALUMINUM AND ALUMINUM ALLOYS

12·16 Aluminum The best-known characteristic of aluminum is its light weight, the density being about one-third that of steel or copper alloys. Certain aluminum alloys have a better strength-to-weight ratio than that of high-strength steels. Aluminum has good malleability and formability, high corrosion resistance, and high electrical and thermal conductivity. An ultra-pure form of aluminum is used for photographic reflectors to take advantage of its high light reflectivity and nontarnishing characteristics.

Aluminum is nontoxic, nonmagnetic, and nonsparking. The nonmagnetic characteristic makes aluminum useful for electrical shielding purposes such as bus-bar housings or enclosures for other electrical equipment.

Although the electrical conductivity of electric-conductor (EC) grade aluminum is about 62 percent that of copper, its light weight makes it more suitable as an electrical conductor for many industrial applications.

Pure aluminum has a tensile strength of about 13,000 psi. However, substantial increases in strength are obtained by cold working or alloying. Some alloys, properly heat-treated, approach tensile strengths of 100,000 psi.

One of the most important characteristics of aluminum is its machinability and workability. It can be cast by any known method, rolled to any desired thickness, stamped, drawn, spun, hammered, forged, and extruded to almost any conceivable shape.

Commercially pure aluminum, 1100 alloy (99.0+ percent Al), is suitable for applications where good formability or very good resistance to corrosion (or both) are required and where high strength is not necessary. It has been used extensively for cooking utensils, various architectural components, food and chemical handling and storage equipment, and welded assemblies.

12·17 Alloy Designation System The designation of wrought aluminum and wrought aluminum alloys was standardized by The Aluminum Association in 1954. It follows a four-digit numbering system. The first digit indicates the alloy group (Table 12·5). The second digit indicates modification of the original alloy or impurity limits; zero is used for the original alloy, and integers 1 through 9 indicate alloy modifications. In the 1xxx group for minimum aluminum purities of 99.00 percent and greater, the last two digits are the same as the two digits to the right of the decimal point in the minimum aluminum percentage when it is expressed to the nearest 0.01 percent. Thus 1060 indicates a material of 99.60 minimum percent aluminum purity and no special control on individual impurities.

In the 2xxx through 8xxx alloy groups, the last two digits serve only to identify the different aluminum alloys in the group.

TABLE 12·4 Chemical Composition and Typical Mechanical Properties of Copper and Some Copper Alloys*

WROUGHT ALLOYS													
MATERIAL	FORM	COMPOSITION, %				TENSILE STRENGTH, 1,000 PSI		YIELD STRENGTH 0.5% OFFSET 1,000 PSI		ELONGATION, % IN 2 IN.		HARDNESS, ROCKWELL B	
		Cu	Zn	Sn	OTHERS	HARD	SOFT	HARD	SOFT	HARD	SOFT	HARD	SOFT
Copper	Sheet	99.9+	. . .	. . .	. . .	55	32	48	. . .	4	50	58	
Gilding metal	Sheet	95.0	5.0	. . .	. . .	55	35	45	11	5	38	68	7
Commercial bronze	Sheet	90.0	10.0	. . .	. . .	67	37	53	11	3	40	75	10
Red brass	Sheet	85.0	15.0	. . .	. . .	80	45	55	15	4	43	85	10
Low brass	Sheet	80.0	20.0	. . .	. . .	85	43	65	15	4	50	86	11
	Rod	80.0	20.0	. . .	. . .	80	45	60	15	5	50		
Spring brass	Sheet	75.0	25.0	. . .	. . .	80	47	60	15	5	45	87	15
Brass	Sheet	70.0	30.0	. . .	. . .	86	45	65	15	4	50	87	
Cartridge brass	Sheet	69.0	31.0	. . .	. . .	85	46	65	15	4	58	87	22
Yellow brass	Sheet	65.0	35.0	. . .	. . .	90	45	70	15	5	60	85	30
Muntz metal	Sheet	60.0	40.0	. . .	. . .	80	57	60	15	9	48	87	42
Phosphor bronze	Sheet	96.0	. . .	4.0	0.25 P	90	45	75	18	4	50	90	30
	Sheet	92.0	. . .	8.0	+P	110	60	85	25	3	55	99	45
Aluminum brass	Tube	76.0	22.0	. . .	2 Al	83	62	75	16	17	52	86	33
Aluminum bronze	. . .	92.0	. . .	. . .	8 Al	134	76	100	30	13	55	99	69
Manganese bronze	Rod	68.0	29.0	. . .	1 Al, 1 Mn, 1 Fe	85	60	50	25	20	45	90	25
Admiralty metal	Tube	71.0	21.0	1.0	. . .	100	53	98	18	3	60	95	13
Naval brass	Rod	60.0	39.0	0.75	0.25 Pb	62	54	39	15	30	40	55	
Silicon brass	. . .	78.0	20.0	. . .	2.0 Si	110	55	83	12	4	60		
Tin brass	. . .	88.0	10.0	2.0	. . .	85	. . .	. . .	. . .	3	. . .	86	

CASTING ALLOYS

MATERIAL	COMPOSITION, %				TENSILE STRENGTH, 1,000 PSI	YIELD STRENGTH, 1,000 PSI	ELONGATION, %	BHN, 500 kg, 10 mm
	Cu	Zn	Sn	OTHERS				
Cond. copper	99.85	. . .	. . .	. . .	30	17	45	40
Brass	70.0	30.0	. . .	. . .	28	16	22	65
Tin brass	63.0	36.0	1.0	. . .	30	18	18	75
Silicon brass	81.0	15.0	. . .	4 Si	90	45	16	
Aluminum brass	63.0	32.5	. . .	2.5 Al	62	35	18	120
Bronze	88.0	. . .	12.0	+P	40	22	11	70
	81.0	. . .	19.0	+P	35	25	12	135
Gear bronze	88.0	4.0	5.5	2.5 Ni	42	17	32	75
Leaded red brass	85.0	5.0	5.0	5 Pb	34	17	25	60
	80.0	. . .	10.0	10 Pb	35	17	20	65
Silicon bronze	95.0	1.0	. . .	4 Si	55	22	35	90
	93.0	4.0	. . .	2.5 Si, 0.5 Fe	50	18	20	
Aluminum bronze	89.0	. . .	. . .	10 Al, 1 Fe	67	32	15	140
	88.0	. . .	. . .	9 Al, 3 Fe	80	35	25	140
Manganese bronze	68.0	20.0	. . .	4 Mn, 5 Al, 2.5 Fe	110	70	15	
	64.0	24.0	. . .	4 Mn, 5 Al, 3 Fe	115	70	15	210
Nickel silver	60.0	20.0	. . .	20 Ni	45	20	35	55
Cupronickel	70.0	. . .	. . .	30 Ni	64	33	35	120

*From S. L. Hoyt, "Metal Data," Van Nostrand Reinhold Company, New York, 1952.

TABLE 12·5 Designation for Alloy Groups*

	ALUMINUM ASSOCIATION NO.
Aluminum, 99.00% and greater, major alloying element	1xxx
Copper	2xxx
Manganese	3xxx
Silicon	4xxx
Magnesium	5xxx
Magnesium and silicon	6xxx
Zinc	7xxx
Other element	8xxx
Unused series	9xxx

*The Aluminum Association.

12·18 Temper Designation The temper designation follows the alloy designation and is separated from it by a dash. The Aluminum Association Temper Designation System, adopted in 1948, is used for wrought and cast aluminum and aluminum alloys. It is based on the sequences of basic treatments used to produce the various tempers.

The standard temper designation system consists of a letter indicating the basic temper. Except for the annealed and as-fabricated tempers, it is more specifically defined by the addition of one or more digits. There are four basic tempers: F as fabricated, O annealed, H strain-hardened, and T heat-treated.

-F: As Fabricated Applied to products which acquire some temper as the result of normal manufacturing operations. There is no guarantee of mechanical properties.

-O: Annealed, Recrystallized This is the softest temper of wrought alloy products.

-H: Strain-hardened This applies to products which have their mechanical properties increased by cold working only. The -H is always followed by two or more digits. The first digit indicates the specific combination of basic operations as follows:

-H1: Strain-hardened Only The second digit designates the amount of cold work performed, with the numeral 8 representing the full-hard condition. Therefore, half hard is -H14, quarter hard is -H12, etc. Extra hard tempers are designated by numeral 9. A third digit is often used to indicate the degree of control of temper or to identify a special set of mechanical properties.

-H2: Strain-hardened Then Partially Annealed Applied to products that are cold-worked to a harder temper and then have their strength reduced to the desired level by partial annealing. The residual amount of cold work is designated by the same method as the -H1 series.

-H3: Strain-hardened and Then Stabilized Applied only to alloys contraining magnesium which are given a low-temperature heating to stabilize their properties. The degree of strain hardening remaining after the stabilizing treatment is indicated in the usual way by one or more digits.

-W: Solution Heat-treated An unstable temper applicable only to alloys which spontaneously age at room temperature after solution heat treatment. This designation, because of natural aging, is specific only when the period of aging is indicated: for example, 2024-W (1/2 h).

-T: Thermally Treated Applies to products thermally treated, with or without supplementary strain hardening, to produce stable tempers. The -T is followed by the numerals 2 through 10, inclusive, designating a specific combination of basic operations. Deliberate variations of the conditions, resulting in significantly different characteristics for the product, are indicated by adding one or more digits to the basic designation.

-T2: Annealed (cast products only)

-T3: Solution heat-treated and then cold-worked

-T4: Solution heat-treated and naturally aged to a substantially stable condition

-T5: Artificially aged only. Applies to products which are artificially aged after an elevated-temperature rapid-cool fabrication process, such as casting or extrusion

-T6: Solution heat-treated and then artificially aged

-T7: Solution heat-treated and then stabilized: applies to products where the temperature and time conditions for stabilizing are such that the alloy is carried beyond the point of maximum hardness, providing control of growth and/or residual stress

-T8: Solution heat-treated, cold-worked, and then artificially aged

-T9: Solution heat-treated, artificially aged, and then cold-worked

-T10: Artificially aged and then cold-worked, the same as -T5 but followed by cold working to improve strength

12·19 Aluminum-Copper Alloys (2xxx Series) The aluminum-rich end of the aluminum copper equilibrium diagram is shown in Figure 12·19. The maximum solubility of copper in aluminum is 5.65 percent at 1018°F, and the solubility decreases to 0.45 percent at 572°F. Therefore, alloys containing between 2.5 and 5 percent copper will respond to heat treatment by age hardening. The theta (θ) phase is an intermediate alloy phase whose composition corresponds closely to the compound $CuAl_2$. Solution treatment is carried out by heating the alloy into the kappa (κ) single-phase region followed by rapid cooling. Subsequent aging, either natural or artificial, will allow precipitation of the θ phase, thus increasing the strength of the alloy. These alloys may contain smaller amounts of silicon, iron, magnesium, manganese, chromium, and zinc.

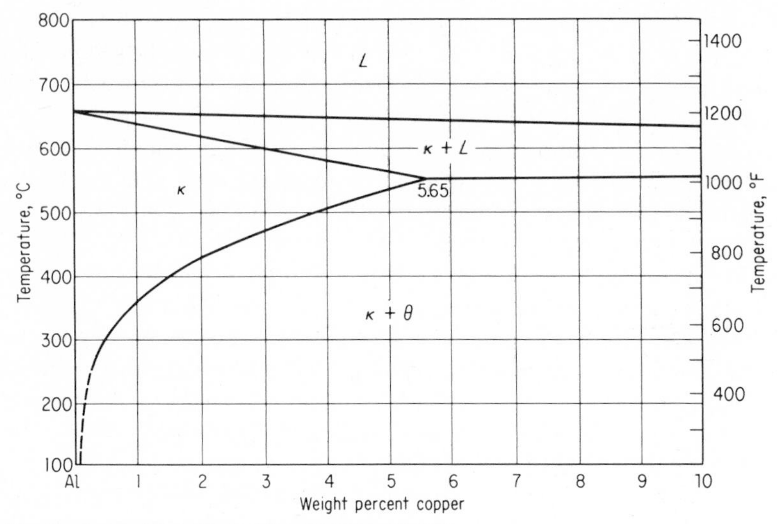

Fig. 12·19 Aluminum-rich portion of the copper-aluminum alloy system. (From Metals Handbook, 1948 ed., p. 1160, American Society for Metals, Metals Park, Ohio.)

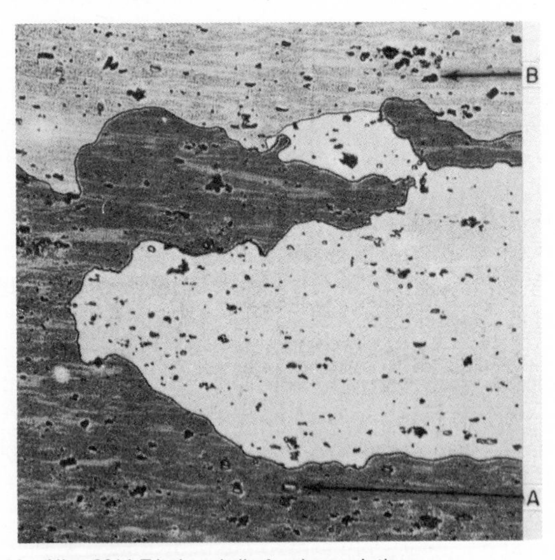

Fig. 12·20 Alloy 2014-T4 closed-die forging, solution-treated at 935°F for 2 h and quenched in water at 150°F. Longitudinal section. Structure consists mainly of dark, insoluble particles (*A*) of a complex compound of Fe, Mn, Si, and Al, and a few particles (*B*) of $CuAl_2$ (white, outlined) in a matrix of α solid solution (*C*). Etched in Keller's reagent, 100×. (From Metals Handbook, vol. 7, "Atlas of Microstructures," American Society for Metals, 1972.)

Figure 12·20 shows the structure of alloy 2014-T4 closed-die forging after solution heat treatment and quenching in warm water. It consists mainly of black insoluble particles of a complex compound of Fe, Mn, Si, and Al in a matrix of aluminum-rich solid solution. There are just a few particles of $CuAl_2$ (white, outlined) since most of this compound is in solution as a result of the heat treatment.

The three most widely used wrought aluminum-copper alloys are 2014, 2017, and 2024. The oldest of all the heat-treatable aluminum alloys is duralumin (2017) containing 4 percent copper. This alloy is widely used for rivets in aircraft construction. Since this is a natural-aging alloy, after solution treatment it is refrigerated to prevent aging. As a single phase, in the solution-treated condition, it has good ductility so that the rivethead may be easily formed. Subsequent return of the material to room temperature causes precipitation of the θ phase as small submicroscopic particles, increasing the hardness and strength.

Alloy 2014 has higher copper and manganese content than 2017 and is susceptible to artificial aging. In the artificially aged temper, 2014 has a higher tensile strength, much higher yield strength, and lower elongation than 2017. This alloy is used for heavy-duty forgings, aircraft fittings, and truck frames. Figure 12·21 shows the relationship of conductivity, strength, and thermal transformation typical of a 2014 aluminum alloy. Alloy 2024, containing 4.5 percent copper and 1.5 percent magnesium, develops the highest strengths of any naturally aged aluminum-copper type of alloy.

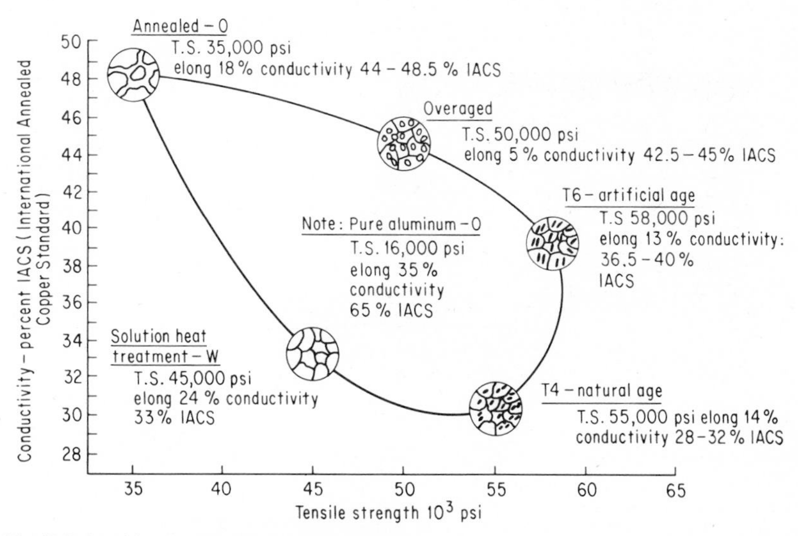

Fig. 12·21 Relationship of conductivity, strength, and thermal transformation typical of a 2014 aluminum alloy.

The higher magnesium content, compared with 2017, makes it more difficult to fabricate. A combination of strain hardening and aging will develop the maximum yield strength attainable in high-strength alloy sheet. Typical uses of 2024 alloy are aircraft structures, rivets, hardware, truck wheels, and screw-machine products. An aluminum-copper alloy containing 2 percent nickel (2218) has been developed for applications involving elevated temperatures such as forged cylinder heads and pistons.

The only binary aluminum-copper casting alloy is 195, containing 4 percent copper. When properly heat-treated, this alloy has an excellent combination of strength and ductility. Alloy 195, sand-cast, is used for flywheel and rear-axle housings, bus wheels, aircraft wheels, and crankcases.

Several casting alloys are produced that contain approximately 8 percent copper. These alloys, 112, 113, and 212, may contain substantial controlled additions of silicon, as well as iron and zinc. The presence of silicon increases fluidity, so that alloys 113 and 212 are preferred for thin-sectioned castings such as housings, cover plates, and hydraulic brake pistons.

A series of casting alloys such as 85, 108, 319, and 380, classed as aluminum-copper-silicon alloys, have been developed containing less than 5 percent copper and from 3 to 8 percent silicon. Figure 12·22 shows the typical structure of a 380 alloy, die-cast, that has desirable properties. The copper provides higher strength and better machining properties than the straight aluminum-silicon alloys, while the silicon provides better casting and pressure-tightness properties than the aluminum-copper alloys. Typical applications include brackets, typewriter frames, manifolds, valve bodies, oil pans, and gasoline and oil tanks.

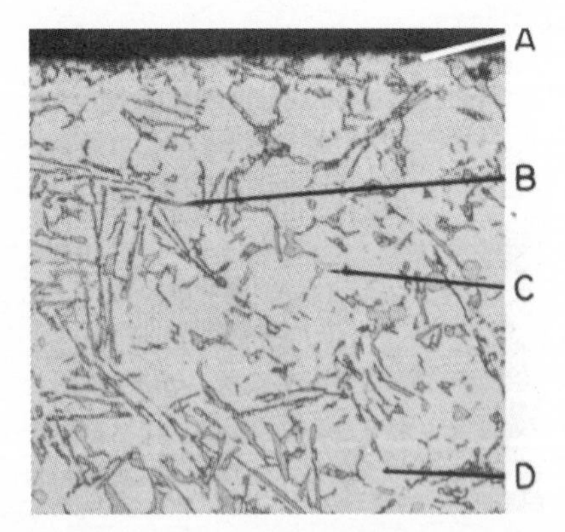

Fig. 12·22 Alloy 380 die casting. Area near a machined surface (*A*) shows structure typical of a casting that has desirable properties: interdendritic particles of eutectic silicon (*B*) and $CuAl_2$ (*C*) in a matrix of aluminum solid solution (*D*). Etched in 0.5 HF, 260×. (From Metals Handbook, vol. 7, "Atlas of Microstructures," American Society for Metals, 1972.)

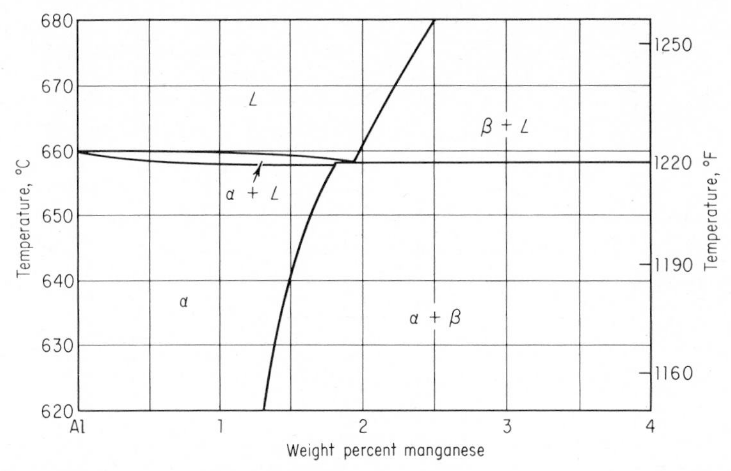

Fig. 12·23 Aluminum-rich portion of the aluminum-manganese alloy system. (From "Metals Handbook," 1948 ed., p. 1163, American Society for Metals, Metals Park, Ohio.)

12·20 Aluminum-Manganese Alloy 5 (3xxx Series) The aluminum-rich portion of the aluminum-manganese alloy system is shown in Figure 12·23. The maximum solubility of manganese in the α solid solution is 1.82 at the eutectic temperature of 1216°F. Although the solubility decreases with decreasing temperature, alloys in this group are generally not age-hardenable. Because of the limited solubility, manganese is not used as a major alloying element in any casting alloys and is used in only a few wrought alloys. One of the alloys in this group is the popular 3003 alloy, which has good formability, very good resistance to corrosion, and good weldability. Typical applications are utensils, food and chemical handling and storage equipment, gasoline and oil tanks, pressure vessels, and piping.

12·21 Aluminum-Silicon Alloys (4xxx Series) The aluminum-rich portion of the aluminum-silicon alloy system is shown in Figure 12·24. The maximum solubility of silicon in the α solid solution is 1.65 percent at the eutectic temperature of 1071°F. Although the solvus line shows lower solubility at lower temperatures, these alloys are generally not heat-treatable. Wrought alloy 4032, containing 12.5 percent silicon, has good forgeability and low coefficient of thermal expansion. It is used for forged automotive pistons.

Aluminum-silicon casting alloys have excellent castability and resistance to corrosion. Alloy 13 (12 percent silicon) and alloy 43 (5 percent silicon) are used for intricate castings, food-handling equipment, and marine fittings.

12·22 Aluminum-Magnesium Alloys (5xxx Series) The aluminum-rich portion of the aluminum-magnesium system is shown in Figure 12·25. Although the solvus line shows a considerable drop in the solubility of magnesium

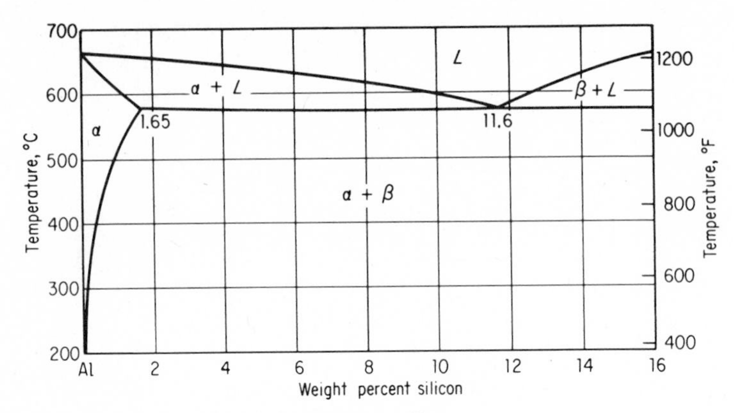

Fig. 12·24 Aluminum-rich portion of the aluminum-silicon alloy system. (From "Metals Handbook," 1948 ed., p. 1166, American Society for Metals, Metals Park, Ohio.)

in aluminum with decreasing temperature, most commercial wrought alloys in this group contain less than 5 percent magnesium, and, with low silicon content, they are not heat-treatable.

The wrought alloys are characterized by good weldability, good corrosion resistance, and moderate strength. Alloy 5005 (0.8 percent magnesium) is used for architectural extrusions; alloy 5050 (1.2 percent magnesium) for tubing and automotive gas and oil lines; alloy 5052 (2.5 percent

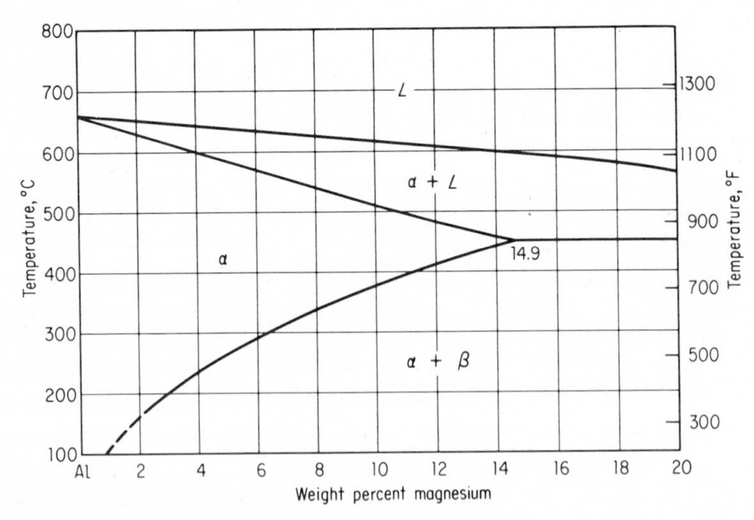

Fig. 12·25 Aluminum-rich portion of the aluminum-magnesium alloy system. (From "Metals Handbook," 1948 ed., p. 1163, American Society for Metals, Metals Park, Ohio.)

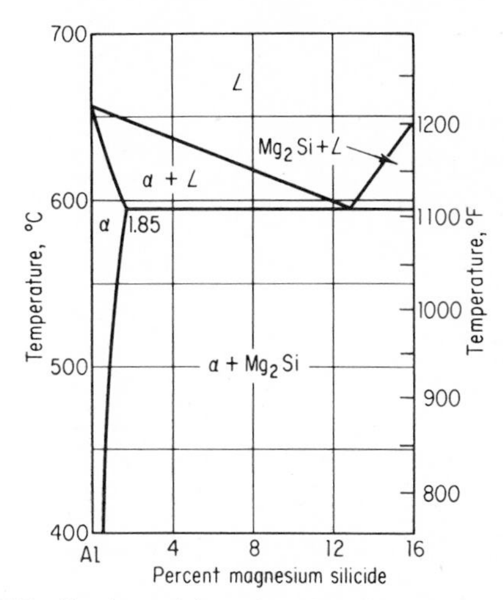

Fig. 12·26 Aluminum-rich portion of the aluminum-magnesium silicide system. (From "Metals Handbook," 1948 ed., p. 1246, American Society for Metals, Metals Park, Ohio.)

magnesium) for aircraft fuel and oil lines; alloy 5083 (4.5 percent magnesium) for marine and welded structural applications; and alloy 5056 (5.2 percent magnesium) for insect screens, cable sheathing, and rivets for use with magnesium alloys.

The aluminum-magnesium casting alloys include alloy 214 (3.8 percent magnesium), alloy 218 (8 percent magnesium), and alloy 220 (10 percent magnesium). The first two are used for dairy and food handling equipment, fittings for chemical and sewage use, fittings for marine use, and aircraft brake shoes. Alloy 220 is the only one in this group which is age-hardenable, resulting in the highest mechanical properties of any of the aluminum casting alloys. The casting properties of alloys in this group are poor, and they require careful foundry practice.

12·23 Aluminum-Silicon-Magnesium Alloys (6xxx Series) Magnesium and silicon combine to form a compound magnesium silicide (Mg_2Si), which in turn forms a simple eutectic system with aluminum. The aluminum-rich portion of the Al-Mg_2Si system is shown in Figure 12·26. It is precipitation of the Mg_2Si after artificial aging (temper T6) which allows these alloys to reach their full strength. The wrought alloys include 6053, 6061, and 6063. Magnesium and silicon are usually present in the ratio to form magnesium silicide. The structure of alloy 6061 plate, hot-rolled, shows particles of Fe_3SiAl_{12} (gray, scriptlike) and Mg_2Si (black) in an aluminum-rich solid-solution matrix, Figure 12·27. These alloys are characterized by excellent corrosion resistance and are more workable than other heat-treatable al-

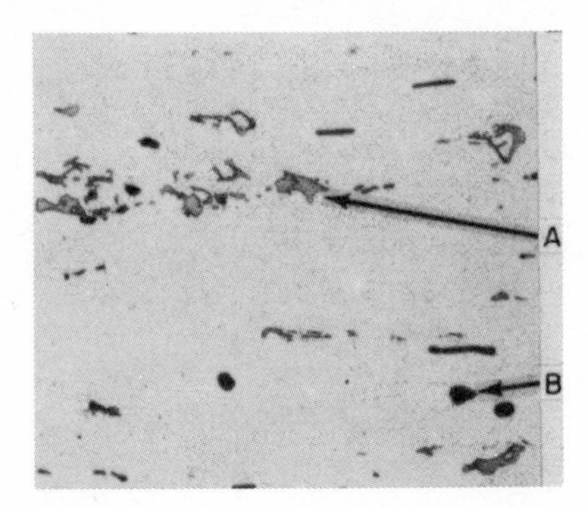

Fig. 12·27 Alloy 6061 plate, as hot-rolled, longitudinal section. Particles (*A*) of Fe_3SiAl_{12} (gray, scriptlike) and Mg_2Si (black) (*B*) in a matrix of aluminum-rich solid solution. Etched in 0.5 percent HF, 250×. (From Metals Handbook, vol. 7, "Atlas of Microstructures," American Society for Metals, 1972.)

loys. Typical applications include aircraft landing mats, canoes, furniture, vacuum-cleaner tubing, bridge railings, and architectural applications.

The aluminum-silicon-magnesium casting alloys 355, 356, and 360 provide a desirable combination of castability, pressure-tightness, strength, and corrosion resistance. In the heat-treated condition their mechanical properties approach those of the aluminum-copper alloys. They are widely used in aircraft applications, machine-tool parts, and general-purpose castings.

12·24 Aluminum-Zinc Alloys (7xxx Series) The aluminum-rich portion of the aluminum-zinc alloy system is shown in Figure 12·28. The solubility of zinc

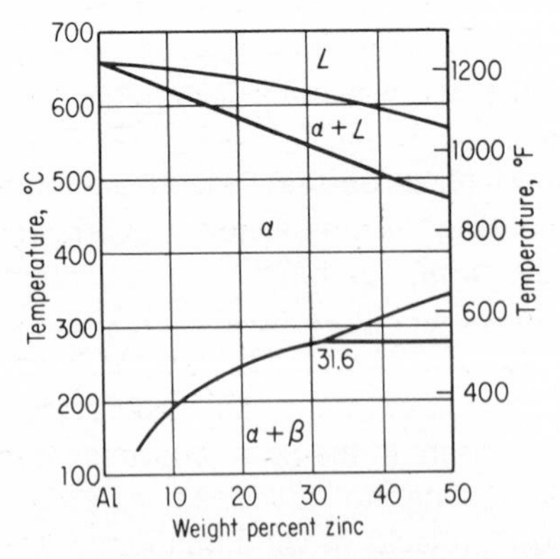

Fig. 12·28 Aluminum-rich portion of the aluminum-zinc alloy system. (From "Metals Handbook," 1948 ed., p. 1167, American Society for Metals, Metals Park, Ohio.)

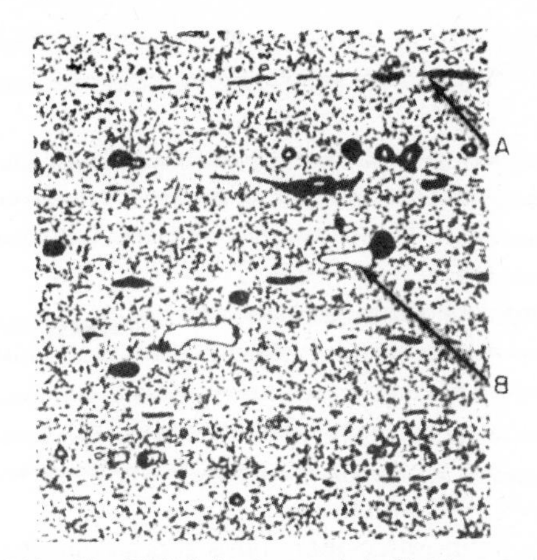

Fig. 12·29 Alloy 7075-0 sheet, annealed. Structure consists of fine and coarse particles (*A*) of $MgZn_2$ (black) and a few insoluble particles (*B*) of $FeAl_3$ (light gray, outlined) in a matrix of aluminum-rich solid solution. Etched in 25 percent nitric acid, 500×. (From Metals Handbook, vol. 7, "Atlas of Microstructures," American Society for Metals, 1972.)

in aluminum is 31.6 percent at 527°F, decreasing to 5.6 percent at 257°F. Commercial wrought alloys contain zinc, magnesium, and copper with smaller additions of manganese and chromium. Alloy 7075 (5.5 percent zinc, 2.5 percent magnesium, 1.5 percent copper), alloy 7079 (4.3 percent zinc, 3.3 percent magnesium, 0.6 percent copper), and alloy 7178 (6.8 percent zinc, 2.7 percent magnesium, 2.0 percent copper) develop the highest tensile strengths obtainable in aluminum alloys. Figure 12·29 shows the structure of alloy 7075-0 sheet, annealed. It consists of fine and coarse particles of $MgZn_2$ (black) and a few insoluble particles of $FeAl_3$ (light gray, outlined) in a matrix of α (aluminum-rich) solid solution.

The susceptibility of these alloys to stress corrosion has been minimized by the addition of chromium and by proper heat treatment. They are used in applications requiring high strength and good corrosion resistance, such as aircraft structural parts.

The aluminum-zinc casting alloy known as 40E, containing 5.5 percent zinc, 0.6 percent magnesium, 0.5 percent chromium, and 0.2 percent titanium, provides high mechanical properties without solution treatment. This alloy also has fair casting characteristics, good corrosion resistance, and very good machinability. It is used for aircraft fittings, turret housings, and radio equipment.

12·25 Corrosion Resistance of Aluminum and Aluminum Alloys The high corrosion resistance of aluminum is due to the self-protecting, thin, invisible oxide film that forms immediately on exposing surfaces to the atmosphere.

This film protects the metal from further corrosion. If the oxide film is removed, in many environments, a new film will form immediately and the metal remains fully protected.

In certain strongly acid or alkaline solutions, or in contact with moist corrosive materials that prevent access of oxygen to the aluminum surface, the protective film does not form readily. Therefore, the aluminum should be adequately protected or not used at all.

A relatively thick oxide coating on aluminum and aluminum alloys may be produced by placing the metal into an aqueous solution containing 15 to 25 percent sulfuric acid. This process, known as *anodizing,* produces a clear, transparent coating containing submicroscopic pores that are usually sealed before use to prevent absorption and staining. Sealing may be accomplished by suitable heating in hot water.

The corrosion resistance of aluminum-copper alloys and aluminum-zinc alloys is satisfactory for most applications but is generally lower than that of the other aluminum alloys. Under certain corrosive conditions they are subject to intergranular corrosion. Therefore, these alloys in the form of

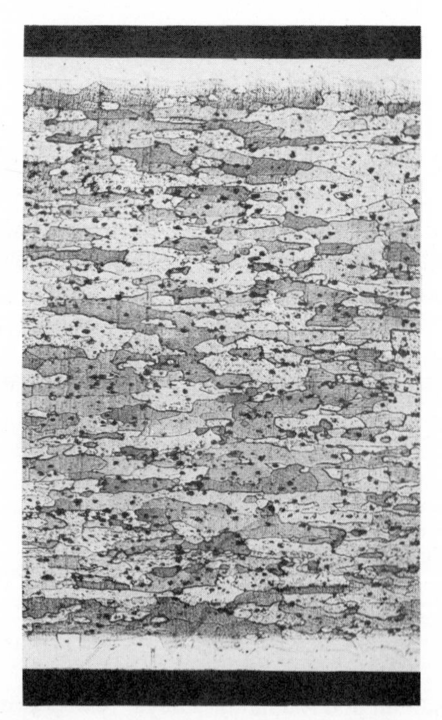

Fig. 12·30 Full cross section of Alclad 2024-T4 sheet showing white cladding layer of commercially pure aluminum on top and bottom of sheet, 125×. (Research Laboratory, Aluminum Company of America.)

TABLE 12·6 Nominal Composition and Typical Mechanical Properties of Some Wrought-aluminum Alloys*

ALLOY AND TEMPER	NOMINAL COMPOSITION, %							TYPICAL MECHANICAL PROPERTIES			
	Si	Cu	Mn	Mg	Cr	Zn	Ni	TENSILE STRENGTH, 1,000 PSI	YIELD STRENGTH, 1,000 PSI	ELONGATION, % IN 2 IN., ½-IN.-DIAM SPECIMEN	BHN
EC	(99.45 + % aluminum)										
-O								12	4		
-H14								16	14		
1100	(99.0 + % aluminum)										
-O								13	5	45	23
-H14								18	17	20	105
-H18								24	22	15	44
2014	0.8	4.4	0.8	0.4							
-O	...	...	...	...	...	...	...	27	14	18	45
-T4	...	...	...	...	...	...	...	62	42	20	105
-T6	...	...	...	...	...	...	...	70	60	13	135
2017	0.8	4.0	0.5	0.5	0.1						
-O	...	...	...	...	...	...	...	26	10	22	45
-T4	...	...	...	...	...	...	...	62	40	22	105
2024	0.5	4.5	0.6	1.5	0.1						
-O	...	...	...	...	...	...	...	27	11	22	47
-T4	...	...	...	...	...	...	...	68	47	19	120
2218	0.2	4.0	...	1.5	...	...	2.0				
-T61	...	...	...	...	...	...	...	59	44	13	
3003	0.6	...	1.2								
-O	...	...	...	...	...	...	...	16	6	40	28
-H14	...	...	...	...	...	...	...	22	21	16	40
4032	12.5	0.9	...	1.0	...	...	0.9				
-T6	...	...	...	...	...	...	...	55	46	9	120
5005	0.4	...	...	0.8							
-O	...	...	...	...	...	...	...	18	6	30†	28
-H34	...	...	...	...	...	...	...	23	20	8†	41
5050	0.4	...	...	1.2							
-O	...	...	...	...	...	...	...	21	8	24†	36
-H34	...	...	...	...	...	...	...	28	24	8†	53
5052	...	...	...	2.5	0.25						
-O	...	...	...	...	...	...	...	28	13	30	47
-H34	...	...	...	...	...	...	...	38	31	14	68
5056	...	...	0.1	5.2	0.1						
-O	...	...	...	...	...	...	...	42	22	35	65
-H18	...	...	...	...	...	...	...	63	59	10	105
5083	...	...	0.7	4.5							
-O	...	...	...	...	...	...	...	42	21	22	
6061	0.6	0.25	...	1.0	0.25						
-O	...	...	...	...	...	...	...	18	8	30	30
-T6	...	...	...	...	...	...	...	45	40	17	95
6063	0.4	0.1	0.1	0.7	0.1	0.1					
-O	...	...	...	...	...	...	...	13	7	...	25
-T6	...	...	...	...	...	...	...	35	31	12†	73
7075	0.5	1.5	...	2.5	0.3	5.5					
-O	...	...	...	...	...	...	...	33	15	16	60
-T6	...	...	...	...	...	...	...	83	73	11	150
7178	0.5	2.0	...	2.7	0.3	6.8					
-O	...	...	...	...	...	...	...	33	15	16	
-T6	...	...	...	...	...	...	...	88	78	11	

*Compiled from information in "Metals Handbook," 1961 edition, American Society for Metals, Metals Park, Ohio.
†Sheet specimens 1/16 in. thick.

TABLE 12·7 Nominal Composition and Typical Mechanical Properties of Some Cast-aluminum Alloys*

ALLOY AND TEMPER†	METHOD OF CASTING‡	NOMINAL COMPOSITION, %					TYPICAL MECHANICAL PROPERTIES			
		Si	Cu	Mg	Zn	Ni	TENSILE STRENGTH, 1,000 PSI	YIELD STRENGTH, 1,000 PSI	ELONGATION, % IN 2 IN.	BHN, 500 KG, 10 MM
13	DC	12.0	...	...	...	...	39	21	2.0	
43	SC	...	...	...	...	...	19	8	8.0	40
	PM	5.0	...	...	...	...	23	9	10.0	45
	DC	...	...	...	...	...	30	16	9.0	
85	DC	5.0	4.0	...	...	...	40	24	5.0	
108	SC	3.0	4.0	...	...	...	21	14	2.5	55
A108	PM	5.5	4.5	...	...	...	28	16	2.0	70
112	SC	...	7.0	...	1.7	...	24	15	1.5	70
113	SC	2.0	7.0	...	1.7	...	24	15	1.5	70
122-T61	SC	...	10.0	0.2	...	...	41	40	...	115
-T65	PM	...	...	...	...	...	48	36	...	140
A132-T65	PM	12.0	0.8	1.2	...	2.5	47	43	0.5	125
D132-T5	PM	9.0	3.5	0.8	...	0.8	36	28	1.0	105
195-T4	SC	0.8	4.5	...	...	...	32	16	8.5	60
-T6	SC	...	...	...	...	...	36	24	5.0	75
B195-T4	PM	2.5	4.5	...	...	...	37	19	9.0	75
B195-T6	PM	...	...	...	...	...	40	26	5.0	90
212	SC	1.2	8.0	...	...	...	23	14	2.0	65
214	SC	...	...	3.8	...	...	25	12	9.0	50
A214	PM	...	...	3.8	1.8	...	27	16	7.0	60
218	DC	...	...	8.0	...	...	45	27	8.0	
220-T4	SC	...	...	10.0	...	...	46	25	14.0	75
319-F	SC	...	...	...	...	...	27	18	2.0	70
-T6	SC	6.3	3.5	...	...	...	36	24	2.0	80
355-T6	SC	5.0	1.3	0.5	...	...	35	25	3.0	80
-T6	PM	...	...	...	...	...	43	27	4.0	90
356-T6	SC	7.0	...	0.3	...	...	33	24	3.5	70
-T6	PM	...	...	...	...	...	40	27	5.0	90
360	DC	9.5	...	...	...	...	44	27	3.0	
380	DC	9.0	3.5	...	...	...	45	26	2.0	

* Compiled from information in "Casting Alcoa Alloys," Aluminum Co. of America.
† Alcoa (Aluminum Co. of America) designation.

sheet are usually clad with a high-purity alloy such as commercial aluminum (1100) or a magnesium-silicon alloy of the 6000 series. The coating slabs are mechanically attached to the alloy core ingot, and the bonding is accomplished by hot rolling. The nominal cladding thickness per side is usually 1½ or 2½ percent of the thickness of the base material. Figure 12·30 shows the full cross section of Alclad 2024 sheet. The clear white cladding layer is visible at the edges of the sheet. Alclad alloys are extensively used for aircraft applications because of the excellent combination of high strength and high resistance to corrosion.

The nominal chemical composition and typical mechanical properties of some wrought- and cast-aluminum alloys are given in Tables 12·6 and 12·7. Table 12·8 shows the relationship of electrical conductivity and hardness for some aluminum alloys.

TABLE 12·8 Relationship of Electrical Conductivity and Hardness for Some Aluminum Alloys*

ALLOY	TEMPER	CONDUCTIVITY, % IACS†	BHN 500 KG, 10MM
EC	0	62	19
1100	0	59	23
	H18	57	44
2014	0	50	45
	T4	30	105
	T6	40	135
2024	0	50	47
	T3,T4	30	120
3003	0	50	28
	H12	42	35
	H18	40	55
5052	0	35	47
	H38	35	77
6061	0	45	30
	T4	40	65
	T6	40	95
7075	T6	30	150
B195	T4	33	75
	T6	33	90
355	T51	43	65
	T6	36	80
	T7	42	85
356	T51	43	60
	T6	39	70
	T7	40	75

*Compiled from "Metals Handbook," vol. 1, American Society for Metals, 1961.
†International Annealed Copper Standard.

MAGNESIUM AND MAGNESIUM ALLOYS

12·26 Magnesium The chief advantages of magnesium are its light weight, ease of machinability, and the high strength-to-weight ratio obtainable with its alloys.

On the basis of equal volumes, aluminum weighs 1½ times more, iron and steel weigh 4 times more, and copper and nickel alloys weigh 5 times more than magnesium.

Magnesium has a c.p.h. (close-packed hexagonal) crystal structure, and plastic deformation takes place at room temperature by slip along the basal planes. The ductility of magnesium is lower than that of f.c.c. metals since there are fewer slip systems available for plastic deformation. Above 400°F, however, additional planes become active, and the plasticity of magnesium and many of its alloys is improved.

Commercially pure magnesium or primary magnesium has a minimum purity of 99.8 percent and usually contains small amounts of aluminum, iron, manganese, silicon, and copper.

Approximately half the magnesium produced is used in alloy form for structural purposes, primarily in the aircraft and missile industries. Magnesium is used as an alloying element in aluminum, zinc, lead, and other nonferrous alloys. It has found increasing use in photo-engraving because of its light weight and rapid but controlled etching characteristics.

Magnesium has a great affinity for oxygen and other chemical oxidizing agents. It is used as a deoxidizer and desulfurizer in the manufacture of nickel and copper alloys, also as a "getter" in the manufacture of vacuum tubes. Because of its high chemical activity, it finds use in the production of uranium and zirconium by thermal reduction with magnesium.

Magnesium anodes provide effective corrosion protection for water heaters, underground pipelines, ship hulls, and ballast tanks.

12·27 Alloy Designation and Temper The American Society for Testing Materials (ASTM) has published a system of alloy nomenclature (Specification B275-61) and temper designation (Specification B296-61) for light metals and alloys. This system has been officially adopted by The Magnesium Association for all magnesium alloys. The temper designation is the same as that adopted by The Aluminum Association for aluminum alloys and is covered in Sec. 12·18. The designation for alloys and unalloyed metals is based on their chemical-composition limits as follows (from ASTM Designation B275-61 by permission of the American Society for Testing Materials):

ALLOYS

a Designations for alloys consist of not more than two letters representing the alloying elements (Note 1) specified in the greatest amount, arranged in order of decreasing percentages, or in alphabetical order if of equal percentages, followed by the respective percentages rounded off to whole numbers and a serial letter (Notes 2 and 3). The full name of the base metal precedes the designation, but it is omitted for brevity when the base metal being referred to is obvious.

b The letters used to represent alloying elements should be those listed in Table 12·9.

c In rounding-off percentages, the nearest whole number shall be used. If the decimal is followed by a 5, the nearest even whole number shall be used.

d When a range is specified for the alloying element, the rounded-off mean should be used in the designation.

e When only a minimum percentage is specified for the alloying element, the rounded-off minimum percentage should be used in the designation.

NOTE 1 For codification, an alloying element is defined as an element (other than the base metal) having a minimum content greater than zero either directly specified or computed in accordance with the percentages specified for other elements. The amount present is the mean of the range (or the minimum percentage if only that is specified) before rounding off.

NOTE 2 The serial letter is arbitrarily assigned in alphabetical sequence starting with A (omitting I and O) and serves to differentiate otherwise identical designations. A serial letter is necessary to complete each designation.

NOTE 3 The designation of a casting alloy in ingot form is derived from the composition specified for the corresponding alloy in the form of castings. Thus, a casting ingot designation may consist of an alloy designation having one or more serial letters, one for each product composition, or it may consist of one or more alloy designations.

UNALLOYED METALS

Designations for unalloyed metals consist of the specified minimum purity, all digits retained but dropping the decimal point, followed by a serial number (Note 2). The full name of the base metal precedes the designation, but it is omitted for brevity when the base metal being referred to is obvious.

As an example, for magnesium alloy AZ92A, "A" represents aluminum, the alloying element specified in the greatest amount; "Z" represents zinc, the alloying element specified in the second greatest amount; "9" indicates that the rounded-off mean aluminum percentage lies between 8.6 and 9.4; "2" signifies that the rounded-off mean zinc percentage lies between 1.5 and 2.5; and "A" as the final letter indicates that this is the first alloy whose composition qualified assignment of the designation AZ92.

TABLE 12·9 Letters Representing Alloy Elements

A	Aluminum	M	Manganese
B	Bismuth	N	Nickel
C	Copper	P	Lead
D	Cadmium	Q	Silver
E	Rare earths	R	Chromium
F	Iron	S	Silicon
G	Magnesium	T	Tin
H	Thorium	Y	Antimony
K	Zirconium	Z	Zinc
L	Beryllium		

12·28 Magnesium Alloys Although most magnesium alloys are ternary alloys, they may be considered as based upon four binary-alloy systems. These are magnesium-aluminum, magnesium-zinc, magnesium–rare earths, and magnesium-thorium. In each case the solvus line on the magnesium-rich side shows a decrease in solubility of the alloying element in solid magnesium as the temperature decreases. This indicates that certain alloy compositions may be strengthened by age hardening.

For example, in Fig. 12·31 showing the magnesium-rich portion of the aluminum-magnesium alloy system, the maximum solubility of aluminum in magnesium is 12.7 percent at 818°F, decreasing to 3.2 percent at 400°F. Alloys containing over 6 percent aluminum, which include all the Mg-Al casting alloys, are therefore heat-treatable.

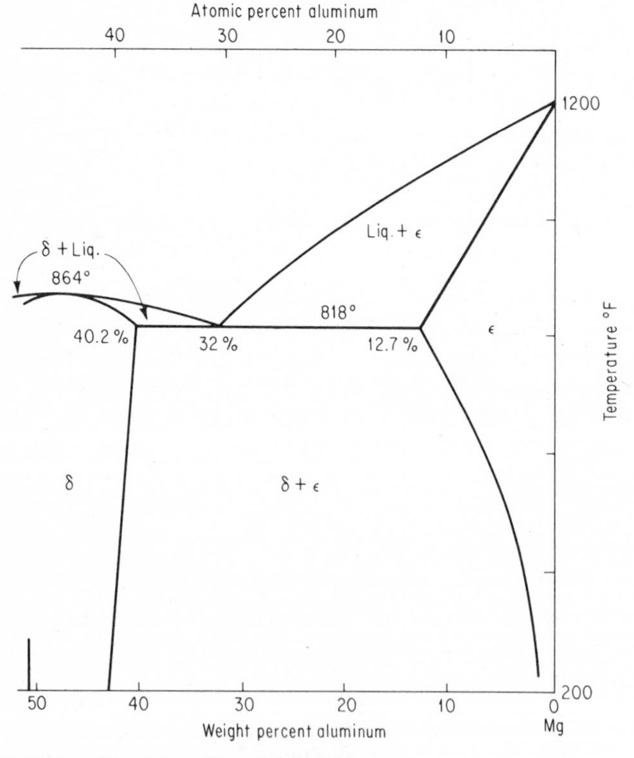

Fig. 12·31 Magnesium-rich portion of the aluminum-magnesium alloy system. The δ phase is also known as the compound $Mg_{17}Al_{12}$, a designation based on crystallographic evidence rather than composition, since Mg_3Al_2 would be simpler and within the δ homogeneity field. (From "Metals Handbook," 1948 ed., p. 1163, American Society for Metals, Metals Park, Ohio.)

The solubility of zinc in solid magnesium varies from 8.4 percent at 644°F to 1.7 percent at 300°F. Alloys in the composition range of 4 to 8 percent zinc show the most potent precipitation-hardening effects of any of the magnesium-based binary systems.

Magnesium-Aluminum-based Alloys This group includes the magnesium-aluminum-manganese (AM) and the magnesium-aluminum-zinc (AZ) casting alloys.

The AM100A alloy is popular for pressure-tight sand and permanent-mold castings with a good combination of tensile strength, yield strength, and elongation.

The sand-casting alloys AZ63A and AZ92A are used for normal-temperature applications. If the operating stresses are not too high, they may give satisfactory service at temperatures as high as 350°F. AZ63A is used where maximum toughness or ductility along with moderately high yield strength are required. AZ92A is used where maximum yield strength plus good pressure-tightness are required. The microstructures of AZ92A alloy are shown in Fig. 12·32. The structure in the sand-cast condition (Fig. 12·32*a*) shows a network of massive δ ($Mg_{17}Al_{12}$). The eutectic was originally composed of δ and ϵ solid solution rich in magnesium. However, most of the eutectic has "separated"; the ϵ phase has joined the primary ϵ grains, leaving behind the compound as a white network. The compound may be retained in solution by the relatively rapid cooling after solution treatment (Fig. 12·32*b*). Subsequent artificial aging causes precipitation of the compound as fine particles within the grains (Fig. 12·32*c*). Because of the more rapid cooling in permanent-mold casting as compared with sand casting, the structure (Fig. 12·32*d*) shows finer particles of the compound and no lamellar constituent. AZ91C and AZ81A are gradually replacing AZ63A for applications requiring good ductility and moderate yield strength.

AZ91A and AZ91B alloys are especially suited for die casting. The castings are pressure-tight with good yield strength and ductility.

The wrought alloy M1A (containing 1.2 percent manganese) is a low-cost, relatively low-strength magnesium alloy. It has excellent weldability, corrosion resistance, and hot formability.

The AZ31 alloys are widely used as general-purpose extrusion alloys having good strength and formability. Since they contain only 3 percent aluminum, they are not heat-treatable and attain their properties by strain hardening. AZ31 with low calcium content, known as PE alloy, is widely used for magnesium photoengraving sheet in the printing industry.

AZ61A has excellent strength and ductility and is used mainly as an extrusion and forging alloy.

AZ80A alloy, containing 8.5 percent aluminum, is a heat-treatable alloy

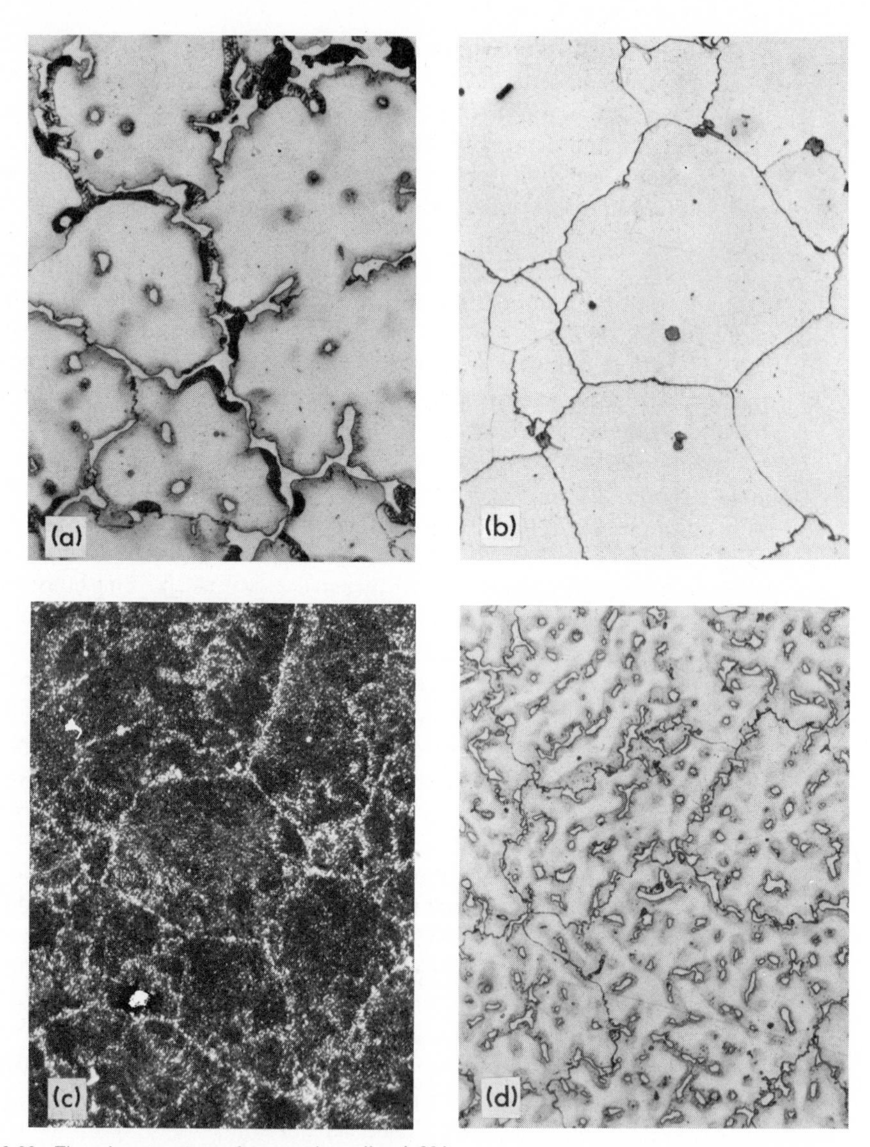

Fig. 12·32 The microstructure of magnesium alloy A-92A, 250×. (*a*) Sand-cast, etched in phosphopicral; (*b*) sand-cast, solution-treated, etched in glycol; (*c*) sand-cast, solution-treated, artificially aged, etched in acetic glycol; (*d*) permanent-mold cast, etched in phosphopicral. (The Dow Metal Products Company.)

used for extruded products and press forgings. It offers an excellent combination of high strength and moderate elongation.

Magnesium-Zinc-based Alloys These are essentially magnesium-zinc-zirconium (ZK) and magnesium-zinc-thorium (ZH) casting alloys. The effect of zirconium additions, up to about 0.7 percent, to magnesium alloys is one of grain refinement. Zirconium completely eliminates the coarse-grained, columnar cast structure, thus increasing the mechanical properties.

The casting alloys ZK51A and ZK61A attain the highest combination of tensile strength and ductility of any magnesium casting alloys. This is due to the strong age-hardening effect of the magnesium-zinc binary system and the fine-grain effect of zirconium. Unfortunately, the high zinc content requires very careful foundry control to produce sound castings free of microporosity and hot cracks. The addition of thorium (ZH62A) helps to reduce these problems with little effect on mechanical properties, while the addition of a rare-earth metal (ZE41A) also reduces the above problems with some reduction in mechanical properties. Figure 12·33 shows the characteristic lamellar form of eutectic Mg-Th-Zn compound at the grain boundaries of magnesium-rich solid solution in a ZH62A-T5 alloy, sand cast.

The wrought alloy ZK60A is the highest-strength extrusion alloy for solid and hollow shapes. It has good toughness and may be heat-treated to improve its properties further. The replacement of zirconium by a rare-earth metal (ZE10A) results in a cheaper and tougher alloy.

Magnesium-Rare Earth-based Alloys This group includes the magnesium-rare earth-zirconium (EK) and magnesium-rare earth-zinc (EZ)

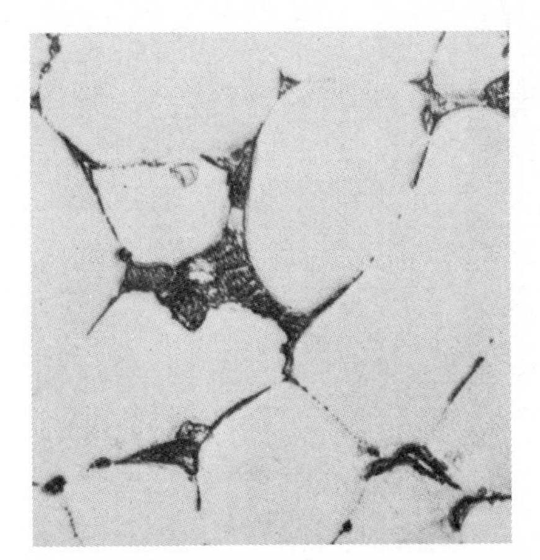

Fig. 12·33 Alloy ZH62A-T5 sand casting. Lamellar eutectic Mg-Th-Zn compound at the grain boundaries of magnesium-rich solid solution. Etched in 2 percent nital, 250×. (From Metals Handbook, vol. 7, "Atlas of Microstructures," American Society for Metals, 1972.)

casting alloys. The rare-earth elements are atomic numbers 58–71 (see Table 2.3).

Improvement in elevated-temperature properties is obtained by a high recrystallization temperature and by precipitates at the grain boundaries that are stable at high temperature to minimize creep. The addition of rare-earth elements to magnesium satisfies both requirements, so that these alloys are suitable for use up to 500°F.

Casting alloys EK30A, EK41A, and EZ33A all have similar mechanical properties, but EZ33A shows poorer corrosion resistance because of the presence of zinc. The structure of EZ33A-T5 alloy, sand cast, is shown in Fig. 12·34. It consists of a network of massive magnesium–rare earth compound (dark) in a magnesium-rich solid solution (light).

When the rare-earth addition contains 50 percent cerium, it is known as *mischmetal* and is used as a low-cost commercial alloy.

The rare-earth elements are not used as the principal addition in any wrought-magnesium alloys.

Magnesium-Thorium–based Alloys The group includes the magnesium-thorium-zirconium (HK) and magnesium-thorium-zinc (HZ) casting alloys. Thorium, like the rare-earth elements, greatly improves the elevated-temperature properties of magnesium.

Casting alloys HK31A and HZ32A are used for applications in the range of 350 to 700°F where properties better than those of the rare earth–containing alloys are needed. HK31A is particularly good for short-time elevated-

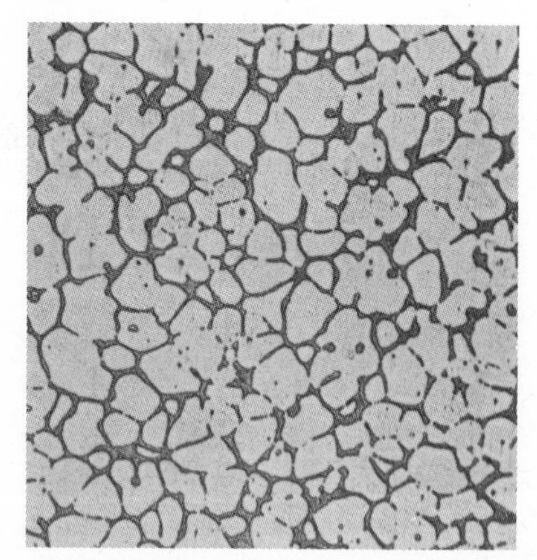

Fig. 12·34 Alloy EZ33A-T5 sand casting. Structure consists of a network of massive magnesium–rare earth compound (dark) in a magnesium-rich solid solution (white). Etched in glycol, 100×. (From Metals Handbook, vol. 7, "Atlas of Microstructures," American Society for Metals, 1972.)

temperature uses where high stresses are encountered, while HZ32A is preferred where long-time lower-stress properties are important.

HK31A is also used as a wrought sheet alloy for high-temperature applications. Other wrought alloys in this group are the extrusion alloy HM31XA and the sheet alloy HM21A, both of which are magnesium-thorium-manganese alloys. These have the best high-temperature properties of any wrought-magnesium alloy.

The nominal composition and typical mechanical properties of some cast- and wrought-magnesium alloys are given in Table 12·10.

12·29 Corrosion Resistance of Magnesium Alloys The resistance of magnesium alloys to atmospheric corrosion depends upon the alloying element, the amount present, indoor or outdoor exposure, and humidity. At low values of humidity (below 10 percent), magnesium alloys show good corrosion resistance, but the resistance decreases with increasing humidity. During indoor exposure, the magnesium alloys show better corrosion resistance then those containing aluminum, but the reverse is true under outdoor exposure conditions. With aluminum content above 9 percent and zinc content above 5 percent, corrosion resistance is decreased. Experimental results indicate that the corrosion resistance of magnesium compares favorably with that of aluminum alloys and is superior to that of low-carbon steel in industrial atmospheres. Under marine exposure, the purity of the alloy seems to be a controlling factor in determining corrosion resistance. Controlled-purity cast alloys show a far superior corrosion resistance under marine conditions than uncontrolled-purity alloys.

The corrosion resistance of magnesium alloys in aqueous salt solutions is dependent upon the presence of impurities such as iron, copper, nickel, and cobalt.

The effect of heat treatment on the corrosion resistance of magnesium alloys in salt solutions varies with the heat-treating temperature, rate of cooling, alloying elements, and impurity content. In the magnesium-manganese alloy M1, heat treatment in the range of 500 to 900°F greatly increases the corrosion rate and the tendency to pit. However, heat treatment at about 1050°F, followed by rapid cooling, completely eliminates pitting and increases corrosion resistance.

12·30 Joining Magnesium Alloys Magnesium and magnesium alloys may be joined by most of the common fusion and mechanical fastening methods. These include shielded-metal arc welding, gas welding, electric resistance seam and spot welding, riveting, bolting, self-fastening devices, and adhesive bonding. Arc welding, spot welding, and riveting are the most commonly used methods.

In all magnesium alloys, the solidification range increases and the melting point and shrinkage decrease with an increase in alloy content. Alloys con-

TABLE 12·10 Nominal Composition and Typical Mechanical Properties of Magnesium Alloys at Room Temperature*

ALLOY AND TEMPER	NOMINAL COMPOSITION, %						TENSILE STRENGTH, 1,000 PSI	TENSILE YIELD STRENGTH, 1,000 PSI	ELONGATION, %	BHN, 500 KG, 10 MM	COMPRESSIVE YIELD STRENGTH, 1,000 PSI
	Al	Mn	Zn	Th	Zr	RARE EARTHS					
SHEET, ANNEALED	99.8% Mg MINIMUM						23–28	13–15	3–15	40–41	10–12
SHEET, HARD TEMPER	...						26–32	17–20	2–10	45–47	15–17
SAND AND PERMANENT-MOLD CASTING ALLOYS											
AM100-T6	10.0	0.1	...	...	...	...	40	22	1	69	19
AZ63A-T6	6.0	0.2	3.0	...	...	...	40	19	5	73	19
AZ81A-T4	7.5	0.15	0.7	...	...	...	40	12	15	55	12
AZ91C-T6	9.0	0.2	0.7	...	...	...	40	19	5	70	19
AZ92A-T6	9.0	0.1	2.0	...	...	...	40	22	3	81	22
ZK51A-T5	...	...	4.5	...	0.7	...	40	24	8	65	24
ZK61A-T5	...	...	6.0	...	0.7	...	...	27	...	68	27
ZK61A-T6	...	...	6.0	...	0.7	...	...	28	...	70	28
ZH62A-T5	...	...	5.7	1.8	0.7	...	39	26	4	70	
ZE41A-T5	...	...	4.25	...	0.5	1.25	30	20	3.5	62	20
EK30A-T6	...	...	...	...	0.6	4.0	23	16	3	45	16
EK41A-T6	...	...	...	...	0.2	3.0	25	18	3	50	18
EZ33A-T5	...	...	2.7	...	0.5	3.0	23	16	3	50	16
HK31A-T6	...	...	...	3.25	0.7	...	32	15	8	66	15
HZ32A-T5	...	...	2.0	3.25	0.75	...	27	13	4	55	16
DIE-CASTING ALLOYS											
AZ91A and B-F	9.0	0.2	0.7	...	...	...	33	22	3	63	24
EXTRUDED BARS AND SHAPES											
MIA-F	...	1.2	...	...	...	...	37	26	12	44	12
AZ31B-F	3.0	0.2	1.0	...	...	...	38	29	15	49	14
AZ61A-F	5.5	0.2	1.0	...	...	...	45	33	16	60	19
AZ80A-T5	8.5	...	0.5	...	...	...	55	40	7	80	35
ZK60A-T5	...	...	5.5	...	0.5	...	53	44	11	82	36
SHEET AND PLATE ALLOYS											
AZ31B-H24	3.0	0.2	1.0	...	...	...	42	32	15	73	26
HK31A-H24	...	...	...	3.25	0.7	...	37	29	8	...	23
HM21A-T8	...	0.5	...	2.0	...	...	34	25	10	...	15
ZE10A-H24	...	...	1.25	...	...	0.17	38	28	13	...	24

*Compiled from information in "Metals Handbook," 1961 ed., American Society for Metals, Metals Park, Ohio.

taining up to 10 percent aluminum aid weldability by refining the grain structure. Alloys containing more than 1 percent zinc have a tendency to crack when hot (*hot-short*) and may result in weld cracking. Thorium alloys have no tendency toward hot-shortness.

Failure of welded joints usually occurs in the heat-affected zone next to the weld rather than the weld itself. This is most likely due to some grain growth of the base metal.

Welds in some magnesium alloys, particularly the Mg-Al-Zn series, are susceptible to stress-corrosion cracking. This is due to the high residual stresses set up in the welding process. These residual stresses may be relieved by a suitable stress-relief treatment.

Adhesive bonding is a comparatively new method of joining. Since no drilling is required, there is less stress concentration and better fatigue strength in adhesive-bonded joints as compared with other types. The adhesive fills the spaces between contacting surfaces and acts as an insulator between any dissimilar metals in the joint. Adhesive bonding allows the use of thinner materials and takes better advantage of the weight saving possible with magnesium. This method of joining is particularly suited for joining stiffeners to sheet and is used in aircraft and radar applications.

Riveting is the most common method of joining magnesium, since it gives joints of good strength and efficiency, is a fairly simple method, and does not require highly skilled labor.

NICKEL AND NICKEL ALLOYS

12·31 Nickel Nickel is characterized by good resistance to corrosion and oxidation. It is white in color and has good workability and good mechanical properties. It forms tough, ductile solid-solution alloys with many of the common metals. Approximately 60 percent of the nickel produced is used in stainless and nickel-alloy steels. Most of the remainder is used in high-nickel alloys and for electroplating. Because of its high corrosion resistance and hardness, nickel makes an ideal coating for parts subjected to corrosion and wear. Although nickel is often given a flash coating of chromium to increase wear resistance, most of the corrosion protection is due to the heavy nickel undercoat.

Cast nickel is sometimes used for corrosion-resistant castings, particularly where contamination with copper or iron must be avoided. Small amounts of silicon and manganese are added to facilitate the production of sound, ductile castings.

Wrought nickel is not adversely affected by cold working, welding, or heating. Its mechanical properties are similar to those of mild steel. It retains its strength at elevated temperatures and its ductility and toughness at low temperatures. The electrical conductivity of nickel, while not so

TABLE 12·11 Nominal Composition and Typical Mechanical Properties of Commercial Nickels*

MATERIAL	NOMINAL COMPOSITION (ESSENTIAL ELEMENTS), %	CONDITION	TENSILE STRENGTH, 1,000 PSI	YIELD STRENGTH, 0.2% OFFSET, 1,000 PSI	ELONGATION, % IN 2 IN.	BHN
Nickel (pure)	Ni 99.99	Annealed	46	8.5	30	
A nickel (wrought)	99.40 Ni(+Co), 0.06 C, 0.25 Mn, 0.15 Fe, 0.05 Si, 0.05 Cu, 0.005 S	Annealed	70	20	40	100
		Hot-rolled	75	25	40	110
		Cold-drawn	95	70	25	170
		Cold-rolled	105	95	5	210
D nickel	95.00 Ni(+Co), 0.10 C, 4.75 Mn, 0.05 Fe, 0.05 Si, 0.02 Cu, 0.005 S,	Annealed	75	35	40	140
		Hot-rolled	90	50	35	150
		Cold-drawn	100	80	25	190
Nickel (cast)	95.6 Ni, 0.5 Cu, 0.5 Fe, 0.8 Mn, 1.5 Si, 0.8 C	As cast	57	25	22	110
Duranickel	93.90 Ni(+Co), 0.15 C, 0.25 Mn, 0.15 Fe, 0.005 S, 0.55 Si, 0.05 Cu, 4.50 Al, 0.45 Ti	Annealed	100	45	40	160
		Annealed, age-hardened	170	125	25	330
		Spring temper	175	. . .	5	320
		Spring, age-hardened	205	. . .	10	370
Permanickel	98.65 Ni(+Co), 0.25 C, 0.10 Mn, 0.10 Fe, 0.005 S, 0.06 Si, 0.02 Cu, 0.45 Ti, 0.35 Mg	Annealed	105	45	45	160
		Annealed, age-hardened	175	125	25	325
		Spring temper	180	. . .	5	
		Spring, age-hardened	210	195	10	

* "Properties of Some Metals and Alloys," The International Nickel Co.

high as that of copper or aluminum, is satisfactory for current-carrying leads and terminals in many electronic applications.

The most important commercial grades of nickel are A nickel, D nickel, E nickel, permanickel, and duranickel.

A nickel is the basic material, containing 99 percent minimum nickel including cobalt. Cast commercial nickel contains approximately 2 percent silicon to improve fluidity and castability. A nickel is used where strength combined with resistance to corrosion and oxidation is required. Rolled nickel is used by the chemical and soap industries for the construction of evaporators, jacketed kettles, heating coils, and other processing equipment.

D nickel and *E nickel* conform generally to the composition of A nickel, the important difference being the inclusion of about 4.5 and 2.0 percent, respectively, of manganese replacing a like amount of nickel. The addition of manganese improves the resistance to atmospheric attack at elevated temperatures. The mechanical strength of D nickel, both at normal and at elevated temperatures, is somewhat greater than that of A nickel, and it has better resistance to attack by sulfur. D nickel is used extensively for spark-plug electrodes, ignition tubes, radio-tube grid wires, and marine-boiler refractory bolts. Since E nickel has a lower manganese content than D nickel, its mechanical properties are intermediate between A and D nickel. Typical uses are for spark-plug wires and as electrical lead-in wires for furnaces.

Duranickel is a wrought, age-hardenable, corrosion-resisting, nickel-aluminum alloy. It offers a combination of high strength (comparable to that of heat-treated steels) and the excellent corrosion resistance of nickel. Duranickel springs are used as laundry clips, jewelry parts, and optical frames. This alloy is also used for instrument parts such as diaphragms, bellows, snap-switch blades, and in the sports field for fish-hooks and parts of fishing tackle.

Permanickel is an age-hardenable, high-nickel alloy having mechanical properties and corrosion resistance similar to those of duranickel. In addition, good electrical and thermal conductivity is present. Its resistance to softening at elevated temperatures is somewhat inferior to that of duranickel, and it should be used in place of duranickel only in applications where higher electrical conductivity and better magnetic properties are essential.

The chemical composition and typical mechanical properties of the commercial grades of nickel are given in Table 12·11.

12·32 Nickel Alloys The most common alloying elements with nickel are copper, iron, chromium, silicon, molybdenum, manganese, and aluminum.

Nickel-Copper–based Alloys Copper is completely soluble in nickel and is added to increase formability, decrease price, and still retain the corrosion resistance of nickel. *Monel* is the most important of the nickel-copper alloys, containing approximately two-thirds nickel and one-third copper. Monel has high corrosion resistance to acids, alkalies, brines, waters, food products, and the atmosphere. It has mechanical properties higher than those of the brasses and bronzes, but lower than those of alloy steels. It also has good toughness and fatigue strength and finds considerable use in elevated-temperature applications. It does not oxidize at a destructive rate below approximately 1000°F in sulfur-free atmospheres and for some applications may be used at temperatures up to 1500°F. Monel has widespread use in the chemical, pharmaceutical, marine, power, electrical, laundry, textile, and paper-equipment fields. The microstructure of Monel annealed and cold-drawn was shown in Fig. 3·16.

R Monel is a nickel-copper alloy which contains high sulfur to improve machinability. It is produced primarily for automatic screw-machine work.

K Monel contains approximately 3 percent aluminum, which makes the alloy age-hardenable. Figure 12·35 shows the fine precipitate of $Ni_3(Al,Ti)$ as black spots in Monel K-500 alloy after aging for 4 h at 1300°F. Since the precipitate is resolvable with the optical microscope, it indicates that the alloy is overaged. Thus it is possible to obtain a nonmagnetic corrosion-resistant material with extra strength and hardness. Some typical applications of K Monel are marine pump shafts, springs, aircraft instruments, ball bearings, and safety tools.

H Monel and *S Monel*, containing 3 and 4 percent silicon, respectively, are casting alloys that combine high strength, pressure-tightness, nongall-

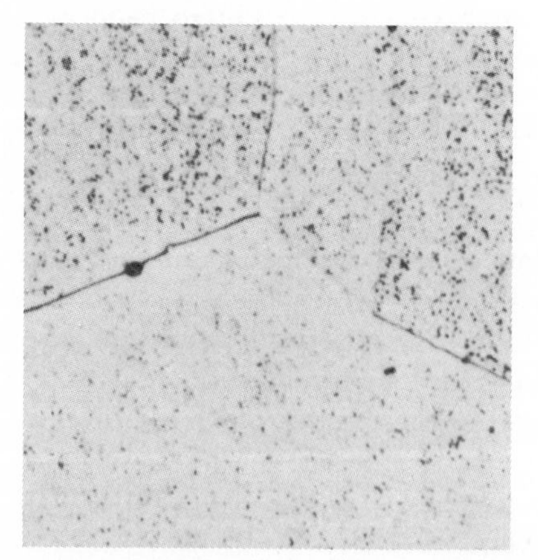

Fig. 12·35 Monel K-500, held 1 h at 2200°F. transferred to a furnace at 1300°F and aged 4 h, water-quenched. Black spots are $Ni_3(Al,Ti)$ precipitate. Etched in NaCN, $(NH_4)_2S_2O_8$, 1,000×. (From Metals Handbook, vol. 7, "Atlas of Microstructures," American Society for Metals, 1972.)

ing, and antiseizing characteristics along with resistance to corrosive attack. Both alloys have similar mechanical properties, but H Monel, containing less silicon, has better machinability. Typical applications include valve seats, pump liners, and impellers.

Constantan (45 percent nickel, 55 percent copper) has the highest electrical resistivity, the lowest temperature coefficient of resistance, and the highest thermal emf against platinum of any of the copper-nickel alloys. The first two properties are important for electrical resistors, while the last property is desirable for thermocouples. The copper-constantan and iron-constantan thermocouples were discussed in Chap. 1.

Nickel-Silicon-Copper–based Alloys The best-known commercial alloy in this group is Hastelloy D. It contains 10 percent silicon and 3 percent copper. It is a casting alloy which is strong, tough, and extremely hard. It can be machined only with difficulty and is generally finished by grinding. Its most important characteristic is its excellent corrosion resistance to concentrated sulfuric acid at elevated temperatures. It is used for evaporators, reaction vessels, pipelines, and fittings in the chemical industry.

Nickel-Chromium-Iron–based Alloys A variety of binary nickel-chromium and ternary nickel-chromium-iron alloys are used as electrical-resistance alloys. Some nominal compositions are 80Ni-20Cr (*Chromel A, Nichrome V,* and others) used as electric heating elements for household appliances and industrial furnaces; 60Ni-16Cr-24Fe (*Chromel C, Nichrome,* and others) used as electrical heating elements for toasters, percolators, waffle irons, heater pads, hair driers, and hot-water heaters, also in high-resistance rheostats for electronic equipment and as dipping baskets for acid pickling; and 35Ni-20Cr-45Fe used for heavy-duty rheostats. Many of the above alloys show good resistance to oxidation, heat fatigue, and carburizing gases. They are widely used in cast and wrought form for heat-treating equipment, furnace parts, carburizing and nitriding containers, cyaniding pots, and other equipment that must withstand temperatures up to about 1800°F.

Inconel, with a nominal composition of 76Ni-16Cr-8Fe, combines the inherent corrosion resistance, strength, and toughness of nickel with the extra resistance to high-temperature oxidation of chromium. The first applications for Inconel were in food-processing equipment such as heaters, coolers, regenerators, pasteurizers, and holding tanks for pasteurizing milk. Inconel is outstanding in its ability to withstand repeated heating and cooling in the range of zero to 1600°F without becoming brittle and is used for exhaust manifolds and heaters of airplane engines. It is used extensively in the furnace and heat-treating field for nitriding containers, carburizing boxes, retorts, muffles, and thermocouple-protection tubes.

Inconel X is an age-hardenable Inconel. Hardening is secured by additions of titanium (2.25 to 2.75 percent) and aluminum (0.4 to 1 percent). A

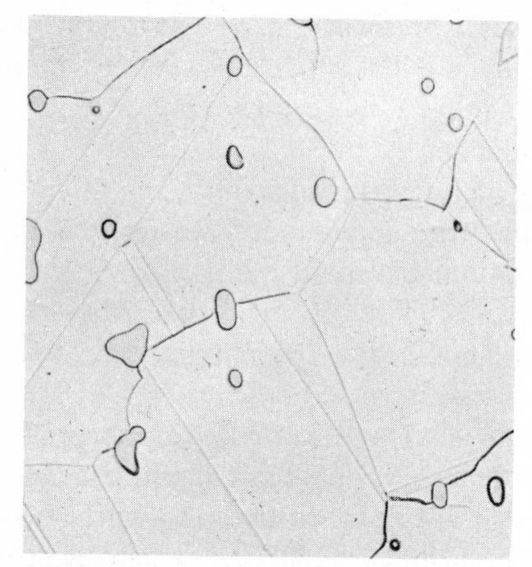

Fig. 12·36 Hastelloy B, solution-annealed at 2150°F and water-quenched. Globular constituent is carbide; matrix is f.c.c. γ solid solution. Etched in chrome regia, 500×. (From Metals Handbook, vol. 7, "Atlas of Microstructures," American Society for Metals, 1972.)

considerable portion of its high room-temperature strength is retained at temperatures up to 1500°F. Typical applications include parts that require high strength and low plastic-flow rate at temperatures up to 1500°F, such as gas turbine supercharger, and jet-propulsion parts, and springs for temperatures up to 1000°F.

Nickel-Molybdenum-Iron–based Alloys *Hastelloy A* (57Ni-20Mo-20Fe) and *Hastelloy B* (62Ni-28Mo-5Fe) are the two best-known alloys in this group. Figure 12·36 shows the structure of Hastelloy B after solution annealing and water quenching. It consists of globular carbides in a matrix of f.c.c.γ solid solution. These alloys are austenitic and therefore do not respond to age hardening. By cold working, it is possible to obtain strength and ductility comparable to those of alloy steel. These alloys are noted for their high resistance to corrosion by hydrochloric, phosphoric, and other nonoxidizing acids. They are used in the chemical industry for equipment to handle, transport, and store acids and other corrosive materials.

Nickel-Chromium-Molybdenum-Iron–based Alloys The remainder of the Hastelloy alloys fall into this group, the best-known one being *Hastelloy C* (54Ni-17Mo-15Cr-5Fe-4W). These alloys are characterized by their high corrosion resistance to oxidizing acids such as nitric, chromic, and sulfuric acids. They generally have good high-temperature properties and are resistant to oxidizing and reducing atmospheres up to 2000°F. They are used in the chemical industry, when dealing with strong oxidizing acids, for pump and valve parts, spray nozzles, and similar applications. *Hastelloy*

X (47Ni-9Mo-22Cr-18Fe) has outstanding strength and oxidation resistance up to 2200°F. It is used for many industrial-furnace applications and for aircraft parts such as jet-engine tail pipes, afterburners, turbine blades, and vanes.

Nickel-Chromium-Molybdenum-Copper-based Alloys The alloys in this group were originally developed as materials resistant to both sulfuric and nitric acids over a wide range of concentration and exposure conditions. Two casting alloys are *Illium B* (50Ni-28Cr-8.5Mo-5.5Cu) and *Illium G* (56Ni-22.5Cr-6.5Mo-6.5Cu). They provide superior corrosion resistance in machinable high-strength casting alloys. Typical applications are thrust and rotary bearings and pump and valve parts where high hardness is required in corrosive environments. *Illium R* (68Ni-21Cr-5Mo-3Cu) is a machinable wrought alloy that provides heat and corrosion resistance. It is used for pump and valve shafting, hardware items, tubing, sheet, and wire.

The nominal chemical composition and typical mechanical properties of some nickel alloys are given in Table 12·12.

12·33 Nickel-Iron Alloys The nickel-iron alloy system is shown in Fig. 12·37. Accurate phase boundaries below about 1100°F have not yet been established because of the sluggishness of structural changes at low temperatures. Nickel and iron are completely soluble in the liquid state and solidify as solid solutions. Nickel lowers progressively the γ to α transformation in

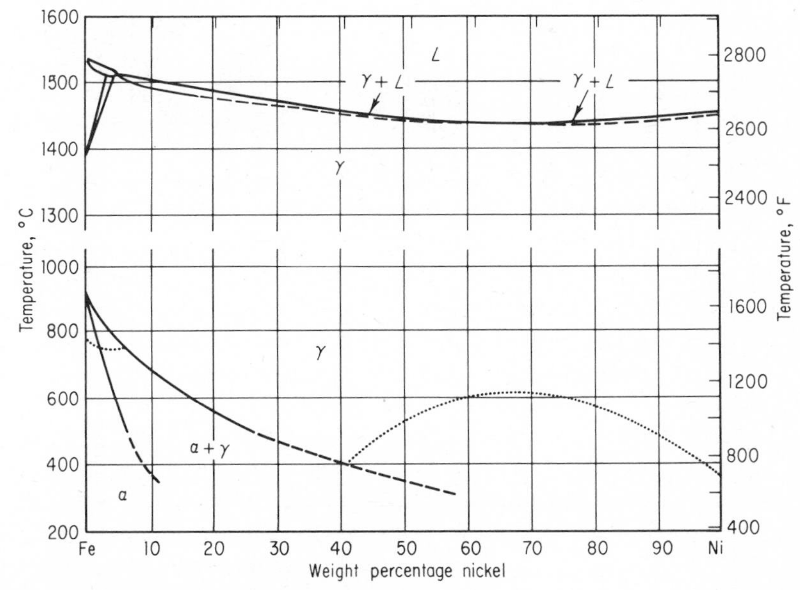

Fig. 12·37 The iron-nickel alloy system. (From Metals Handbook, 1948 ed., p. 1211, American Society for Metals.)

TABLE 12·12 Nominal Composition and Typical Mechanical Properties of Some Nickel Alloys[a]

MATERIAL	NOMINAL COMPOSITION (ESSENTIAL ELEMENTS), %	CONDITION	TENSILE STRENGTH, 1,000 PSI	YIELD STRENGTH, 0.2% OFFSET, 1,000 PSI	ELONGATION, % IN 2 IN.	BHN
Monel (wrought)	66.15 Ni(+Co), 31.30 Cu, 1.35 Fe, 0.90 Mn	Annealed	75	35	40	125
		Hot-rolled	90	50	35	150
		Cold-drawn	100	80	25	190
		Cold-rolled	110	100	5	240
Monel (cast)	64.0 Ni(+Co), 31.5 Cu, 1.5 Si	As cast	75	35	35	140
K Monel	65.25 Ni(+Co), 29.60 Cu, 2.75 Al, 0.45 Ti	Annealed	100	45	40	155
		Annealed[d]	155	100	25	270
		Spring temper	150	140	5	300
		Spring temper[d]	185	160	10	335
H Monel	63.0 Ni(+Co), 30.5 Cu, 3.2 Si	As cast	115	70[b]	10	265
Inconel (wrought)	77.0 Ni(+Co), 15.0 Cr, 7.0 Fe	Annealed	85	35	45	150
		Hot-rolled	100	60	35	180
		Cold-rolled[c]	135	110	5	260
Inconel X	72.85 Ni(+Co), 15.15 Cr, 6.80 Fe, 2.50 Ti, 0.75 Al	Annealed	115	50	50	150
		Annealed[d]	175	115	25	300

[a] "Properties of Some Metals and Alloys," The International Nickel Co.; "Metals Handbook," 1961 ed., American Society for Metals, Metals Park, Ohio.
[b] 0.5 percent extension.
[c] Hard temper.
[d] Age-hardened.

TABLE 12·12 (Continued)

MATERIAL	NOMINAL COMPOSITION (ESSENTIAL ELEMENTS), %	CONDITION	TENSILE STRENGTH, 1,000 PSI	YIELD STRENGTH, 0.2% OFFSET, 1,000 PSI	ELONGATION, % IN 2 IN.	BHN
Hastelloy alloy A	Bal. Ni, 22 Mo, 22 Fe, Mn, Si	As cast	73	44	10	180
		Rolled[e]	115	50	44	210
Hastelloy alloy B	Bal. Ni, 28 Mo, 5 Fe, Mn, Si	Sand-cast[e]	80	50	9	199
		Rolled[e]	120	56	50	215
Hastelloy alloy C	Bal. Ni, 16 Mo, 16 Cr, 5 Fe, 4 W, Mn, Si	Sand-cast[e]	78	50	5	199
		Rolled[e]	130	71	45	204
Hastelloy alloy D	Bal. Ni, 10 Si, 3 Cu, Mn	Sand-cast[e]	118	118	0–2[f]	321
Hastelloy alloy X	47 Ni, 9 Mo, 22 Cr, 8 Fe, C, W	Wrought sheet	114	52	43	90[g]
		Sand-cast	65	42	11[f]	89[g]
Illium B	50 Ni, 28 Cr, 8.5 Mo, 5.5 Cu	Grade B1 (2.5–4.5% Si)[h]	61–67	50–62[i]	1.0–4.5	200–240
		Grade B4 (6.1–6.3% Si)[h]	45–51	45–51[i]	0.5 max	325–360
Illium G	56 Ni, 22.5 Cr, 6.5 Mo, 6.5 Cu	Cast	68	50	32	200
Illium R	68 Ni, 21 Cr, 5 Mo, 3 Cu	Solution-treated	112.8	50.2	45.7	162
		20% cold-worked	142.3	128.1	11.5	238

[e] Annealed.
[f] In 1 in.
[g] Rockwell B.
[h] Grade defined by silicon content.
[i] Elastic limit.

iron. Alloys that contain up to 6 percent nickel are ferritic. As the nickel content increases, the alloys have an increasing tendency to air-harden on slow cooling. Alloys with 6 to about 30 percent nickel are martensitic after fast cooling. After slow cooling or reheating, they decompose into α plus γ phases. The amount of each phase present is dependent upon the nickel content, heat treatment, and amount of cold working. Alloys containing more than 30 percent nickel are predominantly austenitic and nonmagnetic. Alloys of iron and nickel, containing 20 to 90 percent nickel, have wide application because of their useful thermal expansion and magnetic and thermoelastic properties.

As the nickel content is increased above 25 percent, thermal expansion falls off sharply, becoming almost invariable, for ordinary ranges in temperature, at 36 percent nickel. Further additions of nickel result in an increase in thermal expansion. Figure 12·38 shows the effect of nickel on the coefficient of linear thermal expansion of iron-nickel alloys at room temperature. The 35 percent nickel alloy is known as *Invar*, meaning invariable, and is used where very little change in size with change in temperature is desirable. Typical applications include length standards, measuring tapes, instrument parts, variable condensers, tuning forks, and special springs.

In the range of 30 to 60 percent nickel, it is possible to select alloys of appropriate expansion characteristics to fit particular applications. Alloys containing 68 percent iron, 27 percent nickel, and 5 percent molybdenum, or 53 percent iron, 42 percent nickel, and 5 percent molybdenum have high coefficients of thermal expansion. They are used in combination with a low-expansion alloy to produce movement. Applications include thermostatic bimetals, thermoswitches, and other temperature-regulating devices.

Alloys containing approximately 28 percent nickel, 18 percent cobalt, and 54 percent iron have coefficients of expansion closely matching those of standard types of glass. They are used for matched glass-to-metal seals under the trade names of *Kovar* and *Fernico.* A 46 percent nickel alloy, called *Platinite,* has the same coefficient of expansion as platinum. A 42

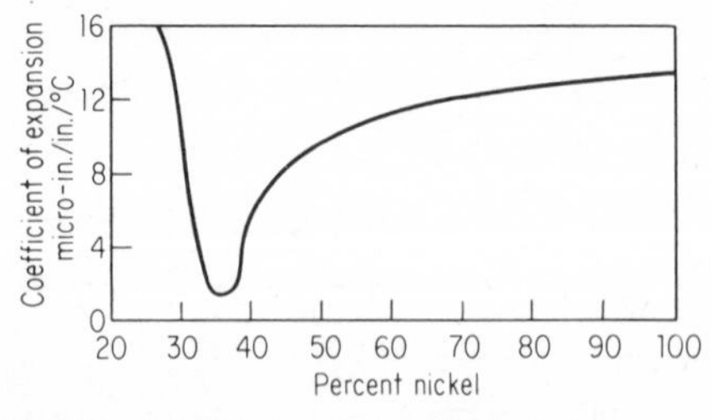

Fig. 12·38 Effect of nickel on the coefficient of linear thermal expansion of iron-nickel alloys at room temperature. (After Guillaume from "Metals Handbook," vol. 1, American Society for Metals, Metals Park, Ohio, 1961.)

percent nickel alloy covered with an oxidized copper plating is known as *Dumet wire* and is used to replace platinum as the "seal-in" wire in vacuum tubes.

An alloy containing 36 percent nickel and 12 percent chromium known as *Elinvar* has a zero thermoelastic coefficient; that is, the modulus of elasticity is almost invariable over a considerable range in temperature. It is used for hair springs and balance wheels in watches and for similar parts in precision instruments.

Permalloys include several nickel-iron alloys in the range of 78 percent nickel that have high magnetic permeability under the influence of very weak magnetizing forces. They also have low hysteresis losses and low electrical resistivity. Permalloy parts are used as loading coils in electrical communication circuits.

The aluminum-nickel-cobalt-iron alloys, commonly called *Alnico*, containing 8 to 12 percent aluminum, 14 to 28 percent nickel, 5 to 35 percent cobalt, have outstanding magnetic properties. These are widely used as permanent magnets in motors, generators, radio speakers, telephone receivers, microphones, and galvanometers.

By variation in the percentage of nickel and proper additions of cobalt, chromium, copper, and molybdenum, different combinations of magnetic properties can be secured.

LEAD AND LEAD ALLOYS

12·34 Lead The major properties of lead include heavy weight, high density, softness, malleability, low melting point, and low strength. It has lubricating properties, low electrical conductivity, high coefficient of expansion, and high corrosion resistance.

By far the largest tonnage of lead is used in the manufacture of storage batteries, followed by the use of tetraethyl lead as the antiknock ingredient in high-test gasoline. Lead compounds are used in the manufacture of many high-grade paints.

The high weight of lead makes it suitable for use as weights and counterbalances, its high density for shielding against β rays and γ rays, its softness for gaskets and for calked joints in cast-iron pipe, and its flexibility for cable sheathing. As a coating on wire, lead acts as a drawing lubricant. Advantage is taken of the high corrosion resistance of lead by its use for equipment in the chemical industry, as a roofing material, and in the plumbing industry as pipe for transporting water and chemicals. Lead is used to improve the machinability of bronzes, brasses, and free-machining steels.

12·35 Lead Alloys Antimony and tin are the most common alloying elements of lead. The antimony-lead phase diagram is shown in Fig. 6·29. This is a simple eutectic system with the eutectic composition at 11.2 percent anti-

TABLE 12·13 Properties of Cast Lead-Antimony Alloys*

ANTIMONY, %	TENSILE STRENGTH, PSI	BHN
0	2,500	4.0
1	3,400	7.0
2	4,200	8.0
3	4,700	9.1
4	5,660	10.0
5	6,360	11.0
6	6,840	11.8
7	7,180	12.5
8	7,420	13.3
9	7,580	14.0
10	7,670	14.6
11	7,620	14.8
12	7,480	15.0
13	7,380	15.2
14	7,000	15.3

*From "Lead in Modern Industry," Lead Industries Association, New York, 1952.

mony. Antimony is generally added to lead to raise the recrystallization temperature and to increase hardness and strength, as shown by the values in Table 12·13. Lead-antimony alloys contain from 1 to 12 percent antimony and are used for storage-battery plates, cable sheathing, collapsible tubes, and for building construction.

The lead-tin alloy system is shown in Fig. 6·30. Like the lead-antimony

TABLE 12·14 Properties of Lead-Tin Alloys*

TIN, %	TENSILE STRENGTH, PSI	BHN
5	3,200	8.0
10	4,100	11.5
15	4,900	12.0
20	5,400	11.7
30	6,200	12.4
40	6,600	13.0
50	7,000	14.3
60	7,200	10.7

*From "Lead in Modern Industry," Lead Industries Association, New York, 1952.

diagram, it is also a simple eutectic system with the eutectic point located at 61.9 percent tin and 361°F. Although lead-tin alloys are most commonly used for their melting characteristics, as in solder, tin also increases hardness and strength, as shown by the values in Table 12·14. The most widely used solders are those containing about 40 percent tin and 60 percent lead, or 50 percent each, with or without small percentages of antimony. *Terne metal*, a lead-tin alloy containing from 10 to 25 percent tin, is used to coat steel sheets for roofing and automotive fuel-tank applications.

Lead alloys containing bismuth, tin, and cadmium form a low-melting-point eutectic. These alloys are useful in electric fuses, sprinkler systems, and boiler plugs.

Lead-tin-antimony alloys are widely used in the printing industry as type metals. The lead base provides low cost, low melting point, and ease in casting; additions of antimony provide hardness and wear resistance and also lower the casting temperature; additions of tin increase fluidity, reduce brittleness, and impart a finer structure. Electrotype metal, being used only as a backing material for the electroformed copper shell and not required to resist wear, contains the lowest percentages of tin and antimony. Foundry type metal, on the other hand, is used exclusively to cast type for hand composition. Since the cast type is used over and over again, it requires the hardest, most wear-resistant alloy that is practical to use. Found-

Fig. 12·39 Linotype metal: 12 percent antimony, 4 percent tin, 84 percent lead. Almost entirely a ternary eutectic structure. (American Smelting and Refining Company.)

TABLE 12.15 Nominal Composition and Typical Mechanical Properties of Some Lead Alloys*

ALLOYS	NOMINAL COMPOSITION, %				CONDITION
	Pb	Sb	Sn	OTHERS	
Chemical lead	99.9+	. . .	. . .	. . .	Rolled sheet
Corroding lead	99.73+	. . .	. . .	. . .	Sand-cast Chill-cast
Arsenical lead	Bal.	. . .	0.10	0.15 As, 0.10 Bi	Extruded sheath
Calcium lead	Bal.	. . .	. . .	0.028 Ca	Extruded and aged
Soft solders	97.5	. . .	1	1.5 Ag	. . .
	95	. . .	5	. . .	. . .
	80	. . .	20	. . .	. . .
	50	. . .	50	. . .	. . .
Antimonial lead	99	1	. . .	. . .	Extruded and aged
	91	9	. . .	. . .	Chill-cast
Hard lead	96	4	. . .	. . .	Cold-rolled 95%
	94	6	. . .	. . .	Cold-rolled 95%
Type metal	95	2.5	2.5	. . .	. . .
	80	14	6	. . .	. . .
	86	11	3	. . .	Cast
	78	15	7	. . .	. . .
	61	25	2 Cu		. . .
Lead-base babbitt:					
SAE 13	85	10	5	. . .	. . .
SAE 14	75	15	10	. . .	Chill-cast
SAE 15	83	15	1	1 As	. . .
G	83.5	12.75	0.75	3 As	. . .

*"Metals Handbook," 1961 ed., American Society for Metals, Metals Park, Ohio.

ry type metal contains the largest amounts of tin and antimony, and generally includes up to 2 percent copper as an additional hardener. Linotype and Intertype casting machines die-cast an entire line of type characters at each casting. It is important that the alloy used should have a low melting point and a short temperature range during solidification. Therefore, the ternary eutectic alloy or compositions near this are preferred. Figure 12·39 shows the microstructure of linotype metal containing 12 percent antimony, 4 percent tin, and 84 percent lead. This alloy has a liquidus temperature of 463°F and a solidus temperature of 462°F, and the structure is almost entirely a ternary eutectic mixture.

Lead-base bearing alloys are known commercially as *babbitts* or *white metal* alloys. One group includes the alloys of lead-tin-antimony and usually arsenic, while the other group includes alloys of lead and tin with small percentages of calcium, barium, magnesium, and sodium. Figure 12·40 shows the microstructure of a lead-base bearing alloy. It consists of cubes of primary antimony-tin compound in a binary eutectic mixture of lead and tin solid solutions. These alloys are used for automotive connecting rods, main and camshaft bearings, diesel-engine bearings, railroad-car journal bearings, and many electric-motor bearings.

Table 12.15 (Continued)

TENSILE STRENGTH, PSI	YIELD STRENGTH, PSI	ELONGATION, %	BHN	TYPICAL USES
2,385	1,180	29	...	Material of construction in the chemical industry
1,800 2,000	800 ...	30 47	3.2–4.5 4.2	Storage batteries, cable sheathing, paint, calking, antiknock fluid, liquid metal for heat treating
2,500	...	40	4.9	Cable sheathing
4,500	...	25	...	Cable sheath and creep-resistant pipe
... 3,400 5,800 6,100	... 1,500 3,650 4,800	... 50 16 60	13 8 11.3 14.5	Coating and joining metals, body solder
3,000 7,500	... ...	50 17	7 15.4	Cable sheathing Storage-battery grids
4,020 4,100	... ...	48.3 47	... ...	Rolled sheet and extruded pipe
... ...	... ...	... ...	12.4 23 19 24 ...	Electrotype Stereotype Linotype Monotype Foundry type
10,000 10,500 10,350 9,800	... ...	5 4 2 1.5	19 22 20 22	Light loads, car journal bearings Moderate loads: blowers, pumps High loads; diesel-engine bearings Elevated-temperature bearing; trucks

The nominal composition and typical mechanical properties of some lead alloys are given in Table 12·15.

TIN AND TIN ALLOYS

12·36 Tin Tin is a white, soft metal that has good corrosion resistance and good lubricating properties. It undergoes a polymorphic transformation from the normal tetragonal structure (white tin) to a cubic form (gray tin) at a temperature of 55.8°F. This transformation is accompanied by a change in density from 7.30 to 5.75, and the resulting expansion causes disintegration of the metal to coarse powder known as *tin pest.* However, the transformation is very sluggish, and considerable undercooling is necessary to initiate it. Common impurities in tin tend to delay or inhibit the change so that, under ordinary conditions, the transformation is of no practical importance.

Over half the primary tin used in this country goes into the coating of other metals, primarily steel in the manufacture of tin cans. Tin-coated copper tubing is useful for handling fresh waters that contain large percentages of carbon dioxide and oxygen. The use of tin as an alloying element in copper, aluminum, and lead has been discussed in preceding sections.

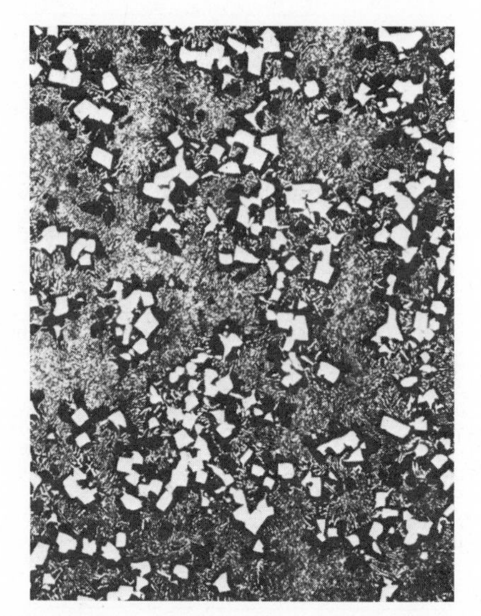

Fig. 12·40 Lead-base bearing alloy: 15 percent antimony, 5 percent tin, 80 percent lead. Cubes of primary antimony-tin compound (white) in a binary eutectic mixture of lead and tin solid solutions, 75×. (American Smelting and Refining Company.)

12·37 Tin Alloys Lead is alloyed with tin to produce several soft solders that have higher strength than the lead-base solders. Tin solders containing 5 percent antimony or 5 percent silver are preferred for electrical equipment because these solders have higher electrical conductivity than the high-lead alloys.

TABLE 12·16 Nominal Composition and Typical Mechanical Properties of Some Lead Alloys*

ALLOY	NOMINAL COMPOSITION, %					CONDITION
	Sn	Sb	Cu	Pb	Ag	
Tin (pure)	99.8 min	...	...	...	...	Cast
Hard tin	99.6	...	0.4	...	...	80% reduction
Antimonial tin solder	95	5	...	...	...	Cast
Tin-silver solder	95	...	...	...	5	Sheet
Soft solder	70	...	...	30	...	Cast
	63	...	...	37	...	Cast
Tin babbitt	91	4.5	4.5	...	...	Chill-cast
	89	7.5	3.5	...	...	Chill-cast
	84	8	8	...	...	Die-cast
	65	15	2	18	...	Die-cast
White metal	92	8	...	...	...	Chill-cast
Pewter	91	7	2	...	...	Annealed sheet

*"Metals Handbook," 1961 ed., American Society for Metals, Metals Park, Ohio.
†In 4 in.
‡Compressive yield (0.125 percent offset).

The most common alloying elements for tin are antimony and copper to produce pewter and the tin-base babbitts that are used for high-grade bearing applications. Figure 12·41 shows a typical microstructure of tin-base babbitt. There are CuSn rods arranged in a star-shaped pattern and large cubes of SnSb compound, all in a ductile tin-rich ternary eutectic. The SnSb cubes are extremely hard and contribute to the excellent wear resistance of babbitt.

The nominal composition and typical mechanical properties of some tin alloys are given in Table 12·16.

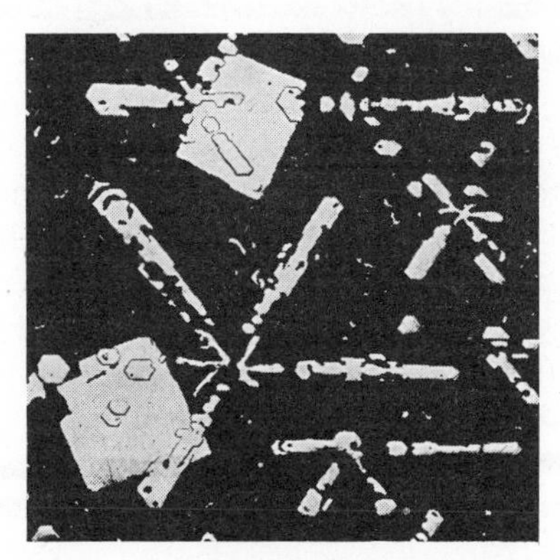

Fig. 12·41 Tin-base hard babbitt of 84 percent tin, 7 percent copper, and 9 percent antimony, 50×. Star-shaped CuSn compound and rectangular crystals of SnSb compound in a ductile ternary eutectic matrix. (By permission from R. M. Brick, R. B. Gordon, and A. Phillips, "Structure and Properties of Alloys," 3d ed., McGraw-Hill Book Company, New York, 1965.)

Table 12.16 (Continued)

TENSILE STRENGTH, PSI	YIELD STRENGTH, PSI	ELONGATION, % IN 2 IN.	BHN	TYPICAL USES
3,100	. . .	55†	5.3	Electrotinning, alloying
4,000	. . .	. . .	. . .	Collapsible tubes and foil
5,900	. . .	38†	. . .	Solder for electrical equipment
4,600	3,600	49	. . .	Solder for electrical equipment
6,800	. . .	. . .	12	For joining and coating of metals
7,500	. . .	32†	14	
9,300	4,400‡	2	17	Automotive applications, better
11,200	6,100‡	. . .	24	corrosion and wear resistance than
10,000	. . .	1	30	lead-base bearing alloys
7,800	. . .	1.5	23	
7,200	. . .	. . .	20	Castings for costume jewelry
8,600	. . .	40	9.5	Vases, candlesticks, book ends

TITANIUM AND TITANIUM ALLOYS

12·38 Titanium The process of producing titanium sponge by the magnesium reduction of titanium tetrachloride was discovered by W. J. Kroll in 1938. Shortly afterward the United States armed services became interested in the metal primarily because of its high melting point (3035°F). There was the possibility of developing titanium alloys with strength at elevated temperatures which might substitute in military equipment for nickel-base and cobalt-base alloys. Titanium has a density of about 0.16 lb/cu in. compared with steel at 0.28. Therefore, titanium alloy structures have a high strength-weight ratio and are particularly useful for aircraft parts. Titanium has excellent corrosion resistance up to approximately 1000°F. While titanium is the fourth most abundant element in the earth's crust, it is relatively expensive to obtain from its ores.

Titanium has a strong affinity for the gases hydrogen, nitrogen, and oxygen, all of which form interstitial solid solutions with titanium. They all have a marked strengthening effect, as illustrated for nitrogen in Fig. 12·42. When the amount of absorbed oxygen, nitrogen, or hydrogen exceed specified limits, they embrittle titanium, reducing impact strength, and cause brittle failure under sustained loads at low stresses.

Titanium metal has a close-packed hexagonal crystal structure, called alpha, at room temperature. This structure transforms to body-centered cubic beta at 1625°F(882°C).

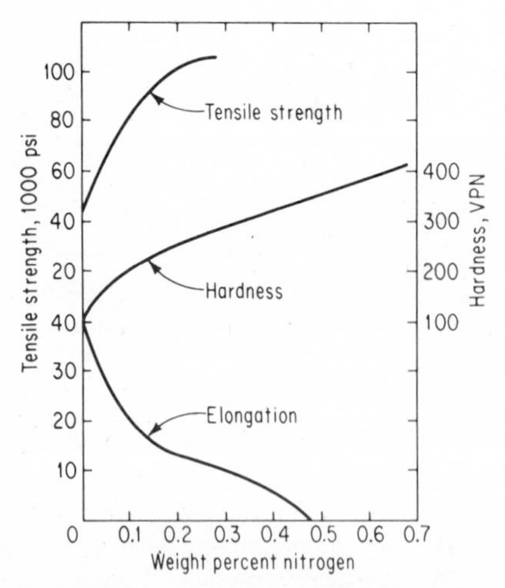

Fig. 12·42 Increase in strength resulting from the presence of nitrogen in otherwise pure titanium. (From Brick, Gordon, and Phillips, "Structure and Properties of Alloys," 3d ed., McGraw-Hill Book Company, New York, 1965.)

Commercially pure titanium is lower in strength, more corrosion-resistant, and less expensive than titanium alloys. It is used for applications requiring high ductility for fabrication but little strength, such as chemical process piping, valves and tanks, aircraft firewalls, tailpipes, and compressor cases.

12·39 Titanium Alloys The addition of alloying elements to titanium will influence the alpha to beta transformation temperature. It is common practice to refer to alloying elements as alpha or beta stabilizers. An alpha stabilizer

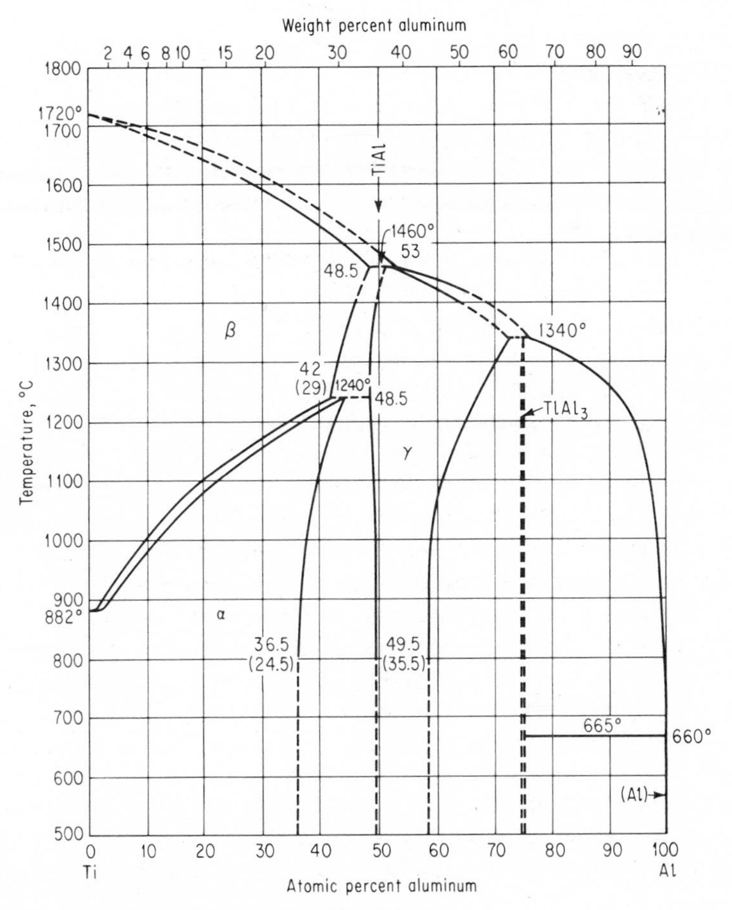

Fig. 12·43 The Ti-Al phase diagram. (From Max Hansen, "Constitution of Binary Alloys," 2d ed., McGraw-Hill Book Company, New York 1958.)

means that as solute is added, the alpha to beta transformation temperature is raised; similarly, a beta stabilizer lowers the transformation temperature. Aluminum is an alpha stabilizer, as may be seen from the Ti-Al phase diagram (Fig. 12·43). Important beta stabilizers are chromium, molybdenum, vanadium, manganese, and iron. Ti-Mo and Ti-V alloy systems (see Fig. 12·44) show complete solid solubility, forming the beta solid solution over the entire range. The alpha-phase field is severely restricted, the maximum extent being 1.8 percent Mo and 3.5 percent V. The Ti-Mn phase diagram, shown in Fig. 12·45, illustrates a beta stabilizer by means of a eutectoid reaction. In eutectoid systems intermetallic compounds always occur. Tin is substantially neutral in the amount present in commercial alloys. The relative amounts of alpha and beta stabilizers in an alloy, and the heat treatment, determine whether its microstructure is predominantly one-phase alpha, a mixture of alpha and beta, or the single-phase beta over its useful temperature range.

Properties are directly related to microstructure. Single-phase alloys are weldable with good ductility; some two-phase alloys are also weldable, but their welds are less ductile. Two-phase alpha-beta alloys are stronger than the one-phase alpha alloys, primarily because b.c.c. beta is stronger than c.p.h. alpha. Most important, two-phase alloys can be strengthened by heat treatment because the microstructure can be manipulated by controlling heating, quenching, and aging cycles.

Alpha Alloys Most of the alpha alloys contain some beta-stabilizing alloying elements. The compositions of these alloys are balanced by high aluminum content so that the alloys are essentially one-phase alpha. Figure 12·46 shows coarse, platelike alpha in a Ti-5Al-2.5Sn alloy after hot working and annealing. The alpha alloys have two main attributes: weldability and retension of strength at high temperatures. The first results from the one-phase microstructure, the second from the presence of aluminum. Alloying elements in solution strengthen the alpha-phase alloys, and aluminum is the most effective strengthener of alpha alloys. Especially important, its effect persists to high temperatures. Hot working of alpha alloys containing more than about 6 percent aluminum is difficult. Hot workability of high-aluminum alpha is improved by additions of beta-stabilizing alloying elements in amounts small enough so that the beta phase is present in small quantities in the annealed microstructure. Some applications of Ti-5Al-2.5Sn alloy include aircraft tailpipe assemblies and other formed sheet components operating up to 900°F, and missile fuel tanks and structural parts operating for short times up to 1100°F.

Alpha-Beta Alloys These contain enough beta-stabilizing elements to cause the beta phase to persist down to room temperature, and they are stronger than alpha alloys. The beta phase, as strengthened by beta-alloying additions in solution, is stronger than the alpha phase. If the alpha phase in alpha-beta alloys is strengthened by aluminum, the alpha-beta alloy is stronger yet, especially at elevated temperatures. Figure 12·47 shows the alpha-beta microstructure in an annealed Ti-8Mn alloy. The alpha-beta alloys can be further strengthened by heat treatment. Essen-

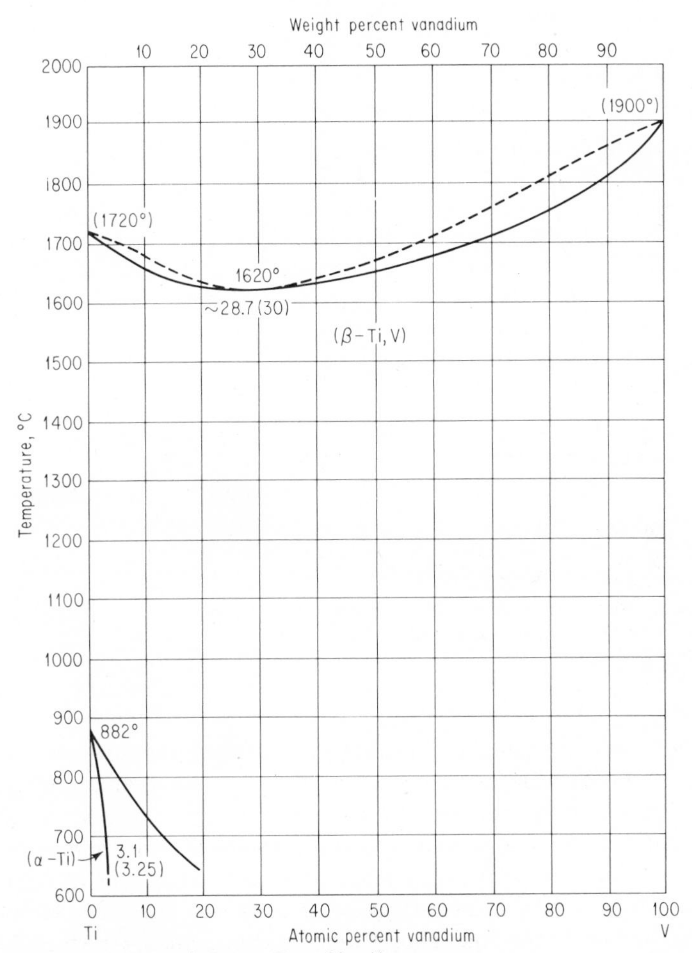

Fig. 12·44 The Ti-V phase diagram. (From Max Hansen, "Constitution of Binary Alloys," 2d ed,. McGraw-Hill Book Company, New York, 1958.)

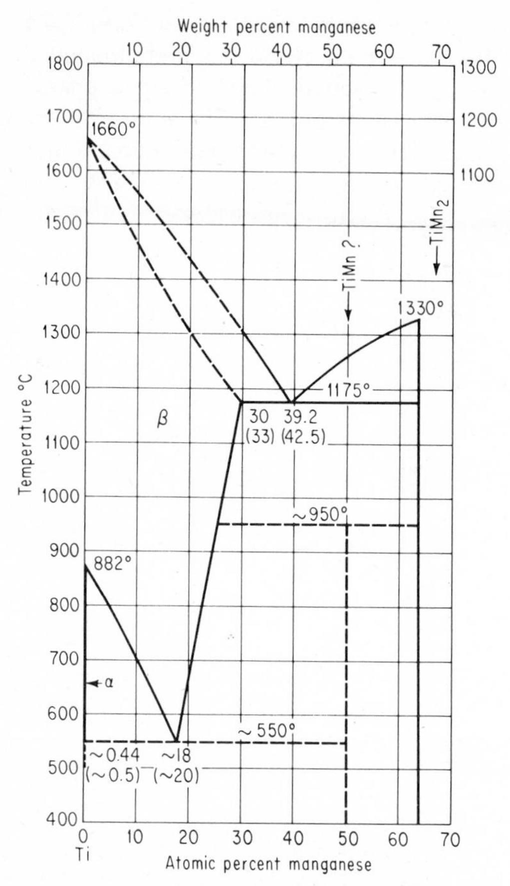

Fig. 12·45 The Ti-Mn phase diagram. (From Max Hansen, "Constitution of Binary Alloys," 2d ed., McGraw-Hill Book Company, New York, 1958.)

tially, this is accomplished by quenching from a temperature in the alpha-beta field followed by aging at moderately elevated temperature. In contrast to the usual age-hardening procedure, a homogeneous beta solid solution is not formed in the first step. If an all-beta structure were formed, the beta grain size would be excessively large, and the subsequent formation of alpha would be mainly at the beta grain boundaries. These two factors reduce the ductility of the aged alloy. Quenching suppresses the transformation of the elevated-temperature beta phase that would occur on slow cooling. Aging at elevated temperature causes precipitation of fine particles of alpha in the volumes that were beta grains prior to quenching. This fine structure is stronger than the coarse, annealed alpha-beta structure. In

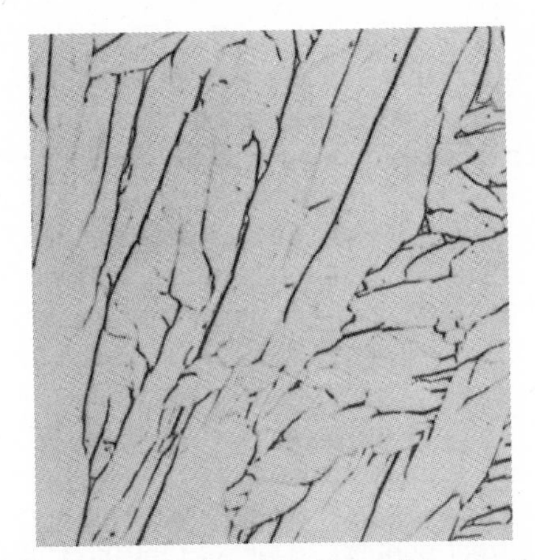

Fig. 12·46 Ti-5Al-2.5Sn hot-worked below the α transformation temperature, annealed 30 min at 2150°F (1177°C), which is above the β transformation temperature, and furnace-cooled. Structure is coarse, platelike α. Etched in Kroll's reagent, 100×. (From Metals Handbook, vol. 7, "Atlas of Microstructures," American Society for Metals, 1972.)

some cases, quenched titanium alloy structures may be of an unstable form of alpha designated alpha prime and called *titanium martensite.* This designation originally was borrowed from steel metallurgy, where martensite is

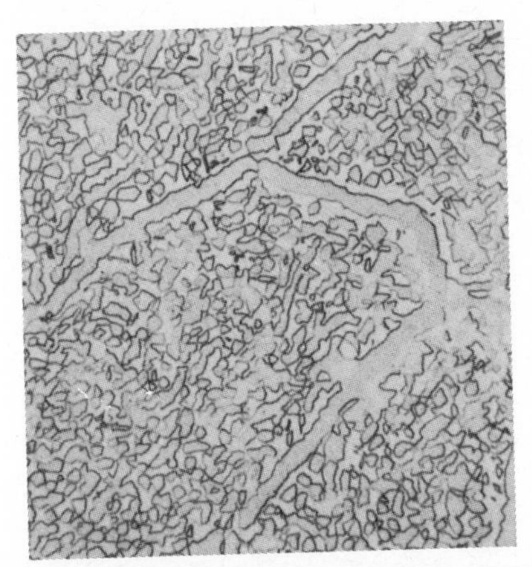

Fig. 12·47 Ti-8Mn alloy, annealed 2 h at 1300°F (704°C), furnace-cooled to 1100°F (593°C), and held 1 h. α grains (gray) in β matrix (light); also α at the prior beta grain boundaries. Etched in Kroll's reagent, 500×. (From Metals Handbook, vol. 7, "Atlas of Microstructures," American Society for Metals, 1972.)

a metastable structure formed by a diffusionless phase transformation when steel is quenched from a high temperature. In modern terminology, however, martensite is a word for any needle-like metallic structure formed by diffusionless shear, usually upon rapid cooling. Figure 12·48 shows the titanium martensite structure in a Ti-6Al-4V alloy which was water-quenched from 1950°F. Strength of alpha-beta alloys can be increased about 35 percent by heat treating, compared with the properties of annealed material. Typical applications of the Ti-6Al-4V alloy include aircraft gas turbine compressor blades and disks; forged airframe fittings; and sheet metal airframe parts. The Ti-8Mn alloy has been used for aircraft skins and primary structural parts subject to temperatures in the range of 200 to 600°F.

Beta Alloys Unlike the alpha alloys, beta alloys can be strengthened by heat treatment. Figure 12·49*a* shows the all-beta microstructure in a Ti-3Al-13V-11Cr alloy after solution treatment for 10 min at 1450°F. This alloy is weldable in both the annealed and heat-treated conditions. Aging at elevated temperature after solution treatment results in the precipitation of fine particles of alpha and $TiCr_2$ compound. The microstructure after aging for 48 h at 900°F shows dark particles of precipitated alpha in beta grains (see Fig. 12·49*b*). Ultimate strengths up to 215,000 psi with 5 percent elongation are possible after heat treatment. This is an increase over the annealed strength of at least 50 percent. Beta alloys have been used for high-strength fasteners and for aerospace components requiring high strength at moderate temperatures.

Fig. 12·48 Ti-6Al-4V alloy, held for 1 h at 1950°F (1066°C), which is above the β transformation temperature, and water-quenched. Structure is α' (titanium martensite); prior β grain boundaries are also visible. Etched in 10HF, $5HNO_3$, 250×. (From Metals Handbook, vol. 7, "Atlas of Microstructures," American Society for Metals, 1972.)

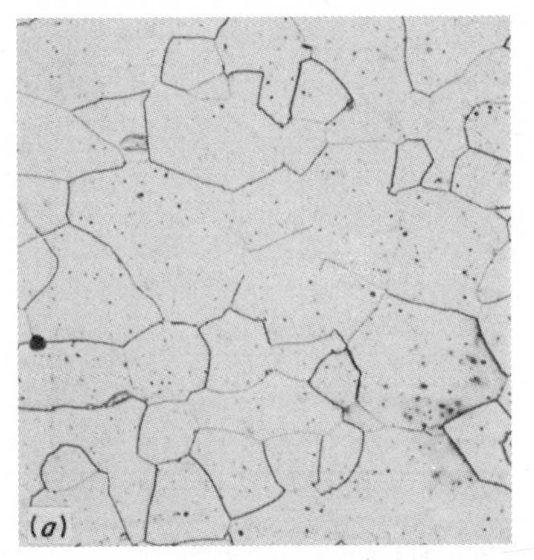

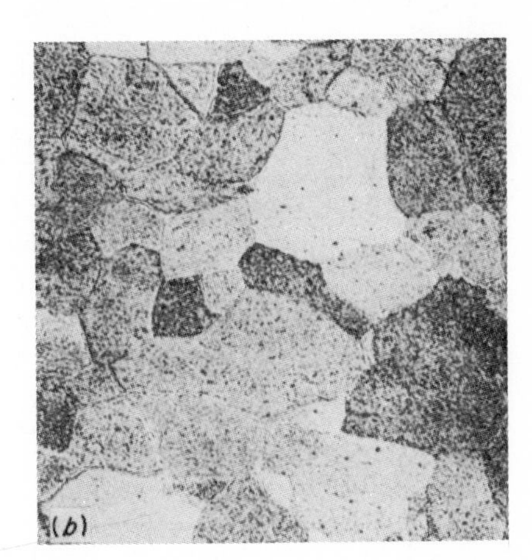

Fig. 12·49 Ti-13V-11Cr-3Al sheet. (*a*) Solution-treated for 10 min at 1450°F (788°C) and air-cooled. Structure consists of equiaxed grains of metastable beta. (*b*) Same as (*a*), except aged for 48 h at 900°F (482°C). Structure consists of dark particles of precipitated α in β grains. Etched in 2HF, $10HNO_3$, $88H_2O$, 250×. (From Metals Handbook, vol. 7, "Atlas of Microstructures," American Society for Metals, 1972.)

Typical mechanical properties of some titanium alloys are given in Table 12·17.

ZINC AND ZINC ALLOYS

12·40 The principal use of zinc is as a coating for steel to prevent corrosion. It is more highly anodic than steel, and in a corrosive atmosphere the zinc coating acts as the sacrificial anode. Thus the zinc is consumed while protecting the steel from any attack. Metallic zinc coatings may be applied by various methods such as hot-dip galvanizing, electrogalvanizing, painting, metallizing or spraying of molten metal, and by sherardizing or cementation. Steel products that are galvanized include bolts, chains, fencing, hardware, pipe and tubing, screws, sheets, tanks, wire, and wire cloth.

In *sherardizing*, parts to be coated are tightly packed with zinc dust in an airtight container, which is then revolved and heated to a temperature slightly below the melting point of zinc. In the presence of heat, the zinc impregnates the surface and diffuses into the steel, providing a thin uniform coating.

Zinc oxide is used in the manufacture of dental cement, enamels, floor tilting, glass, glazes, matches, paint, pottery, rubber goods, tires, and tubes.

Zinc can be easily worked into various shapes and forms by common fab-

TABLE 12·17 Typical Mechanical Properties of Some Titanium Alloys*

ALLOY	CONDITION	ROOM TEMPERATURE TENSILE STRENGTH, PSI	ROOM TEMPERATURE YIELD STRENGTH, PSI	ROOM TEMPERATURE ELONGATION, %
COMMERCIALLY PURE TITANIUM				
Comm. purity (99.0%)	Annealed	79,000	63,000	27
ALPHA TITANIUM ALLOYS				
Ti-5Al-2.5Sn	Annealed	125,000	120,000	18
Ti-6Al-4Zr-1V	Annealed	143,000	138,000	17
Ti-8Al-1Mo-1V	HT†	147,000	135,000	16
ALPHA-BETA TITANIUM ALLOYS				
Ti-8Mn	Annealed	138,000	125,000	15
Ti-4Al-4Mn	Annealed	148,000	133,000	16
	HT†	162,000	140,000	9
Ti-6Al-4V	Annealed	135,000	120,000	11
	HT†	170,000	150,000	7
Ti-7Al-4Mo	Annealed	160,000	150,000	15
	HT†	190,000	175,000	12
BETA TITANIUM ALLOY				
Ti-3Al-13V-11Cr	HT†	180,000	170,000	6

*Compiled from data in Metals Handbook, vol. 1, American Society for Metals, 1961.
†HT is usually after age-hardening.

ricating methods. Pure zinc has a recrystallization temperature below room temperature, so that it "self-anneals" and cannot be work-hardened at room temperature. The presence of natural impurities or added elements raises the recrystallization temperature. Therefore, the less pure grades of wrought zinc will show an increase in hardness and strength with working.

For deep-drawing purposes a relatively pure zinc should be used. Typical applications include drawn and extruded battery cans, eyelets, grommets, laundry tags, and address plates. The addition of lead and cadmium results in higher hardness, stiffness, and uniform etching quality. It is used for weatherstrips, soldered battery cans, and photoengraver's plate. For added stiffness, good creep resistance, and easy work hardening, alloys containing from 0.85 to 1.25 percent copper are recommended. A wrought-zinc alloy containing from 0.50 to 1.50 percent copper and 0.12 to 1.50 percent titanium has outstanding creep resistance and is used for corrugated roofing, leaders, and gutters.

The 4 percent aluminum, 0.04 percent magnesium, and up to 3.5 percent copper alloy has been used in the form of heavy rolled plate in the aircraft industry and for dies in the blanking of aluminum-alloy sheet and thin steel.

The major use of zinc as a structural material is in the form of alloys for die castings. The zinc die-casting alloys are low in cost and easy to cast and have greater strength than all die-casting metals except the copper alloys. They can be cast to close dimensional limits and are machined at minimum cost; their resistance to surface corrosion is adequate for a wide range of applications. They are usually limited to service temperatures below 200°F, since above this temperature their tensile strength is reduced 30 percent and their hardness 40 percent.

The Al-Zn phase diagram is shown in Fig. 12·50. A lamellar eutectic forms at 720°F (382°C) and 5 percent aluminum, containing α' and β solid solutions. The α' constituent of the eutectic is stable only at temperatures above 527°F (275°C). At that temperature, it transforms by a eutectoid reaction into α and β phases. Commercial die-casting alloys are cooled fast enough to prevent the eutectoid transformation and retain the eutectic mixture of α' and β. Figure 12·51*a* shows the microstructure of alloy AG40A (Zamak-3), die-cast. The structure consists of white grains of primary β (zinc-rich solid solution) surrounded by the dark eutectic mixture. Slower cooling in a permanent mold will cause the lamellar eutectic to be much coarser (Fig. 12·51*b*).

When die castings are aged at room temperature or slightly elevated temperature, a precipitation reaction occurs in the zinc-rich β solid solution. The β may contain about 0.35 percent aluminum in solution in a freshly made die casting. During a 5-week room-temperature aging period, this will decrease to about 0.05 percent, the excess appearing as minute particles of α within the β structure (Fig. 12·51*c*).

The two die-casting alloys in general use are known as *Zamak*-3 (ASTM AG40A, SAE 903) and *Zamak*-5 (ASTM AC41A, SAE 925). They both contain

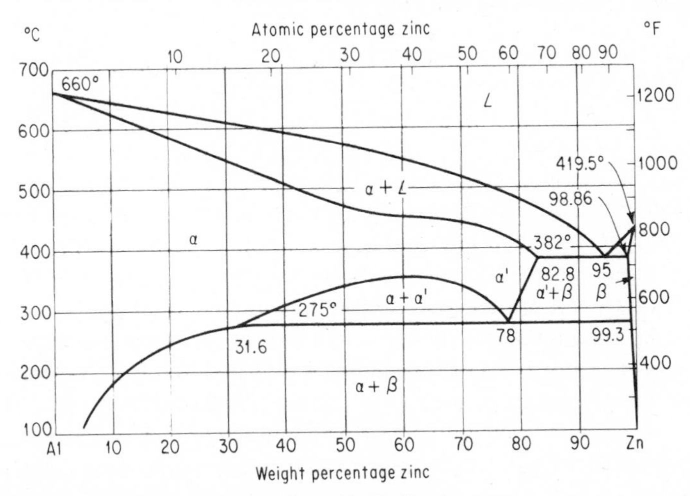

Fig. 12·50 The Al-Zn phase diagram. (From Metals Handbook, 1948 ed., p. 1167, American Society for Metals.)

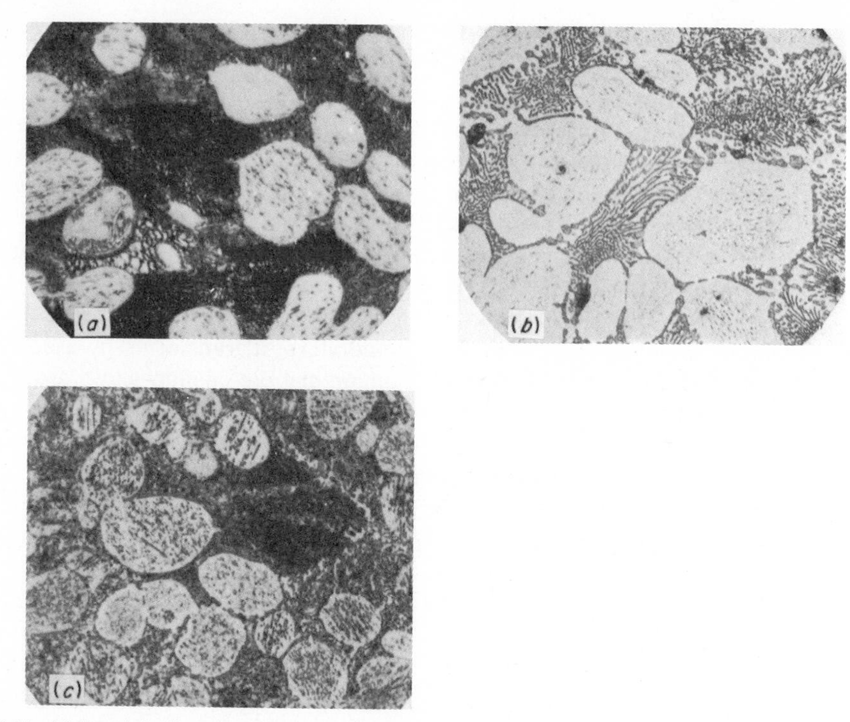

Fig. 12·51 (*a*) Zamak-3 alloy, as die-cast. Structure consists of white grains of primary β (zinc-rich solid solution) surrounded by the dark eutectic mixture; 1,000×. (*b*) Zamak-5 alloy, permanent-mold-cast. Same structure as (*a*), but the lamellar eutectic is coarser due to the slower cooling; 250×. (*c*) Zamak-3 alloy, die-cast then aged for 10 days at 203°F (95°C). Aging has caused additional precipitate of α (black dots) in the light grains of β (zinc-rich) solid solution; 1,000×. All samples etched in 50g CrO_3, 4g Na_2SO_4, 1 liter H_2O. (From Metals Handbook, vol. 7, "Atlas of Microstructures," American Society for Metals, 1972.)

about 4 percent aluminum and 0.04 percent magnesium. Zamak-3 has slightly higher ductility and retains its impact strength better at slightly elevated temperature. Zamak-5, containing about 1 percent copper, is somewhat harder and stronger and has slightly better castability. They are used for automotive parts, household utensils, building hardware, padlocks, toys, and novelties. The maximum composition limit of certain impurities such as lead (0.007 percent), cadmium (0.005 percent), and tin (0.005 percent) must be strictly observed to minimize intergranular corrosion.

The composition and typical mechanical properties of some zinc alloys are given in Table 12·18.

TABLE 12·18 Typical Mechanical Properties of Some Zinc Alloys*

MATERIAL COMPOSITIONS, APPROX. %	TREATMENT	TENSILE STRENGTH, PSI	ELONGATION, % IN 2 IN.	BHN, 10-MM BALL
		WROUGHT ALLOYS		
Commercial rolled zinc (deep drawing), 0.08 max Pb, bal. Zn	Hot-rolled†	19,500	65	38
	Hot-rolled‡	23,000	50	38
	Cold-rolled†	21,000	40	
	Cold-rolled‡	27,000	40	
Commercial rolled zinc, 0.05–0.10 Pb, 0.05–0.08 Cd, bal. Zn	Hot-rolled†	21,000	52	43
	Hot-rolled‡	25,000	30	43
	Cold-rolled†	22,000	40	
	Cold-rolled‡	29,000	30	
Commercial rolled zinc, 0.25–0.50 Pb, 0.25–0.45 Cd, bal. Zn	Hot-rolled†	23,000	50	47
	Hot-rolled‡	29,000	32	47
	Cold-rolled†	25,000	45	
	Cold-rolled‡	31,000	28	
Copper-hardened rolled zinc alloy, 0.85–1.25 Cu, bal. Zn	Hot-rolled†	24,000	20	52
	Hot-rolled‡	32,000	15	60
	Cold-rolled†	32,000	5	
	Cold-rolled‡	40,000	3	
Rolled zinc alloy, 0.85–1.25 Cu, 0.006–0.016 Mg, bal. Zn	Hot-rolled†	28,000	20	61
	Hot-rolled‡	36,000	10	80
	Cold-rolled†	37,000	20	
	Cold-rolled‡	48,000	2	
		CASTING ALLOYS		
Zamak-3, SAE 903, ASTM AG40A (XX111), 3.5–4.3 Al, 0.03–0.08 Mg, 0–0.15 max Cu, bal. Zn (99.99%)	Die-cast	41,000	10	82
Zamak-5,SAE 925, ASTM AC41A (XXV), 3.5–4.3 Al, 0.03–0.08 Mg, 0.75–1.25 Cu, bal. Zn (99.99%)	Die-cast	47,600	7	91
Zamak-2, SAE 921, ASTM (XX1), 3.5–4.5 Al, 0.02–0.10 Mg, 2.5–3.5 Cu, bal. Zn (99.99%)	Sand-cast	20,000–30,000	. . .	70–100
Zamak-5 (same as above)	Sand-cast	20,000–30,000	. . .	70–100

*American Zinc Institute, New York.
†Longitudinal direction.
‡Transverse direction.

PRECIOUS METALS

The precious-metals group includes silver, gold, and the six platinum metals platinum, palladium, iridium, rhodium, ruthenium, and osmium. This group is characterized by softness, good electrical conductivity, and very high corrosion resistance to common acids and chemicals.

12·41 Silver and Silver Alloys The photosensitivity of silver and certain silver salts, coupled with their ease of reduction, forms the basis for photography. Silver-clad copper, brass, nickel, and iron are used for electrical conductors, contacts, and chemical equipment. A recent development is the use of silver coatings on glass, ceramics, and mica to provide a conducting base for subsequent electroplating of electronic devices. The high reflectivity and ease of electroplating make silver useful in reflectors, silverware, and jewelry.

Silver-Copper Alloys The silver-copper alloy system is shown in Fig. 12·52. It is a simple eutectic-type system, with the eutectic point located at 28.1 percent copper and 1435°F. The maximum solubility of copper in silver is 8.8 percent, and the slope of the solvus line indicates the possibility of age-hardening certain alloy compositions. *Sterling silver* (7.5 percent copper) and *coin silver* (10 percent copper) are age-hardenable alloys, but little commercial use is made of this heat treatment because of the close temperature control required. The 28 percent copper eutectic alloy finds

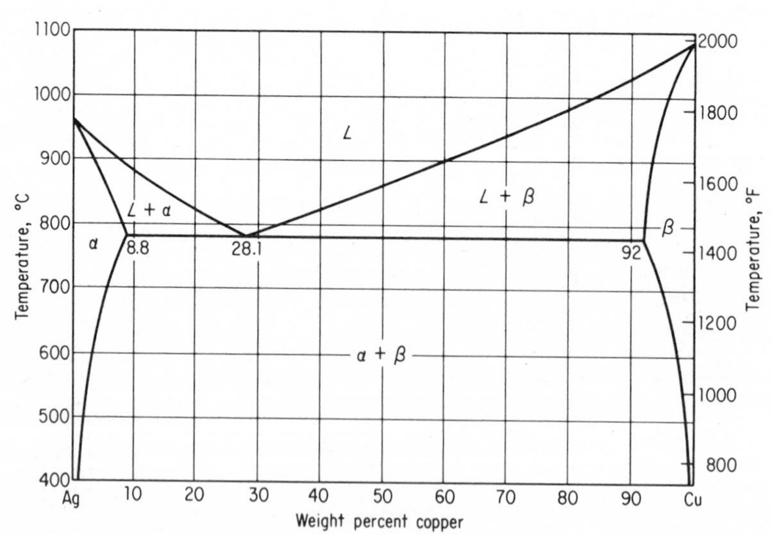

Fig. 12·52 The silver-copper alloy system. (From "Metals Handbook," 1948 ed., p. 1148, American Society for Metals, Metals Park, Ohio.)

some use as a brazing solder. Coin silver is used for United States silver coins and for electrical contacts.

Silver-Copper-Zinc Silver alloys in this group are known as *silver solders* or *silver brazing alloys.* In addition to silver, copper, and zinc, they often contain cadmium and tin. In brazing, the physical mechanism of bonding is similar to that of soft soldering, except that it takes place at a higher temperature. There is no melting of the material being joined, and the bond is achieved by interfacial penetration of the brazing alloy. The important property of these alloys is the temperature at which they melt and flow freely into a joint. By suitable variation of composition, it is possible to obtain brazing alloys that melt anywhere from 1100 to 1550°F. Silver brazing alloys are used for many applications in the joining of ferrous and nonferrous materials.

12·42 Gold and Gold Alloys Aside from the use of gold alloys for coinage, jewelry, and dental products, they have many industrial applications. The very high corrosion resistance, nontarnishing characteristics, good electrical conductivity, and ease of electroplating make gold coating suitable for electrical applications. Electroplated gold is used in wave guides, on grid wires, on contacts, on vibrating components, and as a thin film on glass for selective light filters. Other industrial applications of high-purity gold include thermal-limit fuses to protect electrical furnaces, as a target in x-ray apparatus, as a freezing-point standard, for the lining of chemical equipment, and as a high-melting solder for vacuum-tight pressure welds.

A 70 percent gold–30 percent platinum alloy, with a solidus temperature of 2242°F, is used as a high-melting-point platinum solder.

Gold-palladium-iron alloys develop very high resistivity after proper heat treatment and are used primarily for potentiometer wire. The alloy having the highest resistivity contains 49.5 percent gold, 40.5 percent palladium, and 10 percent iron.

12·43 Platinum and Platinum Alloys Platinum is the most important and most abundant metal in the platinum group. The important properties of platinum are high corrosion resistance, high melting point, white color, and ductility. It forms extensive ductile solid solutions with other metals. In the unalloyed form, platinum is used for thermocouple and resistance-thermometer elements, electrical contacts, crucibles and laboratory ware, dental foil, electrodes, heat- and corrosion-resistant equipment, and for jewelry. It is also used as a catalyst in the production of sulfuric acid, vitamins, and high-octane gasolines.

Most of the important binary platinum alloy systems show complete solid solubility, so that the increase in hardness and strength obtained by alloying is due to solid-solution hardening. The hardening effect of alloying additions upon platinum is shown in Fig. 12·53. Of the metals shown,

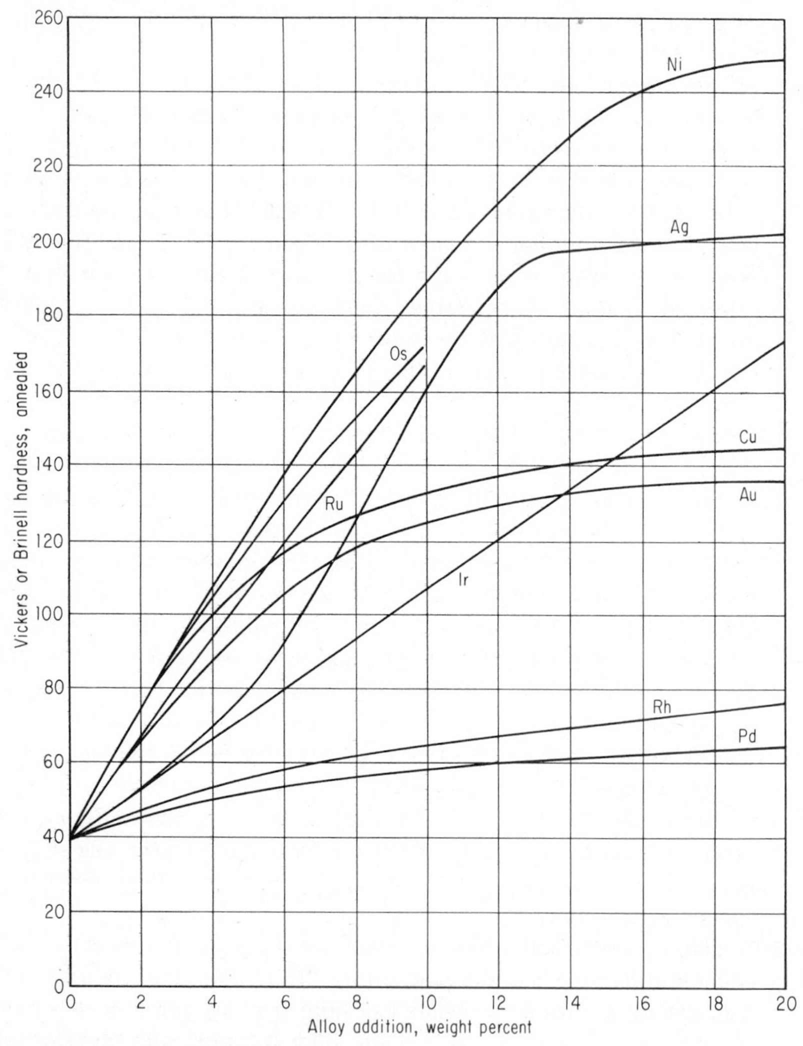

Fig. 12·53 Hardening effect of alloying additions on platinum. (The International Nickel Company.)

nickel produces the greatest hardening effect and palladium the least. In general, the effect of alloying additions on the tensile strength parallels the effect on hardness.

Platinum-Rhodium Alloys These contain between 3.5 and 40 percent rhodium. Rhodium is the preferred alloying element to platinum for most applications at high temperatures under oxidizing conditions. The 10 percent rhodium alloy is the most popular one in this group. It is the standard catalyst for the oxidation of ammonia in the manufacture of nitric acid. This alloy shows excellent resistance to molten glass and is used for nozzles in glassworking equipment. The 10 percent rhodium–90 percent platinum alloy, with its composition carefully controlled, serves as the positive side of the widely used rhodium-platinum vs. platinum thermocouple. The 3.5 percent rhodium alloy is used for crucibles as an alternative to pure platinum. Platinum-rhodium alloys containing between 10 and 40 percent rhodium are used as windings in furnaces operating between 2800 and 3275°F.

Platinum-Iridium Alloys Platinum alloyed with 0.4 to 0.6 percent iridium is employed for crucibles and other laboratory ware. The rich color, high mechanical properties, and excellent corrosion and tarnish resistance of the 5 to 15 percent iridium alloys make them the preferred metal for jewelry. Electrical contacts for dependable service in magnetos, relays, and thermostats generally contain between 10 and 25 percent iridium. The 25 or 30 percent iridium alloys are used for hypodermic needles and as electrodes for aircraft spark plugs.

Platinum-Ruthenium Alloys The alloys in this group have properties and applications similar to those of the platinum-iridium group. Ruthenium is harder to work than iridium, the practical limit of workability being about 15 percent ruthenium. The 5 percent ruthenium alloy is used in jewelry and in medium-duty electrical contacts. The 10 percent alloy is used for contacts in aircraft magnetos and the 14 percent alloy in heavy-duty contacts. Platinum-ruthenium alloys are also employed for aircraft spark-plug electrodes, hypodermic needles, and pen nibs.

Platinum-Nickel Alloys This group of alloys, containing up to 20 percent nickel, has good strength at elevated temperature. The 5 percent nickel alloy is used for long-life oxide-coated cathode wires in electron tubes.

Platinum-Tungsten Alloys The most popular alloys in this group contain 4 and 8 percent tungsten. Typical applications include aircraft spark-plug electrodes, electrical contacts, grids in power tubes for radar, potentiometer wire, strain gauges, and hard corrosion-resisting instrument bearings.

A platinum 23 percent cobalt alloy has unusual magnetic properties and is used as a permanent magnet in small instruments where a very short magnet is essential.

12·44 Palladium and Palladium Alloys Palladium resembles platinum in many respects and is second to it in importance. The principal advantage of palladium as compared with platinum is its lower cost. The major application of palladium is in telephone relay contacts. It is also used as a catalyst to remove oxygen from heat-treating atmospheres and as a filter for purification of hydrogen gas. Palladium leaf is used for decorative effects in bookbinding, glass signs, and trim.

Like platinum, palladium forms complete solid solutions with almost all

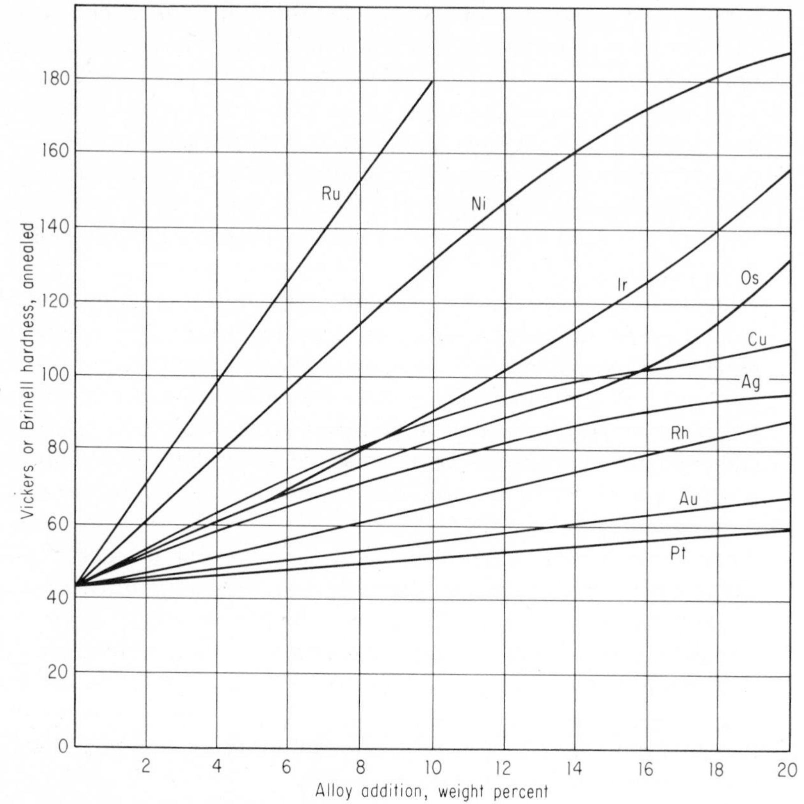

Fig. 12·54 Hardening effect of alloying additions on palladium. (The International Nickel Company.)

alloying elements. The hardening effect of alloying additions on palladium is shown in Fig. 12·54. Of the metals shown, ruthenium and nickel are very effective hardeners, while platinum is least effective.

Palladium-Silver Alloys Alloys containing 1, 3, 10, 40, 50, and 60 percent palladium are widely used for electrical contacts. The 60 percent palladium alloy is employed for electrical contacts operating at reasonably high currents and for precision resistance wires. The lower-palladium-content alloys are used for contacts in low-voltage relays and regulators. Palladium-silver alloys are used for brazing stainless steel and other heat-resistant alloys.

Palladium-silver alloys that have additions of copper, gold, zinc, and platinum are age-hardenable and yield high mechanical properties after heat treatment.

12·45 Iridium Iridium is the most corrosion-resistant element known. Pure iridium has been used for crucibles in studying slag reactions at very high temperatures, and as extrusion dies for very high-melting glasses.

The main use of iridium is as a hardening addition to platinum. Small amounts of iridium, up to about 0.1 percent, are used for refining the grain size and improving the mechanical properties of gold- and silver-base casting alloys.

12·46 Osmium Osmium has a high melting point and cannot be worked even at very high temperatures. Osmium and its alloys have high hardness, high wear resistance, and good corrosion resistance. Typical applications include fountain-pen nibs, phonograph needles, electrical contacts, and instrument pivots.

12·47 Rhodium Rhodium is similar to platinum in color and has considerably higher reflectivity. It has exceptionally high corrosion resistance, almost equal to that of iridium. Rhodium provides a nontarnishing electroplate with high reflectivity. It is used as a finishing plate in the jewelry field and for reflectors for motion-picture projectors and aircraft searchlights. A thin rhodium plate is sometimes used on the surfaces of sliding electrical contacts. The main use of rhodium is as an alloying addition to platinum and palladium.

12·48 Ruthenium This element cannot be cold-worked but may be forged at temperatures above 2800°F. The general corrosion resistance of ruthenium approaches that of iridium. The metal is rarely used in the pure form except as a catalyst for the synthesis of certain hydrocarbons. It is employed mainly as a hardener for platinum and palladium.

Typical mechanical properties of some metals and their alloys are given in Table 12·19.

12·49 Electrical Contacts Since many of the precious metals and their alloys are widely used for electrical contacts, this particular application will be discussed in greater detail (The H. A. Wilson Company, Union, N.J.).

The properties of an ideal contact material are:

High electrical conductivity for maximum current-carrying capacity

Low temperature coefficient of resistance to keep contact resistance as nearly uniform as possible

High thermal conductivity to decrease temperature rise of contact and reduce tendency toward oxidation

Low surface contact resistance to utilize minimum contact pressure

High melting point to prevent the formation of molten bridges, loss of material, and surface roughening

High boiling point to prevent local vaporization and loss of material during arcing

High corrosion resistance to prevent an increase in contact resistance

High nonwelding and nonsticking characteristics

TABLE 12·19 Properties of Some Metals and Alloys*

METAL OR ALLOY AND NOMINAL COMPOSITION	CONDITION	TENSILE STRENGTH, 1,000 PSI	YIELD STRENGTH, 1,000 PSI	ELONGATION, % IN 2 IN.	BHN	MODULUS OF ELASTICITY, 10^6 PSI
Silver (99.9+)	Annealed	23	12	45	30	10.5
	Cold-rolled	43	48	6	90	
Sterling silver,	Hard	64	50	4	125	10.5
92.5 Ag, 7.5 Cu	Annealed	41	20	26	65	
Silver brazing alloy,	Annealed	58	45	38	110	13
60 Ag, 25 Cu, 15 Zn	Cold-drawn	89	75	9	160	
Gold (pure)	Hard	30	...	2	49	
	Annealed	17.5	...	40	28	10.8
Platinum	Hard	65	...	2	101	
(commercial)	Annealed	27	...	28	65	22
Platinum-iridium,	Hard	80	...	2	169	
90 Pt, 10 Ir	Annealed	53	34	23	104	25
Platinum-rhodium,	Hard	93	...	3	169	
90 Pt, 10 Rh	Annealed	50	18.3	36	79	21.2
Platinum-ruthenium,	Hard	145	...	2	210	
90 Pt, 10 Ru	Annealed	91	...	28	156	31.5
Palladium	Hard	55	...	...	91	
(commercial)	Annealed	30	7.6	30	47	16.3
Palladium-silver,	Hard	100	94	...	176	
60 Pd, 40 Ag	Annealed	47	15	40	87	22.4
Iridium (pure)	Annealed	90	34	...	175	75
Rhodium (pure)	Annealed	80	...	...	119	50
Osmium (pure)	Arc-melted button	...	...	...	400†	81
Ruthenium (pure)	Forged	74	...	...	390†	60
Molybdenum	As-rolled	100	75	30	250	46
(99.9+)	Recrystallized	70	50	45	190	
Tantalum	Annealed sheet	60	45	37	55	27
(99.9+)	Unannealed sheet	110	100	3	123	
Tungsten (pure)	Hard wire	600	540	0.8	...	53
Titanium (99.8+)	Annealed	90	70	2.3	200	16.5
Zirconium (pure)	Annealed	36	16	31	77	11

*From "Properties of Some Metals and Alloys," The International Nickel Co.
†Vickers pyramid hardness.

High hardness and toughness to prevent mechanical wear and failure, particularly in parts that operate at high frequencies or under high contact pressures

It is apparent, from the properties listed, that no one metal or alloy can be a universal contact material. The practical selection of a contact material is based upon combining two or more desirable properties while minimizing the less advantageous properties for a particular application.

The contact materials may be classified according to contact properties in the following groups:

High conductivity: silver and silver alloys

Corrosion- and oxidation-resistant: platinum and related alloys

Refractory and arc-resistant: tungsten and molybdenum

High conductivity plus arc-resistant: powder-metallurgy compacts

Silver and Silver Alloys Silver has the highest electrical and thermal conductivity of all the contact materials. The oxides of silver decompose at relatively low temperatures because of arcing, reverting to metallic silver, thus maintaining low contact resistance. Silver is used for sensitive contacts under light and intermediate pressure. They operate satisfactorily if current and voltage do not become excessive. The principal disadvantages of silver are its low hardness, low melting point, tendency to form sulfide films, and tendency to build up on one electrode under excessive current conditions. These disadvantages are minimized by the addition of alloying elements, principally copper, cadmium, zinc, nickel, manganese, platinum, palladium, and iron. The effect of alloying is to increase hardness, raise the melting point, reduce material loss or transfer, increase resistance to welding or sticking, increase resistance to erosion by arcing, and increase corrosion resistance.

Platinum, Palladium, and Gold The outstanding property of the alloys in this group is their corrosion resistance to surrounding atmospheres and the sureness of making contact under light pressures. The metals are able to maintain low contact resistance over long periods because of their resistance to oxidation at the high local temperatures reached in arcing. Since these metals are soft, they are rarely used in the pure form. They are alloyed to obtain higher hardness without undue sacrifice of corrosion resistance or surface-contact resistance. The most common alloying elements are iridium, ruthenium, osmium, rhodium, and silver. Copper, nickel, and iron are also used as alloying elements.

Tungsten and Molybdenum Tungsten and molybdenum have the unique property of high resistance to arc erosion along with very high melting points, high boiling points, and high resistance to welding and pitting. They are used in applications requiring high contact pressures and where the contacts have to operate frequently or continuously. The chief disadvantage of tungsten and molybdenum is their tendency to form oxides, particularly where severe arcing takes place. In some applications these disadvantages may be overcome by using very high contact pressures, by incorporating a wiping action in the contacts, or by use of protective circuits to suppress excessive arcing.

Powder-metallurgy Compacts Silver and tungsten do not alloy with each other by the ordinary process of melting and casting. These metals, however, may be combined by using the techniques of powder metallurgy (discussed in Chap. 16) to produce alloys which are homogeneous and have high physical properties. It is thus possible to take advantage of the high electrical conductivity of silver or copper and the high arc resistance of tungsten or molybdenum. These materials are designed to operate under conditions of heavy current and voltage and are also often used in low-current high-frequency applications.

Typical combinations in this group are silver-tungsten, silver-molybdenum, silver-tungsten carbide, silver-molybdenum carbide, copper-tungsten carbide, and copper-tungsten. Other compositions are manufactured by powder metallurgy to take advantage of the high electrical conductivity of silver and copper and the semirefractory properties of cadmium oxide, iron, graphite, and nickel.

Many of these compositions can be produced in strip and wire form and may be drilled, rolled, swaged, drawn, formed, bent, and extruded without difficulty. They may be used as replacements for fine silver or silver alloys in some applications, since they give greater resistance to sticking and welding, have improved mechanical properties, and possess greater resistance to electrical erosion.

QUESTIONS

12·1 What is the most important property of copper?
12·2 Explain why copper is a suitable material for automobile radiators.
12·3 Explain the reasons for the difference in microstructure of Figs. 12·1 and 12·2.
12·4 What would be the temper of 0.25-in.-diameter copper wire cold-drawn from 0.50-in.-diameter soft wire?
12·5 What is *season cracking*? How may it be minimized?
12·6 What is *dezincification*? How may it be minimized?
12·7 How does the addition of lead to brass improve its machinability?
12·8 Which copper alloy would be best for the tubes in a marine heat exchanger? Why?
12·9 Why are most copper-zinc alloys not age-hardenable?
12·10 Discuss the effect on corrosion resistance of copper by increasing additions of (*a*) zinc, (*b*) tin, (*c*) nickel.
12·11 Why is muntz metal heat-treatable? Describe a typical heat treatment and the resulting microstructure.
12·12 What properties would be important in the choice of a copper-alloy spring?
12·13 Why is "manganese bronze" a misnomer?

12·14 Differentiate between the terms *brass* and *bronze*.
12·15 Why are tin bronzes suitable for use as bearings?
12·16 Why is beryllium suitable for tools in the petroleum industry?
12·17 What are the outstanding properties of cupronickel alloys?
12·18 What are the outstanding properties of aluminum?
12·19 Why do long-range electrical transmission lines use a steel core and an aluminum shell?
12·20 Explain the meaning of the digits in the following aluminum specifications: 2107-T4, 5056-H16, 7075-T6, 6061-0.
12·21 Why do many aluminum alloys respond to age hardening? Give some examples.
12·22 Why do aluminum alloy 2017 rivets have to be refrigerated until used?
12·23 What outstanding properties are generally given an alloy by the addition of nickel?
12·24 What outstanding properties do aluminum-silicon alloys have? Give some typical applications.
12·25 Which aluminum casting alloy develops the highest mechanical properties? Why?
12·26 What is meant by *anodizing* aluminum?
12·27 What are the outstanding properties of magnesium? Of nickel?
12·28 What is the effect of zirconium additions to magnesium alloys?
12·29 What is the effect of a rare-earth metal addition to magnesium? Of thorium addition?
12·30 Compare aluminum and magnesium with regard to corrosion resistance.
12·31 Discuss the methods used to join magnesium alloys.
12·32 Give one application and the reasons for selection of the following nickel alloys: duranickel, permanickel, Monel, K Monel, constantan, Inconel, Inconel X, Hastelloy C, and Illium G.
12·33 Give the composition, special properties, and one application of the following iron-nickel alloys: Invar, Kovar, Platinite, and Elinvar.
12·34 What are the outstanding properties of lead? Give one application for each property.
12·35 What property is important for fusible plugs? Give the composition of a suitable alloy for this application.
12·36 What properties are important in alloys for type metals?
12·37 Why is "white metal" suitable for bearing applications?
12·38 Which type of soft solder is preferred for electrical equipment and why?
12·39 Compare the lead-base and tin-base babbitts with regard to properties and applications.
12·40 Why is it difficult to work-harden lead, tin, or zinc at room temperature?
12·41 Why is tin-coated steel used for tin cans to hold food and not zinc-coated steel?
12·42 What are the outstanding properties of (*a*) silver, (*b*) gold, and (*c*) platinum?
12·43 Which of the platinum metals has the highest corrosion resistance? Which has the highest modulus of elasticity?
12·44 Name two important properties of titanium.
12·45 What is one disadvantage of titanium?
12·46 What are the two crystal structures of titanium, and what is the transformation temperature?
12·47 Explain what is meant by an alpha or beta stabilizer.
12·48 Name one alpha stabilizer and three beta stabilizers.

12·49 Explain why the two-phase titanium alloys are stronger than the single-phase alpha alloys.
12·50 What are the two main attributes of alpha alloys?
12·51 How may alpha-beta alloys be strengthened?
12·52 How may beta alloys be strengthened?
12·53 Give at least two applications for alpha alloys, alpha-beta alloys, and beta alloys.

REFERENCES

Alice, J.: "Introduction to Magnesium and Its Alloys," Ziff-Davis Publishing, Chicago, 1945.
Aluminum Company of America: "Casting Alcoa Alloys," Pittsburgh, Pa., 1951.
American Society for Metals: "Magnesium," Metals Park, Ohio, 1946.
———: "Metals Handbook," Metals Park, Ohio, 1948, 1961, 1972.
———: "Physical Metallurgy of Aluminum Alloys," Metals Park, Ohio, 1949.
Bunn, E.S., and R. A. Wilkins: "Copper and Copper-base Alloys," McGraw-Hill Book Company, New York, 1943.
Dow Chemical Company: "Magnesium Alloys and Products," Midland, Mich., 1955.
Ellis, O. W.: "Copper and Copper Alloys," American Society for Metals, Metals Park, Ohio, 1948.
Gibbs, L. E.: "Cold Working of Brass," American Society for Metals, Metals Park, Ohio, 1946.
International Nickel Co.: "Nickel and Its Alloys," "Nickel and Nickel-base Alloys," "Age-hardening Inco Nickel Alloys," "Engineering Properties of Duranickel," "Engineering Properties of Inconel," "Engineering Properties of Monel," "Engineering Properties of Nickel," New York.
Lead Industries Association: "Lead in Modern Industry," New York, 1952.
Liddell, D. M.: "Handbook of Nonferrous Metallurgy," McGraw-Hill Book Company, New York, 1945.
Mathewson, C. H. (ed.): "Modern Uses of Nonferrous Metals," 2d ed., American Institute of Mining and Metallurgical Engineers, New York, 1958.
Mondolfo, L. F.: "Metallography of Aluminum Alloys," John Wiley & Sons, Inc., New York, 1943.
National Bureau of Standards (U.S.): Zinc and Its Alloys, Natl. Bur. Std. (U.S.) Circ. 395, 1931.
Raudebaugh, R. J.: "Nonferrous Physical Metallurgy," Pitman Publishing Corporation, New York, 1952.
Raynor, G. V.: "Physical Metallurgy of Mg and Its Alloys," Pergamon Press, New York, 1959.
Reynolds Metals Company: "Heat Treating Aluminum Alloys," 1948; "Aluminum Forming," 1952; "Finishes for Aluminum," 1951, Louisville, Ky.
Roberts, C. S.: "Magnesium and Its Alloys," John Wiley & Sons, Inc., New York, 1960.
Samans, C. H.: "Engineering Metals and Their Alloys," The Macmillan Company, New York, 1964.
Van Horn, Kent R.: "Aluminum," 3 vols., American Society for Metals, Metals Park, Ohio, 1967.
Vines, R. F.: "The Platinum Metals and Their Alloys," The International Nickel Co., New York, 1941.

13 METALS AT HIGH AND LOW TEMPERATURES

13·1 Introduction The terms *high* and *low* temperature are entirely relative to our own natural environment. What is considered a high temperature for low-melting metals such as tin and lead may be considered a low temperature for a high-melting metal such as tungsten. Therefore, lower-melting metals will exhibit characteristics at low temperatures that will require relatively higher temperatures for other metals. For example, a temperature of about 1000°F may be necessary to recrystallize iron after cold working, but tin and lead will recrystallize at or near room temperature. The properties of metals are usually determined at room temperature, and our thinking about metals is based on their behavior at normal temperatures. When the temperature is changed, becoming either higher or lower, changes in the behavior of metals often occur which may seriously affect their usefulness in a particular application.

METALS AT HIGH TEMPERATURES

13·2 Elevated-temperature Tests The behavior of metals observed by stressing at elevated temperatures depends upon the length of the test period. Since life expectancy of machine parts is usually high, it is not possible to run a test for many years to determine what to use in current construction. It is necessary to extrapolate from shorter-time tests. This extrapolation, however, must be done with great care from tests that will provide useful data. This is especially difficult for high-temperature applications, since changes in behavior will occur with time at temperature.

In high-temperature tests it is necessary to determine the dependence of ultimate strength (rupture strength) and yield strength (creep strength) on the time application of the stress. In investigating the plastic behavior of metals at elevated temperatures, it is convenient to apply a tensile load on the test specimen. While this procedure, in many cases, does not duplicate service conditions, it is possible, by careful interpretation, for the data to provide useful information which can be applied to combined stress conditions.

Many tests have been developed for high-temperature studies, but the three most widely used ones are:

1 Creep tests at small deformations: low stresses and low strain rates for long time periods

2 Stress-rupture (creep-rupture) tests at larger deformations: higher stresses and larger strain rates for shorter periods of time

3 Short-time tensile tests at large deformations: high stresses and high strain rates available with the usual tension-testing equipment

13·3 Creep Tests Creep is a property of great importance in materials used for high-temperature applications. It may be defined as a continuing slow plastic flow under constant conditions of load or stress. Creep is generally associated with a time rate of deformation continuing under stresses well below the yield strength for the particular temperature to which the metal is subjected. It occurs at any temperature, though its importance depends upon the material and the degree to which freedom from continuing deformation is desired.

A creep test is simply a tension test run at constant load and constant temperature. There is a means of measuring the elongation of the specimen very accurately and a means of heating the specimen under closely controlled conditions. The total creep or percent elongation is plotted against time for the entire duration of the test. Two typical creep curves are shown in Fig. 13·1.

The various stages of creep are illustrated by curve *A*. When the load is first applied, there is an instantaneous elastic elongation, then a primary stage of a transient nature during which slip and work hardening take place in the most favorably oriented grains. The creep rate (tangent to the curve) is initially high and gradually slows to a minimum. This is followed by a secondary stage of steady-state creep, during which the deformation con-

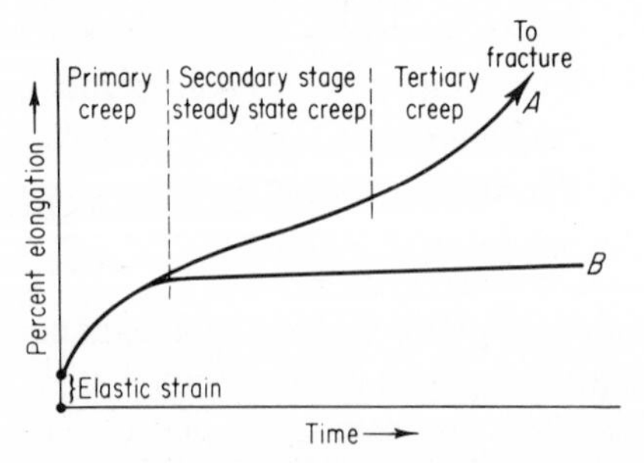

Fig. 13·1 Typical creep curves illustrating the stages of creep.

tinues at an approximately constant rate. During this stage a balance exists between the rate of work hardening and the rate of softening because of recovery or recrystallization. In some cases, under moderate stresses, the creep rate may continue to decrease at a very slow rate, and the secondary stage may continue for a very long time (curve *B*, Fig. 13·1). However, if the stress is sufficiently high, there is a tertiary stage in which the creep rate accelerates until fracture occurs.

There is little or no correlation between the room-temperature mechanical properties of a material and its creep properties. Creep seems to be greatly affected by small variations in microstructure and prior history. The grain size of the metal is an important factor in determining its creep characteristics. Whereas at room temperature fine-grained materials show higher yield and ultimate strengths than coarse-grained materials, the reverse is true at elevated temperatures. It is believed that at high temperatures the grain boundaries may act as centers for the generation of dislocations which cause creep.

The presence of solute atoms, even in minor amounts, tends to retard creep by interfering with the motion of dislocations through the crystal. A more potent factor in retarding creep is the presence of a strong, stable, finely dispersed second phase.

13·4 Stress-rupture Tests These tests are conducted to determine the ability of a material to resist fracture at elevated temperatures. In stress-rupture tests the loads are high enough to cause comparatively rapid rupture. The time involved is usually between 10 and 400 h, although some tests may run as long as 1,000 h.

A series of specimens are broken at each temperature of interest, under constant load, the stresses being selected so fractures will occur from a few minutes to several hundred hours. The results are usually plotted in log-log coordinates, and if no structural changes occur during the test period, the relationship of rupture stress and time to rupture is linear. Typical stress-rupture data for S-590 (cobalt-chromium-nickel–base) alloy are shown in Fig. 13·2. Discontinuities in the straight lines are associated with changes in the alloy and indicate a change in the method of failure from transcrystalline low-temperature type to intercrystalline high-temperature type.

The principal differences between the stress-rupture and creep test are the testing time, the stress or strain rate level, and the sensitivity of control and measurement of temperature, load, and strain. It is possible before fracture to determine the elongation as a function of the time, as in an ordinary creep test. From these data, the steady-state or minimum creep rate at very high stresses may also be determined. The log-log plot of stress vs. minimum creep rate for S-590 alloy is shown in Fig. 13·3.

For some applications, such as superheater tubes, still tubes, piping, pipe

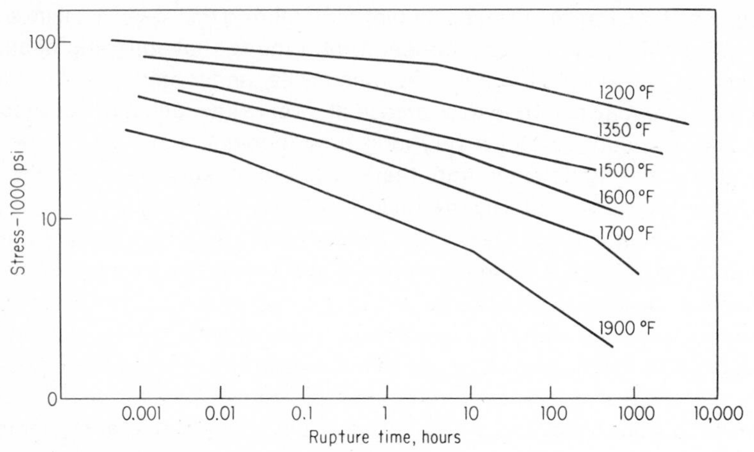

Fig. 13·2 Log-log plot of stress vs. rupture time for S-590 alloy. (From "High Temperature Properties of Metals," American Society for Metals, Metals Park, Ohio, 1951.)

fittings, sheet-metal parts, nozzle guide vanes, and boilers, only rupture data are important. For other applications, such as bolts, steam valves, steam-turbine blading, turbine rotors, turbine casings, and valve stems, creep data are considered most important.

Considering the fact that the load in the stress-rupture test is much higher than design values and that the test continues to fracture, there is some doubt as to the usefulness of the data obtained. Metal behavior at high loads, high rates of deformation, and short life may not be indicative of that

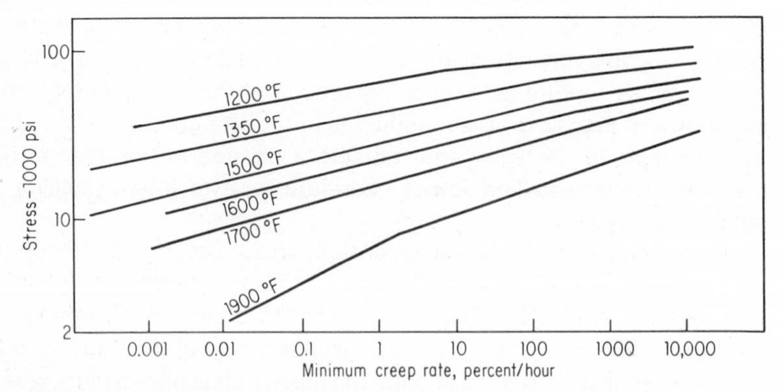

Fig. 13·3 Log-log plot of stress vs. minimum creep rate for S-590 alloy. (From "High Temperature Properties of Metals," American Society for Metals, Metals Park, Ohio, 1951.)

at lower loads. As a method of rating different alloys or in comparison of two different lots of the same alloy, the stress-rupture test seems to show a correlation with creep tests at usable loads.

In some applications the design life is short. Guided missiles are designed for a life of 1 h, and turbojet engines are frequently designed for 1,000-h life. In these cases the test period can be as long as the design life, and stress-rupture data can sometimes be used directly in such designs. For applications of longer design life—for example, steam or gas turbines with a design life of 100,000 h (13 years)—it is necessary to extrapolate data obtained during shorter time periods.

Materials for high-temperature use are usually designed for a certain minimum creep after a stated period. The creep strength, which is the quantity used in design, is the stress required to produce a definite percent deformation in a stated time. Some gas-turbine designers have set a standard of 0.1 percent creep in 10,000 h. This is equivalent to 0.01 percent in 1,000 h.

13·5 Short-time Tension Tests These tests are used to study the effect of heating a sample and testing under strain rates that are available in the ordinary tensile-testing machine. Elastic properties at elevated temperatures are not real, since their values depend upon the time between load applications, and their accuracy depends on the sensitivity of the extensometer. The duration of testing is usually only a few minutes, and the important effects of time at temperature are not measured. The short-time tension test fails to predict what will happen in a shorter or longer period of time and therefore has very little application. The test is sometimes used for rapid estimation of materials which may warrant further study, and the short-time tensile strength is frequently used as the 0.1-h point on a rupture curve. The variation of the short-time yield and tensile strength of Inconel X with temperature is shown in Fig. 13·4.

13·6 Creep Properties of Various Alloys Plain-carbon and low-alloy steels are widely used for moderate-temperature applications, particularly below 900°F. An increase in carbon content improves the creep strength at lower temperatures where the carbides are present in lamellar form. The reverse may be true at higher temperatures where the carbides are spheroidized. The recommended structure of plain-carbon steels for high-temperature service is the normalized one. The annealed structure appears to be less stable and tends to spheroidize more rapidly, reducing creep strength. The use of aluminum as a deoxidizer in the manufacture of steel tends to produce fine grain, which lowers creep strength. Aluminum additions should be kept low, and their effect is considerably reduced by the presence of manganese and molybdenum.

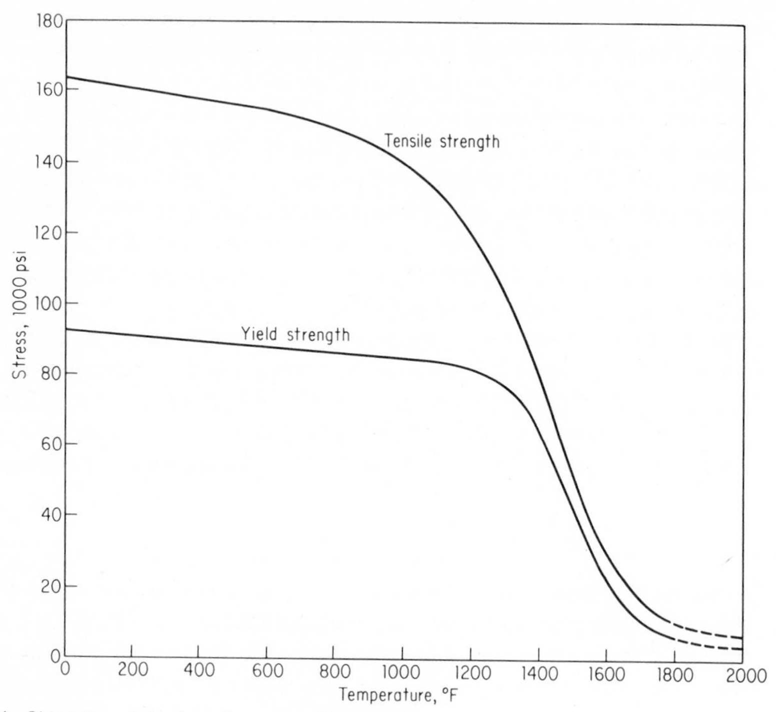

Fig. 13·4 Short-time yield strength and tensile strength of Inconel X at elevated temperatures. (From "Metals Handbook," 1954 supplement, p. 42, American Society for Metals, Metals Park, Ohio.)

In low-alloy steels, containing less than 10 percent alloy, molybdenum and vanadium are most effective in raising the creep resistance. The carbon content is usually kept below 0.15 percent. The 0.5 percent molybdenum steel is used for piping and superheater tubes up to 850°F. Above this temperature, spheroidization and graphitization tend to take place, with a reduction in creep strength. The addition of 1 percent chromium to this steel increases the resistance to graphitization, and the steel is now used for piping and boiler tubes at temperatures to 1000°F.

The chromium-molybdenum-vanadium steels containing up to 0.50 percent carbon are used in the normalized and tempered, or quenched and tempered condition. They have relatively high yield and creep strengths and are suitable for bolts, steam-turbine rotors, and other parts operating at temperatures up to 1000°F.

The straight-chromium (4xx series) stainless steels are used for elevated-temperature applications that require increased corrosion and oxidation resistance. Type 410 is used for bolts, steam valves, pump shafts, and other parts up to 1000°F. Type 422 is used for similar applications up to 1200°F. Both the above types are hardenable to a martensitic structure and are tempered about 100° above the operating temperature to promote stability of structure.

Type 430 (16 percent Cr) and 446 (25 percent Cr) are nonhardenable and ferritic. These grades, generally used in the annealed condition, have lower creep strength than the hardenable types but show greater oxidation resistance. Type 430 is used for heat-exchange equipment, condensers, piping, and furnace parts operating at temperatures to 1550°F. Type 446 is used for similar applications up to 2000°F.

The austenitic stainless steels (3xx series) show better creep properties than the 4xx series and better corrosion and oxidation resistance. Type 310 (25 percent Cr, 20 percent Ni) is used for furnace linings, boiler baffles, thermocouple wells, aircraft-cabin heaters, and jet-engine burner liners. Type 347 (18 percent Cr, 11 percent Ni, + Cb and Ta) is used for steam lines, superheater tubes, and exhaust systems in reciprocating engines and gas turbines operating up to 1600°F.

Another group of alloys that have good creep properties in the temperature range of 1200 to 1400°F are essentially chromium-nickel-molybdenum-iron alloys. Many contain small amounts of titanium and aluminum. Some trade names for alloys in this group are A-286, Discaloy, Incoloy 901, 16-25-6, and D-979. These alloys are used as forgings for turbine wheels, various other components of gas turbines, sheet-metal casings, housings, and exhaust equipment.

The nickel-base alloys such as M-252, Waspaloy, Rene 41, Hastelloy R-235, Inconel 700, Udimet 500, and Unitemp 1753 are widely used for aircraft applications. They are intended for use in the temperature range of 1400 to 1800°F. These alloys contain 50 to 70 percent nickel, about 20 percent chromium, up to 10 percent molybdenum or tungsten, up to 20 percent cobalt, and titanium and aluminum. They are used for manifolds, collector rings, and exhaust valves of reciprocating engines, and in sheet form for combustion liners, tail pipes, and casings of gas turbines and jet engines.

The cobalt-chromium-nickel–base alloys such as S-816, S-590, L-605, and N-155 are suitable for applications in the same range as the nickel-base alloys but have lower rupture strength. They are used for wheels and buckets of gas turbines.

All the commercial alloys mentioned above tend to lose their strength rapidly when heated above about 1700°F. It seems unlikely that alloys using these metals will raise the allowable operating temperature much above the present limit. This limit is related to a great extent by the melting points of the base metals. Most promising base metals for future high-temperature alloys are molybdenum (melting point 4730°F) and tungsten (melting point

TABLE 13·1 Nominal Compositions of Some Wrought Superalloys*

ALLOY	C	Cr	Ni	Co	Mo	W	Cb	Ti	Al	Fe	OTHER
	IRON-NICKEL-CHROMIUM-MOLYBDENUM ALLOYS										
A-286	0.08	15	26	. . .	1.25	. . .	. . .	2.0	0.25	bal	0.3V
DISCALOY	0.06	14	26	. . .	3.00	. . .	. . .	1.7	0.25	bal	. . .
16-25-6	0.06	16	25	. . .	6.00	. . .	. . .	. . .	. . .	bal	0.15N
D-979	0.05	15	45	. . .	4.00	4.00	. . .	3.0	1.00	bal	0.01B
	NICKEL-BASE ALLOYS										
Nimonic 80	0.1 max	20	bal	2 max	. . .	. . .	. . .	2.25	1.25	5 max	. . .
Nimonic 90	0.1 max	20	bal	18	. . .	. . .	. . .	2.50	1.50	5 max	. . .
Hastelloy C	0.15 max	16.5	bal	. . .	17	4.5	. . .	. . .	. . .	6	. . .
Hastelloy R-235	0.15	15.5	bal	. . .	5	. . .	. . .	2.50	2.00	10	0.6 max V
Hastelloy X	0.15	22	bal	. . .	9	. . .	. . .	. . .	. . .	20	. . .
Inconel X	0.04	15	73	. . .	. . .	. . .	1	2.50	0.9	7	. . .
Waspaloy	0.1	19	bal	14	4	. . .	. . .	3	1.3	1	tr B, Zr
Udimet 500	0.1	19	bal	19	4	. . .	. . .	3	2.9	4 max	tr B
Unitemp 1753	0.24	16	bal	7	1.5	8	. . .	3	2.0	9.5	tr B, Zr
René 41	0.1	19	bal	11	10	. . .	. . .	3	1.5	3	tr B
René 63	0.1	14	bal	15	6	3	. . .	2.50	3.8	. . .	0.015 B
	COBALT-CHROMIUM-NICKEL-BASE ALLOYS										
S-816	0.40	20	20	bal	4	4	4	. . .	. . .	4	. . .
V-36	0.25	25	20	bal	4	2	2	. . .	. . .	3	. . .
L-605	0.15	20	10	bal	. . .	15	. . .	. . .	. . .	2	. . .
S-590	0.40	20	20	20	4	4	4	. . .	. . .	bal	. . .
N-155	0.15	21	20	20	3	3	1	. . .	. . .	bal	0.15N

*From Metals Handbook, vol. 1, American Society for Metals, Metals Park, Ohio, 1961.

6170°F). These metals are relatively abundant and are available in high-purity form. A recently developed molybdenum alloy containing 0.5 percent titanium has higher rupture strength in the range of 1600 to 2000°F than any other commercial alloy. Two disadvantages of these metals are their high density and their great susceptibility to oxidation. This latter disadvantage may be overcome by the development of a suitable oxidation-resistant coating.

The nominal composition of some wrought heat-resistant superalloys is given in Table 13·1, while some typical high-temperature strength values for several of these alloys are given in Table 13·2.

METALS AT LOW TEMPERATURES

13·7 Effect of Low Temperature on Properties As the temperature is decreased below normal room temperature, the hardness, yield strength, and, with

TABLE 13·2 High-temperature Strength Values for Several Alloys

ALLOY	TEMP, °F	SHORT-TIME TENSILE STRENGTH, PSI	STRESS-RUPTURE STRENGTH, PSI, 1,000 H	CREEP STRENGTH, PSI	
				0.1% PER 1,000 H	0.01% PER 1,000 H
Carbon steel, 0.15C, annealed	800	55,000	. . .	26,800	18,500
	1000	36,500	12,000	5,750	2,700
	1200	20,000	2,200	620	290
Carbon-molybdenum steel, 0.15 C, 0.55 Mo	800	60,000	. . .	30,000	21,000
	1000	45,000	25,000	10,800	6,700
	1200	27,000	4,400	2,000	700
Type 410 stainless steel, 13 Cr	1000	48,000	19,000	12,000	
	1200	25,000	7,000	2,200	
	1300	17,000			
Type 304 stainless steel, 19 Cr, 9 Ni	1000	60,000	35,000	17,000	12,000
	1200	46,000	14,000	7,000	4,000
	1300	37,000	9,000	3,900	
N-155 (low C), 0.15 C, 21 Cr, 20 Ni, 20 Co, 3 Mo, 3 W, 1 Cb, bal. Fe	1200	83,000	37,000	19,000	16,000
	1350	60,000	22,000	14,500	10,500
	1500	40,000	13,000	8,000	5,000
S-816, 0.4 C, 20 Cr, 20 Ni, 4 Mo, 4 W, 4 Cb, 4 Fe, bal. Co	1350	99,000	30,000	18,000	12,000
	1500	78,000	17,000	11,500	8,100
	1600	60,000	9,500	6,500	5,000
Inconel X, 0.04 C, 15 Cr, 73 Ni, 1 Cb, 2.5 Ti, 0.9 Al, 7 Fe	1200	120,000	69,000	60,000	48,000
	1350	93,000	42,000	37,500	30,000
	1500	60,000	18,000	18,000	15,000

few exceptions, the ultimate strength and modulus of elasticity of all metals and alloys increase. The variation of yield and tensile strengths of iron, nickel, and copper with temperature is shown in Fig. 13·5.

In regard to the effect of temperature on ductility, metals fall into two distinct groups, those which remain ductile at low temperatures and those which become brittle. An indication of the amount of ductility, or plastic deformation, before fracture may be obtained from a study of the fracture surface. A cup-cone type of fracture is typical of a ductile material which has failed in shear after plastic deformation when tested in tension. A brittle material fails by cleavage with no evidence of plastic deformation. As the temperature is decreased, face-centered-cubic metals fracture only by shear and show a gradual and continuous decrease in ductility. Metals with other crystal structures may fail by shear at room temperature, but with decreasing temperature the mode of fracture changes from shear (ductile) to cleavage (brittle). The change in fracture often appears as a sharp drop in ductility. The effect of temperature on the ductility of iron, copper, and nickel is shown in Fig. 13·6. The tensile properties of some steels and nonferrous materials at low temperatures are given in Table 13·3. Cleavage fractures of structural members are often sudden and unexpected and usually result in catastrophic brittle failure of the part. Great interest

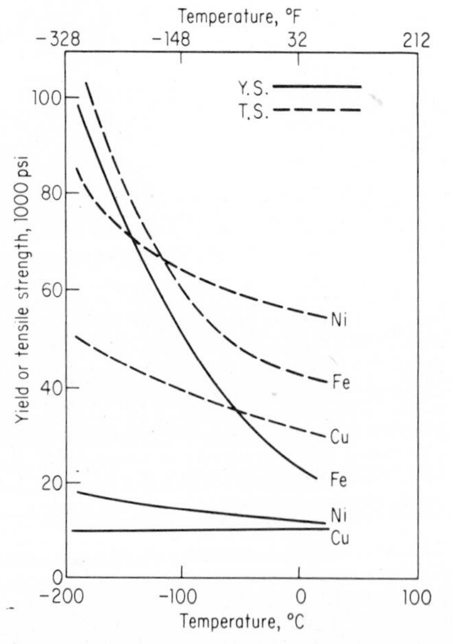

Fig. 13·5 Variation of yield and tensile strengths of iron, copper, and nickel with temperature. (From "Behavior of Metals at Low Temperatures," American Society for Metals, Metals Park, Ohio, 1953.)

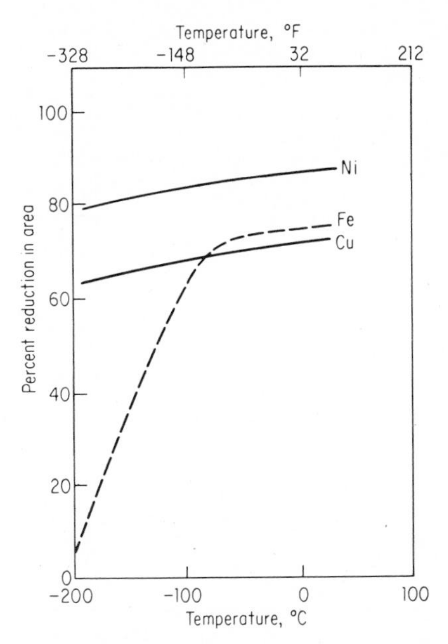

Fig. 13·6 Variation of ductility of iron, copper, and nickel with temperature. (From "Behavior of Metals at Low Temperatures," American Society for Metals, Metals Park, Ohio, 1953.)

in this problem developed during World War II when a number of welded ships failed in a brittle manner with almost explosive rapidity. In some cases, the ship was split in two. The steel used for ship plate was ductile at normal temperatures, yet the failure was of a brittle nature with little plastic deformation.

The tendency of steel to fail in a brittle manner is increased by stress concentration, increased speed of load application, and decrease of temperature. These three factors are interrelated, and the effect of lowering the temperature is the easiest one to measure quantitatively. It is often possible to study the change from ductile to brittle fracture with decreasing temperature, provided that stress concentration and speed of load application are held constant. These conditions are satisfied in the ordinary Charpy or Izod notched-bar impact test (see Sec. 1·33). A rough approach to some degree of correlation with room-temperature and low-temperature toughness behavior is shown by the notched-bar impact test. This test has become so common that most of the information on toughness has been obtained by it.

13·8 Effect of Temperature on Notched-bar Test If tests on steel are made at many temperatures, a plot of energy absorbed vs. temperature will usually

TABLE 13·3 Tensile Properties of Some Steels and Nonferrous Materials at Low Temperature*

MATERIAL AND COMPOSITION	CONDITION	TEMP, °F	YIELD STRENGTH, PSI	TENSILE STRENGTH, PSI	ELONGATION, % IN 2 IN.	REDUCTION IN AREA, %
Low-carbon steel,	As-rolled	70	54,700	66,300	29.7	71.8
0.13–0.14 C	As-rolled	−85	67,700	80,700	33.6	70.3
	As-rolled	−292	...	121,300	26.5	55
	Annealed	70	42,700	45,700	27.5	77.5
	Annealed	−296		137,000	7.5	
	Annealed	−423	155,000	155,000	0.3	2.5
Ni steel, 0.13 C,	Oil-quenched	68	...	103,000	25	74
5.13 Ni, 0.41 Mn,	and	−242	...	153,000	25	57
0.19 Cr, 0.15 Si	tempered	−319	...	175,000	21	50
	1200°F					
Ni-Cr-Mo steel,	Oil-quenched	70	137,700	152,000	14	65
0.33 C, 0.67 Cr,	and	6	141,000	154,500	15.6	64
2.45 Ni, 0.64 Mo	tempered	−90	145,000	163,000	15.6	62
	1185°F	−292	183,500	201,500	17	63
Fe-Ni alloy,	Water-	70	52,400	81,100	32	58
0.16 C, 35.8 Ni,	quenched	−423	127,000	144,000	20	60
0.86 Mn						
Commercially	Annealed	70	24,600	65,500	42	78
pure Ni		−112	27,600	76,400	43	73
		−292	27,600	97,000	53	74
Commercially	Hard-rolled	75	19,700	23,500	16	
pure Al		−112	21,350	24,700	18	

TABLE 13·3 (continued)

MATERIAL AND COMPOSITION	CONDITION	TEMP, °F	YIELD STRENGTH, PSI	TENSILE STRENGTH, PSI	ELONGATION, % IN 2 IN.	REDUCTION IN AREA, %
Al alloy 2017, 4 Cu, 0.5 Mn, 0.5 Mg	Solution-treated	75	45,500	68,000	15	
		−112	46,500	70,000	16	
Al alloy 2052, 2.5 Mg, 0.25 Cr	Hard-rolled	75	38,600	43,500	14	
		−112	39,200	45,600	18.5	
Pure copper	Annealed	75	8,600	31,400	48	76
		−112	10,100	38,500	47	74
		−292	15,000	50,800	58	77
Cu-Be, 2.56 Be	Water-quenched and aged	75	125,000	187,000	2.6	5
		−112	147,000	202,000	4	5
		−292	155,000	214,000	3	6
Cu-Zn, 30.5 Zn, 0.10 Fe	Annealed	75	28,200	51,100	49	77
		−112	27,300	57,100	60	79
		−292	29,600	73,500	75	73
Magnesium alloy MI, 1 Mn	Extruded	77	28,000	39,000	3.5	
		−110	30,900	52,000	3	
Magnesium alloy AZ63, 6 Al, 3 Zn	Cast, heat-treated, and aged	77	21,500	38,900	4.5	5.8
		−110	24,400	35,400	2.5	3.4
Ni alloy Monel, 28.86 Cu, 0.28 Mn	Annealed	75	20,900	70,800	41	75
		−112	27,100	85,300	40	74
		−292	29,600	113,000	51	72

*Compiled from data in "Metals Handbook," 1948 ed., American Society for Metals, Metals Park, Ohio.

show a temperature range in which the impact values drop sharply as the temperature is lowered. At the same time, the mode of fracture changes from a predominantly fibrous-shear type to a crystalline-cleavage type. This is shown graphically in Fig. 13·7. Values in the transition range are often erratic since slight changes in conditions will affect the values.

The temperature at which some specified level of energy absorption or fracture appears is defined as the *transition temperature*. In ASTM specifications it is defined as the temperature at which specimens show a fracture of 50 percent shear and 50 percent cleavage. The lower the transition temperature the better is the steel able to resist the embrittling effect of stress concentration, high loading rate, or low temperature. A study of available data for iron and steel indicates that their low-temperature behavior is affected by two classes of variables, namely, metallurgical factors and mechanical factors.

13·9 Metallurgical Factors The important interrelated metallurgical factors affecting the low-temperature behavior of iron and steel are composition, deoxidation, heat treatment and microstructure, surface condition, and grain size.

Increasing carbon content decreases the notched impact strength at room temperature and raises the transition temperature (Fig. 13·8). The physical form of the carbon is also important. When the cementite is spheroidized, it seems to be less harmful to low-temperature properties.

A manganese content of up to 1.5 percent lowers the transition temperature, as shown in Fig. 13·9, but does not change the shape of the transition curve.

Silicon, in amounts up to 0.30 percent used to deoxidize steels, lowers transition temperature and improves notch toughness because a cleaner steel and a more uniform ferritic grain are produced. Larger amounts have the reverse effect, and the presence of 4 percent silicon results in a brittle structure even at room temperature.

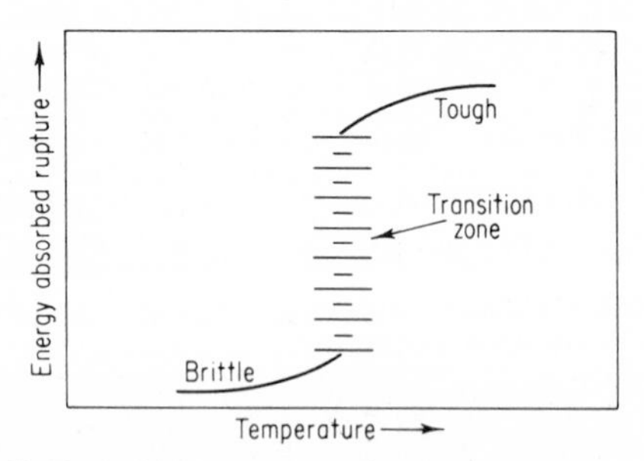

Fig. 13·7 Typical curve of impact strength vs. testing temperature for a ferritic steel, showing transition temperature zone in which erratic values may be expected. (The International Nickel Company.)

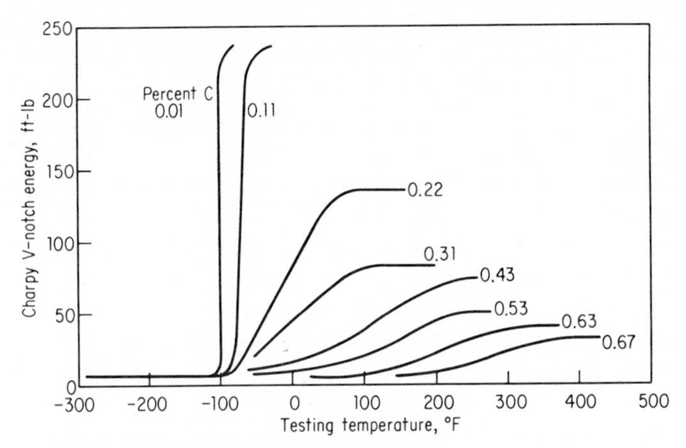

Fig. 13·8 Effect of carbon on the shape of the transition curve. (From "Metals Handbook," 1961 ed., p. 227, American Society for Metals, Metals Park, Ohio.)

The use of aluminum in addition to silicon for the deoxidation of steel seems to have a beneficial effect on the notch toughness of medium-carbon steels. The room-temperature impact resistance is improved and the transition temperature is lowered as the amount of aluminum is increased up to about 0.10 percent. The relatively fine ferritic grain size usually found in aluminum-treated cast and wrought steels contributes largely to the improved toughness.

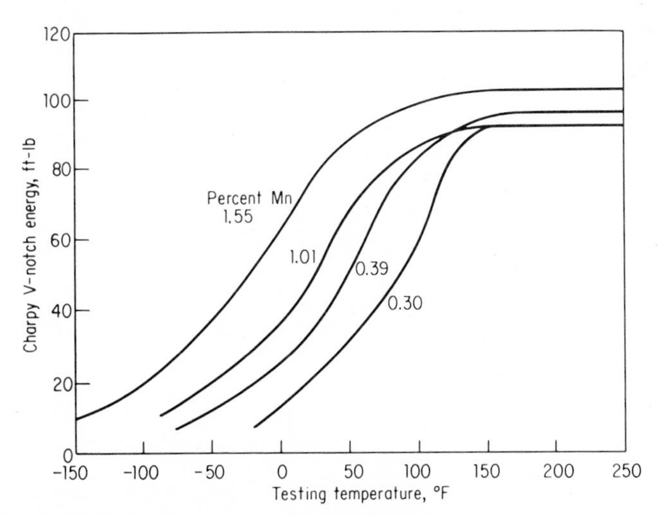

Fig. 13·9 Effect of manganese on Charpy V-notch values of a 0.30 percent carbon steel. (From "Metals Handbook," 1961 ed., p. 227, American Society for Metals, Metals Park, Ohio.)

For a particular type of steel and strength level, fine-grained steels have higher notch toughness than coarse-grained steels. The transition temperature is lowered as the grain size decreases. This is illustrated by Fig. 13·10. The fine-grained condition is usually due to a deoxidation practice that uses silicon, aluminum, or vanadium.

Nickel is the most effective alloying element for increasing resistance to low-temperature embrittlement in steel and is one of the few alloying elements which improve the low-temperature ductility of iron. Nickel additions to steel increase room-temperature toughness, lower the transition temperature, and widen the transition-temperature range. The effects of variation in nickel content and temperature on the Charpy keyhole impact values of low-carbon steels are shown in Fig. 13·11. The curves indicate that the transition temperature for 1020 killed steel is slightly below 0°F. The 2 percent nickel steel retains considerable toughness down to −100°F. The 3½ and 5 percent nickel steels seem to be best between −100 and −200°F. The 8½ percent nickel steel shows only a gradual decrease in toughness with decrease in temperature. The 13 percent nickel steel shows no transition temperature, and its room-temperature toughness remains at almost the same value over the entire test range.

Molybdenum, vanadium, and titanium have a similar effect on notch toughness. Small amounts tend to raise the transition temperature, but as the amount is increased, transition temperature is lowered.

The best microstructure for low-temperature toughness is that of tem-

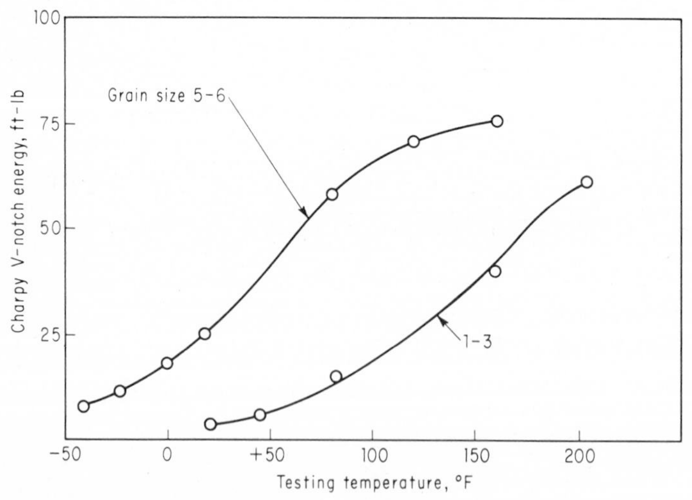

Fig. 13·10 Effect of grain size on notch toughness of 1030 steel. (From "Metals Handbook," 1961 ed., p. 236, American Society for Metals, Metals Park, Ohio.)

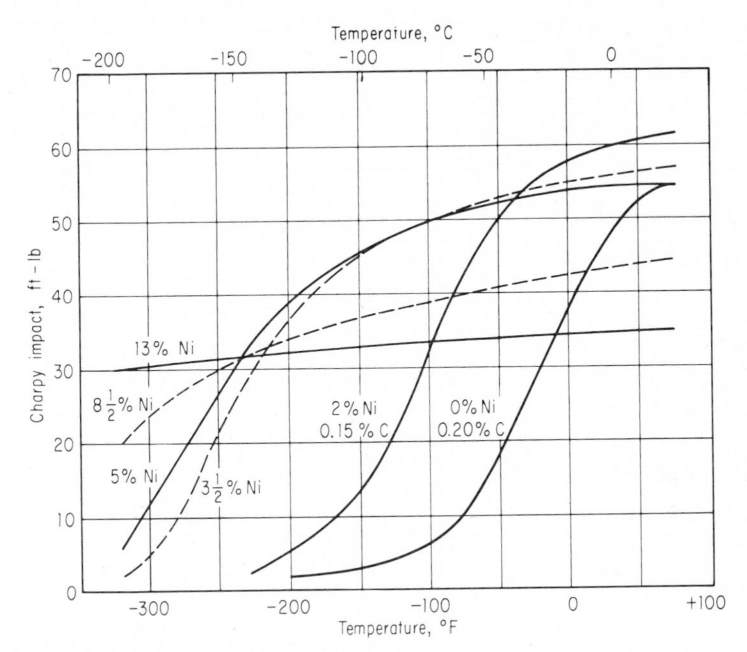

Fig. 13·11 Effect of nickel content on the resistance to low-temperature embrittlement of normalized low-carbon steels (keyhole notch). (The International Nickel Company.)

pered martensite. This structure gives the highest toughness and the lowest transition temperature compared with other microstructures of a specific steel (Fig. 13·12). The notch toughness of martensite decreases with increasing amounts of bainite. Retained austenite has only a slight effect on transition temperature.

Carburized or nitrided surfaces tend to lower the notch toughness of carbon and alloy steels. This is due to the hard, less ductile surface layers that resist plastic bending under shock loads. Decarburization favors plastic bending and may slightly increase notch toughness; however, it will reduce fatigue strength.

Notched-bar impact values of some steels and nonferrous materials at low temperatures are given in Table 13·4.

13·10 Mechanical Factors The mechanical factors that affect notched-bar impact results are the stress concentration and the strain rate. The stress concentration is determined by the sharpness of the notch. Decreasing the radius of the notch increases the stress concentration, which tends to produce brittle behavior at higher temperatures. The strain rate is determined by the striking velocity of the pendulum, and the energy absorbed is very sensitive to the striking velocity when the steel is near the transition tem-

TABLE 13·4 Toughness of Some Steels and Nonferrous Materials at Low Temperatures*

MATERIAL AND COMPOSITION	CONDITION	ROOM-TEMP. TENSILE STRENGTH, PSI OR BHN	NOTCHED-BAR IMPACT VALUES							
			FT-LB	TEMP, °F	FT-LB	TEMP, °F	FT-LB	TEMP, °F	FT-LB	TEMP, °F
Bessemer structural steel, 0.10 C	As-rolled	55,000	37	148	26	68	4	32	3	0
Ni steel, 0.31 C, 3.47 Ni	Normalized 1675°F	192	49	75	32	0	17	−40	9	−100
	Oil-quenched, tempered 1050°F	217	106	75	106	0	105	−40	100	−100
Ni-Cr-Mo steel, 0.39 C, 0.71 Mn, 1.71 Ni, 0.75 Cr, 0.30 Mo	Normalized 1630°F	440	13	75	7	0	6	−40	5	−100
	Oil-quenched, tempered 1100°F	302	82	75	82	0	82	−40	77	−100
Fe-Ni alloy, 35.3 Ni	Annealed	...	28	75	19	−112	11.3	−310		
Ni-Fe alloy, 1.3 Fe	Annealed	...	27.6	75	27	−112	29.3	−310		
Commercially pure Al	Annealed	13,200	19	75	20	−112	21	−184	27	−292
Al alloy 17S, 3.72 Cu, 0.6 Mn, 0.6 Mg	Solution-treated	59,500	15	75	18	−105				
Al alloy 52S, 2.35 Mg, 0.12 Si, 0.14 Fe	As-rolled	31,500	58	75	58	−105				
Pure Cu, 99.98 Cu	Annealed	31,400	43	75	44	−112	44	−184	50	−290
Cu-Be, 2.56 Be	Water-quenched and aged	187,000	2	75	3	−112	3	−292		
Cu-Zn, 30.5 Zn, 0.10 Fe	Annealed	51,100	66	75	69	−112	78	−290		
Mg alloy M1, 1 Mn	Extruded	39,000	3.9	77	2.6	−110				
Mg alloy AZ63, 6 Al, 3 Zn	Cast, heat-treated, and aged	38,900	1.6	77	1.5	−110				
Ni alloy Monel, 28.86 Cu, 0.28 Mn	Annealed	70,800	90	75	90	−112	97	−292		

*Compiled from data in "Metals Handbook," 1948 edition, American Society for Metals, Metals Park, Ohio.

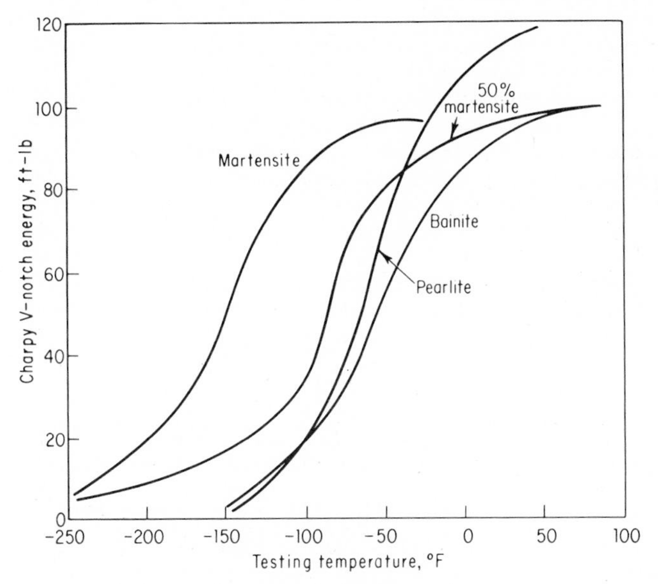

Fig. 13·12 Effect of microstructure on notch toughness of a 0.17 percent carbon, Cr-Mo steel. (From "Metals Handbook," 1961 ed., p. 235, American Society for Metals, Metals Park, Ohio.)

perature. High striking velocity has the same effect as lowering the test temperature and tends to exaggerate brittle behavior.

To summarize, the best notch-toughness properties are obtained by selecting a fine-grained, low-carbon, fully-killed nickel-alloy steel which has been quenched to a fully martensitic structure and tempered to the desired hardness level.

It must be emphasized that notched-bar impact values apply only to the conditions under which the test was run. The results cannot be used in design and serve only to compare different factors under those conditions. Any attempt to use these values for structures of different size or different strain rates must be done with extreme care. The values become significant for design only when correlated with a particular structure in a particular kind of service.

QUESTIONS

13·1 What factors should be considered when testing at elevated temperatures as compared with testing at room temperature?

13·2 Define *creep*. Why is this properly important for high-temperature application?

13·3 Give a specific application where creep properties would be important in design.

13·4 Draw a typical creep curve and explain the stages of creep.
13·5 What metallurgical factors affect the creep characteristics of metals?
13·6 What are the limitations on the use of stress-rupture data?
13·7 Which alloying elements are most effective in raising the creep strength of steels?
13·8 What is the effect of low temperatures on the mechanical properties of metals?
13·9 What is meant by *transition temperature*? How is it measured?
13·10 What factors influence brittle failure?
13·11 What are the limitations on the use of notched-bar data for the performance of actual parts?
13·12 Differentiate between a shear fracture and a cleavage fracture.
13·13 What alloying element is most effective in improving the low-temperature toughness of steel?

REFERENCES

American Society for Metals: "Behavior of Metals at Low Temperatures," Metals Park, Ohio, 1953.
———: "Creep and Recovery," Metals Park, Ohio, 1957.
———: "High Temperature Properties of Metals," Metals Park, Ohio, 1961.
———: "Metals Handbook," 1948 and 1961 editions, Metals Park, Ohio.
American Society for Testing Materials: "Evaluation of Metallic Materials in Design for Low-temperature Service," Special Technical Publication no. 302, Philadelphia, 1962.
Clark, F. H.: "Metals at High Temperatures," Van Nostrand Reinhold Company, New York, 1950.
Clauss, F. J.: "Engineer's Guide to High-temperature Materials," Addison-Wesley Publishing Company, Inc., Reading, Mass., 1969.
Conway, S. B.: "Numerical Methods for Creep and Rupture," Gordon & Breach Science Publishers, New York, 1967.
Dorn, J. E. (ed.): "Mechanical Behavior of Materials at Elevated Temperatures," McGraw-Hill Book Company, New York, 1961.
Heheman and Ault (eds.): "High Temperature Materials," John Wiley & Sons, Inc., New York, 1959.
International Nickel Company: "Nickel Alloy Steels," 2d ed., New York, 1949.
Parker, E. R.: "Brittle Behavior of Engineering Structures," John Wiley & Sons, Inc., New York, 1957.
Savitsky, E. M.: "The Influence of Temperature on the Mechanical Properties of Metals and Alloys," Stanford University Press, Stanford, Calif., 1961.
Seigle, L., and R. M. Brick: "Mechanical Properties of Metals at Low Temperatures," A Survey, *Trans. ASM.*, vol. 40, p. 813, 1948.
Sully, A. H.: "Metallic Creep," Interscience Publishers, Inc., New York, 1949.
Wigley, D. A.: "Mechanical Properties of Materials at Low Temperatures," Plenum Press, New York, 1971.

14 WEAR OF METALS

14·1 Introduction The quality of most metal products depends on the condition of their surfaces and on surface deterioration due to use. Surface deterioration is also important in engineering practice; it is often the major factor limiting the life and the performance of machine components. Wear may be defined as unintentional deterioration resulting from use or environment. It may be considered essentially a surface phenomenon. Wear is one of the most destructive influences to which metals are exposed, and the importance of wear resistance needs no amplification.

The displacement and detachment of metallic particles from a metallic surface may be caused by contact with (1) another metal (adhesive or metallic wear), (2) a metallic or a nonmetallic abrasive (abrasion), or (3) moving liquids or gases (erosion). Erosion is usually accompanied by some form of corrosion. The above three types of wear may be subdivided into wear under rolling friction or sliding friction and, further, according to whether lubrication can or cannot be used.

Wear involving a single type is rare, and in most cases both abrasive and adhesive wear occur. Each form of wear is affected by a variety of conditions, including environment, type of loading, relative speeds of mating parts, lubricant, temperature, hardness, surface finish, presence of foreign particles, and composition and compatibility of the mating parts involved. Since in most machinery applications wear can rarely be avoided completely even with the best lubrication, it is common practice to use a hard metal and a relatively soft one together. The softer material is used (as in a bearing) for the part which is most economical to replace.

14·2 Mechanism of Wear In adhesive wear, also called scoring, galling, seizing and scuffing, tiny projections produce friction by mechanical interference, with the relative motion of contacting surfaces increasing resistance to further movement. If the driving force is sufficient to maintain movement, the interlocked particles are deformed. If they are of a brittle material, they

Fig. 14·1 Surface of a truck drive unit showing galling caused by inadequate lubrication. (From R. D. Barer and B. F. Peters, "Why Metals Fail," Gordon and Breach Science Publishers, New York, 1970.)

may be torn off. This leads to the conclusion that wear resistance will be improved by preventing metal-to-metal contact and by increasing the hardness to resist initial indentation, increasing the toughness to resist tearing out of metallic particles, and increasing the surface smoothness to eliminate the projections. Figure 14·1 shows galling or gouging of the surface of a truck drive unit. Investigation indicated that the galling was due to inadequate lubrication because of leaky oil seals. Figure 14·2 shows severe

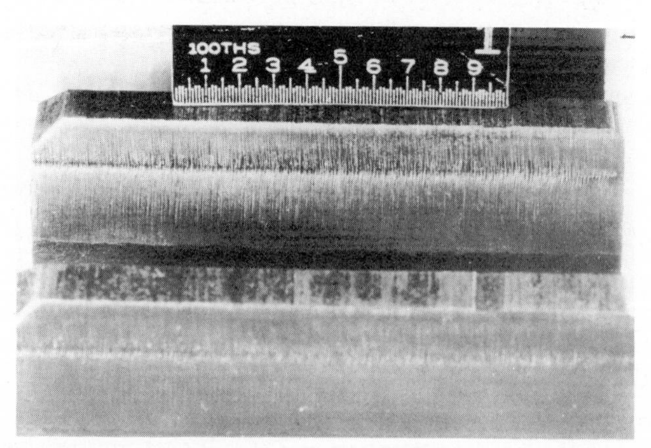

Fig. 14·2 Adhesive wear on the pinion gear of a tractor due to inadequate lubrication. (Courtesy of D. J. Wulpi, International Harvester Company.)

adhesive wear on the surface of an induction-hardened gear tooth operated under heavy load. Lubrication with a light engine oil was inadequate, resulting in metal-to-metal wear.

Abrasive wear occurs when hard particles slide or roll under pressure across a surface, or when a hard surface rubs across another surface. The abrading particles from the harder object tend to scratch or gouge the softer material. These hard particles may also penetrate the softer metal and cause the tearing off of metallic particles. Figure 14·3 shows the main journal bearing of a crankshaft which was damaged by dirty oil. Hard dirt particles with sharp projections broke through the lubricant film, cutting and scratching the journal bearing's relatively soft surface. An obvious solution is to improve the efficiency of the oil filter.

The ease with which the deformed metal may be torn off depends upon the toughness. Therefore, hardness and toughness, the same properties that influence adhesive wear, also determine abrasive wear. Of these two factors, hardness is probably the more important one.

14·3 Factors Influencing Wear Lubrication is an important contributing factor to wear resistance, particularly in adhesive wear. In "thick-film" lubrication, a sufficiently thick lubricating film completely eliminates metallic contact, and metallic wear is reduced to a negligible amount. This is, how-

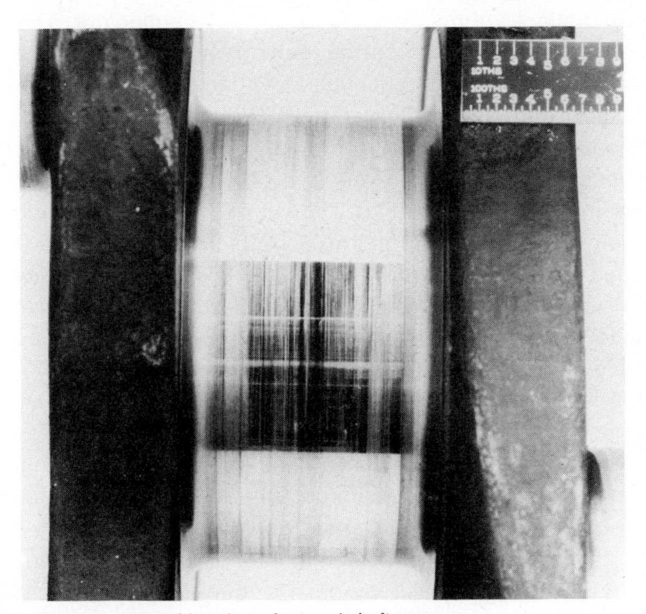

Fig. 14·3 The main journal bearing of a crankshaft was badly scratched by dirt particles which contaminated the lubricant. (Courtesy of D. J. Wulpi, International Harvester Company.)

ever, the ideal condition, and more frequently "boundary lubrication" occurs. This is the condition of intermittent metallic contact that occurs when the oil film cannot be continuously maintained. Under boundary conditions, the amount of wear depends upon speed, pressure, nature of the mating surfaces, and efficiency of the residual oil film. In many cases, however, lubrication is impractical or is not wanted, as in braking.

Although actual melting of the metal occurs only in rare instances, the effect of heat produced by dry wear can reduce wear resistance in several ways. It may temper hardened structures, cause phase changes that increase hardness and brittleness, and decrease mechanical properties, and it accelerates corrosion reactions.

The dominant frictional factor for metallic materials is believed to be welding. Atoms of the same or crystallographically similar metals have very strong forces of cohesion. When two clean surfaces of the same metal actually touch each other, they will weld together because of atomic attraction. If, by friction, sufficient pressure is applied to break through any residual separating material such as oil, dirt, or adsorbed moisture, and the surfaces are in sufficient contact to have elastic or plastic deformation occur, then seizing or welding takes place. The softening of metals by high temperatures increases the ease of plastic deformation and facilitates welding. Seizing may cause complete stoppage, or if relative motion is not prevented, pieces of the opposite face may be pulled out. The resultant projection then may cause scoring, galling, and excessive local wear.

Many methods may be used to minimize the danger of seizing. One is to use thin layers of hard surfacing material. The use of at least one metal that forms some sort of lubricating film or thin, tightly adherent oxide, sulfide, or phosphide coating is frequently helpful. Aluminum oxide is very effective in preventing welding. For parts that operate under such high pressures that elastic deformation permits intimate contact, the best preventive method is a lubricant that combines with the metal surface to form a "corrosion" product of sufficient strength to keep the surfaces separated. The use of materials of high elastic limit will minimize seizure due to intimate contact produced by plastic deformation.

Impact is a factor in wear, since the suddenly applied load may cause plastic flow and a change in shape. Proper design should provide a surface compressive yield strength above the compressive stress produced by impact and sufficient support so that subsurface flow does not occur.

Fatigue failure is included in a discussion of wear since it is a gradual deterioration due to use. Proper design to eliminate stress concentrations at notches and sharp angles will increase fatigue strength. Since fatigue failures are always due to tensile stress, residual compressive stress at the surface will provide additional protection. This may be accomplished by case hardening, such as carburizing, and by shot peening.

14·4 Methods of Testing for Wear Resistance Since wear is not a simple phenomenon, wear resistance is represented by fewer standardized tests than other engineering properties. It is generally accepted that a "universal" wear test is not feasible. Therefore, equipment for wear testing must be designed to simulate actual service conditions. These tests should have proved reproducibility, should be able to rank various materials under consideration, and most important, should be validated by correlation with service data.

14·5 Protection against Wear Many materials and methods are available for protection against wear. The selection of a particular material and process requires a thorough analysis of the actual service conditions, a knowledge of applicability and limitations of the particular material and process, and data concerning the cost involved. The lack of engineering data available for comparisons imposes a need for good judgment on the engineer or technician who selects materials to withstand wear.

Various techniques for providing surface protection to wear are as follows:

Electroplating
Anodizing
Diffusion
Metal spraying
Hard facing
Selective heat treatment

14·6 Electroplating The wear resistance of a metal part can be improved by electroplating a harder metal on its surface. The metals most often plated on base materials are chromium, nickel, and rhodium. Indium plating has been used to reduce the wear of lead bearings.

Two types of chromium plating used industrially are known as *hard chromium* and *porous chromium*. The hard-chromium plate is the same as that used for decorative purposes but much thicker, usually from 0.0001 to 0.010 in. Porous chromium plate has on its surface carefully controlled pits or channels to hold lubricants. The term is misleading, since below the specially prepared surface, it is no more porous than ordinary hard-chromium plate. The hardness of chromium plate is equivalent to 950 to 1050 Vickers. Another factor contributing to the reduction of wear is the low coefficient of friction of chromium plate. Chromium plating is used in the cylinders and piston rings of internal-combustion engines. Nongalling is another useful property of chromium plate. Force-fitted chromium-plated steel parts may be assembled and disassembled many times without seizing or galling. The high corrosion resistance of chromium is helpful in reducing wear under corrosive conditions.

The hardness of nickel plate is from 140 to 425 Vickers depending upon the nickel-plating solution used. Nickel plate is a good deal softer than chromium plate, but in many cases it is hard enough for the purpose and more economical. A nickel plate may be finished by machining, while a chromium plate must be ground. The better throwing power of nickel-plating solutions as compared with chromium-plating solutions is an advantage in plating parts that have recesses.

The hardness of rhodium plate is from 540 to 640 Vickers, and its wear resistance is between those of nickel plate and chromium plate. Rhodium plate has high reflectivity, high heat resistance, and nontarnishing properties along with good hardness and wear resistance. The use of rhodium plate for reflectors of high-intensity light sources, for electrical contacts, and for slip rings and commutators has been mentioned in Chap. 12.

14·7 Anodizing The formation of an oxide coating by anodizing may be used to improve the wear resistance of certain metals. The anodizing process is usually applied to aluminum, magnesium, zinc, and their alloys. In anodizing, the work is the anode, and oxide layers are built up on the base metal. Since the newest oxide layer always forms next to the base metal, in order for the process to continue, the previously formed oxide layers must be porous enough to allow the oxygen ions to pass through them.

Anodizing aluminum is simply a method of building up a much thicker oxide coating than may be obtained by exposure to air. The Alumilite process developed by the Aluminum Company of America uses sulfuric acid as an electrolyte for anodizing. The films produced are transparent, thicker, and more porous than those produced by other electrolytes. Continued development resulted in Alumilite hard coatings, which are thicker and harder than ordinary anodic coatings. Aircraft parts such as hydraulic pistons, guide tracks, gears, cams, screws, swivel joints, and friction locks are made of hard-coated aluminum alloys.

The production of a hard wear-resistant surface by anodizing has greatly extended the uses of magnesium and its alloys. Flash anodic coatings are often used as a base for paint adherence.

Anodizing zinc produces a coating which has greater resistance to wear than chromate films. Anodic zinc coatings are used for cartridge cases, airplane propeller blades, wire-screen cloth, and refrigerator shelves.

14·8 Diffusion Several processes improve wear resistance by diffusion of some element into the surface layers. These are:

Carburizing
Cyaniding
Carbonitriding
Nitriding
Chromizing
Siliconizing

The first four of these processes were discussed in detail in Chap. 8, and only the last two will be discussed here.

Chromizing consists of the introduction of chromium into the surface layers of the base metal. The process is not restricted to ferrous materials and may be applied to nickel, cobalt, molybdenum, and tungsten to improve corrosion resistance and heat resistance.

When it is applied to iron or steel, it converts the surface layer into a stainless-steel case. If the steel contains appreciable amounts of carbon (above 0.60 percent), chromium carbides will precipitate, increasing wear resistance. The chromizing process most widely used employs the principle of transfer of chromium through the gas phase at elevated temperatures. The temperatures used range from 1650 to 2000°F. These high temperatures may produce some distortion and grain growth during treatment. Chromized high-carbon steels have a hardness of 800 to 1000 Vickers and a low coefficient of friction. Chromizing is used on drop-forging dies, tools, hydraulic rams, pistons, and pump shafts.

Siliconizing, or Ihrigizing, consists of impregnation of an iron-base material with silicon. The process is carried out in the temperature range of 1700 to 1850°F. The work is heated in contact with a silicon-bearing material such as silicon carbide, and chlorine gas is used as a catalyst. The case depth ranges from 0.005 to 0.1 in., depending mainly on the carbon content of the base material. The case produced contains approximately 14 percent silicon and is essentially an iron-silicon solid solution. Siliconized cases are difficult to machine, although the hardness is only Rockwell B 80 to 85. The increase in wear resistance by siliconizing is due to a low coefficient of friction and nongalling properties. Siliconized cases have been used on pump shafts, conveyor chain links, cylinder liners, valve guides, valves, and fittings for the chemical and oil industries.

14·9 Metal Spraying Metal spraying or flame spraying has been used for many years in production salvage to build up dimensions that are undersize and to repair worn surfaces. It has found increased use for wear-resistant applications.

Sprayed coatings can be applied by several methods: metallizing; flame-plating, which is used to deposit tungsten carbide and aluminum oxide; plasma arc spraying, which can deposit almost all inorganic materials.

Metallizing is usually done by automatically feeding a metal wire at a controlled rate of speed through the metallizing tool or "gun" (Fig. 14·4). Air, oxygen, and a combustible gas are supplied to the gun by means of hoses and form a high-temperature high-velocity flame around the wire tip. The wire tip is continuously melted off, and the liquid-metal particles are directed at the work by the high-velocity flame. When they strike the surface, these particles flatten out to form irregularly shaped disks. At the same time they are forced into surface pores and irregularities to provide some mechanical interlock with previously deposited material. Cooling is

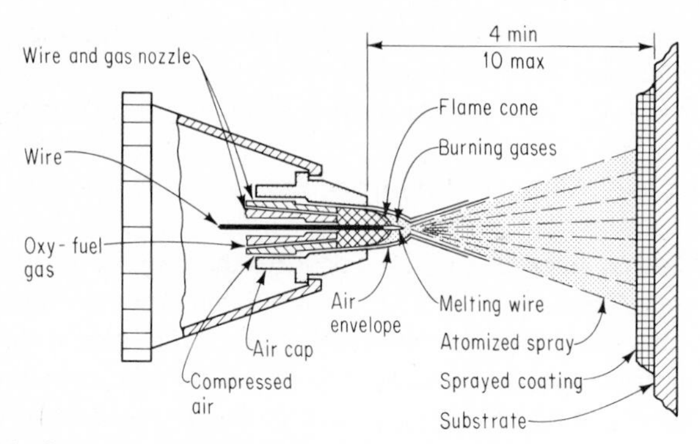

Fig. 14·4 Cross-section diagram of a wire gun for metal spraying. (From "Metals Handbook," vol. 2, 1964, American Society for Metals, Metals Park, Ohio.)

very rapid, and a thin oxide film forms on the exposed surfaces of the deposited particles. The nature of the oxides formed under metallizing conditions determines to a large degree the physical properties of the deposit. Metals that form dense, tenacious oxides having good physical properties, such as chromium steel, aluminum bronze, or silicon-aluminum alloys, show relatively high strength in the sprayed form. Metals which form loose, friable oxides, such as the brasses and copper, produce coatings of low strength.

Another method of metallizing employs an oxyacetylene powder gun, Fig. 14·5. The gun can spray metal powder in place or over almost any contoured surface. It is necessary to postheat the coated surface to fuse the sprayed deposit. The powder gun can apply thin coatings and is useful

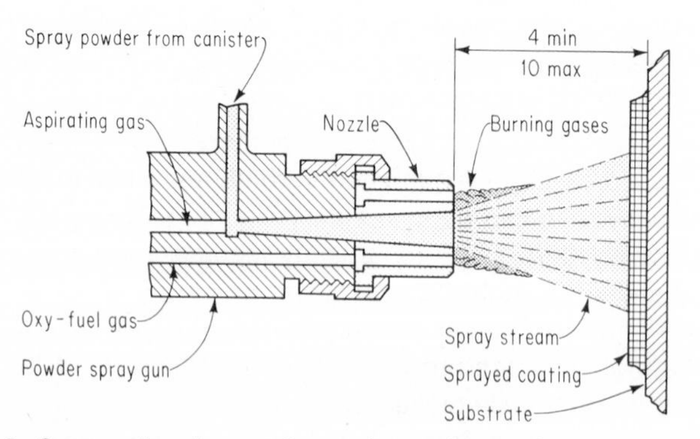

Fig. 14·5 Cross-section diagram of a powder gun for metal spraying. (From "Metals Handbook," vol. 2, 1964, American Society for Metals, Metals Park, Ohio.)

for spraying hard, corrosion-resistant alloys. Most of these alloys are nickel-base or cobalt-base materials containing chromium, boron, and silicon. Aluminum, tin, and zinc powders can also be sprayed by the powder gun.

Plasma is a luminous stream of ionized gas produced by passing a gas through an electric arc. Temperatures up to 30,000°F are economically obtainable; thus the use of plasma flame permits deposition of the highest-melting metals. An additional advantage of the plasma process is that oxygen and combustion gas, as well as their combustion products, are absent from the gas stream. Details of a plasma spray gun are shown schematically in Fig. 14·6. The disadvantage of plasma-arc spraying is the very careful safety measures that must be employed due to the added hazards of high temperatures, high noise level, toxic waste products, and radiation.

Various methods have been developed for bonding sprayed metals to the base material. One of the most widely used methods is to spray a bonding coat of molybdenum 0.001 to 0.003 in. thick which forms a thin alloy layer. Molybdenum adheres to nearly all steels and many other metals and alloys but will not bond to copper, brass, bronze, or nitrided steels. Other bonding methods prepare the surfaces by abrasive blasting or by rough threading. The Fusebond method deposits rough, porous nickel on the surface to provide a good strong anchorage for sprayed coatings, but it cannot be used on brass, bronze, or copper.

In general, a sprayed metal coating is harder and more brittle and porous than equivalent cast or wrought metal. Table 14·1 shows a comparison of hardness by wrought wires and sprayed particles of various metals by metallizing, while Table 14·2 gives the Vickers hardness of typical plasma-arc sprayed coatings.

High-chromium (13.5 percent) stainless steel can be used in many applications where a hard coating is required, even where corrosion resistance is not necessary. The material has high strength and elongation, low shrinkage, and little tendency to crack on shafts. Typical applications are armature shafts, cylinder liners, pistons, valve stems, and hydraulic rams.

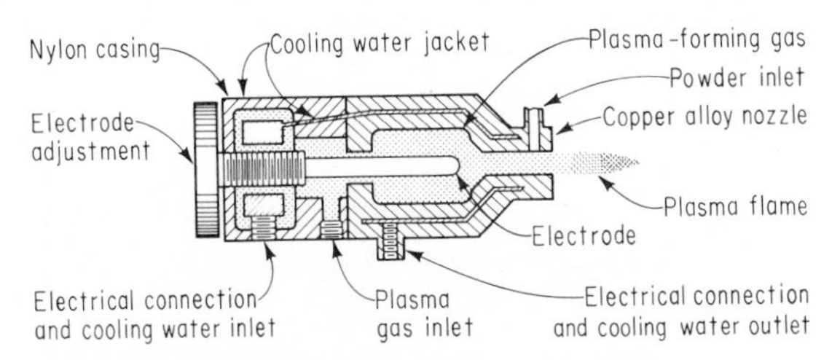

Fig. 14·6 Cross-section diagram of a plasma gun for metal spraying. (From "Metals Handbook," vol. 2, 1964, American Society for Metals, Metals Park, Ohio.)

TABLE 14·1 Comparison of Hardness of Wrought Wires and Sprayed Particles of Various Metals*

METAL	KNOOP HARDNESS VALUES (50-g LOAD) WROUGHT WIRE	SPRAYED PARTICLES
Aluminum bronze	229	408
Molybdenum	404	1535
Monel	338	326
Steels:		
1010	307	445
1025	370	504
1080	398	664
Type 304	360	381
Type 420	372	757

*From "Metals Handbook," vol. 1, 1964, American Society for Metals, Metals Park, Ohio.

Molybdenum coatings combine a hard wearing surface with good adhesion. They have good wear and abrasion resistance and have been used successfully to build up the worn surfaces of aluminum pulleys and aluminum and iron brake drums used on elevators and presses. They also provide an excellent bearing surface and are particularly good when running against bronze bearings under severe abrasion conditions. Copper alloys can be used for general-purpose wear applications. Aluminum bronze is very wear-resistant and machines easily. It performs well on such parts as pump impellers, bronze castings, split motor bearings, and air-brake valves. Monel and nickel coatings are used where wear resistance combined with corrosion resistance is needed, as in pump plungers, shafts and hydraulic pumps.

TABLE 14·2 Vickers Hardness of Typical Plasma-sprayed Coatings*

METALS		CARBIDES AND OXIDES	
Molybdenum wire		Hafnium carbide	1000
As sprayed	168-183	Tantalum carbide	1000
Molybdenum powder		Osmium oxide	168-205
As sprayed	321-368	Chromic oxide	1000
Tantalum powder		Chromium-aluminum oxide	287-455
As sprayed	443	Hafnium oxide	251-313
Tungsten wire		Zirconium oxide	638-805
As sprayed	246-263		
Tungsten powder			
As sprayed	330-390		

*From Materials Engineering, Manual no. 201, Van Nostrand Reinhold Company, New York, 1962.

The development of higher-temperature flame-spraying processes has opened the door to many new coating materials, many of which have outstanding resistance to wear and abrasion. Several tungsten carbide compositions, applied by flame-plating, can be used for applications where either wear resistance alone or wear resistance plus shock, heat, or corrosion resistance is needed. The basic tungsten carbide composition containing 6 to 8 percent cobalt is recommended for general-wear applications such as seals, valve plates, bearings, and shafts. Alloys containing higher cobalt (13 to 17 percent) wear about 30 to 40 percent faster but provide better resistance to mechanical shock. Compositions of tungsten carbide, nickel, and mixed tungsten-chromium carbides are also available where wear resistance and resistance to heat or corrosion (or both) are needed. As shown in Table 14·3, a chromium carbide nickel-chromium composition can also be used for the same kind of service. Aluminum oxide coatings may be deposited by flame-plating. They are superior to tungsten carbides under corrosive or high-temperature oxidizing conditions but have only about one-half the wear resistance to wet abrasion.

The wear resistance of sprayed metals is generally very good, but metallizing is not suitable for service involving heavy impact or extreme abrasion or where highly localized loads are applied. Metallizing is best suited for applications involving light abrasive wear, for conditions where boundary lubrication exists, and to provide a wear-resistant or corrosion-resistant surface to inexpensive base materials.

14·10 Hard Facing The production of a hard wear-resistant surface layer on metals by welding is known as *hard facing.* This method is relatively easy to apply, requiring only the hard-facing alloys in the form of welding rods and an oxyacetylene flame or electric arc. The advantages of hard facing are that (1) it may be applied to localized areas subjected to wear, (2) hard wear-resistant compounds are available, and (3) it provides effective use of expensive alloys and protection in depth.

The hard-facing material is provided in the form of an electrode or welding rod. These rods are generally used bare for oxyacetylene-gas welding and are flux-coated for electric-arc welding. The flux coating contains materials for arc stability, oxidation protection of the molten weld, fluxing of impurities, thermal and electrical insulation, and control of metal transfer. Oxyacetylene-gas welding produces smoother deposits that can be positioned more precisely, while the heating and cooling rates are slower. Electric-arc welding is less expensive, may be faster, and lends itself better to automatic equipment. Arc deposits are generally rougher and more likely to be porous; they tend to develop cracks because of the sharp temperature gradients due to rapid heating and cooling. The hard-facing material is melted and spread over the base metal and bonds with it to form a new surface ranging from 1/16 to 1/4 in. thick, depending on the application.

TABLE 14·3 Properties of Flame-plated Coatings*

COATING ➧	TUNGSTEN CARBIDES					
	WC + 6-8% Co	WC + 13-15% Co	WC + 15-17% Co	25% WC + 7% Ni + Mixed W-Cr Carbides	85% Chromium Carbide (Cr_3C_2) + 15% Ni-Cr	99+% Aluminum Oxide (gamma)
Mechanical properties						
Vickers Hardness	1200-1450	1100-1250	1050-1200	1000-1200	1000-1200	1000-1200
Mod of Rupture, psi ...	67,000-73,000	80,000-90,000	90,000-100,000	40,000	75,000	22,000
Mod of Elast, 10^6 psi ...	40	36-38	34-36	17	22	16
Porosity, %	0.5	0.5-1	0.7-1.2	0.5	0.5	1
Main features	Extreme wear resistance	Excellent wear resistance plus increased resistance to mechanical and thermal shock	Excellent wear resistance plus greatest resistance to mechanical and thermal shock	Excellent wear resistance at higher temperatures. Improved corrosion resistance	Good wear resistance at high temperatures or in corrosive media. Resists flame impingement	Excellent resistance to wear, chemical attack and high temperature oxidation

*From Materials Engineering, Manual no. 201, Van Nostrand Reinhold Company, New York, 1962.

Only the surface of the base metal is brought to melting temperature; this prevents mixing of the alloy with too much of the base metal and thus changing the properties of both the coating and the base metal.

Hard facings can be applied to most ferrous metals, but with a few exceptions it is not advisable to face nonferrous alloys having melting points below 2,000°F. Carbon steels are relatively easy to hard-face, particularly below 0.35 percent carbon. Welding becomes more difficult with increasing carbon content, and the high-carbon and alloy steels must be preheated before and postheated after hard facing. Stainless steels, cast iron, ductile iron, and high-speed steels can also be hard-faced provided that appropriate welding practices are observed. Monel can easily be hard-faced, but copper, brass, and bronze are relatively difficult to do because of their low melting points and high conductivity.

Hard facing is most extensively used where systematic lubrication against abrasion is impossible, as on oil-well drilling tools, agricultural and earth-moving equipment, mining tools, engine valves, and refinery or chemical processing equipment. Hard facing also extends the life of lubricated parts such as metalworking dies and areas of machine parts that have a high wear rate. Hard-faced surfaces are usually more wear-, heat-, and corrosion-resistant than case-hardened or flame-hardened surfaces. Also, critical local areas of large components can be hard-faced where it would be impractical or impossible to harden the component by heat treatment. Since hard facing is a welding process, it may be used to repair parts on location without dismantling heavy equipment. Hard facing serves to increase operating efficiency by extending the life of equipment, by decreasing replacement cost and loss of production time, and by permitting the use of a low-cost base metal for parts that wear or corrode.

There are more than 150 different compositions of hard-facing materials commercially available, ranging from steels containing only about 2 percent total alloy content to nickel-base and cobalt-base alloys and tungsten carbide. A simplified classification of wear-resistant hard-facing alloys is given in Table 14·4.

Tungsten carbide hard facings have the highest hardness and best wear resistance. Although it is relatively costly, this may not be an important factor, because the material has a long life and can be applied in the form of inserts only where needed, as in the case of rock drill bits. High-chromium irons containing 17 to 32 percent chromium are available in many different alloy compositions. The austenitic types are relatively cheap and are best for metal-to-metal wear or low-stress abrasion applications, such as farm equipment used in sandy soil. The hardenable grades are tougher than the austenitic grades and have excellent wear resistance. Additions of tungsten, molybdenum, and vanadium are sometimes used to help increase hot-hardness and add to abrasion resistance. Martensitic

TABLE 14·4 A graded series of wear-resistant alloys*

(Increasing toughness ↓; Increasing abrasion resistance ↑)

1. Tungsten carbide	Maximum abrasion resistance; worn surfaces become rough
2. High-chromium irons	Excellent erosion resistance; oxidation resistance
3. Martensitic irons	Excellent abrasion resistance; high compressive strength
4. Cobalt-base alloys	Oxidation resistance, corrosion resistance, hot strength, and creep resistance
5. Nickel-base alloys	Corrosion resistance; may have oxidation and creep resistance
6. Martensitic steels	Good combinations of abrasion and impact resistance; good compressive strength
7. Pearlitic steels	Inexpensive; fair abrasion and impact resistance
8. Austenitic steels	Work hardening
Stainless steels	Corrosion resistance
Manganese steel	Maximum toughness with fair abrasion resistance; good metal-to-metal wear resistance under impact

*"Surface Protection against Wear," American Society for Metals, Metals Park, Ohio 1954.

irons are mainly chromium-nickel, chromium-molybdenum, or chromium-tungsten alloys. The combination of martensite and a carbide matrix provides a hard composite structure with good abrasion resistance.

The cobalt-base alloys contain from 45 to 63 percent cobalt, 24 to 29 percent chromium, 5.50 to 13.50 percent tungsten, and 1.10 to 3.20 percent carbon. These alloys are used where wear and abrasion resistance combined with resistance to heat and oxidation is required. They have been used as hard-facing materials for ladle linings and pouring spouts to resist hot gases and liquids. The nickel-base alloys contain 70 to 80 percent nickel, 11 to 17 percent chromium, 2.5 to 3.7 percent boron, and smaller amounts of cobalt and silicon. They have slightly better wear and oxidation resistance than the cobalt-base alloys and have been used for applications up to 1800°F. Some typical applications are for hard facing of hot-heading dies, piercing mandrels, and shear blades exposed to hot solid metals.

Martensitic steels have the advantage of low initial cost, good hardness, strength, abrasion resistance, and toughness. They have been used mainly as buildup layers under other alloy compositions that have better abrasion resistance or higher toughness. Pearlitic steels have relatively low hardness and wear resistance and are therefore infrequently used as hard facings. They are used mainly as build-up metal for welding or as the base for hard surfacing. The austenitic steels, particularly the high-manganese grade, have been used for moderate-service conditions such as the crushing and grinding of coal, limestone, and aggregates. Austenitic manganese steels are very tough and work-harden rapidly under impact (see Sec. 9·10). This alloy is used as a base for hard facing because of its toughness, as well as for overlays.

14·11 Selective Heat Treatment The methods used for selective heat treatment are induction hardening and flame hardening. These are essentially shallow-hardening methods to produce a hardened case and relatively tough core. These methods were discussed in Chap. 8.

QUESTIONS

14·1 Differentiate between metallic wear, abrasive wear, and erosion.
14·2 Differentiate between "thick-film" lubrication and boundary lubrication.
14·3 Explain methods that may be used to minimize seizing.
14·4 What factors should be considered to set up a good wear test?
14·5 What would be a good method of reducing wear in a drawing die? Why?
14·6 Aside from lubrication, how may wear be reduced in an automotive cylinder?
14·7 What are the advantages and disadvantages of nickel plating vs. chromium plating for wear resistance?
14·8 List the various diffusion processes for increasing wear resistance, and give a practical application of each.
14·9 What is the principle of metallizing to increase wear resistance? Give some applications.
14·10 Give some applications of hard facing to improve wear resistance.

REFERENCES

American Society for Metals: "Metals Handbook," 7th ed., 1948; 8th ed., vol. 1, 1961, vol. 2, 1964, Metals Park, Ohio.
———: "Surface Protection against Wear and Corrosion," Metals Park, Ohio, 1954.
———: "Surface Treatment of Metals," Metals Park, Ohio, 1941.
Avery, H. S.: "Hard Surfacing by Fusion Welding," American Brake Shoe Co., New York, 1947.
———: Hard Facing for Impact, *Welding J. (N.Y.)*, vol. 31, no. 2, pp. 116-143, 1952.
Burns and Bradley: "Protective Coatings for Metals," Van Nostrand Reinhold Company, New York, 1967.
Burwell, J. T. (ed.): "Mechanical Wear," American Society for Metals, Metals Park, Ohio, 1950.
Gabe, D. R.: "Principles of Metal Surface Treatment and Protection," Pergamon Press, New York, 1972.
Ingham and Shepard: "Metallizing Handbook," Metco, Inc., Westbury, N.Y., 1965.
Riddihough, M.: "Head Facing by Welding," Iliffe & Sons, Ltd., London, 1948.
Stoody Company: "Stoody Hard Facing Guidebook," 2d ed., Whitter, Calif., 1966.

15 CORROSION OF METALS

15·1 Introduction In the broad sense, corrosion may be defined as "the destruction of a material by chemical, electrochemical, or metallurgical interaction between the environment and the material." Generally it is slow but persistent in character. In some instances the corrosion products exist as a thin adherent film which merely stains or tarnishes the metal and may act as a retardant to further corrosive action. In other cases, the products of corrosion are bulky and porous in character, offering no protection.

One of the most serious problems of industry, corrosion causes damage in the billions of dollars each year. It is a complex problem about which a great deal is known; yet despite extensive research and experimentation, there is still a lot to learn. In some cases, such as direct chemical attack, corrosion is highly obvious, but in other cases, such as intergranular corrosion, it is less obvious but just as damaging.

The basic cause of corrosion is the instability of metals in their refined forms. The metals tend to revert to their natural states through the processes of corrosion.

15·2 Electrochemical Principles Corrosion is essentially an electrochemical process resulting in part or all of the metal being transformed from the metallic to the ionic state. Corrosion requires a flow of electricity between certain areas of a metal surface through an electrolyte. An electrolyte is any solution that contains ions. Ions are electrically charged atoms or groups of atoms. Pure water, for example, contains positively charged hydrogen ions (H^+) and negatively charged hydroxyl ions (OH^-) in equal amounts. The electrolyte, therefore, may be plain water, salt water, or acid or alkaline solutions of any concentration. To complete the electric circuit, there must be two electrodes, an anode and a cathode, and they must be connected. The electrodes may be two different kinds of metals, or they may be different areas on the same piece of metal. The connection between the anode and the cathode may be by a metallic bridge, but in cor-

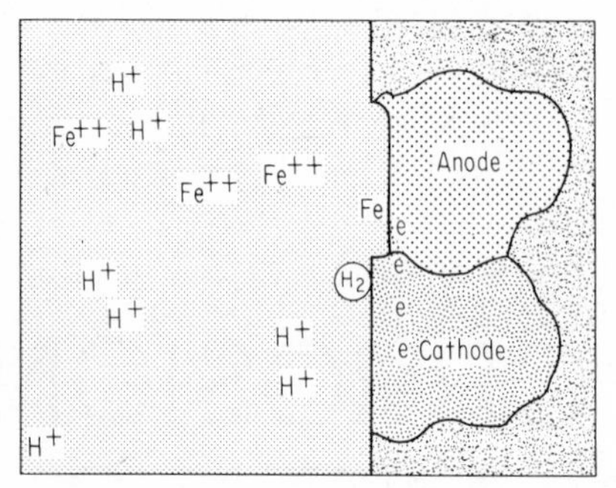

Fig. 15·1 Illustration of the formation of ions at the anode and hydrogen at the cathode in local cell action. (The International Nickel Company.)

rosion it is usually achieved simply by contact. In order for electricity to flow there must be a potential difference between the electrodes.

If a piece of ordinary iron is placed in a solution of hydrochloric acid, vigorous bubbling of hydrogen gas is observed. On the surface of the metal there are numerous tiny anode and cathode areas caused by inclusions in the metal, surface imperfections, localized stresses, orientation of the grains, or perhaps variations in the environment. This condition is shown schematically in Fig. 15·1. At the anode, positive-charged iron atoms detach themselves from the solid surface and enter the solution as positive ions, while the negative charges, in the form of electrons, are left behind in the metal. At the cathode the electrons meet and neutralize some positively charged hydrogen ions which have arrived at the surface through the electrolyte. In losing their charge, the positive ions become neutral atoms again and combine to form hydrogen gas. So, as this process continues, oxidation and corrosion of iron occurs at the anodes, and plating out of hydrogen occurs at the cathodes. The amount of metal which dissolves is proportional to the number of electrons flowing, which in turn is dependent upon the potential and the resistance of the metal.

In order for corrosion to continue it is necessary to remove the corrosion products from the anode and the cathode. In some cases, the evolution of the hydrogen gas at the cathode is very slow, and the accumulation of a layer of hydrogen on the metal slows down the reaction. This is known as *cathodic polarization* (Fig. 15·2). However, oxygen dissolved in the electrolyte can react with accumulated hydrogen to form water, thus allowing corrosion to proceed. For iron and water, the rate of film removal depends on the effective concentration of dissolved oxygen in water adjacent to the cathode. This effective concentration, in turn, depends upon the degree of aeration, amount of motion, temperature, presence of dissolved salts, and other factors.

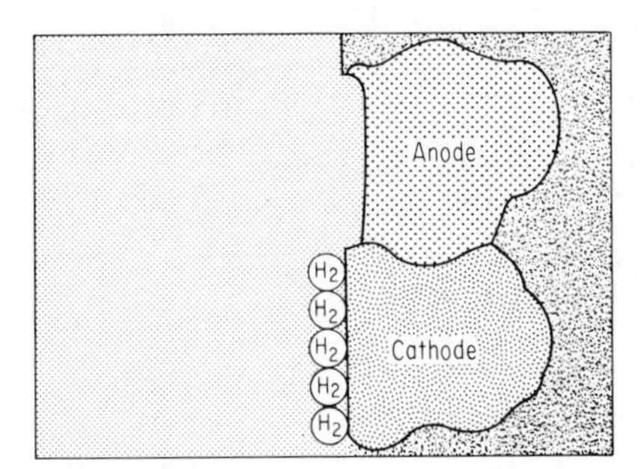

Fig. 15·2 Polarization of the local cathode by a film of hydrogen. (The International Nickel Company.)

The products of anode and cathode processes frequently meet and enter into further reactions that yield many of our common visible corrosion products. For example, with iron in water the hydroxyl ions from the cathodic reaction in their migration through the electrolyte toward the anode encounter ferrous ions moving in the opposite direction. These ions combine to form ferrous hydroxide (Fig. 15·3). This soon becomes oxidized by oxygen in solution to form ferric hydroxide, which precipitates as a form of iron rust. Depending on the alkalinity, oxygen content, and agitation of the solution, this rust may form either away from the iron surface or right next to it, where it can exert more of an influence on the further progress of corrosion.

15·3 Factors Influencing Corrosion One of the most important factors in influencing corrosion is the difference in electrical potential of dissimilar metals when coupled together and immersed in an electrolyte. This potential is due to the chemical natures of the anodic and cathodic regions. Some indication of which metals may be anodic as compared with hydrogen is given by the standard electromotive-force series (Table 15·1). The stan-

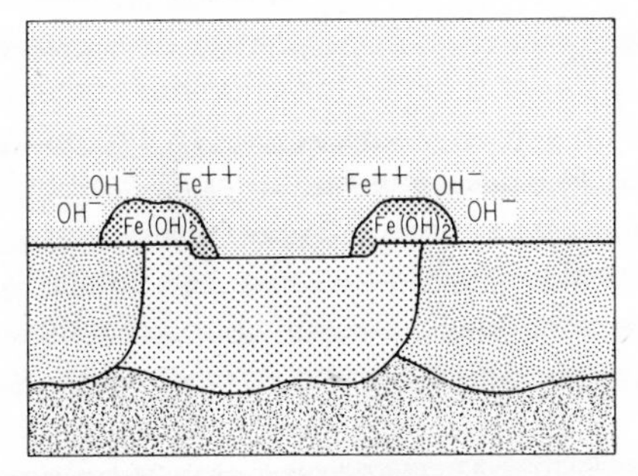

Fig. 15·3 Formation of ferrous hydroxide in the rusting of iron. (The International Nickel Company.)

TABLE 15·1 Electromotive-force Series*

ELECTRODE REACTION	STANDARD ELECTRODE POTENTIAL $E°$, VOLTS, 25°C	ELECTRODE REACTION	STANDARD ELECTRODE POTENTIAL $E°$, VOLTS, 25°C
$K = K^+ + e^-$	–2.922	$Co = Co^{++} + 2e^-$	–0.277
$Ca = Ca^{++} + 2e^-$	–2.87	$Ni = Ni^{++} + 2e^-$	–0.250
$Na = Na^+ + e^-$	–2.712	$Sn = Sn^{++} + 2e^-$	–0.136
$Mg = Mg^{++} + 2e^-$	–2.34	$Pb = Pb^{++} + 2e^-$	–0.126
$Be = Be^{++} + 2e^-$	–1.70	$\frac{1}{2}H_2 = H^+ + e^-$	–0.000
$Al = Al^{3+} + 3e^-$	–1.67	$Cu = Cu^{++} + 2e^-$	0.345
$Mn = Mn^{++} + 2e^-$	–1.05	$Cu = Cu^+ + e^-$	0.522
$Zn = Zn^{++} + 2e^-$	–0.762	$Ag = Ag^+ + e^-$	0.800
$Cr = Cr^{3+} + 3e^-$	–0.71	$Pd = Pd^{++} + 2e^-$	0.83
$Ga = Ga^{3+} + 3e^-$	–0.52	$Hg = Hg^{++} + 2e^-$	0.854
$Fe = Fe^{++} + 2e^-$	–0.440	$Pt = Pt^{++} + 2e^-$	1.2
$Cd = Cd^{++} + 2e^-$	–0.402	$Au = Au^{3+} + 3e^-$	1.42
$In = In^{3+} + 3e^-$	–0.340	$Au = Au^+ + e^-$	1.68
$Tl = Tl^+ + e^-$	–0.336		

*Courtesy of The International Nickel Company.

dard hydrogen cell is assigned a value of zero, and the potential developed by a half cell of the metal in question coupled to a standard half cell is compared with that of the hydrogen cell. This listing in Table 15·1 is in decreasing order of activity. The more active metals at the top of the list exhibit a stronger tendency to dissolve than those at the bottom. A metal higher in the series will displace a metal lower in the series from solution.

The electromotive series holds only for metals under conditions for which the series was determined. The electrolytes contained particular concentrations of salts of the same metal that was being studied. Under actual conditions, in other electrolytes, their behavior may be different. Instead of the electromotive series, a somewhat similar galvanic series is used which is based on experience with combinations of metals in a great variety of environments. Table 15·2 gives such a series for a number of metals and alloys in sea water moving at high velocity. In any couple, the metal near the top of this series will be anodic and suffer corrosion, while the one nearer the bottom will be cathodic and receive some galvanic protection. The difference in electrical potential between two metals is related to distance between them in the galvanic series. A metal coupled with another close to it on this list will usually corrode more slowly than when coupled with one further below it.

The relative concentration of both ions involved in the reaction has a definite influence on the electrical potential. If the metallic-ion concentration is increased relative to the reducible ion concentration, there will be a reduction in potential.

If the metallic ion is removed by the formation of an insoluble compound which is precipitated on the anode, and this film is adherent and impervious to the corroding solution, complete insulation results and corrosion stops. Oxide films of this type are formed on aluminum and chromium, which accounts for their superior corrosion resistance. A porous oxide or metallic coating tends to increase corrosion, especially when the part is exposed to alternating periods of immersion and drying.

The effect of dissolved oxygen on the corrosion rate is twofold: it acts in the formation of oxides and as a cathodic depolarizer. If the oxide formation removes metallic ions from the metal, corrosion will be increased. The effect of oxide film on the metal was mentioned previously. If the oxygen acts to remove hydrogen from around the cathode, corrosion will be increased. The effectiveness of oxygen in removing hydrogen is influenced by the amount of cathode area. With a large cathode the hydro-

TABLE 15·2 Galvanic Series of Metals and Alloys in Sea Water*

Anodic (Corroded) End
Magnesium
Zinc
Aluminum
Cadmium
Aluminum alloys
Low steel
Alloy steel
Cast iron
Stainless steel (active)
Muntz metal
Yellow brass
Aluminum brass
Red brass
Copper
Aluminum bronze
Copper-nickel alloys
Monel
Nickel (passive)
Inconel
Silver
Stainless steel (passive)
Titanium
Gold
Platinum
Catholic (Protected) End

*Courtesy of The International Nickel Company.

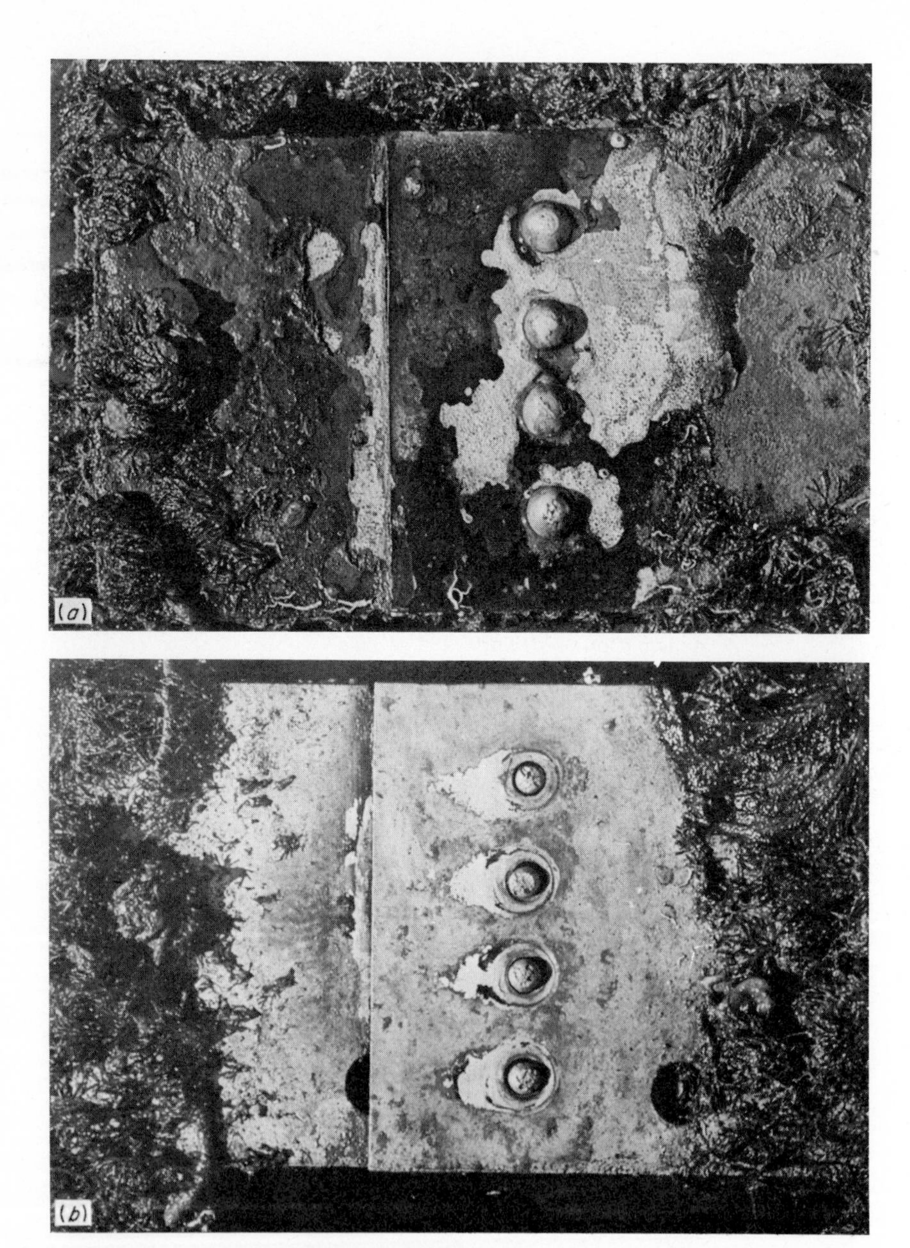

Fig. 15·4 Influence of area relationship between cathode and anode illustrated by copper-steel couples after immersion in sea water. (*a*) Copper rivets with small area in steel plates of large area have caused only slight increase in corrosion of steel. (*b*) Steel rivets with small area in copper plates of large area have caused severe corrosion of steel rivets. (The International Nickel Company.)

gen that reaches it will spread out and will be more accessible for removal by reaction with oxygen. This is why it is poor practice to couple a large cathode with a small anode. This may be illustrated by the following example. If steel plates are joined by copper rivets and immersed in sea water for several months, the copper rivets will remain in good condition and there will be no significant acceleration of the corrosion of the steel near the rivets. If, however, copper plates are joined by steel rivets and immersed under the same conditions, the steel rivets will be attacked very severely (Fig. 15·4).

Agitation acts to increase the corrosion rate by bringing fresh corroding solution into contact with the metal.

Differences in potential from point to point on a single metal surface cause corrosion known as *local action* and may be due to impurities on the surface or differences in surface structure or environment. A difference in environment, such as the difference in concentration of metal ions in the corroding solution at one point on the metal surface as compared with another point on the metal surface, will cause corrosion by local action. This difference in metal-ion concentration can be set up when a metal is in contact with a solution where the velocity is greater at one point than at another. This situation can be created by spinning a metal

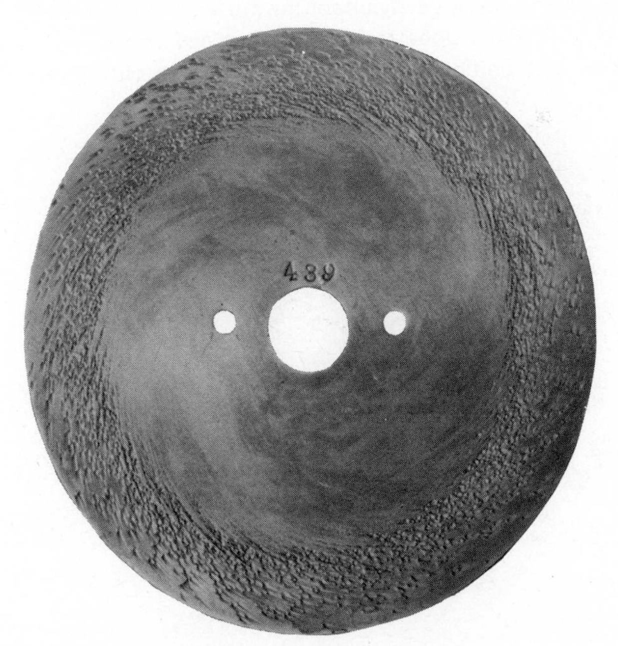

Fig. 15·5 Severe corrosion in region of high velocity on an admiralty-brass disk after rotation in sea water. (The International Nickel Company.)

disk through salt water. Since the metal nearer the center of the disk moves more slowly than at the edge, this allows metal ions to accumulate near the center and be swept away near the edge. At the edge, the region of highest velocity, metal-ion concentration will be the least, and severe corrosion will take place in this region (Fig. 15·5). For this application, a metal must be chosen that will be able to hold its protective film right up to the outer edge under these conditions.

Other factors—such as the presence of other ions in solution, the temperature of the solution, and the existence of stray electric currents—may materially affect the corrosion rate.

15·4 Specific Corrosion Types Specific descriptions are generally used for certain types of industrially important corrosion. When the entire surface of the metal is attacked to the same degree, it is known as *uniform corrosion*. This type is unusual in metals, since they are rarely so homogeneous that the surface will be evenly corroded.

Pitting corrosion is an example of nonuniform corrosion resulting from inhomogeneities in metal due to inclusions, coring, and distorted zones. These inhomogeneities set up differences of potential at localized spots to cause deep isolated holes. Figure 15·6 shows an electrolytically formed pit. The large pit formed when the surface was penetrated in a small area then grew rapidly into a large cavity under the surface. Progressive growth of the cavity caused further penetration of the surface from below. Part of the roof of the cavity then collapsed. Figure 15·7 shows pitting on a metal

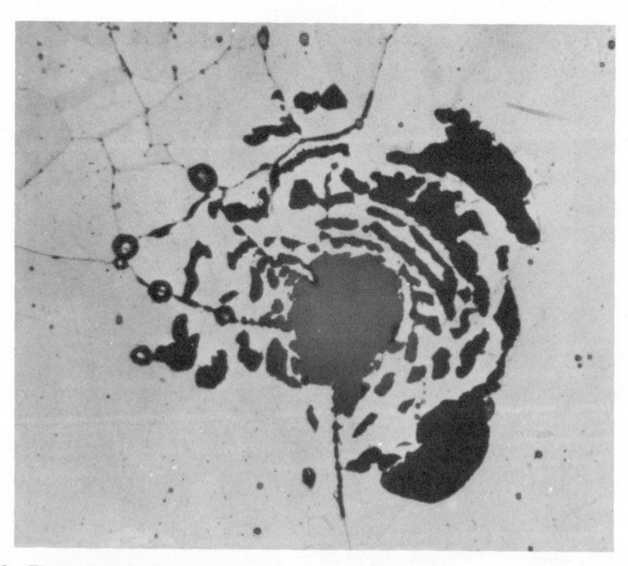

Fig. 15·6 Electrolitically formed pit, 350×. (M. A. Streicher, Du Pont Research Laboratory.)

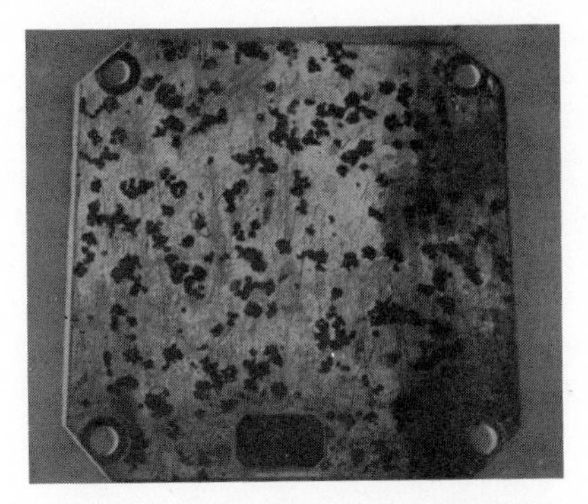

Fig. 15·7 Pitting corrosion due to marine organisms while metal was immersed in sea water. (The International Nickel Company.)

surface resulting from localized corrosion under marine organisms that became attached to the surface while immersed in sea water. Pitting of the base metal occurs when there is a break in the protecting layer or film. For example, when the chromium plate in a steel auto bumper is broken, pitting of the exposed steel takes place.

Cavitation corrosion, illustrated schematically in Fig. 15·8, is caused by the collapse of bubbles and cavities within a liquid. Vibrating motion between a surface and a liquid is such that repeated loads are applied to the surface, causing very high stresses when these bubbles form and collapse regularly. These collapses produce high stress impacts which gradually remove particles of the surface, eventually forming deep pits, depressions, and pockmarks. Figure 15·9 shows numerous small pits formed by cavitation corrosion on the surface of a cast-iron sleeve. This type of corrosion may be minimized or eliminated by switching to a more resistant material or by using a protective coating. In general, aluminum bronzes, Stellite, some stainless steels, and certain tool steels have good resistance to cavi-

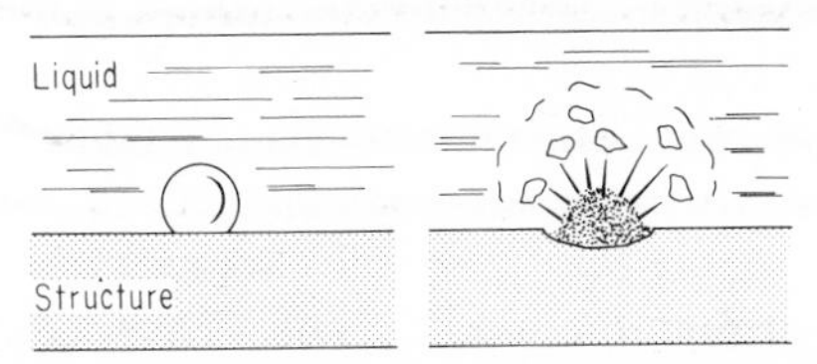

Fig. 15·8 Cavitation corrosion caused by collapsing action of bubbles which form at points where the local pressure is equal to or below the vapor pressure of the liquid. (From Materials Engineering, Special Report No. 202, January 1963.)

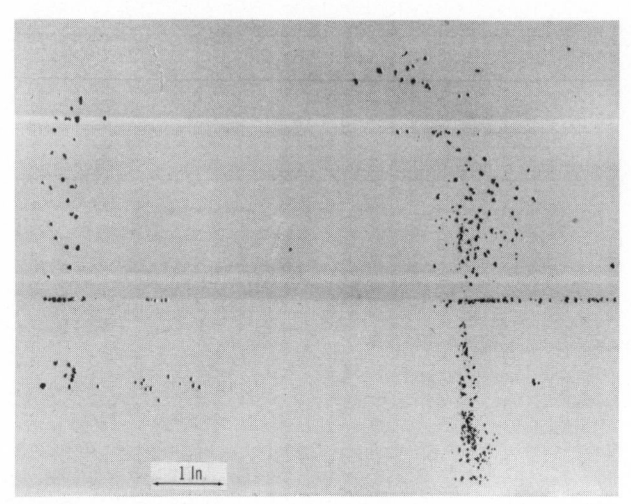

Fig. 15·9 Cavitation corrosion in the form of numerous small pits in local areas on the surface of a cylinder sleeve of gray cast iron. (Courtesy of D. J. Wulpi, International Harvester Company.)

tation damage. Materials such as cast iron, bronze and steel castings, and steel plate have relatively poor resistance to cavitation damage, but they can be protected by welded layers, sprayed-metal coatings, and some non-metallic coatings.

Crevice corrosion is a general term including accelerated attack at the junction of two metals exposed to a corrosive environment. We know from experience that corrosion is more likely to occur in crevices which retain solutions and take longer to dry out. It is also possible for corrosion to occur at crevices even when they are completely immersed. Accelerated attack can occur because of a differential in oxygen concentration (Fig. 15·10*a*). Oxygen has relatively easy access to the outside of the joint, which is cathodic. The metal in the joint is relatively anodic. The deposit of insoluble corrosion product around the anodic center tends to more completely exclude oxygen, resulting in a low oxygen concentration area and increased electrical potential. If the action continues, a pit forms in the center. Corrosion always occurs in the region of oxygen deficiency. Crevices can also lead to differences in metal-ion concentrations at different locations. For example, it is possible for an area in a crevice to have a higher metallic-ion concentration than the area outside. Thus, corrosion can take place in the region of metal-ion deficiency at the edge of a mechanical joint, as illustrated in Fig. 15·10*b*. The best way to avoid this type of corrosion is to eliminate crevices entirely by changing the design or by filling in joints that are liable to cause trouble.

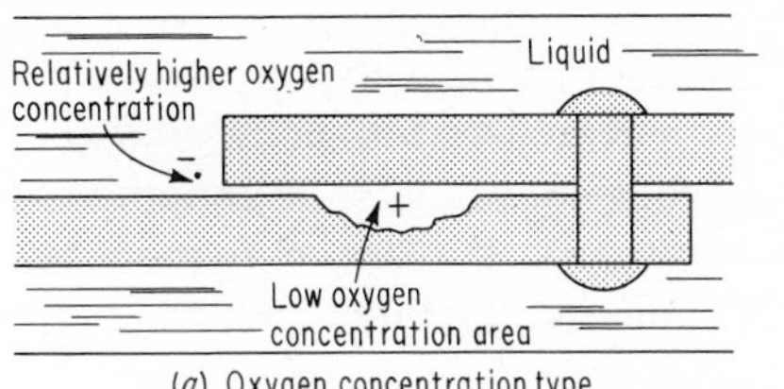
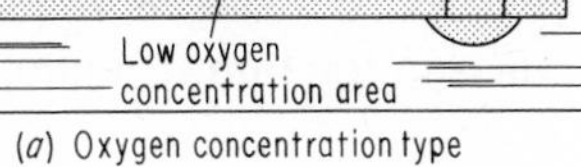

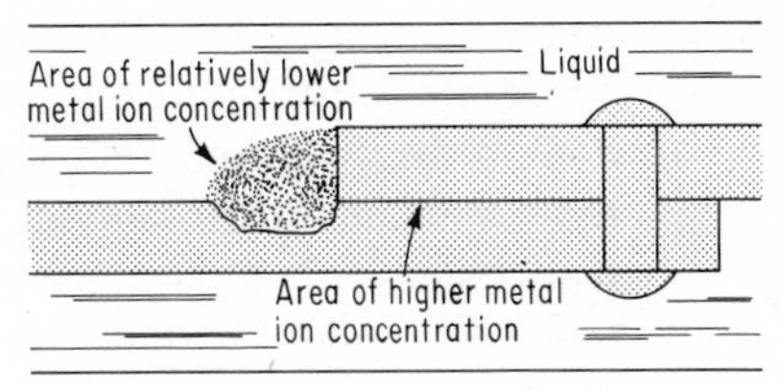

Fig. 15·10 Crevice corrosion caused by: (*a*) difference in oxygen concentration and (*b*) difference in metal-ion concentration. Notice that in each case corrosion takes place where there is a deficiency of either oxygen or metal ions. (From Materials Engineering, Special Report No. 202, January 1963.)

Fretting corrosion is a common type of surface damage produced by vibration which results in striking or rubbing at the interface of close-fitting, highly loaded surfaces. Such corrosion is common at surfaces of clamped or press fits, splines, keyways, and other close-fitting parts subject to minute relative movement. Fretting corrosion ruins bearings, destroys dimensions, and reduces fatigue strength. This type of corrosion is a mechanical-chemical phenomenon. When two components rub together, adhesive forces cause small particles of the surface to weld. With continued slight motion, the welded particles tear away from the opposing surfaces and react chemically with the atmosphere, forming debris or powder in the joint. Figure 15·11 shows fretting corrosion on the shaft of an oil pump gear during fatigue testing. There are several ways in which fretting corrosion may be overcome. The most obvious way is to remove the source of vibration by tighter clamping or more rigid mounting. Other

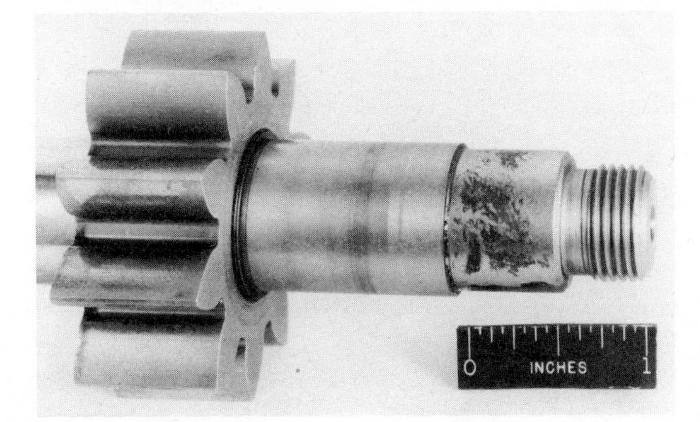

Figure. 15·11 Fretting corrosion of the shaft of an oil-pump drive gear during fatigue testing. (Courtesy of D. J. Wulpi, International Harvester Company.)

methods include raising hardnesses of mating surfaces, inserting rubber gaskets in joints (to absorb motion), lubricating with a dry medium (molybdenum disulfide held in a binder), and sealing the entire area with a material such as rubber cement to exclude atmospheres. Strengthening treatments such as nitriding, shot peening, surface rolling, chromium plating, or flame and induction hardening also lessen fracture resulting from fretting in shafts.

Intergranular corrosion is another example of nonuniform corrosion when a potential difference exists between the grain boundaries and the rest of the alloy. This type of corrosion usually takes place when precipitation of a phase from a solid solution occurs. Since precipitation usually takes place faster at the grain boundaries, the material in the vicinity of the grain boundary becomes depleted of the dissolved element, creating a difference of potential, and the grain boundary will dissolve preferentially (Fig. 15·12). Often a visual examination of the part will not reveal the extent of the damage, and in most cases there is an appreciable loss in mechanical properties.

Stress corrosion is acceleration of corrosion in certain environments when metals are externally stressed or contain internal tensile stresses due to cold working (see Sec. 12·6). The cracks may be transgranular or intergranular or a combination of both. The magnitude of stress necessary to cause failure depends on the corrosive medium and on the structure of

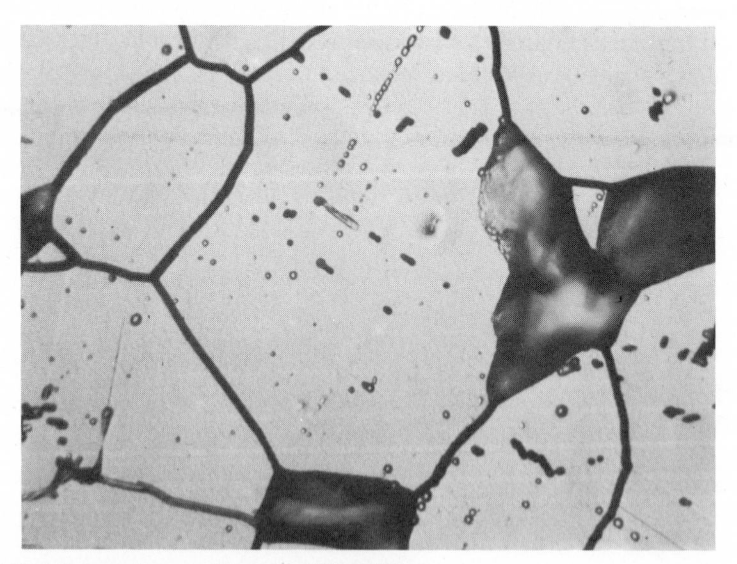

Fig. 15·12 Intergranular corrosion in type 316 stainless steel after a 27-h exposure to boiling sulfate–sulfuric acid solution, 500×. (M. A. Streicher, Du Pont Research Laboratory.)

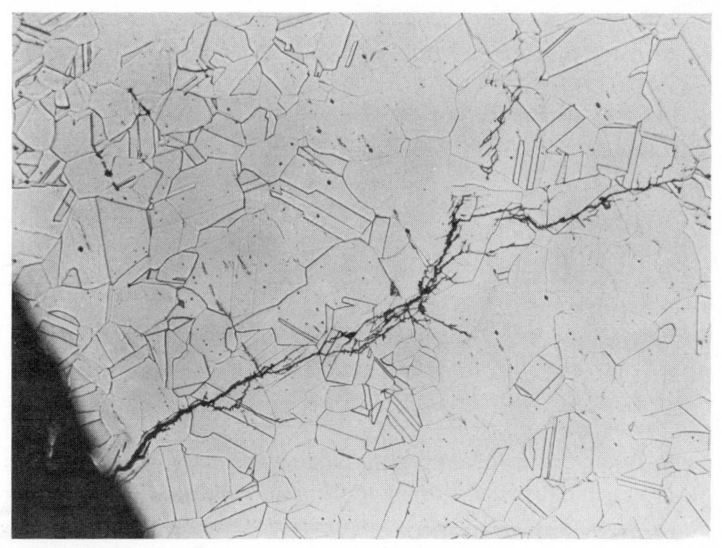

Fig. 15·13 Stress-corrosion cracking in type 304 austenitic stainless steel. A bend was exposed to chloride-containing water, 250×. (M. A. Streicher, Du Pont Research Laboratory.)

the base metal. Stress corrosion is one of the most important types of corrosion because it can occur in so many metals. Almost any metal can be attacked in certain environments, yet the conditions that cause cracking in one metal will not cause cracking in another. Thus, it is difficult to predict where attack will occur. The presence of nitrogen in iron and steel tends to make them more prone to stress-corrosion cracking in some nitrate solutions. Steels containing aluminum have better resistance to stress corrosion because the aluminum combines with nitrogen to form aluminum nitride. Some stainless steels are susceptible to stress-corrosion cracking in the presence of chlorides, e.g., in solutions of sodium chloride, calcium chloride, and several others. Stress-corrosion cracking has been recognized as the major cause of austenitic stainless steel failure in processes involving a chloride-containing environment. Figure 15·13 shows a stress-corrosion crack in type 304 austenitic stainless steel. Although they will pit badly in the presence of chlorides, the ferritic stainless steels are less likely to fail from stress corrosion than the austenitic martensitic grades. Stress-corrosion cracking can also occur in stressed copper alloys when they are exposed to ammonia and its compounds, especially in the presence of oxygen and carbon dioxide. The danger of cracking can be minimized by avoiding residual stresses, by using protective coatings, and, in the case of brasses, by keeping zinc content below

TABLE 15·3 Effect of Corrosion Type on Properties*

TYPE	WEIGHT LOSS, %	DEPTH OF PENETRATION, %	LOSS OF PROPERTIES, %		
			TENSILE STRENGTH	YIELD STRENGTH	ELONGATION
Uniform	1	1	1	1	1
Pitting	0.7	5	7	5	15
Intergranular	0.2	15	25	20	80
Stress	0.1	100	100	100	100

*By permission from L. F. Mondolfo and O. Zmeskal, "Engineering Metallurgy," McGraw-Hill Book Company, New York, 1955.

15 percent. Brasses containing 20 to 40 percent zinc are highly susceptible to attack.

Intergranular corrosion and stress corrosion have a very serious effect on the mechanical properties of the metal. The reduction in strength is not due to the amount of metal removed but rather to the stress concentration produced by the fine cracks. Table 15·3 gives an idea of the effect of the type of corrosion on properties.

Preferential corrosion of one of the components may even occur in single-phase solid-solution alloys. Dezincification in brass (see Sec. 12·6) is an example of this kind of corrosion (Fig. 12·5).

Galvanic corrosion occurs at the interface where two metals are in contact in a corroding medium. This type of corrosion, illustrated in Fig. 15·14, was discussed in Sec. 15·2.

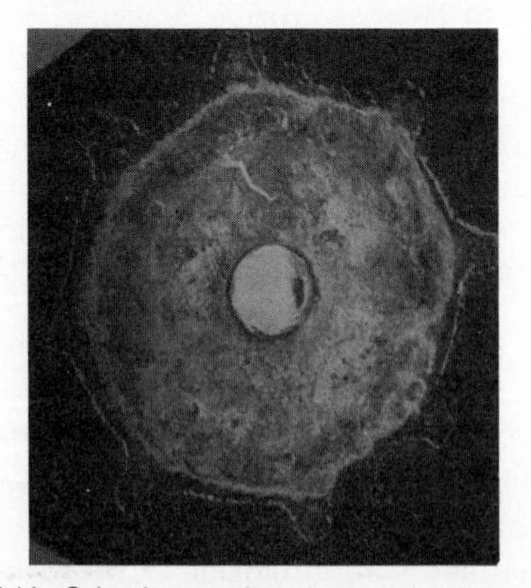

Fig. 15·14 Galvanic corrosion of magnesium where it is in close contact with a steel core around which the magnesium was cast. (The International Nickel Company.)

A type of corrosion that has become increasingly important is *liquid-metal corrosion.* In certain types of nuclear reactors for the production of atomic power, liquid metals such as bismuth and sodium are used as the heat-transfer medium. The path of the liquid metal is a closed loop with one leg at high temperature in the reactor core and the other leg at lower temperature in a heat exchanger. The solubility of a solid in the liquid usually increases with temperature. Therefore, there is a tendency for the solid to dissolve up to its solubility limit in the high-temperature leg and be deposited, because of the lower solubility limit, in the cooler leg. The hot leg is continually corroded, and the cold leg becomes plugged with the deposited corrosion products. This phenomenon is essentially one of mass transfer which leads to gradual deterioration of the metal in the hot zone. The most effective method of controlling this type of corrosion is by use of inhibitors in the liquid metal. Zirconium has been used as an effective inhibitor in liquid bismuth to decrease the liquid-metal corrosion of iron by mass transfer.

15·5 Methods for Combating Corrosion Many methods are used industrially to prevent corrosion by selection of the proper alloy and structure, or by surface protection of a given material. The most important are:

1. Use of high-purity metals
2. Use of alloy additions
3. Use of special heat treatments
4. Proper design
5. Cathodic protection
6. Use of inhibitors
7. Surface coatings

In most cases, the use of high-purity metals tends to reduce pitting corrosion by minimizing inhomogeneities, thereby improving corrosion resistance.

Alloy additions may reduce corrosion by several methods. For example, austenitic stainless steels, when cooled through a temperature range of about 900 to 1400°F, precipitate chromium carbides at the grain boundaries. This precipitation depletes the boundaries of chromium and makes them more susceptible to intergranular corrosion. This type of corrosion may be avoided either by reducing the carbon to a low value (below 0.03 percent) or by converting the carbide to a more stable form. The latter method is more widely used and involves the addition of titanium or columbium. These elements have a great affinity for carbon, producing very stable carbides that are not soluble in austenite at elevated temperature. This leaves very little carbon available for combination with chromium and results in what is known as a *stabilized* stainless steel. Some alloy additions improve corrosion resistance by forming, or helping the formation of, adherent, nonporous surface oxide films. This is particularly true of manganese and aluminum additions to copper alloys, molybdenum additions to stainless steels, and magnesium additions to aluminum.

Heat treatment which leads to homogenization of solid solutions, especially in cast alloys that are subject to coring, tends to improve corrosion resistance. Stress-relief treatments following cold working are widely used to improve the resistance of alloys susceptible to stress corrosion.

Proper design should keep contact with the corroding agent to a minimum. Joints should be properly designed to reduce the tendency for liquids to enter and be retained. Contact between materials far apart in the electromotive series should be avoided. If this cannot be done, they should be separated by rubber or plastic to reduce the possibility of galvanic corrosion. For example, Fig. 15·15*a* shows two cases of galvanic corrosion when using dissimilar metals. Since aluminum is the anode with respect to steel, corrosion of an aluminum rivet can be expected when it is used to fasten steel sheets together. Similarly, if a steel rivet is used to fasten aluminum sheets, then undercutting galvanic corrosion of the aluminum sheet will result in loose rivets, slipping, and possible structural damage. This type of corrosion can be prevented by applying a nonhardening insulating joint compound in the area where the sheet and the rivet or bolt are in contact, or by applying a zinc chromate primer to all contacting surfaces and then coating the primed area with an aluminum paint. Where the fasteners are not subject to high stresses, the contact points can be insulated with plastics or other nonmetallic sleeves, shims, washers, and similar parts, as shown in Fig. 15·15*b*.

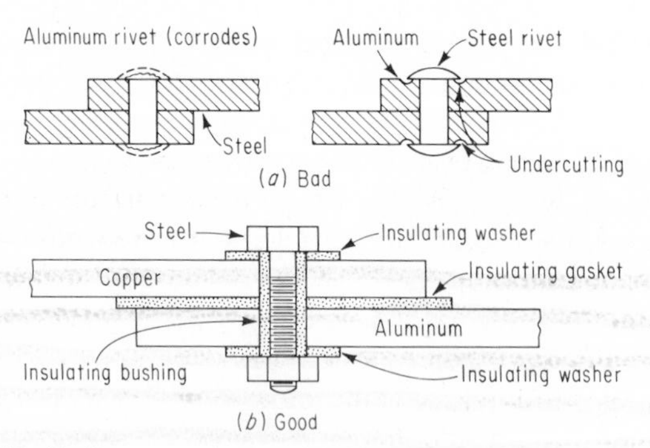

Fig. 15·15 Design suggestion to minimize corrosion when fastening two dissimilar metals. (*a*) Corrosion of aluminum rivets in steel plates or aluminum plates when fastened with a steel rivet. (*b*) Recommended practice of using insulating materials between the steel bolt and copper and aluminum plates. (From Materials Engineering, Special Report No. 202, January 1963.)

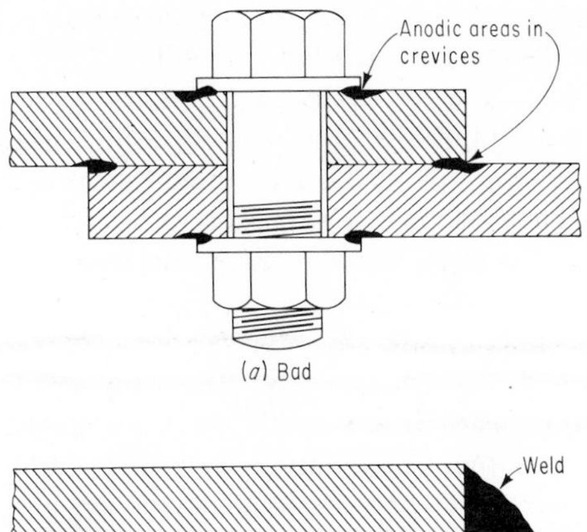

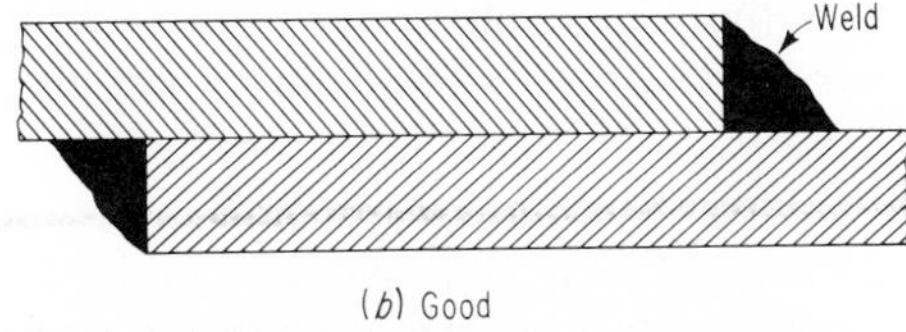

Fig. 15·16 (*a*) Crevice corrosion between plates when bolted together. (*b*) Recommended practice, if possible, is to fasten the plates by welding. (From Materials Engineering, Special Report No. 202, January 1963.)

It was pointed out earlier that crevices are a potential source of concentration-cell corrosion. They are frequently encountered in sections such as shown in Fig. 15·16*a*, where two plates are bolted together in a corrosive solution. No matter how much torque is applied to the bolt, it is practically impossible to eliminate crevices into which the solution gradually penetrates and becomes stagnant. Crevices can be avoided by using welds instead of mechanical fasteners, as shown in Fig. 15·16*b*, or by using insulating gaskets between surfaces that are machined parallel.

Cathodic protection is obtained by placing the metal that would normally corrode in electrical contact with one that is above it in the galvanic series. The more active metal thus becomes the anode. This is essentially a galvanic battery in which the corroding metal is made to function as the cathode. The metals generally used to provide this type of protection are zinc and magnesium. In some cases, the protective direct current is obtained by an external voltage source. The anode in this case usually consists of a relatively inert material such as carbon, graphite, or platinum. The structures most frequently protected by this method are underground pipelines, hulls of ships, and boilers. For the protection of underground pipe, anodes are buried some 8 to 10 ft from the pipe. The depth of the hole should be sufficient to locate the anode in permanently moist soil.

Individual anodes are connected to a *collector* wire which in turn is brazed to the pipeline. The current discharges from the anode to the soil, collects on the pipeline, and returns to the anode through the connecting wire. For the cathodic protection of ship hulls, zinc or magnesium anodes are fastened to the rudder and to the hull itself in the region around the propeller. Magnesium anodes have become widely used to provide cathodic protection in domestic and industrial water heaters and elevated water-storage tanks.

Inhibitors are chemicals which, when added to the corrosive solution, reduce or eliminate its corrosive effect. In most cases, the inhibitor will form a protective layer on the metal surface. Inhibitors are added to the antifreeze mixtures used in automobile radiators. Oxidizing agents when added to the corrosive solution will produce oxide films on aluminum, chromium, and manganese.

Surface coatings include paints, salt and oxide films, and metallic coatings.

Paints and other organic coatings are primarily used to improve the appearance of the surfaces and structures. The use of paint for corrosion protection only is secondary and of little economic importance to the paint industry. Paint provides a protective film to the metal and is effective only as long as the film is unbroken.

Salt and oxide films are obtained by reacting the metal with a solution which produces the desired film. Some examples are: A chromate pickle protects magnesium by forming a film of magnesium chromate, *Parkerizing* or *Bonderizing* for ferrous alloys protects by forming a phosphate film, anodizing for aluminum and magnesium forms a thick oxide film, and passivating for stainless steels also forms an oxide film.

Metallic coatings may be obtained by a variety of methods such as metallizing, hot dipping, electroplating, diffusion, and cladding.

Metallizing was discussed in Chap. 14 in regard to surface coating for wear resistance. Practically all metallized coatings for corrosion use are either zinc or aluminum. These metals are used primarily for corrosion work on iron and steel because steel is cathodic to them and hence is protected electrochemically in spite of any porosity or minor voids in the coating. Grit or sand blasting is almost universally used for cleaning and preparing the surface prior to spraying. Most coating systems employ supplementary organic sealers or top coats. The sealers are usually vinyl chlorinated rubber applied over inhibiting primers.

Hot dipping is used mainly to apply a coating of zinc, tin, cadmium, aluminum, or lead to steel. The hot-dip process has a wide range of applicability, but the coating applied must contend with a brittle diffusion layer of intermetallic compounds at the interface. This may result in poor adhesion and a tendency to flake on bending, unless the thickness of the diffusion layer is properly controlled. The hot-dip process usually includes the operations of pickling with inhibited acid to produce a clean surface; fluxing to facilitate wetting; dipping the article in a molten bath of controlled composition; and wiping, shaking, or centrifuging the dipped piece to regulate the thickness and uniformity of the coating.

Zinc coating, or *galvanizing,* is applied to a greater tonnage of steel products than any other method of metallic coating. The major applications of zinc coating on steel products include roofing and siding sheets; wire and wire products for all outdoor exposure; articles fabricated from sheet steel such as boilers, pails, cans, and tanks; hardware for outdoor use; pipe and conduit; and exposed structural steel. Tin plate is widely used in the familiar tin can. Terne metal is sheet steel plated in a lead bath to which tin has been added to ensure bonding. It is more expensive than zinc, but much cheaper than tin. It is used for gasoline tanks and as a roofing material. Aluminizing, hot-dipped aluminum coating on steel, is used for applications that require a combination of resistance to corrosion and heat, such as mufflers and tailpipes of automotive engines.

Electroplated coatings are used extensively for decorative purposes, but for industrial applications the most important single function is for corrosion protection. In addition to corrosion protection and appearance, electroplated coatings are applied to obtain other surface properties such as wear resistance, high electrical conductivity, good solderability, and high or low light-reflecting ability. Nickel and chromium plates provide wear resistance (see Chap. 14); silver and copper are best for electrical conductivity; silver, copper, tin, and cadmium improve solderability; chromium and rhodium have high resistance to tarnish.

Zinc coatings are used on iron and steel products which require primarily corrosion protection. These include nuts, bolts, screws, nails, hardware, and electrical conduit. Cadmium coatings are used as substitutes for zinc coatings but are not quite so good for outdoor exposure. Cadmium plating is also used for radio chassis and electronic equipment since it is readily solderable. Chromium coatings are used on automobile trim, plumbing fixtures, hardware, and appliances. These coatings are very thin and rather porous and normally offer little protection against corrosion. Therefore, they are usually applied over thicker coatings of nickel or copper. Nickel coatings are used principally as an underplate for chromium, silver, gold, and rhodium plating. Nickel has good corrosion resistance but is tarnished by the atmosphere. Copper coatings are used as an undercoat for nickel and chromium coatings, particularly on zinc-base die castings. The greatest use of electroplated tin coatings is for food containers. These coatings may be made thinner than those made by the conventional hot-dipping method.

The use of chromium and silicon as diffusion coatings for wear resistance has been discussed in Chap. 14. In addition to the above two metals, aluminum and zinc diffusion coatings are also used to provide corrosion protection. All diffusion-coating processes follow essentially the same procedure and are based on the same principles. The part to be coated is placed in contact with a powder containing the metal to provide the coating. At elevated temperature there is a transfer of the metal to the base material, through the vapor phase, usually by means of a suitable catalyst. Holding at temperature or reheating without the powder after the initial penetration will allow further diffusion to the desired depth. The process of alloying steel with aluminum by diffusion is called *calorizing.* Calorized steel is highly resistant to oxidation and corrosion by hot gases, particularly sulfurous gases. Calorized parts are used in furnaces that employ fuels high in sulfur; bolts for use up to 1400°F; salt, cyanide, and lead pots; and oil refineries. Zinc impregnation is obtained by a process known as *sherardizing.* The principal application of sherardizing is for small steel parts, such as bolts, nuts, and washers, or for castings exposed to the atmosphere.

Cladding is a method by which the coating becomes an integral part of the material. This may be accomplished by casting or hot working. Casting is best suited when there is a considerable difference in the melting points of the cladding material and the base material. Hot rolling is the more widely used method for cladding. Slabs or sheets of the cladding material are strapped to an ingot of the base material. After heating to the rolling temperature, the straps are removed and the entire assembly is rolled. The heat and pressure during rolling weld the two materials together. The cladding may be of the same base material as the core. Alclad is the name applied to aluminum alloys which are clad with pure aluminum to improve corrosion resistance (see Fig. 12·30). Steel may be clad with nickel, nickel-chromium, or nickel-copper alloys. Aside from corrosion resistance, cladding is sometimes done to obtain a combination of properties that are not available by any other method. Copper-clad steel wire is a good example. The copper exterior provides high electrical conductivity and good corrosion resistance, while the steel core provides high tensile strength. This wire is produced by pouring molten copper into a mold containing a round, heated steel billet. A weld that is able to withstand cold working is produced between the two metals. After solidification, the composite billet is hot-rolled to a rod and then drawn to the desired size.

QUESTIONS

15·1 Define corrosion.
15·2 How does corrosion differ from erosion?
15·3 Explain the mechanism of electrochemical corrosion.
15·4 Why does corrosion generally occur at the anode?

15·5 What determines whether a given area is anodic or cathodic?
15·6 What is the limitation on the use of the electromotive series (Table 15·1) in the corrosion of metals?
15·7 How does the distance between metals in the galvanic series (Table 15·2) affect corrosion rate?
15·8 Why do aluminum and chromium show superior corrosion resistance?
15·9 What is meant by anodizing? How does it affect corrosion resistance?
15·10 What three factors are necessary to form a galvanic cell?
15·11 Explain and describe the mechanism of dezincification.
15·12 Discuss the differences and similarities between intergranular corrosion and stress corrosion.
15·13 Describe each of the following: pitting corrosion, crevice corrosion, cavitation corrosion, fretting corrosion.
15·14 Explain cathodic protection.
15·15 List seven methods of corrosion protection and explain why each method is effective.

REFERENCES

Ailor, William H. (ed): "Handbook on Corrosion Testing and Evaluation," John Wiley & Sons, Inc., New York, 1971.
American Society for Metals: "Corrosion of Metals," Metals Park, Ohio, 1946.
———: "Metals Handbook," 1948, 1961, and 1964 ed., Metals Park, Ohio.
———: "Surface Protection against Wear and Corrosion," Metals Park, Ohio, 1954.
Bosich, Joseph F.: "Corrosion Prevention for Practicing Engineers," Barnes and Noble, New York, 1970.
Burns, R. M., and W. W. Bradley: "Protective Coatings for Metals," American Chemical Society Monograph 129, Van Nostrand Reinhold Company, New York, 1955.
Evans, U. R.: "An Introduction to Metallic Corrosion," 2d ed., St. Martin's Press, New York, 1963.
———: "Metallic Corrosion Passivity and Protection," Edward Arnold (Publishers) Ltd., London, 1946.
Fontana and Greene: "Corrosion Engineering," McGraw-Hill Book Company, New York, 1967.
International Nickel Company: "Corrosion in Action," New York, 1955.
LaQue, F. L., and H. R. Copson: "Corrosion Resistance of Metals and Alloys," 2d ed., Van Nostrand Reinhold Company, New York, 1963.
Scully, J. C.: "Fundamentals of Corrosion," Pergamon Press, New York, 1966.
Speller, F. N.: "Corrosion," 3d ed., McGraw-Hill Book Company, New York, 1951.
Uhlig, H. H.: "Corrosion and Corrosion Control," 2d ed., John Wiley & Sons, Inc., New York, 1971.
———: "The Corrosion Handbook," John Wiley & Sons, Inc., New York, 1948.

16 POWDER METALLURGY

16·1 Introduction Powder metallurgy may be defined as the art of producing metal powders and using them to make serviceable objects. Powder metallurgy principles were used as far back as 3000 B.C. by the Egyptians to make iron implements. The use of gold, silver, copper, brass, and tin powders for ornamental purposes was commonplace during the middle ages.

In 1829, Woolaston published a paper which described a process for producing compact platinum from platinum sponge powder. Considered the first scientific work in the field of powder metallurgy, this laid the foundations for modern techniques.

It is interesting to note that in the nineteenth century more metallic elements were produced in powder form than in any other form.

The invention of the incandescent electric light by Edison required the development of a suitable filament material. Powders of osmium, tantalum, and tungsten were used, but the filaments were very brittle. It had long been evident that tungsten would make an ideal filament for the electric lamp, but to work tungsten into the necessary fine wire was beyond conventional metallurgy at the beginning of the twentieth century. It remained for Coolidge, in 1909, to make the important discovery that tungsten can be worked in a certain temperature range and will retain its ductility at room temperature. Finely divided tungsten powder was compressed into small ingots which were sintered at temperatures below the melting point of tungsten. These sintered ingots were brittle at room temperature but could be worked at elevated temperatures near the sintering temperature. Subsequent working at the elevated temperature improved its ductility until a stage was reached where the metal was ductile at room temperature and could be drawn into wire with tensile strength approaching 600,000 psi.

The Coolidge method led to a new method of fabrication for refractory metals such as molybdenum, tantalum, and columbium. It also led to the development of cemented carbides and composite metals. At about the same time, porous-metal bearings were manufactured using the technique of powder metallurgy. These and other applications will be discussed later in this chapter.

Initially all P/M (symbol used to represent powder metallurgy) parts were small, and the mechanical properties were more or less comparable with conventional materials. Today, however, parts greater than 1 ft in diameter and weighing more than 50 lb are being produced in large quantities. Figure 16·1 shows a picture of a molybdenum-powder billet which is believed to be the heaviest made to date. It weighed 3,000 lb, and in the green condition it was 18½ in. in diameter by 44 in. long. An important development has been the increasing use of large P/M parts by the automotive industry. Materials with mechanical properties far exceeding those of conventional materials have been developed by improving heat treatments, powder compositions, and processing methods to achieve higher densities. High strength, ductility, and toughness may be obtained in P/M parts, so that the old idea of brittle fragile parts is no longer valid.

16·2 Powder Metallurgy Processes The two main operations of the powder metal process are compacting and sintering.

Compacting, or pressing, consists in subjecting the suitably prepared powder mixtures, at normal or elevated temperature, to considerable pres-

Fig. 16·1 Molybdenum-powder forging billet, shown in the green condition, measuring 18½ in. in diameter by 44 in. long. (Courtesy of GTE Sylvania.)

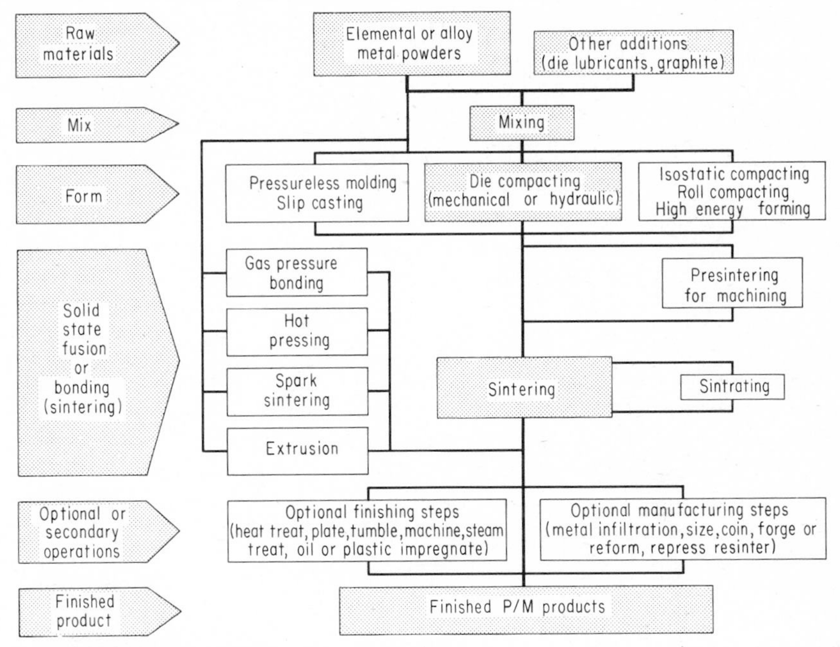

Fig. 16·2 Powder metallurgy processes. (Courtesy of the Metal Powder Industries Federation.)

sure. The resulting powder compact is known as a *briquette* and is said, in this form, to be "green." It can be handled, but it is relatively brittle.

Sintering is an operation in which the green briquettes are subjected to heat, usually in an inert atmosphere, at a temperature below the melting point of the solid metal. Sintering will give the required mechanical strength as well as other desired properties.

In addition to compacting and sintering, and depending upon application, other accessory operations may be added to the process. These include presintering, sizing, machining, and impregnation. A summary of the powder metallurgy processes is shown in Fig. 16·2. Typical flow sheets showing steps in the manufacture of powder metal parts are shown in Fig. 16·3.

16·3 Preparation of Metal Powders There is a definite relation between a particular method of powder production and desired properties of powder metallurgy products. Many mechanical and chemical methods are used to produce powders for specific applications, but the three most important methods are atomization, reduction of oxides, and electrolytic deposition.

Atomization is the method most frequently used for metals having low melting points, such as tin, lead, zinc, cadmium, and aluminum. As the

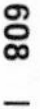

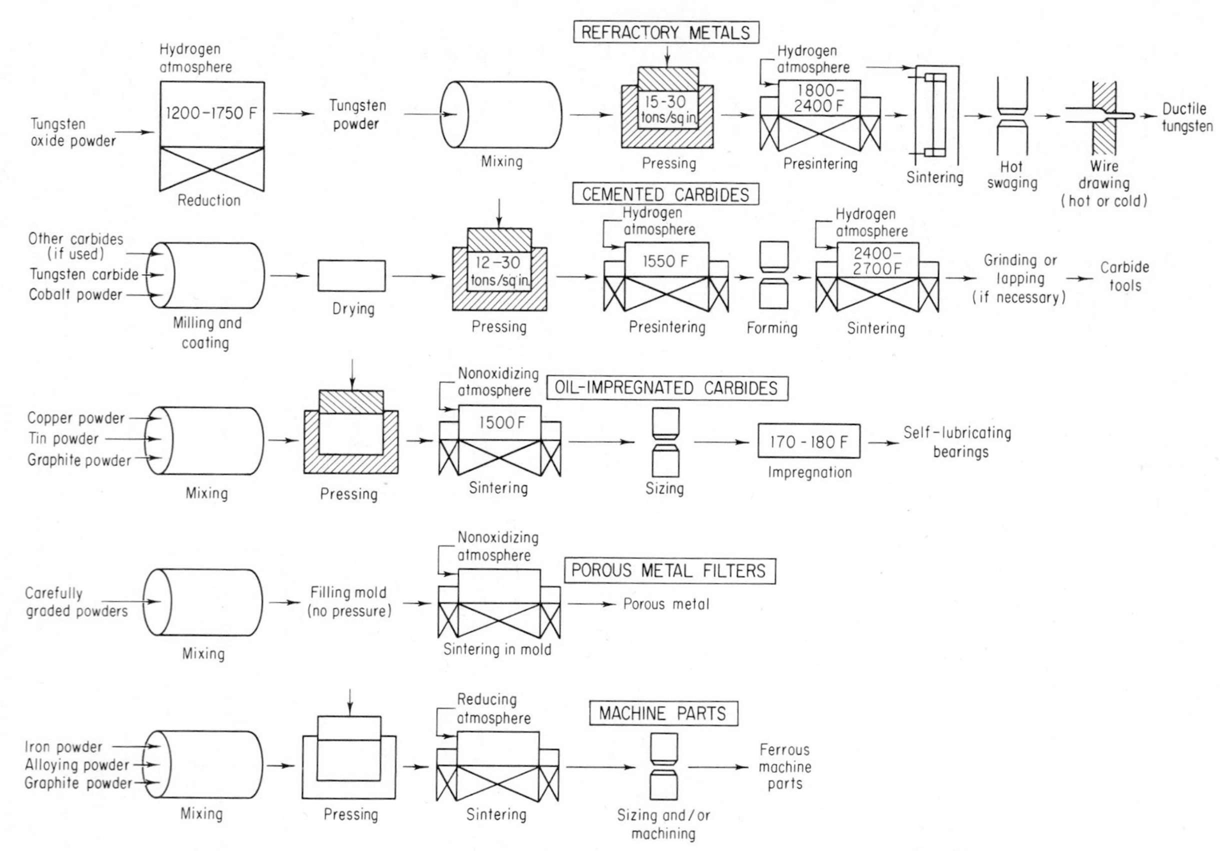

Fig. 16·3 Typical flow sheets showing steps in the manufacture of articles from metal powders. (From "Metals Handbook," 1948 ed., p. 48, American Society for Metals, Metals Park, Ohio.)

liquid metal is forced through a small orifice, a stream of compressed air causes the metal to disintegrate and solidify into finely divided particles. Atomized products are generally in the form of sphere-shaped particles (Fig. 16·4*a*). A wide range of particle-size distributions may be obtained by varying the temperature of the metal, pressure and temperature of the atomizing gas, rate of flow of metal through the orifice, and the design of the orifice and nozzle. The principal advantage of the atomization process is its flexibility. It will produce powders of different degrees of fineness, and in the production of a given fineness, uniformity of particle-size distribution can be closely maintained.

The reduction of compounds of the metals (usually an oxide) provides a convenient, economical, and flexible method of producing powders. The largest volume of metallurgical powder is made by the process of oxide reduction. Mill scale or chemically produced oxides are reduced with carbon monoxide or hydrogen, and the reduced powder is subsequently ground. The nature, particle size, and distribution of the raw material and the conditions of reduction greatly influence the form of the deposited particles. If the oxide powder is graded before reduction, a high degree of size uniformity can be obtained in the reduced powder. The particles produced by oxide reduction are spongelike in structure and are ideal for molding. The shape is generally jagged and irregular (Fig. 16·4*b*), and the particles are porous. This is the only practical method available for producing powders of the refractory metals such as tungsten and molybdenum. Oxide reduction is also an economical method of producing powders of iron, nickel, cobalt, and copper.

The method of electrolytic deposition is most suitable for the production of extremely pure powders of principally copper and iron. This process is essentially an adaptation of electroplating. By regulation of current density, temperature, circulation of the electrolyte, and proper choice of electrolyte, the powder may be directly deposited from the electrolyte. The deposit may be a soft spongy substance which is subsequently ground to powder, or the deposit may be a hard, brittle metal. Powders obtained from hard, brittle electrodeposits are generally not suitable for molding purposes. Most of the powder produced by electrolytic deposition for commercial applications is of the spongy type. The shape of electrolytic powder is generally dendritic (Fig. 16·4*c*). Although the resulting powder has low apparent density, the dendritic structure tends to give good molding properties because of interlocking of the particles during compacting.

16·4 Characteristics of Metal Powders In all cases, the performance of the material during processing as well as the properties of the finished product depends to a large extent upon the basic characteristics of the powder material. Aside from the chemical composition and purity, the basic char-

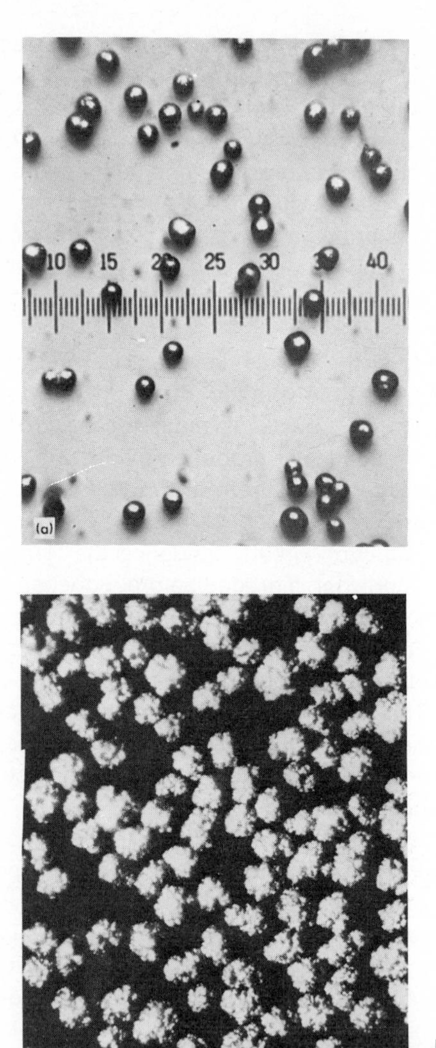

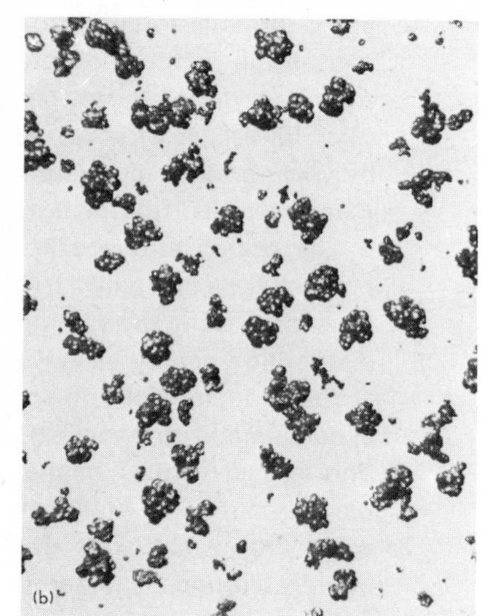

Fig. 16·4 Shape of powders produced by different methods. (*a*) Atomizing; (*b*) oxide reduction; (*c*) electrodeposition. (Courtesy of the Metal Powder Industries Federation.)

acteristics of a metal powder are particle size and size distribution, particle shape, apparent density, and particle microstructure.

Metal powders may be divided into sieve and subsieve size ranges. Those in the sieve-size class are usually designated according to the finest mesh through which all the powder will pass. If all the powder passes through a 200-mesh screen, it is designated as a minus 200-mesh powder, etc. The subsieve-size powders all pass through a 325-mesh sieve used in practice.

The size of these powders may be specified by averaging the actual dimensions as determined by microscopic examination.

Particle-size distribution is important in the packing of the powder and will influence its behavior during molding and sintering. For practical purposes, the selection of a desirable size distribution for a specific application is usually based upon experience. In general, a finer powder is preferred over a coarser powder, since finer powders have smaller pore size and larger contact areas, which usually results in better physical properties after sintering. Particle-size distribution is specified in terms of a sieve analysis, that is, the amount of powder passing through 100-, 200-, etc. mesh sieves. It should be apparent that sieve analysis will yield significant results regarding particle size and distribution only when the particles are spherical in shape. Inaccurate information will be obtained if the particles are irregular or flaky.

The nature of the surface of individual particles is also an important powder characteristic. Powders produced by chemical reduction of oxides (Fig. 16·4*b*) usually have a highly roughened surface which is easily observed, whereas atomized particles (Fig. 16·4*a*) have a much finer degree of surface roughness. The character of the surface will influence the frictional forces between particles, which is important when powder is flowing or settling or during compaction. Since any reaction between particles or between the powder and its environment is initiated at the surface, the amount of surface area per unit of powder can be significant. The surface area is very high for powders made by reduction techniques, as shown by the data in Table 16·1.

Particle shape is important in influencing the packing and flow characteristics of powders. Spherical-shaped particles have excellent sintering qualities and result in uniform physical characteristics of the end product; however, irregular-shaped particles have been found superior for practical

TABLE 16·1 Typical Specific Surface Areas of Commercial Powders*

POWDER	SPECIFIC SURFACE AREA (cm^2/gm)
Reduced Fe	
Fine, 79% – 325 mesh	5160
Normal blend	1500
Coarse, 1% – 325 mesh	516
Sponge Fe – normal blend	800
Atomized Fe – normal blend	525
Electrolytic Fe – normal blend	400
Reduced Tungsten 0.6 micron	5000

*From J. S. Hirschhorn, "An Introduction to Powder Metallurgy," American Powder Metallurgy Institute, New York, 1969.

TABLE 16·2 Apparent Densities of Commercial Metal Powders*

MATERIAL	SPECIFIC GRAVITY	APPARENT DENSITY	MATERIAL	SPECIFIC GRAVITY	APPARENT DENSITY
Aluminum	2.70	0.7–1	Molybdenum	10.2	3–6.5
Antimony	6.68	2–2.5	Nickel	8.9	2.5–3.5
Cadmium	8.65	3	Silicon	2.42	0.5–0.8
Chromium	7.1	2.5–3.5	Silver	10.50	1.2–1.7
Cobalt	8.9	1.5–3	Tin	5.75	1–3
Copper	8.93	0.7–4	Tungsten	19.3	5–10
Lead	11.3	4–6	Zinc	7.14	2.5–3
Magnesium	1.74	0.3–0.7	Iron and steel	7.85	1–4

*By permission from C. G. Goetzel, "Treatise on Powder Metallurgy," vol. 1, Interscience Publishers, Inc., New York, 1949.

molding. The mechanism of packing involves three processes: the filling of gaps between larger particles by smaller ones, breakdown of bridges or arches, and mutual sliding and rotation of particles. These processes are important in loading die cavities with metal powders.

Apparent density may be defined as the weight of a loosely heaped quantity of powder necessary to fill a given die cavity completely. A list of apparent-density ranges of a number of metal powders is given in Table 16·2. Naturally, increasing the specific gravity or density of the solid material increases the apparent density of the powder. As was just pointed out, the packing of powder particles is greatly influenced by particle size and shape. For example, a given space can be completely filled only by the same size of cubes exactly aligned. Any particle shapes that are curved or irregular cannot completely fill a space, and this leads to the presence of porosity. The importance of packing of spheres was discussed in Chap. 2 under crystal structures, where it was shown that face-centered-cubic and close-packed-hexagonal structures had high packing factors. An effective way to increase the apparent density is to fill the spaces among particles with smaller ones. This leads to a filling arrangement known as interstitial packing (illustrated in Fig. 16·5); however, even the smaller particles cannot

Fig. 16·5 An example of interstitial packing, that is, voids among large particles being filled by smaller ones.

Fig. 16·6 Model illustration of the "bridging" effect caused by small particles.

completely fill the pores. It is even possible for the addition of small particles to decrease the apparent density by an effect known as "bridging," illustrated in the model of Fig. 16·6. As shown in Fig. 16·7, the shape of the powder influenced the effect on apparent density. Fine spherical powder is very effective in increasing the apparent density, while flake additions reduce the apparent density sharply. The apparent density of a powder is a property of great importance for both molding and sintering operations. Powders with low apparent density require a longer compression stroke and deeper cavities to produce a briquette of given size and density. The tendency of the compact to shrink during sintering seems to decrease with increasing apparent density.

16·5 Mixing Proper blending and mixing of the powders are essential for uniformity of the finished product. Desired particle-size distribution is obtained by blending in advance the different types of powders used. Alloying powders, lubricants, and volatilizing agents to give a desired amount of porosity are added to the blended powders during mixing. The time for

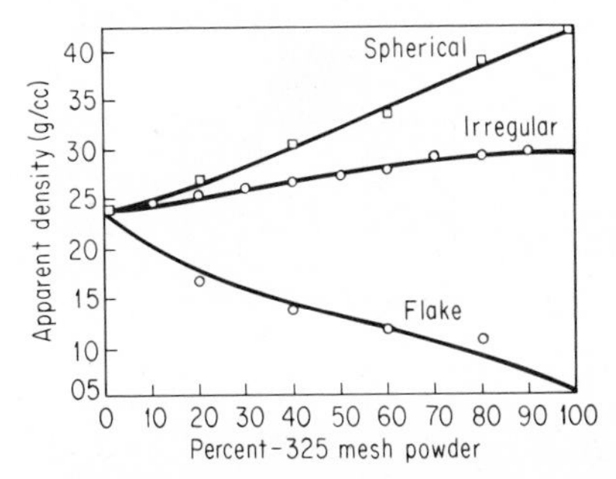

Fig. 16·7 The effect of −325 mesh additions to a +325 mesh distribution of 316 stainless-steel powder on apparent density for three different shapes of the addition. (From J. S. Hirschhorn, "Introduction to Powder Metallurgy," American Powder Metallurgy Institute, New York, 1969.)

mixing may vary from a few minutes to several days, depending upon experience and the results desired. Overmixing should be avoided in many cases, since it may decrease particle size and work-harden the particles.

16·6 Compacting The most important operation in powder metallurgy is compacting or pressing. The ability to obtain a satisfactory pressed density often determines the feasibility of manufacture by powder metallurgy. Most compacting is done cold, although there are some applications for which compacts are *hot-pressed.*

The purpose of compacting is to consolidate the powder into the desired shape and as closely as possible to final dimensions, taking into account any dimensional changes that result from sintering; compacting is also designed to impart the desired level and type of porosity and to provide adequate strength for handling.

Compacting techniques may be classified into two types: (1) pressure techniques, such as die, isostatic, high-energy-rate forming, forging, extrusion, vibratory, and continuous, and (2) pressureless techniques such as slip casting, gravity, and continuous.

Die compaction is the most widely used method. The usual sequence of operations in die compacting consists in filling the die cavity with a definite volume of powder; application of the required pressure by movement of the upper and lower punches toward each other; and finally, ejection of the green compact by the lower punch. These operations are shown schematically in Fig. 16·8.

The pressures commonly employed range from 19 to 50 tons/sq in. The

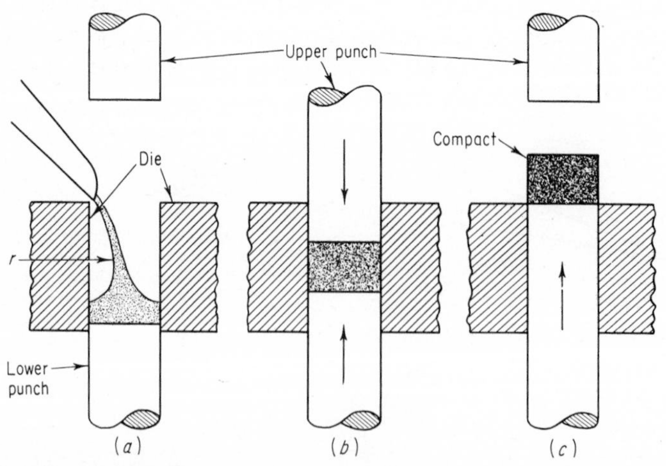

Fig. 16·8 Schematic operations for the die compacting to form powdered parts. (By permission from L. F. Mondolfo and O. Zmeskal, "Engineering Metallurgy," McGraw-Hill Book Company, New York, 1955.)

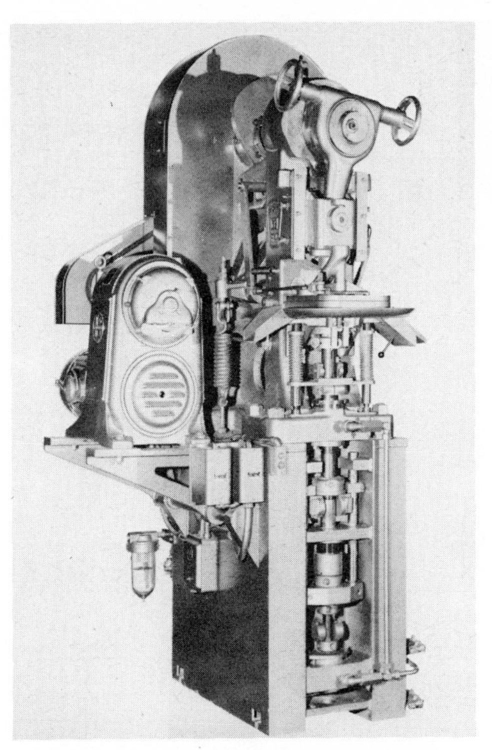

Fig. 16·9 Multiple-action 20-ton powder metal mechanical press. (F. J. Stokes Corporation.)

pressure may be obtained by either mechanical or hydraulic presses. Mechanical presses are available with pressure ratings of 10 to 150 tons and speeds of 6 to 150 strokes/min. The important features of mechanical presses are high-speed production rates, flexibility in design, simplicity and economy in operation, and relatively low investment and maintenance costs. A mechanical press is shown in Fig. 16·9. This is a 20-ton multiple-motion press combining one motion from above with two independent mechanically linked motions from below (Fig. 16·10). The secondary lower motion allows the production of multilevel compacts with simplified tooling and can also be used as a movable core rod to assist in the production of thin-walled pieces.

Hydraulic presses have higher pressure ratings, up to 5,000 tons, but slower stroke speeds, generally less than 20/min. These presses are used for higher-pressure, more complicated powder metal parts.

Dies are usually made of hardened, ground, and lapped tool steels. When the powder to be compacted consists of hard abrasive particles, the die is generally constructed of two parts. The tough outer section supports the

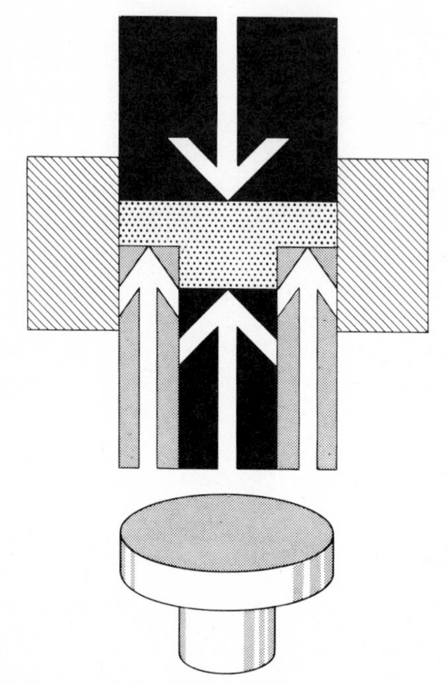

Fig. 16·10 The multiple actions of the press shown in Fig. 16·9. (F. J. Stokes Corporation.)

hardened, polished, wear-resistant insert which is the working surface of the die. These replaceable liners are discarded when worn and reduce the cost of die upkeep.

The punches are made of die steel heat-treated to be slightly softer than the die, since they are usually easier to replace than the die. They must be perfectly aligned and very closely fitted.

In *isostatic compacting,* pressure is applied simultaneously and equally in all directions. The powder is placed in a rubber mold which is immersed in a fluid bath within a pressure vessel, so that the fluid may be placed under high pressures. Since pressure is applied uniformly, it is possible to obtain a very uniform green density and a high degree of uniformity in properties. This method has been used extensively for ceramic materials rather than metals.

High-energy-rate techniques may be either mechanical, pneumatic, or explosive- or spark-discharge methods applied in a closed die. The advantage of these methods is the short time and high pressures that can be attained. It is also possible to use low-grade and very cheap powders, and some parts, due to increased strength of the green compacts, may be used without subsequent sintering. Disadvantages include high punch and die wear, limited tolerances, and high cost.

Forging and *extrusion techniques* have been used only to a limited extent. In either case the powder is "canned" or placed in some kind of metal container. The sealed container is heated or evacuated and then forged or extruded. Figure 16·11 shows the method of extruding wire or rod from powdered metal. After forging or extruding, the container material is removed either mechanically or chemically. Both techniques yield compacts of extremely high density and usually do not require sintering.

In *vibratory compaction,* pressure and vibration are applied simultaneously to a mass of powder in a rigid die. Compared with ordinary die compaction, this method allows the use of much lower pressures to achieve a given level of densification. A major problem is equipment design to apply vibration to practical tooling and presses.

Continuous compaction is applied primarily for simple shapes such as rod, sheet, tube, and plate. Most of the commercial techniques involve flowing loose powder between a set of vertically oriented rolls. The roll gap is adjusted to give a compact of desired properties. In general, the speed of powder rolling is much less than that of conventional rolling operations.

Slip casting is widely used for ceramics but only to a limited extent for metals. The process consists of first preparing a "slip" containing the powder suspended in a liquid vehicle and additives to prevent particle settling. The slip is then placed in a mold made of a fluid-absorbing material (such as plaster of paris) to form the slip casting. After removal from the mold, the slip casting is dried and sintered. This technique is attractive for materials that are relatively incompressible by conventional die compaction, but the process does not lend itself to high production rates because of the long time required for the liquid to be removed through the porosity of the mold. Figure 16·12 shows a molybdenum crucible made by slip-casting molybdenum powder.

In *gravity compaction,* the die is filled with loose powder which is then sintered in the die. The die is usually made of an inert material such as graphite. Since pressure is not used, parts are generally more porous. Commercially this method is used for the production of P/M filters.

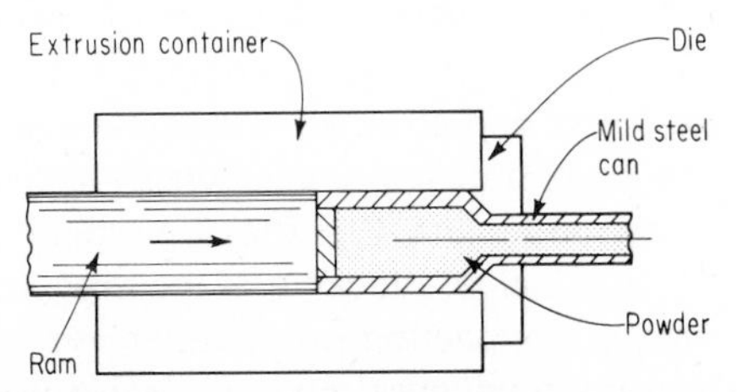

Fig. 16·11 Method of extruding wire and rods from powdered metal.

Fig. 16·12 Molybdenum crucible made by slip-casting molybdenum powder of 2 to 3 μ particle size, followed by air drying and sintering 73 h at 3270°F in hydrogen; unetched, 3×. (Courtesy of T. I. Jones, Los Alamos Scientific Laboratory.)

Continuous pressureless compaction is used to produce porous sheet for electrodes in nickel-cadmium rechargeable batteries. The powder may be applied in the form of a slurry (similar to the slip in slip casting) to be coated on a metal screen or solid metal sheet to produce unusual composites.

The green compact density is a most useful property, since it is an indication of the effectiveness of compaction and also determines the behavior of the material during subsequent sintering. Green density seems to increase with increasing compaction pressure, increasing particle size or apparent density, decreasing hardness and strength of particles, and decreasing compaction speed. The dependence of green density at 30 tons/sq in. on apparent density is shown in Fig. 16·13. The strength of green compacts results chiefly from mechanical interlocking of the irregularities on the particle surfaces. Increasing compaction pressure (Fig. 16·14), or green density, increases green strength because more particle movement and deformation will promote greater interlocking.

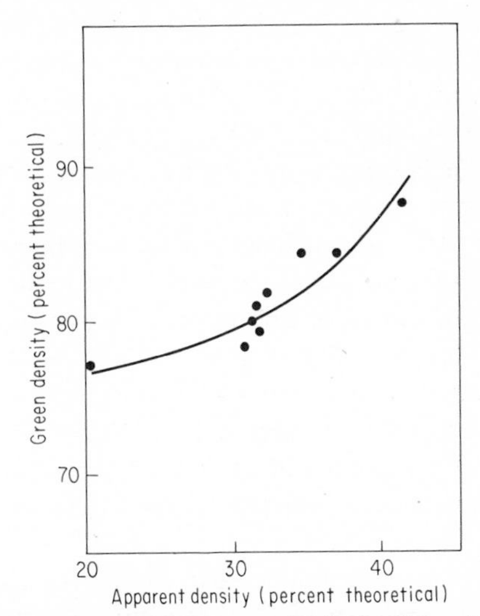

Fig. 16·13 Dependence of green density (at 30 tons/sq in.) on apparent density for different iron powders. (From J. S. Hirschhorn, "Introduction to Powder Metallurgy," American Powder Metallurgy Institute, New York, 1969.)

16·7 Sintering The sintering process is usually carried out at a temperature below the highest melting constituent. In some cases the temperature is high enough to form a liquid constituent, such as in the manufacture of cemented carbides, where sintering is done above the melting point of the binder metal. In other cases, no melting of any constituent takes place.

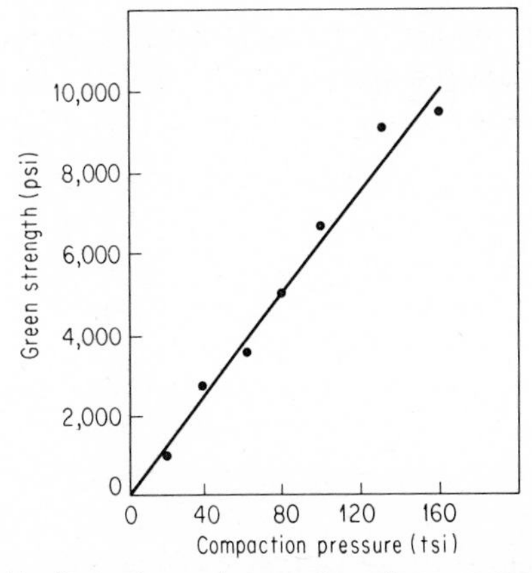

Fig. 16·14 Dependence of green strength on compaction pressure for electrolytic iron powder. (From H. H. Hausner, "Powder Metallurgy," Chemical Publishing Co., New York, 1974.)

Sintering furnaces may be either the electric-resistance type or gas- or oil-fired type. Close control of temperature is necessary to minimize variations in final dimensions. The very uniform and accurate temperature of the electric furnace makes it most suitable for this type of work.

Since bonding between particles is greatly affected by surface films, the formation of undesirable surface films, such as oxides, must be avoided. This may be accomplished by the use of a controlled protective atmosphere. Another function of the atmosphere is to reduce such films if they are present on the powders before mixing and briquetting. The protective atmosphere should not contain any free oxygen and should be neutral or reducing to the metal being sintered. A dry hydrogen atmosphere is used in the sintering of refractory carbides and electrical contacts, but most commercial sintering atmospheres are produced by the partial combustion of various hydrocarbons. Natural gas or propane is often used for this purpose.

Sintering is essentially a process of bonding solid bodies by atomic forces. Sintering forces tend to decrease with increasing temperature, but obstructions to sintering—such as incomplete surface contact, presence of surface films, and lack of plasticity—all decrease more rapidly with increasing temperature. Thus elevated temperatures tend to favor the sintering process. The longer the time of heating or the higher the temperature, the greater will be the bonding between particles and the resulting tensile strength.

Despite a great deal of experimental and theoretical work on the fundamental aspects of sintering, there is still much of the process that is not understood. The sintering process starts with bonding among particles as the material heats up. Bonding involves diffusion of atoms where there is intimate contact between adjacent particles leading to the development of grain boundaries. This stage results in a relatively large increase in strength and hardness, even after short exposures to an elevated temperature. During the next stage, the newly formed bond areas, called "necks,"

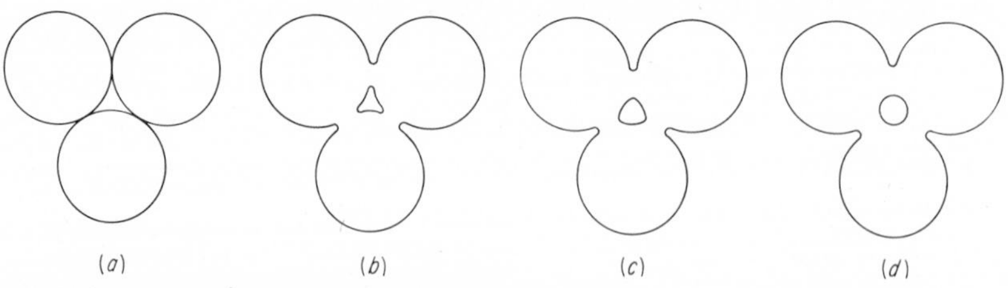

Fig. 16·15 Schematic illustration of a three-sphere sintering model: (*a*) original point contacts, (*b*) neck growth, (*c*) and (*d*) pore rounding. (From J. S. Hirschhorn, "Introduction to Powder Metallurgy," American Powder Metallurgy Institute, New York, 1969.)

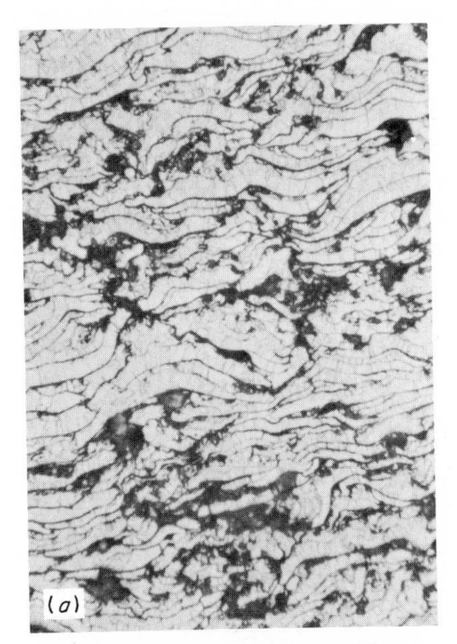

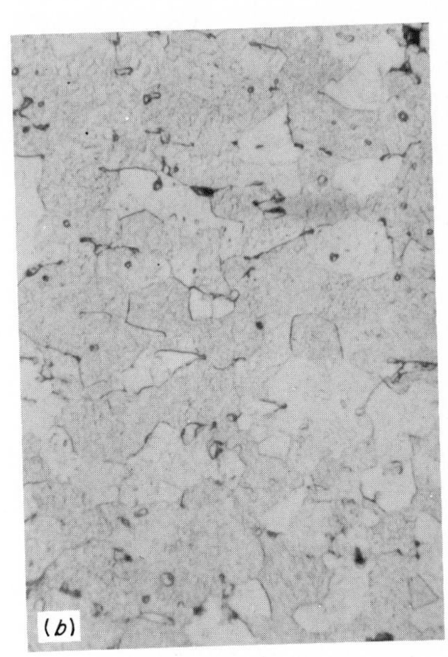

Fig. 16·16 (*a*) Tungsten powder of 10 μ particle size plasma-arc-sprayed on a copper mandrel. Note distorted structure of molten tungsten drops as "splashed" on each other, with appreciable voids (black) between each. Etch 10 g $CuSO_4 \cdot 5H_2O$, 20 ml NH_4OH, 40 ml H_2O, 500×. (*b*) The same material after sintering 24 h at 3460°F in hydrogen. Note growth of equiaxed tungsten grains and substantial reduction in porosity. Same etch; 500×. (Courtesy of T. I. Jones, Los Alamos Scientific Laboratory.)

grow in size, followed by pore rounding. This is shown schematically in terms of a three-sphere model in Fig. 16·15. The last stage is pore shrinkage and eventual elimination. This stage is rarely complete, since the temperatures and times necessary are too impractical. Figure 16·16*a* shows the distorted as-sprayed structure of tungsten powder with considerable voids, whereas Fig. 16·16*b* shows the substantial reduction in porosity after sintering for 24 h at 3460°F.

16·8 Hot Pressing This method consists in applying pressure and temperature simultaneously. Molding and sintering take place at the same time, which results in higher densities and greater productions. The advantages of hot pressing as compared with cold compacting and sintering are a reduction in gas content and shrinkage effects, along with higher strength, hardness, elongation, and density. Hot pressing is used only to a limited extent, primarily for the production of very hard cemented-carbide parts. Figure 16·17 shows the equiaxed single-phase structure and black pores of hot-pressed titanium carbide powder made in a graphite die at 4150°F and a pressure

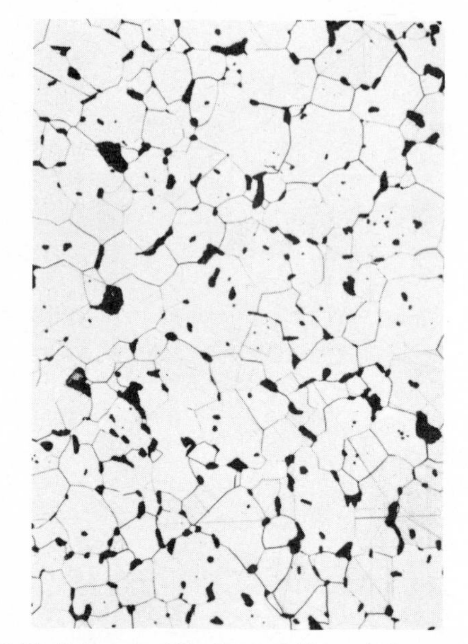

Fig. 16·17 Hot-pressed titanium carbide made from powder in a graphite die at 4150°F and at a pressure of 2,000 psi, showing equiaxed single-phase structure with black voids. Etch one part HF and four parts HNO_3; 100×. (Courtesy of T. I. Jones, Los Alamos Scientific Laboratory.)

of 2,000 psi. The principal disadvantage of this method is the high cost of dies to stand up under pressure at elevated temperatures.

16·9 Supplemental Operations For applications that require higher density or close dimensional tolerances, sintering is followed by a cold-working operation known as *coining* or *repressing*. Coining serves the purpose of condensing the sintered compact. It is possible to obtain considerable plastic deformation within the die, resulting in more complex shapes than may be made directly from powder. The restricted plastic deformation within the die also allows, in many cases, close dimensional tolerances to be held without the necessity of costly machining.

Heating for sintering may be interrupted at some intermediate temperature. This is known as *presintering*. At this point, the compact may have good machinability or be sufficiently soft to allow the use of operations that are not feasible after sintering.

In some cases, resintering after repressing will increase the mechanical properties considerably. As compared with straight-sintered metal, resintering may increase the tensile strength of copper by about 60 percent and of iron by about 30 percent. Despite the increase in strength, resintering may result in large grain size and loss of dimensions due to shrinkage.

This may require another sizing operation, which will increase the cost of the part.

Depending upon the particular application, the sintered compact may be heat-treated to obtain certain desirable properties. The heat treatment may be a stress-relief or annealing treatment. Suitable nonferrous alloy compositions may be age-hardened. Steels may be quench-hardened or case-hardened by carburizing, cyaniding, or nitriding.

Various finishing operations may be carried out to complete the manufacture of powder metal parts. These include machining, shearing, broaching, burnishing, straightening, deburring, grinding, and sandblasting.

Protective surface coatings may be applied by electroplating, metal spraying, and many of the other methods described in Chap. 14.

Of the various joining methods, only brazing has been used extensively for powder metallurgy products.

Impregnation is the means used to fill the internal pores in the sintered compact. This is carried out primarily to improve antifriction properties, as in the self-lubricating bearings. Oil impregnation may be accomplished by dipping the parts in a container of hot oil, or by first drawing the air out of the pores by vacuum and then forcing the oil into the pores under pressure. Waxes and greases may also be used as impregnants. The use of a low-melting metal impregnant, such as tin and lead babbitt alloys in a spongy matrix of nonferrous alloys, tends to improve the bearing properties of the metal. Impregnation with liquid lead has been used to increase the specific gravity of iron-base parts.

16·10 Design of P/M Parts The configurations possible with P/M parts depend largely on the compacting operation. Since metal powders do not flow freely, sharp corners, long, thin sections, and large variations in cross sections that are difficult to fill before pressure is applied can have lower density after pressing. Also, since press action is vertical rather than horizontal, grooves and cutouts perpendicular to the direction of pressing cannot be formed conventionally and must be made by machining. Designs are also limited by press capacity, length of stroke, and platen work area.

Tolerances of P/M parts are influenced by many variables. Alloy, density, sintering temperature, sintering time, and coining or resizing operations all have marked effects on tolerances. P/M parts manufactured to regular tolerances cost considerably less than those manufactured to precision tolerances. Precision-tolerance parts are more demanding in tool accuracy, maintenance, and replacement, so that specified tolerances should not exceed what is absolutely required.

Figure 16·18 shows several design tips for P/M parts. Powder metallurgy is most ideally suited for the production of cylindrical or rectangular shapes, that do not have large variations in cross-sectional dimensions. Surface indentations or projections can easily be formed on the tops or

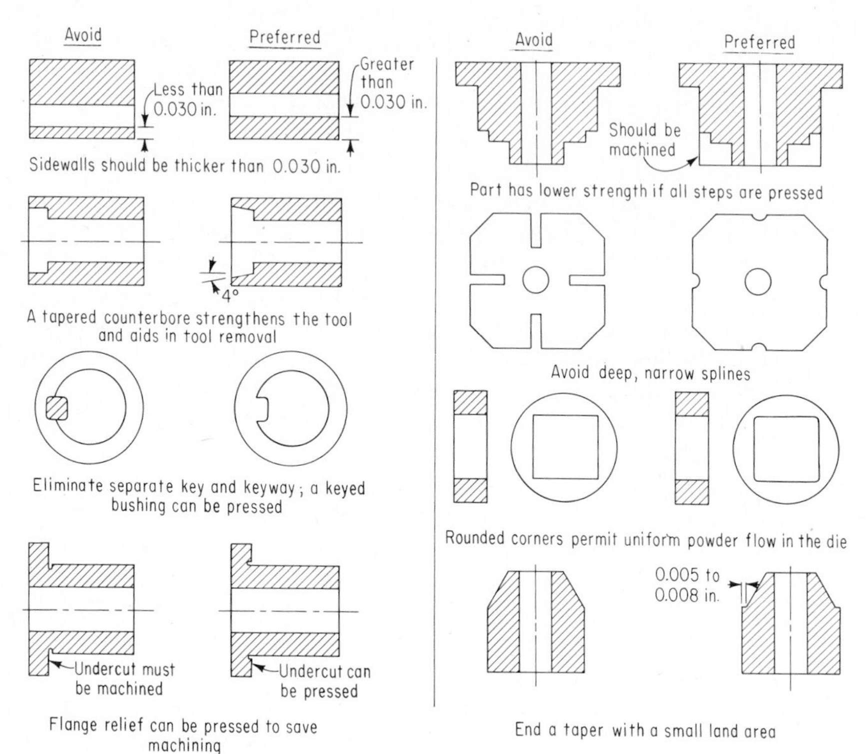

Fig. 16·18 Design tips for P/M parts. (From Machine Design, Metals Reference Issue, The Penton Publishing Co., Cleveland, 1970.)

bottoms of parts, and flanges can be formed at either end. Splines, gear teeth, axial holes, counterbores, straight knurls, slots, and keyways are easily formed. However, undercuts, holes at right angles to the direction of pressing, reverse tapers, reentrant angles, and threads cannot be pressed. In many cases parts requiring features that cannot be pressed directly can still be produced economically by pressing to semifinished shapes and machining the desired details. Thin walls, narrow splines, and sharp corners should be avoided. Side walls of varying thickness should not be less than 0.030 in., and uniform walls should not be less than about 0.060 in. for parts of any appreciable length. Also, abrupt changes in wall thickness and feather edges should be avoided, and corners should be rounded. Small, long holes should not be used, and a minimum flange overhang of about 1/16 in. at heads or shoulders is preferred to no overhang. These precautions will facilitate production as well as minimize tool costs. Sharp edges and long, thin sections on punches or core pins not only are costly to make, but have greater tendency to break and require higher maintenance costs.

16·11 Applications of Powder Metallurgy Powder metallurgy techniques are used for the production of refractory metals, composite metals, porous metals, and metal-nonmetal combinations, and as a more efficient production method for certain parts. Table 16·3 on pages 626–629 shows the composition, properties, and applications of some powdered metals.

The high melting points of the refractory metals make it impossible to use the conventional melting and casting techniques. The use of the powder method for the manufacture of tungsten filaments was described at the beginning of this chapter. This technique offers the only practical method of producing molybdenum, tantalum, and other metals of the same group.

One of the outstanding uses of powder metallurgy is the combination of hard materials in a metallic matrix, which serves as the basis for cemented-carbide products. In the production of cemented-carbide cutting tools, a suitable mixture of the carbides of tungsten, tantalum, and titanium with cobalt as a binder is compacted and presintered. In this condition, the materials can be cut, machined, and ground to the final shape. The compact is then subjected to a high-temperature (about 2750°F) sintering operation during which the liquid cobalt binds the hard carbide particles into a solid piece. Cemented-carbide tools are noted for high compressive strength, red-hardness, and wear resistance. Since they are relatively brittle, they are usually employed as brazed-on tips to a steel tool. They are also used as liners for wear-resistant applications.

Other examples in this classification are diamond-impregnated grinding wheels, drill-core bits, and dressing tools. These consist of diamonds embedded in cemented carbides or more plastic metals and alloys.

Metal-nonmetal combinations have found wide use in the manufacture of friction materials such as clutch facings and brake linings. These materials contain a metallic matrix of copper or bronze for heat conductivity, lead or graphite to form a smoothly engaging lining during operation, and silica or emery for frictional purposes. Iron is sometimes added to increase friction and prevent seizing. Copper-graphite combinations are used as current-collector brushes and in porous bronze and iron bearings.

Composite metals are metal combinations that retain the characteristics of each metal for particular applications. Powder metallurgy is particularly useful for alloys of metals that are not soluble in the liquid state or form monotectics. Casting tends to produce a two-layer alloy unless special techniques are used, whereas homogeneous mixtures are easily produced from powders. The electrical industry makes use of composite metals in the production of heavy-duty contacts which combine the high resistance to abrasion and arcing of a refractory metal such as tungsten with the high conductivity of silver or copper. Similarly, the lubricating qualities of lead

TABLE 16·3 Properties and Applications of Some Powder Metals*

MATERIAL AND SPECIFICATION NUMBERS	CONDITION	DENSITY		TENSILE STRENGTH, PSI	COMPRESSIVE YIELD STRENGTH, PSI	TYPICAL HARDNESS		ELASTIC MODULUS (TENSION), PSI	ELONGATION, %	APPLICATIONS
		G PER Cu Cm	PCT. THEO.			SURFACE	CORE			
Unalloyed iron, 99.9% Fe MPIF 35: F-0000 N through T SAE: 850, 851, 852, 853, 855; J471C ASTM: B310, B439		6.0	78	20,000	15,000	Rf 38	Rf 38	12,700,000	3.0	Structural (lightly loaded gears), magnetic (motor pole pieces), projectile driving bands, self-lubricating bearings—may be steam treated for wear resistance
	As sintered	6.5	83	30,000	25,000	Rf 55	Rf 55	15,500,000	10	
		7.5	96	40,000	35,000	Rf 95	Rf 95	25,500,000	20	
	Carbonitrided 0.040 in. case 0.80 C	7.5	96	100,000 to 140,000	35,000 (core)	Rc 55 to 60	Rf 95	25,500,000	0.5 case 20 core	Structural, wear-resisting (small levers and cams)
Iron-carbon sintered steel MPIF 35: F-0001-P through F-0008-P SAE: 852 ASTM: B310	As sintered 0.8 C	6.2	79	31,000	24,000	Rb 15	Rb 15	13,300,000	0.5	Structural (moderately loaded gears, levers and cams)
	As sintered 0.8C	6.8	87	44,000	30,000	Rb 65	Rb 65	18,300,000	1.0	
	Heat-treated (large parts)	6.8	87	64,000	–	Rc 35 to 45	Rb 65	18,300,000	0.5	Structural (moderately loaded gears, levers, and cams requiring wear resistance)
Low-carbon iron-copper alloy, 0.25 C max, 7.0 to 25.0 Cu, 0 to 1.0 Mn, 0 to 1.0 Zn, bal Fe MPIF 35: FC0300 N through T, FX2000-T (infiltrated) SAE: 870, 872 ASTM: B303A (infiltrated), B222	As sintered 10.5 Cu	5.8	74	29,500	20,000	Rh 80	Rh 80	11,600,000	0.5	Structural (pump housings, support brackets, lightly loaded gears), self-lubricating bearings—may be steam-treated for wear resistance
	As sintered 10.5 Cu	6.2	80	34,000	21,000	Rh 90	Rh 90	13,800,000	1.0	
	Copper-infiltrated, 25 Cu up to 4 Ni	7.5	96	60,000	55,000	Rb 60	Rb 60	25,500,000	10	High-impact structural parts, socket adaptors—may be carbonitrided for wear resistance

TABLE 16·3 (Continued)

MATERIAL AND SPECIFICATION NUMBERS	CONDITION	DENSITY		TENSILE STRENGTH, PSI	COMPRESSIVE YIELD STRENGTH, PSI	TYPICAL HARDNESS		ELASTIC MODULUS (TENSION), PSI	ELONGATION, %	APPLICATIONS
		G PER Cu Cm	PCT. THEO.			SURFACE	CORE			
Iron-copper-carbon alloy, 3.0 to 25.0 Cu, 0.3 to 1.0 C, bal Fe plus up to 0.40 Mn or 1.5 Zn, or 1.5 Co. MPIF 35: FC0308, FC0508, FX2010-T (infiltrated) SAE: 864-B, 866-A, 867-B, 872 ASTM: B303, B, C (infiltrated), B426	As sintered 3.0 Cu, 0.8 C	6.3	81	45,000	—	Rb 50	Rb 50	14,400,000	1.5	Structural (medium loads including gears, cams, support brackets, levers, and ratchets)—can be heat-treated to a high degree of wear resistance
	As sintered 5.0 Cu, 0.8 C	6.2	80	60,000	54,000	—	Rb 70	13,800,000	0.8	
	Sintered and heat-treated 5.0 Cu, 0.8 C	6.2	80	65,000	80,000	Rc 30	Rb 100	13,800,000	0.5	
	Infiltrated 20.0 Cu, 0.8 C	7.4	95	80,000	70,000	Rb 90	Rb 90	24,600,000	0.7	
	Infiltrated, heat treated 20.0 Cu, 0.8 C	7.4	95	100,000	80,000	Rc 42	Rb 100	24,600,000	0.5	
Low-alloy steel, chemically AISI 4630, 0.30 C (combined), 0.25 Mo, 1.85 Ni (prealloy type)	Coined, resintered, heat-treated	7.2	92	110,000	—	Rc 45	Rc 20	22,500,000	0.7	Structural, wear-resisting (oil pump gears)
Low-alloy steel, chemically AISI 4650, 0.45 C (combined), 0.50 Mo, 2.0 Ni (diffusional alloy type) MPIF 35: FM-0205-S and T ASTM: B484	As sintered	7.2	92	60,000	—	Rb 68	—	—	3.0	Structural (couplings)
	Heat-treated	7.2	92	145,000	—	Rc 44	—	22,500,000	1.1	Structural, wear-resisting (oil pump gears and heavily loaded support brackets)
	Coined resintered, heat-treated	7.6	97	200,000	—	Rc 56	—	—	1.5	Structural, wear- and impact-resisting (oil pump gears to 3,000 psi and heavily loaded transmission gears)

*Metal Progress Data Sheet, American Society for Metals, Metals Park, Ohio, April 1971.

TABLE 16·3 **(Continued)**

MATERIAL AND SPECIFICATION NUMBERS	CONDITION	DENSITY		TENSILE STRENGTH, PSI	COMPRESSIVE YIELD STRENGTH, PSI	TYPICAL HARDNESS		ELASTIC MODULUS (TENSION), PSI	ELONGATION, %	APPLICATIONS
		G PER Cu Cm	PCT. THEO.			SURFACE	CORE			
Nickel alloy steel, 4.0 to 7.0 Ni, 0.0 to 2.0 Cu, 0.0 to 0.80 C Nearest equivalent: AISI 2517 (diffusional alloy type) MPIF 35: FN-0408-S and T, FN-0705-S and T ASTM: B484	Heat-treated 4.0 Ni, 1.0 Cu, 0.70 C	6.8	87	100,000	65,000	Rc 35	Rc 20	18,300,000	0.5	Structural, wear-resisting, high-stress (planetary differential and transmission gears up to 6 hp)
	Heat-treated 7.0 Ni, 2.0 Cu, 0.5 C	7.4	95	175,000	125,000	Rc 45	Rc 30	24,600,000	2.5	Structural, high-stress, impact resisting (shifter lugs and clutches)
	Carbonitrided 0.040 in case 4.0 Ni, 1.0 Cu, 0.0 C (core)	6.9	89	90,000	—	Rc 45	Rb 65	19,600,000	3.5 (core)	Structural, wear-resisting, high-stress, and requiring welded assembly (welded assembly of pinion and sprocket)
Sintered austenitic stainless steel, type 316L MPIF 35: SS-316L, P through R ASTM: B525	Prealloyed, as sintered (dis. NH_3)	6.2	80	50,000	20,000	Rb 42	Rb 42	—	4.0	Structural, corrosion-resisting, nonmagnetic (small gears, levers, cams, and other parts for exposure to salt water and specific industrial acids)
	Prealloyed, as sintered (dis. NH_3)	6.6	86	60,000	45,000	Rb 60	Rb 60	—	10	
	Prealloyed, as sintered (dry hydrogen)	6.6	86	45,000	—	—	—	—	12.5	
Sintered martensitic stainless steel, type 410L MPIF 35: SS-410N through P	Prealloyed, as sintered (dis. NH_3)	6.2	80	65,000	35,000	Rb 90	Rb 90	—	3.5	Structural, corrosion-resisting, nonmagnetic (small gears, levers, cams, and other parts for exposure to salt water and specific industrial acids where applications require heat treating for wear resistance)
	Heat-treated	6.2	80	70,000	45,000	Rc 30	—	—	0.8	
	Heat-treated	6.8	88	100,000	80,000	Rc 30	—	—	0.8	

TABLE 16·3 (Continued)

MATERIAL AND SPECIFICATION NUMBERS	CONDITION	DENSITY		TENSILE STRENGTH, PSI	COMPRESSIVE YIELD STRENGTH, PSI	TYPICAL HARDNESS		ELASTIC MODULUS (TENSION), PSI	ELONGATION, %	APPLICATIONS
		G PER Cu Cm	PCT. THEO.			SURFACE	CORE			
Copper-nickel-zinc alloy (nickel silver), 62.0 Cu, 18.0 Zn, 18.0 Ni, 2.0 Sn MPIF 35: BZN-1818 U, W ASTM: B458	Single-pressed as sintered	7.3	82	28,000	20,000	Rh 74	Rh 74	—	7	Structural, nonacid corrosion-resisting (gears, levers, chuck jaws, parts for marine exposure)
		7.9	89	35,000	40,000	Rh 84	Rh 84	—	15	
Sintered brass, 77 to 80 Cu, 1.0 to 2.0 Pb, bal Zn MPIF 35: BZ0218-T and U SAE: 890, 891 ASTM: B282	Prealloyed, as sintered	7.6	87.5	29,600	—	Rh 61	Rh 61	—	10	Mechanical components, atmospheric-corrosion-resisting (builders hardware, mechanism housings, lock parts, pump housings)
	Coined and resintered	8.0	92	31,000	—	—	Re 49	—	24	
Sintered bronze, 86 to 90 Cu, 9.5 to 10.5 Sn, 1.0 max Fe, 1.75 max C MPIF 35: BT0010 N through R ASTM: B255	Blended copper-tin powders, single-pressed as sintered	6.8	76	16,000	—	—	Rh 50	—	2.5	Structural, atmospheric-corrosion-resisting, bearings (journal bearings, thrust bearings, load-carrying bearing plates)
		7.2	81	30,000 (8,000 graphited)	—	—	Rh 65	—	3.0	
Alcoa Type 201 AB—4.4 Cu, 0.8 Si, 0.4 Mg, bal Al	As sintered	2.59	—	28,300	—	Rh 57	Rh 57	—	2.5	Lightly loaded gears and ratchets, camera parts, circuit board, heat sink, cabinet hardware
	Sintered and coined	2.63	—	32,200	—	—	—	—	3.0	
	Sintered, coined, heat-treated	2.63	—	49,400	—	Re 87	Re 87	—	2.0	

are combined with the load-carrying ability of copper in the copper-lead bearings.

Controlled porosity of powder metal parts has led to the production of porous bearings, gears, and filters. Self-lubricating bearings are made of bronze powder with controlled porosity after sintering. The pores are subsequently filled with oil. In operation, the load on the bearing and the increased heat set up by the moving part within the bearing force the oil out of the pores to provide automatic and uniform lubrication. Self-lubricating bearings are used extensively in the automotive industry and in washing machines, refrigerators, electric clocks, and many other types of equipment. Porous-metal gears are used in oil pumps for their lubricating properties. Metal filters, used in the chemical industry, are similar to the ceramic type but have higher mechanical strength and resistance to both mechanical and thermal shock.

Finally, in many applications the use of powder metallurgy techniques results in more economical manufacture of the part. Where load conditions are not severe, small gears, cams, levers, sprockets, and other parts of iron, steel, brass, or bronze may be molded from powders to reduce greatly or completely eliminate expensive and time-consuming machining and other forming operations. For example, the gears of a gear-type oil pump must have accurately formed involute teeth or the pump will be inefficient. The machined gear is cut from a cast blank by a skilled machinist with about 64 percent of the metal lost in chips. On the other hand, any semiskilled man can fill a hopper and operate a press which can turn out hundreds of these gears with dimensional accuracy and with less than 1 percent of the metal as waste.

Small Alnico permanent magnets containing aluminum, nickel, cobalt, and iron may be made from powders or by casting. The cast alloy is difficult to machine, and finishing to dimensions must be done by tedious grinding. These magnets may be molded of powders directly to size and shape and their dimensions held to acceptable tolerances during sintering. In addition, a finer grain size and greater mechanical strength are obtained in the sintered magnets.

Some typical parts produced by powder metallurgy techniques are shown in Fig. 16·19.

QUESTIONS

16·1 Why is particle-size distribution important in the packing of powders?

16·2 Discuss the importance of particle shape on the properties of sintered compacts?

16·3 List the three common methods of powder production and discuss their influences on the properties of the final product.

Fig. 16·19 Typical parts produced from powder metals. (F. J. Stokes Corporation.)

16·4 Contrast mechanical and hydraulic compacting presses with regard to advantages, disadvantages, and applications.
16·5 Why is sintering carried out in a controlled-atmosphere furnace?
16·6 Why do elevated temperatures tend to favor the sintering process although sintering forces tend to decrease with increasing temperature?
16·7 What are the advantages and disadvantages of hot pressing as compared with cold compacting and sintering?
16·8 Give three specific applications of powder metallurgy parts. Describe how these parts may be manufactured by other methods, and give the advantages of the powder metallurgy method.
16·9 Why is pore size important in the manufacture of self-lubricating bearings? How may pore size be controlled?
16·10 Why are canned powders often evacuated in powder extrusion and forging techniques?

REFERENCES

American Society for Metals: "Metals Handbook," 1948 ed., Metals Park, Ohio.
———: "Powder Metallurgy in Nuclear Engineering," Metals Park, Ohio, 1958.
American Society for Testing Materials: "Testing Metal Powders and Metal Powder Products," Philadelphia, 1953.
Clark, F. H.: "Advanced Techniques in Powder Metallurgy," Rowman and Littlefield, New York, 1963.

Goetzel, C. G.: "Treatise on Powder Metallurgy," vols. 1 to 3, Interscience Publishers, Inc., New York, 1949–1952.

Hausner, H. H.: "Powder Metallurgy," Chemical Publishing Company, Inc., New York, 1947.

——— (ed.): "Modern Developments in Powder Metallurgy," Plenum Press, New York, 1966.

———, K. H. Roll, and P. K. Johnson (eds.): "Iron Powder Metallurgy," Plenum Press, New York, 1968.

Hirschhorn, J. S.: "Introduction to Powder Metallurgy," American Powder Metallurgy Institute, New York, 1969.

The Iron and Steel Institute: "Symposium on Powder Metallurgy," Special Report 38, London, 1947.

Jones, W. D.: "Fundamental Principles of Powder Metallurgy," Edward Arnold Publishers Ltd., London, 1960.

Leszynski, W. J. (ed.): "Powder Metallurgy," Interscience Publishers, Inc., New York, 1961.

Metal Powder Industries Federation: "Powder Metallurgy Equipment Manual," New York, 1968.

Poster, A. R. (ed.): "Handbook of Metal Powders," Van Nostrand Reinhold Company, New York, 1966.

Schwartzkopf, P.: "Powder Metallurgy," The Macmillan Company, New York, 1947.

17 FAILURE ANALYSIS

17·1 Introduction When one considers the many millions of metallic parts that are fabricated and placed in service, it is not unusual that some will fail prematurely. Simply from a statistical viewpoint it is not reasonable, with present engineering practice, to expect no failures. However, even though the number of failures of a particular component may be small, they are important because they may affect the manufacturer's reputation for reliability. In some cases, particularly when the failure results in personal injury or death, it will lead to expensive lawsuits. It is not unusual for automotive manufacturers under prodding and publicity from consumer watchdogs to recall millions of cars to correct a design or heat-treating defect even though the actual number of failures was very small.

The purpose of this chapter is to briefly explain the basic causes for metal failure and to illustrate some of the failures by case histories. Most of the illustrations in this chapter were taken from two excellent books on metal failure—"How Components Fail" by Donald J. Wulpi (American Society for Metals, 1966) and "Why Metals Fail" by R. D. Barer and B. F. Peters (Gordon and Breach Science Publishers, 1970).

17·2 Procedure In any failure analysis it is important to get as much information as possible from the failed part itself along with an investigation of the conditions at the time of failure. Some of the questions to be asked are:

1. How long was the part in service?
2. What was the nature of the stresses at the time of failure?
3. Was the part subjected to an overload?
4. Was the part properly installed?
5. Was it subjected to service abuse?
6. Were there any changes in the environment?
7. Was the part properly maintained?

A study of the fractured surface should answer the following questions:

1. Was the fracture ductile, brittle, or a combination of the two?
2. Did failure start at or below the surface?

3. Did the failure start at one point, or did it originate at several points?
4. Did the crack start recently or had it been growing for a long time?

It should be apparent that no suitable solution may be prescribed unless information regarding how the part performed and failed is available.

Laboratory and field testing permit the evaluation of the effects of material, design, and fabrication variables on performance of the part under controlled conditions. Failure analysis, on the other hand, is concerned with parts returned from service and thus gives results of actual operating conditions. By combining the information from tests with the results from analysis, a clear picture of the causes of failure can be obtained. Rarely are failures assigned to a single cause. Usually they result from the combined effects of two or more factors that are deterimental to the life of the part or structure.

When studying a failure, care must be used to avoid destroying important evidence. Detailed studies usually require documentation of the service history (time, temperature, loading, environment, etc.) along with chemical analysis, photomicrographs, and the like. Further study of the sequence of events leading up to the failure, plus knowledge of the location, markings, and condition of all adjacent parts at the time of failure, is necessary to confirm analysis. There always exists the possibility of unforeseen loading, unreported collision, or unanticipated vibration that may have contributed to premature failure.

The procedure for investigating a failure covers four areas as follows:

1 Initial observations. A detailed visual study of the actual component that failed should be made as soon after the failure as possible. Record all details by many photographs for later review. Interpretation must be made of deformation markings, fracture appearance, deterioration, contaminants, and other factors.

2 Background data. Collect all available data concerned with specifications and drawings, component design, fabrication, repairs, maintenance, and service use.

3 Laboratory studies. Verify that the chemical composition of the material is within specified limits. Check dimensions and properties of the component. Supplementary tests may be made as needed—for example, hardness and determination of microstructure to check heat treatment, nondestructive tests to check for processing defects or existing cracks, composition of corrosion products, a free-bend test to check ductility, etc. Very often, examination of a fracture surface with a low-power binocular microscope can reveal the type and cause of failure.

4 Synthesis of failure. Study all the facts and evidence, both positive and negative, and answers to the typical questions given earlier. This, combined with theoretical analysis, should indicate a solution to the problem of failure.

Extensive studies of carburized and hardened gears for heavy-duty trucks, machine tools, mining machines, diesel engines, etc. showed that 38 percent of the failures resulted from surface problems (pitting, spalling, crushing, and scoring), 24 percent from bending fatigue, 15 percent from impact, and 23 percent from miscellaneous causes. From a detailed analysis of failures by steel companies, auto manufacturers, and electrical equipment manufacturers, nearly 50 percent of all failures can be attributed to faulty design, the rest being distributed between production and service problems.

17·3 Modes of Fracture As was pointed out earlier, proper analysis of the fracture often yields much information on the contributing factors and helps to identify the type of failure. Ductile and brittle fractures were discussed in Sec. 3·7, but it will be useful to review the fracture modes.

Ductile fractures are the result of shear forces that produce plastic deformation (slip or twinning) along certain crystallographic planes, whereas brittle fractures are due to tensile forces that produce cleavage. In most fractures, both types are present in varying degrees. Identification of the basic mechanism often determines the type of load that initiated fracture. By the same token, a knowledge of load application can help in determining whether a particular failure was ductile (shear) or brittle (cleavage) in nature.

Figure 17·1 shows two bolts pulled to fracture in tension to illustrate ductile and brittle behavior. The one on the right was soft (Rockwell C 15); it failed in a ductile manner by shear, resulting in extensive plastic deformation. The bolt on the left was relatively hard (Rockwell C 57) and failed in a brittle fashion, with no apparent plastic flow. Shear fractures caused by a single load are dull gray and fibrous, with edges which are usually deformed plastically. Small cavities are initially formed by slip. They join together and eventually grow to form a crack under continued loading. The crack spreads with the aid of stress concentration at the tip of the crack, generally moving perpendicular to the tensile force and eventually forming a "shear lip" at the surface (see Fig. 3·14).

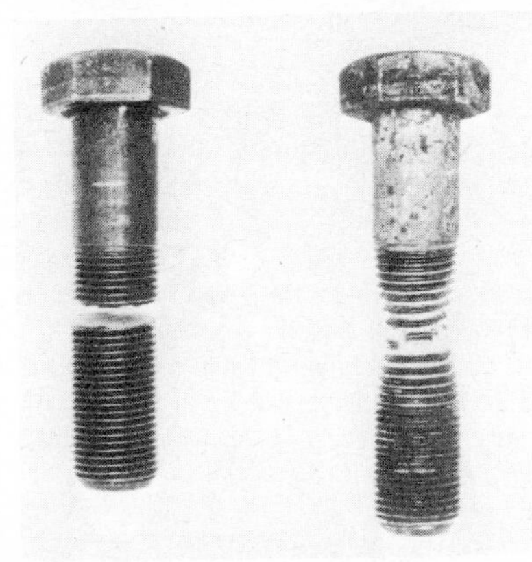

Fig. 17·1 Two bolts intentionally pulled to failure in tension to demonstrate brittle and ductile behavior. The brittle bolt, left, was hard, Rockwell C 57; the ductile bolt was soft, Rockwell C 15. (Courtesy of D. J. Wulpi, International Harvester Company.)

Brittle (cleavage) fractures generally appear bright and crystalline. Each crystal tends to fracture on a single cleavage plane, and this plane varies only slightly from one crystal to the next in the aggregate. For this reason it follows that a cleavage fracture in a polycrystalline specimen will generally sparkle in the light when rotated in the hand. Surfaces of brittle fractures sometimes have distinctive appearances. From the origin of fracture, a characteristic "chevron" or "herringbone" pattern is formed which points to the fracture origin (Fig. 17·2). Since (as pointed out in Chap. 3) slip and cleavage occur on a different set of crystallographic planes, the nature of individual fractures can often be determined by metallographic examination in the laboratory.

Fractures are rarely either cleavage or shear. The variable stresses that usually exist in a structure, the changing of stress patterns during the progress of fracture, or the microscopic differences in orientation of grains produce fractures composed of both shear and cleavage areas. Consideration of combinations of fracture modes can often give information regarding the nature of the fracture. Figure 17·3 shows three samples of the same material as they reacted to notched-impact tests at different temperatures. On the left, the fracture surface is mainly dull gray and fibrous; the edges are curved, indicating plastic deformation, so that the fracture mode is mostly shear. In the center the mode was mixed shear and cleavage, since the surface is both shiny and dull with some evidence of plastic deformation at the edges. The fracture on the right is by cleavage alone.

Fig. 17·2 "Chevron" pattern points to the origin of the brittle fracture (arrow) in this specimen. A fatigue fracture is also apparent in the upper right-hand corner. (Courtesy of D. J. Wulpi, International Harvester Company.)

Fig. 17·3 Combinations of fracture modes are shown by fracture surfaces of three impact test specimens which were broken at different temperatures. On the left, fracture is mostly shear; in the center, combined shear and cleavage; and on the right, cleavage. (Courtesy of D. J. Wulpi, International Harvester Company.)

The entire surface is bright and the edges are straight, showing no evidence of plastic deformation.

17·4 Stress and Strength The solution to failure problems resulting from overstressing of parts depends on the determination of two factors: the stress on the part and the strength required to support that stress. Depending on the type of load and the geometry of the part, there may be simple axial stress or a complex system of multiaxial stresses. The total stress can include internal residual stresses from fabrication or heat treatment as well as stresses from external loads.

The basic stresses in a part under external load were discussed in Sec. 3·2. The most important are the normal stresses (those perpendicular to the plane of the cross section) and shear stresses (those in the plane of the cross section). Normal stresses tend to produce separation, while shear stresses tend to produce plastic flow. It was pointed out that the maximum shear stress occurs at a 45° angle to the initiating tensile stress. When a part is under load, yielding will occur when the shear stress is greater than the shear yield strength; ductile or shear fractures develop when the shear strength is overcome by the shear stress; and brittle fractures occur when the tensile (cohesive) strength is exceeded by the tensile stress.

Consideration should be given to the significant stresses when investigating a particular mode of failure. For example, if failure is due to a fatigue fracture at a gear tooth root, the significant stress would be the repeated bending stress at that location. Contact stress acting on the gear face would not be significant in this case. For a pitting or wear-type failure of the gear tooth, the reverse would be true.

17·5 Types of Loading In many cases, the type of load is a contributing factor to failure. There are essentially five types of loads illustrated in Table 17·1—

TABLE 17·1 Types of Loading*

LOAD		STRESS DISTRIBUTION	EXAMPLES
Axial	 Tension	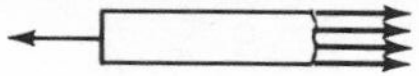	tensile test bars cables
	 Compression	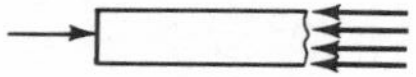	short columns
Bending	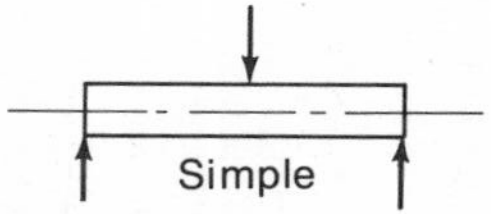 Simple	compression 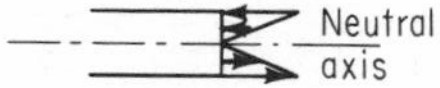 Neutral axis tension	Beams
	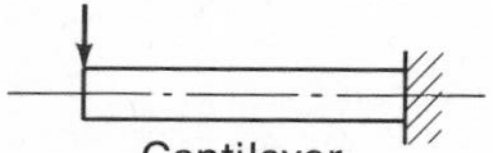 Cantilever	tension  Neutral axis compression	Root of gear teeth
Torsional	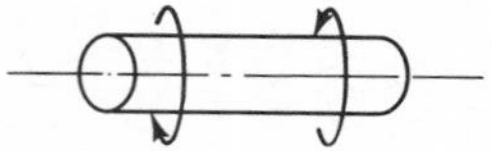	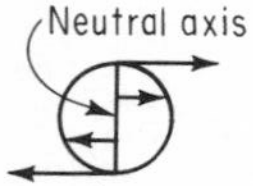 Neutral axis shear	shafts coil springs
Direct shear	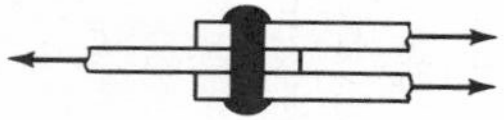	 shear	rivets bolts
Contact	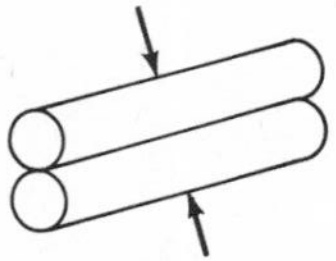	varies with depth and force direction	roller bearings gear teeth

*From D. J. Wulpi, "How Components Fail," American Society for Metals, Metals Park, Ohio, 1966.

axial, bending, torsional, direct shear, and contact. In axial loading, the load is applied coincident with the center line of the part and the stress is uniform across the cross section, as in tensile test bars and supporting cables. Bending loads are produced by couples of forces coincident with the center line. Across the cross section, the stress varies from maximum at the outermost fibers to zero at the neutral axis, as in beams and the root of gear teeth. Torsional loading involves the application of a force couple in a plane normal to the center line. The shear stress varies from a maximum at the surface to zero at the neutral axis. Examples of parts subjected to torsion are shafts and coil springs. Direct shear loads act on closely spaced parallel planes and tend to move part of the material with respect to the rest, similar to a cutting action, as in rivets and bolts. The shear stress distribution is uniform across the cross section. Contact loads are compressive loads perpendicular to two surfaces, combined with sliding forces between the surfaces. The stress distribution varies with depth and force direction. Examples of contact loading are roller bearings and gear teeth.

All these types of loads induce normal and shear stresses which must be balanced by the material's cohesive and shear strengths. It is possible for overload fractures to occur when the applied load reaches excessive values. The failed main bearing of an air compressor was found during an overhaul. The cage was broken, one ball was split in two, several other balls were heavily scored, and both inner and outer races showed signs of

Fig. 17·4 Both halves of inner ball race showing damage on one side only due to misalignment. (From R. D. Barer and B. F. Peters, "Why Metals Fail," Gordon and Breach Science Publishers, New York, 1970.)

being badly overheated. Figure 17·4 shows both halves of the inner race. The failed surface is to one side of the central track and extends about half way around the race. Subsequent examination indicated that the bearing was subjected to misalignment in its casting. This misalignment caused overloading and overheating of the bearing along with an end thrust which resulted in the failure on one side of the central track. Figure 17·5 shows a failed phosphor-bronze spring with numerous cracks on the inside surface outlined by fluorescent penetrant. The maximum tensile stress for phosphor bronze is about 50,000 psi. Calculations indicated that the spring was subjected to a stress of 76,900 psi in service; therefore, the failure was the result of overload. Figure 17·6 shows a gear with two broken teeth. Tooth A failed in service, while tooth B was fractured in the laboratory in an overload test. The similarity in the appearance of both fracture surfaces indicates that a high overload probably caused tooth A to break.

17·6 Fatigue Fractures Fatigue failures are the most common types of fracture in machines and probably constitute about 90 percent of all fractures. Such fractures develop after a large number of load applications, generally at a stress level below the yield strength of the material. Fatigue testing was discussed briefly in Sec. 1·34.

Fatigue stresses develop in three principal ways, as shown in Fig. 17·7. The upper diagram illustrates the stress pattern under reversed loading, typical of a rotating shaft under a bending load, where tension, compression, or shear stresses of the same magnitude alternate. The middle diagram shows the stress variation under unidirectional loading where the load varies from zero to a maximum either in tension, compression, or shear

Fig. 17·5 Failed phosphor-bronze hatch spring showing numerous cracks on the inside surface outlined by dye penetrant. (From R. D. Barer and B. F. Peters, "Why Metals Fail," Gordon and Breach Science Publishers, New York, 1970.)

Fig. 17·6 In this gear, tooth A broke in service, and tooth B was fractured by a single blow in the laboratory. The similarity in fracture surfaces indicates that tooth A was also broken by a sharply applied load. (Courtesy of D. J. Wulpi, International Harvester Company.)

typical of a punch or gear teeth. The diagram on page 642 shows the condition under unidirectional loading with a preload. In this case the stress varies from minimum to maximum without reaching zero, as in cylinder-head bolts and connecting-rod bolts.

Since a fatigue fracture is progressive, developing over a long period of time, the fracture surface usually shows characteristic "beach" or "clamshell" markings. Figure 17·8 shows a bending fatigue fracture of a large threaded axle; an arrow indicates the origin of fracture. Generally, failure originates at the surface of the part where the shear stresses first exceed the shear strength. In this case the fracture started at the indicated discontinuity on the surface and, as shown by the beach marks, proceeded nearly across the section before final separation.

Fatigue fractures initiate in shear by a mechanism involving slip and work hardening, eventually forming microscopic discontinuities which develop into cracks (see Fig. 3·15). Once a crack is formed, its rate of growth depends upon the stress magnitude, stress gradient, endurance limit of the material, notch sensitivity, and the presence or absence of structural flaws and inclusions. Figure 17·9 shows that if, after the crack has formed, the stress is reduced to below the value necessary to initiate it, the crack may not propagate any further. This is probably caused by an increase in strength due to strain hardening at the crack tip. If the applied load is large enough, the crack will advance perpendicular to the maximum tensile stress. Variation in the cyclic load causes small ridges or beach marks to develop on the fracture surface. They indicate the position of

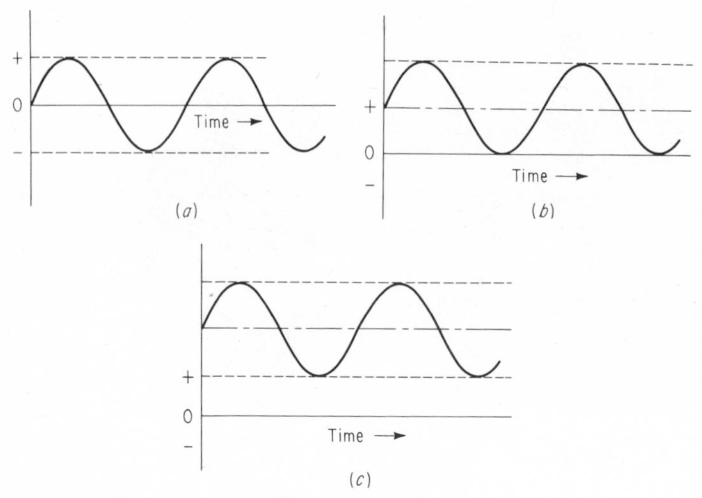

Fig. 17·7 The basic fatigue stress conditions. (*a*) Reversed stress, (*b*) unidirectional stress, (*c*) unidirectional stress with a preload.

root of the advancing crack at a given time. As the section gradually weakens, the crack grows faster, and the clam-shell markings get further apart, larger, and more distinct. Therefore, when these markings are present, they provide a means of locating the origin of fracture accurately.

Fig. 17·8 The presence of "beach marks" usually indicates that failure was caused by fatigue. Here fracture began at a discontinuity (arrow). (Courtesy of D. J. Wulpi, International Harvester Company.)

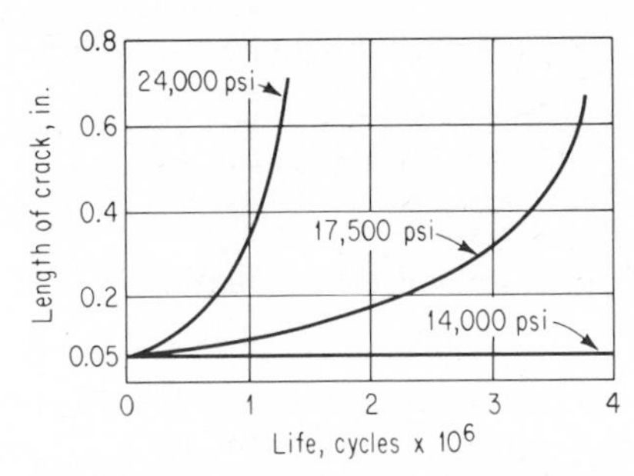

Fig. 17·9 Relationship of part life to crack length. (From Machine Design, The Penton Publishing Co., Cleveland, November 13, 1969.)

When fatigue originates in several locations of a filleted shaft in rotating bending, the progressing cracks run into each other, usually resulting in "ratchet marks" as shown in Fig. 17·10.

17·7 Effect of Stress Raisers In machine and structural members, the highest stresses occur most often at fillets, holes, and similar geometrical irregularities that concentrate and increase surface stress. These are called stress raisers.

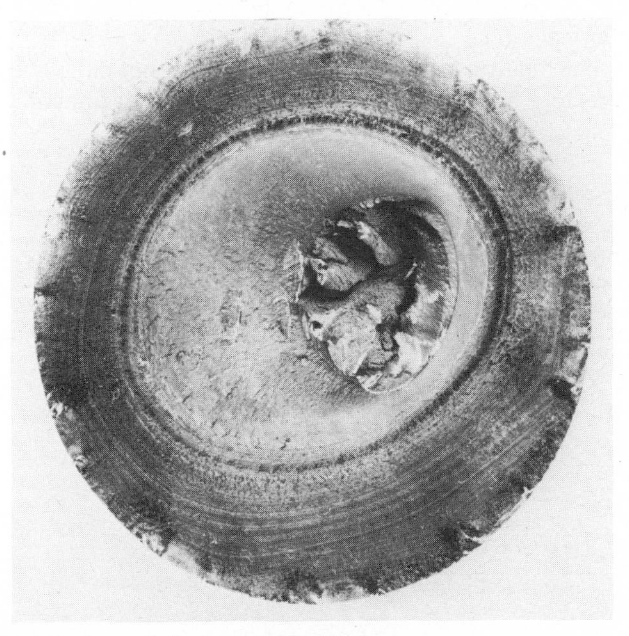

Fig. 17·10 "Ratchet marks" around edges of fatigue failures indicate that fracture began at several points. (Courtesy of D. J. Wulpi, International Harvester Company.)

The majority of stress raisers fall into one of the following broad groups:

1. Those caused by changes in the geometry of a part, such as holes, keyways, threads, steps, or changes in diameter in shafts and bolt heads, etc.
2. Surface discontinuities, such as nicks, notches, machining marks, pitting, corrosion, etc.
3. Defects inherent in the material, such as nonmetallic inclusions, minute cracks, voids, etc.

Primary stress raisers are usually of the first group, although those of the second and third groups may play secondary, related roles. Even ordinary tool marks act as notches which tend to concentrate stresses, particularly at the root of the notch. They are especially damaging when they occur at section discontinuities such as fillets.

Under a static load, the highly stressed metal yields plastically at a notch root or hole edge, thereby passing the high stresses to other sections until fracture occurs. However, under fatigue, or repeated loads, where the stress is below the elastic limit, yielding is more localized, and a crack may start before the stress pattern changes to relieve the stress concentration. Figure 17·11 shows the effect of severe notches in steel specimens. The importance of stress raisers can be shown in a single example. If a small hole is drilled in a wide strip of elastic material and the strip is subjected to axial tension, the stress at the edge of the hole reaches a maximum of 3 times the normal stress.

In rotating machine parts, the final rupture area is not directly opposite the start of fracture but is slightly offset by the effect of rotation. This is

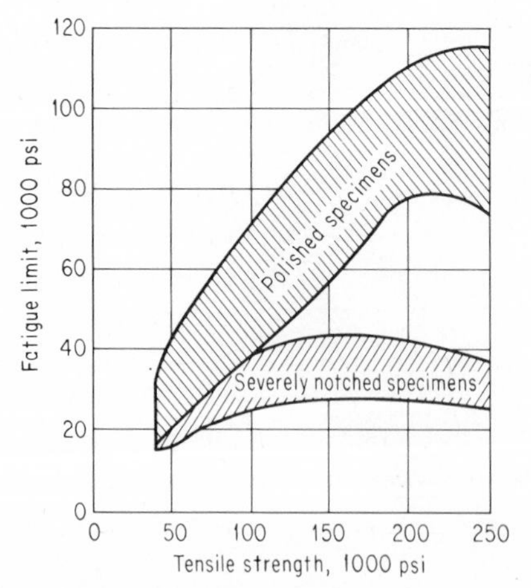

Fig. 17·11 In specimens with severe notches, fatigue limits level out at 25,000 to 45,000 psi and drop off slightly as tensile strengths rise above 180,000 psi.

Fig. 17·12 The offsetting effect of rotation on the zone of final fracture reveals the direction that the shaft rotated during operation. (Courtesy of D. J. Wulpi, International Harvester Company.)

illustrated by Fig. 17·12, which shows the fracture surface of a broken shaft. Fracture originated due to stress concentration at a corner of the keyway, and the beach marks swing around counterclockwise toward the final rupture because of the clockwise rotation.

Sharp corners are always stress raisers and should be avoided when possible. The valve spindle on the left in Fig. 17·13 shows a crack at the sharp corner, whereas the one on the right has a generous radius in the corner and no sign of failure. Figure 17·14 shows a crack which originated at the sharp corner of a valve bonnet. Cracking could be avoided by providing a generous radius at the inner corner. Figure 17·15*a* shows a steel stud which failed by fatigue at the first thread. The fractured end of the stud, Fig. 17·15*b*, shows the characteristic fatigue markings, and the arrow indicates the spot where fatigue started. Examination of the thread root, Fig. 17·15*c*, indicates very rough machining by the presence of torn metal on the thread face, and the radius at the thread root is very small.

In general, hard materials show a higher notch sensitivity than soft materials, and this property will affect the appearance of a fracture under cyclic loading. In a high notch-sensitive material, the crack tends to grow

Fig. 17·13 Valve spindle at left is cracked at the corner (arrow). The other spindle has a generous radius at this location. (From R. D. Barer and B. F. Peters, "Why Metals Fail," Gordon and Breach Science Publishers, New York, 1970.)

more rapidly along the highly stressed surface than toward the center. Therefore, the beach marks curve away from the origin of fracture, as illustrated in Fig. 17·16*a*. In a less notch-sensitive material, such as annealed steel, the crack moves more rapidly toward the center than along the surface and will produce concave beach marks around the origin of fracture, as illustrated in Fig. 17·16*b*.

Fig. 17·14 Sectional valve bonnet showing crack which originated at the sharply machined corner. (From R. D. Barer and B. F. Peters, "Why Metals Fail," Gordon and Breach Science Publishers, New York, 1970.)

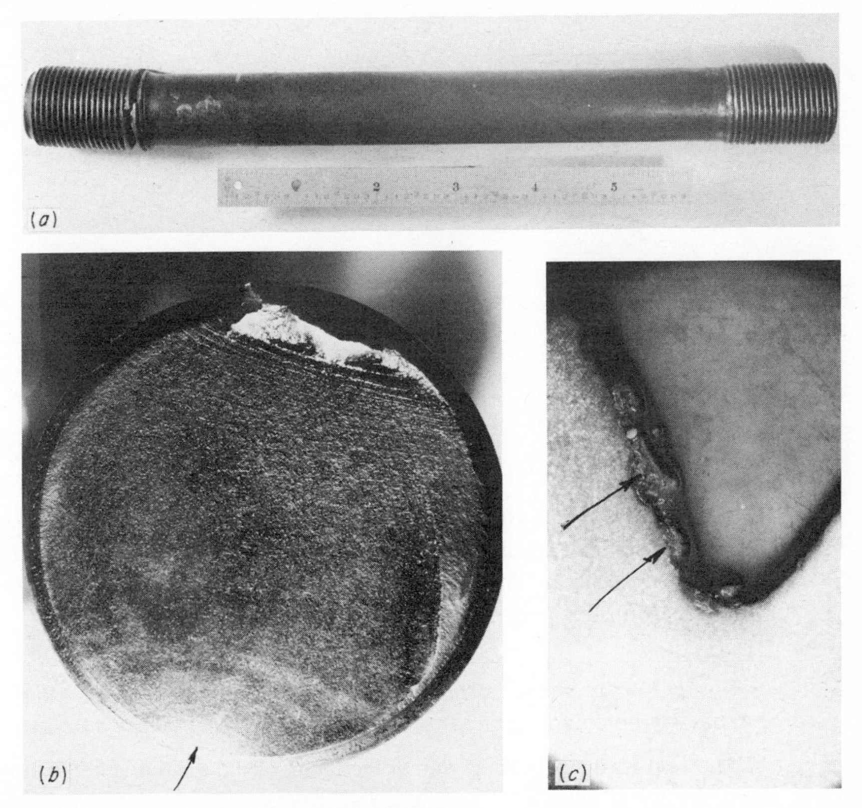

Fig. 17·15 (*a*) Steel stud which failed as a result of fatigue at the first thread. (*b*) Fractured end of the stud, showing fatigue markings. Fatigue initiated at arrow. (*c*) Root of thread of stud. Note the torn metal, from machining, at arrows; 200×. (From R. D. Barer and B. F. Peters, "Why Metals Fail," Gordon and Breach Science Publishers, New York, 1970.)

Internal corners in longitudinal grooves, such as splines and keyways, act as stress raisers. Fatigue cracks that develop follow the paths of maximum stress. In spline shafts, multiple cracks form (Fig. 17·17*a*) and grow together, producing a "starry" fracture appearance (Fig. 17·18). A typical crack path from the inner corner of a keyway is shown in Fig. 17·17*b*. In some cases, particularly when the key is loosely fitted, nearly all the torque is transmitted through the key. Fracture starts at the bottom corner of the keyway and progresses in shear parallel to the surface, resulting in a "peeling" fracture. Sometimes the peeling action goes entirely around the shaft, forming a separated shell (Fig. 17·19).

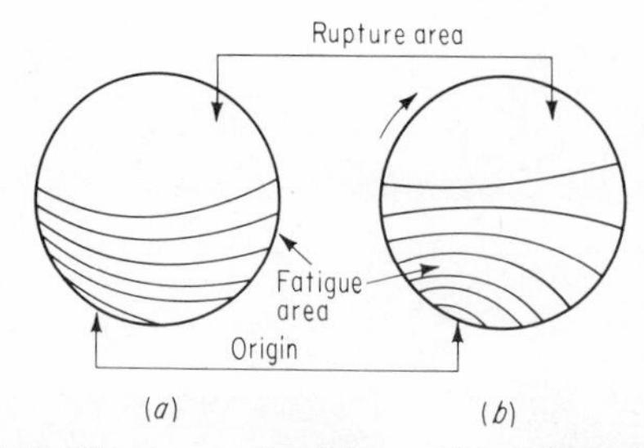

Fig. 17·16 The degree of notch sensitivity affects the manner in which beach marks develop. In notch-sensitive alloys, such as high-strength steel, these marks curve away from the source of failure (left). The reverse is true in notch-insensitive material (right). (From D. J. Wulpi, "How Components Fail," American Society for Metals, Metals Park, Ohio, 1966.)

17·8 Effect of Strength Reducers In addition to stress raisers, certain metallurgical conditions may act to lower the strength of the metal and lead to fracture. Such conditions include overheating, grinding burns, poor heat treating, and poor casting practice.

Figure 17·20*a* shows a ruptured tube from a marine boiler. Visual examination indicated that:

1 The split was at the center line of the tube—facing the fire, and therefore at a high heat input zone.
2 The thin edges of the split and the "stretcher" marks on the inner surface at the break are indicative of plastic flow of the metal.
3 Only a small amount of internal deposit was noted in the immediate area of the split.

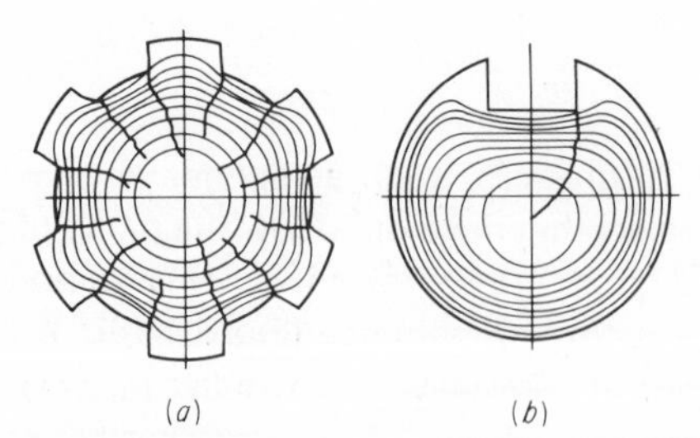

Fig. 17·17 Fatigue cracks tend to follow paths of maximum stress concentration. Circular lines indicate stresses. In splines and keyways, the stresses concentrate at inner corners. (*a*) Spline, (*b*) keyway. (From D. J. Wulpi, "How Components Fail," American Society for Metals, Metals Park, Ohio, 1966.)

Fig. 17·18 When spline shafts fail in fatigue (from reversed torsional loading), they generally develop "starry" fracture surfaces. (Courtesy of D. J. Wulpi, International Harvester Company.)

Microscopic examination of the material some distance from the rupture showed the normal microstructure of ferrite and pearlite typical of a 1025 steel (Fig. 17·20*b*). A sample taken from the lip of the fracture showed martensite (Fig. 17·20*c*). This structure could only arise by heating the steel to 1,700 to 1,800°F and then water quenching. Apparently there was a temporary cause of poor circulation in the tube, thus reducing heat transfer through the wall. This caused the metal of the tube wall to reach 1,700 to 1,800°F in a few minutes. At this temperature the metal is weak but ductile, so that a gentle bulging of the wall took place until it became too thin for the internal pressure. At rupture, the flow of water from the tube effectively quenched the overheated steel to give the martensitic structure.

Figure 17·21*a* shows the inner surface of a failed copper-nickel-iron cooling pipe with pits at *a* and a crack at *b*. The normal fine-grained microstructure some distance from the failure is shown in Fig. 17·21*b*. The

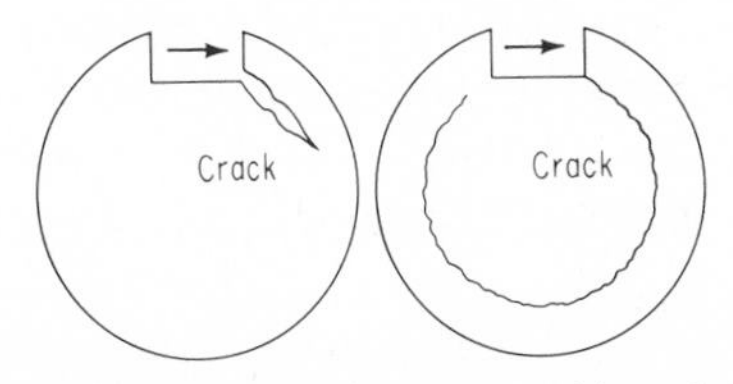

Fig. 17·19 Fatigue cracks in keyways of loosely fitting members may peel around the shaft under the surface. (From D. J. Wulpi, "How Components Fail," American Society for Metals, Metals Park, Ohio, 1966.)

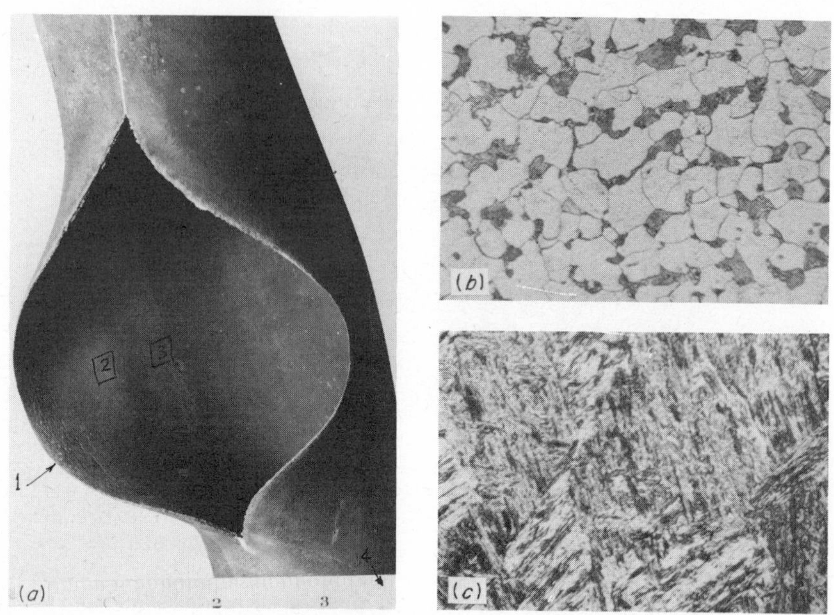

Fig. 17·20 (*a*) A burst boiler tube. Note the thin edges and the "stretcher" marks on the inside surface. (*b*) Normal microstructure of a 0.25 percent carbon steel tube showing a mixture of pearlite (dark) and ferrite (white), 100×. (*c*) Microstructure at the lip of the burst specimens. This shows only martensite and would only result from a drastic quench of the tube; 200×. (From R. D. Barer and B. F. Peters, "Why Metals Fail," Gordon and Breach Science Publishers, New York, 1970.)

microstructure at a section near the pitted area is shown in Fig. 17·21*c*. Notice the relatively large grain size and the presence of cracks. The large grain size was due to overheating in that area, probably when the pipe was bent to shape. The pits were confined to this zone of overheated and enlarged grains, since the corrosion resistance of this alloy is considerably reduced if it is heated to temperatures much above 1000°F.

Hardened steel can be severely damaged by improper grinding. Excessive heat causes damaging residual stresses, but more significantly surface areas may transform to hard, brittle martensite which may result in a pattern of fine hairlike cracks easily revealed by magnetic particle inspection (see Fig. 1·33*a*).

Faulty heat treatment may often be the cause of failure. The possibility of cracking during heat treatment of shallow-hardened and through-hardened steels was discussed in Sec. 8·31. Figure 17·22*a* shows a band-saw blade which cracked after being in use 30 min. Microscopic examination showed that in hardening the teeth of the saw, the manufacturer also extended the hardened zone into the blade. The extent of hardening is shown as the white area through the tooth and below in Fig. 17·22*b*. Notice the crack which started near the root of a tooth in the brittle, hardened area. Figure 17·22*c* shows a properly heat-treated blade from another

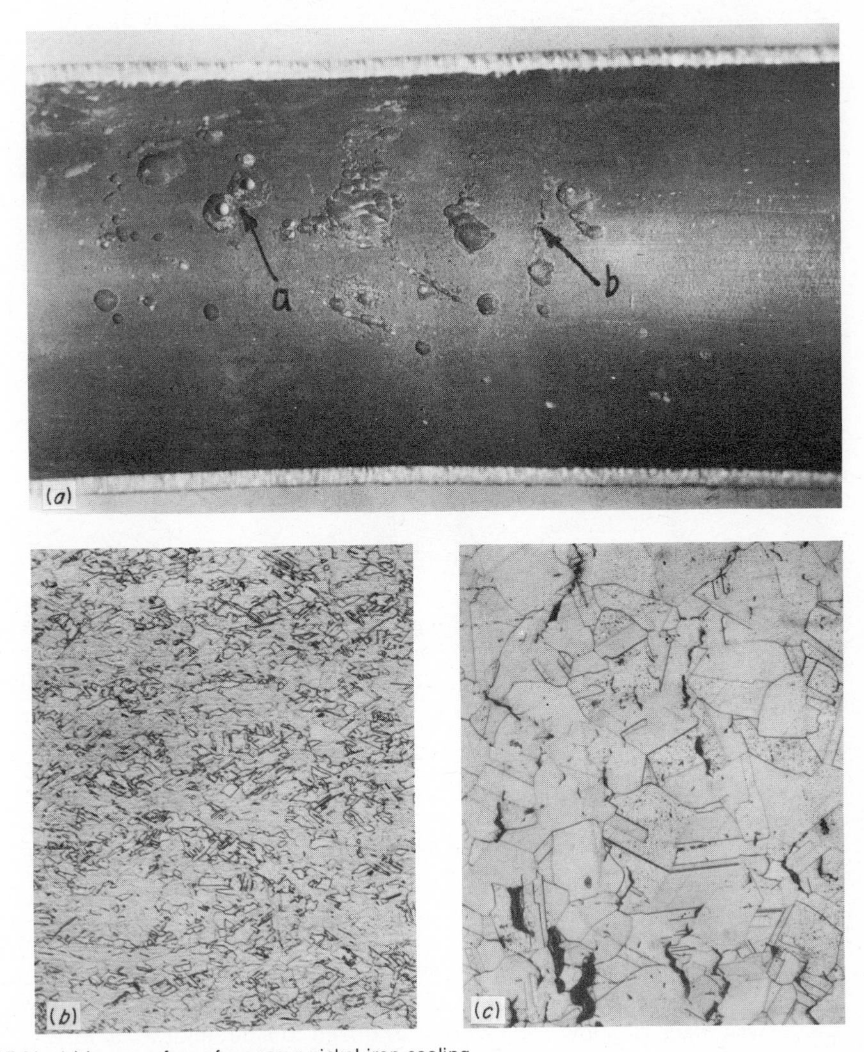

Fig. 17·21 (*a*) Inner surface of a copper-nickel-iron cooling pipe showing pits at *a* and a crack at *b*. (*b*) Normal fine-grained microstructure at some distance from the heated area, 150×. (*c*) Section cut through the pipe near the pitted area. The large grain size confirms exposure to high temperature. Black lines are cracks; 150×. (From R. D. Barer and B. F. Peters, "Why Metals Fail," Gordon and Breach Science Publishers, New York, 1970.)

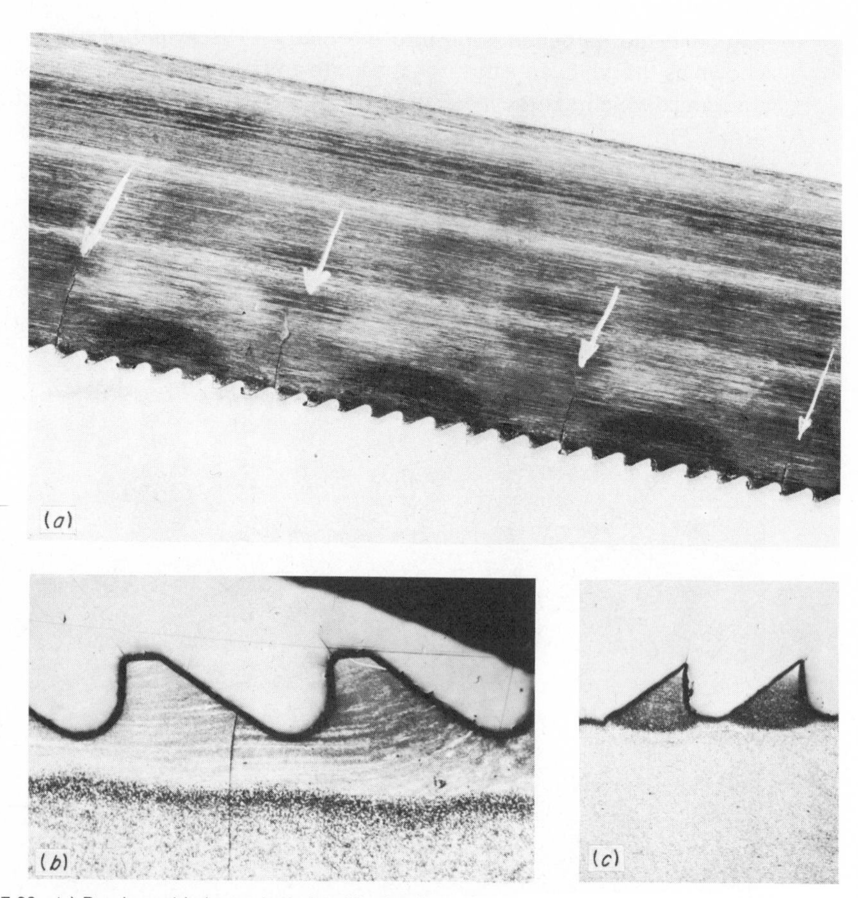

Fig. 17·22 (*a*) Band-saw blade cracked after 30 min of use. (*b*) Cracked blade showing that teeth were hardened well past the teeth roots. Notice the vertical crack in the hardened zone; 12×. (*c*) Properly hardened blade—only the teeth have been hardened; 12×. (From R. D. Barer and B. F. Peters, "Why Metals Fail," Gordon and Breach Science Publishers, New York, 1970.)

supplier in which the root of the thread is not hardened, thereby giving the blade greater toughness and flexibility.

Figure 17·23*a* shows a fractured carburized steel shell support part. It was made of a low-alloy steel, carburized to a depth of 0.030 in. and hardened to Rockwell C 52. Microscopic examination of the case, Fig. 17·23*b*, shows a white carbide network around tempered martensite. The hard, brittle carbide network is a frequent cause of failure in carburized parts. It could have been avoided by use of a proper diffusion cycle during carburizing to reduce the surface carbon content, or by proper heat treatment after carburizing to break up the carbide network (see Sec. 8·26).

Faulty foundry practice may often cause failure. Figure 17·24*a* shows cracks at the bolthole of a cast, leaded gun-metal waveguide flange. Microscopic examination, Fig. 17·24*b*, revealed the presence of excessive coarse shrinkage (black areas), which is due to inadequate foundry practice.

A double-ended piston from a condensate pump was leaking water from a crack in the wall of one of the pistons. Radiographic investigation showed the presence of a very thin wall at the crack location. This was confirmed by sectioning (Fig. 17·25). During casting the core shifted, which accounted for the unusually thin wall.

17·9 Effect of Residual Stresses Residual stresses are stresses that exist in a part independent of any external force. Nearly every manufacturing operation will result in residual stresses in varying degrees (see Table 17·2). This table presents the tendencies to be expected for surface residual stresses

Fig. 17·23 (*a*) Fractured carburized steel shell support part, 2×. (*b*) Microstructure of the carburized area showing white, brittle carbide network around tempered martensite; 400×. (From R. D. Barer and B. F. Peters, "Why Metals Fail," Gordon and Breach Science Publishers, New York, 1970.)

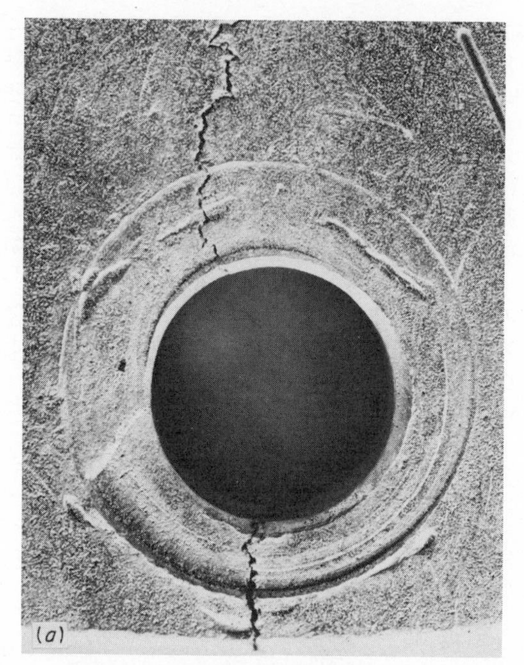

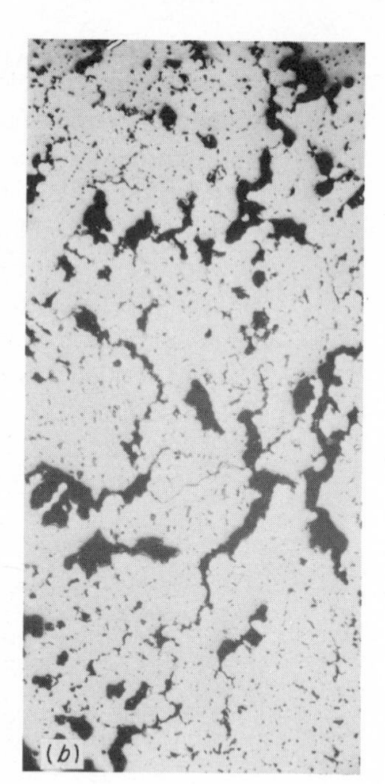

Fig. 17·24 (*a*) Cracks at bolthole of a waveguide flange. (*b*) Microstructure near cracks showing gross shrinkage voids (black areas), 23×. (From R. D. Barer and B. F. Peters, "Why Metals Fail," Gordon and Breach Science Publishers, New York, 1970.)

only. The origin of residual stresses due to heat treatment was discussed in Sec. 8·31. In general, residual stresses are beneficial when they are opposite to the applied load. Since cracks are propagated only by tensile stresses, surface residual compressive stress would be most desirable. Heat-treating processes that usually produce compressive residual stress are the shallow-hardening ones such as nitriding, flame hardening, induction hardening, and usually carburizing. Welding usually produces residual tensile stresses due to the contraction of the weld metal during cooling from the weld temperature.

The effect of residual stresses varies with material hardness and with the presence of stress raisers. In general, soft materials with no stress raisers experience almost complete fading of residual stress while operating under reversing loads, while notched parts made from very hard materials retain almost all of their residual stress (Fig. 17·26).

17·10 Other Variables Aside from stress raisers, strength reducers, and residual stresses, other variables may have to be considered when investigating

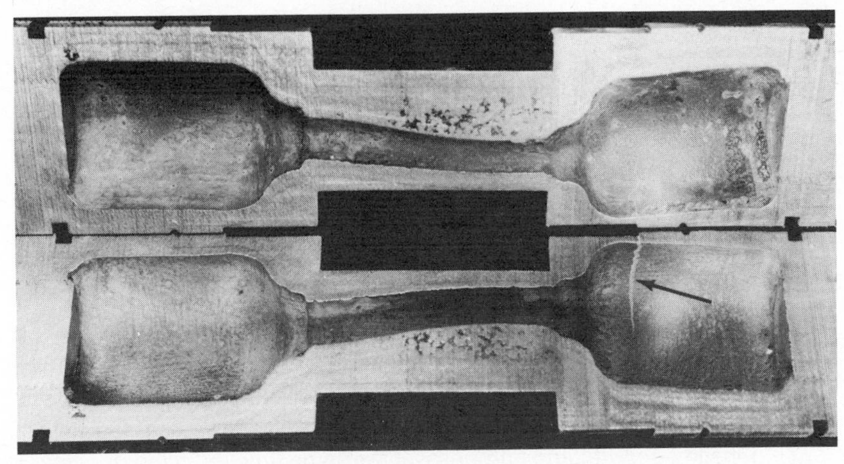

Fig. 17·25 Sectioned piston rod showing off-center core and crack (arrow) outlined by magnetic particle inspection. (From R. D. Barer and B. F. Peters, "Why Metals Fail," Gordon and Breach Science Publishers, New York, 1970.)

failures. The behavior of metals at low and high temperatures was discussed in Chapter 13. Heating a metal above room temperature tends to lower its yield strength, tensile strength, and hardness with a corresponding increase in ductility. Failure resulting from elevated temperature may be related to excessive creep, stress rupture, or thermal fatigue. Unfortunately, components serving at high temperatures often deteriorate by

TABLE 17·2 Residual Stresses Caused by Manufacturing Operations*

TENSILE STRESSES	COMPRESSIVE STRESSES	EITHER
Welding	Nitriding	Carburizing
Grinding	Shot peening	Rolling
Straightening	Flame and induction hardening	Casting
	Heat and quenching	Abrasive metal cutting (tensile stresses most common)
	Single-phase materials	Nonabrasive metal cutting
		Heat and quenching materials that undergo phase transformation (tensile stresses most common)

*From "Machine Design," The Penton Publishing Co., Cleveland, Oct. 16, 1969.

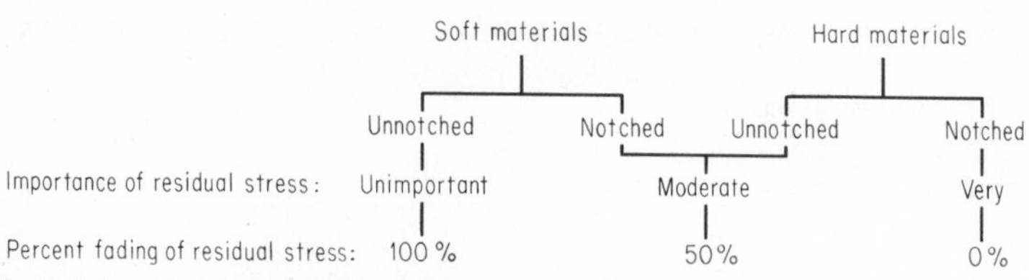

Fig. 17·26 Relationship of material hardness and residual stress fading. (From Machine Design, The Penton Publishing Co., Cleveland, Oct. 16, 1969.)

some form of hot corrosion or instability. Performance depends more upon resistance to this type of attack than upon the material's basic properties.

Generally, decreasing the temperature of a metal raises the yield strength, tensile strength, and hardness, at the same time reducing ductility. While at ordinary temperatures a material may show a ductile, shear fracture, below the "transition temperature" the basic mode of fracture changes to brittle, low-energy cleavage.

In some cases, the rate of loading may determine whether a part will fail and the type of failure. Under extremely low rates of loading, ductile metals show a large drop in strength, but stronger steels show little change. Under rapid loading rates, the apparent strength is somewhat higher. When the rate of loading approaches what may be considered impact loads, brittle fracture may be induced in a normally ductile material because of the lack of time for flow to occur.

The composition and microstructure of a material will also influence the type of fracture. Generally, in heat-treated steel, the best combination of mechanical properties is obtained with a tempered-martensite structure (see Fig. 8·20). As with the rate of loading, the effect of composition and microstructure is determined primarily by their influence on the transition temperature. Above the transition temperature, the fractures are usually ductile shear, whereas below it, brittle cleavage fractures predominate. Transition temperatures appear to be lowered by the addition of nickel and molybdenum and raised by carbon, manganese, and chromium.

17·11 Bending Fractures Bending is one of the common causes of fracture in machine and structural parts. Failure may be from a single application of a load greater than the overall strength of a part or can be due to a reversing load that results in a bending fatigue fracture.

The surface appearance of fatigue fractures was described in Sec. 17·6. In many cases, the pattern on the fracture surface indicates the forces that caused bending fracture. Typical fracture appearances of bending-fatigue failures are shown in Fig. 17·27, with final fracture zones as cross-hatched areas. Ordinarily, bending fatigue cracks are perpendicular to the tensile

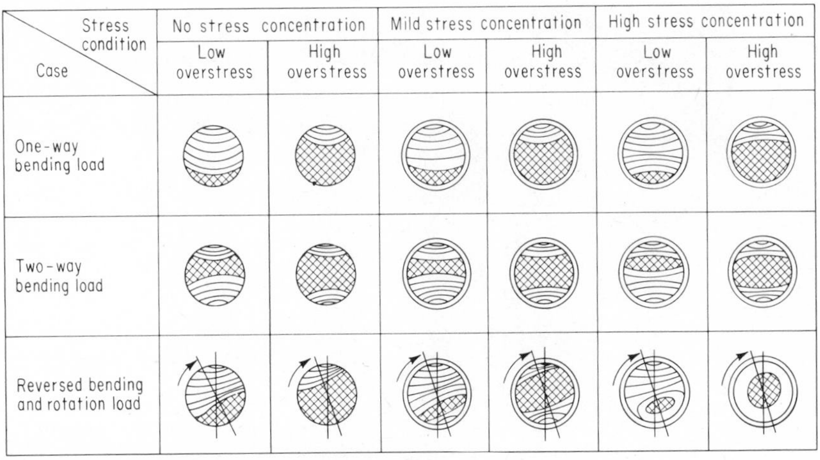

Stress condition / Case	No stress concentration		Mild stress concentration		High stress concentration	
	Low overstress	High overstress	Low overstress	High overstress	Low overstress	High overstress
One-way bending load						
Two-way bending load						
Reversed bending and rotation load						

Fig. 17·27 Fracture appearances of bending-fatigue failures. Final fracture zones are shown as crosshatched areas. (From Machine Design, The Penton Publishing Co., Cleveland, Nov. 27, 1969.)

stress which occurs on one side of the bend and originate at the surface where the maximum applied stress is located. Figure 17·28 shows how cracks are oriented in cylindrical and filleted shafts overloaded due to a one-way bending load. As discussed earlier, stress raisers such as sharp fillets and tool marks tend to initiate cracks because the stress is highest there.

Reversed bending fatigue without rotation will usually cause cracks on

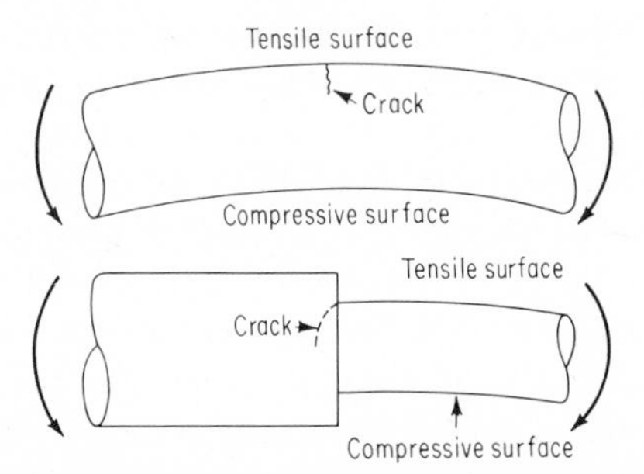

Fig. 17·28 Bending fractures usually develop on surfaces and normal to the stress direction. Sharp fillets concentrate bending stresses, causing cracks to develop more rapidly. Arrows indicate bending direction. (From D. J. Wulpi, "How Components Fail," American Society for Metals, Metals Park, Ohio, 1966.)

Fig. 17·29 This 1050 shaft, 1.94 in. in diameter, broke in reversed bending fatigue. A sharp fillet concentrated the bending stresses, causing a crack to develop on opposite sides with final fracture in the middle. (Courtesy of D. J. Wulpi, International Harvester Company.)

opposite sides of the shaft since each side undergoes alternate tensile and compressive stresses. Figure 17·29 shows a 1050 shaft which broke in reversed bending fatigue. A sharp fillet concentrated the bending stresses, causing failure to start at opposite sides of the shaft with final fracture in the middle.

Under reversed bending with a rotational load, the final fracture area tends to be offset from the initial crack due to the effect of rotation (see Fig. 17·12).

While most bending fatigue cracks originate at the surface, it is possible under certain conditions for the crack to originate below the surface. This condition may arise due to the presence of a microcrack or other metallurgical discontinuity, usually arising from fabrication processes that cause the strength in its immediate vicinity to be considerably lower than the surface strength. Figure 17-30 shows an induction-hardened axle shaft of 1041 steel which failed in rotating bending fatigue. Fracture *A* started below the hardened zone and moved nearly half-way across the cross sec-

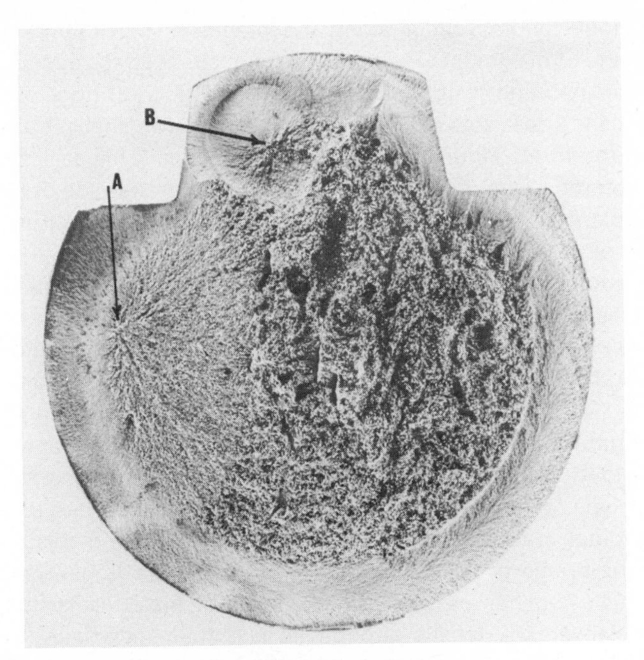

Fig. 17·30 Some rotating bending-fatigue failures begin beneath surfaces. In this induction-hardened axle shaft, fracture started at *A* and moved into the cross section, meeting another subsurface crack that started at *B*, resulting in final fracture. (Courtesy of D. J. Wulpi, International Harvester Company.)

tion before meeting another small subsurface fracture (*B*). Chevron marks point to both origins, indicating brittle, sudden final failure.

17·12 Torsional Failures Torsional failures are most common in shafts, including crankshafts, torsion bars, and axles. The appearance of a torsion-fatigue fracture is quite different from that caused by bending fatigue.

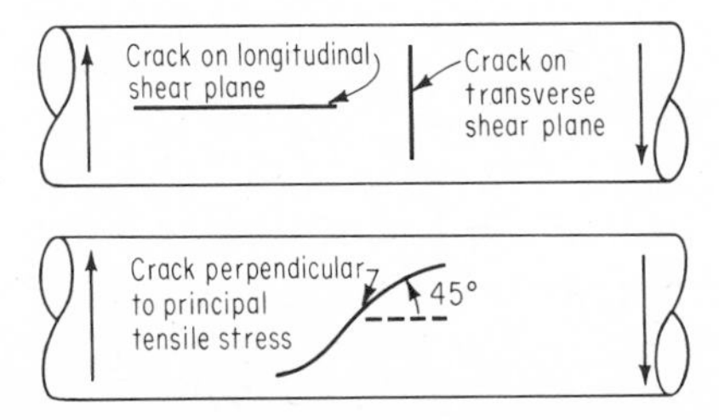

Fig. 17·31 Torsional fatigue can develop parallel to the principal shear stresses (top), or perpendicular to the principal tensile stresses (bottom). (From D. J. Wulpi, "How Components Fail," American Society for Metals, Metals Park, Ohio, 1966.)

Torsional-fatigue failures occur along the planes of maximum shear or along the plane of maximum tension. Maximum shear stress occurs along the axis of the shaft and at right angles to it, as in Fig. 17·31 (top), while the maximum tensile stress acts at an angle of 45° to the two shear stresses, as in Fig. 17·31 (bottom). Figure 17·32 shows schematically the basic types of torsional fractures. Torsional cracks may follow the transverse or longitudinal shear planes, the diagonal planes of maximum tensile stress, or a combination of these.

In a shaft subjected to torsion, the maximum shear stress is equal to the maximum tensile stress. In a part without stress raisers, which fracture will occur will depend upon the relative values of the shear strength and tensile strength. The strength values are a function of the material and its condition. In steel, the shear strength is approximately one-half the tensile strength. Therefore, the shear stress will reach the shear strength of the steel long before the tensile stress will reach the tensile strength, and a shear-type failure will result. Transverse cracks are more prevalent than longitudinal cracks, because grinding or machining marks are oriented in a transverse direction. Figure 17·33*a* shows a transverse shear failure in a 1045 steel bar tested in torsion. In brittle materials such as gray cast iron, however, the tensile strength is less than the shear strength. Therefore, the tensile stress will reach the tensile strength of the cast iron before the shear stress will reach the shear strength, and a tensile-type

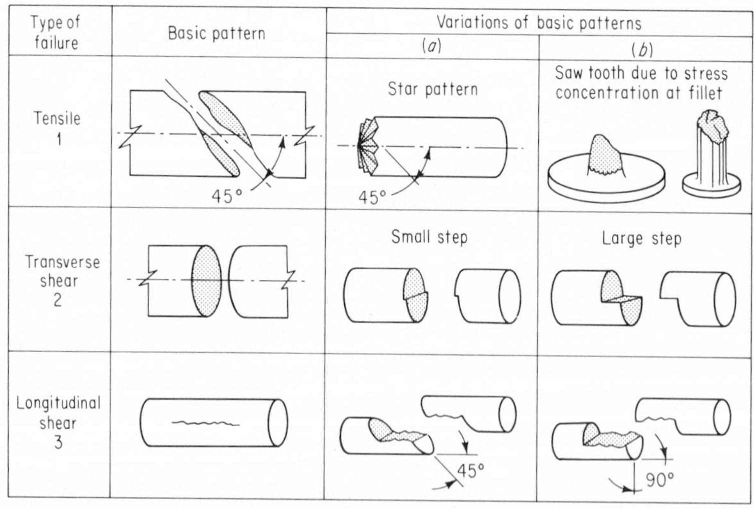

Fig. 17·32 Basic torsional fractures. (From Machine Design, The Penton Publishing Co., Cleveland, Dec. 11,1969.)

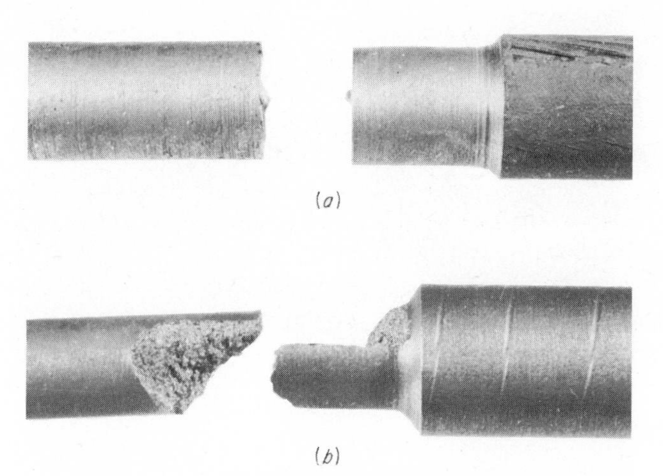

Fig. 17·33 (*a*) Transverse shear failure in a 1045 steel bar tested in torsion. (*b*) Tensile-type failure along a spiral angle in a gray cast-iron bar tested in torsion.

failure will result. Figure 17·33*b* shows the failure in a gray cast-iron bar by a combination of tension and longitudinal shear when tested in torsion.

In torsional fatigue, stress raisers are nearly as serious as they are in bending fatigue. Oil holes, fillets, and grooves in a shaft tend to concentrate the stresses and produce a tensile-type fracture. In the case of a transverse hole, the stress concentration factor may be above 3. The effect is to raise the tensile stress to more than 3 times its normal value, while the shear stress remains essentially the same. Although the tensile strength is still twice the shear strength, the applied tensile stress is now more than 3 times the shear stress. Therefore, the tensile stress will reach the tensile strength before the shear stress will reach the shear strength, and a tensile-type fracture along a 45° spiral angle will result (see Fig. 17·34).

17·13 Summary Other types of failure include surface damage due to wear, covered in Chap. 14, and surface damage due to corrosion, covered in Chap. 15.

To determine the true cause of a failure, the investigator must give full consideration to the interplay of design, fabrication, material properties, environment, and service loads. The cause will usually be classified in one of the categories outlined in the following list. Appropriate solutions may involve redesign, change of material or processing (or both), quality control, protection against environment, changes in maintenance schedules, or restrictions on service loads or service life.

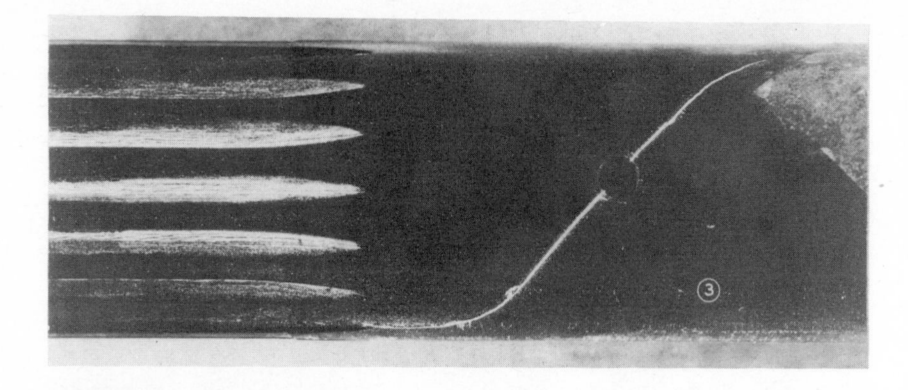

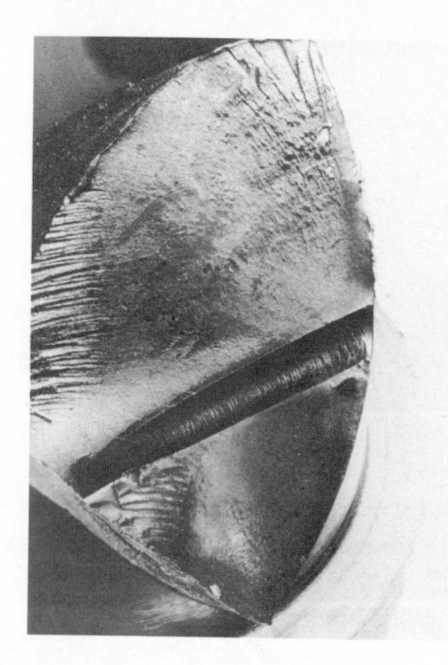

Fig. 17·34 Holes in shafts sometimes concentrate stresses, resulting in torsional-fatigue failures. The splined shaft (top) is carburized alloy steel at Rockwell C 60, and the crankshaft (bottom) is of 1045 steel, induction-hardened and tempered to Rockwell C 55. (Courtesy of D. J. Wulpi, International Harvester Company.)

CLASSIFICATION OF FAILURE CAUSES*

I. *Failures Due to Faulty Processing*

1 Flaws due to faulty composition (inclusions, embrittling impurities, wrong material).
2 Defects originating in ingot making and casting (segregation, unsoundness, porosity, pipes, nonmetallic inclusions).
3 Defects due to working (laps, seams, shatter cracks, hot-short splits, delamination, and excess local deformation).
4 Irregularities and mistakes due to machining, grinding, or stamping (gouges, burns, tearing, fins, cracks, embrittlement).

*From T. J. Dolan, "Analyzing Failures of Metal Components." *Metals Engineering Quarterly,* American Society for Metals, November, 1972.

5 Defects due to welding (porosity, undercuts, cracks, residual stress, lack of penetration, underbead cracking, heat affected zone).
6 Abnormalities due to heat treating (overheating, burning, quench cracking, grain growth, excessive retained austenite, decarburization, precipitation).
7 Flaws due to case hardening (intergranular carbides, soft core, wrong heat cycles).
8 Defects due to surface treatments (cleaning, plating, coating, chemical diffusion, hydrogen embrittlement).
9 Careless assembly (mismatch of mating parts, entrained dirt or abrasive, residual stress, gouges or injury to parts, and the like).
10 Parting line failures in forging due to poor transverse properties.

II. *Failures due to Faulty Design Considerations or Misapplication of Material*

1 Ductile failure (excess deformation, elastic or plastic; tearing or shear fracture).
2 Brittle fracture (from flaw or stress raiser of critical size).
3 Fatigue failure (load cycling, strain cycling, thermal cycling, corrosion fatigue, rolling contact fatigue, fretting fatigue).
4 High-temperature failure (creep, oxidation, local melting, warping).
5 Static delayed fractures (hydrogen embrittlement, caustic embrittlement, environmentally stimulated slow growth of flaws).
6 Excessively severe stress raisers inherent in the design.
7 Inadequate stress analysis, or impossibility of a rational stress calculation in a complex part.
8 Mistake in designing on basis of static tensile properties, instead of the significant material properties that measure the resistance of the material to each possible failure mode.

III. *Failure Due to Deterioration During Service Conditions*

1 Overload or unforeseen loading conditions.
2 Wear (erosion, galling, seizing, gouging, cavitation).
3 Corrosion (including chemical attack, stress corrosion, corrosion fatigue), dezincification, graphitization of cast iron, contamination by atmosphere.
4 Inadequate or misdirected maintenance or improper repair (welding, grinding, punching holes, cold straightening, and so forth).
5 Disintegration due to chemical attack or attack by liquid metals or platings at elevated temperatures.
6 Radiation damage (sometimes must decontaminate for examination which may destroy vital evidence of cause of failure), varies with time, temperature, environment, and dosage.
7 Accidental conditions (abnormal operating temperatures, severe vibration, sonic vibrations, impact or unforeseen collisions, ablation, thermal shock, and so forth).

It is obvious from this chapter that the analysis of failure and proper assignment of primary and secondary causes of failure are often a very complex problem. Knowledge of each type of failure is important to avoid or minimize future problems. Every service failure should be carefully studied to obtain the maximum amount of information concerning its failure. Metallurgical and visual examination, stress analysis, and intelligent questioning will also add a great deal of knowledge regarding the failure. Application of this accumulated knowledge to the prevention of future failures is the goal of failure analysis.

QUESTIONS

17·1 Suppose you were given a gear with several broken teeth and asked to investigate the reason for failure. Make a list of the questions you would ask.
17·2 What would you look for in a visual examination of the above gear?
17·3 What tests would you perform on the above gear? Explain the reasons for selecting the tests.
17·4 Explain the goal of failure analysis.
17·5 Explain the difference in appearance between ductile and brittle fractures.
17·6 Describe the most important basic stresses in a part under external load.
17·7 Explain why brittle fractures occur when the tensile strength is exceeded, and why ductile fractures occur when the shear strength is exceeded.
17·8 Describe the five types of loading and the type of stresses produced.
17·9 What is the most common type of fracture in machine parts?
17·10 Describe the three principal ways in which fatigue stresses develop.
17·11 Describe the development of "beach marks" on the surface of a fatigue failure.
17·12 How may the origin of a fatigue fracture be determined?
17·13 Once a fatigue crack forms, what does its rate of growth depend upon?
17·14 Explain the relation between the origin of fracture and the final fracture zone of a part subjected to a one-way bending load and one subjected to a rotating bending load.
17·15 Give three examples of stress raisers.
17·16 How will the beach marks curve with respect to the origin of failure in a hardened steel and annealed steel under cyclic loading? Explain.
17·17 Describe the kind of fracture which may occur as a result of a loose-fitting key on a shaft.
17·18 Explain the effect of three strength reducers.
17·19 Explain why residual stresses are important in failure analysis.
17·20 Under which conditions are residual stresses most significant with regard to the material?
17·21 What will be the difference in fracture surface appearance between a low overstress and a high overstress on a material?
17·22 What is the difference in appearance between bending fatigue and torsion fatigue?
17·23 What will be the difference in fracture between steel and cast iron subjected to torsion?

REFERENCES

Alban, Lester E.: Why Gears Fail, *Metal Progress,* American Society for Metals, Metals Park, Ohio, November 1970.

Barer, R. D., and B. F. Peters: "Why Metals Fail," Gordon and Breach Science Publishers, New York, 1970.

Bennett, J. A., and G. W. Quick: Mechanical Failures of Metals in Service, *Nat. Bur. Std. Circ.* 550, Washington, D.C., 1954.

Forrest, Peter R.: "Fatigue of Metals," Addison-Wesley Publishing Co., Inc., Reading, Mass., 1962.

Grover, H. J., S. A. Gordon, and L. R. Jackson: "Fatigue of Metals and Structures," Bureau of Aeronautics, Navy Dept., Washington, D.C., 1954.

Heywood, R. B.: "Designing Against Fatigue of Metals," Barnes & Noble, New York, 1962.

Larson, F. R., and F. L. Carr: How Failures Occur, *Metal Progress,* American Society for Metals, Metals Park, Ohio, March 1964.

Lipson, Charles: Basic Course in Failure Analysis, *Machine Design,* The Penton Publishing Co., Cleveland, Ohio, Oct. 16, 1969.

Parker, E. R.: "Brittle Behavior of Engineering Structures," John Wiley & Sons, Inc., New York, 1957.

Polushkin, E. P.: "Defects and Failures of Metals," Elsevier Publishing Co., New York, 1956.

Republic Steel Corporation: "Analysis of Service Failures," Cleveland, Ohio, 1961.

Ruskin, A. M.: "Materials Considerations in Design," Prentice-Hall, Inc., Englewood Cliffs, N.J., 1967.

Wulpi, D. J.: "How Components Fail," American Society for Metals, Metals Park, Ohio, 1966.

APPENDIX

Temperature-conversion Table

°C	°F	°C	°F	°C	°F	°C	°F	°C	°F
−273	−459	400	752	1000	1832	1600	2912	−268	−450
−250	−418	410	770	1010	1850	1610	2930	−240	−400
−200	−328	420	788	1020	1868	1620	2948	−212	−350
−150	−238	430	806	1030	1886	1630	2966	−184	−300
−100	−148	440	824	1040	1904	1640	2984	−157	−250
− 50	− 58	450	842	1050	1922	1650	3002	−129	−200
− 40	− 40	460	860	1060	1940	1660	3020	−101	−150
− 30	− 22	470	878	1070	1958	1670	3038	− 73	−100
− 20	− 4	480	896	1080	1976	1680	3056	− 46	− 50
− 10	+ 14	490	914	1090	1994	1690	3074	− 40	− 40
0	32	500	932	1100	2012	1700	3092	− 34	− 30
5	41	510	950	1110	2030	1710	3110	− 29	− 20
10	50	520	968	1120	2048	1720	3128	− 23	− 10
15	59	530	986	1130	2066	1730	3146	− 18	0
20	68	540	1004	1140	2084	1740	3164	− 15	5
25	77	550	1022	1150	2102	1750	3182	− 12	10
30	86	560	1040	1160	2120	1760	3200	− 9	15
35	95	570	1058	1170	2138	1770	3218	− 7	20
40	104	580	1076	1180	2156	1780	3236	− 4	25
45	113	590	1094	1190	2174	1790	3254	− 1	30
50	122	600	1112	1200	2192	1800	3272	+ 2	35
55	131	610	1130	1210	2210	1810	3290	4	40
60	140	620	1148	1220	2228	1820	3308	7	45
65	149	630	1166	1230	2246	1830	3326	10	50
70	158	640	1184	1240	2264	1840	3344	13	55
75	167	650	1202	1250	2282	1850	3362	16	60
80	176	660	1220	1260	2300	1860	3380	18	65
85	185	670	1238	1270	2318	1870	3398	21	70
90	194	680	1256	1280	2336	1880	3416	24	75
95	203	690	1274	1290	2354	1890	3434	27	80
100	212	700	1292	1300	2372	1900	3452	29	85
110	230	710	1310	1310	2390	1910	3470	32	90
120	248	720	1328	1320	2408	1920	3488	35	95
130	266	730	1346	1330	2426	1930	3506	38	100
140	284	740	1364	1340	2444	1940	3524	43	110
150	302	750	1382	1350	2462	1950	3542	49	120
160	320	760	1400	1360	2480	1960	3560	54	130
170	338	770	1418	1370	2498	1970	3578	60	140
180	356	780	1436	1380	2516	1980	3596	65	150
190	374	790	1454	1390	2534	1990	3614	71	160
200	392	800	1472	1400	2552	2000	3632	76	170
210	410	810	1490	1410	2570	2050	3722	83	180
220	428	820	1508	1420	2588	2100	3812	88	190
230	446	830	1526	1430	2606	2150	3902	93	200
240	464	840	1544	1440	2624	2200	3992	121	250
250	482	850	1562	1450	2642	2250	4082	149	300
260	500	860	1580	1460	2660	2300	4172	177	350
270	518	870	1598	1470	2678	2350	4262	204	400
280	536	880	1616	1480	2696	2400	4352	232	450
290	554	890	1634	1490	2714	2450	4442	260	500
300	572	900	1652	1500	2732	2500	4532	288	550
310	590	910	1670	1510	2750	2550	4622	316	600
320	608	920	1688	1520	2768	2600	4712	343	650
330	626	930	1706	1530	2786	2650	4802	371	700
340	644	940	1724	1540	2804	2700	4892	399	750
350	662	950	1742	1550	2822	2750	4982	427	800
360	680	960	1760	1560	2840	2800	5072	454	850
370	698	970	1778	1570	2858	2850	5162	482	900
380	716	980	1796	1580	2876	2900	5252	510	950
390	734	990	1814	1590	2894	3000	5432	538	1000

GLOSSARY OF TERMS RELATED TO PHYSICAL METALLURGY*

abrasion. The process of rubbing, grinding, or wearing away by friction.
abrasive. A substance used for grinding, honing, lapping, superfinishing, polishing, pressure blasting, or barrel finishing. It includes natural materials such as garnet, emery, corundum, and diamond and electric-furnace products like aluminum oxide, silicon carbide, and boron carbide.
activation energy. The energy required for initiating a metallurgical reaction; for example, plastic flow, diffusion, chemical reaction.
adhesion. Force of attraction between the molecules (or atoms) of two different phases, such as liquid brazing filler metal and solid copper, or plated metal and basis metal. Contrast with *cohesion*.
age hardening. Hardening by aging, usually after rapid cooling or cold working. See *aging*.
aging. In a metal or alloy, a change in properties that generally occurs slowly at room temperature and more rapidly at higher temperatures.
air-hardening. Term describing a steel containing sufficient carbon and other alloying elements to harden fully during cooling in air or other gaseous mediums from a temperature above its transformation range. The term should be restricted to steels that are capable of being hardened by cooling in air in fairly large sections, about 2 in. or more in diameter.
Alclad. Composite sheet produced by bonding either corrosion-resistant aluminum alloy or aluminum of high purity to base metal of structurally stronger aluminum alloy.
allotropy. The reversible phenomenon by which certain metals may exist in more than one crystal structure. If not reversible, the phenomenon is termed "polymorphism."
alloy. A substance having metallic properties and being composed of two or more chemical elements of which at least one is an elemental metal.
alloying element. An element which is added to a metal to effect changes in properties and which remains within the metal.
alloy steel. Steel containing significant quantities of alloying elements (other than carbon and the commonly accepted amounts of manganese, silicon, sulfur, and phosphorus) added to effect changes in the mechanical or physical properties.
alpha iron. The body-centered-cubic form of pure iron, stable below 1670°F.
Alumel. A nickel-base alloy containing about 2.5 percent Mn, 2 percent Al, and 1 percent Si; used chiefly as a component of pyrometric thermocouples.
aluminizing. Forming an aluminum or aluminum alloy coating on a metal by hot dipping, hot spraying, or diffusion.
amorphous. Not having a crystal structure; noncrystalline.
angstrom (A, AU). 10^{-8} cm.
annealing. Heating to and holding at a suitable temperature and then cooling at a suitable rate, for such purposes as reducing hardness, improving machinability, facilitating cold

*From Metals Handbook, 8th ed., vol. 1, American Society for Metals, 1961.

working, producing a desired microstructure, or obtaining desired mechanical, physical, or other properties.

When applied to ferrous alloys, the term *annealing,* without qualification, implies full annealing.

When applied to nonferrous alloys, the term *annealing* implies a heat treatment designed to soften a cold-worked structure by recrystallization or subsequent grain growth or to soften an age-hardened alloy by causing a nearly complete precipitation of the second phase in relatively coarse form.

Any process of annealing will usually reduce stresses, but if the treatment is applied for the sole purpose of such relief, it should be designated *stress relieving.*

anode. The electrode where electrons leave (current enters) an operating system such as a battery, an electrolytic cell, an x-ray tube, or a vacuum tube.

anode corrosion. The dissolution of a metal acting as an anode.

anodizing. Forming a *conversion coating* on a metal surface by anodic oxidation; most frequently applied to aluminum.

apparent density. *(Powd. met.)* The weight of a unit volume of powder, determined by a specified method of loading and usually expressed in grams per cubic centimeter.

artificial aging. Aging above room temperature. See *aging* and *precipitation heat treatment.*

athermal transformation. A reaction that proceeds without benefit of thermal fluctuations; that is, thermal activation is not required. Such reactions are diffusionless and can take place with great speed when the driving force is sufficiently high. For example, many martensitic transformations occur athermally on cooling, even at relatively low temperatures, because of the progressively increasing driving force. In contrast, a reaction that occurs at constant temperature is an isothermal transformation; thermal activation is necessary in this case, and the reaction proceeds as a function of time.

atomic percent. The number of atoms of an element in a total of 100 representative atoms of a substance; often written a/o.

atomization. *(Powd. met.)* The dispersion of a molten metal into particles by a rapidly moving stream of gas or liquid.

austempering. Quenching a ferrous alloy from a temperature above the transformation range, in a medium having a rate of heat abstraction high enough to prevent the formation of high-temperature transformation products, and then holding the alloy, until transformation is complete, at a temperature below that of pearlite formation and above that of martensite formation.

austenite. A solid solution of one or more elements in face-centered-cubic iron. The solute is generally assumed to be carbon.

austenitic steel. An alloy steel whose structure is normally austenitic at room temperature.

austenitizing. Forming austenite by heating a ferrous alloy above the transformation range.

bainite. A decomposition product of austenite consisting of an aggregate of ferrite and carbide. In general, it forms at temperatures lower than those where very fine pearlite forms and higher than those where martensite begins to form on cooling. Its appearance is feathery if formed in the upper part of the temperature range and acicular, resembling tempered martensite, if formed in the lower part.

banded structure. A segregated structure of nearly parallel bands aligned in the direction of working.

basal plane. A plane perpendicular to the *c*, or principal, axis in a tetragonal or hexagonal structure.

base metal. (1) The metal present in the largest proportion in an alloy; brass, for example, is a copper-base alloy. (2) The metal to be brazed, cut, or welded. (3) After welding, that part of the metal which was not melted.

bend test. A test for determining relative ductility of metal that is to be formed, usually sheet, strip, plate, or wire, and for determining soundness and toughness of metal. The specimen is usually bent over a specified diameter through a specified angle for a specified number of cycles.

binary alloy. An alloy containing two component elements.

binder. (1) In founding, a material, other than water, added to foundry sand to bind the particles together, sometimes with the use of heat. (2) *(Powd. met.)* A cementing medium: either a material added to the powder to increase the green strength of the compact that is expelled during sintering, or a material (usually of relatively low melting point) added to a powder mixture for the specific purpose of cementing together powder particles which alone would not sinter into a strong body.

black light. Electromagnetic radiation not visible to the human eye. The portion of the spectrum generally used in fluorescent inspection falls in the ultraviolet region between 3300 and 4000 Å, with the peak at 3650 Å.

blowhole. A hole in a casting or a weld caused by gas entrapped during solidification.

blue brittleness. Brittleness exhibited by some steels after being heated to some temperature within the range of 300 to 650°F, and more especially if the steel is worked at the elevated temperature.

brale. A diamond penetrator of specified sphero-conical shape used with a Rockwell hardness tester for hard metals. This penetrator is used for the A, C, D, and N scales.

brass. An alloy consisting mainly of copper (over 50 percent) and zinc, to which smaller amounts of other elements may be added.

brazing. Joining metals by flowing a thin layer, capillary thickness, of nonferrous filler metal into the space between them. Bonding results from the intimate contact produced by the dissolution of a small amount of *base metal* in the molten *filler metal,* without fusion of the base metal. Sometimes the filler metal is put in place as a thin solid sheet or as a clad layer, and the composite is heated, as in *furnace brazing.* The term *brazing* is used where the temperature exceeds some arbitrary value, such as 800°F; the term *soldering* is used for temperatures lower than the arbitrary value.

Brinell hardness test. A test for determining the hardness of a material by forcing a hard steel or carbide ball of specified diameter into it under a specified load. The result is expressed as the Brinell hardness number, which is the value obtained by dividing the applied load in kilograms by the surface area of the resulting impression in square millimeters.

brittle crack propagation. A very sudden propagation of a crack with the absorption of no energy except that stored elastically in the body. Microscopic examination may reveal some deformation even though it is not noticeable to the unaided eye.

brittle fracture. Fracture with little or no plastic deformation.

brittleness. The quality of a material that leads to crack propagation without appreciable plastic deformation.

bronze. A copper-rich copper-tin alloy with or without small proportions of other elements such as zinc and phosphorus. Also, certain other essentially binary copper-base alloys containing no tin, such as aluminum bronze (copper-aluminum), silicon bronze (copper-silicon), and beryllium bronze (copper-beryllium).

burning. (1) Permanently damaging a metal or alloy by heating to cause either incipient melting or intergranular oxidation. (2) In grinding, getting the work hot enough to cause discoloration or to change the microstructure by tempering or hardening.

carbide. A compound of carbon with one or more metallic elements.

carbide tools. Cutting tools, made of tungsten carbide, titanium carbide, tantalum carbide, or combinations of them, in a matrix of cobalt or nickel, having sufficient wear resistance and heat resistance to permit high machining speeds.

carbonitriding. Introducing carbon and nitrogen into a solid ferrous alloy by holding above Ac_1 in an atmosphere that contains suitable gases such as hydrocarbons, carbon monoxide, and ammonia. The carbonitrided alloy is usually quench-hardened.

carbon steel. Steel containing carbon up to about 2 percent and only residual quantities of other elements, except those added for deoxidation, with silicon usually limited to 0.60 percent and manganese to about 1.65 percent. Also termed *plain-carbon steel, ordinary steel,* and *straight-carbon steel.*

carburizing. Introducing carbon into a solid ferrous alloy by holding above Ac_1 in contact with a suitable carbonaceous material, which may be a solid, liquid, or gas. The carburized alloy is usually quench-hardened.
case. In a ferrous alloy, the outer portion that has been made harder than the inner portion, or core, by *case hardening*.
case hardening. Hardening a ferrous alloy so that the outer portion, or case, is made substantially harder than the inner portion, or core. Typical processes used for case hardening are *carburizing, cyaniding, carbonitriding, nitriding, induction hardening*, and *flame hardening*.
casting. (1) An object at or near finished shape obtained by solidification of a substance in a mold. (2) Pouring molten metal into a mold to produce an object of desired shape.
cast iron. An iron containing carbon in excess of the solubility in the austenite that exists in the alloy at the eutectic temperature. For the various forms *gray cast iron, white cast iron, malleable cast iron*, and *nodular cast iron*, the word "cast" is often left out, resulting in "gray iron," "white iron," "malleable iron," and "nodular iron," respectively.
cast structure. The internal physical structure of a casting evidenced by shape and orientation of crystals and segregation of impurities.
cathode. The electrode where electrons enter (current leaves) an operating system such as a battery, an electrolytic cell, an x-ray tube, or a vacuum tube.
cathodic protection. Partial or complete protection of a metal from corrosion by making it a cathode, using either a galvanic or impressed current.
cavitation. The formation and instantaneous collapse of innumerable tiny voids or cavities within a liquid subjected to rapid and intense pressure changes.
cavitation damage. Wearing away of metal through the formation and collapse of cavities in a liquid.
cemented carbide. A solid and coherent mass made by pressing and sintering a mixture of powders of one or more metallic carbides and a much smaller amount of a metal, such as cobalt, to serve as a binder.
cementite. A compound of iron and carbon, known chemically as iron carbide, and having the approximate chemical formula Fe_3C. It is characterized by an orthorhombic crystal structure.
ceramic tools. Cutting tools made from fused, sintered, or cemented metallic oxides.
cermet. A body consisting of ceramic particles bonded with a metal.
Charpy test. A pendulum-type single-bow impact test in which the specimen, usually notched, is supported at both ends as a simple beam and broken by a falling pendulum. The energy absorbed, as determined by the subsequent rise of the pendulum, is a measure of impact strength or notch toughness.
Chromel. (1) A 90Ni-10Cr alloy used in thermocouples. (2) A series of nickel-chromium alloys, some with iron, used for heat-resistant applications.
chromizing. A surface treatment at elevated temperature, generally carried out in pack, vapor, or salt bath, in which an alloy is formed by the inward diffusion of chromium into the base metal.
clad metal. A composite metal containing two or three layers that have been bonded together. The bonding may have been accomplished by co-rolling, welding, casting, heavy chemical deposition, or heavy electroplating.
cleavage. The splitting (fracture) of a crystal on a crystallographic plane of low index.
cleavage fracture. A fracture, usually of a polycrystalline metal, in which most of the grains have failed by cleavage, resulting in bright reflecting facets.
cleavage plane. A characteristic crystallographic plane or set of planes on which cleavage fracture easily occurs.
coherency. The continuity of lattice of precipitate and parent phase (solvent) maintained by mutual strain and not separated by a phase boundary.
coherent precipitate. A precipitate in a stage intermediate between a solute and a distinct phase, formed from a supersaturated solid solution but still retaining the lattice of the solvent even though the lattice may be distorted. Such a precipitate has no phase boundary.
cohesion. Force of attraction between the molecules (or atoms) within a single phase. Contrast with *adhesion*.

cohesive strength. (*1*) The hypothetical stress in an unnotched bar causing tensile fracture without plastic deformation. (*2*) The stress corresponding to the forces between atoms.

cold-short. A condition of brittleness existing in some metals at temperatures below the recrystallization temperature.

cold treatment. Cooling to a low temperature, often near −100°F, for the purpose of obtaining desired conditions or properties, such as dimensional or structural stability.

cold working. Deforming metal plastically at a temperature lower than the recrystallization temperature.

columnar structure. A coarse structure of parallel columns of grains, having the long axis perpendicular to the casting surface.

combined carbon. The part of the total carbon in steel or cast iron that is present as other than free carbon.

congruent melting. An isothermal melting in which both the solid and liquid phases have the same composition throughout the transformation.

constantan. A group of copper-nickel alloys containing 45 to 60 percent Cu with minor amounts of iron and manganese, and characterized by relatively constant electrical resistivity irrespective of temperature; used in resistors and thermocouples.

constitution diagram. A graphical representation of the temperature and composition limits of phase fields in an alloy system as they actually exist under the specific conditions of heating or cooling (synonymous with *phase diagram*).

continuous phase. In an alloy or portion of an alloy containing more than one phase, the phase that forms the background or matrix in which the other phase or phases are present as isolated units.

cooling stresses. Residual stresses resulting from nonuniform distribution of temperature during cooling.

coring. A variable composition between the center and surface of a unit of structure (such as a dendrite, grain, carbide particle) resulting from nonequilibrium growth which occurs over a range of temperature.

corrosion. The deterioration of a metal by chemical or electrochemical reaction with its environment.

corrosion embrittlement. The severe loss of ductility of a metal resulting from corrosive attack, usually intergranular and often not visually apparent.

corrosion fatigue. Effect of the application of repeated or fluctuating stresses in a corrosive environment characterized by shorter life than would be encountered as a result of either the repeated or fluctuating stresses alone or the corrosive environment alone.

covalent bond. A bond between two or more atoms resulting from the completion of shells by the sharing of electrons.

creep. Time-dependent strain occurring under stress. The creep strain occurring at a diminishing rate is called primary creep; that occurring at a minimum and almost constant rate, secondary creep; that occurring at an accelerating rate, tertiary creep.

creep strength. (1) The constant nominal stress that will cause a specified quantity of creep in a given time at constant temperature. (2) The constant nominal stress that will cause a specified creep rate at constant temperature.

crevice corrosion. A type of concentration-cell corrosion; corrosion of a metal that is caused by the concentration of dissolved salts, metal ions, oxygen or other gases, and such, in crevices or pockets remote from the principal fluid stream, with a resultant building up of differential cells that ultimately cause deep pitting.

critical cooling rate. The minimum rate of continuous cooling just sufficient to prevent undesired transformations. For steel, the slowest rate at which it can be cooled from above the upper critical temperature to prevent the decomposition of austenite at any temperature above the M_s.

critical point. (1) The temperature or pressure at which a change in crystal structure, phase, or physical properties occurs. Same as *transformation temperature*. (2) In an equilibrium diagram, that specific value of composition, temperature and pressure, or combinations thereof, at which the phases of a heterogeneous system are in equilibrium.

critical strain. The strain just sufficient to cause the growth of very large grains during heating where no phase transformations take place.

crystal. A solid composed of atoms, ions, or molecules arranged in a pattern which is repetitive in three dimensions.
crystalline fracture. A fracture of a polycrystalline metal characterized by a grainy appearance.
crystallization. The separation, usually from a liquid phase on cooling, of a solid crystalline phase.
cubic plane. A plane perpendicular to any one of three crystallographic axes of the cubic (isometric) system; the Miller indices are {100}.
cup fracture (cup-and-cone fracture). Fracture, frequently seen in tensile test pieces of a ductile material, in which the surface of failure on one portion shows a central flat area of failure in tension, with an exterior extended rim of failure in shear.
cyaniding. Introducing carbon and nitrogen into a solid ferrous alloy by holding above Ac_1, in contact with molten cyanide of suitable composition. The cyanided alloy is usually quench-hardened.

decalescence. A phenomenon, associated with the transformation of alpha iron to gamma iron on the heating (superheating) of iron or steel, revealed by the darkening of the metal surface owing to the sudden decrease in temperature caused by the fast absorption of that latent heat of transformation.
decarburization. The loss of carbon from the surface of a ferrous alloy as a result of heating in a medium that reacts with the carbon at the surface.
deep etching. Severe etching of a metallic surface for examination at a magnification of ten diameters or less to reveal gross features such as segregation, cracks, porosity, or grain flow.
defect. A condition that impairs the usefulness of an object or of a part.
deformation bands. Parts of a crystal which have rotated differently during deformation to produce bands of varied orientation within individual grains.
dendrite. A crystal that has a tree-like branching pattern, being most evident in cast metals slowly cooled through the solidification range.
deoxidizer. A substance that can be added to molten metal to remove either free or combined oxygen.
dezincification. Corrosion of some copper-zinc alloys involving loss of zinc and the formation of a spongy porous copper.
diamond-pyramid hardness test. An indentation hardness test employing a diamond-pyramid indenter and variable loads enabling the use of one hardness scale for all ranges of hardness from very soft lead to tungsten carbide. See *Knoop hardness* and *Vickers hardness.*
dichromate treatment. A chromate conversion coating produced on magnesium alloys in a boiling solution of sodium dichromate.
die casting. (1) A casting made in a die. (2) A casting process where molten metal is forced under high pressure into the cavity of a metal mold.
diffusion. (1) Spreading of a constituent in a gas, liquid, or solid, tending to make the composition of all parts uniform. (2) The spontaneous movement of atoms or molecules to new sites within a material.
diffusion coating. (1) An alloy coating produced at high temperatures by the inward diffusion of the coating material into the base metal. (2) Composite electrodeposited coatings which are subsequently interdiffused by thermal treatment.
diffusion coefficient. A factor of proportionality representing the amount of substance in grams diffusing across an area of 1 sq cm through a unit concentration gradient in 1 s of time.
dilatometer. An instrument for measuring the expansion or contraction in a metal resulting from changes in such factors as temperature or allotropy.
dip brazing. Brazing by immersion in a molten salt or metal bath. Where a metal bath is employed, it may provide the filler metal.
direct quenching. Quenching carburized parts directly from the carburizing operation.
discontinuous precipitation. Precipitation, mainly at the grain boundaries of a super-

saturated solid solution, accompanied by the appearance of two *lattice parameters:* one, of the solute atoms still in solution, the other, of the precipitate.

dislocation. A linear defect in the structure of a crystal. Two basic types are recognized, but combinations and partial dislocations are most prevalent. An "edge dislocation" corresponds to the row of mismatched atoms along a straight edge formed by an extra, partial plane of atoms within the body of the crystal, that is, by a plane of smaller area than any other parallel section through the crystal. A "screw dislocation" corresponds to the highly distorted lattice adjacent to the axis of a spiral structure in a crystal, the spiral structure being characterized by a distortion that has joined normally parallel planes together to form a continuous helical ramp winding about the dislocation as an axis with a pitch of one interplanar distance.

distortion. Any deviation from the desired shape or contour.

drawability. A measure of the workability of a metal subject to a drawing process. A term usually expressed to indicate a metal's ability to be deep-drawn.

drawing. (1) Forming recessed parts by forcing the plastic flow of metal in dies. (2) Reducing the cross section of wire or tubing by pulling it through a die. (3) A misnomer for *tempering.*

ductile crack propagation. Slow crack propagation that is accompanied by noticeable plastic deformation and requires energy to be supplied from outside the body.

ductility. The ability of a material to deform plastically without fracturing, being measured by elongation or reduction of area in a tensile test.

duralumin. (obsolete) A term formerly applied to the class of age-hardenable aluminum-copper alloys containing manganese, magnesium, or silicon.

dye penetrant. Penetrant with dye added to make it more readily visible under normal lighting conditions.

eddy-current testing. Nondestructive testing method in which eddy-current flow is induced in the test object. Changes in the flow caused by variations in the object are reflected into a nearby coil or coils for subsequent analysis by suitable instrumentation and techniques.

elastic deformation. Change of dimensions accompanying stress in the elastic range, original dimensions being restored upon release of stress.

elasticity. That property of a material by virtue of which it tends to recover its original size and shape after deformation.

elastic limit. The maximum stress to which a material may be subjected without any permanent strain remaining upon complete release of stress.

electrochemical corrosion. Corrosion which occurs when current flows between cathodic and anodic areas on metallic surfaces.

electrogalvanizing. The electroplating of zinc upon iron or steel.

electrolyte. (1) An ionic conductor. (2) A liquid, most often a solution, that will conduct an electric current.

electromotive series. A list of elements arranged according to their *standard electrode potentials.* In corrosion studies, the analogous but more practical *galvanic series* of metals is generally used. The relative position of a given metal is not necessarily the same in the two series.

electron compound. A term used to describe intermediate phases of metal systems that have both a common crystal structure and a common ratio of valence electrons to atoms. Thus, CuZn, Cu_3Al, Cu_5Sn, and FeAl all have the body-centered-cubic structure and an electron-to-atom ratio of 3 to 2.

electroplating. Electrodepositing metal (may be an alloy) in an adherent form upon an object serving as a cathode.

elongation. In tensile testing, the increase in the gage length, measured after fracture of the specimen within the gage length, usually expressed as a percentage of the original gage length.

emulsifier. (1) Same as *emulsifying agent.* (2) In penetrant inspection, a material that is added to some penetrants, after the penetrant is applied, to make a water-washable mixture.

emulsion. A suspension of one liquid phase in another.

equilibrium. A dynamic condition of balance between atomic movements where the resultant is zero and the condition appears to be one of rest rather than change.

equilibrium diagram. A graphical representation of the temperature, pressure, and composition limits of phase fields in an alloy system as they exist under condtions of complete equilibrium. In metal systems, pressure is usually considered constant (synonymous with phase diagram).

erosion. Destruction of metals or other materials by the abrasive action of moving fluids, usually accelerated by the presence of solid particles or matter in suspension. When *corrosion* occurs simultaneously, the term erosion-corrosion is often used.

etching. Subjecting the surface of a metal to preferential chemical or electrolytic attack in order to reveal structural details.

eutectic. (1) An isothermal reversible reaction in which a liquid solution is converted into two or more intimately mixed solids on cooling, the number of solids formed being the same as the number of components in the system. (2) An alloy having the composition indicated by the eutectic point on an equilibrium diagram. (3) An alloy structure of intermixed solid constituents formed by a eutectic reaction.

eutectoid. (1) An isothermal reversible reaction in which a solid phase (usually a solid solution) is converted into two or more intimately mixed solids on cooling, the number of solids formed being the same as the number of components in the system. (2) An alloy having the composition indicated by the eutectoid point on an equilibrium diagram. (3) An alloy structure of intermixed solid constituents formed by a eutectoid reaction.

extensometer. An instrument for measuring changes caused by stress in a linear dimension of a body.

fatigue. The phenomenon leading to fracture under repeated or fluctuating stresses having a maximum value less than the tensile strength of the material. Fatigue fractures are progressive, beginning as minute cracks that grow under the action of the fluctuating stress.

fatigue life. The number of cycles of stress that can be sustained prior to failure for a stated test condition.

fatigue limit. The maximum stress below which a material can presumably endure an infinite number of stress cycles. If the stress is not completely reversed, the value of the mean stress, the minimum stress, or the stress ratio should be stated.

fatigue strength. The maximum stress that can be sustained for a specified number of cycles without failure, the stress being completely reversed within each cycle unless otherwise stated.

ferrite. (1) A solid solution of one or more elements in body-centered-cubic iron. Unless otherwise designated, the solute is generally assumed to be carbon. On some equilibrium diagrams there are two ferrite regions separated by an austenite area. The lower area is alpha ferrite; the upper, delta ferrite. If there is no designation, alpha ferrite is assumed.

ferrite banding. Parallel bands of free ferrite aligned in the direction of working. Sometimes referred to as ferrite streaks.

fiber (fibre). (1) The characteristic of wrought metal that indicates *directional properties* and is revealed by the etching of a longitudinal section or is manifested by the fibrous or woody appearance of a fracture. It is caused chiefly by the extension of the constituents of the metal, both metallic and nonmetallic, in the direction of working. (2) The pattern of preferred orientation of metal crystals after a given deformation process, usually wiredrawing.

fibrous fracture. A fracture where the surface is characterized by a dull gray or silky appearance. Contrast with crystalline fracture.

fibrous structure. (1) In forgings, a structure revealed as laminations, not necessarily detrimental, on an etched section or as a ropy appearance on a fracture. It is not to be confused with the "silky" or "ductile" fracture of a clean metal. (2) In wrought iron, a structure consisting of slag fibers embedded in ferrite.

file hardness. Hardness as determined by the use of a file of standardized hardness on the assumption that a material which cannot be cut with the file is as hard as, or harder than, the file. Files covering a range of hardnesses may be employed.

flame hardening. Quench hardening in which the heat is applied directly by a flame.

fluidity. The ability of liquid metal to run into and fill a mold cavity.

fluorescent magnetic-particle inspection. Inspection with either dry magnetic particles or those in a liquid suspension, the particles being coated with a fluorescent substance to increase the visibility of the indications.

fluoroscopy. An inspection procedure in which the radiographic image of the subject is viewed on a fluorescent screen, normally limited to low-density materials or thin sections of metals because of the low light output of the fluorescent screen at safe levels of radiation.

focal spot. The area on the anode of an x-ray tube where the electron beam impinges, thus producing x-rays.

forging. Plastically deforming metal, usually hot, into desired shapes with compressive force, with or without dies.

formability. The relative ease with which a metal can be shaped through plastic deformation.

fractography. Descriptive treatment of fracture, especially in metals, with specific reference to photographs of the fracture surface. Macrofractography involves photographs at low magnification; microfractography, at high magnification.

fracture stress. (1) The maximum principal true stress at fracture. Usually refers to unnotched tensile specimens. (2) The (hypothetical) true stress which will cause fracture without further deformation at any given strain.

fracture test. Breaking a specimen and examining the fractured surface with the unaided eye or with a low-power microscope to determine such things as composition, grain size, case depth, soundness, or presence of defects.

fragmentation. The subdivision of a grain into small discrete crystallites outlined by a heavily deformed network of intersecting slip as a result of cold working. These small crystals or fragments differ from one another in orientation and tend to rotate to a stable orientation determined by the slip systems.

free carbon. The part of the total carbon in steel or cast iron that is present in the elemental form as graphite or temper carbon.

free ferrite. Ferrite that is structurally separate and distinct, as may be formed without the simultaneous formation of carbide when cooling hypoeutectoid austenite into the critical temperature range. Also *proeutectoid ferrite*.

fretting (fretting corrosion). Action that results in surface damage, especially in a corrosive environment, when there is relative motion between solid surfaces in contact under pressure.

full annealing. Annealing a ferrous alloy by austenitizing and then cooling slowly through the transformation range. The austenitizing temperature for hypoeutectoid steel is usually above Ac_3; and for hypereutectoid steel, usually between $Ac_{1,3}$ and Ac_{cm}.

gage length. The original length of that portion of the specimen over which strain, change of length, and other characteristics are measured.

galling. Developing a condition on the rubbing surface of one or both mating parts where excessive friction between high spots results in localized welding with subsequent spalling and a further roughening of the surface.

galvanic corrosion. Corrosion associated with the current of a galvanic cell consisting of two dissimilar conductors in an electrolyte or two similar conductors in dissimilar electrolytes. Where the two dissimilar metals are in contact, the resulting reaction is referred to as *couple action*.

galvanic series. A series of metals and alloys arranged according to their relative electrode potentials in a specified environment.

gamma iron. The face-centered-cubic form of pure iron, stable from 1670 to 2550°F.

grain. An individual crystal in a polycrystalline metal or alloy.

grain growth (coarsening). An increase in the size of grains in polycrystalline metal, usually effected during heating at elevated temperatures. The increase may be gradual or abrupt, resulting in either uniform or nonuniform grains after growth has ceased. A mixture of nonuniform grains is sometimes termed *duplexed*. Abnormal grain growth (exaggerated grain growth) implies the formation of excessively large grains, uniform or nonuniform.

grain refiner. A material added to a molten metal to attain finer grains in the final structure.

grain size. For metals, a measure of the areas or volumes of grains in a polycrystalline material, usually expressed as an average when the individual sizes are fairly uniform. Grain sizes are reported in terms of number of grains per unit area or volume, in terms of average diameter, or as a grain-size number derived from area measurements.

granular fracture. A type of irregular surface produced when metal is broken; characterized by a rough, grain-like appearance as differentiated from a smooth and silky or a fibrous type. It can be subclassified into transgranular and intergranular forms. This type of fracture is frequently called crystalline fracture, but the inference that the metal has crystallized is not justified.

graphitic carbon. Free carbon in steel or cast iron.

graphite steel. Alloy steel made so that part of the carbon is present as graphite.

graphitization. Formation of graphite in iron or steel. Where graphite is formed during solidification, the phenomenon is called *primary graphitization*; where formed later by heat treatment, *secondary graphitization*.

graphitizing. Annealing a ferrous alloy in such a way that some or all of the carbon is precipitated as graphite.

grey cast iron. A cast iron that gives a gray fracture due to the presence of flake graphite. Often called gray iron.

grinding cracks. Shallow cracks formed in the surface of relatively hard materials because of excessive grinding heat or the high sensitivity of the material.

hard chromium. Chromium deposited for engineering purposes, such as increasing the wear resistance of sliding metal surfaces, rather than as a decorative coating. It is usually applied directly to basis metal and is customarily thicker than a decorative deposit.

hard-drawn. Temper of copper or copper-alloy tubing drawn in excess of 25 percent reduction in area.

hardenability. In a ferrous alloy, the property that determines the depth and distribution of hardness induced by quenching.

hardening. Increasing the hardness by suitable treatment, usually involving heating and cooling. When applicable, the following more specific terms should be used: *age hardening, case hardening, flame hardening, induction hardening, precipitation hardening,* and *quench hardening*.

hard facing. Depositing filler metal on a surface by welding, spraying, or braze welding for the purpose of resisting abrasion, erosion, wear, galling, and impact.

hardness. Resistance of metal to plastic deformation usually by identation. However, the term may also refer to stiffness or temper, or to resistance to scratching, abrasion, or cutting. Indentation hardness may be measured by various hardness tests, such as *Brinell, Rockwell*, and *microhardness*.

heat check. A pattern of parallel surface cracks that are formed by alternate rapid heating and cooling of the extreme surface metal, sometimes found on forging dies and piercing punches. There may be two sets of parallel cracks, one set perpendicular to the other.

heat treatment. Heating and cooling a solid metal or alloy in such a way as to obtain desired conditions or properties. Heating for the sole purpose of hot working is excluded from the meaning of this definition.

homogenizing. Holding at high temperature to eliminate or decrease chemical segregation by diffusion.

homogenizing treatment. A heat treatment of an alloy intended to make it uniform in composition by eliminating *coring* and concentration gradients.

homopolar crystal. A crystal characterized by covalent bonding, the type of atomic bonding resulting from the sharing of electrons by neighboring atoms.

hot forming. Working operations such as bending, drawing, forging, piercing, pressing, and heading performed above the recrystallization temperature of the metal.

hot-shortness. Brittleness in metal in the hot-forming range.
hot tear. A fracture formed in a metal during solidification because of hindered contraction.
hot working. Deforming metal plastically at such a temperature and rate that strain hardening does not occur. The low limit of temperature is the recrystallization temperature.
hypereutectic alloy. Any binary alloy whose composition lies to the right of the eutectic on an equilibrium diagram and which contains some eutectic structure.
hypereutectoid. Analogous to hypereutectic.
hypoeutectic alloy. Any binary alloy whose composition lies to the left of the eutectic on an equilibrium diagram and which contains some eutectic structure.
hypoeutectoid. Analogous to hypoeutectic.

immersion plating (dip plating). Depositing a metallic coating on a metal immersed in a liquid solution, without the aid of an external electric current.
impact energy (impact value). The amount of energy required to fracture a material, usually measured by means of an *Izod* or *Charpy* test. The type of specimen and testing conditions affect the values and therefore should be specified.
impact test. A test to determine the behavior of materials when subjected to high rates of loading, usually in bending, tension, or torsion. The quantity measured is the energy absorbed in breaking the specimen by a single blow, as in the *Charpy* or *Izod* tests.
inclusions. Nonmetallic materials in a solid metallic matrix.
indentation hardness. The resistance of a material to indentation. This is the usual type of hardness test, in which a pointed or rounded indenter is pressed into a surface under a substantially static load.
indication. In nondestructive inspection, a response, or evidence of a response, that requires interpretation to determine its significance.
induction furnace. An a-c electric furnace in which the primary conductor is coiled and generates, by electromagnetic induction, a secondary current which develops heat within the metal charge.
induction hardening. Quench hardening in which the heat is generated by electrical induction.
inoculation. The addition of a material to molten metal to form nuclei for crystallization.
intercept method. A method of determining grain size by counting the number of grains per unit length intersected by straight lines.
intercrystalline. Between the crystals, or grains, of a metal.
interdendritic corrosion. Corrosive attack which progresses preferentially along interdendritic paths. This type of attack results from local differences in composition, such as coring commonly encountered in alloy castings.
intergranular corrosion. Corrosion occuring preferentially at grain boundaries.
intermediate annealing. Annealing wrought metals at one or more stages during manufacture and before final treatment.
intermediate phase. In an alloy or a chemical system, a distinguishable homogeneous phase whose composition range does not extend to any of the pure components of the system.
intermetallic compound. An intermediate phase in an alloy system, having a narrow range of homogeneity and relatively simple stoichiometric proportions, in which the nature of the atomic binding can vary from metallic to ionic.
interrupted aging. Aging at two or more temperatures, by steps, and cooling to room temperature after each step.
interrupted quenching. Quenching in which the metal object being quenched is removed from the quenching medium while the object is at a temperature substantially higher than that of the quenching medium. See also time quenching.
interstitial solid solution. A solid solution in which the solute atoms occupy positions within the lattice of the solvent. See also *substitutional solid solution.*
intracrystalline. Within or across the crystals or grains of a metal; same as transcrystalline and transgranular.
ion. An atom, or group of atoms, that has gained or lost one or more outer electrons and

thus carries an electric charge. Positive ions, or cations, are deficient in outer electrons. Negative ions, or anions, have an excess of outer electrons.

ionic crystal. A crystal in which atomic bonding results from the electrostatic attractive forces between positively and negatively charged ions; for example, sodium chloride. This type of atomic linkage, also known as *polar bonding,* is characteristic of many compounds.

isothermal annealing. Austenitizing a ferrous alloy and then cooling to and holding at a temperature at which austenite transforms to a relatively soft ferrite carbide aggregate.

isothermal transformation. A change in phase at any constant temperature.

isotropy. Quality of having identical properties in all directions.

Izod test. A pendulum type of single-blow impact test in which the specimen, usually notched, is fixed at one end and broken by a falling pendulum. The energy absorbed, as measured by the subsequent rise of the pendulum, is a measure of impact strength or notch toughness.

Knoop hardness. Microhardness determined from the resistance of metal to indentation by a pyramidal diamond indenter, having edge angles of 172° 30′ and 130°, making a rhombohedral impression with one long and one short diagonal. The long diagonal is measured microscopically to determine the KHN.

lead screen. In radiography, a screen used (1) to filter out soft-wave or scattered radiation and (2) to increase the intensity of the remaining radiation so that the exposure time can be decreased.

ledeburite. The eutectic of the iron-carbon system, the constituents being austenite and cementite. The austenite decomposes into ferrite and cementite on cooling below the Ar_1.

liquidus. In a constitution or equilibrium diagram, the locus of points representing the temperatures at which the various compositions in the system begin to freeze on cooling or to finish melting on heating.

Lüder bands. Surface markings or depressions resulting from localized plastic deformation in metals which show discontinuous yielding. See *stretcher strains.*

macro-etch. Etching of a metal surface for accentuation of gross structural details and defects for observation by the unaided eye or at magnifications not exceeding ten diameters.

macrograph. A graphic reproduction of the surface of a prepared specimen at a magnification not exceeding ten diameters. When photographed, the reproduction is known as a photomacrograph.

macroscopic. Visible at magnifications from one to ten diameters.

macroshrinkage. A casting defect, detectable at magnifications not exceeding ten diameters, consisting of voids in the form of stringers shorter than *shrinkage cracks.* This defect results from contraction during solidification where there is not an adequate opportunity to supply filler material to compensate for the shrinkage. It is usually associated with abrupt changes in section size.

macrostructure. The structure of metals as revealed by examination of the etched surface of polished specimen at a magnification not exceeding ten diameters.

magnetic-particle inspection. A nondestructive method of inspection for determining the existence and extent of possible defects in ferromagnetic materials. Finely divided magnetic particles, applied to the magnetized part, are attracted to and outline the pattern of any magnetic-leakage fields created by discontinuities.

malleability. The characteristic of metals which permits plastic deformation in compression without rupture.

malleable cast iron. A cast iron made by a prolonged anneal of *white cast iron* in which decarburization or graphitization, or both, take place to eliminate some or all of the cementite. The graphite is in the form of temper carbon. If decarburization is the predominant reaction, the product will have a light fracture, hence, "white-heart malleable"; otherwise, the fracture will be dark, hence, "black-heart malleable." "Pearlitic malleable" is a black-heart variety having a pearlitic matrix along with perhaps some free ferrite.

malleableizing. Annealing white cast iron in such a way that some or all of the combined carbon is transformed to graphite, or, in some instances, part of the carbon is removed completely.

martempering. Quenching an austenitized ferrous alloy in a medium at a temperature in the upper part of the martensite range, or slightly above that range, and holding it in the medium until the temperature throughout the alloy is substantially uniform. The alloy is then allowed to cool in air through the martensite range.

martensite. (1) In an alloy, a metastable transitional structure intermediate between two allotropic modifications whose abilities to dissolve a given solute differ considerably, the high-temperature phase having the greater solubility. The amount of the high-temperature phase transformed to martensite depends to a large extent upon the temperature attained in cooling, there being a rather distinct beginning temperature. (2) A metastable phase of steel, formed by a transformation of austenite below the M_5 (or Ar'') temperature. It is an interstitial supersaturated solid solution of carbon in iron having a body-centered-tetragonal lattice. Its microstructure is characterized by an acicular, or needlelike, pattern.

martensitic transformation. A reaction which takes place in some metals on cooling, with the formation of an acicular structure called martensite.

matrix. The principal phase or aggregate in which another constituent is embedded.

mechanical metallurgy. The technology dealing with the manner in which metals react to their mechanical environment.

mechanical properties. The properties of a material that reveal its elastic and inelastic behavior where force is applied, thereby indicating its suitability for mechanical applications; for example, modulus of elasticity, tensile strength, elongation, hardness, and fatigue limit.

mechanical twin. A twin formed in a metal during plastic deformation by a simple shear of the lattice.

melting point. The temperature at which a pure metal, compound, or eutectic changes from solid to liquid; the temperature at which the liquid and the solid are in equilibrium.

metal. (1) An opaque lustrous elemental chemical substance that is a good conductor of heat and electricity and, when polished, a good reflector of light. Most elemental metals are malleable and ductile and are, in general, heavier than the other elemental substances. (2) As to structure, metals may be distinguished from nonmetals by their atomic binding and electron availability. Metallic atoms tend to lose electrons from the outer shells, the positive ions thus formed being held together by the electron gas produced by the separation. The ability of these free electrons to carry an electric current, and the fact that the conducting power decreases as temperature increases, establish one of the prime distinctions of a metallic solid.

metallizing (spray metallizing). Forming a metallic coating by atomized spraying with molten metal or by vacuum deposition.

metallograph. An optical instrument designed for both visual observation and photomicrography of prepared surfaces of opaque materials at magnifications ranging from about 25 to about 1,500 diameters. The instrument consists of a high-intensity illuminating source, a microscope, and a camera bellows. On some instruments provisions are made for examination of specimen surfaces with polarized light, phase contrast, oblique illumination, dark-field illumination, and customary bright-field illumination.

metallography. The science dealing with the constitution and structure of metals and alloys as revealed by the unaided eye or by such tools as low-powered magnification, optical microscope, electron microscope, and diffraction or x-ray techniques.

metallurgy. The science and technology of metals. Process (chemical) metallurgy is concerned with the extraction of metals from their ores and with the refining of metals; physical metallurgy deals with the physical and mechanical properties of metals as affected by composition, mechanical working, and heat treatment.

metal spraying. Coating metal objects by spraying molten metal upon the surface with gas pressure.

metastable. In a state of pseudoequilibrium which has a higher free energy than has the true equilibrium state yet does not change spontaneously.

micrograph. A graphic reproduction of the surface of a prepared specimen, usually etched, at a magnification greater than ten diameters. If produced by photographic means it is called a photomicrograph (not a microphotograph).

microhardness. Hardness determined by using a microscope to measure the diagonal of the impression left by a Knoop or Vickers indenter. The hardness is the load divided by the area of the impression.

microsegregation. Segregation within a grain, crystal, or small particle. See *coring*.

microshrinkage. A casting defect, not detectable at magnifications lower than ten diameters, consisting of interdendritic voids. This defect results from contraction during solidification where there is not an adequate opportunity to supply filler material to compensate for shrinkage. Alloys with a wide range in solidification temperature are particularly susceptible.

microstructure. The structure of polished and etched metals as revealed by a microscope at a magnification greater than ten diameters.

Miller indices, plane. Indices which identify a family of planes in crystal structure. The intercepts m, n, and p for any plane within a crystal give the reciprocals l/m, l/n, and l/p, which may be changed to a common denominator, resulting in the numerators h, k, and l, respectively. These numerators when written as (hkl) identify the family of planes to which the specific plane belongs.

modulus of elasticity. A measure of the stiffness of metal. Ratio of stress, within proportional limit, to corresponding strain.

monotectic. An isothermal reversible reaction in a binary system, in which a liquid on cooling decomposes into a second liquid of a different composition and a solid. It differs from a eutectic in that only one of the two products of the reaction is below its freezing range.

M_s temperature. The temperature at which martensite begins to form in an alloy system on cooling. Specifically for steel, it is the temperature at which austenite begins to change into martensite on cooling.

natural aging. Spontaneous aging of a supersaturated solid solution at room temperature. See *aging*, and compare with *artificial aging*.

necking. (1) Reducing the cross-sectional area of metal in a localized area by stretching. (2) Reducing the diameter of a portion of the length of a cylindrical shell or tube.

network structure. A structure in which one constituent occurs primarily at the grain boundaries, thus partially or completely enveloping the grains of the other constituents.

neutron. Elementary nuclear particle with a mass approximately the same as that of a hydrogen atom and electrically neutral.

nitriding. Introducing nitrogen into a solid ferrous alloy by holding at a suitable temperature (below *Ac*, for ferritic steels) in contact with a nitrogenous material, usually ammonia or molten cyanide of appropriate composition. Quenching is not required to produce a hard case.

noble metal. (1) A metal whose potential is highly positive relative to the hydrogen electrode. (2) A metal with marked resistance to chemical reaction, particularly to oxidation and to solution by inorganic acids. The term as often used is synonymous with *precious metal*.

nodular cast iron. A *cast iron* that has been treated while molten with a master alloy containing an element such as magnesium or cerium to give primary graphite in the spherulitic form.

nondestructive inspection. Inspection by methods that do not destroy the part to determine its suitability for use.

normalizing. Heating a ferrous alloy to a suitable temperature above the transformation range and then cooling in air to a temperature substantially below the transformation range.

nucleation. The initiation of a phase transformation at discrete sites, the new phase growing on nuclei.

nucleus. (1) The first structurally stable particle capable of initiating recrystallization of a phase or the growth of a new phase, and possessing an interface with the parent matrix. The term is also applied to a foreign particle which initiates such action. (2) The heavy central core of an atom, in which most of the mass and the total positive electric charge are concentrated.

octahedral plane. In cubic crystals, a plane with equal intercepts on all three axes.

offset. The distance along the strain coordinate between the initial portion of a stress-strain curve and a parallel line that intersects the stress-strain curve at a value of stress which is used as a measure of the *yield strength.* It is used for materials that have no obvious *yield point.* A value of 0.2 percent is commonly used.

optical pyrometer. An instrument for measuring the temperature of heated material by comparing the intensity of light emitted with a known intensity of an incandescent lamp filament.

orange peel. A pebble-grained surface which develops in forming of metals having coarse grains.

ordering. Forming a *superlattice.*

overaging. Aging under conditions of time and temperature greater than those required to obtain maximum change in a certain property, so that the property is altered in the direction of the initial value. See *aging.*

overheating. Heating a metal or alloy to such a high temperature that its properties are impaired. When the original properties cannot be restored by further heat treating, by mechanical working, or a combination of working and heat treating, the overheating is known as *burning.*

oxidation. A reaction in which there is an increase in valence resulting from a loss of electrons.

parameter (lattice). In a crystal, the length, usually in angstrom units, of the unit cell along one of its axes or edges; also called *lattice constant.*

particle size. *(Powd. met.)* The controlling lineal dimension of an individual particle, as determined by analysis with screens or other suitable instruments.

patenting. In wire making, a heat treatment applied to medium-carbon or high-carbon steel before the drawing of wire or between drafts. This process consists in heating to a temperature above the transformation range and then cooling to a temperature below Ae_1 in air or in a bath of molten lead or salt.

pearlite. A lamellar aggregate of *ferrite* and *cementite*, often occurring in steel and cast iron.

penetrant. A liquid with low surface tension used in *penetrant* inspection to flow into surface openings of parts being inspected.

penetrant inspection. A method of nondestructive testing for determining the existence and extent of discontinuities that are open to the surface in the part being inspected. The *indications* are made visible through the use of a dye or fluorescent chemical in the liquid employed as the inspection medium.

peritectic. An isothermal reversible reaction in which a liquid phase reacts with a solid phase to produce another solid phase on cooling.

peritectoid. An isothermal reversible reaction in which a solid phase reacts with a second solid phase to produce yet a third solid phase on cooling.

permanent mold. A metal mold (other than an ingot mold) of two or more parts that is used repeatedly for the production of many castings of the same form. Liquid metal is poured in by gravity.

permanent set. Plastic deformation that remains upon releasing the stress that produces the deformation.

phase. A physically homogeneous and distinct portion of a material system.

phase diagram. Same as *constitution diagram* or *equilibrium diagram*.

phosphatizing. Forming an adherent phosphate coating on a metal immersed in a suitable aqueous phosphate solution.

pitting. Forming small sharp cavities in a metal surface by nonuniform electrodeposition or by corrosion.

plastic deformation. Deformation that does or will remain permanent after removal of the load which caused it.

plasticity. The ability of a metal to deform nonelastically without rupture.

polycrystalline. Pertaining to a solid composed of many crystals.

polymorphism. See *allotropy*.

porosity. Fine holes or pores within a metal.

powder metallurgy. *(Powd.met.)* The art of producing metal powders and of utilizing metal powders for the production of massive materials and shaped objects.

precipitation hardening. Hardening caused by the precipitation of a constituent from a supersaturated solid solution. See also *age hardening* and *aging*.

precipitation heat treatment. Artificial aging in which a constituent precipitates from a supersaturated solid solution.

precision casting. A metal casting of reproducible accurate dimensions, regardless of how it is made.

preferred orientation. A condition of a polycrystalline aggregate in which the crystal orientations are not random.

preheating. Heating before some further thermal or mechanical treatment. For tool steel, heating to an intermediate temperature immediately before final austenitizing. For some nonferrous alloys, heating to a high temperature for a long time, in order to homogenize the structure before working.

process annealing. In the sheet and wire industries, heating a ferrous alloy to a temperature close to, but below, the lower limit of the transformation range and then cooling, in order to soften the alloy for further cold working.

progressive aging. Aging by increasing the temperature in steps or continuously during the aging cycle. See *aging* and compare with *interrupted aging*.

proportional limit. The maximum stress at which strain remains directly proportional to stress.

pyrometer. A device for measuring temperatures above the range of liquid thermometers.

quench hardening. Hardening a ferrous alloy by austenitizing and then cooling rapidly enough so that some or all of the austenite transforms to martensite. The austenitizing temperature for hypoeutectoid steels is usually above Ac_3 and for hypereutectoid steels usually between $Ac_{1,3}$ and Ac_{cm}.

quenching. Rapid cooling. When applicable, the following more specific terms should be used: direct quenching, interrupted quenching, selective quenching, spray quenching, and time quenching.

radiograph. A photographic shadow image resulting from uneven absorption of radiation in the object being subjected to penetrating radiation.

radiography. A nondestructive method of internal examination in which metal or other objects are exposed to a beam of x-ray or gamma radiation. Differences in thickness, density, or absorption caused by internal discontinuities are apparent in the shadow image either on a fluorescent screen or on photographic film placed behind the object.

recalescence. A phenomenon associated with the transformation of gamma iron to alpha iron on the cooling (supercooling) of iron or steel; it is revealed by the brightening (reglowing) of the metal surface owing to the sudden increase in temperature caused by the fast liberation of the latent heat of transformation.

recarburize. (1) To increase the carbon content of molten cast iron or steel by adding carbonaceous material, high-carbon pig iron or a high-carbon alloy. (2) To carburize a metal part to return surface carbon lost in processin .

recovery. Reduction or removal of work-hardening effects, without motion of large-angle grain boundaries.

recrystallization. (1) The change from one crystal structure to another, as occurs on heating or cooling through a critical temperature. (2) The formation of a new, strain-free grain structure from that existing in cold-worked metal, usually accomplished by heating.

recrystallization annealing. Annealing cold-worked metal to produce a new grain structure without phase change.

recrystallization temperature. The approximate minimum temperature at which complete recrystallization of a highly cold-worked metal occurs within a specified time, usually 1 h.

reduction of area. (1) Commonly, the difference, expressed as a percentage of original area, between the original cross-sectional area of a tensile test specimen and the minimum cross-sectional area measured after complete separation. (2) The difference, expressed as a percentage of original area, between original cross-sectional area and that after straining the specimen.

refractory metal. A metal having an extremely high melting point. In the broad sense, it refers to metals having melting points above the range of iron, cobalt, and nickel.

residual method. Method of *magnetic-particle inspection* in which the particles are applied after the magnetizing force has been removed.

residual stress. Stress present in a body that is free of external forces or thermal gradients.

resilience. (1) The amount of energy per unit volume released upon unloading. (2) The capacity of a metal, by virtue of high yield strength and low elastic modulus, to exhibit considerable elastic recovery upon release of load.

Rockwell hardness test. A test for determining the hardness of a material based upon the depth of penetration of a specified penetrator into the specimen under certain arbitrarily fixed conditions of test.

rolling. Reducing the cross-sectional area of metal stock, or otherwise shaping metal products, through the use of rotating rolls.

rubber forming. Forming where rubber is used as a functional die part. Processes in which rubber is employed only to contain the hydraulic fluid should not be classified as rubber forming.

rust. A corrosion product consisting of hydrated oxides of iron. Applied only to ferrous alloys.

scale. Formation of a layer of iron oxide on the surface of steel when heated at high temperature in a furnace.

scleroscope test. A hardness test where the loss in kinetic energy of a falling diamond-tipped metal "tup," absorbed by indentation upon impact of the tup on the metal being tested, is indicated by the height of rebound.

scoring. Marring or scratching of any formed part by metal pickup on the punch or die.

scratch hardness. The hardness of a metal determined by the width of a scratch made by a cutting point drawn across the surface under a given pressure.

seam. On the surface of metal, an unwelded fold or lap which appears as a crack, usually resulting from a defect obtained in casting or in working.

season cracking. Cracking resulting from the combined effects of corrosion and internal stress. A term usually applied to stress-corrosion cracking of brass.

secondary hardening. Tempering certain alloy steels at certain temperatures so that the resulting hardness is greater than that obtained by tempering the same steel at some lower temperature for the same time.

seizing. Stopping of a moving part by a mating surface as a result of excessive friction caused by galling.

selective heating. Heating only certain portions of an object so that they have the desired properties after cooling.

selective quenching. Quenching only certain portions of an object.

self-diffusion. The spontaneous movement of an atom to a new site in a crystal of its own species, as, for example, a copper atom within a crystal of copper.

shear. (1) That type of force which causes or tends to cause two contiguous parts of the same body to slide relative to each other in a direction parallel to their plane of contact. (2) A type of cutting tool with which a material in the form of wire, sheet, plate, or rod is cut between two opposing blades.

shear angle. The angle that the *shear plane*, in metal cutting, makes with the work surface.

shear fracture. A fracture in which a crystal (or a polycrystalline mass) has separated by sliding or tearing under the action of shear stresses.

shear strength. The stress required to produce fracture in the plane of cross section, the conditions of loading being such that the directions of force and of resistance are parallel and opposite, although their paths are offset a specified minimum amount.

shortness. A form of brittleness in metal. It is designated as "cold," "hot," and "red," to indicate the temperature range in which the brittleness occurs.

shot peening. Cold working the surface of a metal by metal-shot impingement.

shrinkage cavity. A void left in cast metals as a result of solidification shrinkage.

shrinkage cracks. Hot tears associated with shrinkage cavities.

sieve analysis. (*Powd. met.*) Particle-size distribution; usually expressed as the weight percentage retained upon each of a series of standard sieves of decreasing size and the percentage passed by the sieve of finest size.

siliconizing. Diffusing silicon into solid metal, usually steel, at an elevated temperature.

silky fracture. A metal fracture in which the broken metal surface has a fine texture usually dull in appearance. Characteristic of tough and strong metals.

sinter. To heat a mass of fine particles for a prolonged time below the melting point, usually to cause agglomeration.

sintering. (*Powd. met.*) (1) The bonding of adjacent surfaces of particles in a mass of metal powders or a compact, by heating. (2) A shaped body composed of metal powders and produced by sintering with or without prior compacting.

skin. A thin outside metal layer, not formed by bonding (as in cladding or electroplating), that differs in composition, structure, or other characteristic from the main mass of metal.

slack quenching. The process of hardening steel by quenching from the austenitizing temperature at a rate slower than the *critical cooling rate* for the particular steel, resulting in incomplete hardening and the formation of one or more transformation products in addition to or instead of martensite.

slag. A nonmetallic product resulting from the mutual dissolution of flux and nonmetallic impurities in smelting and refining operations.

slip. Plastic deformation by the irreversible shear displacement (translation) of one part of a crystal relative to another in a definite crystallographic direction and usually on a specific crystallographic plane.

slip band. A group of parallel slip lines so closely spaced as to appear as a single line when observed under an optical microscope. See *slip line*.

slip direction. The crystallographic direction in which the translation of slip takes place.

slip line. The trace of the slip plane on the viewing surface; the trace is (usually) observable only if the surface has been polished before deformation. The usual observations on metal crystals (under the light microscope) are of a cluster of slip lines.

slip plane. The crystallographic plane in which slip occurs in a crystal.

S-N diagram. A plot showing the relationship of stress S and the number of cycles N before failure in fatigue testing.

soldering. Similar to *brazing*, with the filler metal having a melting temperature range below an arbitrary value, generally 800°F. Soft solders are usually lead-tin alloys.

solidification shrinkage. The decrease in volume of a metal during solidification.

solid solution. A single solid homogeneous crystalline phase containing two or more chemical species.

solidus. In a constitution or equilibrium diagram, the locus of points representing the temperatures at which various compositions finish freezing on cooling or begin to melt on heating.

solute. The component of either a liquid or solid solution that is present to a lesser or minor extent; the component that is dissolved in the *solvent*.

solution heat treatment. Heating an alloy to a suitable temperature, holding at that temperature long enough to allow one or more constituents to enter into solid solution, and then cooling rapidly enough to hold the constituents in solution. The alloy is left in a supersaturated, unstable state.

solvent. The component of either a liquid or solid solution that is present to a greater or major extent; the component that dissolves the *solute*.

solvus. In a phase or equilibrium diagram, the locus of points representing the temperatures at which the various compositions of the solid phases coexist with other solid phases, that is, the limits of solid solubility.

sorbite. (obsolete) A fine mixture of ferrite and cementite produced either by regulating the rate of cooling of steel or tempering steel after hardening. The first type is very fine pearlite difficult to resolve under the microscope; the second type is tempered martensite.

spheroidal-graphite cast iron. Same as *nodular cast iron*.

spheroidite. An aggregate of iron or alloy carbides of essentially spherical shape dispersed throughout a matrix of ferrite.

spheroidizing. Heating and cooling to produce a spheroidal or globular form of carbide in steel. Spheroidizing methods frequently used are:

1. Prolonged holding at a temperature just below Ae_1.
2. Heating and cooling alternately between temperatures that are just above and just below Ae_1.
3. Heating to a temperature above Ae_1 or Ae_3 and then cooling very slowly in the furnace or holding at a temperature just below Ae_1.
4. Cooling at a suitable rate from the minimum temperature at which all carbide is dissolved, to prevent the reformation of a carbide network, and then reheating in accordance with methods 1 or 2 above. (Applicable to hypereutectoid steel containing a carbide network.)

spray quenching. Quenching in a spray of liquid.

stabilizing treatment. Any treatment intended to stabilize the structure of an alloy or the dimensions of a part. (1) Heating austenitic stainless steels that contain titanium, columbium, or tantalum to a suitable temperature below that of a full anneal in order to inactivate the maximum amount of carbon by precipitation as a carbide of titanium, columbium, or tantalum. (2) Transforming retained austenite in parts made from tool steel. (3) Precipitating a constituent from a nonferrous solid solution in order to improve the workability, to decrease the tendency of certain alloys to age harden at room temperature, or to obtain dimensional stability.

steadite. A hard structural constituent of cast iron that consists of a binary eutectic of ferrite (containing some phosphorous in solution) and iron phosphide (Fe_3P). The composition of the eutectic is 10.2 percent P and 89.8 percent Fe, and the melting temperature is 1920°F.

steel. An iron-base alloy, malleable in some temperature range as initially cast, containing manganese, usually carbon, and often other alloying elements. In carbon steel and low-alloy steel, the maximum carbon is about 2.0 percent; in high-alloy steel, about 2.5 percent. The dividing line between low-alloy and high-alloy steels is generally regarded as being at about 5 percent metallic alloying elements.

Steel is to be differentiated from two general classes of "irons": the cast irons, on the high-carbon side, and the relatively pure irons such as ingot iron, carbonyl iron, and electrolytic iron, on the low-carbon side. In some steels containing extremely low carbon, the manganese content is the principal differentiating factor, steel usually containing at least 0.25 percent; ingot iron contains considerably less.

stiffness. The ability of a metal or shape to resist elastic deflection. For identical shapes, the stiffness is proportional to the modulus of elasticity.

strain. A measure of the change in the size or shape of a body, referred to its original size or shape. *Linear strain* is the change per unit length of a linear dimension. *Conventional strain* is the linear strain referred to the original gage length. When the term strain is used alone, it usually refers to the linear strain in the direction of the applied stress.

strain hardening. An increase in hardness and strength caused by plastic deformation at temperatures lower than the recrystallization range.

stress. Force per unit area, often thought of as force acting through a small area within a plane. It can be divided into components, normal and parallel to the plane, called *normal stress* and *shear stress*, respectively.

stress-corrosion cracking. Failure by cracking under combined action of corrosion and stress, either external (applied) or internal (residual). Cracking may be either intergranular or transgranular, depending on metal and corrosive medium.

stress raisers. Changes in contour or discontinuities in structure that cause local increases in stress.

stress relieving. Heating to a suitable temperature, holding long enough to reduce residual stresses, and then cooling slowly enough to minimize the development of new residual stresses.

stress-rupture test. A tension test performed at constant load and constant temperature, the load being held at such a level as to cause rupture. Also known as *creep-rupture test*.

stretcher strains. Elongated markings that appear on the surface of some materials when deformed just past the yield point. These markings lie approximately parallel to the direction of maximum shear stress and are the result of localized yielding. Same as *Lüders lines*.

stringer. In wrought materials, an elongated configuration of microconstituents or foreign material aligned in the direction of working. Commonly, the term is associated with elongated oxide or sulfide inclusions in steel.

substitutional solid solution. A solid alloy in which the solute atoms are located at some of the lattice points of the solvent, the distribution being random.

substrate. Layer of metal underlying a coating, regardless of whether the layer is basis metal.

subsurface corrosion. Formation of isolated particles of corrosion products beneath the metal surface. This results from the preferential reaction of certain alloy constituents by inward diffusion of oxygen, nitrogen, and sulfur.

superalloy. An alloy developed for very high-temperature service where relatively high stresses (tensile, thermal, vibratory, and shock) are encountered and where oxidation resistance is frequently required.

supercooling. Cooling below the temperature at which an equilibrium-phase transformation can take place without actually obtaining the transformation.

superficial Rockwell hardness test. Form of Rockwell hardness test using relatively light loads which produce minimum penetration. Used for determining surface hardness or hardness of thin sections or small parts, or where a large hardness impression might be harmful.

superlattice. An ordered arrangement of atoms in a solid solution to form a lattice superimposed on the normal solid-solution lattice.

temper. (1) In heat treatment, to reheat hardened steel or hardened cast iron to some temperature below the eutectoid temperature for the purpose of decreasing the hardness and increasing the toughness. The process is also sometimes applied to normalized steel. (2) In nonferrous alloys and in some ferrous alloys (steels that cannot be hardened by heat treatment), the hardness and strength produced by mechanical or thermal treatment, or both, and characterized by a certain structure, mechanical properties, or reduction in area during cold working.

temper brittleness. Brittleness that results when certain steels are held within, or are cooled slowly through, a certain range of temperature below the transformation range. The brittleness is revealed by notched-bar impact tests at or below room temperature.

tempering. Reheating a quench-hardened or normalized ferrous alloy to a temperature below the transformation range and then cooling at any rate desired.

tensile strength. In tensile testing, the ratio of maximum load to original cross-sectional area. Also called *ultimate strength*.

terminal phase. A solid solution, represented by an area, at either extreme side of a binary diagram.

thermal analysis. A method for determining transformations in a metal by noting the temperatures at which thermal arrests occur. These arrests are manifested by changes in slope of the plotted or mechanically traced heating and cooling curves. When such data are secured under nearly equilibrium conditions of heating and cooling, the method is commonly used for determining certain critical temperatures required for the construction of equilibrium diagrams.

thermal fatigue. Fracture resulting from the presence of temperature gradients which vary with time in such a manner as to produce cyclic stresses in a structure.

thermocouple. A device for measuring temperatures consisting of two dissimilar metals which produce an electromotive force roughly proportional to the temperature difference between their hot and cold junction ends.

time quenching. Interrupted quenching in which the time in the quenching medium is controlled.

tinning. Coating metal with a very thin layer of molten filler metal.

torsion. A twisting action resulting in shear stresses and strains.

total carbon. The sum of the free and combined carbon (including carbon in solution) in a ferrous alloy.

toughness. Ability of a metal to absorb energy and deform plastically before fracturing. It is usually measured by the energy absorbed in a notch-impact test, but the area under the stress-strain curve in tensile testing is also a measure of toughness.

transformation temperature. The temperature at which a change in phase occurs. The term is sometimes used to denote the limiting temperature of a transformation range. The following symbols are used for iron and steels:

Ac_{cm}. In hypereutectoid steel, the temperature at which the solution of cementite in austenite is completed during heating.

Ac_1. The temperature at which austenite begins to form during heating.

Ac_3. The temperature at which transformation of ferrite to austenite is completed during heating.

Ac_4. The temperature at which austenite transforms to delta ferrite during heating.

Ae_{cm}, Ae_1, Ae_3, Ae_4. The temperatures of phase changes at equilibrium.

Ar_{cm}. In hypereutectoid steel, the temperature at which precipitation of cementite starts during cooling.

Ar_1. The temperature at which transformation of austenite to ferrite or to ferrite plus cementite is completed during cooling.

Ar_3. The temperature at which austenite begins to transform to ferrite during cooling.

Ar_4. The temperature at which delta ferrite transforms to austenite during cooling.

M_s (or Ar''). The temperature at which transformation of austenite to martensite starts during cooling.

M_f. The temperature at which martensite formation finishes during cooling.

Note: All these changes, except the formation of martensite, occur at lower temperatures during cooling than during heating, and depend on the rate of change of temperature.

transition lattice. An intermediate unstable crystallographic configuration that forms during solid-state reactions such as precipitation from solid solution and eutectoid decomposition.

transition temperature. (1) An arbitrarily defined temperature within the temperature range in which metal-fracture characteristics, determined usually by notched tests, are changing rapidly, for example from primarily fibrous (shear) to primarily crystalline (cleavage) fracture. Commonly used definitions are *transition temperature for 50 percent cleavage fracture, 10-ft-lb transition temperature,* and *transition temperature for half maximum energy.* (2) Sometimes also used to denote the arbitrarily defined temperature in a range in which the ductility changes rapidly with temperature.

troostite. (obsolete) A previously unresolvable, rapidly etching fine aggregate of carbide and ferrite produced either by tempering martensite at low temperature or by quenching a steel at a rate slower than the critical cooling rate. Preferred terminology for the first product is tempered martensite; for the latter, fine pearlite.

twin. Two portions of a crystal having a definite crystallographic relationship; one may be regarded as the parent, the other as the twin. The orientation of the twin is either a mirror image of the orientation of the parent about a "twinning plane" or an orientation that can be derived by rotating the twin portion about a "twinning axis."

twin band. On a polished and etched surface, the section through a twin and the parent crystal.

ultimate strength. The maximum conventional stress—tensile, compressive, or shear—that a material can withstand.

upsetting. Working metal so that the cross-sectional area of a portion or all of the stock is increased.

vacancy. A type of lattice imperfection in which an individual atom site is temporarily unoccupied. Diffusion (of other than interstitial solutes) is generally visualized as the shifting of vacancies.

Vickers hardness. Microhardness determined from the resistance of a metal to indentation by a 136° diamond-pyramid indenter making a square impression.

white cast iron. *Cast iron* that gives a white fracture because the carbon is in combined form.

white metal. A general term covering a group of white-colored metals of relatively low melting points (lead, antimony, bismuth, tin, cadmium, and zinc) and of the alloys based on these metals.

Widmanstätten structure. A structure characterized by a geometrical pattern resulting from the formation of a new phase along certain crystallographic planes of the parent solid solution. The orientation of the lattice in the new phase is related crystallographically to the orientation of the lattice in the parent phase. The structure was originally observed in meteorites but is readily produced in many other alloys with certain heat treatment.

wire drawing. Reducing the cross section of wire by pulling it through a die.

wrought iron. A commercial iron consisting of slag (iron silicate) fibers entrained in a ferrite matrix.

x-ray. Electromagnetic radiation, of wavelength less than about 500 angstrom units, emitted as the result of deceleration of fast-moving electrons or decay of atomic electrons from excited orbital states; specifically, the radiation produced when an electron beam of sufficient energy impinges upon a target of suitable material.

yield point. The first stress in a material, usually less than the maximum attainable stress, at which an increase in strain occurs without an increase in stress. Only certain metals exhibit a yield point. If there is a decrease in stress after yielding, a distinction may be made between upper and lower yield points.

yield strength. The stress at which a material exhibits a specified deviation from proportionality of stress and strain. An offset of 0.2 percent is used for many metals.

Young's modulus. See *modulus of elasticity*.

zone melting. Highly localized melting, usually by induction heating, of a small volume of an otherwise solid piece. By moving the induction coil along the rod, the melted zone can be transferred from one end to the other. In a binary mixture where there is a large difference in composition on the liquidus and solidus lines, high purity can be attained by concentrating one of the constituents in the liquid as it moves along the rod.

INDEX

Abrasion of metals, 567
Actual cooling rate, 283–287
  determination of, 283
  effect on microstructure, 290–292
  effect of size and mass of piece, 289–296
  quenching medium, 284–287
  surface area to mass ratio, 289–290
  surface condition of piece, 288–289
Admiralty metal, 468
Age hardening, 190–194
  aging process, 191–194
  property changes due to, 193–194
  solution treatment, 190–191
AISI (American Iron and Steel Institute) classification system for steels, 244–245
Alclad, 497, 602
Allotropy, 85, 208–209
Alloy cast iron, 454–458
  chromium, 457
  copper, 457
  molybdenum, 457
  nickel, 458
  vanadium, 458
Alloy steels, 349–383
  chromium, 358
  definition of, 349
  manganese, 359–360
  molybdenum, 360–361
  nickel, 355–358
  nickel-chromium, 358–359
  silicon, 361
  stainless, 361–376
  tungsten, 361
  vanadium, 361
Alloy systems:
  aluminum-copper, 485–486
  aluminum-lead, 206
  aluminum-magnesium, 490
  aluminum-magnesium-silicide, 491
  aluminum-manganese, 489
  aluminum-silicon, 179, 490
  aluminum-zinc, 492, 533
  cobalt-tungsten, 219
  copper-antimony, 222
  copper-beryllium, 478
  copper-lead, 205
  copper-nickel, 168
  copper-palladium, 211
  copper-silicon, 474
  copper-tin, 472
  copper-zinc, 465
  gold-copper, 210
  iron-chromium-carbon, 362-364
  iron-nickel, 209
  iron-nickel-chromium-carbon, 366
  iron-silicon, 222
  iron-tin, 222
  lead-antimony, 187
  lead-tin, 188
  magnesium-aluminum, 500
  nickel-iron, 513
  silver-copper, 536
  silver-platinum, 197
  titanium-aluminum, 526
Alloy systems:
  titanium-manganese, 528
  titanium-vanadium, 527
Alnico, 517, 630
Aluminizing, 601
Aluminum, 481–485
  anodizing, 494
  corrosion resistance, 493–497
Aluminum alloys, 485–497
  aluminum-copper, 485–488
  aluminum-magnesium, 489–491
  aluminum-manganese, 489
  aluminum-silicon, 489
  aluminum-silicon-magnesium, 491–492
  aluminum-zinc, 492–493
  composition and properties of, table, 496
  corrosion resistance, 493–497
  designation system, 481, 484
  temper designation, 484–485
Aluminum Association:
  alloy designation, 481, 484
  temper designation, 484–485
Aluminum brass, 468
Aluminum bronze, 476
Annealing, 129–144
  effect on properties, 137
  grain growth, 135–136
  grain size, 136–137
  recovery, 129–130
  recrystallization, 131–135
    temperature, 132–135
  of steel: full, 249–252
    process, 254
    properties, table, 252
    spheroidizing, 252–254
    stress-relief, 254
Annealing twins, 118
Anodizing, 572
Apparent density of metal powders, table, 612
Architectural bronze, 470
Arsenical copper, 462
Atom binding, 76–78
  covalent, 77
  ionic, 76–77
  metallic, 77–78
  van der Waals, 78
Atomic diameter, 78–79
  table, 79
Atomic numbers and weights, table, 68
Atomic structure, 65–70
  electrons, 65
  neutrons, 65
  protons, 65
  quantum numbers, 69
Atomization, 607–609
Ausforming, 383
Austempering, 313–315
Austenite:
  definition of, 234
  homogeneity, effect of, 282–283
  microstructure of, 235
Austenite transformation:
  on continuous cooling, 273–276

Austenite transformation:
isothermal, 261–273
slow cooling, 237–241
Austenitic stainless steels, 371–374
Austenitizing temperature, 281–282
Axial ratio, 84

Babbitt:
lead-base, 520–521
tin-base, 521
Bain, E. C., 260
Bainite:
austempering, 313–315
microstructure, 268–269
Bakelite, 15
Bending fractures, 656–659
Beryllium bronze, 476–479
age hardening, effect of, 478–479
microstructure, 479
phase diagram, 478
Bonderizing, 600
Bragg equation, 90
Brasses, 464–472
alpha, 466–469
red, 468–469
yellow, 466–468
alpha plus beta, 469–471
cast, 471–472
Breaking strength, 40–41
Brinell hardness test, 26–27
Briquette, 607
Bronzes, 472–478
aluminum, 476
beryllium, 476–478
silicon, 475–476
tin, 472–475
Bulls-eye structure, 432

Calorizing, 602
Carbon-concentration gradient, 318–319
Carbon solubility in iron, 234–236
Carbonitriding, 326–328
Carburizing, 317–326
case depth by, table, 321
equation, 319
gas, 322–323
heat treatment after, 325–326
liquid, 323–325
pack, 319–321
Cartridge brass, 467
Case hardening, 315–336
carbonitriding, 326–328
carburizing, 316–326
cyaniding, 326–328
flame hardening, 332–333
gas cyaniding, 327
induction hardening, 333–336
nitriding, 328–332
Cast iron, 423–459
alloy, 453–458
chilled, 448–450
gray, 434–448
graphite flake, 441–445
size chart, 442
types of, 444
malleable, 426–434

Cast iron, malleable:
ferritic, 429–430
manufacture of, 426–429
pearlitic, 430–434
nodular, 450–453
microstructure, 450–454
properties of, 452
types of, 423–424
white, 424–426
Castings, defects in, 98–100
hot tears, 99–100
pipe, 98
porosity, 98–99
shrinkage, 98
Cathodic polarization, 584
Cavitation corrosion, 591–592
Cemented carbide tools, 415–419, 625
Cementite, definition of, 234
Ceramic tools, 420–421
Cermets, 419–420
Charpy impact specimen, 43
Chemical elements:
atomic number, atomic weight, and symbols, table, 68
definition of, 65
periodic table, 71
Chilled cast iron, 448–450
Chromel A, 511
Chromel-alumel, 7
Chromel C, 511
Chromium in cast iron, 453–457
Chromium steels, 358
Chromizing, 573
Classification of failure causes, 662–663
Coherent lattice theory, 192
Coin silver, 536
Coining, 622
Commercial bronze, 468
Comparison method to measure grain size, 101–102
Composite metals, 625
Congruent-melting alloy, 170, 195
Constantan, 511
Cooling rate:
actual (*see* Actual cooling rate)
critical, 260
Copper, 461–464
arsenical, 462
in cast iron, 457
electrolytic tough-pitch, 461
free-cutting, 462
mechanical properties of, table, 482
oxygen-free high-conductivity, 461
silver-bearing, 462
temper designation, 462–463
Copper alloys, 464–480
brasses, 464–472
alpha, 466–469
alpha plus beta, 469–471
effect of zinc on, table, 465
bronzes, 472–480
aluminum, 476
beryllium, 476–479
silicon, 475–476
tin, 472–475
copper-aluminum, 476

Copper alloys:
  copper-beryllium, 478
  copper-nickel, 478–479
  copper-silicon, 474
  copper-tin, 472
  mechanical properties of, table, 482–483
  nickel silvers, 480
  temper designation, 463
Copper-constantan, 7
Copper oxide in copper, 461–462
Coring, 165–166
Corrosion of metals, 583–602
  definition of, 583
  dezincification, 467
  effect on properties, table, 596
  electrochemical principles, 583–585
  electromotive-force series, 586
  factors in, 585–590
  galvanic series, 587
  inhibitors, 600
  methods of combating, 597–602
  season cracking, 466
  specific types, 590–597
Creep, properties of, table, 555
Creep tests, 45, 548–549
Crevice corrosion, 592
Critical cooling rate, definition of, 260
Critical resolved shear stress, table, 115
Crystal imperfections, 94–98
  dislocations, 97–98
  interstitials, 96–97
  vacancies, 96
Crystal structure, 80–85
  body-centered cubic, 80–81
  close-packed hexagonal, 84–85
  face-centered cubic, 81–83
  space lattice, 80
  unit cell, 80
Crystal systems, table, 81
Crystallization, 91–94
Crystallographic planes, 87–89
Cupronickels, 478–479
Cyaniding, 326–328

Davenport, E. S., 260
Decarburization, 319
  in tool steels, 397
Dendrite, 93
Dendritic segregation, 166
Dezincification, 467
Die compaction of powders, 614–616
Diffusion, 163–165
Dislocation, 97–98
  in slip, 113–115
Ductile iron (*see* Nodular cast iron)
Ductility elongation, 41
  reduction in area, 41
Dumet wire, 517

Eddy current inspection, 57–60
Elastic hardness, 24–25
Elastic limit, 38–39
Electrolytic deposition of powders, 609
Electrolytic tough-pitch copper, 461
Electron configuration, table, 75
Electroplating, 571–572
Electrotype metal, 519
Elinvar, 517
End-quench hardenability test, 297–302
  correlation with C-T diagram, 300
Endurance limit, 44
Equilibrium diagrams (*see* Phase diagrams)
Erosion of metals, 567
Etching, 18–19
  reagents, table, 22
Eutectic reaction, 175
  equation, 178, 183, 218
Eutectic systems, 171–179
  microstructures: Al-Si, 180
    Pb-Sb, 187
    Pb-Sn, 188
  properties, 189–190
Eutectoid reaction, 212–214
  equation, 212
Extractive metallurgy, viii

Failure analysis, 633–664
  bending fractures, 656–659
  classification of causes, 662–663
  effect of residual stresses, 653–654
  effect of strength reducers, 648
  effect of stress raisers, 643–647
  fatigue fractures, 640–643
  loading, types of, 637–640
  modes of fracture, 635–636
  torsional fractures, 659–661
Fatigue fractures, 640–643
Fatigue tests, 44–45
Fernico, 516
Ferrite:
  definition of, 234
  microstructure of, 235
Ferritic irons, 451
Ferritic stainless steels, 369–370
Flame hardening, 332–333
Flame-plated coatings, properties of,
    table, 578
Floe process, 330
Fluorescent-penetrant inspection, 51–54
Forging brass, 470
Foundry type metal, 519
Fracture, 118–121
Fracture modes, 635–636
Free-cutting brass, 470
Free-cutting copper, 462
Fretting corrosion, 593–594
Fusebond method of spraying, 575

Galvanic corrosion, 596
Galvanizing, 601
Gas carburizing, 322–323
Gilding metal, 468
Gold, 543
  properties of, 537
Gold alloys, 537
Grain size in castings, 100–101
Grain size measurement, 101–104
Gravity compaction of powders, 617
Gray cast iron, 434–448
  graphite flake, 441–445
    size chart, 442
    types of, 444

Gray cast iron:
heat treatment, 439–441
manganese in, 438
microstructure, 436
phosphorus in, 438–439
properties of, 446
silicon in, 435–438
sulfur in, 438

H Monel, 510
Hadfield manganese steel, 360
Hadfield silicon steel, 361
Hard facing, 577–581
Hardenability, 296–302
definition of, 293
specification for, 301–302
test for, 297–299
Hardenability bands, 301
Hardenable carbon steels, 343–345
Hardness, 24–37
conversion, table, 28–29
cylindrical correction chart, 36
definition of, 24
tests for: applications, table, 34
Brinell, 26–27
file, 26
microhardness, 32–33
Knoop, 32–33
Vickers, 32
Rockwell: scales, table, 30
superficial tester, 30–31
scleroscope, 24–25
scratch, 25–26
Vickers, 31–32
Hardness-penetration diagram, 293
Hastelloy A, 512
Hastelloy B, 512
Hastelloy C, 512
Hastelloy D, 511
Hastelloy X, 512–513
Heat of fusion, 92
Heyn method to measure grain size, 102–103
Holography, 60–62
Homogenization, 166–168
Hot hardness, 395
Hot pressing of powders, 621–622
Hot working, 137–140
Hume-Rothery, William, 150, 152
Hypereutectic alloys, 174
Hypereutectic cast irons, 234
Hypereutectoid alloys, 213
Hypereutectoid steels, 234
Hypoeutectic alloys, 174
Hypoeutectic cast irons, 234
Hypoeutectoid alloys, 213
Hypoeutectoid steels, 234

Ihrigizing, 573
Illium B, 513
Illium G, 513
Illium R, 513
Impact test, 42–44
specimen: Charpy, 42–43
Izod, 43–44
Impregnation, 623
Inconel, 511
Inconel X, 511–512
Induction hardening, 333–336
Ingot iron, 225
Intergranular corrosion, 594
Intermediate alloy phases, 148–150, 195–196
electron compound, 150
intermetallic compound, 149
interstitial compound, 149–150
Intermetallic compound, 149
Interstitial compound, 149–150
Interstitial solid solution, 153–154
Interstitials, 96–97
Invar, 516
Iridium, 541
Iron:
alpha, 225
beta, 225
cooling curve, 226
delta, 225
gamma, 225
Iron-constantan, 7
Iron-iron carbide phase diagram, 230–234
delta region, 231–232
effect of alloying elements on, 352–353
eutectic reaction, 232
eutectoid reaction, 233
peritectic reaction, 231
structures: austenite, 234
cementite, 234
ferrite, 234
ledeburite, 234
pearlite, 234
Isostatic compaction of powders, 616
Isothermal annealing, 270
Isothermal-transformation diagram, 260–266
alloy element, effect of, 277–279
austenitic grain size, effect of, 280–281
determination, 260–265
Isotopes, 74–76
Izod impact specimen, 43

Jeffries method to measure grain size, 103–104
Jominy hardenability test, 297–302

K Monel, 510
Knoop hardness test, 32–33
Kovar, 516

Lead, 517
Lead alloys, 517–521
composition and properties of, table, 520–521
lead-antimony, 517–518
lead-tin, 518–520
terne metal, 519
Ledeburite, definition of, 234
Linotype metal, 520
Liquid carburizing, 323–325
Liquid-metal corrosion, 597
Liquidus line, 158
Low brass, 469
Lucite, 15–16

M_f temperature, 257
effect of carbon, 259

M_s temperature, 257–259
  effect of carbon, 259
  formula for calculation, 259
Magnaflus test, 49–51
Magnaglo test, 49–51
Magnesium, 498–499
Magnesium alloys, 500–507
  composition and properties of, table, 506
  corrosion resistance, 505
  designation and temper, 498–499
  joining, 505, 507
  magnesium-aluminum-based, 501–503
  magnesium-rare earth-based, 503–504
  magnesium-thorium-based, 504–505
  magnesium-zinc-based, 503
Magnetic-particle inspection, 49–51
Malleable cast iron, 426–434
  copper-alloyed, 434
  ferritic, 429–430
  manufacture of, 426–429
  pearlitic, 430–434
Manganese bronze, 470
Manganese steel, 359–360
Maraging steels, 377–383
  composition of, table, 378
  properties of, table, 380
Martens, A., vii
Martensite:
  definition of, 257
  microstructure of, 258–260
Martensite transformation, 257–260
Martensitic stainless steels, 366–368
Metal spraying, 573–577
Metal "whiskers," 95
Metallizing, 600
  (*See also* Metal spraying)
Metallograph, 19
Metallographic sample preparation, 15–19
  etching, 18–19
    reagents, table, 22
  fine polishing, 16–17
  intermediate polishing, 16
  mounting, 15–16
    in Bakelite, 15
    in Lucite, 15–16
  rough grinding, 15
  sampling, 15
Metallography, definition of, 14
Metalloids, 76
Metallurgical microscopes, 19–24
  electron, 19–24
  optical, 19–20
Metals:
  characteristics of, 76
  at high temperatures, 547–555
    elevated-temperature test, 547–548
    properties of, table, 555
  at low temperatures, 555–565
    properties of, table, 558–559
Metastable phase, 230
Miller indices, 87–89
Modules of elasticity, 41
Molybdenum:
  in cast iron, 457
  electrical contacts, 543–544
  properties of, 542
Molybdenum steel, 360–361
Monel, 170, 510
Monotectic reaction, equation, 204
Muntz metal, 469

Naval brass, 470
Navy steel, 361
Nichrome, 511
Nichrome V, 511
Nickel, 507–509
  A nickel, 509
  in cast iron, 458
  composition and properties, table, 508
  D nickel, 509
  duranickel, 509
  E nickel, 509
  effect of low temperature, 562
  effect on transition temperature, 562
  permanickel, 509
Nickel alloys, 509–517
  composition and properties, table, 514–515
  nickel-chromium-iron-based, 511–512
  nickel-chromium-molybdenum-copper-based, 513
  nickel-chromium-molybdenum-iron-based, 512–513
  nickel-copper-based, 509–511
  nickel-iron, 513, 516–517
  nickel-molybdenum-iron-based, 512
  nickel-silicon-copper-based, 511
Nickel-chromium steels, 358–359
Nickel silvers, 480
Nickel steels, 355–358
Nitriding, 328–332
  case depth by, 328
  Floe process, 330
Nodular cast iron, 450–453
  microstructure, 450–452
  properties of, 452
Nondestructive testing, 45–62
  eddy current, 57–60
  elements of, 46
  fluorescent penetrant, 51–54
  holography, 60–62
  magnetic particle, 49–50
  radiography, 46–49
  summary, table, 58–59
  ultrasonic, 54–57
Normalizing, steel, 254–256

Optical pyrometer, 11–14
Order-disorder transformation, 209–212
Osmium, 541
Oxygen-free high-conductivity copper, 461

Pack carburizing, 319–321
Palladium, 543
  properties of, 539
Palladium alloys, 539–540
  properties of, table, 542
Parkerizing, 600
Pauli exclusion principle, 67
Pearlite:
  definition of, 234
  microstructure, 235

Peltier effect, 5
Periodic table of elements, 70–74
Peritectic reaction, 196–202
  equation, 196
Peritectoid reaction, 214–216
  equation, 214
Permalloys, 517
Pewter, 523
Phase:
  definition of, 147
  types, 148–154
    intermediate alloy, 148–150
      electron compound, 150
      intermetallic compound, 149
      interstitial compound, 149–150
    pure metal, 148
    solid solution, 150–154
      interstitial, 153–154
      substitutional, 152–153
Phase diagrams, 155–220
  coordinates of, 156
  determination of, 156–157
  reactions on: eutectic, 171–178
    eutectoid, 212–214
    monotectic, 202–205
    peritectic, 196–202
    peritectoid, 214–216
    table, 218
  rules for: chemical composition of phases, 159
    relative amounts of each phase, 159–161
Phosphor bronze (*see* Tin bronze)
Physical metallurgy, viii
Piezoelectric effect, 54
Pitting corrosion, 591
Plasma spraying, 575
Plastic deformation:
  effect on properties, 124–126
  polycrystalling material, 121–124
  by slip, 107–116
    in different lattice structures, 115–116
    mechanism, 110–115
    resolved shear stress, 110
  by twinning, 116–118
Platinite, 516
Platinum, 537, 543
Platinum alloys, 537–539
  platinum-iridium, 539
  platinum-nickel, 539
  platinum-rhodium, 538–539
  platinum-ruthenium, 539
  platinum-tungsten, 539
  properties of, table, 542
Polymorphism, 85
Powder metallurgy, 605–630
  characteristics of powders, 609–613
  coining, 622
  compacting, 614–619
  electrical contacts, 544
  hot pressing, 621–622
  impregnation, 623
  mixing, 613–614
  particle size, 610–611
  powder preparation, 607–609
  presintering in, 622

Powder metallurgy:
  processes, 606–607
  properties and applications of, table, 626–629
  sintering, 619–621
Precipitation hardening (*see* Age hardening)
Precipitation-hardening stainless steels, 374–376
Preferential corrosion, 596
Preferred orientation, 122
Proportional limit, 38
Pseudoeutectic alloy, 171
Pyrometers, 5–14
  optical, 11–14
  radiation, 10–11
  thermoelectric, 5–10
Pyrometry, 1

Quantum numbers, 69
Quenching, 283–287
  actual cooling rate, 284
  mechanism, 283–284
  medium, 284–286
    circulation, 287
    temperature, 286–287
    type of, 284–285
  stages in, 283–284

R Monel, 510
Radiation pyrometer, 10–11
Radiography of metals, 46–49
Raoult's law, 172
Recrystallization, 131–135
  temperature, 132–135
Red brass, 469
Red hardness, 395–396
Red-short, 438
Residual stresses, 653–654
  from heat treatment, 336–343
Resilience, 42
Rhodium, 541
Rockwell hardness test:
  cylindrical correction chart, 36
  scales, table, 30
Rockwell superficial hardness test, 30–31
Ruthenium, 541

S curves (*see* Isothermal-transformation diagram)
S Monel, 510
Sauveur, Albert, vii
Scale on steel, 288
  protection against, 288–289
Scleroscope hardness test, 24–25
Season cracking in brass, 466
Seeback effect, 5
Self-lubricating bearings, 630
Sherardizing, 531, 602
Short-time tension tests, 551
Silicon bronze, 475–476
Silicon steel, 361
Siliconizing, 573
Silver, properties of, 536
Silver alloys, 536–537
  electrical contacts, 543
  properties of, table, 542

Silver alloys:
silver brazing, 537
silver-copper, 536–537
silver-copper-zinc, 537
Silver-bearing copper, 462
Silver solders, 537
Sintered carbide tools, 415–419
Sintering in powder metallurgy, 619–621
Slip, 107–116
Slip casting, 617
Slip system, 115
Solders, 519, 522
Solid solutions, 150–154
alloys, properties of, 168–170
factors in, 152–153
interstitial, 153–154
microstructures, Cu-Ni, 169
order-disorder, 209–212
substitutional, 152–153
terminal, 182
types of, 152
Solidus line, 158
Solvus line, 182
Sorbite, 309
Sorby, Henry C., vii
Spheroidal graphite iron (*see* Nodular cast iron)
Spheroidite, 254
Spheroidize annealing, 252–254
Spherulitic iron (*see* Nodular cast iron)
Stainless steels, 362–376
austenitic, 371–374
classification of, 362
composition and properties of, table, 370–371
ferritic, 369–370
manganese in, 374
martensitic, 366–368
precipitation-hardening, 374–376
Steadite in cast iron, 438–439
Steels:
alloy, composition of, 247
alloying elements in, effects of, table, 356–357
annealing: full, 249–252
process, 254
properties of, table, 252
spheroidizing, 252–254
stress-relief, 254
austempering, 313–315
austenite, homogeneity of, effect of, 282–283
austenitizing temperature, 281–282
bainite, 268–269
microstructure, 268–269
chromium, 358
classification of, 244–245
cold-short, 248
cooling-transformation diagram, 273–276
critical cooling rate, 260
decarburization, 319
effect: of manganese, 248
of phosphorus, 248
of silicon, 248
of sulfur, 248
hardenability, 296–302
Steels, hardenability: specification, 301–302
test for, 297–299
hardenable carbon, 343–345
hardening temperature, 281–282
heat treatment, 249–345
hot-short, 248
hypereutectoid, 234
critical temperatures, 240–241
microstructures of, 242
slow cooling, 240–241
hypoeutectoid, 234
critical temperatures, 237
microstructures of, 239
slow cooling, 237–240
in iron-iron carbide diagram, 233
isothermal annealing, 270
isothermal-transformation diagram, 260–266
manganese, 359–360
maraging, 377–383
martensite, 257
molybdenum, 360–361
nickel, 355–358
nickel-chromium, 358–359
nitralloys, 330
normalizing, 254–256
properties of, table, 252
plain carbon, composition of, 246
quenching, 283–287
residual stresses, 336–343
scale on, 288
protection against, 288–289
silicon, 361
stainless (*see* Stainless steels)
tempering, 305–313
tool (*see* Tool steels)
tungsten, 361
vanadium, 361
Stellite, 413–414
Sterling silver, 536
Stiffness, 41
Strain, 37
Stress, 37
true, 41–42
Stress corrosion, 594–596
Stress-relief annealing, 254
Stress-rupture tests, 549–552
Substitutional solid solution, 152–153
chemical-affinity factor, 152
crystal-structure factor, 152
relative-size factor, 152
relative-valence factor, 152–153
Superlattice, 209
Surface heat treatment (*see* Case hardening)

Temper brittleness, 306
Temperature versus electromotive force, table, 6
Temperature measurement, 1–14
by color, 1–2
by pyrometers, 5–14
optical, 11–14
radiation, 10–11
recording and controlling, 10

Temperature measurement, by pyrometers:
thermoelectric, 5–10
by thermometers, 2–4
gas or vapor pressure, 3–4
liquid-expansion, 3
metal expansion, 2
resistance, 4
Temperature scales, 1
Tempered martensite, 311
Tempering:
alloying elements, effect of, 353–354
secondary hardness, 354
steel, 305–313
effects: on microstructure, 308–311
on properties, 305–307
of time and temperature, 311
Tensile properties, 38–42
breaking strength, 40–41
ductility: elongation, 41
reduction in area, 41
elastic limit, 38–39
modulus of elasticity, 41
proportional limit, 38
resilience, 42
toughness, 42
true stress-strain, 41–42
ultimate strength, 40
yield point, 39
yield strength, 39–40
Tensile test, 37–42
Terne metal, 519
Testing:
creep, 45
fatigue, 44–45
hardness, 24–37
nondestructive, 45–62
impact, 42–44
Thermocouple, 5–8
construction, 7–8
materials, 7
protecting tubes, 8
Thermometers, 1–4
Thermometry, 1
Thermopile, 11
Thomson effect, 5
Tin, 521–522
Tin alloys, 522–523
composition and properties of, table, 522–523
Tin bronze, 472–475
Tin pest, 521
Titanium, 524–525
Titanium alloys, 525–531
alpha alloys, 526
alpha-beta alloys, 527–530
beta alloys, 530–531
composition and properties of, table, 532
Titanium martensite, 529
Tobin bronze, 470
Tool failures, 410–412
Tool steels, 387–421
AISI classification, 387
brand names, table, 398–399
cold-work, 401–404
comparative properties of, table, 394–395
decarburization, 397
Tool steels:
distortion, 393
failures, 410–412
hardenability, 393–394
heat-treatment, 408–410
high-speed, 405–406
hot-work, 404–405
machinability, table, 396
mold, 406–407
selection of, 392–393
shock-resisting, 400–401
special-purpose, 407–408
toughness, 394–395
water-hardening, 397–400
Torsional fractures, 659–661
Toughness, 42
at low temperatures, table, 564
of tool steels, 394–395
Transducer, 55
Transition carbide, 308
Transition temperature, 560
Troostite, 309
Tungsten:
electrical contacts, 543–544
properties of, 542
steel, 361
Twinning, 116–118
Twins:
annealing, 118
deformation, 118

Ultimate strength, 40
Ultrasonic inspection, 54–57
Uniform corrosion, 590

Vacancies, 96
Vanadium in cast iron, 458
Vanadium steels, 361
Vibratory compaction of powders, 617
Vickers hardness test, 31–32

Wear of metals, 567–581
factors, 569–570
mechanism, 567–569
protection against, 571–581
White cast iron, 424–426
microstructure, 426
White metal, 520–521
Work hardening, 123
Wrought iron, 226–230
composition of, 228
manufacture of, 226–228
microstructure, 229
nickel, 229–230
properties of, table, 229

X-ray diffraction, 89–90
X-rays for radiography, 46–47

Yellow brass, 467
Yield point, 39
Yield strength, 39–40

Zamak-3, 533
Zamak-5, 533
Zinc alloys, 531–535
composition and properties of, table, 535
Zyglo, 51–54